Student's Solutions Manual

to accompany

Applied Calculus

For Business, Economics, and the Social and Life Sciences

Expanded Ninth Edition

Laurence D. Hoffmann
Solomon Smith Barney

Gerald L. Bradley
Claremont McKenna College

Prepared by
Devilyna Nichols
Purdue University

Boston Burr Ridge, IL Dubuque, IA Madison, WI New York San Francisco St. Louis
Bangkok Bogotá Caracas Kuala Lumpur Lisbon London Madrid Mexico City
Milan Montreal New Delhi Santiago Seoul Singapore Sydney Taipei Toronto

The McGraw·Hill Companies

Student's Solutions Manual to accompany
APPLIED CALCULUS: FOR BUSINESS, ECONOMICS, AND THE SOCIAL AND LIFE SCIENCES,
EXPANDED NINTH EDITION
LAURENCE D. HOFFMANN AND GERALD L. BRADLEY

Published by McGraw-Hill Higher Education, an imprint of The McGraw-Hill Companies, Inc., 1221 Avenue of the Americas, New York, NY 10020.

Recycled/acid free paper
This book is printed on recycled, acid-free paper containing 10% postconsumer waste.

6 7 8 9 0 QPD/QPD 0 9 8 7

ISBN 978-0-07-325882-9
MHID 0-07-325882-2

www.mhhe.com

CONTENTS

Chapter 1

Functions, Graphs, and Limits

1.1 Functions

1. $$f(x) = 3x^2 + 5x - 2,$$
$$f(0) = 3(0)^2 + 5(0) - 2 = -2,$$
$$f(-2) = 3(-2)^2 + 5(-2) - 2 = 0,$$
$$f(1) = 3(1)^2 + 5(1) - 2 = 6.$$

3. $$g(x) = x + \frac{1}{x},$$
$$g(-1) = -1 + \frac{1}{-1} = -2,$$
$$g(1) = 1 + \frac{1}{1} = 2,$$
$$g(2) = 2 + \frac{1}{2} = \frac{5}{2}.$$

5. $$h(t) = \sqrt{t^2 + 2t + 4},$$
$$h(2) = \sqrt{2^2 + 2(2) + 4} = 2\sqrt{3},$$
$$h(0) = \sqrt{0^2 + 2(0) + 4} = 2,$$
$$h(-4) = \sqrt{(-4)^2 + 2(-4) + 4} = 2\sqrt{3}$$

7. $$f(t) = (2t - 1)^{-3/2} = \frac{1}{(\sqrt{2t - 1})^3},$$
$$f(1) = \frac{1}{[\sqrt{2(1) - 1}]^3} = 1,$$
$$f(5) = \frac{1}{[\sqrt{2(5) - 1}]^3} = \frac{1}{[\sqrt{9}]^3} = \frac{1}{27},$$
$$f(13) = \frac{1}{[\sqrt{2(13) - 1}]^3} = \frac{1}{[\sqrt{25}]^3} = \frac{1}{125}.$$

9. $$f(x) = x - |x - 2|,$$
$$f(1) = 1 - |1 - 2| = 1 - |-1| = 1 - 1 = 0,$$
$$f(2) = 2 - |2 - 2| = 2 - |0| = 2,$$
$$f(3) = 3 - |3 - 2| = 3 - |1| = 3 - 1 = 2.$$

11. $$f(t) = \begin{cases} 3 & \text{if } t < -5 \\ t + 1 & \text{if } -5 \le t \le 5 \\ \sqrt{t} & \text{if } t > 5 \end{cases}$$

$$f(-6) = 3,$$
$$f(-5) = -5 + 1 = -4,$$
$$f(16) = \sqrt{16} = 4.$$

13. $g(x) = \dfrac{x}{1 + x^2}.$

Since $1 + x^2 \neq 0$ for any real number, the domain is the set of all real numbers.

15. $f(t) = \sqrt{1 - t}.$
Since negative numbers do not have real square roots, the domain is all real numbers such that $1 - t \ge 0$, or $t \le 1$. Therefore, the domain is not the set of all real numbers.

17. $g(x) = \dfrac{x^2 + 5}{x + 2}.$
Since denominators cannot be 0, the domain consists of all real numbers such that $x \neq -2$.

19. $f(x) = \sqrt{2x + 6}.$
Since negative numbers do not have real square roots, the domain is all real numbers such that $2x + 6 \ge 0$, or $x \ge -3$.

21. $f(t) = \dfrac{t+2}{\sqrt{9-t^2}}$.
Since negative numbers do not have real square roots and denominators cannot be zero, the domain is the set of all real numbers such that $9 - t^2 > 0$, namely $-3 < t < 3$.

23. $f(u) = 3u^2 + 2u - 6$ and $g(x) = x + 2$, so

$$\begin{aligned} f(g(x)) &= f(x+2) = 3(x+2)^2 + 2(x+2) - 6 \\ &= 3x^2 + 14x + 10. \end{aligned}$$

25. $f(u) = (u-1)^3 + 2u^2$ and $g(x) = x + 1$, so

$$\begin{aligned} f(g(x)) &= f(x+1) \\ &= [(x+1) - 1]^3 + 2(x+1)^2 \\ &= x^3 + 2x^2 + 4x + 2. \end{aligned}$$

27. $f(u) = \dfrac{1}{u^2}$ and $g(x) = x - 1$, so

$$f(g(x)) = f(x-1) = \frac{1}{(x-1)^2}.$$

29. $f(u) = \sqrt{u+1}$ and $g(x) = x^2 - 1$, so

$$\begin{aligned} f(g(x)) &= f(x^2 - 1) \\ &= \sqrt{(x^2-1)+1} \\ &= \sqrt{x^2} = |x|. \end{aligned}$$

31. $f(x) = x^2$

$$\begin{aligned} \frac{f(x+h) - f(x)}{h} &= \frac{(x+h)^2 - x^2}{h} \\ &= \frac{x^2 + 2xh + h^2 - x^2}{h} = \frac{h(2x+h)}{h} = 2x + h. \end{aligned}$$

33. $f(x) = \dfrac{1}{x}$

$$\begin{aligned} \frac{f(x+h) - f(x)}{h} &= \frac{\frac{1}{x+h} - \frac{1}{x}}{h} \cdot \frac{x(x+h)}{x(x+h)} \\ &= \frac{x - (x+h)}{hx(x+h)} = \frac{x - x - h}{hx(x+h)} = \frac{-h}{hx(x+h)} \\ &= \frac{-1}{x(x+h)}. \end{aligned}$$

35. $f(g(x)) = f(1-3x) = \sqrt{1-3x}$
$g(f(x)) = g(\sqrt{x}) = 1 - 3\sqrt{x}$
To solve $\sqrt{1-3x} = 1 - 3\sqrt{x}$, square both sides, so

$$\begin{aligned} 1 - 3x &= 1 - 6\sqrt{x} + 9x \\ -3x &= -6\sqrt{x} + 9x \\ 6\sqrt{x} &= 12x \\ \sqrt{x} &= 2x \end{aligned}$$

squaring both sides again,

$$\begin{aligned} x &= 4x^2 \\ 0 &= 4x^2 - x \\ 0 &= x(4x-1) \\ x = 0, &\quad x = \frac{1}{4} \end{aligned}$$

Since squaring both sides can introduce extraneous solutions, need to check these values.

$$\begin{aligned} \sqrt{1 - 3(0)} &\stackrel{?}{=} 1 - 3\sqrt{0} \\ 1 &= 1 \\ \sqrt{1 - 3\left(\frac{1}{4}\right)} &\stackrel{?}{=} 1 - 3\sqrt{\frac{1}{4}} \\ \frac{1}{2} &\stackrel{?}{=} 1 - \frac{3}{2} \\ \frac{1}{2} &\neq -\frac{1}{2} \end{aligned}$$

Also check remaining value to see if is in domain of f and g functions. Since $f(0)$ and $g(0)$ are both defined, $f(g(x)) = g(f(x))$ when $x = 0$.

37.

$$f(g(x)) = f\left(\frac{x+3}{x-2}\right) = \frac{2\left(\frac{x+3}{x-2}\right) + 3}{\frac{x+3}{x-2} - 1} = x$$

$$g(f(x)) = g\left(\frac{2x+3}{x-1}\right) = \frac{\frac{2x+3}{x-1} + 3}{\frac{2x+3}{x-1} - 2} = x$$

Answer will be all real #'s for which f and g are defined. So, $f(g(x)) = g(f(x))$ for all real #'s except $x = 1$ and $x = 2$.

39. $$f(x) = 2x^2 - 3x + 1,$$
$$f(x-2) = 2(x-2)^2 - 3(x-2) + 1$$
$$= 2x^2 - 11x + 15.$$

41. $$f(x) = (x+1)^5 - 3x^2,$$
$$f(x-1) = [(x-1)+1]^5 - 3(x-1)^2$$
$$= x^5 - 3x^2 + 6x - 3.$$

43. $f(x) = \sqrt{x}$,
$f(x^2 + 3x - 1) = \sqrt{x^2 + 3x - 1}$.

45. $$f(x) = \frac{x-1}{x},$$
$$f(x+1) = \frac{(x+1)-1}{x+1}$$
$$= \frac{x}{x+1}.$$

47. $f(x) = (x-1)^2 + 2(x-1) + 3$ can be rewritten as $g(h(x))$ with $g(u) = u^2 + 2u + 3$ and $h(x) = x - 1$.

49. $f(x) = \dfrac{1}{x^2+1}$
can be rewritten as $g\ (h(x))$

with $g(u) = \dfrac{1}{u}$

and $h(x) = x^2 + 1$.

51. $f(x) = \sqrt[3]{2-x} + \dfrac{4}{2-x}$
can be rewritten as $g\ (h(x))$ with

$g(u) = \sqrt[3]{u} + \dfrac{4}{u}$

and $h(x) = 2 - x$.

53. $f(x) = -x^3 + 6x^2 + 15x$

(a) $f(2) = -8 + 6(2)^2 + 15(2) = 46$

(b) $f(1) = -1 + 6 + 15 = 20$
$f(2) - f(1) = 46 - 20 = 26$.

55. $P(t) = 20 - \dfrac{6}{t+1}$

(a) $P(9) = 20 - \dfrac{6}{9+1}$ or 19,400 people.

(b) $P(8) = 20 - \dfrac{6}{8+1}$
$P(9) - P(8) = 20 - \dfrac{3}{5} - \left(20 - \frac{2}{3}\right) = \frac{1}{15}$
This accounts for about $\frac{1}{15}$ of 1,000 people, or 67 people.

(c) $P(t)$ approaches 20, or 20,000 people.
Writing exercise–Answers will vary.

57. $$S(r) = C(R^2 - r^2)$$
$$= 1.76 \times 10^5(1.2^2 \times 10^{-4} - r^2).$$

(a) $S(0) = (1.76 \times 10^5)(1.44 \times 10^{-4}) =$ 25.344cm/sec.

(b) $S(0.6 \times 10^{-2})$

$$= 1.76 \times 10^5(1.44 \times 10^4 - 0.6^2 \times 10^{-4})$$
$$= 1.76 \times 10^5(1.08 \times 10^4)$$
$$= 19.008 \text{ cm/sec.}$$

59. $s(A) = 2.9\sqrt[3]{A}$

(a) $s(8) = 2.9\sqrt[3]{8} = 2.9 \times 2 = 5.8$
Since the number of species should be an integer, you would expect to find approximately 6 species.

(b) $s_1 = 2.9\sqrt[3]{A}$ and $s_2 = 2.9\sqrt[3]{2A}$
$s_2 = 2.9\sqrt[3]{2}\sqrt[3]{A} = \sqrt[3]{2}\left(2.9\sqrt[3]{A}\right) = \sqrt[3]{2}s_1$.

(c) $100 = 2.9\sqrt[3]{A}$

$$\frac{100}{2.9} = \sqrt[3]{A}$$
$$\left(\frac{100}{2.9}\right)^3 = \left(\sqrt[3]{A}\right)^3$$
$$\left(\frac{100}{2.9}\right)^3 = A$$

Need an area of approximately 41,002 square miles.

61. $C(x) = \dfrac{150x}{200 - x}$

(a) All real numbers except $x = 200$.

(b) All real numbers for which $0 \le x \le 100$. If $x < 0$ or $x > 200$ then $C(x) < 0$ but cost is non-negative. $x > 100$ means more than 100%.

(c) $C(50) = \dfrac{150(50)}{200 - 50} = 50$ million dollars.

(d) $C(100) = \dfrac{150(100)}{200 - 100} = 150$

$C(100) - C(50) = 100$ million dollars.

(e) $\dfrac{150x}{200 - x} = 37.5$

$$187.5x = 37.5(200),$$

$$x = \frac{7,500}{187.5} = 40\%.$$

63. **(a)** $c(p) = 0.4p + 1$ and $p(t) = 8 + 0.2t^2$

$c(t) = 0.4(8 + 0.2t^2) + 1 = 0.08t^2 + 4.2$ PPM.

(b) $c\,(2) = 0.08(2)^2 + 4.2 = 0.32 + 4.2 =$ 4.52PPM.

(c) $6.2 = 0.08t^2 + 4.2, \quad t^2 = \dfrac{2}{0.08} = 25$, or $t = 5$ years.

65. $D(x) = -0.02x + 29$; $C(x) = 1.43x^2 + 18.3x + 15.6$

(a) $R(x) = xD(x) = x(-0.02x + 29)$

$= -0.02x^2 + 29x$

$P(x) = R(x) - C(x)$

$= \left(-0.02x^2 + 29x\right) - \left(1.43x^2 + 18.3x + 15.6\right)$

$= -1.45x^2 + 10.7x - 15.6$

(b) $P(x) > 0$ when

$-1.45x^2 + 10.7x - 15.6 > 0$

Using the quadratic formula, the zeros of P are

$$x = \frac{-10.7 \pm \sqrt{(10.7)^2 - (4)(-1.45)(-15.6)}}{2(-1.45)}$$

$x = 2, 5.38$

so, $P(x) > 0$ when $2 < x < 5.38$.

67. $D(x) = -0.5x + 39$; $C(x) = 1.5x^2 + 9.2x + 67$

(a) $R(x) = xD(x) = x(-0.5x + 39)$

$= -0.5x^2 + 39x$

$P(x) = R(x) - C(x)$

$= \left(-0.5x^2 + 39x\right) - \left(1.5x^2 + 9.2x + 67\right)$

$= -2x^2 + 29.8x - 67$

(b) $P(x) > 0$ when

$-2x^2 + 29.8x - 67 > 0$

Using the quadratic formula, the zeros of P are

$$x = \frac{-29.8 \pm \sqrt{(29.8)^2 - (4)(-2)(-67)}}{2(-2)}$$

$x \approx 2.76, 12.14$

so, $P(x) > 0$ when $2.76 < x < 12.14$.

69. $Q(p) = \dfrac{4,374}{p^2}$ and

$p(t) = 0.04t^2 + 0.2t + 12$

(a) $Q(t) = \dfrac{4,374}{(0.04t^2 + 0.2t + 12)^2}$

(b) $Q(10) = \dfrac{4,374}{(4 + 2 + 12)^2} = \dfrac{4,374}{324}$

$= 13.5$ kg/week.

(c) $30.375 = \dfrac{4,374}{(0.04t^2 + 0.2t + 12)^2}$

$(0.04t^2 + 0.2t + 12)^2 = \dfrac{4,374}{30.375} = 144 = 12^2$

So $0.04t^2 + 0.2t + 12 = \pm 12$.

The positive root leads to $t(0.04t + 0.2) = 0$ or $t = 0$. (Disregard $t < 0$.) The negative root produces imaginary numbers. $t = 0$ now.

71. To find the domain of $f(x) = \dfrac{4x^2 - 3}{2x^2 + x - 3}$, Press [y =].

Enter $(4x \wedge 2 - 3) \div (2x \wedge 2 + x - 3)$ for $y_1 =$

Press [graph].

For a better view of the vertical asymptotes, press [zoom] and enter Zoom In. Use arrow buttons to move cross-hair to the left-most vertical asymptote. When it appears cross-hair is on the line, zoom in again for a more accurate reading. Move cross-hair again to be on the line. It appears that $x = -1.5$ is not in the domain of f. Zoom out once to move cross-hair

to the rightmost vertical asymptote and repeat the procedure of zoom in to find that $x = 1$ is not the domain of f.
The domain consists of all values except $x = -1.5$ and $x = 1$.

73. For $f(x) = 2\sqrt{x-1}$ and $g(x) = x^3 - 1.2$, to find $f\ (g(2.3))$, we must find $g\ (2.3)$ first and then input that answer into f. Press [y=].
Input $2\sqrt{(x-1)}$ for $y_1 =$ and press [enter].
Input $x \wedge 3 - 1.2$ for $y_2 =$.
Use the window dimensions $[-15, 15]1$ by $[-10, 10]1$. Use the value function under the calc menu, input 2.3, and press [enter].
Use ↑ and ↓ arrows to be sure that $y_2 = x \wedge 3 - 1.2$ is displayed in the upper left corner. The lower right corner display should read $y = 10.967$
Use the value function again and input 10.967. Verify $y_1 = 2\sqrt{(x-1)}$ is displayed in the upper left corner.
The answer of $y = 6.31$ is displayed in lower right corner.

75. **(a)** VE1

Level of Education	Year 1991	1992
No H.S. diploma	$\frac{16,582}{16,582} = 1$	1
H.S. diploma	$\frac{24,007}{16,582} = 1.45$	$\frac{23,908}{16,344} = 1.46$
Some college	$\frac{27,017}{16,582} = 1.63$	$\frac{26,626}{16,344} = 1.63$
Bachelor's degree	$\frac{41,178}{16,582} = 2.48$	$\frac{41,634}{16,344} = 2.55$
Advanced degree	$\frac{60,525}{16,582} = 3.65$	$\frac{62,080}{16,344} = 3.80$

1993	1994	1995
1	1	1
$\frac{24,072}{15,889} = 1.52$	$\frac{24,458}{16,545} = 1.48$	$\frac{25,180}{16,465} = 1.53$
$\frac{26,696}{15,889} = 1.68$	$\frac{26,847}{16,545} = 1.62$	$\frac{28,037}{16,465} = 1.70$
$\frac{43,529}{15,889} = 2.74$	$\frac{44,963}{16,545} = 2.72$	$\frac{43,450}{16,465} = 2.64$
$\frac{69,145}{15,889} = 4.35$	$\frac{67,770}{16,545} = 4.10$	$\frac{66,581}{16,465} = 4.04$

1996	1997	1998
1	1	1
$\frac{25,289}{17,135} = 1.48$	$\frac{25,537}{17,985} = 1.42$	$\frac{25,937}{17,647} = 1.47$
$\frac{28,744}{17,135} = 1.68$	$\frac{29,263}{17,985} = 1.63$	$\frac{30,304}{17,647} = 1.72$
$\frac{43,505}{17,135} = 2.54$	$\frac{45,150}{17,985} = 2.51$	$\frac{48,131}{17,647} = 2.73$
$\frac{69,993}{17,135} = 4.08$	$\frac{70,527}{17,985} = 3.92$	$\frac{69,777}{17,647} = 3.95$

1999	2000
1	1
$\frac{26,439}{17,346} = 1.52$	$\frac{27,097}{18,727} = 1.45$
$\frac{30,561}{17,346} = 1.76$	$\frac{31,212}{18,727} = 1.67$
$\frac{49,149}{17,346} = 2.83$	$\frac{51,653}{18,727} = 2.76$
$\frac{72,841}{17,346} = 4.20$	$\frac{72,175}{18,727} = 3.85$

(b) 1.45, 1.67, 2.76, 3.85.
Writing exercise—Answers will vary.

1.2 The Graph of a Function

1. **(a)** Since of form x^n, where n is non-integer real number, is a power function.

(b) Since of form $a_n x^n + a_{n-1}x^{n-1} + \cdots + a_1 x + a_0$, where n is nonnegative integer, is polynomial function.

(c) Since can multiply out and simplify to form $a_{nx}x^n + a_{n-1}x^{n-1} + \cdots + a_1 x + a_0$, is polynomial function.

(d) Since is quotient of two polynomial functions is a rational function.

3. $f(x) = x$
A function of the form

$$y = f(x) = ax + b$$

is a linear function, and its graph is a line. Two points are sufficient to draw that line. The x-intercept is 0, as is the y-intercept, and $f(1) = 1$.

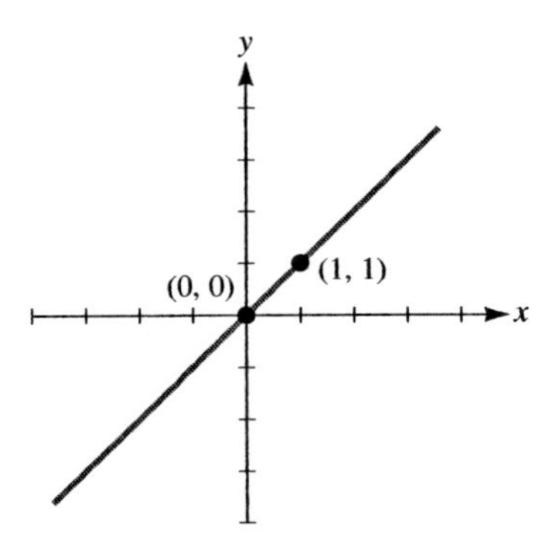

5. $f(x) = x^3$
Note that if $x > 0$ then $f(x) > 0$ and if $x < 0$, then $f(x) < 0$. This means that the curve will only appear in the first and third quadrants. Since x^3 and $(-x)^3$ have the same absolute value, only their signs are opposites, the curve will be symmetric with respect to (wrt) the origin. The x-intercept is 0, as is the y-intercept.

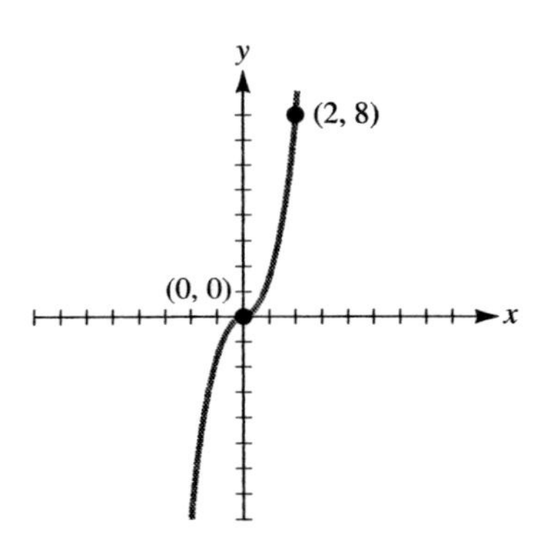

7. $f(x) = 2x - 1$
A function of the form $y = f(x) = ax + b$ is a linear function, and its graph is a line. Two points are sufficient to draw that line. The x-intercept is $\frac{1}{2}$ and the y-intercept is -1.

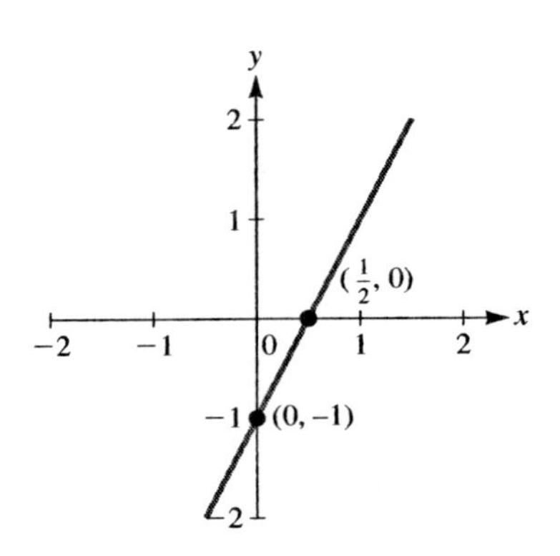

9. Since function is of form $y = Ax^2 + Bx + C$ (where $C = 0$), the graph is a parabola; its vertex is $\left(-\frac{5}{4}, -\frac{25}{8}\right)$, it opens up (A is positive), and its intercepts are (0,0) and $\left(-\frac{5}{2}, 0\right)$.

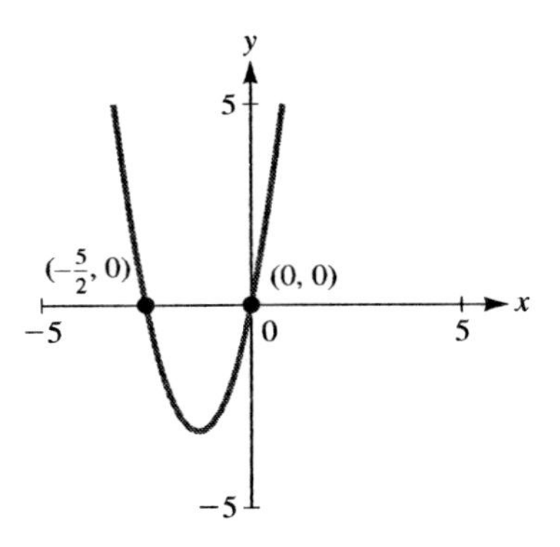

11. Since function is of form $y = Ax^2 + Bx + C$, the graph is a parabola which opens down (A is negative) and its vertex is $(-1, 16)$. Further,

$$\begin{aligned} f(x) &= -x^2 - 2x + 15 \\ &= -(x^2 + 2x - 15) \\ &= -(x + 5)(x - 3). \end{aligned}$$

So the x-intercepts are $(-5, 0)$ and $(3, 0)$, and the y-intercept is $(0, 15)$.

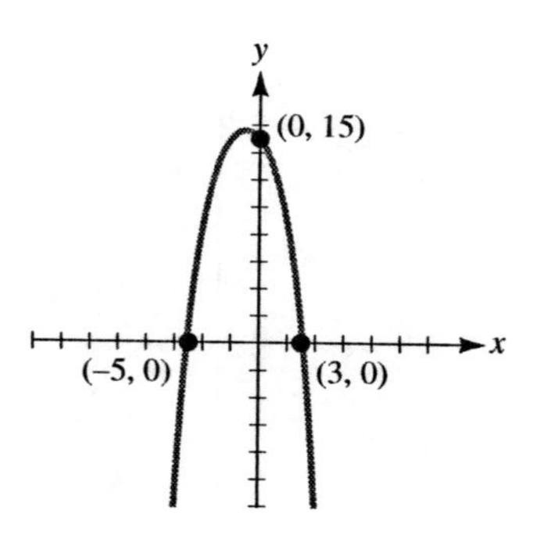

13. $f(x) = \sqrt{x}$
This represents one branch of the graph of $y^2 = x$.
(0,0) is only intercept,$x \geq 0$ and $y \geq 0$.

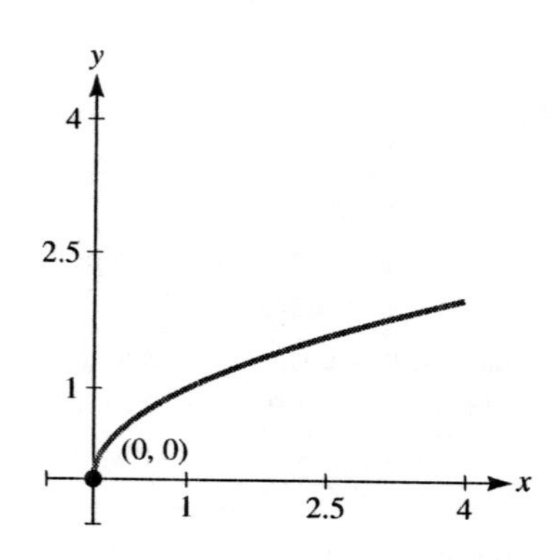

15.

$$f(x) = \begin{cases} x - 1 & \text{if } x \leq 0 \\ x + 1 & \text{if } x > 0 \end{cases}$$

Note that the graph consists of two half lines on either side of $x = 0$. There is no x-intercept for either half line. The half line $y = x - 1$ has a y-intercept of -1, while the half line $y = x + 1$ has no y-intercept.

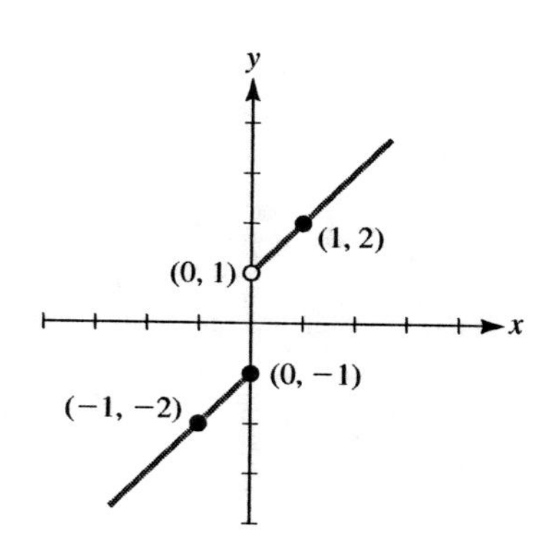

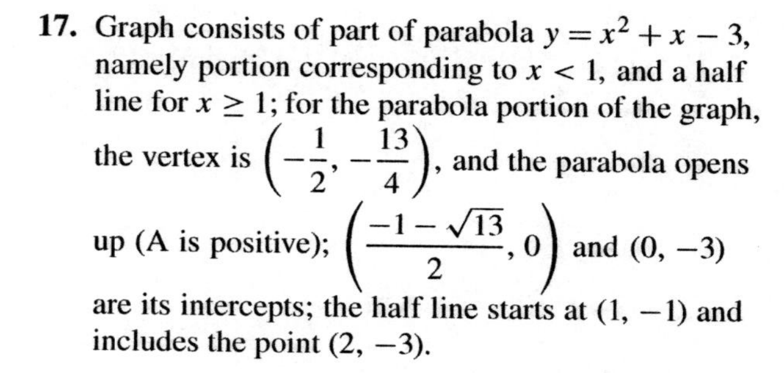

17. Graph consists of part of parabola $y = x^2 + x - 3$, namely portion corresponding to $x < 1$, and a half line for $x \geq 1$; for the parabola portion of the graph, the vertex is $\left(-\frac{1}{2}, -\frac{13}{4}\right)$, and the parabola opens up (A is positive); $\left(\frac{-1-\sqrt{13}}{2}, 0\right)$ and $(0, -3)$ are its intercepts; the half line starts at $(1, -1)$ and includes the point $(2, -3)$.

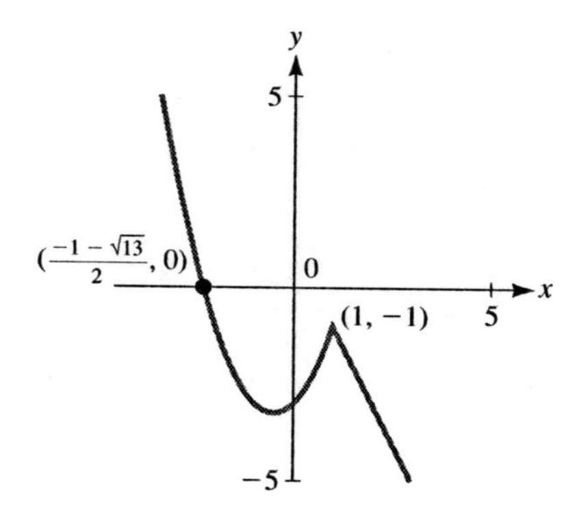

19. $y = 3x + 5$ and $y = -x + 3$
Add 3 times the second equation to the first. Then $4y = 14$ or $y = \frac{7}{2}$. Substitute in the first, then $x = 3 - y = -\frac{1}{2}$. The point of intersection is $P\left(-\frac{1}{2}, \frac{7}{2}\right)$.

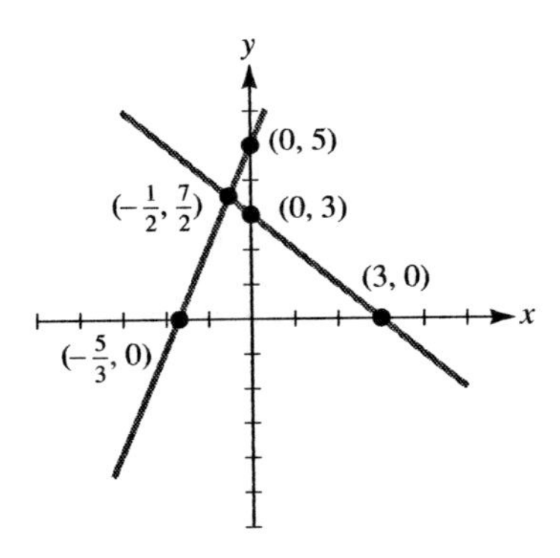

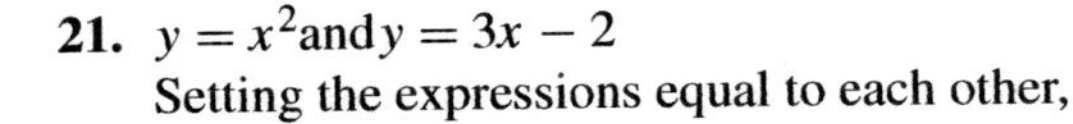

21. $y = x^2$ and $y = 3x - 2$
Setting the expressions equal to each other,

$$x^2 = 3x - 2$$
$$x^2 - 3x + 2 = 0$$
$$(x - 1)(x - 2) = 0$$
$$x = 1, 2$$

So points of intersection are $P_1(1, 1)$ and $P_2(2, 4)$.

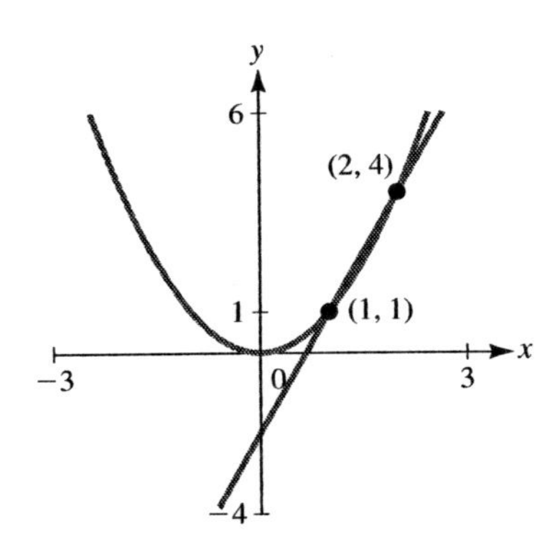

23. $3y - 2x = 5$ and $y + 3x = 9$.
Multiply the second equation by -3 and add it to the first one. Then,

$$-2x - 9x = 5 - 27,$$
$$x = 2, \quad y = 9 - 3(2) = 3.$$

The point of intersection is $P(2, 3)$.

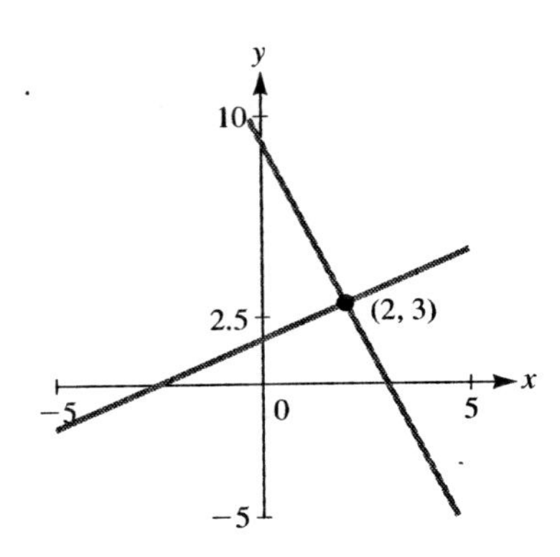

25. **(a)** Crosses y-axis at $y = -1$, y-intercept is $(0, -1)$.
(b) Crosses x-axis at $x = 1$, x-intercept is $(1, 0)$.
(c) Largest value of f is 3 and occurs at $x = 4$ (highest point on graph).
(d) Smallest value of f is -3 and occurs at $x = -2$ (lowest point on graph).

27. **(a)** Crosses y-axis at $y = 2$, y-intercept is $(0, 2)$.
(b) Crosses x-axis at $x = -1$ and 3.5; x-intercepts are $(-1, 0)$ and $(3.5, 0)$.
(c) Largest value of f is 3 and occurs at $x = 2$ (highest point on graph).
(d) Smallest value of f is -3 and occurs at $x = 4$ (lowest point on graph).

29. The monthly profit is

$$\begin{aligned} P(p) &= (\text{number of recorders sold}) \\ &\quad (\text{price} - \text{cost}) \\ &= (120 - p)(p - 40) \end{aligned}$$

So, the intercepts are (40, 0), (120, 0), and (0, −4800). The graph suggests a maximum profit when $p \approx 80$, that is, when 80 recorders are sold.

$$P(80) = (120 - 80)(80 - 40) = 1600$$

So estimated max profit is \$1600.

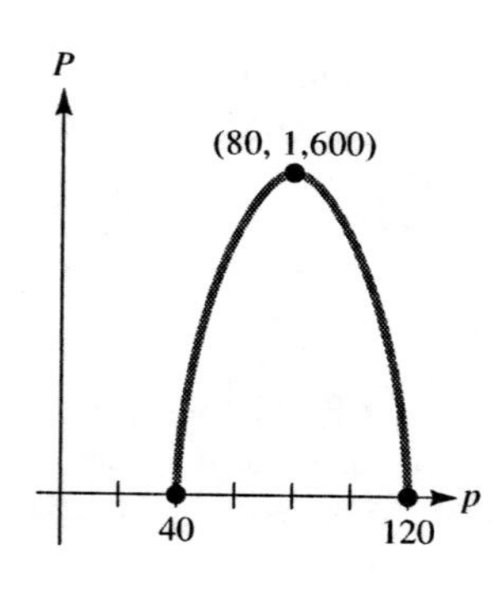

31. (a)

$$E(p) = (\text{price per unit})(\text{demand})$$
$$= -200p(p - 60)$$

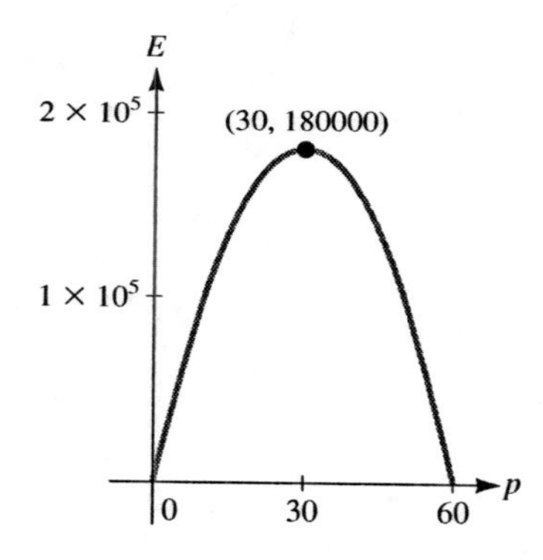

(b) The p intercepts represent prices at which consumers do not buy commodity.

(c) The graph suggests a maximum expenditure when $p \approx 30$.

$$E(30) = -200(30)(30 - 60) = 180,000$$

So estimated max expenditure is $180,000.

33. (a) profit = revenue − cost

$$= (\#\text{sold})(\text{sellingprice}) - \text{cost}$$
$$P(x) = x(-0.05x + 38) - (0.02x^2 + 3x + 574.77)$$
$$= -0.07x^2 + 35x - 574.77 \text{ hundred dollars}$$

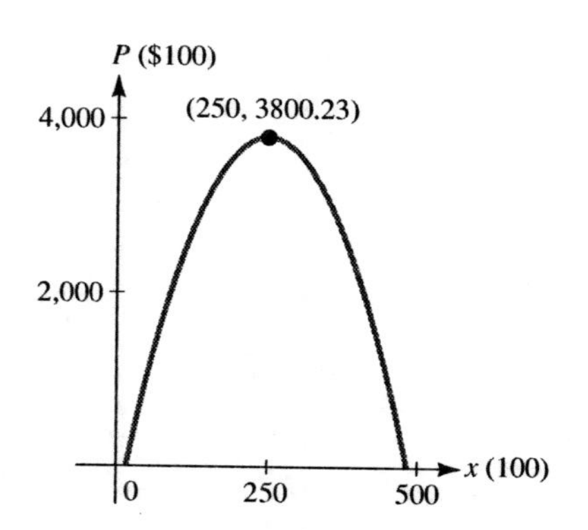

(b) The graph suggests a maximum profit when $x = 250$, that is, when 25,000 units are purchased. Note that the max profit is $P(250) = -0.07(250)^2 + 35(20) - 574.77 \approx 3800.23$ hundred, or $380,023. For the unit price,

$$p = -0.05(250) + 38 = \$25.50$$

35. $D(v) = 0.065v^2 + 0.148v$
For practical domain, graph is part of parabola corresponding to $v \geq 0$.

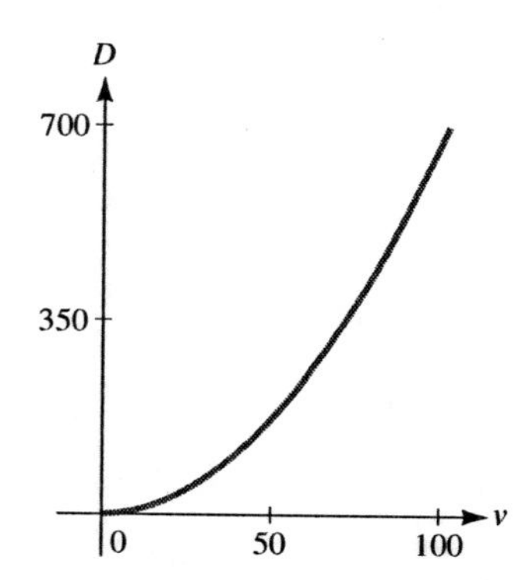

37. (a) revenue = (#apts) (rent per apt)
Since $\dfrac{p - 1200}{100}$ represents the number of $100 increases,

$$150 - 5\left(\frac{p - 1200}{100}\right) = 210 - 0.05p$$

represents the number of apartments that will be leased. So,

$$R(p) = 210p - 0.05p^2$$

(b)

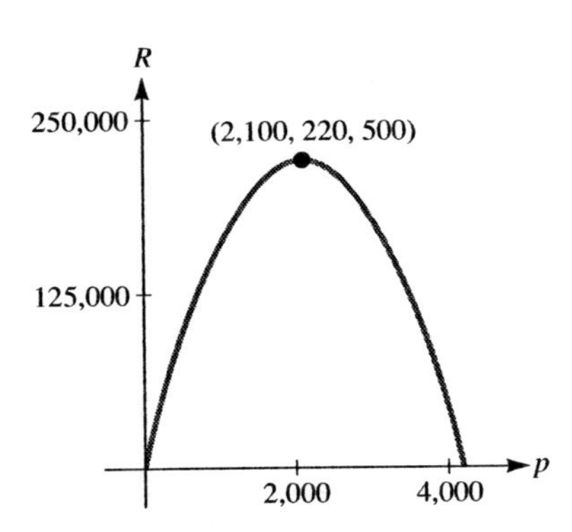

(c) The graph suggests a maximum profit when $p = 2100$; that is, when the rental price is \$2,100. The max profit is $R(2100) = 210(2100) - 0.05(2100)^2 \approx \$220{,}500$.

39. $N(t) = -35t^2 + 299t + 3{,}347$

(a)

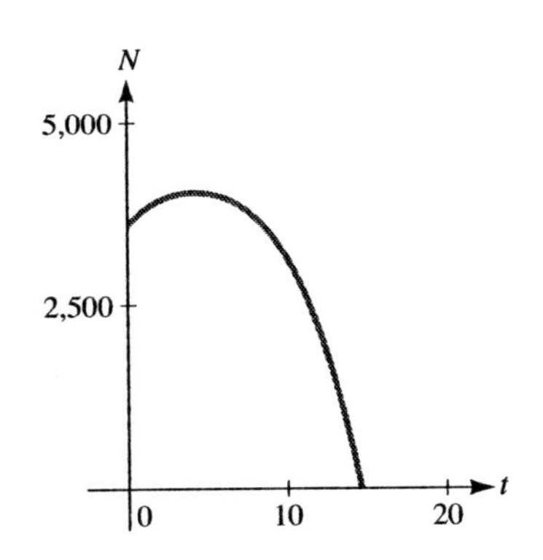

(b) Since the year 1995 is represented by $t = 5$, the amount predicted was $N(5) = -35(5)^2 + 299(5) + 3{,}347 = 3{,}967$ thousand tons.

(c) Based on the formula, the maximum lead emission would occur at the vertex, or when

$$t = -\frac{299}{2(-35)} \approx 4.27 \text{years}$$

This would be during March of the year 1994.

(d) No. From the graph, $N(t) < 0$ when $t \approx 15$, or during the year 2005.

41. The graph is a function because no vertical line intersects the graph more than once.

43. The graph is not a function because there are vertical lines intersecting the graph at more than one point; for example, the y-axis.

45. $f(x) = -9x^2 + 3600x - 358, 200$
Answers will vary, but one viewing window has the following dimensions: [180, 200] 10 by [−500, 1850] 500.

47. **(a)** The graph of $y = x^2 + 3$ is graph of $y = x^2$ translated up 3 units.

(b)

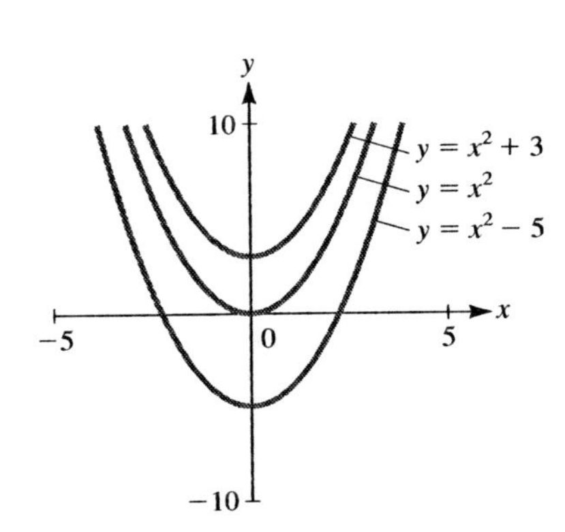

(c) When $c > 0$, the graph of g is the graph of f translated up c units. When $c < 0$, the graph is translated down $|c|$ units.

49. **(a)** The graph of $y = (x - 2)^2$ is the graph of $y = x^2$ translated two units to the right.

(b)

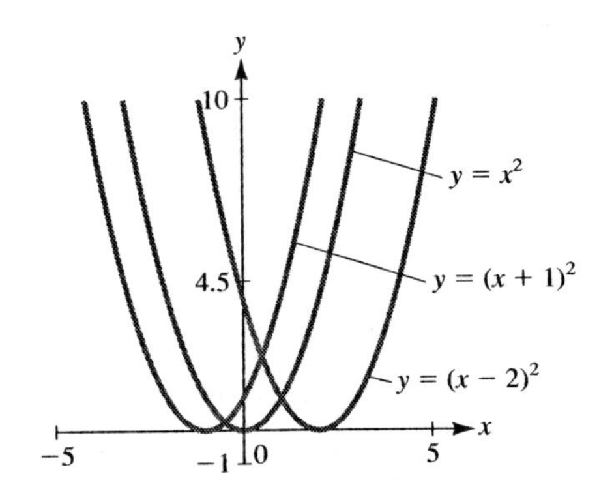

(c) When $c > 0$, the graph of g is the graph of f translated c units to the right. When $c < 0$, the graph is translated $|c|$ units to the left.

51. (a)

Days of Training	Mowers per Day
2	6
3	7.23
5	8.15
10	8.69
50	8.96

(b) The number of mowers per day approaches 9.

(c) To graph $N(t) = \dfrac{45t^2}{5t^2 + t + 8}$, press [y =]
Input $(45x \wedge 2) \div (5x \wedge 2 + x + 8)$ for $Y_1 =$.
Use window dimensions [−10, 10]1 by [−10, 10]1 (z standard).
Press [graph].

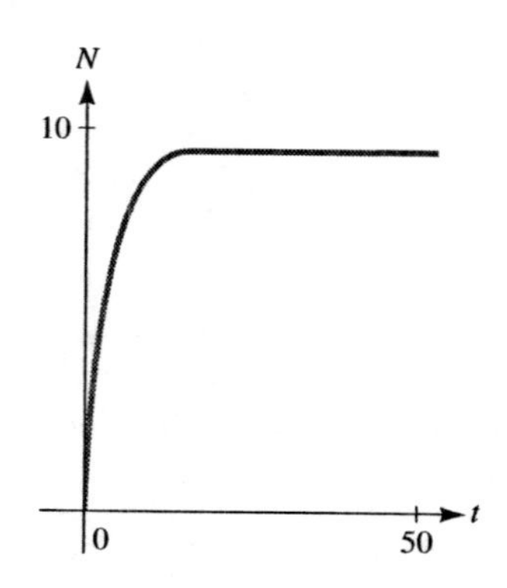

53. To graph $f(x) = \dfrac{-9x^2 - 3x - 4}{4x^2 + x - 1}$,
Press [y =]
Input $(-9x \wedge 2 - 3x - 4) \div (4x \wedge 2 + x - 1)$ for $y_1 =$
Press [graph]
Use the Zoom in function under the Zoom menu to find the vertical asymptotes to be $x_1 \approx -0.65$and$x_2 \approx 0.39$. The function f is defined for all real x except $x_3 \approx -0.65$and$x \approx 0.39$.

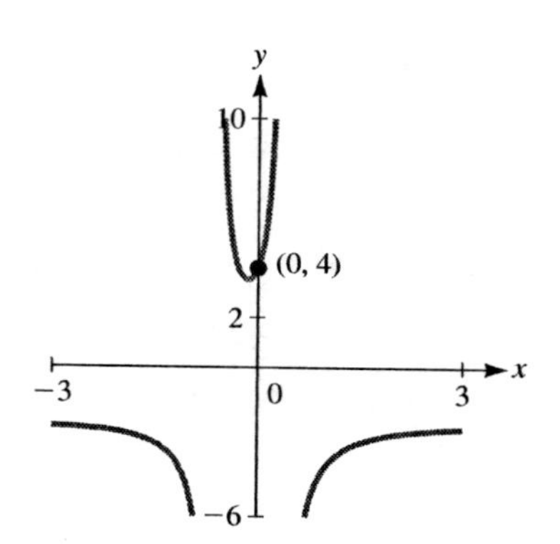

55. To graph $g(x) = -3x^3 + 7x + 4$ and find x-intercepts,
Press [y =]
Input $-3x \wedge 3 + 7x + 4$for$y_1 =$
Press [graph]
Press [trace]
Use left arrow to move cursor to the left most x-intercept. When the cursor appears to be at the x-intercept, use the Zoom In feature under the Zoom menu twice. It can be seen that there are two x-intercepts in close proximity to each other. These x-intercepts appear to be $x_1 \approx -1$ and $x_2 \approx -0.76$. To estimate the third x-intercept, use the z-standard function under the Zoom menu to view the original graph. Use right arrow and zoom in to estimate the third x-intercept to be $x_3 \approx 1.8$.

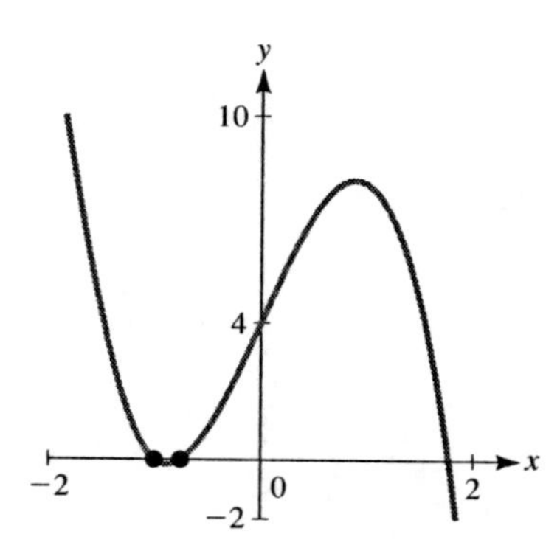

57. From problem #46,

$$d = \sqrt{(x_2 - x_1)^2 + (y_2 - y_1)^2}$$

Let (a, b), the center of circle, correspond to the point (x_2, y_2); then,

$$d = \sqrt{(x-a)^2 + (y-b)^2}$$

Squaring both sides,

$$d^2 = (x-a)^2 + (y-b)^2$$

Since (a, b) is the center of and (x, y) is a point on the circle, the distance is the radius of the circle and

$$R^2 = (x-a)^2 + (y-b)^2$$

1.3 Linear Functions

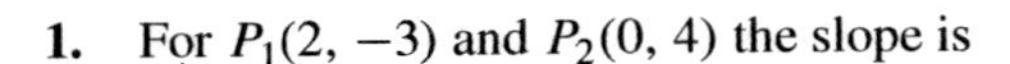

1. For $P_1(2, -3)$ and $P_2(0, 4)$ the slope is

$$m = \frac{4-(-3)}{0-2} = -\frac{7}{2}$$

3. For $P_1(2, 0)$ and $P_2(0, 2)$ the slope is

$$m = \frac{2-0}{0-2} = -1$$

5. For $P_1(2, 6)$ and $P_2(2, -4)$ the slope is

$$m = \frac{6-(-4)}{2-2},$$

which is undefined, since the denominator is 0. The line through the given points is vertical.

7. The line has slope $= 2$ and an intercept of $(0, 0)$. So, the equation of line is $y = 2x + 0$, or $y = 2x$.

9. The line has slope $\frac{2\frac{1}{2}}{4} = \frac{\frac{5}{2}}{4} = \frac{5}{8}$. The x-intercept is $(-4, 0)$ and the y-intercept is $\left(0, \frac{5}{2}\right)$. The equation of line is $y = \frac{5}{8}x + \frac{5}{2}$.

11. The line $x = 3$ is a vertical line that includes all points of the form $(3, y)$. Therefore, the x-intercept is $(3, 0)$ and there is no y-intercept. The slope of the line is undefined, since $x_2 - x_1 = 3 - 3 = 0$.

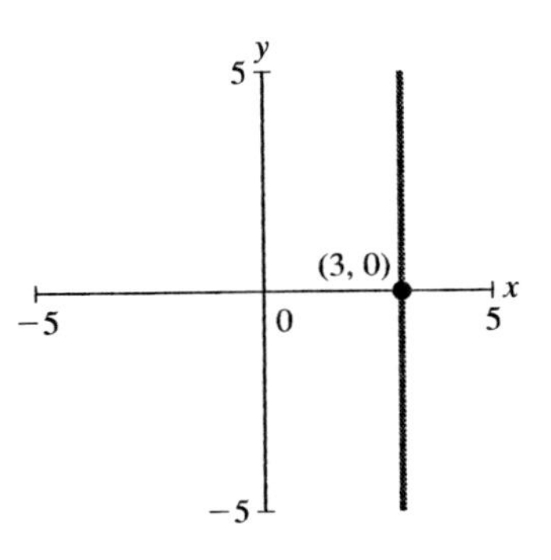

13. $y = 3x$
$m = 3$, y-intercept $b = 0$, and the x-intercept is 0.

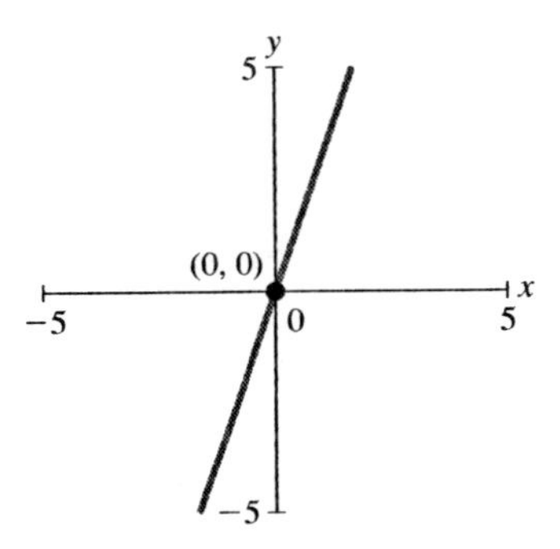

15. $3x + 2y = 6$ or $y = -\frac{3}{2}x + 3$

$$m = -\frac{3}{2},$$

y-intercept $b = 3$, and the x-intercept is 2.

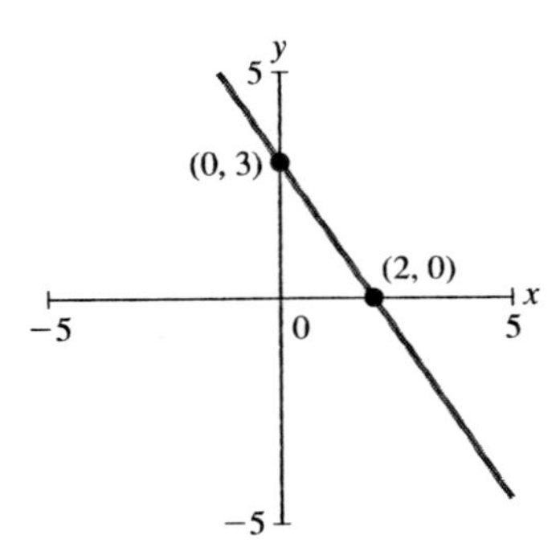

17. $\frac{x}{2} + \frac{y}{5} = 1$ or $y = -\frac{5}{2}x + 5$
$m = -\frac{5}{2}$, y-intercept $b = 5$, and the x-intercept is 2.

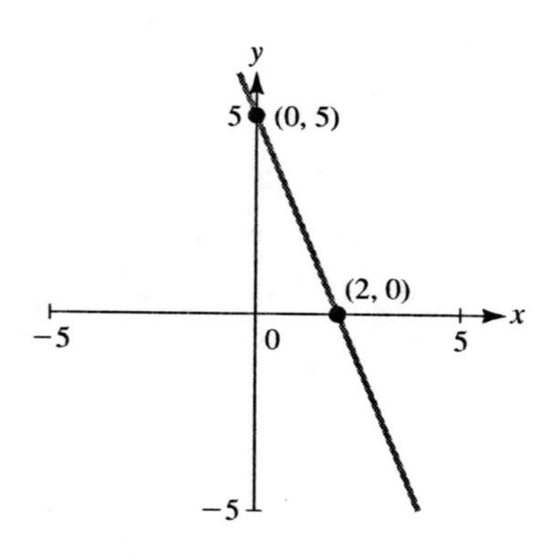

19. $m = 1$ and $P(2, 0)$, so

$$y - 0 = (1)(x - 2), \quad \text{or} \quad y = x - 2$$

21. $m = -\frac{1}{2}$ and $P(5, -2)$, so

$$y - (-2) = -\frac{1}{2}(x - 5), \quad \text{or} \quad y = -\frac{1}{2}x + \frac{1}{2}$$

23. Since the line is parallel to the x-axis, it is horizontal and its slope is 0. For $P(2, 5)$, the line is

$$y - 5 = 0(x - 2), \quad \text{or} \quad y = 5$$

25. $m = \dfrac{1-0}{0-1}$
and for $P(1, 0)$ the equation of the line is

$$y - 0 = -1(x - 1) \quad \text{or} \quad y = -x + 1$$

The equation would be the same if the point (0, 1) had been used.

27.

$$m = \frac{1 - \left(\frac{1}{4}\right)}{-\left(\frac{1}{5}\right) - \left(\frac{2}{3}\right)} = -\frac{45}{52}$$

For $P\left(-\frac{1}{5}, 1\right)$, the line is $y - 1 = -\frac{45}{52}(x + \frac{1}{5})$, or $y = -\frac{45}{52}x + \frac{43}{52}$.

29. The slope is 0 because the y-values are identical. So, $y = 5$.

31. The given line $2x + y = 3$, or$y = -2x + 3$, has a slope of -2. Since parallel lines have the same slope, $m = -2$ for the desired line. Given that the point (4, 1) is on the line, $y - 1 = -2(x - 4)$, or $y = -2x + 9$.

33. The given line $x + y = 4$, or $y = -x + 4$, has a slope of -1. A perpendicular line has slope $m = -\frac{1}{-1} = 1$. Given that the point (3, 5) is on the line, $y - 5 = 1(x - 3)$, or $y = x + 2$.

35. Let x be the number of units manufactured. Then $60x$ is the cost of producing x units, to which the fixed cost must be added.

$$y = 60x + 5,000$$

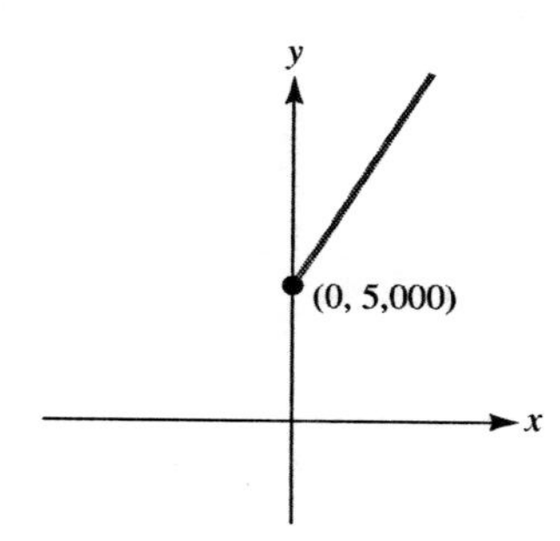

37. **(a)** Let x be the number of hours spent registering students in person. During the first 4 hours $(4)(35) = 140$ students were registered. So,

$$360 - 140 = 220$$

students had pre-registered. Let y be the total number of students who register. Then,

$$y = 35x + 220$$

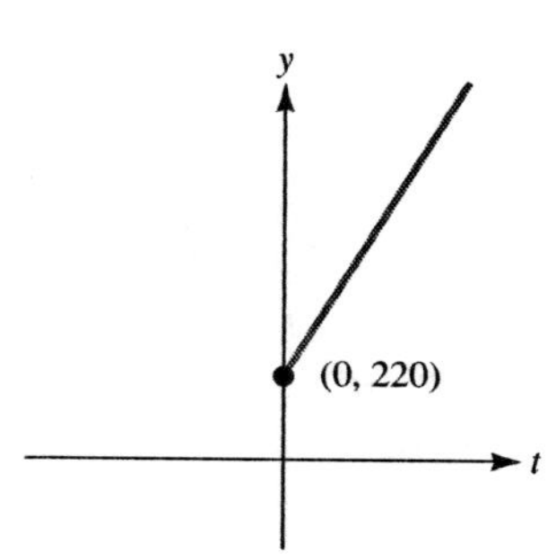

(b) $$y = (3)(35) + 220 = 325$$

(c) From part **(a)**, we see that 220 students had pre-registered.

39. The slope is

$$m = \frac{1,500 - 0}{0 - 10} = -150$$

Originally (when time $x = 0$), the value y of the books is 1500 (this is the y intercept.)

$$y = -150x + 1,500$$

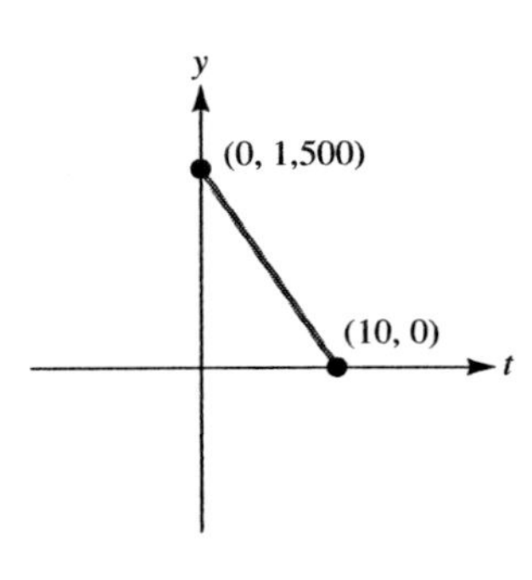

41. **(a)** Let x be the number of days. The slope is

$$m = \frac{200 - 164}{12 - 21} = -4$$

For $P(12, 200)$,
$y - 200 = -4(x - 12)$, or $y = -4x + 248$.

(b) $y = 248 - (4)(8) = 216$ million gallons.

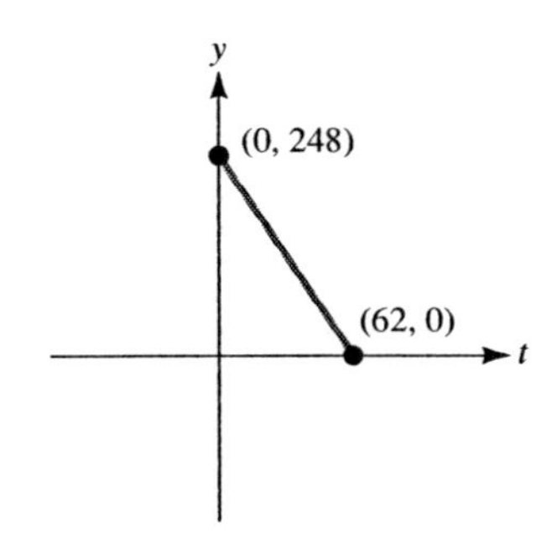

43. Let the x-axis represent time in months and the y-axis represent price per share.

(a)

(b)

(c)

45. **(a)** $H(7) = 6.5(7) + 50 = 95.5$ cm tall.

(b) $150 = 6.5A + 50$, $A = 15.4$ years old

(c) $H(0) = 6.5(0) + 50 = 50$ cm tall. This height ≈ 19.7 inches) seems reasonable.

(d) $H(20) = 6.5(20) + 50 = 180$ cm tall. This height ≈ 5.9 feet) seems reasonable.

47. (a) Let C be the temperature in degrees Celsius and F the temperature in degrees Farenheit. The slope is

$$m = \frac{212 - 32}{100 - 0} = \frac{9}{5}$$

So, $\frac{F - 32}{C - 0} = \frac{9}{5}$, or $F = \frac{9}{5}C + 32$

(b)

$$F = \frac{9}{5}(15) + 32$$
$$= 59 \text{ degrees}$$

(c)

$$68 = \frac{9}{5}C + 32,$$
$$36 = \frac{9}{5}C,$$
$$C = 20 \text{ degrees}$$

(d) Solving $C = \frac{9}{5}C+32$, $C = -40$. So, the temperature $-40°$ C is also $-40°$ F.

49. (a) The original value of the book is \$100 and the value doubles every 10 years. At the end of 30 years, in 1930, the book was worth \$800. At the end of 90 years, in 1990, the book was worth \$51,200. At the end of 100 years, in 2000, the book will be worth \$102,400.

(b) The value of the book is *not* a linear function.

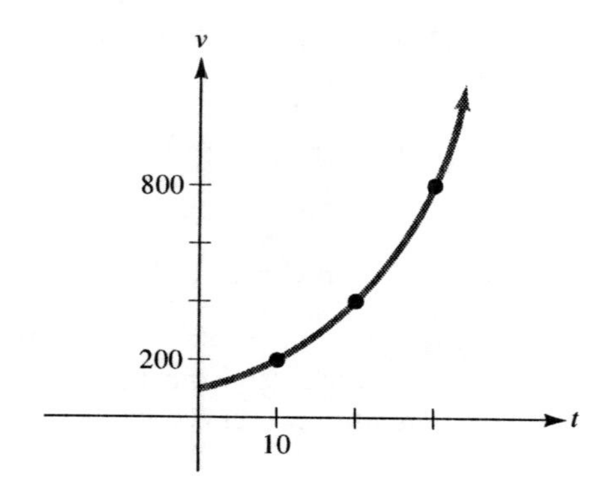

51. (a) Let t represent years after 1995. Using the points (0,575) and (5,545), the slope is $m = \frac{545 - 575}{5 - 0} = -6$. If S represents the average SAT score, $S(t) = -6t + 575$.

(b) $S(10) = -6(10) + 575 = 515$.

(c) $527 = -6t + 575$, $t = 8$, and the year would be 2003.

53. (a) Using the points $(0, V)$ and (N, S), the slope of the line is $\frac{S - V}{N}$. So, the value of an asset after t years is $B(t) = \frac{S - V}{N}t + V$.

(b) For this equipment, $B(t) = -6,400t + 50,000$. So, $B(3) = -6,400(3) + 50,000 = 30,800$. Value after three years is \$30,800.

55. To graph $y = \frac{25}{7}x + \frac{13}{2}$ and $y = \frac{144}{45}x + \frac{630}{229}$ on the same set of axes, Press [y =].
Input $\frac{(25x)}{7} + \frac{13}{2}$ for $y_1 =$ and press [enter].
Input $\frac{(144x)}{45} + \frac{630}{229}$ for $y_2 =$.
Use the window dimensions [0, 4] 0.5 by [0, 14] 2
Press [graph].
It does not appear that the lines are parallel.
To verify this, press [2ND] [quit].
Input $\frac{25}{7} - \frac{144}{45}$ and [enter].
If the lines were parallel the difference in their slopes would equal zero (the slopes would be the same). The difference of these slopes is 0.37 and therefore, the lines are not parallel.

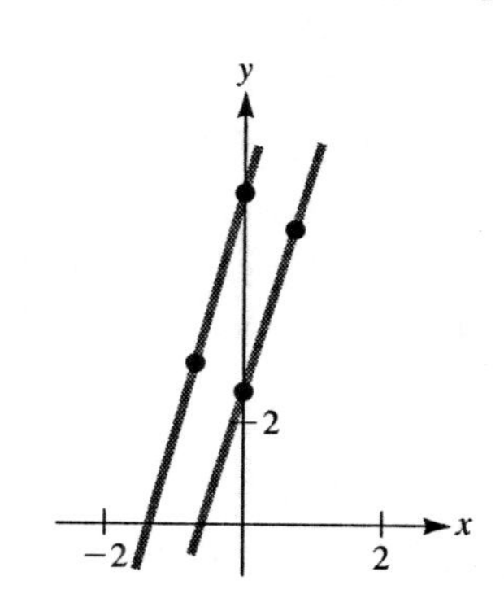

57. A rental company rents a piece of equipment for a \$60.00 flat fee plus an hourly fee of \$5.00 per hour.

(a) Let $y =$ cost of renting the equipment and $t =$ number of hours.

t	2	5	10	t
$y(t)$	70	85	110	$60 + 5t$

(b) $y(t) = 5t + 60$, $t \geq 0$

(c) Press [y=].
Input $5x + 60$ for $y_1 =$.
Use dimensions $[-10, 10]$ 1 by $[-10, 100]$ 10
Press [graph].

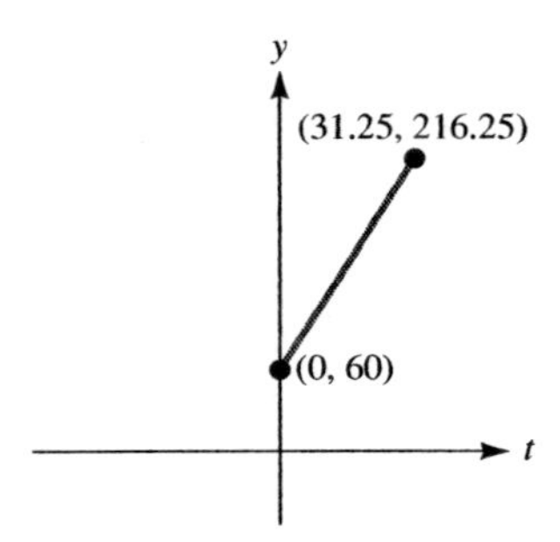

(d) To answer part (d), it may be easiest to use window dimensions $[30, 33]$ 5 by $[200, 230]$ 5.
Press [graph].
Press [trace] and move cross-hairs to be as close to $y = 216.25$ as possible.
When $y = 216.2234$, the x-coordinate is 31.24.
It takes approximately 31.24 hours for the rental charge to be \$216.25. Using algebra, we see it takes exactly 31.25 for the charge to be \$216.25.

59. The slope of -0.389 means the unemployment rate drops by approximately 0.389% from year to year. Writing exercise—Answers will vary.

61. The slope of L_1 is $m_1 = \dfrac{b}{a}$ and that of L_2 is $m_2 = \dfrac{c}{a}$. By hypothesis, $L_1 \perp L_2$.

$$OA = \sqrt{a^2 + b^2} \text{ and}$$
$$OB = \sqrt{a^2 + c^2}$$

Since $AB = b - c$ and by the Pythagorean theorem,

$$(a^2 + b^2) + (a^2 + c^2) = (b - c)^2$$
$$2a^2 + b^2 + c^2 = b^2 - 2bc + c^2$$

from which $2a^2 = -2bc$

$$-1 = \frac{bc}{a^2}$$
$$-1 = \left(\frac{b}{a}\right)\left(\frac{c}{a}\right) = m_1 m_2$$

or $\quad m_1 = -\dfrac{1}{m_2}$.

1.4 Functional Models

1. This problem has two possible forms of the solution. Assume the stream is along the length, say l. Then w is the width and

$$l + 2w = 1,000 \text{ or } l = 1,000 - 2w$$

The area is

$$A = lw = 2w(500 - w) \text{ squarefeet}$$

3. Let x and y be the smaller and larger numbers, respectively. Then

$$x + y = 18 \text{ or } y = 18 - x$$

The product is $P = xy = x(18 - x)$.

5. Revenue = (number of units sold) (price per unit)

$$R = x(35x + 15)$$

7. Let x be the length and y the width of the rectangle. Then

$$2x + 2y = 320 \quad \text{or} \quad y = 160 - x$$

The area is (length)(width) or

$$A(x) = x(160 - x)$$

The length is estimated to be 80 meters from the graph below, which also happens to be the width. So the maximum area seems to correspond to that of a square.

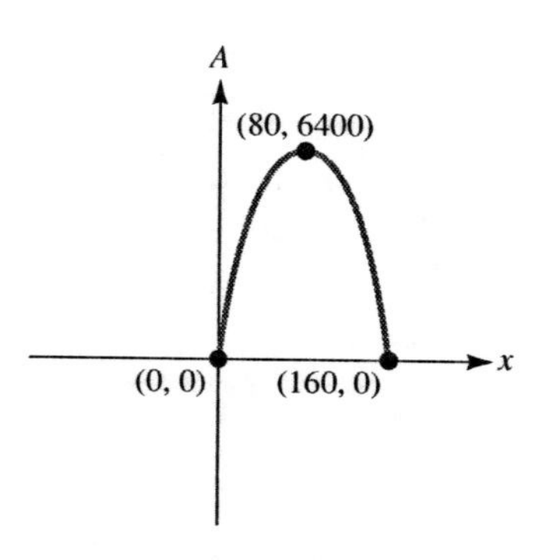

9. Let x be the length of the square base and y the height of the box. The surface area is $2x^2 + 4xy = 4{,}000$ So $y = \dfrac{2{,}000 - x^2}{2x}$ and the volume is

$$V = x^2 y = x\left(1{,}000 - \frac{x^2}{2}\right)$$

11. Let r be the radius and h the height of the cylinder. The surface area of the closed cylinder is

$$S = 120\pi = 2\pi r^2 + 2\pi rh \quad \text{or} \quad h = \frac{60 - r^2}{r}$$

So $V(r) = \pi r^2 h = \pi r(60 - r^2)$

13. Let r be the radius and h the height of the cylinder. Since the volume is

$$V = \pi r^2 h = 4\pi, \quad \text{or} \quad h = \frac{4}{r^2}$$

The cost of the top or bottom is

$$C_t = C_b = 2(0.02)\pi r^2,$$

while the cost of the side is

$$2\pi rh(0.02) = \frac{0.16\pi}{r}$$

The total cost is

$$C(x) = 0.08\pi r^2 + \frac{0.16\pi}{r}$$

15. Let R denote the rate of population growth and p the population size. Since R is directly proportional to p,

$$R(p) = kp,$$

where k is the constant of proportionality.

17. Let R denote the rate at which temperature changes, M the temperature of the medium, and T the temperature of the object. Then $T - M$ is the difference in the temperature between the object and the medium. Since the rate of change is directly proportional to the difference,

$$R(T) = k(T - M),$$

where k is the constant of proportionality.

19. Let R denote the rate at which people are implicated, x the number of people implicated, and n the total number of people involved. Then $n - x$ is the number of people involved but not implicated. Since the rate of change is jointly proportional to those implicated and those not implicated,

$$R(x) = kx(n - x),$$

where k is the constant of proportionality.

21. Let s be the speed of the truck.
The cost due to wages is $\dfrac{k_1}{s}$,
where k_1 is a constant of proportionality, and the cost due to gasoline is $k_2 s$, where k_2 is another constant of proportionality.
If $C(s)$ is the total cost,

$$C(s) = \frac{k_1}{s} + k_2 s$$

23.

$$C = \left(\frac{N+1}{24}\right)(300) = \left(\frac{N+1}{2}\right)(25)$$

$$C = \frac{2N \cdot 300}{25} = 24N$$

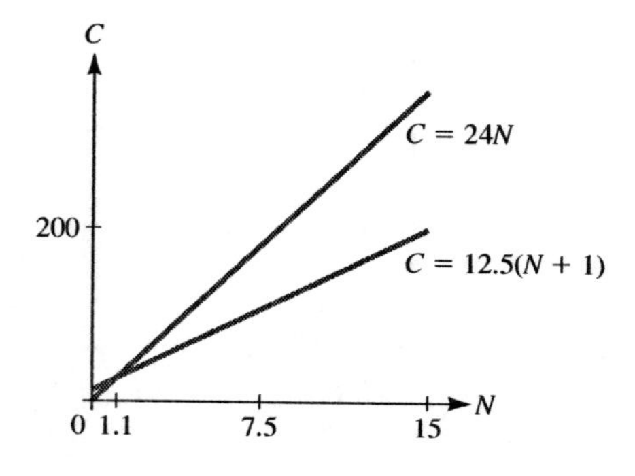

25. **(a)** The estimated surface area of the child is:

$S = 0.0072(18)^{0.425}(91)^{0.725} \approx 0.6473$

so, $C = \dfrac{(0.6473)(250)}{1.7} \approx 95.2\text{mg}$

(b) Using $2H$ and $2W$ for the larger child,

$C = \dfrac{0.0072(2W)^{0.425}(2H)^{0.725}A}{1.7}$

Comparing to drug dosage for the smaller child,

$$\frac{\frac{0.0072(2W)^{0.425}(2H)^{0.725}A}{1.7}}{\frac{0.0072W^{0.425}H^{0.725}A}{1.7}} = (2)^{0.425}(2)^{0.725} \approx 2.22$$

So, drug dosage for larger child is approx. 2.22 times the dosage for the smaller child.

27. Let x be the number of passengers. There will be $x - 40$ passengers between $40 < x \leq 80$ (if the total number is below 80). The price for the second category is

$$60 - 0.5(x - 40) = 80 - 0.5x$$

The revenue generated in this category is

$$80x - 0.5x^2$$

$$R(x) = \begin{cases} 2{,}400 & \text{if } 0 < x \leq 40 \\ 80x - 0.5x^2 & \text{if } 40 < x < 80 \\ 40x & \text{if } x \geq 80 \end{cases}$$

Only the points corresponding to the integers $x = 0, 1, 2, \cdots$ are meaningful in the practical context.

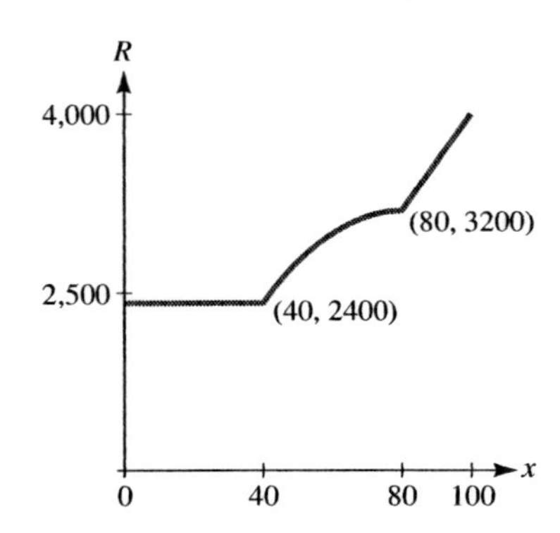

29. (a) Let x denote the number of people in the group and $C(x)$ the corresponding total admission charge for the group. If $x \geq 50$, the group is charged $3x$ dollars and if $0 \leq x < 50$, the group is charged $3.5x$ dollars. So

$$C(x) = \begin{cases} 3.5x & \text{if } 0 \leq x < 50 \\ 3.0x & \text{if } x \geq 50 \end{cases}$$

Only the points corresponding to the integers $x = 0, 1, 2, \ldots$ are meaningful in the practical context.

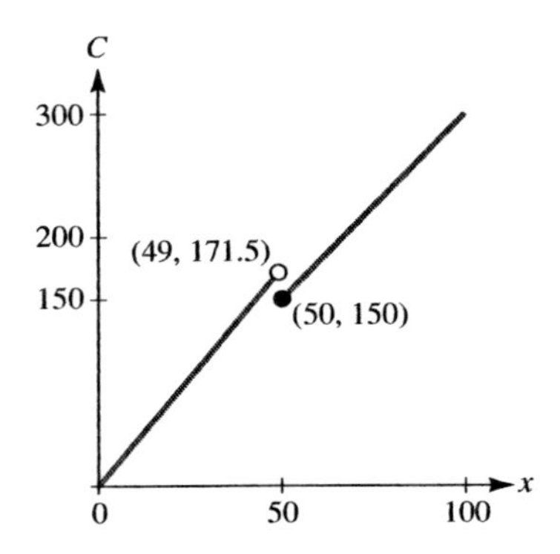

(b) $C(49) = 3.5(49) = 171.50$; $C(50) = 150$. So, by recruiting one additional member, a group of 49 can save $171.50 - 150 = \$21.50$.

31. Let x be the side of the square base and y the height of the open box. The area of the base is x^2 square meters and that of each side is xy square meters. The total cost is

$$4x^2 + 3(4xy) = 48$$

Solving for y in terms of x,

$$12xy = 48 - 4x^2$$
$$3xy = 12 - x^2$$
$$y = \frac{12 - x^2}{3x}$$

The volume of the box is

$$V = x^2y = \frac{x(12 - x^2)}{3} = 4x - \frac{x^3}{3} \text{ cubic meters.}$$

33. Let x denote the width of the printed portion and y the length of the printed portion. Then $x + 4$ is the width of the poster and $y + 8$ is its length.
The area A of the poster is
$A = (x + 4)(y + 8)$
which is a function of two variables.

$A = 25$ leads to $xy = 25$ or $y = \dfrac{25}{x}$.
So

$$A(x) = (x+4)\left(\frac{25}{x} + 8\right) = 8x + 57 + \frac{100}{x}$$

35. Let x be the sales price per lamp. Then, $x - 30$ will be the number of \$1.00 increases over the base price of \$30, and $1{,}000(x - 30)$ is the number of unsold lamps. Therefore the number of lamps sold is $3{,}000 - 1{,}000(x - 30)$. The profit is

$$\begin{aligned} P &= [3{,}000 - 1{,}000(x-30)]x \\ &\quad - 18[3{,}000 - 1{,}000(x-30)] \\ &= [3{,}000 - 1{,}000(x-30)](x-18) \\ &= (33{,}000 - 1{,}000x)(x-18) \end{aligned}$$

The optimal selling price is \$25.50.

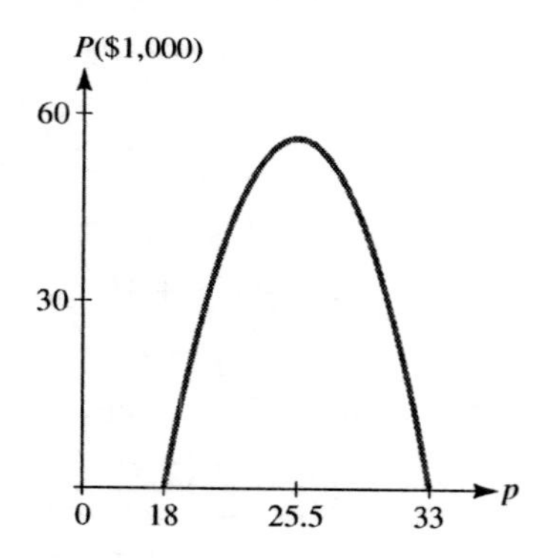

37. Let x be the number of machines used and t the number of hours of production. The number of kickboards produced per machine per hour is $30x$. It costs $20x$ to set up all the machines. The cost of supervision is $19.20t$. The number of kickboards produced by x machines in t hours is $30xt$ which must account for all 8,000 kickboards. Solving $30xt = 8{,}000$ for t leads to

$$t = \frac{800}{3x}$$

Cost of supervision: $19.20\left(\dfrac{800}{3x}\right) = \dfrac{5,120}{x}$

Total cost: $C(x) = 20x + \dfrac{5,120}{x}$

The number of machines which minimize cost is approximately 16. Note that $C(16) = 20(16) + \dfrac{5,120}{16} = 640$. So, the estimated min cost is \$640.

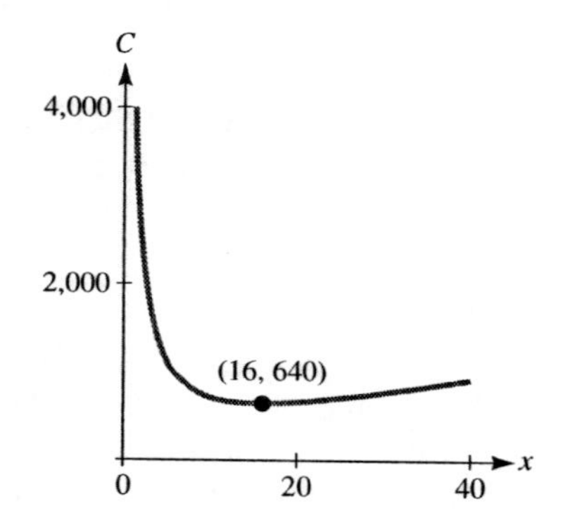

39. Let x denote the number of days after July 1 and $R(x)$ the corresponding revenue (in dollars). Then

$$R(x) = (\text{number of bushels sold})(\text{price per bushel})$$

Since the crop increases at the rate of 1 bushel per day and 80 bushels were available on July 1, the number of bushels sold after x days is $140 + x$. Since the price per bushel decreases by 0.02 dollars per day and was \$3 on July 1, the price per bushel after x days is $3 - 0.02x$ dollars. Putting it all together,

$$R(x) = (140 + x)(3 - 0.02x) = 0.02(150 - x)(140 + x)$$

The number of days to maximize revenue is approximately 5 days after July 1, or July 6. Note that $R(5) = 0.02(150 - 5)(140 + 5) = 420.50$. So, the estimated max revenue is \$420.50.

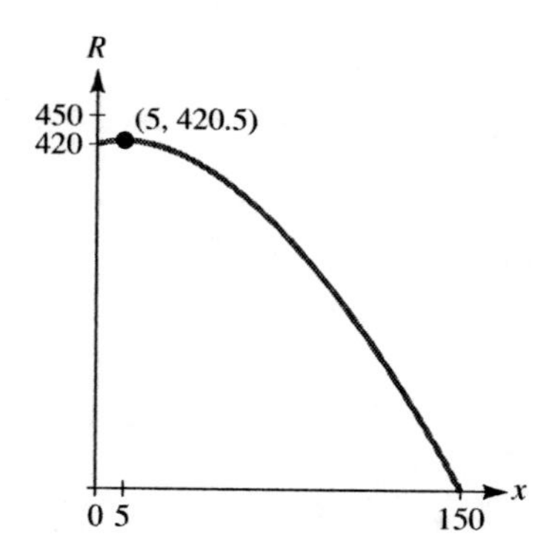

41. (a) Equilibrium occurs when $S(x)(D(x)$, or

$$3x + 150 = -2x + 275$$
$$5x = 125$$
$$x = 25$$

The corresponding equilibrium price is $p = S(x) = D(x)$ or $p = 3(25) + 150 = \$225$.

(b)

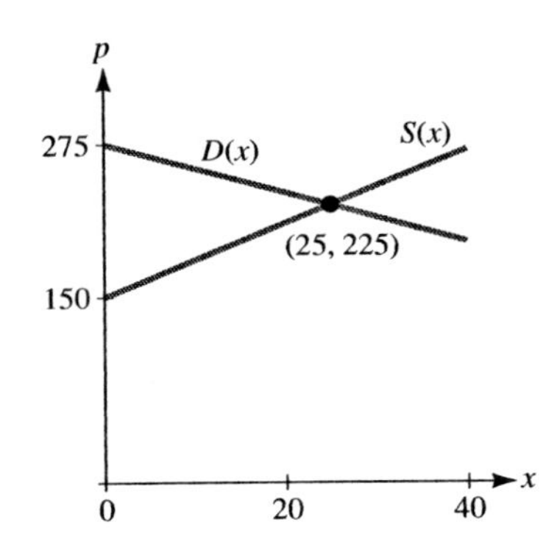

(c) There is a market shortage when demand exceeds supply. Here, a market shortage occurs when $0 < x < 25$. A market surplus occurs when supply exceeds demand. Here, a market surplus occurs when $x > 25$.

43. (a) Equilibrium occurs when $S(x) = D(x)$, or

$$2x + 7.43 = -0.21x^2 - 0.84x + 50$$
$$0.21x^2 + 2.84x - 42.57 = 0$$

Using the quadratic formula,

$$x = \frac{-2.84 \pm \sqrt{(2.84)^2 - 4(0.21)(-42.57)}}{2(0.21)}$$

so $x = 9$ (disregarding the negative root.) The corresponding equilibrium price is

$$p = S(x) = D(x), \text{ or } p = 2(9) + 7.43 = 25.43$$

(b)

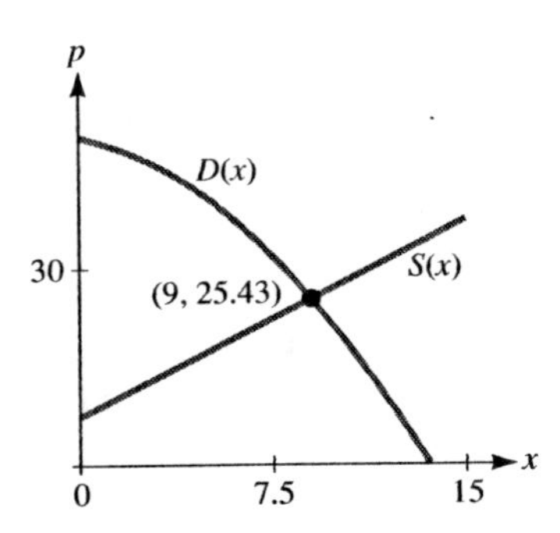

(c) There is a market shortage when demand exceeds supply. Here, a market shortage occurs when $0 < x < 9$. A market surplus occurs when supply exceeds demand. Here, a market surplus occurs when $x > 9$.

45. (a) Equilibrium occurs when $S(x) = D(x)$, or

$$2x + 15 = \frac{385}{x + 1}$$
$$(2x + 15)(x + 1) = 385$$
$$2x^2 + 17x + 15 = 385$$
$$2x^2 + 17x - 370 = 0$$

Using the quadratic formula,

$$x = \frac{-17 \pm \sqrt{(17)^2 - 4(2)(-370)}}{2(2)}$$

so $x = 10$ (disregard the negative root). The corresponding equilibrium price is $p = S(x) = D(x)$, or $p = 2(10) + 15 = 35$

(b)

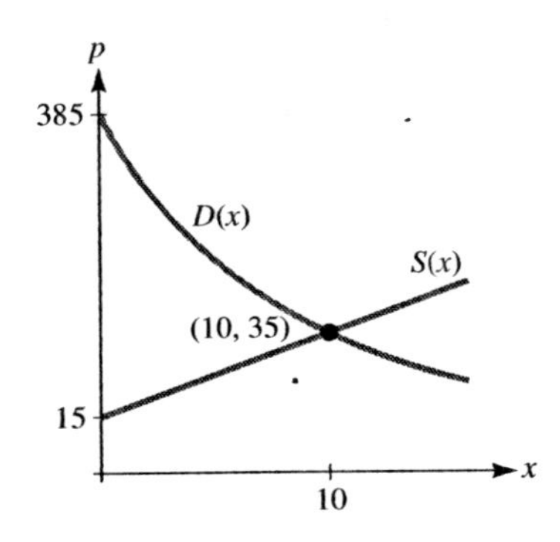

(c) The supply curve intersect the y-axis at $S(0) = 15$. Since this is the price at which producers are willing to supply zero units, it corresponds to their overhead at the start of production.

47. Let t be the number of hours the second plane has been flying. Since distance = (rate)(time), the equation for its distance is

$$d = 650t$$

The first plane has been flying for $t + \frac{1}{2}$ hours, so the equation for its distance is

$$d = 550\left(t + \frac{1}{2}\right)$$

The planes will meet when

$$\begin{aligned} 650t &= 550\left(t + \frac{1}{2}\right) \\ 650t &= 550t + 275 \\ 100t &= 275 \\ t &= 2.75 \end{aligned}$$

Since three-quarters of an hour is 45 minutes, the second plane passes the first plane after it has been flying 2 hours and 45 minutes.

49. Royalties for publisher A are given by

$$R_A(N) = \begin{cases} 0.01(2)(N) & 0 < N \leq 30{,}000 \\ 0.01(2)(30,\ 000) & \\ +0.035(2)(N - 30,\ 000) & N > 30{,}000 \end{cases}$$

Royalties for publisher B are given by

$$R_B(N) = \begin{cases} 0 & N \leq 4{,}000 \\ 0.02(3)(N - 4{,}000) & N > 4{,}000 \end{cases}$$

Clearly, for $N \leq 4{,}000$, publisher A offers the better deal. When $N = 30{,}000$, publisher A pays \$600, but publisher B now pays more, paying \$1,560. Therefore, the plans pay the same amount for some value of $N < 30{,}000$. To find the value,

$$\begin{aligned} 0.01(2)(N) &= 0.02(3)(N - 4{,}000) \\ 0.02N &= 0.06N - 240 \\ 240 &= 0.04N \\ 6{,}000 &= N \end{aligned}$$

So, when $N < 6{,}000$, publisher A offers the better deal. When $N > 6{,}000$, publisher B initially offers the better deal. Then, the plans again pay the same amount when

$$\begin{aligned} 0.01(2)(30{,}000) + 0.035(2)(N - 30{,}000) &= 0.02(3)(N - 4{,}000) \\ 0.07N - 1{,}500 &= 0.06N - 240 \\ 0.01N &= 1{,}260 \\ N &= 126{,}000 \end{aligned}$$

So, when more than 126,000 copies are sold, plan A becomes the better plan.

51. Since I is proportional to the area, A, of the pupil, $I = kA$, where k is a constant of proportionality. Since the pupil of the eye is circular and the area of a circle is $A = \pi r^2$, $I = k\pi r^2$.

53. $R = \dfrac{R_m[S]}{K_m + [S]}$

$$y = \frac{1}{R} = \frac{K_m + [S]}{R_m[S]} = \frac{K_m(1/S) + 1}{R_m} = \frac{K_m}{R_m}x + \frac{1}{R_m}.$$

55. (a) Let x = number of books and C = cost of producing x books

$$C(x) = 5.5x + 74{,}200$$

x	2,000	4,000	6,000	8,000
$C(x)$	85,200	96,200	107,200	118,200

(b) Let x = number of books and R = revenue from the sale of x books

$$R(x) = 19.5x$$

x	2,000	4,000	6,000	8,000
$R(x)$	39,000	78,000	117,000	156,000

(c) $y = 5.5x + 74,\ 200$

(d) $y = 19.5x$

(e) To graph both functions of the same coordinate axes,
Press [y=].
Input $5.5x + 74{,}200$ for $y_1 =$ and press [enter].
Input $19.5x$ for $y_2 =$.

Use window dimensions [5000, 5500] 100 by [100000, 105000] 1250.
Press graph.

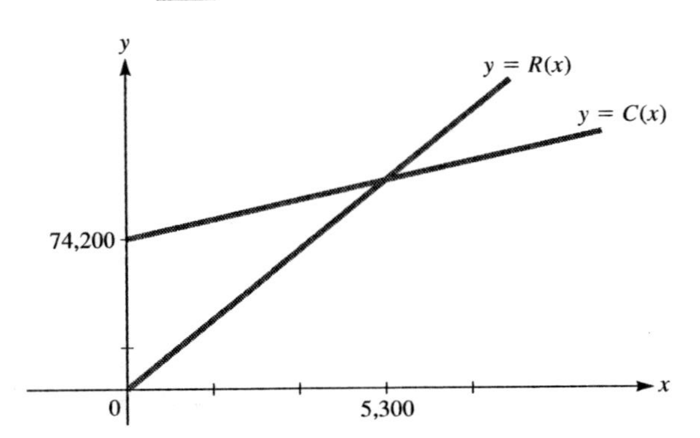

(f) To find where cost = revenue,
Press trace.
Use arrow buttons to move along one of the graphs to the apparent intersection. Use the Zoom in function under the calc menu. Repeat process using arrows buttons and zoom in for a more accurate reading. As an alternative, use the intersect function under the calc menu. Enter a value close to the point of intersection for y_1 and also for y_2. Finally, enter a guess.
The coordinate (5300, 103350) appears to be the point at which cost equals revenue.

(g) To use the graph to determine how many books need to be made to produce revenue of at least \$85,000, use the window settings of (e).
Press graph.
Press trace.
Press ↓ arrow to be sure that $y_2 = 19.5x$ is displayed in the upper left corner.
Use arrow buttons to trace along the revenue graph. It appears that approximately 4,360 books must be sold for a revenue of \$85,000. The profit when 4,360 books are sold is $-13,160$, a loss of \$13,160.

1.5 Limits

1. $\lim_{x\to a} f(x) = b$, even though $f(a)$ is not defined.

3. $\lim_{x\to a} f(x) = b$ even though $f(a) = c$.

5. $\lim_{x\to a} f(x)$ does not exist since as x approaches a from the left, the function becomes unbounded.

7.
$$\lim_{x\to 2}(3x^2 - 5x + 2) = 3\lim_{x\to 2} x^2 - 5\lim_{x\to 2} x + \lim_{x\to 2} 2 = 3(2)^2 - 5(2) + 2 = 4.$$

9.
$$\lim_{x\to 0}(x^5 - 6x^4 + 7) = \lim_{x\to 0} x^5 - 6\lim_{x\to 0} x^4 + \lim_{x\to 0} 7 = 7.$$

11.
$$\lim_{x\to 3}(x-1)^2(x+1) = \lim_{x\to 3}(x-1)^2 \lim_{x\to 3}(x+1) = (3-1)^2(3+1) = 16.$$

13.
$$\lim_{x\to 1/3} \frac{x+1}{x+2} = \frac{\lim_{x\to 1/3} x+1}{\lim_{x\to 1/3} x+2} = \frac{\frac{4}{3}}{\frac{7}{3}} = \frac{4}{7}$$

15. $\lim_{x\to 5} \frac{x+3}{5-x}$ does not exist since the limit of the denominator is zero while the limit of the numerator is not zero.

17.
$$\lim_{x\to 1} \frac{x^2-1}{x-1} = \lim_{x\to 1} \frac{(x+1)(x-1)}{x-1} = \lim_{x\to 1}(x+1) = 2.$$

19.
$$\lim_{x\to 5} \frac{x^2-3x-10}{x-5} = \lim_{x\to 5} \frac{(x-5)(x+2)}{x-5} = \lim_{x\to 5}(x+2) = 7.$$

21.
$$\lim_{x\to 4} \frac{(x+1)(x-4)}{(x-1)(x-4)} = \frac{\lim_{x\to 4}(x+1)}{\lim_{x\to 4}(x-1)} = \frac{5}{3}.$$

23.

$$\lim_{x\to -2} \frac{x^2 - x - 6}{x^2 + 3x + 2} = \lim_{x\to -2} \frac{(x-3)(x+2)}{(x+1)(x+2)} = \frac{\lim_{x\to -2}(x-3)}{\lim_{x\to -2}(x+1)} = \frac{-5}{-1} = 5.$$

25.

$$\lim_{x\to 4} \frac{\sqrt{x}-2}{x-4} = \lim_{x\to 4} \frac{\sqrt{x}-2}{x-4}\frac{\sqrt{x}+2}{\sqrt{x}+2} = \lim_{x\to 4} \frac{x-4}{(x-4)(\sqrt{x}+2)} = \frac{1}{4}.$$

27.

$$f(x) = x^3 - 4x^2 - 4,$$

$$\lim_{x\to +\infty} f(x) = \lim_{x\to +\infty} x^3 = +\infty$$

$$\lim_{x\to -\infty} f(x) = \lim_{x\to -\infty} x^3 = -\infty$$

29.

$$f(x) = (1-2x)(x+5) = -2x^2 - 9x + 5$$

$$\lim_{x\to +\infty} f(x) = \lim_{x\to +\infty} -2x^2 = -\infty$$

$$\lim_{x\to -\infty} f(x) = \lim_{x\to -\infty} -2x^2 = -\infty$$

31.

$$f(x) = \frac{x^2 - 2x + 3}{2x^2 + 5x + 1}$$

$$\lim_{x\to +\infty} f(x) = \lim_{x\to +\infty} \frac{1 - \frac{2}{x} + \frac{3}{x^2}}{2 + \frac{5}{x} + \frac{1}{x^2}} = \frac{1}{2}$$

$$\lim_{x\to -\infty} f(x) = \lim_{x\to -\infty} \frac{1 - \frac{2}{x} + \frac{3}{x^2}}{2 + \frac{5}{x} + \frac{1}{x^2}} = \frac{1}{2}$$

33. $f(x) = \dfrac{2x+1}{3x^2 + 2x - 7}$,

$$\lim_{x\to +\infty} f(x) = \lim_{x\to +\infty} \frac{\frac{2}{x} + \frac{1}{x^2}}{3 + \frac{2}{x} - \frac{7}{x^2}} = 0$$

$$\lim_{x\to -\infty} f(x) = \lim_{x\to -\infty} \frac{\frac{2}{x} + \frac{1}{x^2}}{3 + \frac{2}{x} - \frac{7}{x^2}} = 0$$

35. $f(x) = \dfrac{3x^2 - 6x + 2}{2x - 9}$,

$$\lim_{x\to +\infty} f(x) = \lim_{x\to +\infty} \frac{3x^2 - 6x + 2}{2x - 9} = \lim_{x\to +\infty} \frac{3x - 6 + \frac{2}{x}}{2 - \frac{9}{x}}$$

$$\lim_{x\to +\infty} 3x - 6 + \frac{2}{x} = +\infty \quad \text{and}$$

$$\lim_{x\to +\infty} 2 - \frac{9}{x} = 2$$

So, $\displaystyle\lim_{x\to +\infty} \frac{3x - 6 + \frac{2}{x}}{2 - \frac{9}{x}} = +\infty$

$$\lim_{x\to -\infty} f(x) = \lim_{x\to -\infty} \frac{3x - 6 + \frac{2}{x}}{2 - \frac{9}{x}}$$

$$\lim_{x\to -\infty} 3x - 6 + \frac{2}{x} = -\infty \quad \text{and}$$

$$\lim_{x\to -\infty} 2 - \frac{9}{x} = 2$$

So, $\displaystyle\lim_{x\to -\infty} \frac{3x-6+\frac{2}{x}}{2-\frac{9}{x}} = -\infty$

37. $\displaystyle\lim_{x\to +\infty} f(x) = 1$
and $\displaystyle\lim_{x\to -\infty} f(x) = -1$

39. The corresponding table values are:

$$f(1.9) = (1.9)^2 - 1.9 = 1.71$$
$$f(1.99) = (1.99)^2 - 1.99 = 1.9701$$
$$f(1.999) = (1.999)^2 - 1.999 = 1.997001$$
$$f(2.001) = (2.001)^2 - 2.001 = 2.003001$$
$$f(2.01) = (2.01)^2 - 2.01 = 2.0301$$
$$f(2.1) = (2.1)^2 - 2.1 = 2.31$$

$$\lim_{x\to 2} f(x) = 2$$

41. The corresponding table values are

$$f(0.9)=\frac{(0.9)^3+1}{0.9-1}=-17.29$$

$$f(0.99)=\frac{(0.99)^3+1}{0.99-1}=-197.0299$$

$$f(0.999)=\frac{(0.999)^3+1}{0.999-1}=-1{,}997.002999$$

$$f(1.001)=\frac{(1.001)^3+1}{1.001-1}=2{,}003.003001$$

$$f(1.01)=\frac{(1.01)^3+1}{1.01-1}=203.0301$$

$$f(1.1)=\frac{(1.1)^3+1}{1.1-1}=23.31$$

$\lim_{x\to 1} f(x)$ does not exist

43. $$\lim_{x\to c}[2f(x)-3g(x)]=\lim_{x\to c}2f(x)-\lim_{x\to c}3g(x)$$
$$=2\lim_{x\to c}f(x)-3\lim_{x\to c}g(x)$$
$$=2(5)-3(-2)=16$$

45. $$\lim_{x\to c}\frac{f(x)}{g(x)}=\frac{\lim_{x\to c}f(x)}{\lim_{x\to c}g(x)}$$
$$=\frac{5}{-2}=-\frac{5}{2}$$

47. $$\lim_{x\to c}\sqrt{g(x)}=\sqrt{\lim_{x\to c}g(x)}$$
$$=\sqrt{4}=2$$

49. As the weight approaches 18 lbs., displacement approaches a limit of 1.8 inches.

51. $p=0.2t+1{,}500$; $E(t)=\sqrt{9t^2+0.5t+179}$

(a) Since the units of p are thousands and the units of E are millions, the units of E/p will be thousands. $P(t)=\dfrac{\sqrt{9t^2+0.5t+179}}{0.2t+1500}$ thousand dollars per person

(b) Dividing each term by t (note that each term under the square root will be divided by t^2 since $\sqrt{t^2}=t$),

$$\lim_{t\to\infty}P(t)=\lim_{t\to\infty}\frac{\sqrt{9+\frac{0.5}{t}+\frac{179}{t^2}}}{0.2+\frac{1500}{t}}$$

$$=\frac{\sqrt{\lim_{t\to\infty}\left(9+\frac{0.5}{t}+\frac{179}{t^2}\right)}}{\lim_{t\to\infty}\left(0.2+\frac{1500}{t}\right)}=\frac{\sqrt{9}}{0.2}=15$$

or, $15,000 per person.

53. **(a)** $\displaystyle\lim_{S\to\infty}=\frac{aS}{S+c}=\lim_{S\to\infty}\frac{a}{1+\frac{c}{S}}=a$
As bite size increases indefinitely, intake approaches a limit of a. This signifies that the animal has a limit of how much it can consume, no matter how large its bites become.

(b) Writing exercise—Answers will vary.

55. $\displaystyle\lim_{x\to+\infty}\frac{7.5x+120,000}{x}=\lim_{x\to+\infty}7.5+\frac{120,000}{x}=$ 7.5 As the number of units produced increases indefinitely, the average cost per unit decreases, approaching a minimum of $7.50. The average cost cannot decrease further, as the expense of materials cannot be eliminated completely.

57. $\lim_{x\to 0}f(x)$ does not exist because $f(x)$ oscillates infinitely many times between -1 and 1, regardless how close x gets to 0.

59. $$\lim_{x\to+\infty}=\frac{a_nx^n+a_{n-1}x^{n-1}+\cdots+a_1x+a_0}{b_mx^m+b_{m-1}x^{m-1}+\cdots+b_1x+b_0}$$

(a) When $n<m$,

$$=\lim_{x\to+\infty}\frac{a_n+\frac{a_{n-1}}{x}+\cdots+\frac{a_1}{x^{n-1}}+\frac{a_0}{x^n}}{b_m\frac{x^m}{x^n}+b_{m-1}\frac{x^{m-1}}{x^n}+\cdots+b_1\frac{x}{x^n}+b_0\frac{1}{x^n}}$$

Since

$$\lim_{x\to+\infty}\frac{x^m}{x^n}=+\infty,\quad \lim_{x\to+\infty}f(x)=0$$

(b) When $n<m$,

$$\frac{x^m}{x^n}=\text{land}\lim_{x\to+\infty}f(x)=\frac{a_n}{b_m}$$

(c) When $n>m$,

$$= \lim_{x\to+\infty} \frac{a_n\frac{x^n}{x^m} + a_{n-1}\frac{x^{n-1}}{x^m} + \cdots + a_1\frac{x}{x^m} + a_0\frac{1}{x^m}}{b_m + \frac{b_{m-1}}{x} + \cdots + \frac{b_1}{x^{m-1}} + \frac{b_0}{x^m}}$$

Now,

$$\lim_{x\to+\infty} a_n\frac{x^n}{x^m} + a_{n-1}\frac{x^{n-1}}{x^m} + \cdots + a_1\frac{x}{x^m} + a_0\frac{1}{x^m}$$
$$= \pm\infty,$$

depending on the sign of a_n. Also

$$\lim_{x\to+\infty} b_m + \frac{b_{m-1}}{x} + \cdots + \frac{b_1}{x^{m-1}} + \frac{b_0}{x^m} = b_m$$

So, $\lim_{x\to+\infty} \frac{a_n\frac{x^n}{x^m} + a_{n-1}\frac{x^{n-1}}{x^m} + \cdots}{b_m + \frac{b_{m-1}}{x} + \cdots} = \pm\infty,$
depending on the signs of a_n and b_m. When a_n and b_m have the same sign, the limit is $+\infty$; when they have opposite signs, the limit is $-\infty$.

1.6 One-Sided Limits and Continuity

1. $\lim_{x\to2^-} f(x) = -2$; $\lim_{x\to2^+} f(x) = 1$
Since $-2 \neq 1$, $\lim_{x\to2} f(x)$ does not exist

3. $\lim_{x\to2^-} f(x) = 2$; $\lim_{x\to2^+} f(x) = 2$
Since limits are the same, $\lim_{x\to2} f(x) = 2$.

5. $\lim_{x\to4^+} (3x^2 - 9) = \lim_{x\to4^+} 3x^2 - \lim_{x\to4^+} 9$
$= 3(4)^2 - 9 = 39$

7. $\lim_{x\to3^+} \sqrt{3x-9} = \sqrt{3(3)-9} = 0$

9. $\lim_{x\to0^+} (x - \sqrt{x}) = 0 - 0 = 0$

11.
$$\lim_{x\to3^+} \frac{\sqrt{x+1}-2}{x-3}$$
$$= \lim_{x\to3^+} \frac{\sqrt{x+1}-2}{x-3}\cdot\frac{\sqrt{x+1}+2}{\sqrt{x+1}+2}$$
$$= \lim_{x\to3^+} \frac{x+1-4}{(x-3)(\sqrt{x+1}+2)} = \frac{1}{4}.$$

13. $\lim_{x\to3^-} f(x) = \lim_{x\to3^-} (2x^2 - x) = 2(3)^2 - 3 = 15$
$\lim_{x\to3^+} f(x) = \lim_{x\to3^+} (3 - x) = 3 - 3 = 0.$

15. If $f(x) = 5x^2 - 6x + 1$, then $f(2) = 9$ and $\lim_{x\to2} f(x) = 9$,
So, f is continuous at $x = 2$.

17. If $f(x) = \frac{x+2}{x+1}$,
then $f(1) = \frac{3}{2}$ and

$$\lim_{x\to1} f(x) = \lim_{x\to1} \frac{x+2}{x+1} = \frac{\lim_{x\to1}(x+2)}{\lim_{x\to1}(x+1)} = \frac{3}{2}$$

So, f is continuous at $x = 1$.

19. If $f(x) = \frac{x+1}{x-1}$,
$f(1)$ is undefined since the denominator is zero, and so f is not continuous at $x = 1$.

21. If $f(x) = \frac{\sqrt{x}-2}{x-4}$,
$f(4)$ is undefined since the denominator is zero, and so f is not continuous at $x = 4$.

23. If $f(x) = \begin{cases} x+1 & \text{if } x \le 2 \\ 2 & \text{if } x > 2 \end{cases}$
then $f(2) = 3$ and $\lim_{x\to2} f(x)$ must be determined.
As x approaches 2 from the left,

$$\lim_{x\to2^-} f(x) = \lim_{x\to2^-} (x+1) = 3$$

and as x approaches 2 from the right,

$$\lim_{x\to2^+} f(x) = \lim_{x\to2^+} 2 = 2$$

So the limit does not exist (since different limits are obtained from the left and the right), and f is not continuous at $x = 2$.

25.
$$\text{If } f(x)=\begin{cases} x^2+1 & \text{if } x\le 3\\ 2x+4 & \text{if } x>3\end{cases}$$
then $f(3)=(3)^2+1=10$ and $\lim_{x\to 3} f(x)$ must be determined. As x approaches 3 from the left,
$$\lim_{x\to 3^-} f(x)=\lim_{x\to 3^-}(x^2+1)=(3)^2+1=10$$
and as x approaches 3 from the right,
$$\lim_{x\to 3^+} f(x)=\lim_{x\to 3^+}(2x+4)=2(3)+4=10$$
So $\lim_{x\to 3} f(x)=10$. Since $f(x)=\lim_{x\to 3} f(x)$, f is continuous at $x=3$.

27. $f(a)=3a^2-6a+9$ so f is defined for all real numbers. $\lim_{x\to a} f(a)=3(a)^2-6a+9$, so the limit of f exists for all real numbers. Since $f(a)=\lim_{x\to a} f(a)$, there are no values for which f is not continous.

29. $f(x)=\dfrac{x+1}{x-2}$
is not defined at $x=2$, so f is *not* continuous at $x=2$.

31. $f(x)=\dfrac{3x+3}{x+1}$
is not defined at $x=-1$, so f is *not* continuous at $x=-1$.

33. $f(x)=\dfrac{3x-2}{(x+3)(x-6)}$
is not defined at $x=-3$ and $x=6$, so f is *not* continuous at $x=-3$ and $x=6$.

35. $f(x)=\dfrac{x}{x^2-x}$
is not defined at $x=0$ and $x=1$, so f is *not* continuous at $x=0$ and $x=1$.

37. f is defined for all real numbers. Further,
$$\begin{aligned}\lim_{x\to 1^-} f(x)&=2+3=5\\ =\lim_{x\to 1^+} f(x)&=6-1\\ &=f(1),\end{aligned}$$
so there are no values for which f is not continuous.

39. f is defined for all real numbers. However,
$$\lim_{x\to 0^-} f(x)=\lim_{x\to 0^-} 3x-2=3(0)-2=-2$$
$$\lim_{x\to 0^+} f(x)=\lim_{x\to 0^+} x^2+x=0+0=0$$
So $\lim_{x\to 0} f(x)$ does not exist and therefore f is *not* continuous at $x=0$.

41. **(a)** When $v=20$, the middle expression is used to find $W(v)$.
$$\begin{aligned}W(20)&=1.25(20)-18.67\sqrt{20}+62.3\\ &\approx 3.75°\text{F}\end{aligned}$$
For $v=50$, the bottom expression is used to find $W(v)$, so $W(50)=-7°$F.

(b) If $0\le v\le 4$, $W(v)=30°$F, so v cannot be between 0 and 4 (inclusive). If $v\ge 45$, $W(v)=-7$, so v cannot be 45 or more. If $4<v<45$,
$$W(v)=1.25v-18.67\sqrt{v}+62.3$$
If $W(v)=0$, then
$$0=1.25v-18.67\sqrt{v}+62.3$$
Using the quadratic formula, $v=25$ mph.

(c) When rounded to the nearest degree, for practical purposes,
$$\lim_{v\to 4^-} W(v)=\lim_{v\to 4^-} 30=30$$
$$\begin{aligned}\lim_{v\to 4^+} W(v)&=\lim_{v\to 4^+}(1.25v-18.67\sqrt{v}+62.3)\\ &=1.25(4)-18.67\sqrt{4}+62.3=30\end{aligned}$$
So, W is continuous at $v=4$. Similarly for $v=45$,
$$\begin{aligned}\lim_{v\to 45^-} W(v)&=\lim_{v\to 45^-}(1.25v-18.67\sqrt{v}+62.3)\\ &=1.25(45)-18.67\sqrt{45}+62.3\\ &\approx -7\\ &=\lim_{v\to 45^+} W(v)=\lim_{v\to 45^+} -7=-7\end{aligned}$$
So, W is continuous at $v=45$.

43. $p(x)$ is discontinuous at $x=1$, $x=2$, $x=3$, $x=4$, $x=5$.

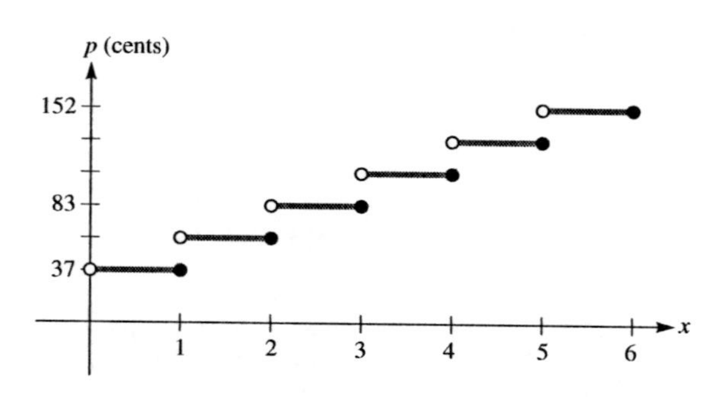

45. The graph is discontinuous at $x = 10$ and $x = 25$. Sue is probably at the gas station replenishing fuel.

47. $C(x) = \dfrac{12x}{100 - x}$

(a) $C(25) = \dfrac{12(25)}{100 - 25} = 4$
or, \$4,000
$C(50) = \dfrac{12(50)}{100 - 50} = 12$
or, \$12,000

(b)

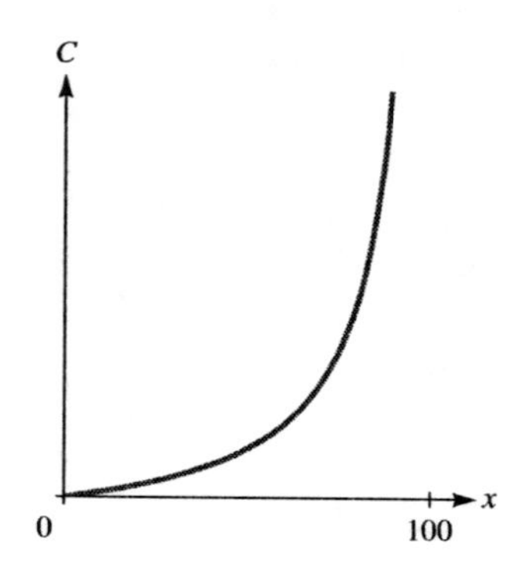

(c) From the graph,

$$\lim_{x \to 100^-} C(x) = \infty$$

So, it is not possible to remove all of the pollution.

49. $C(x) = \dfrac{8x^2 - 636x - 320}{x^2 - 68x - 960}$

(a) $C(0) = \dfrac{-320}{-960} = \dfrac{1}{3} \approx 0.333$

$$C(100) = \frac{8(100)^2 - 636(100) - 320}{(100)^2 - 68(100) - 960} \approx 7.179$$

(b) Since the denominator factors as $(x + 12)(x - 80)$, the function has a vertical asymptote when $x = 80$. This means that C is not continuous on the interval $0 \le x \le 100$, and the intermediate value theorem cannot be used.

51. $f(x) = \begin{cases} Ax - 3 & \text{if } x < 2 \\ 3 - x + 2x^2 & \text{if } 2 \le x \end{cases}$

f is continuous everywhere except possibly at $x = 2$, since $Ax - 3$ and $3 - x + 2x^2$ are polynomials. Since $f(2) = 3 - 2 + 2(2)^2 = 9$, in order that f be continuous at $x = 2$, A must be chosen so that $\lim_{x \to 2} f(x) = 9$.
As x approaches 2 from the right,

$$\begin{aligned} \lim_{x \to 2^+} f(x) &= \lim_{x \to 2^+} (3 - x + 2x^2) \\ &= \lim_{x \to 2^+} 3 - \lim_{x \to 2^+} x + 2 \lim_{x \to 2^+} x^2 \\ &= 3 - 2 + 2(2)^2 = 9 \end{aligned}$$

and as x approaches from the left,

$$\begin{aligned} \lim_{x \to 2^-} f(x) &= \lim_{x \to 2^-} (Ax - 3) \\ &= A \lim_{x \to 2^-} x - \lim_{x \to 2^-} 3 \\ &= 2A - 3 \end{aligned}$$

For $\lim_{x \to 2} f(x) = 9$, $2A - 3$ must equal 9, or $A = 6$.
f is continuous at $x = 2$ only when $A = 6$.

53. On the open interval $0 < x < 1$, since $x \neq 0$,

$$f(x) = x\left(1 + \frac{1}{x}\right) = x + 1$$

So, $f(x)$, a polynomial on $0 < x < 1$, is continuous. On the closed interval $0 \le x \le 1$, the endpoints must now be considered.

$$f(x) = x\left(x + \frac{1}{x}\right)$$

is not continuous at $x = 0$ since $f(0)$ is not defined. However, f is continuous at $x = 1$ since

$f(1) = 1\left(1 + \frac{1}{1}\right) = 2$ and as x approaches 1 from the left,

$$\lim_{x\to 1^-} x\left(x + \frac{1}{x}\right) = \lim_{x\to 1^-} x \cdot \lim_{x\to 1^-}\left(x + \frac{1}{x}\right)$$
$$= 1\left(1 + \frac{1}{1}\right) = 2$$

55. Rewrite as $\sqrt[3]{x-8} + 9x^{2/3} - 29$ and notice that at $x = 0$ this expression is negative and at $x = 8$ it is positive. Therefore, by the intermediate value property, there must be a value of x between 0 and 8 such that this expression is 0 or

$$\sqrt[3]{x-8} + 9x^{2/3} = 29$$

57. To investigate the behavior of

$$f(x) = \frac{2x^2 - 5x + 2}{x^2 - 4},$$

Press [y =].
Input $(2x \wedge 2 - 5x + 2)/(x \wedge 2 - 4)$ for $y_1 =$
Press [graph].

(a) Press [trace]. Use arrows to move cursor to be near $x = 2$ we see that (1.9, 0.72) and (2.1, 0.79) are two points on the graph. By zooming in, we find (1.97, 0.74) and (2.02, 0.76) to be two points on the graph. The $\lim_{x\to 2} f(x) = \frac{3}{4}$, however, the function is not continuous at $x = 2$ since $f(2)$ is undefined. To show this, use the value function under the calc menu and enter $x = 2$. There is no y-value displayed, which indicates the function is undefined for $x = 2$.

(b) Use the z standard function under the Zoom menu to return to the original graph. We see from the graph that there is a vertical asymptote at $x = -2$. The $\lim_{x\to -2^-} f(x) = \infty$ and $\lim_{x\to -2^+} f(x) = -\infty$ and therefore $\lim_{x\to -2} f(x)$ does not exist. So f is not continuous at $x = -2$.

59. Let's assume the hands of a clock move in a continuous fashion. During each hour the minute hand moves from being behind the hour to being ahead of the hour. Therefore, at some time, the hands must be in the same place.

Checkup for Chapter 1

1. Since negative numbers do not have square roots and denominators cannot be zero, the domain of the function $f(x) = \frac{2x-1}{\sqrt{4-x^2}}$ is all real numbers such that $4 - x^2 > 0$ or $(2 + x)(2 - x) > 0$, namely $-2 < x < 2$.

2. $$g(h(x)) = g\left(\frac{x+2}{2x+1}\right) = \frac{1}{2\left(\frac{x+2}{2x+1}\right) + 1} = \frac{1}{\frac{2x+4}{2x+1} + 1}$$
$$= \frac{1}{\frac{2x+4+2x+1}{2x+1}} = \frac{2x+1}{4x+5}, \quad x \neq -\frac{1}{2}$$

3. **(a)** Since $m = -\frac{1}{2}$ and the point ((1,2) is on the line, the equation of the line is

$$y - 2 = -\frac{1}{2}(x - (-1))$$
$$y - 2 = -\frac{1}{2}(x + 1)$$
$$y - 2 = -\frac{1}{2}x - \frac{1}{2}$$
$$y = -\frac{1}{2}x - \frac{1}{2} + 2$$
$$y = -\frac{1}{2}x + \frac{3}{2}$$

(b) Since $m = 2$ and $b = -3$, the equation of the line is $y = 2x - 3$.

4. **(a)** The graph is a line with x-intercept $\frac{5}{3}$ and y-intercept -5.

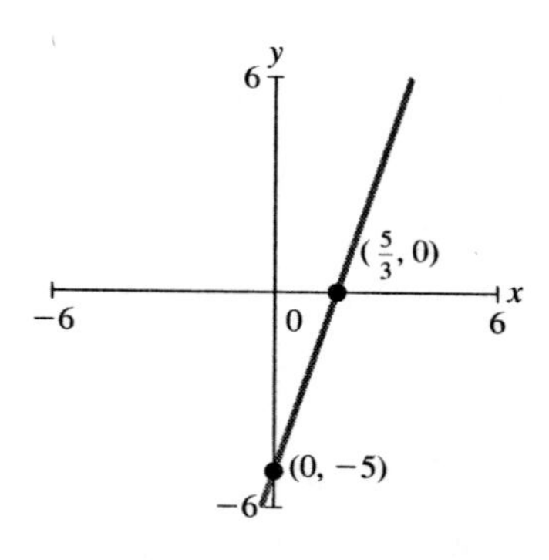

(b) The graph is a parabola which opens down (since $A < 0$). The vertex is

$$\left(-\frac{b}{2a}, f\left(-\frac{b}{2a}\right)\right), \text{ or } \left(\frac{3}{2}, \frac{25}{4}\right).$$

The x-intercepts are

$$\begin{aligned} 0 &= -x^2 + 3x + 4 \\ 0 &= x^2 - 3x - 4 \\ 0 &= (x - 4)(x + 1) \\ x &= 4, -1 \end{aligned}$$

The y-intercept is 4.

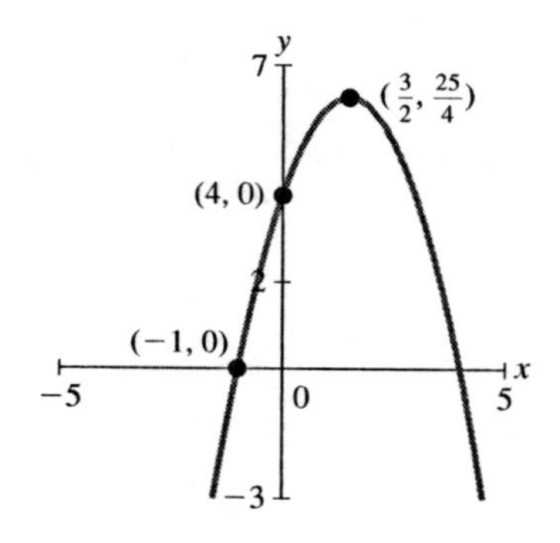

5. (a)

$$\lim_{x \to -1} \frac{x^2 + 2x - 3}{x - 1} = \frac{(-1)^2 + 2(-1) - 3}{-1 - 1} = \frac{1 - 2 - 3}{-2} = 2$$

(b)

$$\lim_{x \to 1} \frac{x^2 + 2x - 3}{x - 1} = \lim_{x \to 1} \frac{(x + 3)(x - 1)}{1 - 2} = \lim_{x \to 1} x + 3 = 4$$

(c) $$\lim_{x \to 1} \frac{x^2 - x - 1}{x - 2} = \frac{(1)^2 - 1 - 1}{x - 1} = \frac{-1}{-1} = 1$$

(d) $$\lim_{x \to +\infty} \frac{2x^3 + 3x - 5}{-x^2 + 2x + 7} = \lim_{x \to +\infty} \frac{2x + 3 - \frac{5}{x^2}}{-1 + \frac{2}{x} + \frac{7}{x^2}}$$

Since $\lim\limits_{x \to +\infty} 2x + 3 - \frac{5}{x^2} = +\infty$ and

$$\lim_{x \to +\infty} -1 + \frac{2}{x} + \frac{7}{x^2} = -1,$$

$$\lim_{x \to +\infty} \frac{2x^3 + 3x - 5}{-x^2 + 2x + 7} = -\infty.$$

6. The function is defined at $x = 1$, and $f(1) = 2(1) + 1 = 3$. If $\lim\limits_{x \to 1} f(x) = 3$, the function will be continuous at $x = 1$. From the left of $x = 1$,

$$\lim_{x \to 1^-} f(x) = \lim_{x \to 1^-} 2x + 1 = 2(1) + 1 = 3.$$

From the right of $x = 1$,

$$\begin{aligned} \lim_{x \to 1^+} f(x) &= \lim_{x \to 1^+} \frac{x^2 + 2x - 3}{x - 1} \\ &= \lim_{x \to 1^+} \frac{(x + 3)(x - 1)}{x - 1} = \lim_{x \to 1^+} (x + 3) \\ &= 1 + 3 = 4. \end{aligned}$$

Since $\lim\limits_{x \to 1^-} f(x) \neq \lim\limits_{x \to 1^+} f(x)$, the limit does not exist and the function is not continuous at $x = 1$.

7. (a) Let t denote the time in months since the beginning of the year and $P(t)$ the corresponding price (in cents) of gasoline. Since the price increases at a constant rate of 2 cents per gallon per month, P is a linear function of t with slope $m = 2$. Since the price on June first (when $t = 5$) is 180 cents, the graph passes through (5, 180). The equation is therefore

$$P - 180 = 2(t - 5)$$

or $P(t) = 2t + 170$ cents,
$P(t) = 0.02t + 1.70$ dollars.

units is $S(10) = (10)^2 + 3 = \$103$, and the demand price is $D(10) = -(10) + 59 = \$49$. The difference is $\$103 - \$49 = \$54$. (Note that for 5 units, the demand price is higher than the supply price. However, for 10 units, the opposite is true.)

9. **(a)** The population is positive and increasing for $0 \le t < 5$. However, for $t \ge 5$, the population decreases. Therefore, the colony dies out when

$$-8t + 72 = 0, \quad \text{or } t = 9$$

(b) $f(1) = 8$ and $f(7) = -56 + 72 = 16$. Since

$$f(5) = \lim_{x \to 5} f(x) = 32,$$

f is continuous. Since $8 < 10 < 16$, by the intermediate value property there exists a value $1 < c < 7$ such that $f(c) = 10$.

10. Since M is a linear function of D, $M = aD + b$, for some constants a and b. Using $M = 7.7$ when $D = 3$, and $M = 12.7$ when $D = 5$, solve the system

$$a \cdot 3 + b = 7.7$$
$$a \cdot 5 + b = 12.7$$

So, $a = 2.5$ and $b = 0.2$. Thus, $M = 2.5D + 0.2$ When $D = 0$, $M = 0.2$, so 0.2% will mutate when no radiation is used.

Review Problems

1. **(a)** The domain of the quadratic function

$$f(x) = x^2 - 2x + 6$$

is all real numbers x.

(b) Since denominators cannot be zero, the domain of the rational function

$$f(x) = \frac{x-3}{x^2 + x - 2} = \frac{x-3}{(x+2)(x-1)}$$

is all real numbers x except $x = -2$ and $x = 1$.

(c) Since negative numbers do not have square roots, the domain of the function

$$f(x) = \sqrt{x^2 - 9} = \sqrt{(x+3)(x-3)}$$

is all real numbers x such that $(x+3)(x-3) \ge 0$, that is for $x \le -3$, or $x \ge 3$, or $|x| \ge 3$.

3. **(a)** If $g(u) = u^2 + 2u + 1$ and $h(x) = 1 - x$

$$\text{then} \quad g(h(x)) = g(1-x) = (1-x)^2 + 2(1-x) + 1 = x^2 - 4x + 4.$$

(b) If $g(u) = \dfrac{1}{2u+1}$ and $h(x) = x + 2$,

$$\text{then} \quad g(h(u)) = g(x+2) = \frac{1}{2(x+2)+1} = \frac{1}{2x+5}.$$

(c) If $g(u) = \sqrt{1-u}$ and $h(x) = 2x + 4$,

$$\text{then} \quad g(h(x)) = g(2x+4) = \sqrt{1-(2x+4)} = \sqrt{-2x-3}.$$

5. **(a)** One of many possible solutions is

$$g(u) = u^5 \quad \text{and} \quad h(x) = x^2 + 3x + 4.$$

Then,

$$g(h(x)) = g(x^2 + 3x + 4) = (x^2 + 3x + 4)^5 = f(x).$$

(b) One of many possible solutions is

$$g(u) = u^2 + \frac{5}{2(u+1)^3} \text{ and } h(x) = 3x + 1.$$

Then,

$$g(h(x)) = g(3x+1) = (3x+1)^2 + \frac{5}{2((3x+1)+1)^3} = (3x+1)^2 + \frac{5}{2(3x+2)^3} = f(x).$$

7. If the graph of

$$y = 3x^2 - 2x + c$$

passes through the point (2, 4),

$$4 = 3(2)^2 - 2(2) + c \quad \text{or} \quad c = -4$$

9. **(a)** If $y = 3x + 2$, $m = 3$ and $b = 2$.

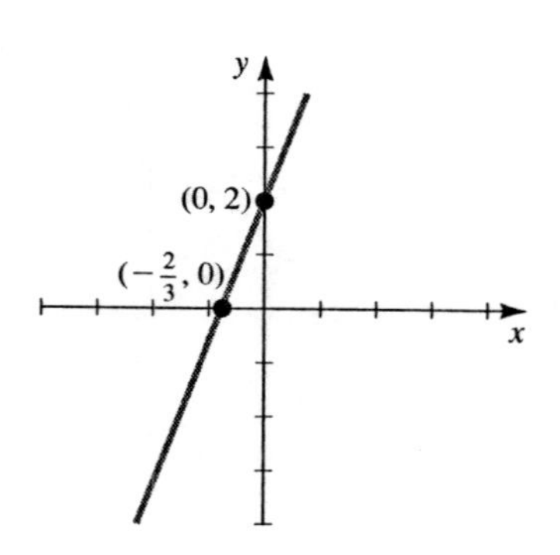

(b) If $5x - 4y = 20$ then

$$y = \frac{5}{4}x - 5$$

and $m = \frac{5}{4}, \quad b = -5.$

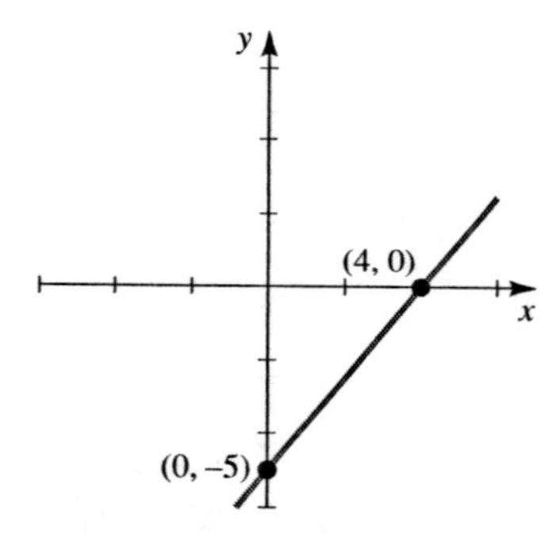

(c) If $2y + 3x = 0$ then

$$y = -\frac{3}{2}x + 0$$

and $m = -\frac{3}{2}, \quad b = 0.$

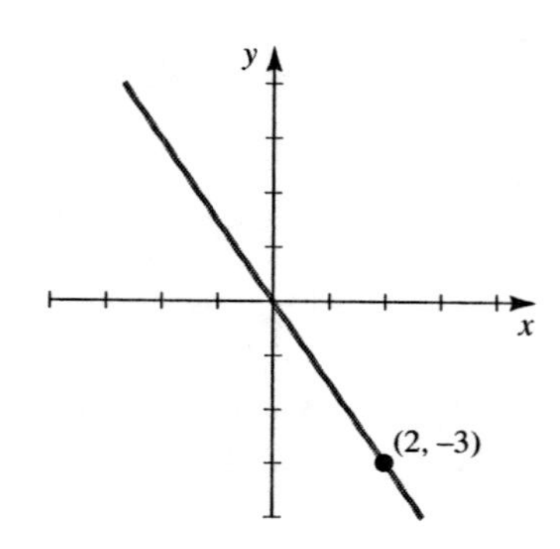

(d) If $\frac{1}{3}x + \frac{1}{2}y = 4$ then

$$y = -\frac{2}{3}x + 8$$

and $m = -\frac{2}{3}, \quad b = 8.$

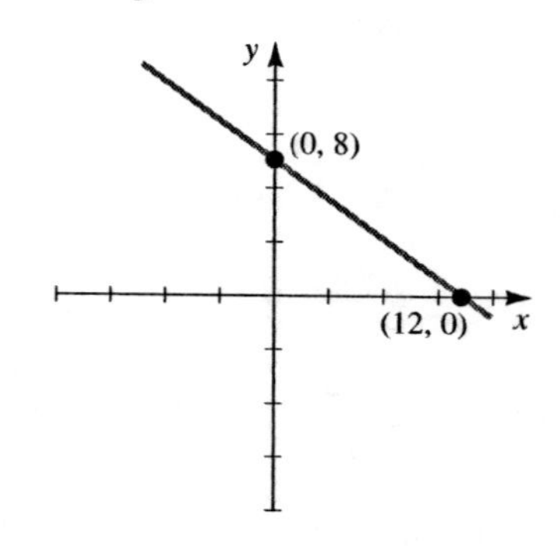

11. The number of weeks needed to reach x percent of the fund raising goal is given by

$$f(x) = \frac{10x}{150 - x}$$

(a) Since x denotes a percentage, the function has a practical interpretation for

$$0 \leq x \leq 100$$

The corresponding portion of the graph is sketched.

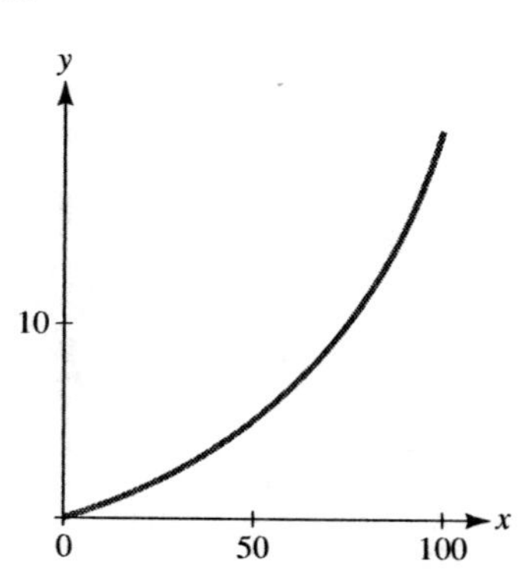

(b) The number of weeks needed to reach 50% of the goal is

$$f(50) = \frac{10(50)}{150 - 50} = 5\text{weeks}$$

(c) The number of weeks needed to reach 100% of the goal is

$$f(100) = \frac{10(100)}{150 - 100} = 20\text{weeks}$$

13.

$$S = 4\pi r^2, \text{ or } r = \sqrt{\frac{S}{4\pi}} = \left(\frac{S}{4\pi}\right)^{1/2}$$

$$V(S) = \frac{4}{3}\pi\left(\left(\frac{S}{4\pi}\right)^{1/2}\right)^3 = \frac{4}{3}\pi\left(\frac{S}{4\pi}\right)^{3/2}$$

$$V(S) = \frac{4}{3}\pi\frac{S^{3/2}}{4^{3/2}\pi^{3/2}} = \frac{4}{3}\pi\frac{S^{3/2}}{8\pi^{3/2}} = \frac{S^{3/2}}{6\pi^{1/2}} = \frac{S^{3/2}}{6\sqrt{\pi}}$$

$$V(2S) = \frac{(2S)^{3/2}}{6\sqrt{\pi}} = \frac{2^{3/2}S^{3/2}}{6\sqrt{\pi}},$$

so volume increased by a factor of $2^{3/2}$, or $2\sqrt{2}$, when S is doubled.

15. (a) The graphs of

$$y = -3x + 5 \quad \text{and} \quad y = 2x - 10$$

intersect when

$$-3x + 5 = 2x - 10, \quad \text{or} \quad x = 3.$$

When $x = 3$, $y = -3(3) + 5 = -4$. So the point of intersection is $(3, -4)$.

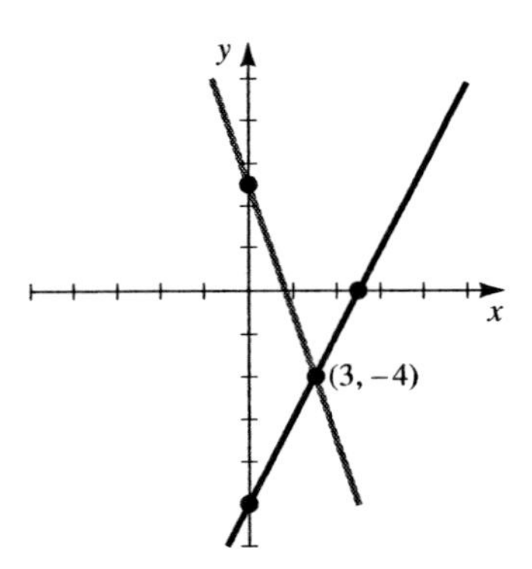

(b) The graphs of

$$y = x + 7 \quad \text{and} \quad y = -2 + x$$

are lines having the same slope, so they are parallel lines and there are no points of intersection.

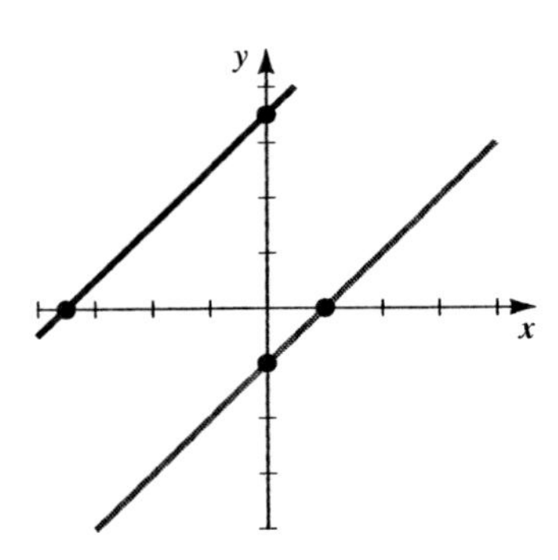

(c) The graphs of

$$y = x^2 - 1 \quad \text{and} \quad y = 1 - x^2$$

intersect when

$$x^2 - 1 = 1 - x^2, \quad 2x^2 = 2, \quad x^2 = 1, \quad \text{or} \quad x = \pm 1$$

When $x = \pm 1$, $y = (\pm 1)^2 - 1 = 0$. So the points of intersection are $(-1, 0)$ and $(1, 0)$.

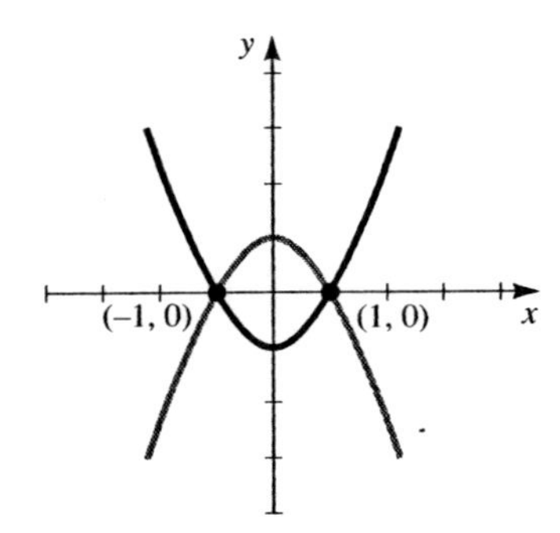

(d) The graphs of

$$y = x^2 \quad \text{and} \quad . \; y = 15 - 2x$$

intersect when

$$x^2 = 15 - 2x$$

$(x + 5)(x - 3) = 0$, or $x = -5$ and $x = 3$. When $x = -5$, $y = 25$, and when $x = 3$, $y = 9$. So, the points of intersection are $(-5, 25)$ and $(3, 9)$.

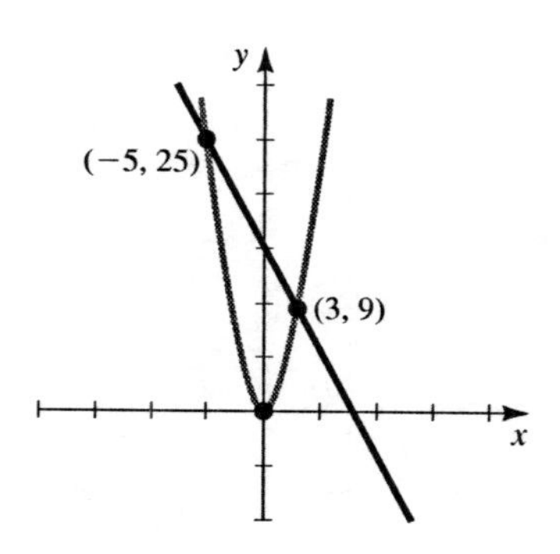

17. If p represents the selling price, the monthly profit is $P(p)$ =(number of cameras sold) (price − cost) Since $\frac{340 - P}{5}$ represents the number of \$5 decreases,

$$40 + 10\left(\frac{340 - p}{5}\right) =$$

represents the number of cameras that will sell. So,

$$\begin{aligned} P(p) &= (720 - 2p)(p - 150) \\ &= 2(360 - p)(p - 150) \end{aligned}$$

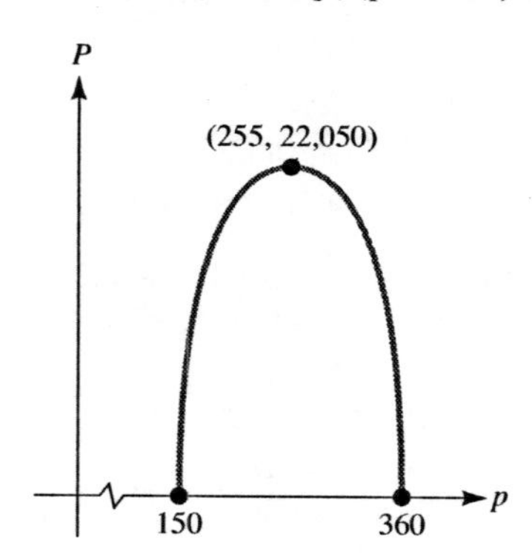

The graph suggests a maximum profit when $p = 255$, that is when the selling price is \$255.

19. Let x denote the number of machines used and $C(x)$ the corresponding cost function. Then,

$$\begin{aligned} C(x) &= \text{(set up cost)} + \text{(operating cost)} \\ &= 80\text{(number of machines)} \\ &\quad + 5.76\text{(number of hours)}. \end{aligned}$$

Since 400,000 medals are to be produced and each of the x machines can produce 200 medals per hour,

$$\text{number of hours} = \frac{400{,}000}{200x} = \frac{2{,}000}{x}$$

So,

$$\begin{aligned} C(x) &= 80x + 5.76\left(\frac{2{,}000}{x}\right) \\ &= 80x + \frac{11{,}520}{x}. \end{aligned}$$

The graph suggests that the cost will be smallest when x is approximately 12.
Note: In chapter 3 you will learn how to use calculus to find the optimal number of machines exactly.

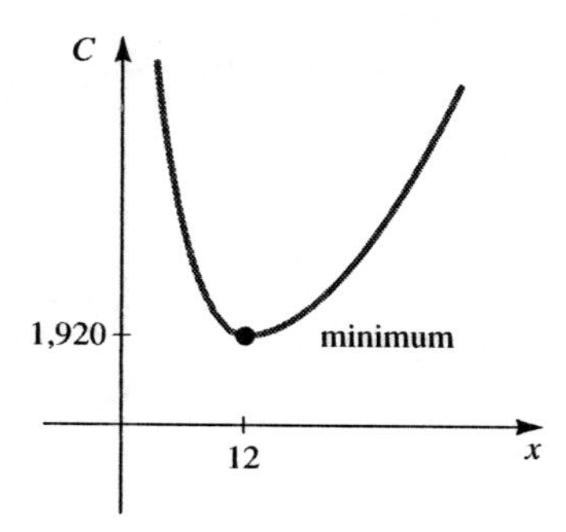

21. **(a)** Let x denote the number of units manufactured and sold. $C(x)$ and $R(x)$ are the corresponding cost and revenue functions, respectively.

$$\begin{aligned} C(x) &= 4{,}500 + 50x \\ R(x) &= 80x \end{aligned}$$

For the manufacturer to break even, since profit = revenue − cost, 0 = revenue − cost, or revenue = cost. That is.

$$4{,}500 + 50x = 80x \quad \text{or} \quad x = 150 \text{ units}$$

(b) Let $P(x)$ denote the profit from the manufacture and sale of x units. Then,

$$\begin{aligned} P(x) &= R(x) - C(x) \\ &= 80x - (4{,}500 + 50x) \\ &= 30x - 4{,}500. \end{aligned}$$

When 200 units are sold, the profit is

$$P(200) = 30(200) - 4{,}500 = \$1,500$$

(c) The profit will be $900 when

$$900 = 30x - 4{,}500 \quad \text{or} \quad x = 180$$

that is, when 180 units are manufactured and sold.

23. Let x denote the number of relevant facts recalled, n the total number of relevant facts in the person's memory, and $R(x)$ the rate of recall.
Then $n - x$ is the number of relevant facts not recalled.
So, $R(x) = k(n - x)$
where k is a constant of proportionality.

25. The cost for the clear glass is (area) (cost per sq ft) $= (2xy)(3)$, and similarly, the cost for the stained glass is

$$\left(\frac{1}{2}\pi x^2\right)(10)$$

So, $C = 6xy + 5\pi x^2$.
Now, the perimeter is $\frac{1}{2}(2\pi x) + 2x + 2y = 20$
so, $\quad \pi x + 2x + 2y = 20$,
or $\quad y = \dfrac{20 - \pi x - 2x}{2}$
Cost as a function of x is

$$\begin{aligned} C(x) &= 6x\left(\frac{20 - \pi x - 2x}{2}\right) + 5\pi x^2 \\ &= 3x(20 - \pi x - 2x) + 5\pi x^2 \\ &= 60x - 3\pi x^2 - 6x^2 + 5\pi x^2 \\ &= 60x - 6x^2 + 2\pi x^2 \end{aligned}$$

27. The fixed cost is $1,500 and the cost per unit is $2, so the cost is $C(x) = 1{,}500 + 2x$, for $0 \le x \le 5{,}000$. As to the question of continuity, the answer is both yes and no. Yes, if (as we normally do) x is any real number. No, if x is discrete ($x = 0, 1, 2, \ldots, 5{,}000$).

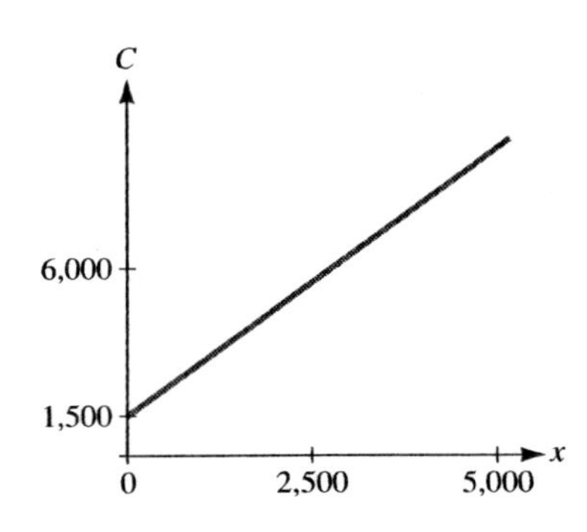

29. Taxes under Proposition A are $100 + .08a$, where a is the assessed value of the home. Taxes under Proposition B are $1{,}900 + .02a$. Taxes are the same when

$$\begin{aligned} 100 + .08a &= 1,900 + .02a \\ .06a &= 1,800 \\ a &= 30,000 \end{aligned}$$

or for an assessed value of $30,000.
Since both tax functions are linear, it is only necessary to test one additional assessed value to determine which proposition is best for all assessed values. For $a = 20{,}000$

$$100 + .08(20{,}000) = \$1{,}700$$
$$1{,}900 + .02(20{,}000) = \$2{,}300$$

So, for $0 < a < 30{,}000$, Proposition A is preferable while for $a > 30{,}000$, Proposition B is preferable.

31.
$$\begin{aligned} &\lim_{x \to 1} \frac{x^2 + x - 2}{x^2 - 1} \\ &= \lim_{x \to 1} \frac{(x+2)(x-1)}{(x+1)(x-1)} = \lim_{x \to 1} \frac{x+2}{x+1} \\ &= \frac{\lim_{x \to 1}(x+2)}{\lim_{x \to 1}(x+1)} = \frac{1+2}{1+1} = \frac{3}{2} \end{aligned}$$

33.
$$\lim_{x \to 2} \frac{x^3 - 8}{2 - x} = \lim_{x \to 2} \frac{(x-2)(x^2 + 2x + 4)}{-(x-2)}$$
$$= \lim_{x \to 2} -(x^2 + 2x + 4) = -(2^2 + 2(2) + 4) = -12.$$

35.
$$\lim_{x \to 0}\left(2 - \frac{1}{x^3}\right) = \lim_{x \to 0} 2 - \lim_{x \to 0} \frac{1}{x^3}$$

Now, $\lim_{x\to 0} 2 = 2$; but $\lim_{x\to 0^+} \frac{1}{x^3} = +\infty$ and $\lim_{x\to 0^-} \frac{1}{x^3} = -\infty$. Since $\lim_{x\to 0} \frac{1}{x^3}$ does not exist, $\lim_{x\to 0}\left(x^3 - \frac{1}{x^3}\right)$ does not exist.

37.
$$\lim_{x\to -\infty} \frac{x}{x^2+5} = \lim_{x\to -\infty} \frac{\frac{1}{x}}{1+\frac{5}{x^2}} = 0$$

39.
$$\lim_{x\to -\infty} \frac{x^4+3x^2-2x+7}{x^3+x+1} = \lim_{x\to -\infty} \frac{x+\frac{3}{x}-\frac{2}{x^2}+\frac{7}{x^3}}{1+\frac{1}{x^2}+\frac{1}{x^3}}$$

Since
$$\lim_{x\to -\infty}\left(x+\frac{3}{x}-\frac{2}{x^2}+\frac{7}{x^3}\right) = -\infty$$

and
$$\lim_{x\to -\infty}\left(1+\frac{1}{x^2}+\frac{1}{x^3}\right) = 1,$$

then
$$\lim_{x\to -\infty} \frac{x^4+3x^2-2x+7}{x^3+x+1} = -\infty$$

41. Since $\lim_{x\to -\infty}\left(1+\frac{1}{x}+\frac{1}{x^2}\right) = 1$ and
$$\lim_{x\to -\infty}(x^3+x+1) = -\infty,$$
$$\lim_{x\to -\infty} \frac{1+\frac{1}{x}+\frac{1}{x^2}}{x^3+x+1} = 0$$

43.
$$\lim_{x\to 0^-} x\sqrt{1-\frac{1}{x}} = \left(\lim_{x\to 0^-} x\right)\left(\lim_{x\to 0^-}\sqrt{1-\frac{1}{x}}\right)$$

Since $\lim_{x\to 0^-} x = 0$, and $\lim_{x\to 0^-}\left(1-\frac{1}{x}\right) = \infty$ implies

$$\lim_{x\to 0^-}\sqrt{1-\frac{1}{x}} = \infty, \quad \text{then } \lim_{x\to 0^-} x\sqrt{1-\frac{1}{x}} = 0.$$

45. $f(x) = \dfrac{x^2-1}{x+3}$
is not continuous at $x = -3$
since $f(-3) = \dfrac{10}{0}$ and division by 0 is undefined.

47.
$$h(x) = \begin{cases} x^3+2x-33 & \text{if } x \le 3 \\ \dfrac{x^2-6x+9}{x-3} & \text{if } x > 3 \end{cases}$$

The denominator in $\frac{x^2-6x+9}{x-3}$ will never be zero, since $x = 3$ is not included in its domain. However, in checking the break point (the only point in question),

$$h(3) = (3)^3 + 2(3) - 33 = 0$$

Further, $\lim_{x\to 3^-} h(x) = \lim_{x\to 3^-}(x^3+2x-33) = 0$ and

$$\lim_{x\to 3^+} h(x) = \lim_{x\to 3^+} \frac{x^2-6x+9}{x-3} = \lim_{x\to 3^+} \frac{(x-3)(x-3)}{x-3}$$
$$= \lim_{x\to 3^+}(x-3) = 3-3 = 0.$$

Since $h(3) = \lim_{x\to 3} h(x)$, h is continuous for all x.

49.
$$w(x) = \begin{cases} Ax & \text{if } x \le 4{,}000 \\ \frac{B}{x^2} & \text{if } x > 4{,}000 \end{cases}$$

For continuity,

$$4{,}000A = \frac{B}{(4{,}000)^2}$$

or $B = A(4,000)^3$

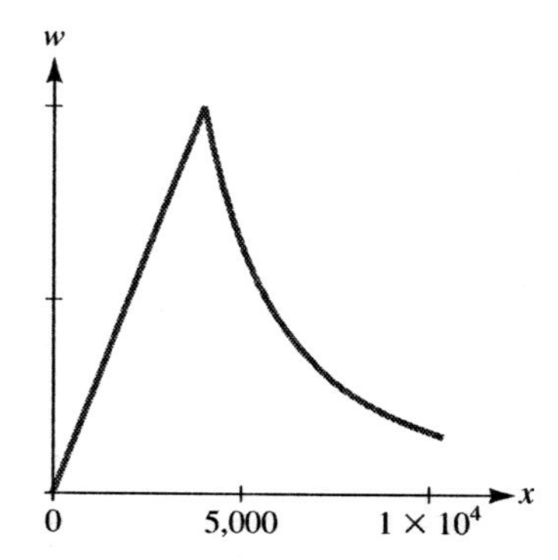

51. To graph $y = \frac{21}{9}x - \frac{84}{35}$ and $y = \frac{654}{279}x - \frac{54}{10}$, press [y =].
Input $(21x)/9 - 84/35$ for $y_1 =$ and press [enter].
Input $(654x)/279 - 54/10$ for $y_2 =$.
Use the z-standard function under the zoom menu to use the window dimensions given.
Press [graph].

It appears from the graph that the two lines are parallel. However, the difference in the slopes is $\frac{21}{9} - \frac{654}{279} = -.01$ which shows that, in fact, the lines are not parallel since they have different slopes.

53. Press [y=].
Input $(x \wedge 2 + 1)/(x \leq 1)$ for $y_1 =$ and press [enter].
(You can obtain the $\leq$ from [2ND] [test] and enter 6: $\leq$).
Input $(x \wedge 2 - 1)/(x > 1)$ for $y_2 =$ and press [enter].
(You can obtain the $>$ from [2ND] [test] and enter 3: $>$). Press [graph].
The graph of y is discontinuous $x = 1$.

55. This limit does exist. The curve is bounded by the lines $y = mx$ and $y = -mx$. Since $-m|x| \leq g(x) \leq m|x|$, as x approaches 0, the bounding values on the right and the left of the inequality also approach 0. The function in the middle $g(x)$, is squeezed or sandwiched between 0 and 0. Its limit has to be 0.
Note: $\lim_{x \to 0^-} |x| \sin(1/x) = \lim_{x \to 0^+} |x| \sin(1/x) = 0$ since $-1 \leq \sin x \leq 1$.

Chapter 2

Differentiation: Basic Concepts

2.1 The Derivative

1. If $f(x) = 5x - 3$, then

$$f(x+h) = 5(x+h) - 3$$

The difference quotient (DQ) is

$$\frac{f(x+h) - f(x)}{h}$$

$$= \frac{[5(x+h) - 3] - [5x - 3]}{h} = \frac{5h}{h} = 5$$

$$f'(x) = \lim_{h \to 0} \frac{f(x+h) - f(x)}{h} = 5$$

The slope is $m = f'(2) = 5$.

3. If $f(x) = 2x^2 - 3x + 5$, then

$$f(x+h) = 2(x+h)^2 - 3(x+h) + 5$$

The difference quotient (DQ) is

$$\frac{f(x+h) - f(x)}{h}$$

$$= \frac{[2(x+h)^2 - 3(x+h) + 5]}{h} - \frac{[2x^2 - 3x + 5]}{h}$$

$$= \frac{4xh + 2(h)^2 - 3h}{h} = 4x + 2h - 3$$

$$f'(x) = \lim_{h \to 0} \frac{f(x+h) - f(x)}{h} = 4x - 3$$

The slope is

$$m = f'(0) = -3$$

5. If $g(t) = \frac{2}{t}$, then

$$g(t+h) = \frac{2}{t+h}$$

The difference quotient (DQ) is

$$\frac{g(t+h) - g(t)}{h}$$

$$= \frac{\frac{2}{t+h} - \frac{2}{t}}{h}$$

$$= \frac{\frac{2}{t+h}}{h} \cdot \frac{t(t+h)}{t(t+h)}$$

$$= \frac{2t - 2(t+h)}{h(t)(t+h)} = \frac{-2}{t(t+h)}$$

$$g'(t) = \lim_{h \to 0} \frac{g(t+h) - g(t)}{h} = -\frac{2}{t^2}$$

The slope is $m = g'\left(\frac{1}{2}\right) = -8$.

7. If $f(x) = \sqrt{x}$, then

$$f(x+h) = \sqrt{x+h}$$

The difference quotient (DQ) is

$$\frac{f(x+h) - f(x)}{h}$$

$$= \frac{\sqrt{x+h} - \sqrt{x}}{h}$$

$$= \frac{\sqrt{x+h} - \sqrt{x}}{h} \cdot \frac{\sqrt{x+h} + \sqrt{x}}{\sqrt{x+h} + \sqrt{x}}$$

$$= \frac{x + h - x}{h\left(\sqrt{x+h} + \sqrt{x}\right)} = \frac{1}{\sqrt{x+h} + \sqrt{x}}$$

$$f'(x) = \lim_{h \to 0} \frac{f(x+h) - f(x)}{h} = \frac{1}{2\sqrt{x}}$$

The slope is $m = f'(9) = \frac{1}{6}$

9. If $f(x) = x^2$, then

$$f(x+h) = (x+h)^2$$

The difference quotient (DQ) is

$$\frac{f(x+h) - f(x)}{h}$$
$$= \frac{(x+h)^2 - x^2}{h}$$
$$= \frac{2xh + h^2}{h}$$
$$= 2x + h$$
$$f'(x) = \lim_{h \to 0} \frac{f(x+h) - f(x)}{h} = 2x$$

The slope of the line is $m = f'(1) = 2$.
Since $f(1) = 1$, $(1, 1)$ is a point on the curve and the equation of the tangent line is

$$y - 1 = 2(x - 1)$$
$$\text{or} \quad y = 2x - 1$$

11. If $f(x) = 7 - 2x$, then

$$f(x+h) = 7 - 2(x+h)$$

The difference quotient (DQ) is

$$\frac{f(x+h) - f(x)}{h}$$
$$= \frac{[7 - 2(x+h)] - [7 - 2x]}{h}$$
$$= -2$$
$$f'(x) = \lim_{h \to 0} \frac{f(x+h) - f(x)}{h} = -2$$

The slope of the line is $m = f'(5) = -2$.
Since $f(5) = -3$, $(5, -3)$ is a point on the curve and the equation of the tangent line is

$$y - (-3) = -2(x - 5)$$
$$\text{or} \quad y = -2x + 7$$

13. If $f(x) = -\frac{2}{x}$, then

$$f(x+h) = \frac{-2}{x+h}$$

The difference quotient (DQ) is

$$\frac{f(x+h) - f(x)}{h}$$
$$= \frac{\frac{-2}{x+h} - \frac{-2}{x}}{h}$$
$$= \frac{\frac{-2}{x+h} + \frac{2}{x}}{h} \cdot \frac{x(x+h)}{x(x+h)}$$
$$= \frac{-2x + 2(x+h)}{h(x)(x+h)} = \frac{2}{x(x+h)}$$
$$f'(x) = \lim_{h \to 0} \frac{f(x+h) - f(x)}{h} = \frac{2}{x^2}$$

The slope of the line is $m = f'(-1) = 2$.
Since $f(-1) = 2$, $(-1, 2)$ is a point on the curve and the equation of the tangent line is

$$y - 2 = 2(x - (-1))$$
$$y = 2x + 4$$

15. Since $\frac{d}{dx} k \cdot f(x) = k \cdot \frac{d}{dx} f(x)$, from problem 7,

$$f'(x) = 2\left(\frac{1}{2\sqrt{x}}\right) = \frac{1}{\sqrt{x}}$$

The slope is $m = f'(4) = \frac{1}{2}$, $f(4) = 4$, the equation of the tangent line is

$$y - 4 = \frac{1}{2}(x - 4), \text{ or}$$
$$y = \frac{1}{2}x + 2$$

17. If $y = f(x) = 3$, then

$$f(x+h) = 3$$

The difference quotient (DQ) is

$$\frac{f(x+h) - f(x)}{h}$$

$$= \frac{3-3}{h} = 0$$

$$\frac{dy}{dx} = \lim_{h\to 0} \frac{f(x+h)-f(x)}{h} = 0$$

$$\frac{dy}{dx} = 0 \text{ when } x = 2.$$

19. If $y = f(x) = x(1-x)$, or $f(x) = x - x^2$, then

$$f(x+h) = (x+h) - (x+h)^2$$

The difference quotient (DQ) is

$$\frac{f(x+h)-f(x)}{h}$$

$$= \frac{[(x+h)-(x+h)^2] - [x - x^2]}{h}$$

$$= \frac{h - 2xh - h^2}{h} = 1 - 2x - h$$

$$\frac{dy}{dx} = \lim_{h\to 0} \frac{f(x+h)-f(x)}{h} = 1 - 2x$$

$$\frac{dy}{dx} = 3 \text{ when } x = -1.$$

21. If $y = f(x) = x^2 - 2x$, then

$$f(x+h) = (x+h)^2 - 2(x+h)$$

The difference quotient (DQ) is

$$\frac{f(x+h)-f(x)}{h}$$

$$= \frac{[(x+h)^2 - 2(x+h)] - [x^2 - 2x]}{h}$$

$$= \frac{2xh + h^2 - 2h}{h} = 2x + h - 2$$

$$\frac{dy}{dx} = \lim_{h\to 0} \frac{f(x+h)-f(x)}{h} = 2x - 2$$

$$\frac{dy}{dx} = 0 \text{ when } x = 1.$$

23. **(a)** If $f(x) = x^3$, then $f(1) = 1$, $f(1.1) = (1.1)^3 = 1.331$.

The slope of the secant line joining the points (1,1) and (1.1,1.331) on the graph of f is

$$m_{\text{sec}} = \frac{y_2 - y_1}{x_2 - x_1} = \frac{1.331 - 1}{1.1 - 1} = 3.31$$

(b) If $f(x) = x^3$, then

$$f(x+h) = (x+h)^3$$

The difference quotient (DQ) is

$$\frac{f(x+h)-f(x)}{h}$$

$$= \frac{(x+h)^3 - x^3}{h}$$

$$= \frac{3x^2h + 3xh^2 + h^3}{h}$$

$$= 3x^2 + 3xh + h^2$$

$$f'(x) = \lim_{h\to 0} \frac{f(x+h)-f(x)}{h} = 3x^2$$

The slope is $m_{\text{tan}} = f'(1) = 3$
Notice that this slope was approximated by the slope of the secant in part **(a)**.

25. **(a)** If $f(x) = 2x - x^2$, then $f(0) = 0$, and
$f\left(\frac{1}{2}\right) = 2\left(\frac{1}{2}\right) - \left(\frac{1}{2}\right)^2 . = 1 - \frac{1}{4} = \frac{3}{4}.$
The slope of the secant line joining the points (0,0) and $\left(\frac{1}{2}, \frac{3}{4}\right)$ on the graph of f is

$$m_{\text{sec}} = \frac{y_2 - y_1}{x_2 - x_1} = \frac{\frac{3}{4} - 0}{\frac{1}{2} - 0} = \frac{3}{2}$$

(b) If $f(x) = 2x - x^2$, then
$f(x+h) = 2(x+h) - (x+h)^2$
The difference quotient (DQ) is
$$\frac{f(x+h)-f(x)}{h}$$

$$= \frac{2(x+h)-(x+h)^2-(2x-x^2)}{h}$$
$$= \frac{2x+2h-x^2-2xh-h^2-2x+x^2}{h}$$
$$= \frac{2h-2xh-h^2}{h}$$
$$= 2-2x-h$$
$$f'(x) = \lim_{h\to 0}(2-2x-h) = 2-2x$$

The slope is $m_{\tan} = f'(0) = 2$
Notice that the estimate given by the slope of the secant in part **(a)** differs somewhat.

27. (a) If $f(x) = 3x^2 - x$, the average rate of change of f is
$$\frac{f(x_2)-f(x_1)}{x_2-x_1}$$
Since $f(0) = 0$ and $f\left(\frac{1}{16}\right) = 3\left(\frac{1}{16}\right)^2 - \frac{1}{16} = -\frac{13}{256}$,
$$\frac{f(x_2)-f(x_1)}{x_2-x_1} = \frac{-\frac{13}{256}-0}{\frac{1}{16}-0} = -\frac{13}{16} = -0.8125$$

(b) If $f(x) = 3x^2 - x$, then
$f(x+h) = 3(x+h)^2 - (x+h)$
The difference quotient (DQ) is
$$\frac{f(x+h)-f(x)}{h}$$
$$= \frac{3(x+h)^2-(x+h)-(3x^2-x)}{h}$$
$$= \frac{3x^2+6xh+3h^2-x-h-3x^2+x}{h}$$
$$= \frac{6xh+3h^2-h}{h} = 6x+3h-1$$
$$f'(x) = \lim_{h\to 0}(6x+3h-1) = 6x-1$$

The instantaneous rate of change at $x = 0$ is $f'(0) = -1$. Notice that this rate is estimated by the average rate in part **(a)**.

29. (a) If $s(t) = \frac{t-1}{t+1}$, the average rate of change of s is
$$\frac{s(t_2)-s(t_1)}{t_2-t_1}$$
Since $s\left(-\frac{1}{2}\right) = \frac{\frac{1}{2}-1}{-\frac{1}{2}+1} = -3$ and $s(0) = \frac{0-1}{0+1} = -1, = \frac{-3+1}{-\frac{1}{2}-0} = 4$

(b) If $s(t) = \frac{t-1}{t+1}$, then
$$s(t+h) = \frac{(t+h)-1}{(t+h)+1}$$
The difference quotient (DQ) is
$$\frac{s(t+h)-s(t)}{h}$$
$$= \frac{\frac{t+h-1}{t+h+1}-\frac{t-1}{t+1}}{h}$$
Multiplying numerator and denominator by $(t+h+1)(t+1)$,
$$= \frac{(t+h-1)(t+1)-(t-1)(t+h+1)}{h(t+h+1)(t+1)}$$
$$= \frac{t^2+th-t+t+h-1-t^2-th-t+t+h+1}{h(t+h+1)(t+1)}$$
$$= \frac{2h}{h(t+h+1)(t+1)} = \frac{2}{(t+h+1)(t+1)}$$
$$s'(t) = \lim_{h\to 0}\frac{2}{(t+h+1)(t+1)} = \frac{2}{(t+1)^2}$$
The instantaneous rate of change when $t = -\frac{1}{2}$ is
$$s'\left(-\frac{1}{2}\right) = \frac{2}{\left(-\frac{1}{2}+1\right)^2} = 8$$
Notice that the estimate given by the average rate in part **(a)** differs significantly.

31. (a) The average rate of temperature change between t_0 and $t_0 + h$ hours after midnight.

The instanteous rate of temperature change t_0 hours after midnight.

(b) The average rate of change in blood alcohol level between t_0 and $t_0 + h$ hours after consumption. The instantaneous rate of change in blood alcohol level t_0 hours after consumption.

(c) The average rate of change of the 30-year fixed mortgage rate between t_0 and $t_0 + h$ years after 2000. The instantaneous rate of change of 30-year fixed mortgage rate t_0 years after 2000.

33. When $t = 30$, $\frac{dV}{dt} \approx \frac{65 - 50}{50 - 30} = \frac{3}{4}$.
In the "long run", the rate at which V is changing with respect to time is getting smaller and smaller, decreasing to zero.

35. When $h = 1{,}000$ meters,

$$\frac{dT}{dh} \approx \frac{-6 - 0}{2{,}000 - 1{,}000} = \frac{-6}{1{,}000} = -0.006\ °\text{C/meter}$$

When $h = 2{,}000$ meters,

$$\frac{dT}{dh} = 0\ °\text{C/meter}$$

Since the line tangent to the graph at $h = 2{,}000$ is horizontal, its slope is zero.

37. **(a)** Profit = (number sold)(profit on each)
Profit on each = selling price − cost to obtain

$$P(p) = (120 - p)(p - 20)$$

Since $q = 120 - p,\ p = 120 - q$

$$P(q) = q\left[(120 - q) - 20\right]$$

or $P(q) = q(100 - q) = 100q - q^2$.

(b) The average rate as q increases from $q = 0$ to $q = 20$ is

$$\frac{P(20) - P(0)}{20} = \frac{\left[100(20) - (20)^2\right] - 0}{20}$$

$= \$80$ per recorder

(c) The rate the profit is changing at $q = 20$ is $P'(20)$.

The difference quotient (DQ) is

$$\frac{P(q+h) - P(q)}{h}$$

$$= \frac{\left[100(q+h) - (q+h)^2\right] - \left[100q - q^2\right]}{h}$$

$$= \frac{100q + 100h - q^2 - 2qh - h^2 - 100q - q^2}{h}$$

$$= \frac{100h - 2qh - h^2}{h} = 100 - 2q - h$$

$$P'(q) = \lim_{h\to 0} \frac{P(q+h) - P(q)}{h} = 100 - 2q$$

$P'(20) = 100 - 2(20) = \$60$ per recorder.
Since $P'(20)$ is positive, profit is increasing.

39. $P(x) = 4{,}000(15 - x)(x - 2)$

(a) The difference quotient (DQ) is

$$\frac{P(x+h) - P(x)}{h}$$

$$= \frac{[4{,}000\,(15 - (x+h))\,((x+h) - 2)]}{h}$$

$$- \frac{[4{,}000(15 - x)(x - 2)]}{h}$$

$$= \frac{4{,}000\,[(15 - x - h)(x + h - 2) - (15 - x)(x - 2)]}{h}$$

$$= \frac{4{,}000(17h - 2xh - h^2)}{h} = 4{,}000(17 - 2x - h)$$

$$P'(x) = \lim_{h\to 0} \frac{P(x+h) - P(x)}{h} = 4{,}000(17 - 2x)$$

(b) $P'(x) = 0$ when

$$4{,}000(17 - 2x) = 0$$

$x = \frac{17}{2} = 8.5$, or 850 units.
When $P'(x) = 0$, the line tangent to the graph of P is horizontal. Since the graph of P is a parabola which opens down, this horizontal tangent indicates a maximum profit.

41. $C(x) = 0.04x^2 + 5.1x + 40$

(a) The average rate of change is

$$\frac{C(x_2) - C(x_1)}{x_2 - x_1}$$

Since $C(10) = 0.04(10)^2 + 5.1(10) + 40 = 95$ and $C(11) = 0.04(11)^2 + 5.1(11) + 40 = 100.94$,

$$\frac{C(x_2) - C(x_1)}{x_2 - x_1} = \frac{100.94 - 95}{11 - 10} = \$5.94 \text{ per unit}$$

(b) $C(x + h) = 0.04(x + h)^2 + 5.1(x + h) + 40$

So, the difference quotient (DQ) is

$$\frac{C(x + h) - C(x)}{h}$$

$$= \frac{0.04(x + h)^2 + 5.1(x + h) + 40 - (0.04x^2 + 5.1x + 40)}{h}$$

$$= \left[0.04x^2 + 0.08xh + 0.04h^2 + 5.1x + 5.1h + 40 - 0.04x^2 - 5.1x - 40\right] / h$$

$$= \frac{0.08xh + 0.04h^2 + 5.1h}{h}$$

$$= 0.08x + 0.04h + 5.1$$

$$C'(x) = \lim_{h \to 0} (0.08x + 0.04h + 5.1) = 0.08x + 5.1$$

The instantaneous rate of change when $x = 10$ is $C'(10) = 0.08(10) + 5.1 = \$5.90$ per unit. Since $C'(10)$ is positive, the cost is increasing when 10 units are being produced.

43. Writing Exercise—Answers will vary.

45. $D(p) = -0.0009p^2 + 0.13p + 17.81$

(a) The average rate of change is

$$\frac{D(p_2) - D(p_1)}{p_2 - p_1}$$

Since $D(60) = -0.0009(60)^2 + 0.13(60) + 17.81 = 22.37$ and $D(61) = -0.0009(61)^2 + 0.13(61) + 17.81 = 22.3911$,

$$= \frac{22.3911 - 22.37}{61 - 60}$$

$$= 0.0211 \text{ mm per mm of mercury}$$

(b) $D(p + h) = -0.0009(p + h)^2 + 0.13 (p + h) + 17.81$

So, the difference quotient (DQ) is

$$\frac{D(p + h) - D(p)}{h}$$

$$= \left[-0.0009(p + h)^2 + 0.13(p + h) + 17.81 - (-0/0009p^2 + 0.13p + 17.81)\right] / h$$

$$= \left[-0.0009p^2 - 0.0018ph - 0.0009h^2 + 0.13p + 0.13h + 17.81 + 0.0009p^2 - 0.13p - 17.81\right] / h$$

$$= \frac{-0.0018ph - 0.0009h^2 + 0.13h}{h}$$

$$= -0.0018p - 0.0009h + 0.13$$

$$D'(x) = \lim_{h \to 0} (-0.0018p - 0.0009h + 0.13)$$

$$= -0.0018p + 0.13$$

The instantaneous rate of change when $p = 60$ is $D'(60) = -0.0018(60) + 0.13 = 0.022$ mm per mm of mercury. Since $D'(60)$ is positive, the pressure is increasing when $p = 60$.

(c) $-0.0018p + 0.13 = 0$

$p \approx 72.22$ mm of mercury

At this pressure, the diameter is neither increasing nor decreasing.

47. **(a)** For $y = f(x) = x^2$, $f(x + h) = (x + h)^2$

The difference quotient (DQ) is

$$\frac{f(x + h) - f(x)}{h} = \frac{(x + h)^2 - x^2}{h}$$

$$= \frac{2xh + h^2}{h} = 2x + h$$

$$\frac{dy}{dx} = f'(x) = \lim_{h \to 0} \frac{f(x + h) - f(x)}{h} = 2x$$

For $y = f(x) = x^2 - 3$, $f(x + h) = (x + h)^2 - 3$

The difference quotient (DQ) is

$$\frac{[(x+h)^2-3]-(x^2-3)}{h}$$

$$=\frac{2xh+h^2}{h}=2x+h$$

$$\frac{dy}{dx}=f'(x)=\lim_{h\to 0}\frac{f(x+h)-f(x)}{h}=2x$$

The graph of $y=x^2-3$ is the graph of $y=x^2$ shifted down 3 units. So the graphs are parallel and their tangent lines have the same slopes for any value of x. This accounts geometrically for the fact that their derivatives are identical.

(b) Since $y=x^2+5$ is the parabola $y=x^2$ shifted up 5 units and the constant appears to have no effect on the derivative, the derivative of the function $y=x^2+5$ is also $2x$.

49. (a) For $y=f(x)=x^2$,
$f(x+h)=(x+h)^2$
The difference quotient (DQ) is

$$\frac{f(x+h)-f(x)}{h}=\frac{(x+h)^2-x^2}{h}$$

$$=\frac{2xh+h^2}{h}=2x+h$$

$$\frac{dy}{dx}=f'(x)=\lim_{h\to 0}\frac{f(x+h)-f(x)}{h}=2x$$

For $y=f(x)=x^3$,
$f(x+h)=(x+h)^3$
The difference quotient (DQ) is

$$\frac{(x+h)^3-x^3}{h}$$

$$=\frac{3x^2h+3xh^2+h^3}{h}=3x^2+3xh+h^2$$

$$\frac{dy}{dx}=f'(x)=\lim_{h\to 0}\frac{f(x+h)-f(x)}{h}=3x^2$$

(b) The pattern seems to be that the derivative of x raised to a power (x^n) is that power times x raised to the power decreased by one (nx^{n-1}). So, the derivative of the function $y=x^4$ is $4x^3$ and the derivative of the function $y=x^{27}$ is $27x^{26}$.

51. When $x<0$, the difference quotient (DQ) is

$$\frac{f(x+h)-f(x)}{h}=\frac{-(x+h)-(-x)}{h}=\frac{-h}{h}=-1$$

So, $f'(x)=\lim_{h\to 0}-1=-1$.
When $x>0$, the difference quotient (DQ) is

$$\frac{f(x+h)-f(x)}{h}=\frac{(x+h)-x}{h}=1$$

So, $f'(x)=\lim_{h\to 0}1=1$.
Since there is a sharp corner at $x=0$ (graph changes from $y=-x$ to $y=x$), the graph makes an abrupt change in direction at $x=0$. So, f is not differentiable at $x=0$.

53. To show that $f(x)=\dfrac{|x^2-1|}{x-1}$ is not differentiable at $x=1$,
Press [y=] and input $(abs(x^2-1))/(x-1)$ for $y_1=$

The abs is under the NUM menu in the math application.
Use window dimensions [−4, 4]1 by [−4, 4]1
Press [Graph]

We see that f is not defined at $x=1$. There can be no point of tangency.

$$\lim_{x\to 1^+}\frac{|x^2-1|}{x-1}=\lim_{x\to 1^+}\frac{|(x-1)(x+1)|}{x-1}=2$$

$$\lim_{x\to 1^-}\frac{|x^2-1|}{x-1}=\lim_{x\to 1^-}\frac{|(x-1)(x+1)|}{x-1}=-2$$

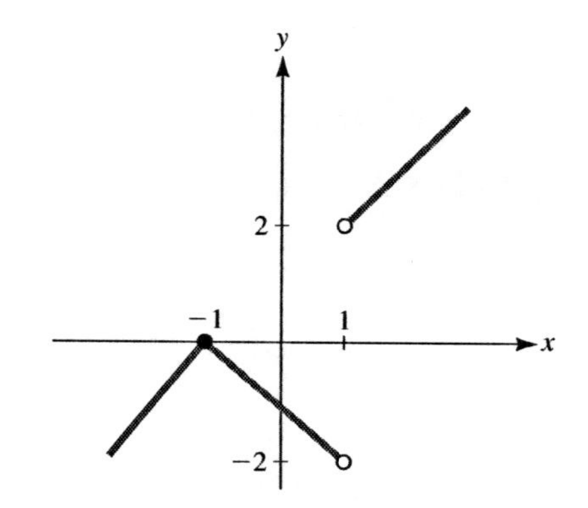

55. To find the slope of line tangent to the graph of $f(x)=\sqrt{x^2+2x-\sqrt{3x}}$ at $x=3.85$, fill in the table below.
The $x+h$ row can be filled in manually.
For $f(x)$, press [y=] and input
$\sqrt{}\left(x\wedge 2+2x-\sqrt{}(3x)\right)$ for $y_1=$
Use window dimensions [−1, 10]1 by [−1, 10]1
Use the value function under the calc menu and enter $x=3.85$ to find $f(x)=4.37310$.
For $f(x+h)$, use the value function under the calc menu and enter $x=3.83$ To find $f(x+h)=4.35192$. Repeat this process for $x=3.84, 3.849, 3.85, 3.851, 3.86$, and 3.87.

The $\dfrac{f(x+h)-f(x)}{h}$ can be filled in manually given that the rest of the table is now complete.
So, slope $=f'(3.85)\approx 1.059$.

h	−0.02	−0.01	−0.001	0	0.001	0.01	0.02
$x+h$	3.83	3.84	3.849	3.85	3.851	3.86	3.87
$f(x)$	4.37310	4.37310	4.37310	4.37310	4.37310	4.37310	4.37310
$f(x+h)$	4.35192	4.36251	4.37204	4.37310	4.37415	4.38368	4.39426
$\dfrac{f(x+h)-f(x)}{h}$	1.059	1.059	1.059	undefined	1.05	1.058	1.058

2.2 Techniques of Differentiation

1. $y=x^{-4}$

$$\frac{dy}{dx}=-4x^{-4-1}=-4x^{-5}=-\frac{4}{x^5}$$

3. Since the derivative of any constant is zero,

$$y=-2$$
$$\frac{dy}{dx}=0$$

(Note: $y=-2$ is a horizontal line and all horizontal lines have a slope of zero, so $\frac{dy}{dx}$ must be zero.)

5. $y=\pi r^2$

$$\frac{dy}{dx}=\pi\left(2r^{2-1}\right)=2\pi r$$

7. $y=\sqrt{2x}=\sqrt{2}\cdot x^{1/2}$

$$\frac{dy}{dx}=\sqrt{2}\left(\frac{1}{2}x^{1/2-1}\right)=\sqrt{2}\left(\frac{1}{2}x^{-1/2}\right)$$
$$=\sqrt{2}\cdot\frac{1}{2x^{1/2}}=\frac{1}{\sqrt{2}x^{1/2}}\text{ or }\frac{1}{\sqrt{2x}}$$

9. $y=\dfrac{9}{\sqrt{t}}=9t^{-1/2}$

$$\frac{dy}{dx}=9\left(-\frac{1}{2}t^{-1/2-1}\right)=9\left(-\frac{1}{2}t^{-3/2}\right)$$
$$=-\frac{9}{2t^{3/2}}\text{ or }-\frac{9}{2\sqrt{t^3}}$$

11.
$$y=x^2+2x+3$$
$$\frac{dy}{dx}=\frac{d}{dx}(x^2)+\frac{d}{dx}(2x)+\frac{d}{dx}(3)$$
$$\frac{dy}{dx}=2x+2$$

13.
$$y=x^9-5x^8+x+12$$
$$\frac{dy}{dx}=\frac{d}{dx}(x^9)-\frac{d}{dx}(5x^8)+\frac{d}{dx}(x)+\frac{d}{dx}(12)$$
$$\frac{dy}{dx}=9x^8-40x^7+1$$

15.
$$f(x)=-0.02x^3+0.3x$$
$$f'(x)=\frac{d}{dx}(-0.02x^3)+(0.3x)$$
$$f'(x)=-0.02(3x^2)+0.3=-0.06x^2+0.3$$

17.
$$y=\frac{1}{t}+\frac{1}{t^2}-\frac{1}{\sqrt{t}}$$
$$=t^{-1}+t^{-2}-t^{-1/2}$$

$$\frac{dy}{dt} = \frac{d}{dt}(t^{-1}) + \frac{d}{dt}(t^{-2}) - \frac{d}{dt}\left(t^{-1/2}\right)$$
$$= -1t^{-1-1} + -2t^{-2-1} - \left(-\frac{1}{2}t^{-1/2-1}\right)$$
$$= -1t^{-2} - 2t^{-3} + \frac{1}{2}t^{-3/2}$$
$$= -\frac{1}{t^2} - \frac{2}{t^3} + \frac{1}{2t^{3/2}}, \text{ or } -\frac{1}{t^2} - \frac{2}{t^3} + \frac{1}{2\sqrt{t^3}}$$

19. $f(x) = \sqrt{x^3} + \dfrac{1}{\sqrt{x^3}} = x^{3/2} + x^{-3/2}$,

$$f'(x) = \frac{d}{dx}(x^{3/2}) + \frac{d}{dx}(x^{-3/2})$$
$$= \frac{3}{2}x^{3/2-1} + \frac{-3}{2}x^{-3/2-1}$$
$$= \frac{3}{2}x^{1/2} - \frac{3}{2}x^{-5/2}$$
$$= \frac{3}{2}x^{1/2} - \frac{3}{2x^{5/2}}, \text{ or } \frac{3}{2}\sqrt{x} - \frac{3}{2\sqrt{x^5}}$$

21.
$$y = -\frac{x^2}{16} + \frac{2}{x} - x^{3/2} + \frac{1}{3x^2} + \frac{x}{3}$$
$$= -\frac{1}{16}x^2 + 2x^{-1} - x^{3/2} + \frac{1}{3}x^{-2} + \frac{1}{3}x,$$
$$\frac{dy}{dx} = \frac{d}{dx}\left(-\frac{1}{16}x^2\right) + \frac{d}{dx}\left(2x^{-1}\right) - \frac{d}{dx}\left(x^{3/2}\right) + \frac{d}{dx}\left(\frac{1}{3}x^{-2}\right) + \frac{d}{dx}\left(\frac{1}{3}x\right)$$
$$= -\frac{1}{16}(2x) + 2(-1x^{-1-1)} - \frac{3}{2}x^{3/2-1} + \frac{1}{3}(-2x^{-2-1}) + \frac{1}{3}$$
$$= -\frac{1}{8}x - 2x^{-2} - \frac{3}{2}x^{1/2} - \frac{2}{3}x^{-3} + \frac{1}{3}$$
$$= -\frac{1}{8}x - \frac{2}{x^2} - \frac{3}{2}x^{1/2} - \frac{2}{3x^3} + \frac{1}{3},$$
$$\text{or } -\frac{1}{8}x - \frac{2}{x^2} + \frac{3}{2}\sqrt{x} - \frac{2}{3x^3} + \frac{1}{3}$$

23. $y = -\dfrac{2}{x^2} + x^{2/3} + \dfrac{1}{2\sqrt{x}} + \dfrac{x^2}{4} + \sqrt{5} + \dfrac{x+2}{3}$

$$= -2x^{-2} + x^{2/3} + \frac{1}{2}x^{-1/2} + \frac{1}{4}x^2 + \sqrt{5} + \frac{1}{3}x + \frac{2}{3}$$
$$\frac{dy}{dx} = \frac{d}{dx}(-2x^{-2}) + \frac{d}{dx}(x^{2/3}) + \frac{d}{dx}\left(\frac{1}{2}x^{-1/2}\right) + \frac{d}{dx}\left(\frac{1}{4}x^2\right) + \frac{d}{dx}\left(\sqrt{5}\right) + \frac{d}{dx}\left(\frac{1}{3}x\right)$$
$$= -2(-2x^{-2-1}) + \frac{2}{3}x^{2/3-1} + \frac{1}{2}\left(-\frac{1}{2}x^{-1/2-1}\right) + \frac{1}{4}(2x) + 0 + \frac{1}{3} + 0$$
$$= 4x^{-3} + \frac{2}{3}x^{-1/3} - \frac{1}{4}x^{-3/2} + \frac{1}{2}x + \frac{1}{3}$$
$$= \frac{4}{x^3} + \frac{2}{3x^{1/3}} - \frac{1}{4x^{3/2}} + \frac{1}{2}x + \frac{1}{3}, \text{ or}$$
$$\frac{4}{x^3} + \frac{2}{3\sqrt[3]{x}} - \frac{1}{4\sqrt{x^3}} + \frac{1}{2}x + \frac{1}{3}$$

25.
$$y = \frac{x^5 - 4x^2}{x^3} = \frac{x^5}{x^3} - \frac{4x^2}{x^3} = x^2 - \frac{4}{x} = x^2 - 4x^{-1}$$
$$\frac{dy}{dx} = \frac{d}{dx}(x^2) - \frac{d}{dx}(4x^{-1}) = 2x - 4(-1x^{-1-1})$$
$$= 2x + 4x^{-2} = 2x + \frac{4}{x^2}$$

27.
$$y = -x^3 - 5x^2 + 3x - 1$$
$$\frac{dy}{dx} = -3x^2 - 10x + 3$$

At $x = -1$, $\dfrac{dy}{dx} = 10$. The equation of the tangent line at $(-1, -8)$ is

$$y + 8 = 10(x + 1),$$
$$\text{or } y = 10x + 2$$

29.
$$y = 1 - \frac{1}{x} + \frac{2}{\sqrt{x}}$$
$$= 1 - x^{-1} + 2x^{-1/2}$$
$$\frac{dy}{dx} = x^{-2} - x^{-3/2} = \frac{1}{x^2} - \frac{1}{x^{3/2}}$$

At $\left(4, \frac{7}{4}\right)$, $\frac{dy}{dx} = -\frac{1}{16}$. The equation of the tangent line is

$$y - \frac{7}{4} = -\frac{1}{16}(x-4), \text{ or}$$
$$y = -\frac{1}{16}x + 2$$

31. $$y = (x^2 - x)(3 + 2x) = 2x^3 + x^2 - 3x$$
$$\frac{dy}{dx} = 6x^2 + 2x - 3$$

At $x = -1$, $\frac{dy}{dx} = 1$. The equation of the tangent line at $(-1, 2)$ is

$$y - 2 = 1(x+1), \text{ or } y = x + 3$$

33. $$f(x) = -2x^3 + \frac{1}{x^2} = -2x^3 + x^{-2}$$
$$f'(x) = -6x^2 - \frac{2}{x^3}$$

At $x = -1$, $f'(-1) = -4$. Further, $y = f(-1) = 3$. The equation of the tangent line at $(-1, 3)$ is

$$y - 3 = -4(x+1), \text{ or } y = -4x - 1$$

35. $$f(x) = x - \frac{1}{x^2} = x - x^{-2}$$
$$f'(x) = 1 + \frac{2}{x^3}$$

At $x = 1$, $f'(1) = 3$. Further, $y = f(1) = 0$. The equation of the tangent line at (1,0) is

$$y - 0 = 3(x-1), \text{ or } y = 3x - 3$$

37. $$f(x) = -\frac{1}{3}x^3 + \sqrt{8x} = -\frac{1}{3}x^3 + \sqrt{8} \cdot x^{1/2}$$
$$f'(x) = -x^2 + \frac{\sqrt{8}}{2x^{1/2}}$$

At $x = 2$, $f'(2) = -4 + \frac{\sqrt{8}}{2\sqrt{2}} = -4 + \frac{1}{2}\sqrt{\frac{8}{2}}$
$= -4 + \frac{1}{2} \cdot 2 = -3$.

Further, $y = f(2) = -\frac{8}{3} + 4 = \frac{4}{3}$. The equation of the tangent line at $\left(2, \frac{4}{3}\right)$ is

$$y - \frac{4}{3} = -3(x-2), \text{ or } y = -3x + \frac{22}{3}$$

39. $$f(x) = 2x^4 + 3x + 1$$
$$f'(x) = 8x^3 + 3$$

The rate of change of f at $x = -1$ is $f'(-1) = -5$.

41. $$f(x) = x - \sqrt{x} + \frac{1}{x^2} = x - x^{1/2} + x^{-2}$$
$$f'(x) = 1 - \frac{1}{2x^{1/2}} - \frac{2}{x^3}$$

The rate of change of f at $x = 1$ is $f'(1) = -\frac{3}{2}$.

43. $$f(x) = \frac{x + \sqrt{x}}{\sqrt{x}} = \frac{x}{\sqrt{x}} + \frac{\sqrt{x}}{\sqrt{x}} = \sqrt{x} + 1 = x^{1/2} + 1$$
$$f'(x) = \frac{1}{2x^{1/2}}$$

The rate of change of f at $x = 1$ is $f'(1) = \frac{1}{2}$.

45. (a) $$Q(t) = 0.05t^2 + 0.1t + 3.4 \text{ PPM}$$
$$Q'(t) = 0.1t + 0.1 \text{ PPM/year}$$

The rate of change of Q is at $t = 1$ is $Q'(1) = 0.2$ PPM/year.

(b) $Q(1) = 3.55$ PPM, $Q(0) = 3.40$, and $Q(1) - Q(0) = 0.15$ PPM.

(c) $Q(2) = 0.2 + 0.2 + 3.4 = 3.8$, $Q(0) = 3.4$, and $Q(2) - Q(0) = 0.4$ PPM.

47. (a) $f(x) = -6x + 582$
The rate of change of SAT scores is $f'(x) = -6$.

(b) The rate of change is constant, so the drop will not vary from year to year. The rate of change is negative, so the scores are declining.

49. (a) $T(x) = 20x^2 + 40x + 600$ dollars
The rate of change of property tax is

$$T'(x) = 40x + 40 \text{ dollars/year}$$

In the year 2000, $x = 0$, $T'(0) = 40$ dollars/year.

(b) In the year 2004, $x = 4$ and $T(4) = \$1{,}080$.
In the year 2000, $x = 0$ and $T(0) = \$600$.
The change in property tax is
$T(4) - T(0) = \$480$.

51. $N(t) = 10t^3 + 5t + \sqrt{t} = 10t^3 + 5t + t^{1/2}$
The rate of change of the infected population is

$$N'(t) = 30t^2 + 5 + \frac{1}{2t^{1/2}} \text{ people/day}$$

On the 9th day, $N'(9) = 2{,}435$ people/day.

53. (a) Costs = cost driver + cost gasoline

$$\text{cost driver } = 20(\#\text{hrs}) = 20\left(\frac{250mi}{x}\right) = \frac{5{,}000}{x}$$

$$\begin{aligned}\text{cost gasoline } &= 1.9(\#\text{gals})\\ &= 1.9(250)\left[\frac{1}{250}\left(\frac{1,200}{x} + x\right)\right]\\ &= \frac{2{,}280}{x} + 1.9x = \frac{2{,}280}{x} + 1.9x \text{ dollars}\end{aligned}$$

So, the cost function is $C(x) = \dfrac{7{,}280}{x} + 1.9x$.

(b) The rate of change of the cost is $C'(x)$.

$$C(x) = 7{,}280x^{-1} + 1.9x$$

$$C'(x) = -\frac{7{,}280}{x^2} + 1.9 \text{ dollars/miles per hr.}$$

When $x = 40$, $C'(40) = -2.65$ dollars/miles per hour. Since $C'(40)$ is negative, the cost is decreasing.

55.

$$f(x) = 2x^3 - 5x^2 + 4$$

$$f'(x) = 6x^2 - 10x$$

The relative rate of change is

$$\frac{f'(x)}{f(x)} = \frac{6x^2 - 10x}{2x^3 - 5x^2 + 4}$$

When $x = 1$,

$$\frac{f'(1)}{f(1)} = \frac{6 - 10}{2 - 5 + 4} = -4$$

57. (a)

$$A(t) = 0.1t^2 + 10t + 20$$

$$A'(t) = 0.2t + 10$$

In the year 2004, the rate of change is

$$A'(4) = 0.8 + 10 \text{ or } \$10{,}800 \text{ per year}$$

(b) $A(4) = (0.1)(16) + 40 + 20 = 61.6$,
so the percentage rate of change is

$$\frac{(100)(10.8)}{61.6} = 17.53\%$$

59. (a) Since your starting salary is \$45,000 and you get a raise of \$2,000 per year, your salary t years from now will be

$$S(t) = 45{,}000 + 2{,}000t \text{ dollars.}$$

The percentage rate of change of this salary t years from now is

$$\begin{aligned}100\left[\frac{S'(t)}{S(t)}\right] &= 100\left(\frac{2{,}000}{45{,}000 + 2{,}000t}\right)\\ &= \frac{200}{45 + 2t} \text{ percent per year.}\end{aligned}$$

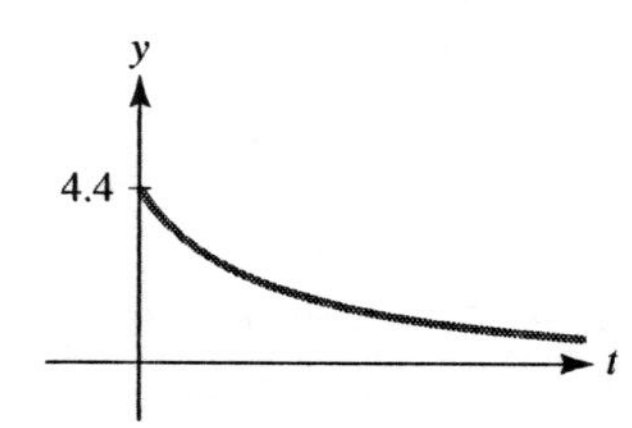

(b) The percentage rate of change after 1 year is

$$\frac{200}{47} \approx 4.26\%$$

(c) In the long run, $\dfrac{200}{45 + 2t}$ approaches 0.
That is, the percentage rate of your salary will approach 0 (even though your salary will continue to increase at a constant rate).

61. (a) $P(x) = 2x + 4x^{3/2} + 5{,}000$ is the population x months from now. The rate of population growth is

$$P'(x) = 2 + 4\left(\frac{3x^{1/2}}{2}\right) = 2 + 6x^{1/2}$$

people per month. Nine months from now, the population will be changing at the rate of

$$P'(9) = 2 + 6(9^{1/2}) = 20 \text{ people per month.}$$

(b) The percentage rate at which the population will be changing 9 months from now is

$$100\frac{P'(9)}{P(9)} = \frac{100(20)}{2(9) + 4(9^{3/2}) + 5{,}000} = \frac{2{,}000}{5{,}126} = 0.39\%$$

63. (a) $T(t) = -68.07t^3 + 30.98t^2 + 12.52t + 37.1$

$$T'(t) = -204.21t^2 + 61.96t + 12.52$$

$T'(t)$ represents the rate at which the bird's temperature is changing after t days, measured in °C per day.

(b) $T'(0) = 12.52$ °C/day
since $T'(0)$ is positive, the bird's temperature is increasing.

$$T'(0.713) \approx -47.12 \text{ °C/day}$$

Since $T'(0.713)$ is negative, the bird's temperature is decreasing.

(c) Find t so that $T'(t) = 0$.

$$0 = -204.21t^2 + 61.96t + 12.52$$

$$t = \frac{-61.96 \pm \sqrt{(61.96)^2 - 4(-204.21)(12.52)}}{2(-204.21)}$$

$t \approx 0.442$ days.
The bird's temperature when $t = 0.442$ is $T(0.442) \approx 42.8$°C.
The bird's temperature starts at $T(0) = 37.1$°C, increases to $T(0.442) = 42.8$°C, and then begins to decrease.

65. (a)

$$s(t) = 3t^2 + 2t - 5 \text{ for } 0 \le t \le 1$$
$$v(t) = 6t + 2 \text{ and } a(t) = 6$$

(b) $6t + 2 = 0$ at $t = -3$. The particle is not stationary between $t = 0$ and $t = 1$.

67. (a) $s(t) = t^4 - 4t^3 + 8t$ for $0 \le t \le 4$
$v(t) = 4t^3 - 12t^2 + 8$ and $a(t) = 12t^2 - 24t$

(b) To find all time in given interval when stationary,

$$4t^3 - 12t^2 + 8 = 0$$
$$4(t^3 - 3t^2 + 2) = 0$$
$$t^3 - 3t^2 + 2 = 0$$

Press [y=] Input $x \wedge 3 - 3x^2 + 2$ for $y_1 =$
Use window dimensions [−4, 4]1 by [−4, 4]1
Use trace and zoom-in to find the x-intercepts or use the zero function under the calc menu.
To use the zero function (for the left-most x-intercept), enter a value to the left of (but close to) the x-intercept for the left bound. Enter a value close to but to the right of the x-intercept for the right bound. Enter $x = -0.7$ for the guess. We see that the left most x-intercept is $x \approx -0.732$.
Repeat this process for the other two x-intercepts to find $x = 1$ and $x \approx 2.73$.
In the interval $0 \le t \le 4$, the particle is stationary when $t = 1$ and $t \approx 2.73$.

69. (a) If after 2 seconds the ball passes you on the way down, then

$$H(2) = H_0$$

where $H(t) = -16t^2 + V_0t + H_0$.
So, $-16(2^2) + (V_0)(2) + H_0 = H_0$,
$-64 + 2V_0 = 0$, or $V_0 = 32\frac{\text{ft}}{\text{sec}}$.

(b) The height of the building is H_0 feet. From part (a) you know that

$$H(t) = -16t^2 + 32t + H_0$$

Moreover, $H(4) = 0$ since the ball hits the ground after 4 seconds.
So, $-16(4^2) + 32(4) + H_0 = 0$, or
$H_0 = 128$ feet.

(c) From parts (a) and (b) you know that

$$H(t) = -16t^2 + 32t + 128$$

and so the speed of the ball is

$$H'(t) = -32t + 32\frac{\text{ft}}{\text{sec}}$$

After 2 seconds, the speed will be $H'(2) = -32$ feet per second, where the minus sign indicates that the direction of motion is down.

(d) The speed at which the ball hits the ground is

$$H'(4) = -96\frac{\text{ft}}{\text{sec}}$$

71. $f(x) = ax^2 + bx + c$
Since $f(0) = 0$, $c = 0$ and $f(x) = ax^2 + bx$.
Since $f(5) = 0$, $0 = 25a + 5b$.
Further, since the slope of the tangent is 1 when $x = 2$, $f'(2) = 1$.

$$f'(x) = 2ax + b$$
$$1 = 2a(2) + b = 4a + b$$

Now, solve the system: $0 = 25a + 5b$ and $1 = 4a + b$.
Since $1 - 4a = b$, using substitution

$$0 = 25a + 5(1 - 4a)$$
$$0 = 25a + 5 - 20a$$
$$0 = 5a + 5$$

or $\quad a = -1$ and
$$b = 1 - 4(-1) = 5$$

So, $f(x) = -x^2 + 5x$.

73. $(f + g)'(x)$

$$= \lim_{h\to 0}\frac{(f+g)(x+h) - (f+g)(x)}{h}$$
$$= \lim_{h\to 0}\frac{f(x+h) + g(x+h) - [f(x) + g(x)]}{h}$$
$$= \lim_{h\to 0}\frac{f(x+h) - f(x) + g(x+h) - g(x)}{h}$$
$$= \lim_{h\to 0}\frac{f(x+h) - f(x)}{h} + \lim_{h\to 0}\frac{g(x+h) - g(x)}{h}$$
$$= f'(x) + g'(x).$$

75. (a) Using the graph, the x-value (tax rate) that appears to correspond to a y-value (percentage reduction) of 50 is 150, or a tax rate of 150 dollars per ton carbon.

(b) Using the points (200,60) and (300,80), from the graph, the rate of change is approximately

$$\frac{dP}{dT} \approx \frac{80 - 60}{300 - 200} = \frac{20}{100} = 0.2\%$$

or increasing at approximately 0.2% per dollar. (Answers will vary depending on the choice of h.)

(c) Writing Exercise—
Answers will vary.

2.3 Product and Quotient Rules; Higher-Order Derivatives

1. $f(x) = (2x + 1)(3x - 2)$,

$$f'(x) = (2x+1)\frac{d}{dx}(3x - 2) + (3x - 2)\frac{d}{dx}(2x + 1)$$
$$= (2x + 1)(3) + (3x - 2)(2)$$
$$= 12x - 1.$$

3. $y = 10(3u + 1)(1 - 5u)$,

$$\frac{dy}{du} = 10\frac{d}{du}(3u + 1)(1 - 5u)$$
$$= 10\left[(3u+1)\frac{d}{du}(1 - 5u) + (1 - 5u)\frac{d}{du}(3u + 1)\right]$$
$$= 10[(3u + 1)(-5) + (1 - 5u)(3)]$$
$$= -300u - 20.$$

5.
$$f'(x) = \frac{1}{3}\left[(x^5 - 2x^3 + 1)\frac{d}{dx}\left(x - \frac{1}{x}\right) + \left(x - \frac{1}{x}\right)\frac{d}{dx}\left(x^5 - 2x^3 + 1\right)\right]$$
$$= \frac{1}{3}\left[\left(x^5 - 2x^3 + 1\right)\left(1 + \frac{1}{x^2}\right) + \left(x - \frac{1}{x}\right)\left(5x^4 - 6x^2\right)\right]$$
$$= 2x^5 - 4x^3 + \frac{4}{3}x + \frac{1}{3x^2} + \frac{1}{3}$$

7.
$$y = \frac{x+1}{x-2},$$
$$\frac{dy}{dx} = \frac{(x-2)\frac{d}{dx}(x+1) - (x+1)\frac{d}{dx}(x-2)}{(x-2)^2}$$
$$= \frac{(x-2)(1) - (x+1)(1)}{(x-2)^2}$$
$$= -\frac{3}{(x-2)^2}.$$

9.
$$f(t) = \frac{t}{t^2-2},$$
$$f'(t) = \frac{(t^2-2)\frac{d}{dt}(t) - t\frac{d}{dt}(t^2-2)}{(t^2-2)^2}$$
$$= \frac{(t^2-2)(1) - (t)(2t)}{(t^2-2)^2}$$
$$= \frac{-t^2-2}{(t^2-2)^2}.$$

11.
$$y = \frac{3}{x+5},$$
$$\frac{dy}{dx} = \frac{(x+5)\frac{d}{dx}(3) - 3\frac{d}{dx}(x+5)}{(x+5)^2}$$
$$= \frac{(x+5)(0) - 3(1)}{(x+5)^2}$$
$$= -\frac{3}{(x+5)^2}.$$

13. $f(x) = \dfrac{x^2-3x+2}{2x^2+5x-1},$

$$f'(x) = \frac{(2x^2+5x-1)\frac{d}{dx}(x^2-3x+2)}{(2x^2+5x-1)^2} - \frac{(x^2-3x+2)\frac{d}{dx}(2x^2+5x-1)}{(2x^2+5x-1)^2}$$
$$= \frac{(2x^2+5x-1)(2x-3)}{(2x^2+5x-1)^2} - \frac{(x^2-3x+2)(4x+5)}{(2x^2+5x-1)^2}$$
$$= \frac{11x^2-10x-7}{(2x^2+5x-1)^2}.$$

15.
$$f(x) = \frac{(2x-1)(x+3)}{x+1} = \frac{2x^2+5x-3}{x+1}$$
$$f'(x) = \frac{(x+1)\frac{d}{dx}(2x^2+5x-3)}{(x+1)^2} - \frac{(2x^2+5x-3)\frac{d}{dx}(x+1)}{(x+1)^2}$$
$$= \frac{(x+1)(4x+5) - (2x^2+5x-3)(1)}{(x+1)^2}$$
$$= \frac{2x^2+4x+8}{(x+1)^2} = \frac{2(x^2+2x+4)}{(x+1)^2}$$

17. $f(x) = (2+5x)^2 = (2+5x)(2+5x)$
$$f'(x) = (2+5x)\frac{d}{dx}(2+5x) + (2+5x)\frac{d}{dx}(2+5x)$$
$$= 2(2+5x)\frac{d}{dx}(2+5x)$$
$$= 2(2+5x)(5)$$
$$= 20 + 50x = 10(2+5x)$$

19. $$g(t) = \frac{t^2 + \sqrt{t}}{2t + 5} = \frac{t^2 + t^{1/2}}{2t + 5}$$

$$g'(t) = \frac{(2t+5)\frac{d}{dt}(t^2 + t^{1/2}) - (t^2 + t^{1/2})\frac{d}{dt}(2t+5)}{(2t+5)^2}$$

$$= \frac{(2t+5)\left(2t + \frac{1}{2t^{1/2}}\right) - (t^2 + t^{1/2})(2)}{(2t+5)^2}$$

$$= \frac{2t^2 + 10t - t^{1/2} + \frac{5}{2t^{1/2}}}{(2t+5)^2} \cdot \frac{2t^{1/2}}{2t^{1/2}}$$

$$= \frac{4t^{5/2} + 20t^{3/2} - 2t + 5}{2t^{1/2}(2t+5)^2}$$

$$= \frac{4\sqrt{t^5} + 20\sqrt{t^3} - 2t + 5}{2\sqrt{t}(2t+5)^2}$$

21. $$y = (5x - 1)(4 + 3x)$$

$$\frac{dy}{dx} = 30x + 17$$

When $x = 0$, $y = -4$ and $\frac{dy}{dx} = 17$. The equation of the tangent line at $(0, -4)$ is

$$y + 4 = 17(x - 0), \text{ or } y = 17x - 4$$

23. $$y = \frac{x}{2x + 3}$$

$$\frac{dy}{dx} = \frac{3}{(2x+3)^2}$$

When $x = -1$, $y = -1$ and $\frac{dy}{dx} = 3$. The equation of the tangent line at $(-1, -1)$ is

$$y + 1 = 3(x + 1), \text{ or } y = 3x + 2$$

25. $$y = (3\sqrt{x} + x)\left(2 - x^2\right) = \left(3x^{1/2} + x\right)\left(2 - x^2\right)$$

$$\frac{dy}{dx} = -3x^2 - \frac{15}{2}x^{3/2} + \frac{3}{x^{1/2}} + 2$$

When $x = 1$, $y = 4$ and $\frac{dy}{dx} = -\frac{11}{2}$

The equation of the tangent line at (1,4) is

$$y - 4 = -\frac{11}{2}(x - 1), \text{ or } y = -\frac{11}{2}x + \frac{19}{2}$$

27. $$f(x) = (x+1)(x^2 - x - 2)$$

$$f'(x) = (x+1)(2x-1) + (x^2 - x - 2)(1)$$

$$= 3x^2 - 3$$

Since $f'(x)$ represents the slope of the tangent line and the slope of a horizontal line is zero, need to solve
$0 = 3x^2 - 3 = 3(x+1)(x-1)$
or $x = -1, 1$.
When $x = -1$, $f(-1) = 0$ and when $x = 1$, $f(1) = -4$. So, the tangent line is horizontal at the points $(-1, 0)$ and $(1, -4)$.

29. $$f(x) = \frac{x + 1}{x^2 + x + 1}$$

$$f'(x) = \frac{-x^2 - 2x}{(x^2 + x + 1)^2}$$

Since $f'(x)$ represents the slope of the tangent line and the slope of a horizontal line is zero, need to solve

$$0 = \frac{-x^2 - 2x}{(x^2 + x + 1)^2}$$

$$0 = -x^2 - 2x = -x(x + 2)$$

or $x = 0, -2$.

When $x = 0$, $f(0) = 1$ and when $x = -2$, $f(-2) = -\frac{1}{3}$. So, the tangent line is horizontal at the points (0,1) and $\left(-2, -\frac{1}{3}\right)$.

31. $$f(x) = x^3(x - 5)^2$$

$$f'(x) = x^3 \cdot 2(x-5)(1) + (x-5)^2(3x^2)$$

$$= x^2(x-5)\,[2x + 3(x-5)]$$

$$= x^2(x-5)(5x - 15)$$

$$= 5x^2(x-5)(x-3)$$

Since $f'(x)$ represents the slope of the tangent line and the slope of a horizontal line is zero, need to solve

$0 = 5x^2(x-5)(x-3)$
or $x = 0, 3, 5$.
When $x = 0$, $f(0) = 0$; when $x = 3$, $f(3) = 108$; and when $x = 5$, $f(5) = 0$. So, the tangent line is horizontal at the points (0, 0), (3, 108) and (5, 0).

33. $$y = (x^2+3)(5-2x^3)$$

$$\frac{dy}{dx} = (x^2+3)(-6x^2) + (5-2x^3)(2x)$$

When $x = 1$,

$$\frac{dy}{dx} = (1+3)(-6) + (5-2)(2) = -18$$

35. $y = x + \dfrac{3}{2-4x}$

$$\frac{dy}{dx} = 1 + \frac{(2-4x)(0) - 3(-4)}{(2-4x)^2}$$

When $x = 0$,

$$\frac{dy}{dx} = 1 + \frac{12}{(2)^2} = 4$$

37. $$y = \frac{2}{x} - \sqrt{x} = 2x^{-1} - x^{1/2}$$

$$\frac{dy}{dx} = \frac{-2}{x^2} - \frac{1}{2x^{1/2}}$$

When $x = 1$,

$$\frac{dy}{dx} = -2 - \frac{1}{2} = -\frac{5}{2}$$

The slope of a line perpendicular to the tangent line at $x = 1$ is $\frac{2}{5}$. The equation of the normal line at (1,1) is

$$y - 1 = \frac{2}{5}(x-1), \text{ or } y = \frac{2}{5}x + \frac{3}{5}$$

39. $y = \dfrac{5x+7}{2-3x}$

$$\frac{dy}{dx} = \frac{(2-3x)(5) - (5x+7)(-3)}{(2-3x)^2}$$

When $x = 1$,

$$\frac{dy}{dx} = \frac{(2-3)(5) - (5+7)(-3)}{(2-3)^2} = 31$$

The slope of a line perpendicular to the tangent line at $x = 1$ is $-\frac{1}{31}$.
The equation of the normal line at $(1, -12)$ is

$$y + 12 = -\frac{1}{31}(x-1), \text{ or } y = -\frac{1}{31}x - \frac{371}{31}$$

41. (a) $$y = \frac{2x-3}{x^3}$$

$$\frac{dy}{dx} = \frac{(x^3)(2) - (2x-3)(3x^2)}{x^6} = \frac{-4x^3+9x^2}{x^6}$$

$$= \frac{-4x+9}{x^4}$$

(b) $y = (2x-3)(x^{-3})$

$$\frac{dy}{dx} = (2x-3)(-3x^{-4}) + (x^{-3})(2)$$

$$= \frac{-3(2x-3)+2x}{x^4}$$

$$= \frac{-4x+9}{x^4}$$

(c) $y = 2x^{-2} - 3x^{-3}$

$$\frac{dy}{dx} = -4x^{-3} + 9x^{-4} = \frac{-4}{x^3} + \frac{9}{x^4} = \frac{-4x+9}{x^4}$$

43. $$f(x) = \frac{2}{5}x^5 - 4x^3 + 9x^2 - 6x - 2$$

$$f'(x) = 2x^4 - 12x^2 + 18x - 6$$

$$f''(x) = 8x^3 - 24x + 18$$

45. $y = \dfrac{2}{3}x^{-1} - \sqrt{2}x^{1/2} + \sqrt{2}x - \dfrac{1}{6}x^{-1/2}$

$$\frac{dy}{dx} = y' = -\frac{2}{3}x^{-2} - \frac{\sqrt{2}}{2}x^{-1/2} + \sqrt{2} + \frac{1}{12}x^{-3/2}$$

$$\frac{d^2y}{dx^2} = y'' = \frac{4}{3}x^{-3} + \frac{\sqrt{2}}{4}x^{-3/2} - \frac{1}{8}x^{-5/2}$$

$$= \frac{4}{3x^3} + \frac{\sqrt{2}}{4x^{3/2}} - \frac{1}{8x^{5/2}}$$

47. $y = (x^3 + 2x - 1)(3x + 5)$

$$\frac{dy}{dx} = y' = (x^3 + 2x - 1)(3) + (3x + 5)(3x^2 + 2)$$

$$= 12x^3 + 15x^2 + 12x + 7$$

$$\frac{d^2y}{dx^2} = y'' = 36x^2 + 30x + 12$$

49. $S(t) = \dfrac{2000t}{4 + 0.3t}$

(a) $S'(t) = \dfrac{(4 + 0.3t)(2000) - (2000t)(0.3)}{(4 + 0.3t)^2}$

The rate of change in the year 2002 is

$S'(2) = \dfrac{(4 + 0.6)(2,000) - (4,000)(0.3)}{(4 + 0.6)^2}$

$\approx$ \$378,070 per year.

(b) Rewrite the function as

$$S(t) = \frac{2,000}{\frac{4}{t} + 0.3}$$

Since $\frac{4}{t} \to 0$ as $t \to +\infty$, sales approach $\frac{2,000}{0.3} \approx 6,666.67$ thousand, or approximately \$6,666,667 in the long run.

51. $P(t) = 100\left[\dfrac{t^2 + 5t + 5}{t^2 + 10t + 30}\right]$

(a) $P'(t) =$

$100\dfrac{(t^2 + 10t + 30)(2t + 5) - (t^2 + 5t + 5)(2t + 10)}{(t^2 + 10t + 30)^2}$

The rate of change after 5 weeks is

$P'(5) =$

$100\dfrac{(25 + 50 + 30)(10 + 5) - (25 + 25 + 5)(10 + 10)}{(25 + 50 + 30)^2}$

$P'(5) = 4.31\%$ per week.

Since $P'(5)$ is positive, the percentage is increasing.

(b) Rewrite the function as

$$p(t) = 100\frac{1 + \frac{5}{t} + \frac{5}{t^2}}{1 + \frac{10}{t} + \frac{30}{t^2}}$$

Since $\frac{5}{t}$, $\frac{5}{t^2}$, $\frac{10}{t}$ and $\frac{30}{t^2}$ all go to zero as $t \to +\infty$, the percentage approaches 100% in the long run, so the rate of change approaches 0.

53. $P(x) = \dfrac{100\sqrt{x}}{0.03x^2 + 9} = 100\dfrac{x^{1/2}}{0.03x^2 + 9}$

(a) $P'(x) =$

$100\dfrac{(0.03x^2 + 9)\left(\frac{1}{2}x^{-1/2}\right) - (x^{1/2})(0.06x)}{(0.03x^2 + 9)^2}$

The rate of change of percentage pollution when 16 million dollars are spent is

$$P'(16) = 100\left(\frac{[0.03(16)^2 + 9]\left[\frac{1}{2}(16)^{-1/2}\right]}{[0.03(16)^2 + 9]^2} - \frac{(16)^{1/2}[0.06(16)]}{[0.03(16)^2 + 9]^2}\right)$$

$= -0.63$ percent

Since $P'(16)$ is negative, the percentage is decreasing.

(b) $P'(x) = 0$ when

$0 = (0.03x^2 + 9)(\frac{1}{2}x^{-1/2}) - (x^{1/2})(0.06x)$

or $x = 10$ million dollars.

Testing one value less than 10 and one value greater than 10 shows $P'(x)$ is increasing when $0 < x < 10$, and decreasing when $x > 10$.

55. (a) $Q(t) = -t^3 + 8t^2 + 15t$

$$R(t) = Q'(t) = -3t^2 + 16t + 15$$

(b) The rate of change of the worker's rate is the second derivative

$$R'(t) = Q''(t) = -6t + 16$$

At 9:00 a.m., $t = 1$ and

$Q''(1) = -6(1) + 16 = 10$ units/hr^2

57. **(a)**
$$s(t) = 3t^5 - 5t^3 - 7$$
$$v(t) = 15t^4 - 15t^2 = 15(t^4 - t^2)$$
$$a(t) = 15(4t^3 - 2t) = 30t(2t^2 - 1)$$

(b) $a(t) = 0$ when $30t(2t^2 - 1) = 0$, or $t = 0$ and $t = \dfrac{\sqrt{2}}{2}$.

59. $s(t) = -t^3 + 7t^2 + t + 2$

(a) $v(t) = -3t^2 + 14t + 1$
$a(t) = -6t + 14$

(b) $a(t) = 0$ when $-6t + 14 = 0$, or $t = \dfrac{7}{3}$

61. $D(t) = 10t + \dfrac{5}{t+1}$

(a) Speed = rate of change of distance with respect to time.

$$\frac{dD}{dt} = 10 + \frac{(t+1)(0) - (5)(1)}{(t+1)^2} = 10 - \frac{5}{(t+1)^2}$$

When $t = 4$,

$$\frac{dD}{dt} = 10 - \frac{5}{25} = \frac{49}{5} \text{ meters/minute.}$$

(b)
$$D(5) = 10(5) + \frac{5}{5+1} = 50 + \frac{5}{6}$$
$$D(4) = 10(4) + \frac{5}{4+1} = 41$$
$$D(5) - D(4) = 9 + \frac{5}{6} = \frac{59}{6} \text{ meters.}$$

63. $F = \dfrac{1}{3}(KM^2 - M^3)$

(a) $S = \dfrac{dF}{dM} = \dfrac{1}{3}(2KM - 3M^2) = \dfrac{2}{3}KM - M^2$

(b) $\dfrac{dS}{dM} = \dfrac{1}{3}(2K - 6M) = \dfrac{2}{3}K - 2M$

is the rate at which the sensitivity is changing.

65. $H(t) = -16t^2 + S_0t + H_0$

(a) $H'(t) = -32t + S_0$ and the acceleration is $H''(t) = -32$.

(b) Since the acceleration is a constant, it does not vary with time.

(c) The only acceleration acting on the object is due to gravity. The negative sign signifies that this acceleration is directed downward.

67. $y = x^{1/2} - \dfrac{1}{2}x^{-1} + \dfrac{1}{\sqrt{2}}x$

$$\frac{dy}{dx} = \frac{1}{2}x^{-1/2} + \frac{1}{2}x^{-2} + \frac{1}{\sqrt{2}}$$
$$\frac{d^2y}{dx^2} = -\frac{1}{4}x^{-3/2} - x^{-3}$$
$$\frac{d^3y}{dx^3} = \frac{3}{8}x^{-5/2} + 3x^{-4} = \frac{3}{8x^{5/2}} + \frac{3}{x^4}$$

69. **(a)**
$$\frac{d}{dx}\left(\frac{fg}{h}\right) = \frac{h\dfrac{d}{dx}(fg) - (fg)\dfrac{d}{dx}h}{h^2} = \frac{h\left(f\dfrac{d}{dx}g + g\dfrac{d}{dx}f\right) - fg\dfrac{d}{dx}h}{h^2}$$

(b) $y = \dfrac{(2x+7)(x^2+3)}{3x+5}$

$$\frac{dy}{dx} = \frac{(3x+5)\left[(2x+7)(2x) + (x^2+3)(2)\right]}{(3x+5)^2} - \frac{(2x+7)(x^2+3)(3)}{(3x+5)^2}$$
$$= \frac{(3x+5)(6x^2+14x+6)}{(3x+5)^2} - \frac{3(2x^3+7x^2+6x+21)}{(3x+5)^2}$$
$$= \frac{12x^3+51x^2+70x-33}{(3x+5)^2}$$

71. For f/g the difference quotient (DQ) is

$$= \frac{(f/g)(x+h) - (f/g)(x)}{h}$$

$$= \frac{1}{h}\left[\frac{f(x+h)}{g(x+h)} - \frac{f(x)}{g(x)}\right]$$

$$= \frac{1}{h}\left[\frac{f(x+h)g(x) - f(x)g(x+h)}{g(x+h)g(x)}\right]$$

$$= \frac{1}{h}\left[\frac{f(x+h)g(x) - f(x)g(x) + f(x)g(x)}{g(x+h)g(x)} - \frac{f(x)g(x+h)}{g(x+h)g(x)}\right]$$

$$= \frac{1}{h}\left[\frac{g(x)[f(x+h) - f(x)]}{g(x+h)g(x)} - \frac{f(x)[g(x+h) - g(x)]}{g(x+h)g(x)}\right]$$

$$= \frac{1}{g(x+h)g(x)} \cdot \left[\frac{g(x)\left[f(x+h) - f(x)\right]}{h} - \frac{f(x)\left[g(x+h) - g(x)\right]}{h}\right]$$

$$\frac{d}{dx}(f/g) = \lim_{h\to 0} \frac{1}{g(x+h)g(x)} \cdot \left[\frac{g(x)[f(x+h) - f(x)]}{h} - \frac{f(x)[g(x+h) - g(x)]}{h}\right]$$

$$= \frac{1}{g(x)g(x)}\left[g(x)f'(x) - f(x)g'(x)\right]$$

$$= \frac{g(x)f'(x) - f(x)g'(x)}{[g(x)]^2}$$

73. To use a graphing utility to sketch $f(x) = x^2(x-1)$ and find where $f'(x) = 0$,
Press [y=]
Input $x^2(x-1)$ for $y_1 =$
Use window dimensions [−2, 3].5 by [−2, 2].5
Press graph
Press [2nd] [Draw] and enter the tangent function
Enter $x = 1$
The calculator draws the line tangent to the graph of f at $x = 1$ and gives $y = 1.000001x - 1.000001$ as the equation of that line. $f'(x) = 0$ when the slope of the line tangent to the graph of f is zero. This happens where the graph of f has a local high or low point. Use the trace button to move cross-hairs to the local low point on the graph of f. Use the zoom-in function under the zoom menu to find $f'(x) = 0$ when $x \approx 0.673$. Repeat this process to find where the local high point occurs. We see $f'(x) = 0$ also for $x = 0$.

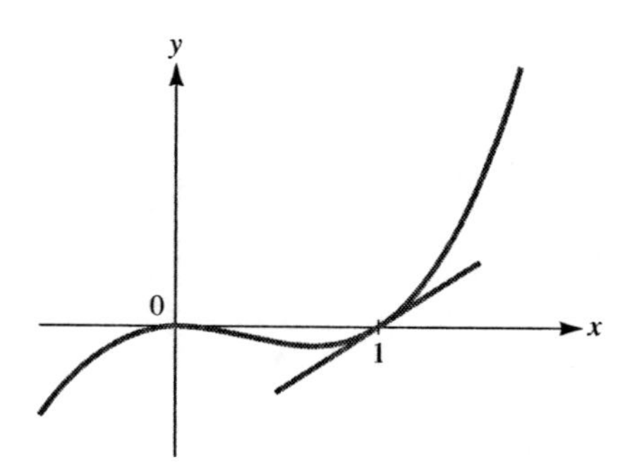

75. To use a graphing utility to graph $f(x) = x^4 + 2x^3 - x + 1$ and to find minima and maxima,
Press [y=] and input $x \wedge 4 + 2x \wedge 3 - x + 1$ for $y_1 =$
Use window dimensions [−5, 5]1 by [0, 2].5
Press [Graph]
We see from the graph that there are two minimums and one maximum.
To find the first minimum, use trace and zoom-in for a more accurate reading.
Alternatively, use the minimum function under the calc menu. Using trace, enter a value to the left of (but close to) the minimum for the left bound.
Enter a value to the right of (but close to) the minimum for the right bound. Finally, enter a guess in between the bounds and the minimum is displayed.
One minimum occurs at (−1.37, 0.75).
Repeat this process for the other minimum and find it to be at (0.366, 0.75)
Repeat again for the maximum (using the maximum function) to find it at (−0.5, 1.31)
$f'(x) = 4x^3 + 6x^2 - 1$
Press [y=] and input $4x \wedge 3 + 6x^2 - 1$ for $y_2 =$
Change window dimensions to [−5, 5]1 by [−2, 2].5
Use trace and zoom-in to find the x-intercepts of $f'(x)$ or use the zero function under the calc menu.

The three x-intercepts of $f'(x)$ are $x \approx -1.37, -0.5$, and 0.366.
The x values extrema occur at the x-intercepts of f' because the tangent line at the corresponding points on the curve are horizontal and so, the slopes are zero.

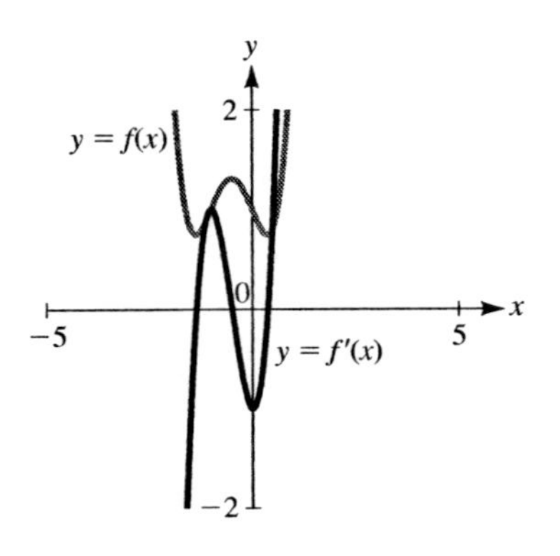

2.4 The Chain Rule

1.
$$y = u^2 + 1,\ u = 3x - 2,$$
$$\frac{dy}{du} = 2u,\ \frac{du}{dx} = 3,$$
$$\frac{dy}{dx} = \frac{dy}{du} \cdot \frac{du}{dx} = (2u)(3) = 6(3x - 2).$$

3.
$$y = \sqrt{u} = u^{1/2},\ u = x^2 + 2x - 3,$$
$$\frac{dy}{du} = \frac{1}{2}u^{-1/2} = \frac{1}{2u^{1/2}},$$
$$\frac{du}{dx} = 2x + 2, = 2(x + 1)$$
$$\frac{dy}{dx} = \frac{dy}{du} \cdot \frac{du}{dx} = \frac{x + 1}{(x^2 + 2x - 3)^{1/2}}.$$

5.
$$y = \frac{1}{u^2} = u^{-2},\ u = x^2 + 1,$$
$$\frac{dy}{du} = -2u^{-3} = -\frac{2}{u^3},\ \frac{du}{dx} = 2x,$$
$$\frac{dy}{dx} = \frac{dy}{du} \cdot \frac{du}{dx} = -\frac{4x}{(x^2 + 1)^3}$$

7.
$$y = \frac{1}{\sqrt{u}} = u^{-1/2},\ u = x^2 - 9,$$
$$\frac{dy}{du} = -\frac{1}{2}u^{-3/2} = -\frac{1}{2u^{3/2}},$$
$$\frac{du}{dx} = 2x,$$
$$\frac{dy}{dx} = \frac{dy}{du} \cdot \frac{du}{dx} = -\frac{x}{(x^2 - 9)^{3/2}}.$$

9.
$$y = \frac{1}{u - 1} = (u - 1)^{-1},\ u = x^2$$
$$\frac{dy}{du} = -(u - 1)^{-2} = -\frac{1}{(u - 1)^2},$$
$$\frac{du}{dx} = 2x,$$
$$\frac{dy}{dx} = \frac{dy}{du} \cdot \frac{du}{dx} = -\frac{2x}{(x^2 - 1)^2}.$$

11.
$$y = 3u^4 - 4u + 5,\ u = x^3 - 2x - 5$$
$$\frac{dy}{du} = 12u^3 - 4,\ \frac{du}{dx} = 3x^2 - 2,$$
$$\frac{dy}{dx} = \frac{dy}{du} \cdot \frac{du}{dx} = (12u^3 - 4)(3x^2 - 2).$$

When $x = 2$, $u = 2^3 - 2(2) - 5 = -1$, so
$$\frac{dy}{dx} = [12(-1)^3 - 4][3(2^2) - 2] = -160$$

13.
$$y = \sqrt{u} = u^{1/2},\ u = x^2 - 2x + 6,$$
$$\frac{dy}{du} = \frac{1}{2}u^{-1/2} = \frac{1}{2u^{1/2}},$$
$$\frac{du}{dx} = 2x - 2,$$
$$\frac{dy}{dx} = \frac{dy}{du} \cdot \frac{du}{dx} = \frac{x - 1}{u^{1/2}}.$$

When $x = 3$, $u = 3^2 - 2(3) + 6 = 9$, so
$$\frac{dy}{dx} = \frac{3 - 1}{9^{1/2}} = \frac{2}{3}$$

15. $$y = \frac{1}{u} = u^{-1},\ u = 3 - \frac{1}{x^2} = 3 - x^{-2},$$

$$\frac{dy}{du} = -u^{-2} = -\frac{1}{u^2},\ \frac{du}{dx} = 2x^{-3} = \frac{2}{x^3}$$

$$\frac{dy}{dx} = \frac{dy}{du}\cdot\frac{du}{dx} = -\frac{1}{u^2}\cdot\frac{2}{x^3}$$

When $x = \frac{1}{2}$, $u = 3 - \frac{1}{(1/2)^2} = 3 - 4 = -1$,

$$\frac{dy}{dx} = \frac{-1}{(-1)^2}\cdot\frac{2}{(1/2)^3} = -16$$

17. $$f(x) = (2x+1)^4,$$

$$f'(x) = 4(2x+1)^3\frac{d}{dx}(2x+1) = 8(2x+1)^3$$

19. $$f(x) = (x^5 - 4x^3 - 7)^8$$

$$\begin{aligned} f'(x) &= 8(x^5 - 4x^3 - 7)^7\frac{d}{dx}(x^5 - 4x^3 - 7) \\ &= 8(x^5 - 4x^3 - 7)^7(5x^4 - 12x^2) \\ &= 8x^2(x^5 - 4x^3 - 7)^7(5x^2 - 12) \end{aligned}$$

21. $$f(t) = \frac{1}{5t^2 - 6t + 2} = (5t^2 - 6t + 2)^{-1},$$

$$\begin{aligned} f'(t) &= -(5t^2 - 6t + 2)^{-2}\frac{d}{dt}(5t^2 - 6t + 2) \\ &= -\frac{10t - 6}{(5t^2 - 6t + 2)^2} = \frac{-2(5t - 3)}{(5t^2 - 6t + 2)^2} \end{aligned}$$

23. $$g(x) = \frac{1}{\sqrt{4x^2+1}} = (4x^2+1)^{-1/2}$$

$$\begin{aligned} g'(x) &= -\frac{1}{2}(4x^2+1)^{-3/2}\frac{d}{dx}(4x^2+1) \\ &= \frac{-8x}{2(4x^2+1)^{3/2}} = \frac{-4x}{(4x^2+1)^{3/2}} \end{aligned}$$

25. $$f(x) = \frac{3}{(1-x^2)^4} = 3(1-x^2)^{-4},$$

$$\begin{aligned} f'(x) &= -12(1-x^2)^{-5}\frac{d}{dx}(1-x^2) \\ &= \frac{24x}{(1-x^2)^5} \end{aligned}$$

27. $$h(s) = \left(1 + \sqrt{3s}\right)^5$$

$$\begin{aligned} h'(s) &= 5\left(1 + \sqrt{3s}\right)^4\frac{d}{ds}\left(1 + \sqrt{3s}\right) \\ &= 5\left(1 + \sqrt{3s}\right)^4\frac{d}{ds}\left(1 + \sqrt{3}s^{1/2}\right) \\ &= 5\left(1 + \sqrt{3s}\right)^4\cdot\frac{\sqrt{3}}{2s^{1/2}} \\ &= \frac{5\sqrt{3}\left(1 + \sqrt{3s}\right)^4}{2\sqrt{s}} \end{aligned}$$

29. $f(x) = (x+2)^3(2x-1)^5$

$$f'(x) = (x+2)^3\frac{d}{dx}(2x-1)^5 + (2x-1)^5\frac{d}{dx}(x+2)^3$$

Now,

$$\begin{aligned} \frac{d}{dx}(2x-1)^5 &= 5(2x-1)^4\frac{d}{dx}(2x-1) \\ &= 10(2x-1)^4 \end{aligned}$$

and

$$\begin{aligned} \frac{d}{dx}(x+2)^3 &= 3(x+2)^2\frac{d}{dx}(x+2) \\ &= 3(x+2)^2 \end{aligned}$$

So,

$$\begin{aligned} f'(x) &= 10(x+2)^3(2x-1)^4 + 3(2x-1)^5(x+2)^2 \\ &= (x+2)^2(2x-1)^4\left[10(x+2) + 3(2x-1)\right] \\ &= (x+2)^2(2x-1)^4(16x+17) \end{aligned}$$

31. $$G(x) = \sqrt{\frac{3x+1}{2x-1}} = \left(\frac{3x+1}{2x-1}\right)^{1/2}$$

$$G'(x) = \frac{1}{2}\left(\frac{3x+1}{2x-1}\right)^{-1/2}\cdot\frac{d}{dx}\left(\frac{3x+1}{2x-1}\right)$$

Now,

$$\begin{aligned} \frac{d}{dx}\left(\frac{3x+1}{2x-1}\right) &= \frac{(2x-1)(3) - (3x+1)(2)}{(2x-1)^2} \\ &= -\frac{5}{(2x-1)^2} \end{aligned}$$

$$\text{So, } G'(x) = \frac{1}{2}\left(\frac{3x+1}{2x-1}\right)^{-1/2} \cdot \frac{-5}{(2x-1)^2}$$
$$= -\frac{5}{2}\left(\frac{2x-1}{3x+1}\right)^{1/2} \cdot \frac{1}{(2x-1)^2}$$
$$= -\frac{5}{2}\frac{(2x-1)^{1/2}}{(3x+1)^{1/2}} \cdot \frac{1}{(2x-1)^2}$$
$$= \frac{-5}{2(3x+1)^{1/2}(2x-1)^{3/2}}$$

33. $$f(x) = \frac{(x+1)^5}{(1-x)^4}$$
$$f'(x) = \frac{(1-x)^4\frac{d}{dx}(x+1)^5 - (x+1)^5\frac{d}{dx}(1-x)^4}{\left[(1-x)^4\right]^2}$$

Now,
$$\frac{d}{dx}(x+1)^5 = 5(x+1)^4\frac{d}{dx}(x+1)$$
$$= 5(x+1)^4$$

and
$$\frac{d}{dx}(1-x)^4 = 4(1-x)^3\frac{d}{dx}(1-x^3)$$
$$= -4(1-x)^3$$

So,
$$f'(x) = \frac{5(1-x)^4(x+1)^4 + 4(x+1)^5(1-x)^3}{(1-x)^8}$$
$$= \frac{(1-x)^3(x+1)^4\,[5(1-x)+4(x+1)]}{(1-x)^8}$$
$$= \frac{(x+1)^4(9-x)}{(1-x)^5}$$

35. $$f(y) = \frac{3y+1}{\sqrt{1-4y}} = \frac{3y+1}{(1-4y)^{1/2}}$$

$f'(y) =$
$$\frac{(1-4y)^{1/2}\frac{d}{dy}(3y+1) - (3y+1)\frac{d}{dy}(1-4y)^{1/2}}{\left[(1-4y)^{1/2}\right]^2}$$

Now,
$$\frac{d}{dy}(3y+1) = 3$$

and
$$\frac{d}{dy}(1-4y)^{1/2} = \frac{1}{2}(1-4y)^{-1/2}\frac{d}{dy}(1-4y)$$
$$= \frac{-2}{(1-4y)^{1/2}}$$

So,

$f'(y) =$
$$\frac{3(1-4y)^{1/2} - (3y+1)\cdot\frac{-2}{(1-4y)^{1/2}}}{1-4y} \cdot \frac{(1-4y)^{1/2}}{(1-4y)^{1/2}}$$
$$= \frac{3(1-4y) + 2(3y+1)}{(1-4y)^{3/2}}$$
$$= \frac{5-6y}{(1-4y)^{3/2}}$$

37. $$f(x) = (3x^2+1)^2$$
$$f'(x) = 2(3x^2+1)(6x)$$

$m = f'(-1) = -48$ and $f(-1) = 16$, so the equation of the tangent line at $(-1, 16)$ is

$$y - 16 = -48(x+1), \text{ or } y = -48x - 32$$

39. $$f(x) = \frac{1}{(2x-1)^6} = (2x-1)^{-6}$$
$$f'(x) = -6(2x-1)^{-5}(2) = -\frac{12}{(2x-1)^5}$$

$m = f'(1) = -12$ and $f(1) = 1$, so the equation of the tangent line at $(1, 1)$ is

$$y - 1 = -12(x-1), \text{ or } y = -12x + 13$$

41. $$f(x) = \sqrt[3]{\frac{x}{x+2}} = \left(\frac{x}{x+2}\right)^{1/3}$$
$$f'(x) = \frac{1}{3}\left(\frac{x}{x+2}\right)^{-2/3} \cdot \frac{(x+2)(1)-(x)(1)}{(x+2)^2}$$
$$= \frac{(x+2)^{2/3}}{3x^{2/3}} \cdot \frac{2}{(x+2)^2}$$
$$= \frac{2}{3x^{2/3}(x+2)^{4/3}}$$

$m = f'(-1) = \frac{2}{3}$ and $f(-1) = -1$, so the equation of the tangent line at $(-1, -1)$ is
$y + 1 = \frac{2}{3}(x + 1)$, or $y = \frac{2}{3}x - \frac{1}{3}$

43. $f(x) = (x^2 + x)^2$

$f'(x) = 2(x^2 + x)(2x + 1) = 2x(x + 1)(2x + 1) = 0$

when $x = -1$, $x = 0$, and $x = -\frac{1}{2}$.

45. $f(x) = \dfrac{x}{(3x - 2)^2}$

$$f'(x) = \frac{(3x-2)^2(1) - (x)\,[2(3x-2)(3)]^2}{\left[(3x-2)^2\right]^2}$$

$$= \frac{(3x-2)\,[(3x-2) - 6x]}{(3x-2)^4}$$

$$= \frac{-3x-2}{(3x-2)^3}$$

$$0 = \frac{-3x-2}{(3x-2)^3} \text{ when } -3x - 2 = 0, \text{ or } x = -\frac{2}{3}.$$

47. $f(x) = \sqrt{x^2 - 4x + 5} = (x^2 - 4x + 5)^{1/2}$

$$f'(x) = \frac{1}{2}(x^2 - 4x + 5)^{-1/2}(2x - 4)$$

$$= \frac{2x - 4}{2(x^2 - 4x + 5)^{1/2}}$$

$$= \frac{x - 2}{(x^2 - 4x + 5)^{1/2}}$$

$$0 = \frac{x-2}{(x^2 - 4x + 5)^{1/2}} \text{ when } x - 2 = 0, \text{ or } x = 2.$$

49. $f(x) = (3x + 5)^2$

(a) $f'(x) = 2(3x + 5)(3) = 6(3x + 5)$

(b) $f(x) = (3x + 5)(3x + 5)$

$f'(x) = (3x + 5)(3) + (3x + 5)(3) = 6(3x + 5)$

51. $f(x) = (3x + 1)^5$

$f'(x) = 5(3x + 1)^4(3) = 15(3x + 1)^4$,

$f''(x) = 60(3x + 1)(3)^3 = 180(3x + 1)^3$

53. $h = (t^2 + 5)^8$

$$\frac{dh}{dt} = 8(t^2 + 5)^7(2t) = 16t(t^2 + 5)^7,$$

$$\frac{d^2h}{dt^2} = 16t[7(t^2 + 5)^6(2t)] + (t^2 + 5)^7(16)$$

$$= 16(t^2 + 5)^6[14t^2 + (t^2 + 5)]$$

$$= 16(t^2 + 5)^6(15t^2 + 5)$$

$$= 80(t^2 + 5)^6(3t^2 + 1)$$

55.

$$f(x) = \sqrt{1 + x^2} = (1 + x^2)^{1/2}$$

$$f'(x) = \frac{1}{2}(1 + x^2)^{-1/2}(2x)$$

$$= \frac{x}{(1 + x^2)^{1/2}}$$

$$f''(x) = \frac{(1 + x^2)^{1/2}(1) - (x)\left[\frac{1}{2}(1 + x^2)^{-1/2}(2x)\right]}{1 + x^2}$$

$$= \frac{(1 + x^2)^{1/2} - \dfrac{x^2}{(1 + x^2)^{1/2}}}{1 + x^2} \cdot \frac{(1 + x^2)^{1/2}}{(1 + x^2)^{1/2}}$$

$$= \frac{1 + x^2 - x^2}{(1 + x^2)^{3/2}} = \frac{1}{(1 + x^2)^{3/2}}$$

57. (a)

$$f(t) = \sqrt{10t^2 + t + 236}$$
$$= (10t^2 + t + 236)^{1/2}$$

The rate at which the earnings are growing is

$$f'(t) = \frac{1}{2}(10t^2 + t + 236)^{-1/2}(20t + 1)$$

$$= \frac{20t + 1}{2(10t^2 + t + 236)^{1/2}}$$

thousand dollars per year.
The rate of growth in 2003 ($t = 5$) is

$$f'(5) = \frac{20(5) + 1}{2(10(5)^2 + 5 + 236)^{1/2}} = 2.279$$

or $2,279 per year.

(b) The percentage rate of the earnings increases in 2003 was

$$100\frac{f'(5)}{f(5)}$$
$$= \frac{100(2.279)}{\sqrt{10(5^2)+5+236}} = 10.285\% \text{ per year.}$$

59. $D(p) = \dfrac{4{,}374}{p^2} = 4{,}374p^{-2}$

(a) $\dfrac{dD}{dp} = -8{,}748p^{-3} = \dfrac{-8{,}784}{p^3}$ When the price is \$9,

$$\frac{dD}{dp} = \frac{-8{,}748}{(9)^3} = -12 \text{ pounds per dollar}$$

(b) $\dfrac{dD}{dt} = \dfrac{dD}{dp} \cdot \dfrac{dp}{dt}$

Now, $p(t) = 0.02t^2 + 0.1t + 6$

$$\frac{dp}{dt} = 0.04t + 0.1 \text{ dollars per week}$$
$$\frac{dD}{dt} = \frac{-8{,}748}{p^3}(0.04t + 0.1) \text{ pounds per week}$$

When $t = 10$,
$p(10) = 0.02(10)^2 + 0.1(10) + 6 = 9$

$$\text{so, } \frac{dD}{dt} = \frac{-8{,}748}{9^3}[0.04(10) + 0.1]$$
$$= -6 \text{ pounds per week}$$

Since the rate is negative, demand will be decreasing.

61.
$$D(p) = \frac{40{,}000}{p}$$
$$p(t) = 0.4t^{3/2} + 6.8$$

Need $100\dfrac{\frac{dD}{dt}}{D(t)}$ when $t = 4$.
When $t = 4$, $p(4) = 0.4(4)^{3/2} + 6.8 = 10$.

$$D(10) = \frac{40{,}000}{10} = 4{,}000$$
$$\frac{dD}{dt} = \frac{dD}{dp} \cdot \frac{dp}{dt}$$

Since $D(p) = \dfrac{40{,}000}{p} = 40{,}000p^{-1}$,

$$\frac{dD}{dp} = -40{,}000p^{-2} = \frac{-40{,}000}{p^2}$$

and $\dfrac{dp}{dt} = 0.6t^{1/2} = 0.6\sqrt{t}$.

$$\frac{dD}{dt} = \frac{-40{,}000}{p^2} \cdot 0.6\sqrt{t}$$

When $t = 4$,

$$\frac{dD}{dt} = \frac{-40{,}000}{(10)^2} \cdot 0.6\sqrt{4} = -480$$
$$100\frac{\frac{dD}{dt}}{D(t)} = 100\frac{-480}{4{,}000} = -12\%$$

Since the percentage rate is negative, demand will be decreasing.

63. $L = 0.25w^{2.6}$; $w = 3 + 0.21A$

(a) $\dfrac{dL}{dw} = 0.65w^{1.6}$ mm per kg
When $w = 60$,
$\dfrac{dL}{dw} = 0.65(60)^{1.6} \approx 455$ mm per kg

(b) When $A = 100$, $w = 3 + 0.21(100) = 24$ and $L(24) = 0.25(24)^{2.6} \approx 969$ mm long.

$$\frac{dL}{dA} = \frac{dL}{dw} \cdot \frac{dw}{dA}$$

Since $\dfrac{dw}{dA} = 0.21$,

$$\frac{dL}{dA} = (0.65w^{1.6})(0.21)$$

When $A = 100$, since $w = 24$,

$$\frac{dL}{dA} = 0.65(24)^{1.6}(0.21) \approx 22.1$$

The tiger's length is increasing at the rate of about 22.1 mm per day

65.
$$Q(K) = 500K^{2/3}$$
$$K(t) = \frac{2t^4 + 3t + 149}{t + 2}$$

(a) $K(3)=\dfrac{2(3)^4+3(3)+149}{3+2}=64$ or \$64,000.

$Q(64)=500(64)^{2/3}=8{,}000$ units

(b) $\dfrac{dQ}{dt}=\dfrac{dQ}{dK}\cdot\dfrac{dK}{dt}$

$$\frac{dQ}{dK}=500\left(\frac{2}{3}K^{-1/3}\right)=\frac{1000}{3K^{1/3}}$$

$$\frac{dK}{dt}=\frac{(t+2)(8t^3+3)-(2t^4+3t+149)(1)}{(t+2)^2}$$

$$=\frac{6t^4+16t^3-143}{(t+2)^2}$$

When $t=5$, $K(5)=\dfrac{2(5)^4+3(5)+149}{5+2}=202$. So,

$$\frac{dQ}{dt}=\frac{1000}{3(202)^{1/3}}\cdot\frac{6(5)^4+16(5)^3-143}{(5+2)^2}$$
$$\approx 6{,}501 \text{ units per month}$$

Since $\dfrac{dQ}{dt}$ is positive when $t=5$, production will be increasing.

67. $V(N)=\left(\dfrac{3N+430}{N+1}\right)^{2/3}$

$N(t)=\sqrt{t^2-10t+45}=(t^2-10t+45)^{1/2}$

(a) $N(9)=\sqrt{(9)^2-10(9)+45}=$ 6 hours per day

$$V(6)=\left[\frac{3(6)+430}{6+1}\right]^{2/3}=16$$

or \$16,000.

(b) $\dfrac{dV}{dt}=\dfrac{dV}{dN}\cdot\dfrac{dN}{dt}$

$$\frac{dV}{dN}=\frac{2}{3}\left(\frac{3N+430}{N+1}\right)^{-1/3}\cdot\frac{(N+1)(3)-(3N+430)(1)}{(N+1)^2}$$

$$=\frac{2(N+1)^{1/3}}{3(3N+430)^{1/3}}\cdot\frac{-427}{(N+1)^2}$$

$$=-\frac{854}{3(3N+430)^{1/3}(N+1)^{5/3}}$$

$$\frac{dN}{dt}=\frac{1}{2}(t^2-10t+45)^{-1/2}(2t-10)=\frac{t-5}{(t^2-10t+45)^{1/2}}$$

Using $t=9$ and $N=6$,

$$\frac{dV}{dt}=\frac{854}{3[3(6)+430]^{1/3}(6+1)^{5/3}}\cdot\frac{9-5}{\left[(9)^2-10(9)+45\right]^{1/2}}$$

≈ -0.968 thousand

or $-$ 968 dollars per month

Since $\dfrac{dV}{dt}$ is negative when $t=9$, the value will be decreasing.

69. $V(T)=0.41(-0.01T^2+0.4T+3.52)$

$$m(V)=\frac{0.39V}{1+0.09V}$$

(a) $\dfrac{dV}{dt}=0.41(-0.02T+0.4)\text{cm}^3$ per °C

(b)

$$\frac{dm}{dV}=\frac{(1+0.09V)(0.39)-(0.39V)(0.09)}{(1+0.09V)^2}$$

$$=\frac{0.39}{(1+0.09V)^2}\text{ gm per cm}^3$$

(c) When $T=10$,

$V(10)=0.41[-0.01(10)^2+0.4(10)+3.52]=2.6732\text{cm}^3$

$$\frac{dm}{dT}=\frac{dm}{dV}\cdot\frac{dV}{dt}$$

$$=\frac{0.39}{(1+0.09V)^2}\cdot 0.41(-0.02T+0.4)$$

When $T=10$,

$$\frac{dm}{dT}=\frac{0.39}{[1+0.09(2.6732)]^2}\cdot 0.41[-0.02(10)+0.4]$$
$$=0.02078 \text{ gm per °C}$$

71. $T=aL\sqrt{L-b}=aL(L-b)^{1/2}$

(a) $\dfrac{dT}{dL}=aL\cdot\dfrac{1}{2}(L-b)^{-1/2}(1)+(L-b)^{1/2}(a)$

$$=\frac{aL}{2(L-b)^{1/2}}+a(L-b)^{1/2}\frac{2(L-b)^{1/2}}{2(L-b)^{1/2}}$$

$$=\frac{aL+2a(L-b)}{2(L-b)^{1/2}}=\frac{3aL-2ab}{2\sqrt{L-b}}=\frac{a(3L-2b)}{2\sqrt{L-b}}$$

$\dfrac{dT}{dL}$ is the rate of change in the time required

with respect to the number of items in the list.

(b) Writing Exercise—Answers will vary.

73. $s(t) = (3 + t - t^2)^{3/2}$, $0 \le t \le 2$

(a) $v(t) = s'(t) = \frac{3}{2}(3 + t - t^2)^{1/2}(1 - 2t)$

$$a(t) = v'(t)$$
$$= \frac{3}{2}\left[(3+t-t^2)^{1/2}(-2) + (1-2t)\frac{1}{2}(3+t-t^2)^{-1/2}(1-2t)\right]$$
$$= \frac{3}{2}\left[-2(3+t-t^2)^{1/2}\frac{2(3+t-t^2)^{1/2}}{2(3+t-t^2)^{1/2}} + \frac{(1-2t)^2}{2(3+t-t^2)^{1/2}}\right]$$
$$= \frac{3}{2}\left[\frac{-4(3+t-t^2)+(1-2t)^2}{2(3+t-t^2)^{1/2}}\right]$$
$$= \frac{3}{2}\left[\frac{-12-4t+4t^2+1-4t+4t^2}{2(3+t-t^2)^{1/2}}\right]$$
$$= \frac{24t^2-24t-33}{4\sqrt{3+t-t^2}}$$

(b) To find when object is stationary for $0 \le t \le 2$,

$$\frac{3}{2}\sqrt{3+t-t^2}(1-2t) = 0$$

Press [y=] and input $1.5\sqrt{}(3+x-x^2) * (1-2x)$ for $y_1 =$
Use window dimensions $[-5, 5]1$ by $[-5, 5]1$
Use the zero function under calc menu to find the only x-intercept occurs at $x = 1/2$.
(Note: algebraically, $\sqrt{3+t-t^2} = 0$ when $t = \frac{1+\sqrt{13}}{2}$, but this value is not in the domain.)
Object is stationary when $t = 1/2$.

$$s\left(\frac{1}{2}\right) = \left[3+\frac{1}{2}-\left(\frac{1}{2}\right)^2\right]^{3/2} = \frac{\sqrt{2197}}{8} \approx 5.859$$

$$a\left(\frac{1}{2}\right) = \frac{24\left(\frac{1}{2}\right)^2 - 24\left(\frac{1}{2}\right) - 33}{4\sqrt{3+\frac{1}{2}-\left(\frac{1}{2}\right)^2}}$$
$$= \frac{-39}{2\sqrt{13}} = \frac{-3\sqrt{13}}{2} \approx -5.4083$$

For $a(1/2)$ you can use the dy/dx function under the calc menu and enter $x = .5$ to find $v'(1/2) = a(1/2) \approx -5.4083$.

(c) To find when the acceleration is zero for $0 \le t \le 2$,

$$\frac{24t^2-24t-33}{4\sqrt{3+t-t^2}} = 0$$

Press [y=] and input $(24x^2 - 24x - 33)/(4\sqrt{}(3+x-x^2))$ for $y_2 =$
Press [Graph]
You may wish to deactivate y_1 so only the graph of y_2 is shown.
Use the zero function under the calc menu to find the x-intercepts are $x \approx -0.775$ and $x \approx 1.77$. (disregard $x = -0.775$.)
The acceleration is zero for $t = 1.77$,
$s(1.77) = (3 + 1.77 - (1.77)^2)^{3/2} \approx 2.09$
Reactivate y_1 and use the value function under the calc menu. Make sure that y_1 is displayed in the upper left corner and enter $x = 1.77$ to find $v(1.77) \approx -4.87$.

(d) We already have $v(t)$ inputted for $y_1 =$ and $a(t)$ inputted for $y_2 =$
Press [y=] and input $(3 + x - x^2) \wedge (3/2)$ for $y_3 =$
Use window dimensions $[0, 2]1$ by $[-5, 5]1$
Press [Graph]

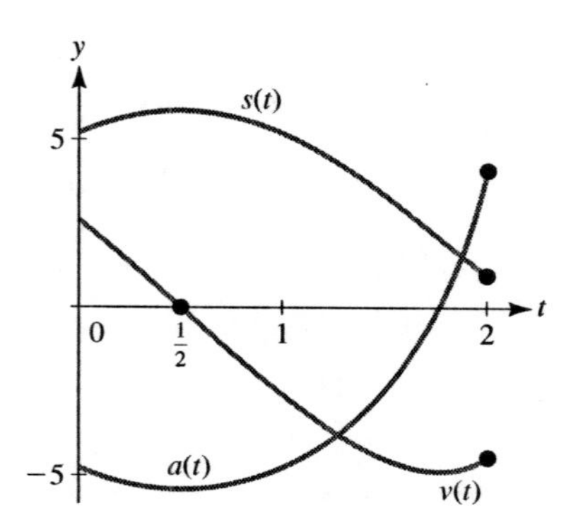

(e) To determine when $v(t)$ and $a(t)$ have opposite signs, press [y=] and deativate $y_3 =$ so only $v(t)$ and $a(t)$ are shown. Press [graph]. We see from the graph, $v(t)$ and $a(t)$ have opposite signs in two intervals. We know the t-intercept of $v(t)$ is $t = 1/2$ and the t-intercept of $a(t)$ is $t = 1.77$. The object is slowing down for $0 \le t < 0.5$ and $1.77 < t \le 2$.

75. To prove that

$$\frac{d}{dx}[h(x)]^2 = 2h(x)h'(x),$$

use the product rule to get

$$\begin{aligned}\frac{d}{dx}[h(x)]^2 &= \frac{d}{dx}[h(x)h(x)]\\ &= h(x)h'(x) + h'(x)h(x)\\ &= 2h(x)h'(x).\end{aligned}$$

77. To use numeric differentiation to calculate $f'(1)$ and $f'(-3)$, press [y=] and input $(3.1x^2 + 19.4) \wedge (1/3)$ for $y_1 =$
Use the window dimensions $[-5, 5]1$ by $[-3, 8]1$
Press [Graph]
Use the dy/dx function under the calc menu and enter $x = 1$ to find $f'(1) \approx 0.2593$
Repeat this for $x = -3$ to find $f'(-3) \approx -0.474$
Since there is only one minimum, we can conclude the graph has only one horizontal tangent.

2.5 Marginal Analysis; Approximations Using Increments

1. $C(x) = \frac{1}{5}x^2 + 4x + 57;$
$p(x) = \frac{1}{4}(36 - x) = 9 - \frac{1}{4}x$

(a) Marginal cost $= C'(x) = \frac{2}{5}x + 4$
Revenue = (# sold)(selling price)

$$R(x) = x\left(9 - \frac{1}{4}x\right) = 9x - \frac{x^2}{4}$$

Marginal revenue $= R'(x) = 9 - \frac{x}{2}$

(b) Estimated cost of 4th unit $= C'(3) = \frac{2}{5}(3) + 4 =$ \$5.20

(c) Actual cost of 4th unit $= C(4) - C(3)$

$$= \left[\frac{1}{5}(4)^2 + 4(4) + 57\right] - \left[\frac{1}{5}(3)^2 + 4(3) + 57\right]$$

$= \$5.40$

(d) Estimated revenue from sale of 4th unit $= R'(3) = 9 - \frac{3}{2} = \$7.50$

(e) Actual revenue from sale of 4th unit

$$= R(4) - R(3)$$

$$= \left[9(4) - \frac{(4)^2}{4}\right] - \left[9(3) - \frac{(3)^2}{4}\right] = \$7.25$$

3. $C(x) = \frac{1}{3}x^2 + 2x + 39;\ p(x) = -x^2 - 4x + 80$

(a) $C'(x) = \frac{2}{3}x + 2$

$R(x) = x(-x^2 - 4x + 80) = -x^3 - 4x^2 + 80x$, so
$R'(x) = -3x^2 - 8x + 80$

(b) $C'(3) = \frac{2}{3}(3) + 2 = \$4.00$

(c) $$C(4) - C(3) = \left[\frac{1}{3}(4)^2 + 2(4) + 39\right] - \left[\frac{1}{3}(3)^2 + 2(3) + 39\right] \approx \$4.33$$

(d) $R'(3) = -3(3)^2 - 8(3) + 80 = \$29$

(e) $$R(4) - R(3) = \left[-(4)^3 - 4(4)^2 + 80(4)\right] - \left[-(3)^3 - 4(3)^2 + 80(3)\right] = \$15$$

5. $C(x) = \frac{1}{4}x^2 + 43;\ p(x) = \frac{3+2x}{1+x}$

(a) $$C'(x) = \frac{1}{2}x$$

$$R(x) = x\left(\frac{3+2x}{1+x}\right) = \frac{3x+2x^2}{1+x}, \text{ so}$$

$$R'(x) = \frac{(1+x)(3+4x) - (3x+2x^2)(1)}{(1+x)^2} = \frac{2x^2+4x+3}{(1+x)^2}$$

(b) $C'(3) = \frac{1}{2}(3) = \$1.50$

(c) $$C(4) - C(3) = \left[\frac{1}{4}(4)^2 + 43\right] - \left[\frac{1}{4}(3)^2 + 43\right] = 47 - 45.25 = \$1.75$$

(d) $R'(3) = \frac{2(3)^2 + 4(3) + 3}{(1+3)^2} = \frac{33}{16} \approx \$2.06$

(e) $R(4) - R(3)$

$$= \frac{3(4) + 2(4)^2}{1+4} - \frac{3(3) + 2(3)^2}{1+3} = \frac{44}{5} - \frac{27}{4} = \$2.05$$

7. $f(x) = x^2 - 3x + 5$; x increases from 5 to 5.3

$$\Delta f \approx f'(x)\Delta x$$
$$f'(x) = 2x - 3$$
$$\Delta x = 5.3 - 5 = 0.3$$
$$\Delta f \approx [2(5) - 3](0.3) = 2.1$$

9. $f(x) = x^2 + 2x - 9$; x increases from 4 to 4.3. Estimated percentage change is

$$100\frac{\Delta f}{f} \text{ where } \Delta f \approx f'(x)\Delta x$$

$$f'(x) = 2x + 2,\ \Delta x = 4.3 - 4 = 0.3$$
$$\Delta f \approx [2(4) + 2](0.3) = 3$$
$$f(4) = (4)^2 + 2(4) - 9 = 15$$
$$100\frac{\Delta f}{f} = 100\frac{3}{15} = 20\%$$

11. $C(q) = 0.1q^3 - 0.5q^2 + 500q + 200$

(a) $$C'(q) = 0.3q^2 - q + 500$$
$$C'(3) = 0.3(3)^2 - 3 + 500 = \$499.70$$

(b) $C(4) - C(3)$

$$= [0.1(4)^3 - 0.5(4)^2 + 500(4) + 200] - [0.1(3)^3 - 0.5(3)^2 500(3) + 200] = \$2198.40 - \$1698.20 = \$500.20$$

13. $C(q) = 3q^2 + q + 500$

(a) $$C'(q) = 6q + 1$$
$$C'(40) = 6(40) + 1 = \$241$$

(b) $C(41) - C(40)$

$$= [3(41)^2 + 41 + 500] - [3(40)^2 + 40 + 500] = \$244$$

15. $C(t) = 100t^2 + 400t + 5{,}000$

$$\Delta C \approx C'(t)\Delta t$$
$$C'(t) = 200t + 400$$

Since t is measured in years, the next six months $= \frac{1}{2}$ year $= \Delta t$

$$\Delta C \approx C'(0)\left(\frac{1}{2}\right) = [200(0) + 400]\left(\frac{1}{2}\right) = 200,$$

or an increase of approximately 200 newspapers.

17. $$R(q) = 240q - 0.05q^2$$
$$\Delta R \approx R'(q)\Delta q$$
$$R'(q) = 240 - 0.1q$$

Since will decrease by 0.65 unit,

$$\Delta q = -0.65$$

$$\Delta R \approx R'(80)(-0.65) = [240 - 0.1(80)](-0.65)$$

$= -150.8$, or a decrease of approximately \$150.80.

19. $Q(K) = 600K^{1/2}$

$$\Delta Q \approx Q'(K)\Delta K$$

$$Q'(K) = 300K^{-1/2} = \frac{300}{\sqrt{K}}$$

Since K is measured in thousands of dollars, the current value of K is 900 and

$$\Delta K = \frac{800}{1000} = 0.8$$

$$\Delta Q \approx Q'(900)(0.8) = \left(\frac{300}{\sqrt{900}}\right)(0.8)$$

$$= 8,$$

or an increase of approximately 8 units.

21. $T(x) = 60x^{3/2} + 40x + 1{,}200$
Estimated percentage change is

$$100\frac{\Delta T}{T} \text{ where } \Delta T \approx T'(x)\Delta x$$

$$T'(x) = 90x^{1/2} + 40 = 90\sqrt{x} + 40$$

The beginning of the year 2008 is 8 years after the beginning of 2000, so the beginning value of t is 8. Measured in years, 6 months = $\frac{1}{2}$ year = Δt.

$$\Delta T \approx T'(8)\left(\frac{1}{2}\right) = \left(90\sqrt{8} + 40\right)\left(\frac{1}{2}\right) = 147.279$$

$$T(8) = 60(8)^{3/2} + 40(8) + 1{,}200 = 2{,}877.645$$

$$100\frac{\Delta T}{T} = 100\frac{147.279}{2877.645} \approx 5.12\%$$

23. $Q = 3{,}000K^{1/2}L^{1/3}$
Since labor force is to remain unchanged, write Q as

$$Q = 3{,}000\sqrt[3]{1331}K^{1/2}$$

Since increase in $K = 1$ (noting that K is measured in thousands of dollars)

$$\Delta Q \approx Q'(K)$$

$$Q'(K) = 1{,}500\sqrt[3]{1{,}331}K^{-1/2} = \frac{1{,}500\sqrt[3]{1{,}331}}{\sqrt{K}}$$

In thousands of dollars, the current value of $K = 400$, so

$$\Delta Q \approx Q'(400) = \frac{1,500\sqrt[3]{1{,}331}}{\sqrt{400}} = 825,$$

or an increase of approximately 825 units.

25. $C(q) = \frac{1}{6}q^3 + 642q + 400$

$$\Delta C \approx C'(q)\Delta q$$

We want to approximate Δq, so

$$\Delta q \approx \frac{\Delta C}{C'(q)}$$

$$C'(q) = \frac{1}{2}q^2 + 642,\ C'(4) = \frac{1}{2}(4)^2 + 642 = 650,$$

and $\Delta C = -130$. So, $\Delta q \approx \frac{-130}{650} = -0.2$, or increase production by 0.2 units.

27. The maximum percentage error in C is

$$100\frac{\Delta C}{C} \text{ where } \Delta C \approx C'(x)\Delta x$$

$$C'(x) = -a(x-b)^{-2}(1) = \frac{-a}{(x-b)^2}$$

$$\Delta C \approx C'(c)(\pm 0.03c) = \frac{-a}{(c-b)^2}(\pm 0.03c)$$

$$= \frac{\pm .03ac}{(c-b)^2}$$

$$C(c) = \frac{a}{c-b}$$

So, $$100\frac{\Delta C}{C} = 100\frac{\frac{\pm .03ac}{(c-b)^2}}{\frac{a}{(c-b)}}$$

$$= \frac{\pm 3c}{|c-b|}\%$$

29. $V = \pi R^2 L$, where L is constant for a given artery. The percentage error in V is

$$100\frac{\Delta V}{V} \text{ where } \Delta V \approx V'(R)\Delta R$$

$V'(R) = 2\pi RL$ so, noting that the radius is decreased by the plaque,

$$\Delta V \approx V'(0.3)(-0.07) = 2\pi(0.3)L(-0.07)$$
$$= -0.042\pi L$$
$$V(0.3) = \pi(0.3)^2 L = 0.09\pi L, \text{ so}$$
$$100\frac{\Delta V}{V} = 100\frac{-0.042\pi L}{0.09\pi L} = -46.67\%,$$

or a blockage in the volume of 46.67%.

31. $\Delta L \approx L'(T)\Delta T$

Since $\sigma = \dfrac{L'(T)}{L(T)}$, $L'(T) = \sigma L(T)$.

Also, $\Delta T = 35 - (-20) = 55$.

$$\text{So, } \Delta L \approx \sigma L(T)\Delta T$$
$$\approx (1.4 \times 10^{-5})(50)(55)$$
$$\approx 3{,}850 \times 10^{-5}$$

or an increase in length of approximately 0.0385 feet.

33. First application of Newton's method:

The equation of the tangent line at $(x_0, f(x_0))$ is

$$y - f(x_0) = f'(x_0)(x - x_0)$$

The x-intercept is when $y = 0$, or when

$$-f(x_0) = f'(x_0)(x - x_0)$$

Solving for $x = x_1$

$$x_1 = x_0 - \frac{f(x_0)}{f'(x_0)}$$

Second application of Newton's method:

Using the point $(x, f(x_1))$,

$$y - f(x_1) = f'(x_1)(x - x_1)$$
$$-f(x_1) = f'(x_1)(x - x_1)$$

Solving for $x = x_2$

$$x_2 = x_1 - \frac{f(x_1)}{f'(x_1)}$$

In general, using the point $(x_{n-1}, f(x_{n-1}))$,

$$y - f(x_{n-1}) = f'(x_{n-1})(x - x_{n-1})$$
$$-f(x_{n-1}) = f'(x_{n-1})(x - x_{n-1})$$

Solving for $x = x_n$,

$$x_n = x_{n-1} - \frac{f(x_{n-1})}{f'(x_{n-1})}$$

35. To use graphing utility to graph f and to estimate each root,

Press [y=] and input $x \wedge 4 - 4x \wedge 3 + 10$ for $y_1 =$

Use window dimensions $[-10, 10]1$ by $[-20, 20]2$

Press [Graph]

Use the zero function under the calc menu to find the zeros (x-intercepts) of f to be $x \approx 1.6$ and $x \approx 3.8$

To use Newton's method, $f(x) = x^4 - 4x^3 + 10$ and $f'(x) = 4x^3 - 12x^2$

$$x - \frac{f(x)}{f'(x)} = x - \frac{x^4 - 4x^3 + 10}{4x^3 - 12x^2} = \frac{3x^4 - 8x^3 - 10}{4x^3 - 12x^2}$$

For $n = 1, 2, 3, \ldots$

$$x_n = \frac{3x_{n-1}^4 - 8x_{n-1}^3 - 10}{4x_{n-1}^3 - 12x_{n-1}^2}$$

Using the graph shown on the calculator, we see one x-intercept is between 1 and 2.

Let $x_0 = 1$, then

$$x_1 = \frac{3x_0^4 - 8x_0^3 - 10}{4x_0^3 - 12x_0^2} = \frac{-15}{-8} = 1.875 \text{ using } x_0 = 1$$

$$x_2 = \frac{3x_1^4 - 8x_1^3 - 10}{4x_1^3 - 12x_1^2} = 1.621 \text{ using } x_1 = 1.875$$

Thus, one x-intercept is $x = 1.6$

The second x-intercept is between 3 and 4.

Let $x_0 = 4$, then

$$x_1 = \frac{3x_0^4 - 8x_0^3 - 10}{4x_0^3 - 12x_0^2} = 3.844 \text{ using } x_0 = 4$$

$$x_2 = \frac{3x_1^4 - 8x_1^3 - 10}{4x_1^3 - 12x_1^2} = 3.821 \text{ using } x_1 = 3.844$$

Thus, the second x-intercept is $x = 3.8$.

Note: Enter $(3x \wedge 4 - 8x \wedge 3 - 10)/(4x \wedge 3 - 12x \wedge 2)$ for $y_2 =$ and use the value function under the calc menu to do all the calculations for Newton's method.

37. $f(x) = \sqrt[3]{x} = x^{1/3}$; $f'(x) = \frac{1}{3}x^{-2/3} = \frac{1}{3x^{2/3}}$

(a)

$$x_{n+1} = x_n - \frac{(x_n)^{1/3}}{\frac{1}{3(x_n)^{2/3}}}$$

$$x_{n+1} = x_n - 3x_n, \text{ or } x_{n+1} = -2x_n$$

So, if x_0 is first guess,

$$x_1 = -2x_0,$$
$$x_2 = -2x_1 = -2(-2x_0) = 4x_0$$
$$x_3 = -2x_2 = -2(4x_0) = -8x_0,$$

etc.

(b) To use the graphing utility to graph f and to draw the tangent lines,
Press [y=] and input $x \wedge (1/3)$ for $y_1 =$
Use window dimensions [−5, 5]1 by [−5, 5]1

Arbitrarily, let's use $x_0 = 1$. Then we will draw tangent lines to the graph of f for $x = 1, -2, 4 \ldots$
Press [2nd] [Draw] and use the tangent function.
Enter $x = 1$ and the tangent line is drawn.
Repeat for $x = -2$ and $x = 4$.
>From the graph, can see that $x = 0$ is the root of $\sqrt[3]{x}$. Any choice besides zero for the first estimate leads to successive approximations on opposite sides of the root, getting farther and farther from the root.

2.6 Implicit Differentiation and Related Rates

1. $2x + 3y = 7$

(a) $2 + 3\frac{dy}{dx} = 0$

$$\frac{dy}{dx} = -\frac{2}{3}$$

(b) Solving for y,

$$y = -\frac{2}{3}x + \frac{7}{3}$$

$$\frac{dy}{dx} = -\frac{2}{3}$$

3. $xy + 2y = 3$

(a) $x\frac{dy}{dx} + y \cdot 1 + 2\frac{dy}{dx} = 0$

$$(x+2)\frac{dy}{dx} = -y$$

$$\frac{dy}{dx} = \frac{-y}{x+2}$$

(b) Solving for y,

$$y = \frac{3}{x+2} = 3(x+2)^{-1}$$

$$\frac{dy}{dx} = -3(x+2)^{-2}(1) = \frac{-3}{(x+2)^2}$$

$$= \frac{3}{x+2} \cdot \frac{-1}{x+2}$$

$$= y \cdot \frac{-1}{x+2} = \frac{-y}{x+2}$$

5. $xy + 2y = x^2$

(a)

$$x\frac{dy}{dx} + y \cdot 1 + 2\frac{dy}{dx} = 2x$$

$$(x+2)\frac{dy}{dx} = 2x - y$$

$$\frac{dy}{dx} = \frac{2x - y}{x+2}$$

(b) Solving for y,

$$y = \frac{x^2}{x+2}$$

$$\frac{dy}{dx} = \frac{(x+2)(2x) - (x^2)(1)}{(x+2)^2} = \frac{x^2 + 4x}{(x+2)^2}$$

Rewriting, $\frac{dy}{dx} = \frac{2x^2 + 4x - x^2}{(x+2)^2}$

$$= \frac{2x(x+2) - x^2}{(x+2)^2}$$

Dividing through by $x+2$,

$$= \frac{2x - \frac{x^2}{x+2}}{x+2} = \frac{2x-y}{x+2}$$

7.
$$x^2 + y^2 = 25$$
$$2x + 2y\frac{dy}{dx} = 0$$
$$\frac{dy}{dx} = -\frac{x}{y}$$

9.
$$x^3 + y^3 = xy,$$
$$3x^2 + 3y^2\frac{dy}{dx} = x\frac{dy}{dx} + y \cdot 1$$
$$(3y^2 - x)\frac{dy}{dx} = y - 3x^2,$$
$$\frac{dy}{dx} = \frac{y-3x^2}{3y^2-x}$$

11. $y^2 + (2x)(y^2) - 3x + 1 = 0$

$$2y\frac{dy}{dx} + (2x)\left(2y\frac{dy}{dx}\right) + (y^2)(2) - 3 + 0 = 0$$
$$(2y + 4xy)\frac{dy}{dx} = 3 - 2y^2$$
$$\frac{dy}{dx} = \frac{3-2y^2}{2y(1+2x)}$$

13. $\sqrt{x} + \sqrt{y} = 1$, or $x^{1/2} + y^{1/2} = 1$

$$\frac{1}{2}x^{-1/2} + \frac{1}{2}y^{-1/2}\frac{dy}{dx} = 0$$
$$x^{-1/2} + y^{-1/2}\frac{dy}{dx} = 0$$
$$\frac{dy}{dx} = \frac{-x^{-1/2}}{y^{-1/2}} = \frac{-\sqrt{y}}{\sqrt{x}}$$

15. $xy - x = y + 2$

$$x\frac{dy}{dx} + y \cdot 1 - 1 = \frac{dy}{dx} + 0$$
$$(x-1)\frac{dy}{dx} = 1 - y$$
$$\frac{dy}{dx} = \frac{1-y}{x-1}$$

17. $(2x+y)^3 = x$,

$$3(2x+y)^2\left(2 + \frac{dy}{dx}\right) = 1,$$
$$2 + \frac{dy}{dx} = \frac{1}{3(2x+y)^2},$$
$$\frac{dy}{dx} = \frac{1}{3(2x+y)^2} - 2$$

19. $(x^2 + 3y^2)^5 = (2x)(y)$

$$5(x^2+3y^2)^4\left(2x + 6y\frac{dy}{dx}\right) = 2x\frac{dy}{dx} + y \cdot 2$$
$$10x(x^2+3y^2)^4 + 30y(x^2+3y^2)^4\frac{dy}{dx} = 2x\frac{dy}{dx} + 2y$$
$$5x(x^2+3y^2)^4 + 15y(x^2+3y^2)^4\frac{dy}{dx} = x\frac{dy}{dx} + y$$
$$\left[15y(x^2+3y^2)^4 - x\right]\frac{dy}{dx} = y - 5x(x^2+3y^2)^4$$
$$\frac{dy}{dx} = \frac{y - 5x(x^2+3y^2)^4}{15y(x^2+3y^2)^4 - x}$$

21.
$$x^2 = y^3$$
$$2x = 3y^2\frac{dy}{dx}$$
$$\frac{2x}{3y^2} = \frac{dy}{dx}$$

The slope of the tangent line at (8,4) is

$$\frac{dy}{dx} = \frac{2(8)}{3(4)^2} = \frac{1}{3}$$

and the equation of the tangent line is

$$y - 4 - \frac{1}{3}(x-8), \text{ or } y = \frac{1}{3}x + \frac{4}{3}$$

23. $xy = 2$

$$x\frac{dy}{dx} + y \cdot 1 = 0$$
$$\frac{dy}{dx} = \frac{-y}{x}$$

The slope of the tangent line at (2,1) is

$$\frac{dy}{dx} = \frac{-1}{2}$$

and the equation of the tangent line is

$$y - 1 = -\frac{1}{2}(x - 2), \text{ or } y = -\frac{1}{2}x + 2$$

25. $(1 - x + y)^3 = x + 7$

$$3(1 - x + y)^2\left(-1 + \frac{dy}{dx}\right) = 1 + 0$$

$$-1 + \frac{dy}{dx} = \frac{1}{3(1 - x + y)^2}$$

$$\frac{dy}{dx} = \frac{1}{3(1 - x + y)^2} + 1$$

When $x = 1$, $(1 - 1 + y)^3 = 1 + 7$, so $y = 2$ and the slope of the tangent line is

$$\frac{dy}{dx} = \frac{1}{3(1 - 1 + 2)^2} + 1 = \frac{13}{12}$$

The equation of the tangent line is

$$y - 2 = \frac{13}{12}(x - 1), \text{ or } y = \frac{13}{12}x + \frac{11}{12}$$

27.

$$x + y^2 = 9$$

$$1 + 2y\frac{dy}{dx} = 0$$

$$\frac{dy}{dx} = \frac{-1}{2y}$$

(a) For horizontal tangent(s), need $\frac{dy}{dx} = 0$, but $-\frac{1}{2y} \neq 0$ for any value of y, so there are no horizontal tangents.

(b) For vertical tangent(s), need the denominator of the slope $2y = 0$, or $y = 0$. When $y = 0$, $x + 0 = 9$, or $x = 9$. There is a vertical tangent at (9,0).

29. $x^2 + xy + y = 3$

$$2x + x\frac{dy}{dx} + y \cdot 1 + \frac{dy}{dx} = 0$$

$$\frac{dy}{dx} = \frac{-2x - y}{x + 1}$$

(a) $\frac{-2x - y}{x + 1} = 0$ when $-2x - y = 0$, or $y = -2x$.

Substituting in the original equation,

$$x^2 - 2x^2 - 2x = 3$$

$$0 = x^2 + 2x + 3$$

Since there are no real solutions, there are no horizontal tangents.

(b) $x + 1 = 0$ when $x = -1$
When $x = -1$, $1 - y + y = 3$
So no such y exists and there are no vertical tangents.

31. $x^2 + xy + y^2 = 3$

$$2x + x\frac{dy}{dx} + y \cdot 1 + 2y\frac{dy}{dx} = 0$$

$$\frac{dy}{dx} = \frac{-2x - y}{x + 2y}$$

(a) $\frac{-2x - y}{x + 2y} = 0$ when $-2x - y = 0$, or $y = -2x$.
Substituting in the original equation,

$$x^2 - 2x^2 + 4x^2 = 3$$

$$3x^2 = 3$$

$$x = \pm 1$$

When $x = -1$, $y = -2(-1) = 2$, and when $x = 1$, $y = -2(1) = -2$. So, there are horizontal tangents at $(-1, 2)$ and $(1, -2)$.

(b) $x + 2y = 0$ when $x = -2y$.
Substituting in the original equation,

$$4y^2 - 2y^2 + y^2 = 3$$

$$3y^2 = 3$$

$$y = \pm 1$$

When $y = -1$, $x = -2(-1) = 2$, and when $y = 1$, $x = -2(1) = -2$. So, there are vertical tangents at $(-2, 1)$ and $(2, -1)$.

33. $x^2 + 3y^2 = 5$

$$2x + 6y\frac{dy}{dx} = 0$$

$$\frac{dy}{dx} = -\frac{x}{3y}$$

$$\frac{d^2y}{dx^2} = \frac{(3y)(-1) - (-x)\left(3\frac{dy}{dx}\right)}{(3y)^2}$$

$$= \frac{-3y + 3x\frac{dy}{dx}}{9y^2}$$

$$= \frac{-3 + 3x\left(\frac{-x}{3y}\right)}{9y^2}.$$

$$= \frac{-3y - \frac{x^2}{y}}{9y^2} \cdot \frac{y}{y}$$

$$= \frac{-3y^2 - x^2}{9y^3}$$

$$= \frac{-(x^2 + 3y^2)}{9y^3} = -\frac{5}{9y^3}$$

35. Need to find

$$\Delta y \approx \frac{dy}{dx}$$

Since Q is to remain constant, let c be the constant value of Q. Then

$$c = 0.08x^2 + 0.12xy + 0.03y^2$$

$$0 = 0.16x + 0.12x\frac{dy}{dx} + y \cdot 0.12 + 0.06y\frac{dy}{dx}$$

$$\frac{dy}{dx} = \frac{-0.16x - 0.12y}{0.12x + 0.06y}$$

Since $x = 80$ and $y = 200$, $\frac{dy}{dx} \approx -1.704$, or a decrease of 1.704 hours of unskilled labor.

37.

$$3p^2 - x^2 = 12$$

$$6p\frac{dp}{dt} - 2x\frac{dx}{dt} = 0$$

When $p = 4$, $48 - x^2 = 12$, $x^2 = 36$, or $x = 6$. Substituting, $6(4)(0.87) - 2(6)\frac{dx}{dt} = 0$

$$20.88 - 12\frac{dx}{dt} = 0$$

$\frac{dx}{dt} = \frac{20.88}{12} = 1.74$ or increasing at a rate of 174 units/month.

39.

$$V = \frac{4}{3}\pi R^3$$

$$\frac{dV}{dt} = 4\pi R^2\frac{dR}{dt}$$

Substituting,

$$\frac{dV}{dt} = 4\pi(0.54)^2(0.13) \approx 0.476$$

or increasing at a rate of 0.476 cm^3 per month.

41.

$$V = \frac{4}{3}\pi r^3$$

$$\frac{dV}{dt} = 4\pi r^2\frac{dr}{dt}$$

Substituting, $0.002\pi = 4\pi(0.005)^2\frac{dr}{dt}$

$$\frac{dr}{dt} = \frac{0.002}{4(0.005)^2} = 20,$$

or increasing at a rate of 20 mm per min.

43.

$$75x^2 + 17p^2 = 5{,}300$$

$$150x\frac{dx}{dt} + 34p\frac{dp}{dt} = 0$$

When $p = 7$, $75x^2 = 5,\ 300 - 833$ or $x = 7.72$. Substituting,

$$150(7.72)\frac{dr}{dt} + 34(7)(-0.75) = 0$$

$$\text{so } \frac{dx}{dt} = \frac{178.5}{1158} = 0.154$$

or increasing at a rate of 15.4 units/month.

45. $M = 70w^{3/4}$

(a) $\frac{dM}{dt} = 52.5w^{-1/4}\frac{dw}{dt}$

Substituting, $\frac{dM}{dt} = \frac{52.5}{(80)^{1/4}}(0.8) \approx 14.04$,

or increasing at a rate of 14.04 kg per day^2.

(b) $\frac{dM}{dt} = \frac{52.5}{(50)^{1/4}}(-0.5) \approx -9.87$, or decreasing at a rate of 9.87 kg per day^2.

47. Since Q is to remain constant, $C = 60K^{1/3}L^{2/3}$
$0 = 60K^{1/3} \cdot \frac{2}{3}L^{-1/3}\frac{dL}{dt} + L^{2/3} \cdot 20K^{-2/3}\frac{dK}{dt}$
Substituting,

$$0 = \frac{40(8)^{1/3}}{(1{,}000)^{1/3}}(25) + \frac{20(1{,}000)^{2/3}}{(8)^{2/3}}\frac{dK}{dt}$$

$$0 = 200 + 500\frac{dK}{dt}$$

$$\frac{dK}{dt} = -0.4,$$

or decreasing at a rate of $400 per week.

49. Let x be the distance between the man and the base of the street light and L the length of the shadow. Because of similar triangles,

$$\frac{L}{6} = \frac{x+L}{12} \text{ or } L = x$$

So, $\frac{dL}{dt} = \frac{dx}{dt} = 4$, or increasing at a rate of 4 feet per second.

51. Need $\Delta y \approx \frac{dy}{dx}\Delta x$
Since Q is to remain constant, let C be the constant value of Q. Then

$$C = 2x^3 + 3x^2y^2 + (1+y)^3$$

$$0 = 6x^2 + (3x^2)\left(2y\frac{dy}{dx}\right) + (y^2)(6x) + 3(1+y)^2\frac{dy}{dx}$$

Substituting,

$$0 = 6(30)^2 + 6(30)^2(20)\frac{dy}{dx} + (20)^2(6 \cdot 30) + 3(1+20)^2\frac{dy}{dx}$$

$$0 = 77{,}400 + 109{,}323\frac{dy}{dx}$$

$$\frac{dy}{dx} \approx -0.7080$$

Since $\Delta x = -0.8$,

$$\Delta y \approx (-0.7080)(-0.8) = 0.5664,$$

or an increase of 0.5664 units in input y.

53. $v = \frac{K}{L}(R^2 - r^2)$
At the center of the vessel, $r = 0$ so

$$v = \frac{K}{L}R^2 = KL^{-1}R^2$$

Using implicit differentiation with t as the variable,

$$\frac{dv}{dt} = K\left[L^{-1}\left(2R\frac{dR}{dt}\right) + R^2\left(-L^{-2}\frac{dL}{dt}\right)\right]$$

Since the speed is unaffected, $\frac{dv}{dt} = 0$ and

$$0 = K\left[\frac{2R}{L} \cdot \frac{dR}{dt} - \frac{R^2}{L^2} \cdot \frac{dL}{dt}\right]$$

$$0 = \frac{2R}{L} \cdot \frac{dR}{dt} - \frac{R^2}{L^2} \cdot \frac{dL}{dt}$$

Solving for the relative rate of change of L,

$$\frac{R^2}{L^2} \cdot \frac{dL}{dt} = \frac{2R}{L} \cdot \frac{dR}{dL}$$

$$\frac{dL/dt}{L} = 2\frac{dR/dL}{R}$$

or double the relative rate of change of R.

55. $\frac{x^2}{a^2} + \frac{y^2}{b^2} = 1$

$$\frac{2x}{a^2} + \frac{2y}{b^2}\frac{dy}{dx} = 0$$

$$\frac{dy}{dx} = \frac{\frac{-2x}{a^2}}{\frac{2y}{b^2}} = \frac{-b^2x}{a^2y}$$

Substituting,

$$\frac{dy}{dx} = \frac{-b^2x_0}{a^2y_0}$$

and the equation of the tangent line is

$$y - y_0 = \frac{-b^2 x_0}{a^2 y_0}(x - x_0)$$

$$\frac{y_0 y}{b^2} - \frac{y_0^2}{b^2} = \frac{-x_0 x}{a^2} + \frac{x_0^2}{a^2}$$

$$\frac{x_0 x}{a^2} + \frac{y_0 y}{b^2} = \frac{x_0^2}{a^2} + \frac{y_0^2}{b^2}$$

So, $\frac{x_0 x}{a^2} + \frac{y_0 y}{b^2} = 1$

57. $y = x^{r/s}$ or $y^s = x^r$

$$s y^{s-1} \frac{dy}{dx} = r x^{r-1}$$

$$\frac{dy}{dx} = \frac{r x^{r-1}}{s y^{s-1}}$$

$$= \frac{r x^{r-1}}{s (x^{r/s})^{s-1}}$$

$$= \frac{r x^{r-1}}{s x^{r - r/s}}$$

$$= \frac{r}{s} x^{(r-1)-(r-r/s)}$$

$$= \frac{r}{s} x^{r/s - 1}$$

59. To use the graphing utility to graph $11x^2 + 4xy + 14y^2 = 21$, we must express y in terms of x.

$$11x^2 + 4xy + 14y^2 = 21$$

$$14y^2 + 4xy + 11x^2 - 21 = 0$$

Using the quadratic formula,

$$y = \frac{-4x \pm \sqrt{16x^2 - 4(14)(11x^2 - 21)}}{28}$$

$$= \frac{-2x \pm \sqrt{294 - 150x^2}}{14}$$

Press [y=] and input $(-2x + \sqrt{}(294 - 150x^2))/14$ for $y_1 =$
Input $(-2x - \sqrt{}(294 - 150x^2))/14$ for $y_2 =$
Use window dimensions $[-1.5, 1.5].5$ by $[-1.5, 1.5].5$
Press [Graph]
Press [2nd] [Draw] and select tangent function. Enter $x = -1$. We see the equation of the line tangent at $(-1, 1)$ is approximately $y = .75x + 1.75$. To find the horizontal tangents,

$$22x + 4x\left(\frac{dy}{dx}\right) + 4y + 28y\left(\frac{dy}{dx}\right) = 0$$

$$\frac{dy}{dx} = \frac{-22x - 4y}{4x + 28y}$$

$\frac{dy}{dx} = 0$ when $y = -\frac{11}{2}x$

$$11x^2 + 4x11x^2 + 4x\left(-\frac{11}{2}x\right) + 14\left(-\frac{11}{2}x\right)^2 = 21$$

Solving yields $x = \pm 0.226$ and $y = \mp 1.241$
The two horizontal tangents are at $y = -1.241$ and $y = 1.241$.

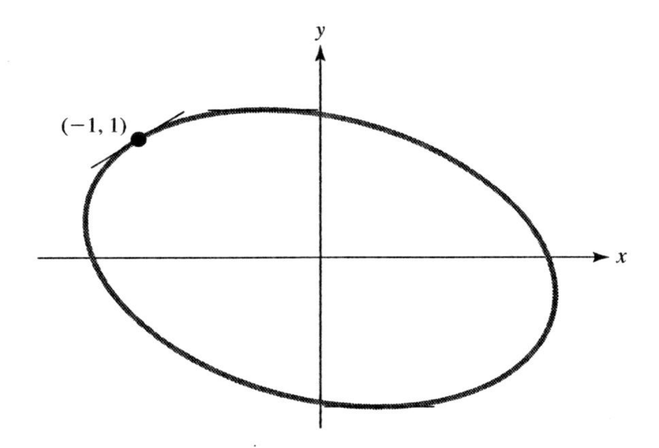

61. To use the graphing utility to graph curve $x^2 + y^2 = \sqrt{x^2 + y^2} + x$,
It is best to graph $x^2 + y^2 = \sqrt{x^2 + y^2} + x$ using polar coordinates.
Given that $r^2 = x^2 + y^2$ and $r \cos\theta = x$, we change the equation to

$$r^2 = r + r\cos\theta$$

$$r(r - 1 - \cos\theta) = 0$$

$r = 0$ gives the origin and thus, we graph $r = 1 + \cos\theta$ using the graphing utility.
Press [2nd] [format] and select Polar Gc
Press [mode] and selct Pol
Press [y=] and input $1 + \cos\theta$

In the viewing window, use θ min $= 0$, θ max $= 2\pi$, θstep $= \pi/24$ and dimensions $[-1, 2].5$ by $[-1.5, 1.5].5$
Using trace and zoom, it appears that a horizontal tangent is approximately $y = \pm 1.23$.

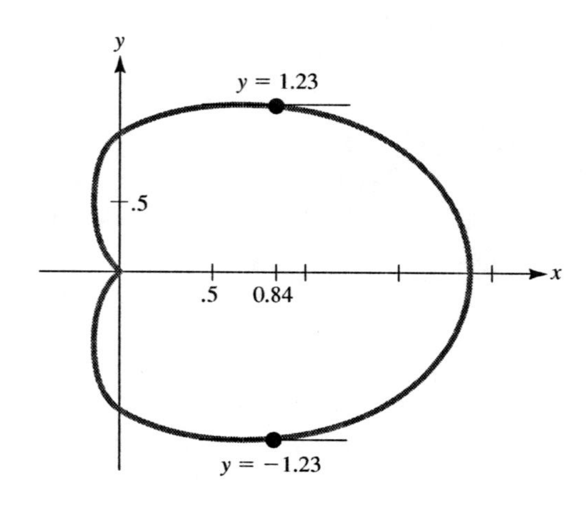

Checkup for Chapter 2

1. **(a)** $y = 3x^4 - 4\sqrt{x} + \dfrac{5}{x^2} - 7$

$$y = 3x^4 - 4x^{1/2} + 5x^{-2} - 7$$
$$\frac{dy}{dx} = 12x^3 - 2x^{-1/2} - 10x^{-3} - 0$$
$$\frac{dy}{dx} = 12x^3 - \frac{2}{\sqrt{x}} - \frac{10}{x^3}$$

(b) $y = (3x^3 - x + 1)(4 - x^2)$

$$\frac{dy}{dx} = (3x^3 - x + 1)(-2x) + (4 - x^2)(9x^2 - 1)$$
$$\frac{dy}{dx} = -6x^4 + 2x^2 - 2x + 36x^2 - 9x^4 - 4 + x^2$$
$$\frac{dy}{dx} = -15x^4 + 39x^2 - 2x - 4$$

(c) $y = \dfrac{5x^2 - 3x + 2}{1 - 2x}$

$$\frac{dy}{dx} = \frac{(1-2x)(10x-3) - (5x^2 - 3x + 2)(-2)}{(1-2x)^2}$$
$$\frac{dy}{dx} = \frac{10x - 20x^2 - 3 + 6x + 10x^2 - 6x + 4}{(1-2x)^2}$$
$$\frac{dy}{dx} = \frac{-10x^2 + 10x + 1}{(1-2x)^2}$$

(d) $y = (3 - 4x + 3x^2)^{3/2}$

$$\frac{dy}{dx} = \frac{3}{2}(3 - 4x + 3x^2)^{1/2}(-4 + 6x)$$
$$\frac{dy}{dx} = (9x - 6)(3 - 4x + 3x^2)^{1/2}$$

2. $f(t) = t(2t + 1)^2$

$$f'(t) = t \cdot 2(2t + 1)(2) + (2t + 1)^2(1)$$
$$f'(t) = (2t + 1)(4t + 2t + 1)$$
$$f'(t) = (2t + 1)(6t + 1) = 12t^2 + 8t + 1$$
$$f''(t) = 24t + 8$$

3. $y = x^2 - 2x + 1$
Slope $= \dfrac{dy}{dx} = 2x - 2$
When $x = -1$, $y = (-1)^2 - 2(-1) + 1 = 4$ and $\dfrac{dy}{dx} = 2(-1) - 2 = -4$. The equation of the tangent line is

$$y - 4 = -4(x + 1), \text{ or } y = -4x$$

4. $f(x) = \dfrac{x + 1}{1 - 5x}$

$$f'(x) = \frac{(1 - 5x)(1) - (x + 1)(-5)}{(1 - 5x)^2}$$
$$f'(x) = \frac{1 - 5x + 5x + 5}{(1 - 5x)^2} = \frac{6}{(1 - 5x)^2}$$
$$f'(1) = \frac{6}{(1 - 5)^2} = \frac{3}{8}$$

5. $T(x) = 3x^2 + 40x + 1800$

(a) $T'(x) = 6x + 40$

In 2003, $x = 3$ and $T'(3) = 6(3) + 40 = \$58$ per year.

(b) Need $100\,\dfrac{T'(3)}{T(3)}$

$$T(3) = 3(3)^2 + 40(3) + 1800 = 1947$$

$$100\frac{T'(3)}{T(3)} = 100\frac{58}{1947} \approx 2.98\%$$

6. $s(t) = 2t^3 - 3t^2 + 2,\ t \geq 0$

(a) $v(t) = s'(t) = 6t^2 - 6t$
$a(t) = s''(t) = 12t - 6$

(b) When stationary, $v(t) = 0$

$$6t^2 - 6t = 0$$

$$6t(t-1) = 0, \quad \text{or } t = 0,\ 1$$

When $0 < t < 1$, $v(t) < 0$, so retreating
$t > 1$, $v(t) > 0$, so advancing.

(c) $|s(1) - s(0)| + |s(2) - s(1)|$
$= 1 + 5 = 6$

7. $C(x) = 0.04x^2 + 5x + 73$

(a) $C'(x) = 0.08x + 5$

$$C'(5) = 0.08(5) + 5 = 5.4, \quad \text{or \$540 per unit}$$

(b) $C(6) - C(5) = 104.44 - 99 = 5.44$, or \$544

8. $Q = 500L^{3/4}$

$$\Delta Q \approx Q'(L)\Delta L$$

$$Q'(L) = 375L^{-1/4} = \frac{375}{L^{1/4}}$$

$$Q'(2401) = \frac{375}{(2401)^{1/4}} = \frac{375}{7}$$

Since $\Delta L = 200$,

$$\Delta Q \approx \frac{375}{7}(200) = \frac{75{,}000}{7},$$

or an increase of approximately 10,714.29 units.

9. $S = 0.2029w^{0.425}$

$$\frac{dS}{dt} = (0.2029)(0.425)w^{-0.575}\frac{dw}{dt}$$

$$= \frac{(0.2029)(0.425)}{(30)^{0.575}}(0.13) \approx 0.001586,$$

or increasing at a rate of 0.001586 m^2 per week.

10. $V = \frac{4}{3}\pi r^3$

Want $100\,\frac{\Delta V}{V} \leq 8$, where $\Delta V \approx V'(r)\Delta r$, $V'(r) = 4\pi r^2$ and $\Delta r = a \cdot r$, where a represents the % error in the measure of r (as a decimal).

$$100\frac{\Delta V}{V} \leq 8$$

$$100\frac{4\pi r^2 \cdot ar}{\frac{4}{3}\pi r^3} \leq 8$$

$$100a \leq \frac{8}{3}$$

or $\frac{8}{3}$% error in the measurement of r.

Review Problems

1. $f(x) = x^2 - 3x + 1$

$$\frac{f(x+h) - f(x)}{h} =$$

$$\frac{\left[(x+h)^2 - 3(x+h) + 1\right] - (x^2 - 3x + 1)}{h}$$

$$= \frac{x^2 + 2xh + h^2 - 3x - 3h + 1 - x^2 + 3x + 1}{h}$$

$$= \frac{2xh + h^2 - 3h}{h} = 2x + h - 3$$

$$f'(x) = \lim_{h \to 0} 2x + h - 3 = 2x - 3$$

3. $f(x) = 6x^4 - 7x^3 + 2x + \sqrt{2}$

$$f'(x) = 24x^3 - 21x^2 + 2$$

5.
$$y = \frac{2 - x^2}{3x^2 + 1}.$$
$$\frac{dy}{dx} = \frac{(3x^2 + 1)(-2x) - (2 - x^2)(6x)}{(3x^2 + 1)^2}$$
$$= \frac{-14x}{(3x^2 + 1)^2}$$

7. $f(x) = (5x^4 - 3x^2 + 2x + 1)^{10}$
$$f'(x) = 10(5x^4 - 3x^2 + 2x + 1)^9(20x^3 - 6x + 2)$$

9.
$$y = \left(x + \frac{1}{x}\right)^2 - \frac{5}{\sqrt{3x}}$$
$$= (x + x^{-1})^2 - \frac{5}{\sqrt{3}}x^{-1/2}$$
$$\frac{dy}{dx} = 2(x + x^{-1})(1 - x^{-2}) + \frac{5}{2\sqrt{3}}x^{-3/2}$$
$$= 2\left(x + \frac{1}{x}\right)\left(1 - \frac{1}{x^2}\right) + \frac{5}{2\sqrt{3}x^{3/2}}$$

11.
$$f(x) = (3x + 1)\sqrt{6x + 5}$$
$$= (3x + 1)(6x + 5)^{1/2}.$$
$$f'(x) = (3x + 1)\left(\frac{1}{2}\right)(6x + 5)^{-1/2}(6) + (6x + 5)^{1/2}(3)$$
$$= \frac{3(3x + 1)}{(6x + 5)^{1/2}} + 3(6x + 5)^{1/2}$$
$$= \frac{3(3x + 1) + 3(6x + 5)}{(6x + 5)^{1/2}}$$
$$= \frac{27x + 18}{(6x + 5)^{1/2}}$$
$$= \frac{9(3x + 2)}{\sqrt{6x + 5}}$$

13. $y = \sqrt{\dfrac{1 - 2x}{3x + 2}} = \left(\dfrac{1 - 2x}{3x + 2}\right)^{1/2}$

$$\frac{dy}{dx} = \frac{1}{2}\left(\frac{1 - 2x}{3x + 2}\right)^{-1/2} \cdot \frac{(3x + 2)(-2) - (1 - 2x)(3)}{(3x + 2)^2}$$
$$= \frac{1}{2}\frac{(3x + 2)^{1/2}}{(1 - 2x)^{1/2}} \cdot \frac{-7}{(3x + 2)^2}$$
$$= \frac{-7}{2(1 - 2x)^{1/2}(3x + 2)^{3/2}}$$

15.
$$f(x) = \frac{4}{x - 3}$$
$$f'(x) = \frac{-4}{(x - 3)^2}$$

$f(1) = -2.$
The slope of the tangent line at $(1, -2)$ is $f'(1) = -1$.
The equation of the tangent line is
$$y + 2 = -(x - 1), \text{ or } y = -x - 1$$

17. $f(x) = \sqrt{x^2 + 5} = (x^2 + 5)^{1/2}$

$$f'(x) = \frac{1}{2}(x^2 + 5)^{-1/2}(2x) = \frac{x}{\sqrt{x^2 + 5}}$$
$f(-2) = 3$. The slope of the tangent line at $(-2, 3)$ is $f'(-2) = -2/3$.
The equation of the tangent line is
$$y - 3 = -\frac{2}{3}(x + 2), \text{ or } y = -\frac{2}{3}x + \frac{5}{3}$$

19. (a) $f(t) = t^2 - 3t + \sqrt{t} = t^2 - 3t + t^{1/2}$
Need $100\dfrac{f'(4)}{f(4)}$

$$f'(t) = 2t - 3 + \frac{1}{2}t^{-1/2} = 2t - 3 + \frac{1}{2\sqrt{t}}$$
$$f'(4) = 2(4) - 3 + \frac{1}{2\sqrt{4}} = \frac{21}{4}$$
$$f(4) = (4)^2 - 3(4) + \sqrt{4} = 6$$
$$100\frac{f'(4)}{f(4)} = 100\frac{\frac{21}{4}}{6} = 87.5\%$$

(b) $f(t) = t^2(3-2t)^3$

$f'(t) = t^2 \cdot 3(3-2t)^2(-2) + (3-2t)^3(2t)$

$f'(1) = 1 \cdot 3(3-2)^2(-2) + (3-2)^3(2) = -4$

$f(1) = 1(3-2)^3 = 1$

$$100\frac{f'(1)}{f(1)} = 100\frac{-4}{1} = -400\%$$

(c)
$$f(t) = \frac{1}{t+1} = (t+1)^{-1}$$

$$f'(t) = -(t+1)^{-2} = \frac{-1}{(t+1)^2}$$

$$f'(0) = \frac{-1}{(0+1)^2} = -1$$

$$f(0) = \frac{1}{0+1} = 1$$

$$100\frac{f'(0)}{f(0)} = 100\frac{-1}{1} = -100\%$$

21. (a) $y = u^3 - 4u^2 + 5u + 2,\ u = x^2 + 1.$

$\frac{dy}{du} = 3u^2 - 8u + 5,\ \frac{du}{dx} = 2x,$

$\frac{dy}{dx} = \frac{dy}{du}\frac{du}{dx}$

When $x = 1$, $u = 2$, and so

$$\frac{dy}{dx} = [3(2^2) - 8(2) + 5][2(1)] = 2$$

(b)
$$y = \sqrt{u} = u^{1/2},$$
$$u = x^2 + 2x - 4,$$
$$\frac{dy}{du} = \frac{1}{2u^{1/2}},$$
$$\frac{du}{dx} = 2x + 2,$$
$$\frac{dy}{dx} = \frac{dy}{du}\cdot\frac{du}{dx}$$

When $x = 2$, $u = 4$, and so

$$\frac{dy}{dx} = \frac{1}{2(4)^{1/2}}\cdot[2(2)+2] = \frac{3}{2}$$

(c)
$$y = \left(\frac{u-1}{u+1}\right)^{1/2},\ u = \sqrt{x-1} = (x-1)^{1/2}$$

$$\frac{dy}{du} = \frac{1}{2}\left(\frac{u-1}{u+1}\right)^{-1/2}\cdot\frac{(u+1)(1)-(u-1)(1)}{(u+1)^2}$$
$$= \frac{(u+1)^{1/2}}{2(u-1)^{1/2}}\cdot\frac{2}{(u+1)^2}$$
$$= \frac{1}{(u-1)^{1/2}(u+1)^{3/2}}$$

$$\frac{du}{dx} = \frac{1}{2}(x-1)^{-1/2}(1) = \frac{1}{2(x-1)^{1/2}}$$

$$\frac{dy}{dx} = \frac{dy}{du}\cdot\frac{du}{dx}$$

When $x = \frac{34}{9}$, $u = \sqrt{\frac{34}{9} - 1} = \frac{5}{3}$, and so

$$\frac{dy}{dx} = \frac{1}{\left(\frac{5}{3}-1\right)^{1/2}\left(\frac{5}{3}+1\right)^{3/2}}\cdot\frac{1}{2\left(\frac{34}{9}-1\right)^{1/2}}$$
$$= \frac{1}{\left(\frac{2}{3}\right)^{1/2}\left(\frac{8}{3}\right)^{3/2}}\cdot\frac{1}{2\left(\frac{5}{3}\right)}$$
$$= \frac{1}{\left(\frac{2}{3}\right)^{1/2}\left(\frac{512}{27}\right)^{1/2}}\cdot\frac{3}{10}$$
$$= \frac{1}{\left(\frac{1024}{81}\right)^{1/2}}\cdot\frac{3}{10}$$
$$= \frac{9}{32}\cdot\frac{3}{10} = \frac{27}{310}$$

23. (a) $5x + 3y = 12,$

$$5 + 3\frac{dy}{dx} = 0, \text{ or } \frac{dy}{dx} = -\frac{5}{3}$$

(b) $x^2 y = 1$,

$$x^2\frac{dy}{dx} + y(2x) = 0$$
$$\frac{dy}{dx} = -\frac{2xy}{x^2} = -\frac{2y}{x}$$

(c) $(2x+3y)^5 = x+1$,

$$5(2x+3y)^4\left(2+3\frac{dy}{dx}\right) = 1$$
$$10(2x+3y)^4 + 15(2x+3y)^4\frac{dy}{dx} = 1$$
$$\frac{dy}{dx} = \frac{1-10(2x+3y)^4}{15(2x+3y)^4}$$

(d) $(1-2xy^3)^5 = x+4y$

$$5(1-2xy^3)^4\left(-2x\cdot 3y^2\frac{dy}{dx} + y^3\cdot -2\right) = 1+4\frac{dy}{dx}$$
$$-30xy^2(1-2xy^3)^4\frac{dy}{dx} - 10y^3(1-2xy^3)^4 = 1+4\frac{dy}{dx}$$
$$\frac{dy}{dx} = \frac{1+10y^3(1-2xy^3)^4}{-30xy^2(1-2xy^3)^4-4}$$

25. $3x^2 - 2y^2 = 6$,

$$6x - 4y\frac{dy}{dx} = 0, \text{ or } \frac{dy}{dx} = \frac{3x}{2y}$$
$$\frac{d^2y}{dx^2} = \frac{2y(3) - 3x\left(2\frac{dy}{dx}\right)}{(2y)^2} = \frac{3y - 3x\frac{dy}{dx}}{2y^2}$$

Since $\dfrac{dy}{dx} = \dfrac{3x}{2y}$

$$\frac{d^2y}{dx^2} = \frac{3y - 3x\left(\frac{3x}{2y}\right)}{2y^2} = \frac{6y^2-9x^2}{4y^3}$$

From the original equation

$$6y^2 - 9x^2 = 3(2y^2-3x^2) = -3(3x^2-2y^2) = -3(6) = -18$$

and so $\dfrac{d^2y}{dx^2} = -\dfrac{18}{4y^3} = -\dfrac{9}{2y^3}$

27. $P(t) = -t^3 + 9t^2 + 48t + 200$

(a) $P'(t) = -3t^2 + 18t + 48$

$P'(3) = -3(3)^2 + 18(3) + 48 = 75$,

or increasing at a rate of 75,000 people per year.

(b) $P''(t) = -6t + 18$

$P''(3) = -6(3) + 18 = 0$ people per year

29. $s(t) = \dfrac{2t+1}{t^2+12}$ for $0 \le t \le 4$

(a)

$$v(t) = \frac{(t^2+12)(2) - (2t+1)(2t)}{(t^2+12)^2} = \frac{-2t^2-2t+24}{(t^2+12)^2} = \frac{-2(t+4)(t-3)}{(t^2+12)^2}$$

$$a(t) = \frac{(t^2+12)^2(-4t-2)}{(t^2+12)^4} - \frac{(-2t^2-2t+24)2(t^2+12)(2t)}{(t^2+12)^4}$$
$$= -2(t^2+12)\left[\frac{(t^2+12)(2t+1)}{(t^2+12)^4} + \frac{(-2t^2-2t+24)(2t)}{(t^2+12)^4}\right]$$
$$= \frac{2(2t^3+3t^2-72t-12)}{(t^2+12)^3}$$

Now, for $0 \le t \le 4$,
$v(t) = 0$ when $t = 3$ and $a(t) \ne 0$.
When $0 \le t < 3$, $v(t) > 0$ and $a(t) < 0$, so the object is advancing and decelerating.
When $3 < t \le 4$, $v(t) < 0$ and $a(t) < 0$, so the object is retreating and decelerating.

(b) The distance for $0 < t < 3$ is

$$|s(3) - s(0)| = \left|\frac{1}{3} - \frac{1}{12}\right| = \frac{1}{4}$$

The distance for $3 < t < 4$ is

$$|s(4) - s(3)| = \left|\frac{9}{28} - \frac{1}{3}\right| = \frac{1}{84}$$

So, the total distance travelled is

$$\frac{1}{4} + \frac{1}{84} = \frac{22}{84} = \frac{11}{42}$$

31. **(a)** $Q(x) = 50x^2 + 9{,}000x$

$\Delta Q \approx Q'(x) = 100x + 9{,}000$

$Q'(30) = 12{,}000$, or an increase 12,000 units.

(b) The actual increase in output is $Q(31) - Q(30) = 12{,}050$ units.

33.

$$Q(L) = 20{,}000L^{1/2}$$
$$\Delta Q \approx Q'(L)\Delta L$$
$$Q'(L) = 10{,}000L^{-1/2} = \frac{10{,}000}{\sqrt{L}}$$
$$Q'(900) = \frac{10{,}000}{\sqrt{900}} = \frac{1{,}000}{3}$$

Since L will decrease to 885,

$$\Delta L = 885 - 900 = -15$$
$$\Delta Q \approx \left(\frac{1{,}000}{3}\right)(-15) = -5{,}000,$$

or a decrease in output of 5,000 units.

35. Let A be the level of air pollution and p be the population.
$A = kp^2$, where k is a constant of proportionality

$$\Delta A \approx A'(p)\Delta p$$
$$A'(p) = 2kp \text{ and } \Delta p = .05p, \text{ so}$$
$$\Delta A \approx (2kp)(0.05p)$$
$$= 0.1kp^2 = 0.1A,$$

or a 10% increase in air pollution.

37. $D = 36m^{-1.14}$

(a) $D = 36(70)^{-1.14} \approx 0.2837$ individuals per square kilometer.

(b) (0.2837 individuals/km^2)(9.2×10^6)km^2 ≈ 2.61 million people.

(c) The ideal population density would be

$$36(30)^{-1.14} \approx 0.7454 \text{ animals/km}^2$$

Since the area of the island is 3,000 km^2, the number of animals on the island for the ideal population density would be (0.7454 animals/km^2)(3,000 km^2) $\approx$ 2,235 animals.
Since the animal population is given by

$$P(t) = 0.43t^2 + 13.37t + 200,$$

this population is reached when

$$2236 = 0.43t^2 + 13.37t + 200$$
$$0 = 0.43t^2 + 13.37t - 2036$$

or, using the quadratic formula, when $t \approx 55$ years. The rate the population is changing at this time is $P'(55)$, where $P'(t) = 0.86t + 13.37$, or $0.86(55) + 13.37 = 60.67$ animals per year.

39. Need $100\frac{\Delta L}{L}$, given that $100\frac{\Delta Q}{Q} = 1\%$, where $\Delta Q \approx Q'(L)\Delta L$. Since,

$$100\frac{Q/(L)\Delta L}{Q(L)} = 1,$$

solving for ΔL yields

$$\Delta L = \frac{Q(L)}{100Q'(L)} \text{ and}$$

$$100\frac{\Delta L}{L} = 100\frac{\frac{Q(L)}{100Q'(L)}}{L}$$
$$= \frac{Q(L)}{Q'(L) \cdot L}$$

Since $Q(L) = 600L^{2/3}$,

$$Q'(L) = 400L^{-1/3} = \frac{400}{L^{1/3}}$$

$$100\frac{\Delta L}{L} = \frac{600L^{2/3}}{\left(\frac{400}{L^{1/3}}\right)(L)} = \frac{3}{2}, \text{ or } 1.5\%$$

Increase labor by approximately 1.5%.

41. $F = kD^2\sqrt{A-C} = kD^2(A-C)^{1/2}$

(a) Treating A and D as constants,

$$\frac{dF}{dC} = \frac{1}{2}kD^2(A-C)^{-1/2}(-1) = \frac{-kD^2}{2\sqrt{A-C}}$$

As C increases, the denominator increases, so F decreases.

(b) We need $100\dfrac{dF/dA}{F}$

Treating C and D as constants,

$$\frac{dF}{dA} = \frac{1}{2}kD^2(A-C)^{-1/2}(1) = \frac{kD^2}{2\sqrt{A-C}}$$

$$100\frac{\frac{dF}{dA}}{F} = 100\frac{\frac{kD^2}{2\sqrt{A-C}}}{kD^2\sqrt{A-C}} = \frac{50}{(A-C)}\%$$

43. Need $\Delta A \approx A'(r)\Delta r$
Since $A = \pi r^2$,

$$A'(r) = 2\pi r$$

When $r = 12$, $A'(12) = 2\pi(12) = 24\pi$
Since $\Delta r = \pm 0.03r$, $\Delta r = \pm 0.03(12) = \pm 0.36$ and

$$\Delta A \approx (24\pi)(\pm 0.36) \approx \pm 27.14\text{cm}^2$$

When $r = 12$, $A = \pi(12)^2 = 144\pi \approx 452.39$ square centimeters. The calculation of area is off by ± 27.14 at most, so

$$425.25 \le A \le 479.53$$

45. $Q = 600K^{1/2}L^{1/3}$
Need $100\dfrac{\Delta Q}{Q}$, where $\Delta Q \approx Q'(L)\Delta L$
Treating K as a constant

$$Q'(L) = 200K^{1/2}L^{-2/3} = \frac{200K^{1/2}}{L^{2/3}}$$

with $\Delta L = 0.02L$

$$100\frac{\Delta Q}{Q} = 100\frac{\left(\frac{200K^{1/2}}{L^{2/3}}\right)(0.02L)}{600K^{1/2}L^{1/3}} \approx 0.67\%$$

47. $\Delta V \approx V'(r)\Delta r$

$V'(r) = 4\pi r^2$, $r = 1.2$, and $\Delta r = (\pm 0.03)(1.2)$

So,

$$\Delta V \approx 4\pi(1.2)^2(\pm 0.03)(1.2) \approx \pm 0.6514$$

When $r = 1.2$, $V = \frac{4}{3}\pi(1.2)^3 \approx 7.2382$ cubic centimeters. The calculation of volume is off by ± 0.6514, at most, so

$$6.5868 \le V \le 7.8897$$

49.

$$D(p) = \frac{32{,}670}{2p+1} = 32{,}670(2p+1)^{-1}$$

$$p(t) = 0.04t^{3/2} + 44$$

Need $\dfrac{dD}{dt}$ when $t = 25$.

$$\frac{dD}{dt} = \frac{dD}{dp}\cdot\frac{dp}{dt}$$

Now,

$$\frac{dD}{dp} = -32{,}670(2p+1)^{-2}(2) = \frac{65{,}340}{(2p+1)^2}$$

$$\frac{dp}{dt} = 0.06t^{1/2}$$

When $t = 25$, $p = 0.04(25)^{3/2} + 44 = 49$, so

$$\frac{dD}{dt} = \frac{65{,}340}{[2(49)+1]^2}\cdot 0.06(25)^{1/2} = 2,$$

or the demand will be increasing at a rate of 2 toasters per month.

51. $P(t) = 20 - \dfrac{6}{t+1} = 20 - 6(t+1)^{-1}$

Need $100\frac{\Delta P}{P}$, where $\Delta P \approx P'(t)\Delta t$

$$P'(t) = 6(t+1)^{-2}(1) = \frac{6}{(t+1)^2}$$

The next quarter year is from $t = 0$ to $t = \frac{1}{4}$, so $P(0) = 14$, $P'(0) = 6$ and $\Delta t = \frac{1}{4}$.

$$100\frac{\Delta P}{P} = 100\frac{(6)\left(\frac{1}{4}\right)}{14} \approx 10.7\%$$

53.

$$s(t) = 88t - 8t^2$$
$$v(t) = s'(t) = 88 - 16t$$

The car is stopped when $v(t) = 0$, so

$$0 = 88 - 16t, \text{ or } t = 5.5 \text{ seconds.}$$

The distance travelled until it stops is

$$s(5.5) = 88(5.5) - 8(5.5)^2 = 242 \text{ feet}$$

55. $P(t) = -t^3 + 7t^2 + 200t + 300$

(a)
$$P'(t) = -3t^2 + 14t + 200$$
$$P'(5) = -3(5)^2 + 14(5) + 200 = 195,$$

or increasing at a rate of \$195 per unit per month.

(b)
$$P''(t) = -6t + 14$$
$$P''(5) = -6(5) + 14 = -16,$$

or decreasing at a rate of \$16 per unit per month per month.

(c) Need $\Delta P' \approx P''(t)\Delta t$
Now, $P''(5) = -16$ and the first six months of the sixth year corresponds to $\Delta t = \frac{1}{2}$.

$$\Delta P' \approx (-16)\left(\frac{1}{2}\right) = -8,$$

or a decrease of \$8 per unit per month.

(d) Need $P'(5.5) - P'(5)$

$$P'(5.5) = -3(5.5)^2 + 14(5.5) + 200 = 186.25$$

The actual change in the rate of price increase is $186.25 - 195 = -8.75$, or decreasing at a rate of \$8.75 per unit per month.

57. $C(x) = 0.06x + 3x^{1/2} + 20$ hundred

$$\frac{dx}{dt} = -11 \text{ when } x = 2{,}500$$
$$\frac{dC}{dt} = \frac{dC}{dx}\cdot\frac{dx}{dt}$$
$$\frac{dC}{dx} = 0.06 + 1.5x^{-1/2} = 0.06 + \frac{1.5}{\sqrt{x}}$$
$$\frac{dC}{dt} = \left(0.06 + \frac{1.5}{\sqrt{2{,}500}}\right)(-11)$$
$$= -0.99 \text{ hundred,}$$

or decreasing at a rate of \$99 per month.

59. Consider the volume of the shell as a change in volume, where $r = \frac{8.5}{2}$ and $\Delta r = \frac{1}{8} = 0.125$.
$\Delta V \approx V'(r)\Delta r$

$$V(r) = \frac{4}{3}\pi r^3$$
$$V'(r) = 4\pi r^2$$
$$V'(4.25) = 4\pi(4.25)^2 = 72.25\pi$$
$$\Delta V = (72.25\pi)(0.125) \approx 28.37 \text{ in}^3$$

61. Let the length of string be the hypotenuse of the right triangle formed by the horizontal and vertical distance of the kite from the child's hand. Then,

$$s^2 = x^2 + (80)^2$$
$$2s\frac{ds}{dt} = 2x\frac{dx}{dt}$$
$$\frac{ds}{dt} = \frac{2x\frac{dx}{dt}}{2s} = \frac{x\frac{dx}{dt}}{s}$$

When $s = 100$, $(100)^2 = x^2 + (80)^2$, or $x = 60$
$\frac{ds}{dt} = \frac{(60)(5)}{100} = 3$, or increasing at a rate of 3 feet per second.

63. Need $\frac{dx}{dt}$.

$$x^2 + y = (10)^2$$

$$2x\frac{dx}{dt} + 2y\frac{dy}{dt} = 0$$

$$\frac{dx}{dt} = \frac{-2y\frac{dy}{dt}}{2x} = \frac{-y\frac{dy}{dt}}{x}$$

When $y = 6$, $x^2 + 36 = 100$, or $x = 8$.
Since $\frac{dy}{dt} = -3$,

$$\frac{dx}{dt} = \frac{(-6)(-3)}{8} = 2.25,$$

or increasing at a rate of 2.25 feet per second.

65. Let x be the distance from the player to third base. Then,

$$s^2 = x^2 + (90)^2$$

$$2s\frac{ds}{dt} = 2x\frac{dx}{dt}$$

$$\frac{ds}{dt} = \frac{2x\frac{dx}{dt}}{2s} = \frac{x\frac{dx}{dt}}{s}$$

When $x = 15$, $s^2 = (15)^2 + (90)^2$, or $s = \sqrt{8325}$.

$$\frac{ds}{dt} = \frac{(15)(-20)}{\sqrt{8325}} \approx -3.29,$$

or decreasing at a rate of 3.29 feet per second.

67. Let x be the distance from point P to the object.

$$V = ktx$$

When $t = 5$ and $x = 20$, $V = 4$, so

$$4 = k(5)(20), \text{ or } k = \frac{1}{25}$$

Since $a = V'$,

$$a = k\left(t\frac{dx}{dt} + x \cdot 1\right)$$

$$a = \frac{1}{25}(5 \cdot 4 + 20) = \frac{8}{5}\text{ft/sec}^2$$

69. Need $100\,\frac{y'}{y}$ as $x \to \infty$.

$$y = mx + b$$

$$y' = m$$

$$100\frac{y'}{y} = 100\frac{m}{mx + b}$$

As x approaches ∞, this value approaches zero.

71. To use a graphing utility to graph f and f',
Press [y=] and input (3x + 5)(2x ∧ 3 − 5x + 4) for $y_1 =$

$$f'(x) = (3x + 5)(6x^2 - 5) + (3)(2x^3 - 5x + 4)$$

Input $f'(x)$ for $y_2 =$
Use window dimensions [−3, 2]1 by [−20, 30]10
Use trace and zoom-in to find the x-intercepts of $f'(x)$ or use the zero function under the calc menu. In either case, make sure that y_2 is displayed in the upper left corner. The three zeros are $x \approx -1.78$, $x \approx -0.35$, and $x \approx 0.88$.

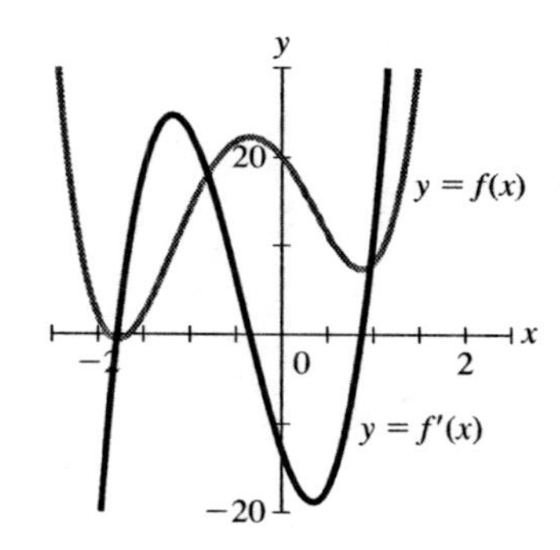

73. **(a)** To graph $y^2(2 - x) = x^3$,

$$y^2 = \frac{x^3}{2 - x}$$

$$y = \pm\sqrt{\frac{x^3}{2 - x}}$$

Press [y=] and input √((x) ÷ (2 − x)) for $y_1 =$ and input $-y_1$ for $y_2 =$ (you can find y_1 by pressing [vars] and selecting function under the y-vars menu). Use window dimensions [−2, 5]1 by [−10, 10]5 and press [graph].

(b) With the graph shown, press [2nd] [Draw] and select the tangent function. Enter $x = 1$ to obtain the equation of the tangent line to be approximately $y = x - 5$.

(c) It can be seen as x approaches 2 from the left the portion of the graph above the x-axis approaches ∞ and the portion below the x-axis approaches $-\infty$.

(d) From the graph, the portion above the graph has a horizontal tangent of $x = 0$, as does the portion below the graph.

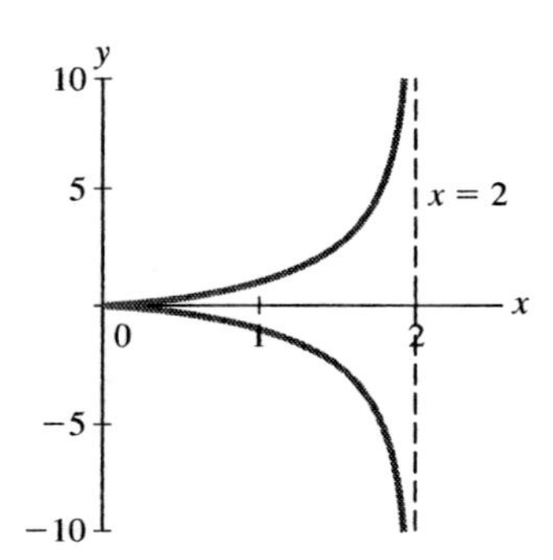

Chapter 3

Additional Applications of the Derivative

3.1 Increasing and Decreasing Functions; Relative Extrema

1. $f'(x) > 0$ when f is increasing, or $-2 < x < 2$
$f'(x) < 0$ when f is decreasing, or $x < -2$ and $x > 2$

3. $f'(x) > 0$ when f is increasing, or $x < -4$ and $0 < x < 2$
$f'(x) < 0$ when f is decreasing, or $-4 < x < -2$, $-2 < x < 0$, and $x > 2$

5. Function is decreasing, so $f'(x) < 0$ and graph of f' is below the x-axis. Function then levels, so $f'(x) = 0$ and graph of f crosses the x-axis. Function next increases for a period of time, so $f'(x) > 0$ and graph of f' is above the x-axis. Function then levels again, so $f'(x) = 0$ and graph of f' crosses the x-axis. Lastly, function decreases, so $f'(x) < 0$ and graph of f' is below the x-axis. Therefore, graph of f' is B.

7. Function is decreasing, so $f'(x) < 0$ and graph of f' is below the x-axis. Function then levels, so $f'(x) = 0$ and graph of f' crosses the x-axis. Function next increases, so $f'(x) > 0$ and graph of f' is above the x-axis. Therefore, graph of f' is D.

9.
$$f(x) = x^2 - 4x + 5$$
$$f'(x) = 2x - 4$$

f is increasing when $f'(x) > 0$

$$2x - 4 > 0, \quad \text{or } x > 2$$

f is decreasing when $f'(x) < 0$

$$2x - 4 < 0, \quad \text{or } x < 2$$

11.
$$f(x) = x^3 - 3x - 4$$
$$f'(x) = 3x^2 - 3 = 3(x + 1)(x - 1)$$
$$f'(x) = 0 \text{ when } x = -1, 1$$
$$\text{When } x < -1, \; f'(x) > 0$$
$$-1 < x < 1, \; f'(x) < 0$$
$$x > 1, \; f'(x) > 0.$$

So, f is increasing when $x < -1$ and $x > 1$; f is decreasing when $-1 < x < 1$.

13.
$$g(t) = t^5 - 5t^4 + 100$$
$$g'(t) = 5t^4 - 20t^3 = 5t^3(t - 4)$$
$$g'(t) = 0 \text{ when } t = 0, 4$$
$$\text{When } t > 0, \; g'(t) > 0$$
$$0 < t < 4, \; g'(t) < 0$$
$$t > 4, \; g'(t) > 0.$$

So, g is increasing when $t < 0$ and $t > 4$; g is decreasing when $0 < t < 4$.

15.
$$f(t) = \frac{1}{4 - t^2} = (4 - t^2)^1, \text{ defined for } t \neq -2, 2$$
$$f'(t) = -(4 - t^2)^{-2}(-2t)$$
$$= \frac{2t}{(4 - t^2)^2} = \frac{2t}{[(2 + t)(2 - t)]^2}$$
$$f'(t) = 0 \text{ when } t = 0$$

When $t < -2,\ f'(t) < 0$

$-2 < t < 0,\ f'(t) < 0$

$0 < t < 2,\ f'(t) > 0$

$t > 2,\ f'(t) > 0.$

So, f is increasing when $0 < t < 2$ and $t > 2$; f is decreasing when $t < -2$ and $-2 < t < 0$.

17. $h(u) = \sqrt{9 - u^2} = (9 - u^2)^{1/2}$

$= [(3 + u)(3 - u)]^{1/2}$, defined for $-3 \leq u \leq 3$

$h'(u) = \frac{1}{2}(9 - u^2)^{-1/2}(-2u) = \frac{-u}{\sqrt{9 - u^2}}$

$h'(u) = 0$ when $u = 0$

When $-3 < u < 0,\ h'(u) > 0$

$0, u < 3,\ h'(u) < 0.$

So, h is increasing when $-3 < u < 0$; h is decreasing when $0 < u < 3$.

19. $F(x) = x + \frac{9}{x} = x + 9x^{-1} = \frac{x^2 + 9}{x}$,

defined when $x \neq 0$

$F'(x) = 1 - 9x^{-2} = 1 - \frac{9}{x^2} = \frac{x^2 - 9}{x^2}$

$= \frac{(x + 3)(x - 3)}{x^2}$

$F'(x) = 0$ when $x = -3, 3$

When $x < -3,\ F'(x) > 0$

$-3 < x < 0,\ F'(x) < 0$

$0 < x < 3,\ F'(x) < 0$

$x > 3,\ F'(x) > 0.$

So, F is increasing when $x < -3$ and $x > 3$; F is decreasing when $-3 < x < 0$ and $0 < x < 3$.

21. $f(x) = \sqrt{x} + \frac{1}{\sqrt{x}} = x^{1/2} + x^{-1/2} = \frac{x + 1}{\sqrt{x}}$,

defined for $x > 0$

$f'(x) = \frac{1}{2}x^{-1/2} - \frac{1}{2}x^{-3/2} = \frac{1}{2x^{1/2}} - \frac{1}{2x^{3/2}}$,

$= \frac{x - 1}{2x^{3/2}}$

$f'(x) = 0$ when $x = 1$

When $0 < x < 1,\ f'(x) < 0$

$x > 1,\ f'(x) > 0.$

So, f is increasing when $x > 1$; f is decreasing when $0 < x < 1$.

23. $f(x) = 3x^4 - 8x^3 + 6x^2 + 2$

$f'(x) = 12x^3 - 24x^2 + 12x = 12x(x - 1)^2$

$f'(x) = 0$ when $x = 0, 1$

When $x < 0,\ f'(x) < 0$ so f decreasing

$0 < x < 1,\ f'(x) > 0$ so f increasing

$x > 1,\ f'(x) > 0$ so f increasing.

When $x = 0,\ f(0) = 2$ and the point (0,2) is a relative minimum. When $x = 1,\ f(1) = 3$, but there is no relative extremum at (1,3).

25. $f(t) = 2t^3 + 6t^2 + 6t + 5$

$f'(t) = 6t^2 + 12t + 6 = 6(t + 1)^2$

$f'(t) = 0$ when $t = -1$

When $t < -1,\ f'(t) > 0$ so f increasing

$t > -1,\ f'(t) > 0$ so f increasing.

When $t = -1,\ f(-1) = 3$, but there is no relative extremum at $(-1, 3)$.

27. $g(x) = (x - 1)^5$

$g'(x) = 5(x - 1)^4(1)$

$g'(x) = 0$ when $x = 1$

When $x < 1,\ g'(x) > 0$ so g increasing

$x > 1,\ g'(x) > 0$ so g increasing.

When $x = 1,\ g(1) = 0$, but there is no relative extremum at (1,0).

29. $f(t) = \dfrac{t}{t^2+3}$

$f'(t) = \dfrac{(t^2+3)(1) - (t)(2t)}{(t^2+3)^2} = \dfrac{3-t^2}{(t^2+3)^2}$

$f'(t) = 0$ when $t = \pm\sqrt{3}$

When $t < -\sqrt{3}$, $f'(t) < 0$ so f decreasing
$-\sqrt{3} < t < \sqrt{3}$, $f'(t) > 0$ so f increasing
$t > \sqrt{3}$. $f'(t) < 0$ so f decreasing.

When $x = -\sqrt{3}$, $f\left(-\sqrt{3}\right) = -\dfrac{\sqrt{3}}{6}$ and the point $\left(-\sqrt{3}, -\dfrac{\sqrt{3}}{6}\right)$ is a relative minimum.

When $x = \sqrt{3}$, $f(\sqrt{3}) = \dfrac{\sqrt{3}}{6}$ and the point $\left(\sqrt{3}, \dfrac{\sqrt{3}}{6}\right)$ is a relative maximum.

31. $h(t) = \dfrac{t^2}{t^2+t-2} = \dfrac{t^2}{(t+2)(t-1)}$

defined for $t \neq -2, 1$

$h'(t) = \dfrac{(t^2+t-2)(2t) - (t^2)(2t+1)}{(t^2+t-2)^2}$

$= \dfrac{t(t-4)}{(t^2+t-2)^2}$

$h'(t) = 0$ when $t = 0, 4$

When $-2 < t < 0$, $h'(t) > 0$ so h increasing
$0 < t < 1$, $h'(t) < 0$ so h decreasing
$1 < t < 4$, $h'(t) < 0$ so h decreasing
$t > 4$, $h'(t) > 0$ so h increasing.

When $t = 0$, $h(0) = 0$ and the point $(0, 0)$ is a relative maximum.
When $t = 4$, $h(4) = \dfrac{8}{9}$ and the point $\left(4, \dfrac{8}{9}\right)$ is a relative minimum.

33. $s(t) = (t^2-1)^4$

$s'(t) = 4(t^2-1)^3(2t) = 8t\,[(t+1)(t-1)]^3$

$s'(t) = 0$ when $t = -1, 0, 1$

When $t < -1$, $s'(t) < 0$ so s decreasing
$-1 < t < 0$, $s'(t) > 0$ so s increasing
$0 < t < 1$, $s'(t) < 0$ so s decreasing
$t > 1$, $s'(t) > 0$ so s increasing.

When $t = -1$, $s(-1) = 0$ and the point $(-1, 0)$ is a relative minimum. When $t = 0$, $s(0) = 1$ and the point $(0, 1)$ is a relative maximum. When $t = 1$, $s(1) = 0$ and the point $(1, 0)$ is a relative minimum.

35. $f(x) = x^3 - 3x^2 = x^2(x-3)$,
intercepts: $(0, 0)$ $(3, 0)$
$f'(x) = 3x^2 - 6x = 3x(x-2)$
$f'(x) = 0$ when $x = 0, 2$

When $x < 0$, $f'(x) > 0$ so f increasing
$x = 0$, $f'(x) = 0$ so f levels
$0 < x < 2$, $f'(x) < 0$ so f decreasing
$x = 2$, $f'(x) = 0$ so f levels
$x > 0$, $f'(x) > 0$ so f increasing.

The point $(0, 0)$ is a relative maximum and the point $(2, -4)$ is a relative minimum.

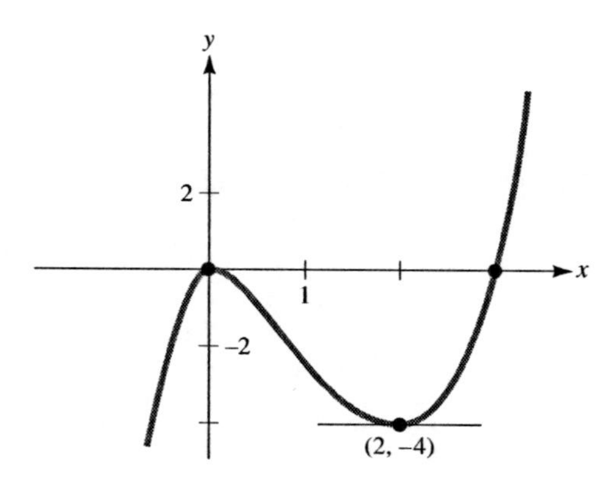

37. $f(x) = 3x^4 - 8x^3 + 6x^2 + 2$

When $x = 0$, $f(0) = 2$ so $(0, 2)$ is an intercept.
$f(x) = 0$ is too difficult to solve.

$f'(x) = 12x^3 - 24x^2 + 12x$
$= 12x(x-1)(x-1)$
$f'(x) = 0$ when $x = 0, 1$

When $x < 0$, $f'(x) < 0$ so f decreasing
$x = 0$, $f'(x) = 0$ so f levels
$0 < x < 1$, $f'(x) > 0$ so f increasing
$x = 1$, $f'(x) = 0$ so f levels
$x > 1$, $f'(x) > 0$ so f increasing.

The point (0, 2) is a relative minimum, but the point (1, 3) is not a relative extremum.

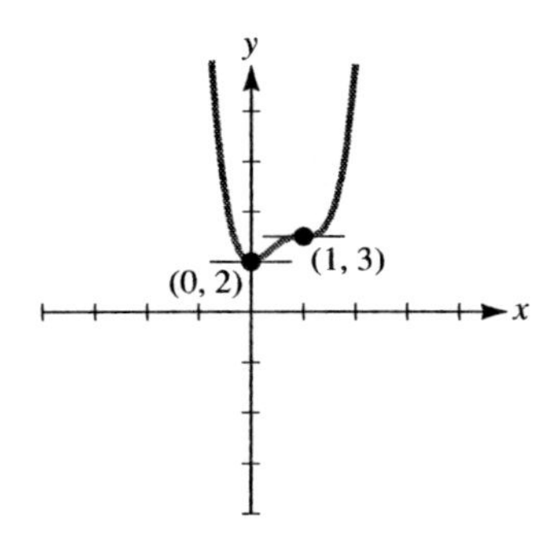

39. $f(t) = 2t^3 + 6t^2 + 6t + 5$
$f'(t) = 6t^2 + 12t + 6 = 6(t + 1)^2$
$f'(t) = 0$ when $t = -1$

When $t < -1$, $f'(t) > 0$ so f increasing
$t = -1$, $f'(t) = 0$ so f levels
$t > -1$, $f'(t) > 0$ so f increasing.

The point (−1, 3) is not a relative extremum.

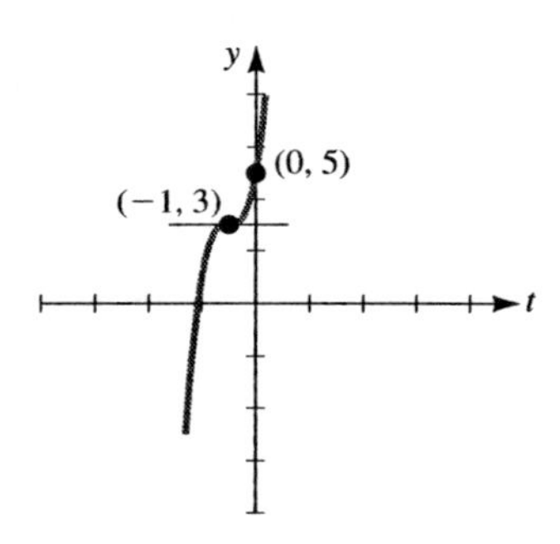

41. $$g(t) = \frac{t}{t^2 + 3}$$

$$g'(t) = \frac{(t^2 + 3)(1) - (t)(2t)}{t^2 + 3} = \frac{3 - t^2}{(t^2 + 3)^2}$$

$g'(t) = 0$ when $t = -\sqrt{3}, \sqrt{3}$

When $t < -\sqrt{3}$, $f'(t) < 0$ so f decreasing
$t = -\sqrt{3}$, $f'(t) = 0$ so f levels
$-\sqrt{3} < t < \sqrt{3}$, $f'(t) > 0$ so f increasing
$t = \sqrt{3}$, $f'(t) = 0$ so f levels
$t > \sqrt{3}$, $f'(t) < 0$ so f decreasing.

The point $\left(-\sqrt{3}, \frac{-\sqrt{3}}{6}\right)$ is a relative minimum and the point $\left(\sqrt{3}, \frac{\sqrt{3}}{6}\right)$ is a relative maximum.

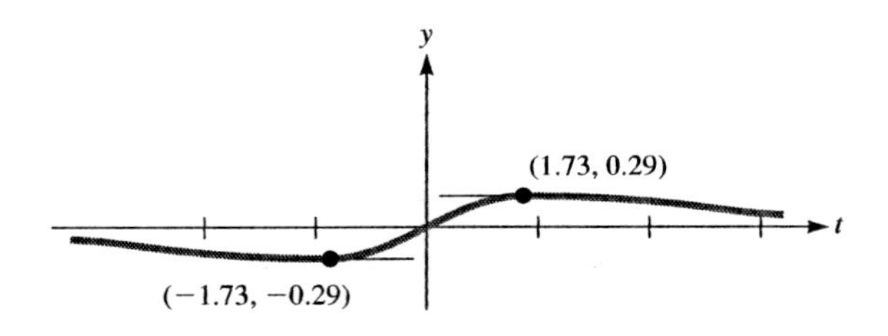

43. $f(x) = 3x^5 - 5x^3 + 4$
$f'(x) = 15x^4 - 15x^2 = 15x^2(x + 1)(x - 1)$
$f'(x) = 0$ when $x = -1, 0, 1$

When $x < -1$, $f'(x) > 0$ so f increasing
$x = -1$, $f'(x) = 0$ so f levels
$-1 < x < 0$, $f'(x) < 0$ so f decreasing
$x = 0$, $f'(x) = 0$ so f levels
$0 < x < 1$, $f'(x) < 0$ so f decreasing
$x = 1$, $f'(x) = 0$ so f levels
$x > 1$, $f'(x) > 0$ so f increasing.

The point (−1, 6) is a relative maximum, the point (0, 4) is not a relative extremum, and the point (1, 2) is a relative minimum.

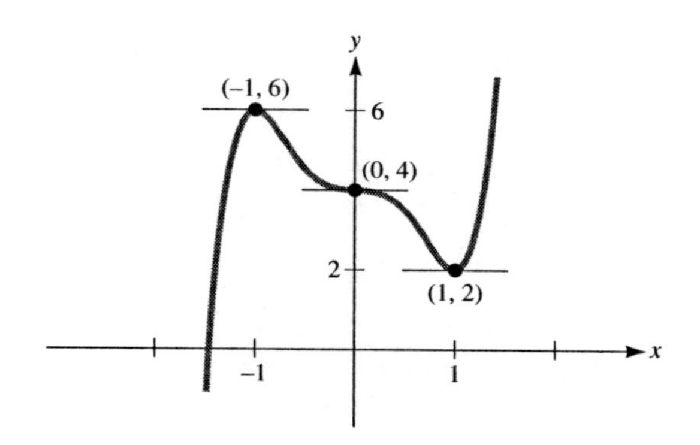

45. $f'(x) = x^2(4 - x^2) = x^2(2 + x)(2 - x)$
$f'(x) = 0$ when $x = -2, 0, 2$

When $x < -2$, $f'(x) < 0$ so f decreasing
$-2 < x < 0$, $f'(x) > 0$ so f increasing
$0 < x < 2$, $f'(x) > 0$ so f increasing
$x > 2$, $f'(x) < 0$ so f decreasing.

When $x = -2$, f has a relative minimum, when $x = 0$, f does not have a relative extremum, and when $x = 2$, f has a relative maximum.

47. $f'(x) = \dfrac{(x+1)^2(4-3x)^3}{(x^2+1)^2}$
$f'(x) = 0$ when $x = -1, \frac{4}{3}$

When $x < -1$, $f'(x) > 0$ so f increasing
$-1 < x < \frac{4}{3}$, $f'(x) > 0$ so f increasing
$x > \frac{4}{3}$, $f'(x) < 0$ so f decreasing.

When $x = -1$, f does not have a relative extremum, and when $x = \frac{4}{3}$, f has a relative maximum.

49. When
$x < 1$, f is decreasing and graph of f' is below x-axis
$x = 1$, f levels and graph of f' crosses the x-axis
$1 < x < 3$, f is increasing and graph of f' is above x-axis
$x = 3$, f levels and graph of f' crosses the x-axis
$x > 3$, f is decreasing and graph of f' is below x-axis.

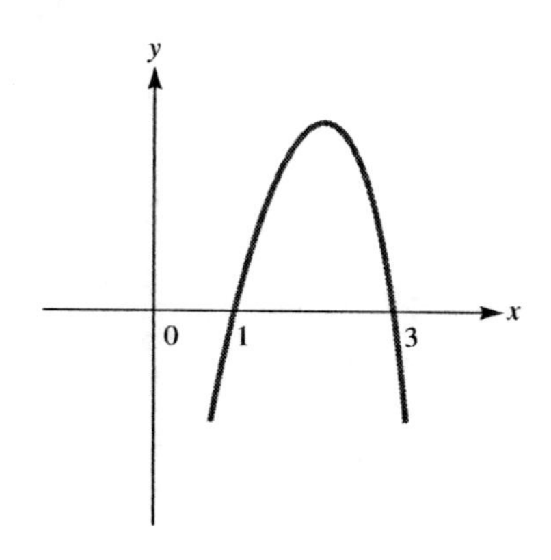

51. When
$x < 2$, f is decreasing and graph of f' is below x-axis
$x = 2$, f levels and graph of f' crosses the x-axis
$2 < x < 5$, f is increasing and graph of f' is above x-axis
$x = 5$, f levels and graph of f' touches x-axis
$x > 5$, f is increasing and graph of f' is above x-axis.

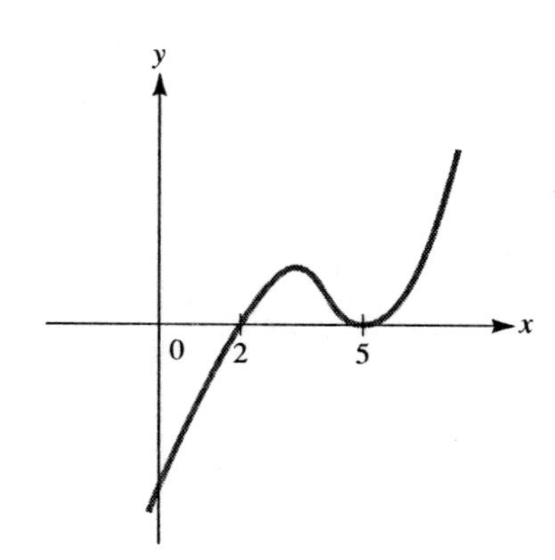

53. $C(x) = x^3 - 20x^2 + 179x + 242$

(a) $$A(x) = \frac{C(x)}{x} = \frac{x^3 - 20x^2 + 179x + 242}{x}$$
$$= x^2 - 20x + 179 + \frac{242}{x}$$
$$= x^2 - 20x + 179 + 242x^{-1}$$
$$A'(x) = 2x - 20 - \frac{242}{x^2}$$

(b) $A'(x) = 0$ when
$0 = 2x - 20 - \frac{242}{x^2}$
$0 = 2x^3 - 20x^2 - 242$
$x^3 - 10x^2 - 121 = 0$
Press [y=] and enter $x^3 - 10x^2 - 121$ for $y_1 =$.
Use window dimensions [−10, 100]10 by [−500, 500]100.
Press [graph].
To find the zero (x-intercept), enter the zero function under the calc menu. Enter a left bound close to the x-intercept, a right bound close to the x-intercept and a guess. The x-intercept or zero is $x = 11$.

When $0 \le x < 11$, $A'(x) < 0$ so A decreasing
$x > 11$, $A'(x) > 0$ so A increasing.

(c) When $x = 11$, A has a relative minimum which is actually an absolute minimum. So the average cost is minimized when 11 units are produced. The corresponding minimum average cost is

$$A(11) = (11)^2 - 20(11) + 179 + \frac{242}{11} = 102$$

or $102,000 per unit.

55. $$R(x) = xp(x) = x(10 - 3x)^2,\ 0 \le x \le 3$$
$$R'(x) = x \cdot 2(10 - 3x)(-3) + (10 - 3x)^2(1)$$
$$= (10 - 3x)(-6x + 10 - 3x)$$
$$= (10 - 3x)(10 - 9x)$$
$$R'(x) = 0 \text{ when } x = \frac{10}{9}, \frac{10}{3}$$

When $0 \le x < \frac{10}{9}$, $R'(x) > 0$ so R increasing
$x = \frac{10}{9}$, $R'(x) = 0$ so R levels
$\frac{10}{9} < x \le 3$, $R'(x) < 0$ so R decreasing.

The point (1.11, 49.38) is a relative maximum, so revenue is maximized when approximately 1.11 hundred, or 111 units are produced.

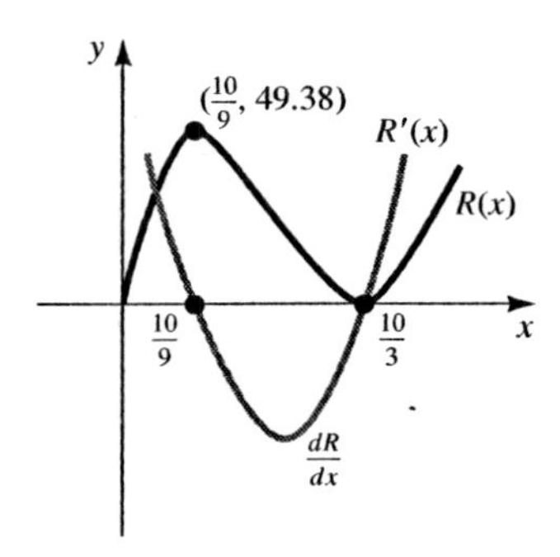

57. $$C(t) = \frac{0.15t}{t^2 + 0.81}$$
Note that, since degree numerator < degree of denominator, $y = 0$ is a horizontal asymptote.

$$C'(t) = \frac{(t^2 + 0.81)(0.15) - (0.15t)(2t)}{(t^2 + 0.81)^2}$$
$$\frac{-0.15t^2 + 0.1215}{(t^2 + 0.81)^2}$$
$C'(t) = 0$ when $t = 0.9$

When $0 < t < 0.9$, $C'(t) > 0$ so C increasing
$t = 0.9$, $C'(t) = 0$ and C levels
$t > 0.9$, $C'(t) < 0$ and C decreasing.

The point (0.9, 0.083) is a relative maximum, so the maximum concentration occurs when $t = 0.9$ hours.

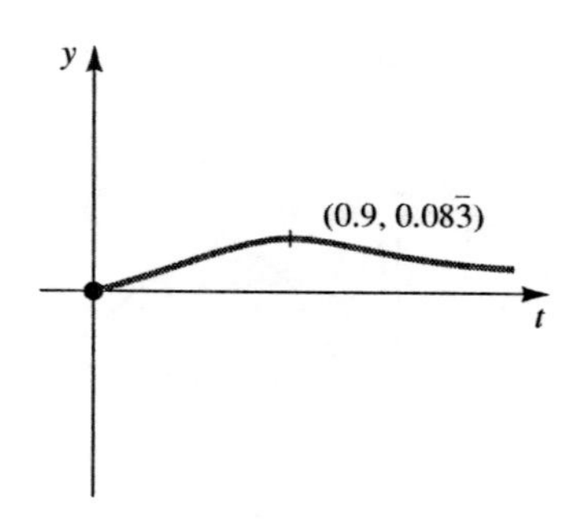

59. $S(x) = -2x^3 + 27x^2 + 132x + 207, 0 \le x \le 17$

(a)
$$S'(x) = -6x^2 + 54x + 132$$
$$= -6(x - 11)(x + 2)$$
$$S'(x) = 0 \text{ when } x = -2, 11$$

When $0 \le x < 11$, $S'(x) > 0$ and S is increasing
$x = 11$, $S'(x) = 0$ and S is levels
$11 < x \le 17$, $S'(x) < 0$ and S is decreasing.

The point (11, 2264) is a relative maximum.

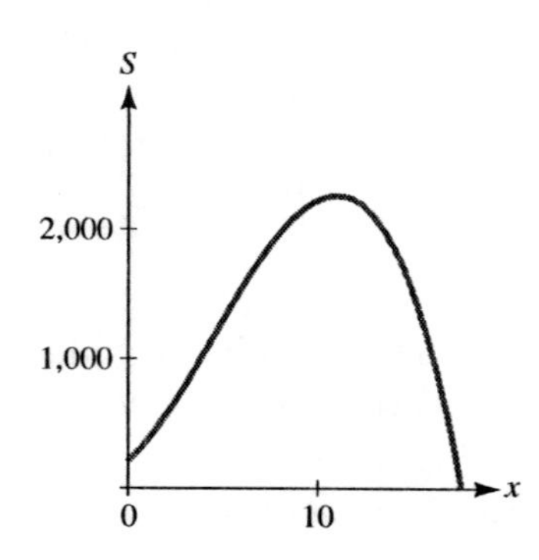

(b) $S(0) = 207$, or 207 units will sell.

(c) Since (11, 2264) is a relative maximum, sales are maximized when 11 thousand, or $11,000 are spent on advertising. The maximum number of units sold is 2,264.

61. $M(r) = \dfrac{1 + 0.05r}{1 + 0.004r^2}$

(a)
$$M'(r) = \frac{(1 + 0.004r^2)(0.005) - (1 + 0.05r)(0.008r)}{(1 + 0.004r^2)^2}$$
$$= \frac{0.05 - 0.008r - 0.0002r^2}{(1 + 0.004r^2)^2}$$
$$= \frac{500 - 80r - 2r^2}{10{,}000(1 + 0.004r^2)^2}$$

Using the quadratic formula, $M'(r) = 0$ When

$$r = \frac{80 \pm \sqrt{(-80)^2 - (4)(-2)(500)}}{2(-2)}$$
$r \approx 5.495$ (rejecting the negative answer)

When $0 \le r < 5.495$, $M'(r) > 0$ so M is increasing
$r > 5.495$, $M'(r) < 0$ so M is decreasing.

(b) When $r \approx 5.495$, M has a relative maximum which is actually an absolute maximum. So, the number of mortages is maximized when the rate is 5.495%. The corresponding maximum number of mortages is

$$M(5.495) = \frac{1 + 0.05(5.495)}{1 + 0.004(5.495)^2} \approx 1.137$$

or 1,137 refinanced mortgages.

63. **(a)** Approximately 1971, 1976, 1980, 1983, 1988, 1994.

(b) Approximately 1973, 1979, 1981, 1985, 1989.

(c) Approximately $\frac{1}{2}$% per year.

(d) Approximately $\frac{1}{2}$% per year.

65. **(a)** Yield $= \left(\begin{smallmatrix}\text{orig}\\ \text{\#fish}\end{smallmatrix}\right)\left(\begin{smallmatrix}\text{proportion}\\ \text{still living}\end{smallmatrix}\right)\left(\begin{smallmatrix}\text{weight}\\ \text{per fish}\end{smallmatrix}\right)$

$$Y(t) = 300\left(\frac{31}{31+t}\right)(3+t-0.05t^2)$$
$$= 9{,}300(31+t)^{-1}(3+t-0.05t^2)$$
$$Y'(t) = 9{,}300\Big[(31+t)^{-1}(1-0.1t)$$
$$+(3+t-0.05t^2)\cdot -(31+t)^{-2}(1)\Big]$$
$$= 9{,}300\left(\frac{1-0.1t}{31+t} - \frac{3+t-0.05t^2}{(31+t)^2}\right)$$
$$= 9{,}300\frac{29-3.1t-0.05t^2}{(31+t)^2}$$

$Y'(t) = 0$ when $t \approx 8.3$

When

$0 \le t < 8.3$, $Y'(t) > 0$ and Y is increasing

$t = 8.3$, $Y'(t) = 0$ and Y levels

$8.3 < t \le 10$, $Y'(t) < 0$ and Y is decreasing.

The point (8.3, 1859) is a relative maximum.

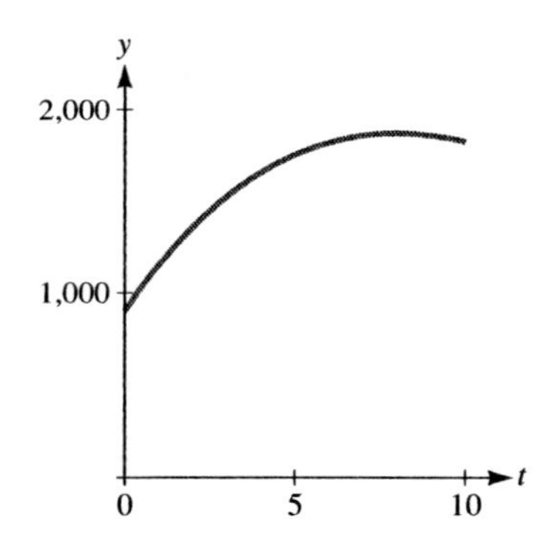

(b) Since (8.3, 1859) is a relative maximum, the yield is maximized after 8.3 weeks and the maximum yield is 1,859 pounds.

67. $H(t) = -053T^2 + 25T - 209,\ 15 \le T \le 30$

$H'(t) = -1.06T + 25$

When $H'(0) = 0$ when $t \approx 23.58$

When $15 \le T < 23.58$, $H'(T) > 0$ so H is increasing

$T = 23.58$, $H'(T) = 0$ so H levels

$23.58 < T \le 30$, $H'(t) < 0$ so H is decreasing.

The point (23.58, 85.81) is a relative maximum.

So, the maximum percentage is 85.81% and it occurs at 23.58°C.

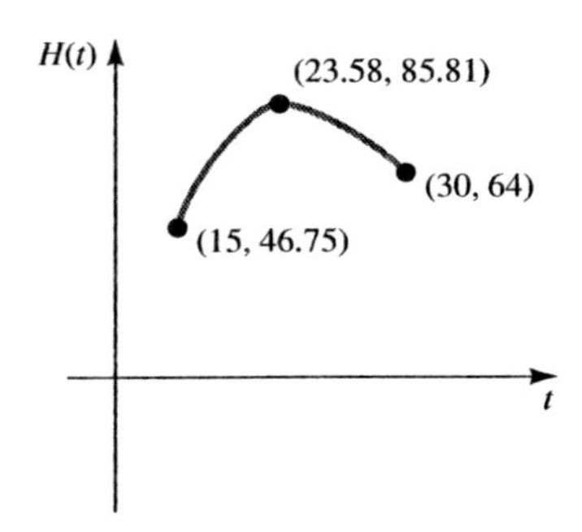

69. (a) Graph levels when $x = 0, 1, 2$.

(b) Graph is decreasing when $0 < x < 1$.

(c) Graph is increasing when $x < 0$, $1 < x < 2$, and $x > 2$.

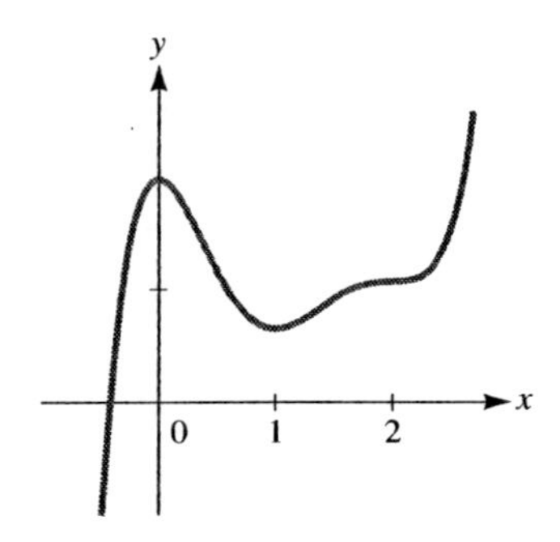

71. (a) Graph is decreasing when $x < -1$.

(b) Graph is increasing when $-1 < x < 3$ and $x > 3$.

(c) Graph levels when $x = -1, 3$.

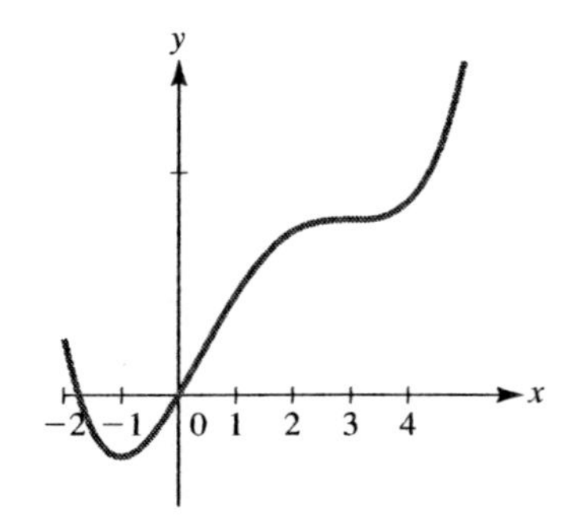

73.
$$f(x) = ax^3 + bx^2 + cx + d$$
$$f'(x) = 3ax^2 + 2bx + c$$
$f'(x) = 0$ when $x = -2$, so
$$0 = 3a(-2)^2 + 2b(-2) + c$$
$$0 = 12a - 4b + c$$
$f'(x) = 0$ when $x = 1$, so
$$0 = 3a(1)^2 + 2b(1) + c$$
$$0 = 3a + 2b + c$$

So, $12a - 4b + c = 3a + 2b + c$, or $b = \frac{3}{2}a$.
Now, $f(-2) = 8$ so
$$8 = a(-2)^3 + \frac{3}{2}a(-2)^2 + c(-2) + d$$
$$8 = -8a + 6a - 2c + d$$
$$8 = -2a - 2c + d$$
or, $d = 8 + 2a + 2c$

Now, $f(1) = -19$ so
$$-19 = a(1)^3 + \frac{3}{2}a(1)^2 + c(1) + d$$
$$-19 = \frac{5}{2}a + c + (8 + 2a + 2c)$$
$$-27 = \frac{9}{2}a + 3c, \text{ or}$$
$$c = \frac{1}{3}\left(-27 - \frac{9}{2}a\right) = -9 - \frac{3}{2}a$$

Using
$$0 = 3a + 2b + c$$
$$0 = 3a + 2\left(\frac{3}{2}a\right) + \left(-9 - \frac{3}{2}a\right)$$
$$0 = \frac{9}{2}a - 9, \text{ or}$$
$$a = 2,\ b = \frac{3}{2}(2) = 3,$$
$$c = -9 - \frac{3}{2}(2) = -12,$$
$$d = 8 + 2(2) + 2(-12) = -12$$

75. $f(x) = 1 - x^{3/5}$

When $x = 0$, $f(0) = 1$ so (0,1) is an intercept.
$f(x) = 0$, $x = 1$ so (1,0) is an intercept.
$$f'(x) = -\frac{3}{5}x^{-2/5}$$
$$= \frac{-3}{5x^{2/5}}$$
When $x < 0$, $f'(x) < 0$ so f is decreasing
$x > 0$, $f'(x) < 0$ so f is decreasing.

f' is undefined when $x = 0$, but f is defined, so this corresponds to a vertical tangent at $x = 0$.

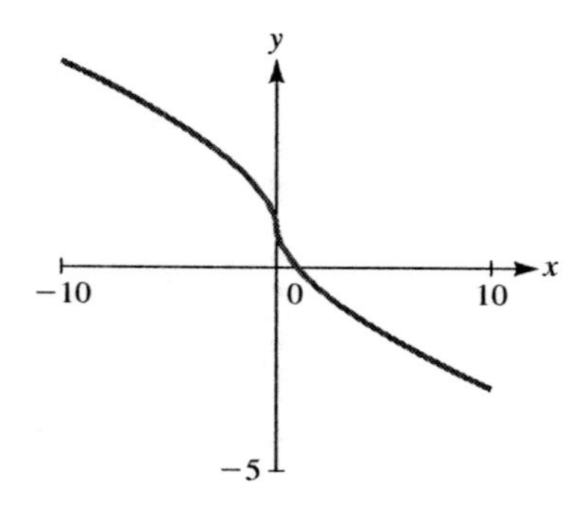

77. $y = (x - p)(x - q)$
$$y' = (x - p)(1) + (x - q)(1)$$
$$= 2x - p - q$$

So, $y' = 0$ when $0 = 2x - p - q$ or, $x = \frac{p + q}{2}$, which is the midpoint of the segment PQ. So any relative extremum occurs midway between its intercepts.

79. $f(x) = (x^2 + x - 1)^3(x + 3)^2$
Press [y=] and input f for $y_1 =$
Use the window dimensions [−4, 2]1 by [−20, 25]5
Press [Graph]

$$f'(x) = 3(x^2 + x - 1)^2(2x + 1)(x + 3)^2 + (x^2 + x - 1)^3(2)(x + 3)$$
$$= (x^2 + x - 1)^2(x + 3)[3(2x + 1)(x + 3) + (x^2 + x - 1)(2)]$$
$$= (x^2 + x - 1)^2(x + 3)(8x^2 + 23x + 7)$$

Press [y=] and input $f'(x)$ for $y_2 =$
Press [Graph]
To find the values of x for which $f'(x) = 0$, it may be easiest to deactivate y_1 so only the graph of $y_2 = f'$ is shown. Use [Trace] and verify $y_2 = (x^2 + x - 1)^2(x + 3)(8x^2 + 23x + 7)$ is shown in the upper left corner. Trace along the graph to move near an x-intercept. Use Zoom in function for more accurate readings. The values of x for which $f'(x) = 0$ are $x_1 \approx -3$, $x_2 \approx -2.5$, $x_3 \approx -1.6$, $x_4 \approx -0.35$, $x_5 \approx 0.62$.

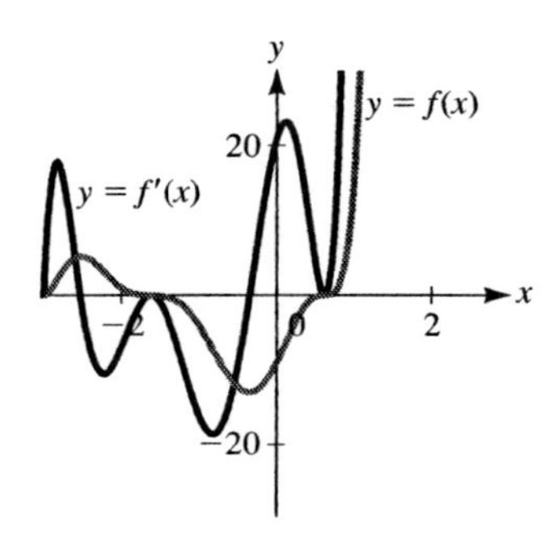

81. $f(x) = (1 - x^{1/2})^{1/2}$
Press [y=] and input f for $y_1 =$
Use the window dimensions [0, 1]0.5 by [−2, 1]1 (the domain of f is $0 < x < 1$)
Press [Graph]

$$f'(x) = \frac{1}{2}(1 - x^{1/2})^{-1/2}\left(-1/2x^{-1/2}\right)$$

$$f'(x) = \frac{-1}{4x^{1/2}(1 - x^{1/2})^{1/2}}$$

Press [y=] and input f' for $y_2 =$
Press [Graph]
We see from the graph that there are no values of x for which $f'(x) = 0$.

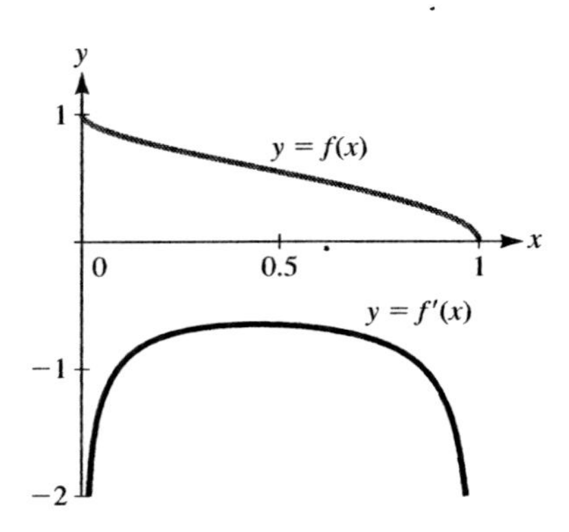

83. Let $f(x) = 4 + \sqrt{9 - 2x - x^2}$. Before graphing, f appears to be the upper half of a circle.

$$y = 4 + \sqrt{9 - 2x - x^2}$$
$$y - 4 = \sqrt{9 - 2x - x^2}$$

By squaring both sides and completing the square we obtain the equation of the whole circle with center (−1, 4) and radius $\sqrt{10}$.

$$(y - 4)^2 = 9 - 2x - x^2$$
$$x^2 + 2x + 1 + (y - 4)^2 = 9 + 1$$
$$(x + 1)^2 + (y - 4)^2 = 10$$

Therefore, $f(x) = 4 + \sqrt{9 - 2x - x^2}$ should be the upper half of this circle.
Press [y=] and input f for $y_1 =$
Use window dimensions [−5, 5] by [−10, 10]
Press [Graph]
Initially, the graph appears to be the upper half of an ellipse but by using the Zsquare function, we see the graph is, in fact, the upper half of the circle.

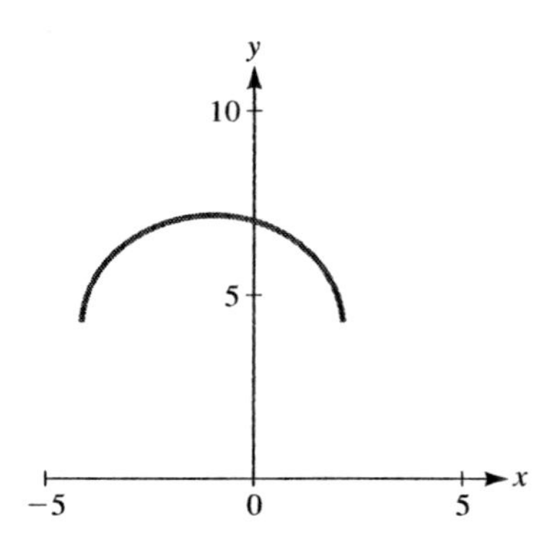

3.2 Concavity and Points of Inflection

1. The graph is:
concave downward ($f''(x) < 0$) for $x < 2$,
and concave upward ($f''(x) > 0$) for $x > 2$.

3. The graph is:
concave downward ($f''(x) < 0$) for $-1 < x < 1$,
and concave upward ($f''(x) > 0$) for $x < -1$ and $x > 1$.

5.
$$f(x) = x^3 + 3x^2 + x + 1$$
$$f'(x) = 3x^2 + 6x + 1$$
$$f''(x) = 6x + 6 = 6(x + 1)$$
$$f''(x) = 0 \text{ when } x = -1$$

When $x < -1$, $f''(x) < 0$ so f is concave down
$x > -1$, $f''(x) > 0$ so f is concave up.

Since the concavity changes at the critical value $x = -1$, the point $(-1, 2)$ is an inflection point.

7. $f(x) = x(2x + 1)^2 = x(4x^2 + 4x + 1) = 4x^3 + 4x^2 + x$
$$f'(x) = 12x^2 + 8x + 1$$
$$f''(x) = 24x + 8 = 8(3x + 1)$$
$$f''(x) = 0 \text{ when } x = -\frac{1}{3}$$

When $x < -\frac{1}{3}$, $f''(x) < 0$ so f is concave down
$x > -\frac{1}{3}$, $f''(x) > 0$ so f is concave up.

Since the concavity changes at the critical value $x = -\frac{1}{3}$, the point $\left(-\frac{1}{3}, -\frac{1}{27}\right)$ is an inflection point.

9.
$$g(t) = t^2 - \frac{1}{t} = t^2 - t^{-1}$$
$$g'(t) = 2t + t^{-2}$$
$$g''(t) = 2 - 2t^{-3} = 2 - \frac{2}{t^3} = \frac{2t^3 - 2}{t^3} = \frac{2(t^3 - 1)}{t^3}$$
$$g''(t) = 0 \text{ when } t = 1$$

(Note that $g''(t)$ and $g(t)$ are undefined for $t = 0$.)

When $t < 0$, $g''(t) > 0$ so g is concave up
$0 < t < 1$, $g''(t) < 0$ so g is concave down
$t > 1$, $g''(t) > 0$ so g is concave up.

Since the concavity changes at the critical value $t = 1$, the point $(1, 0)$ is an inflection point.

11.
$$f(x) = x^4 - 6x^3 + 7x - 5$$
$$f'(x) = 4x^3 - 18x^2 + 7$$
$$f''(x) = 12x^2 - 36x = 12x(x - 3)$$
$$f''(x) = 0 \text{ when } x = 0, 3$$

When $x < 0$, $f''(x) > 0$ so f is concave up
$0 < x < 3$, $f''(x) < 0$ so f is concave down
$x > 3$, $f''(x) > 0$ so f is concave up.

Since the concavity changes at both critical values $x = 0$ and $x = 3$, the points $(0, -5)$ and $(3, -65)$ are inflection points.

13.
$$f(x) = \frac{1}{3}x^3 - 9x + 2$$
$$f'(x) = x^2 - 9 = (x + 3)(x - 3)$$
$$f'(x) = 0 \text{ when } x = -3, 3$$
$$f''(x) = 2x$$
$$f''(x) = 0 \text{ when } x = 0$$

When $x < -3$, $f'(x) > 0$ so f is increasing
$f''(x) < 0$ so f is concave down
$-3 < x < 0$, $f'(x) < 0$ so f is decreasing
$f''(x) < 0$ so f is concave down
$0 < x < 3$, $f'(x) < 0$ so f is decreasing
$f''(x) > 0$ so f is concave up
$x > 3$, $f'(x) > 0$ so f is increasing
$f''(x) > 0$ so f is concave up.

Overall, f is increasing for $x < -3$ and $x > 3$; decreasing for $-3 < x < 3$; concave up for $x > 0$; and concave down for $x < 0$.
The critical value $x = -3$ corresponds to the point $(-3, 20)$, which is a relative maximum.

The critical value $x = 3$ corresponds to the point $(3, -16)$, which is a relative minimum. Since the concavity changes at $x = 0$, the point $(0, 2)$ is an inflection point.

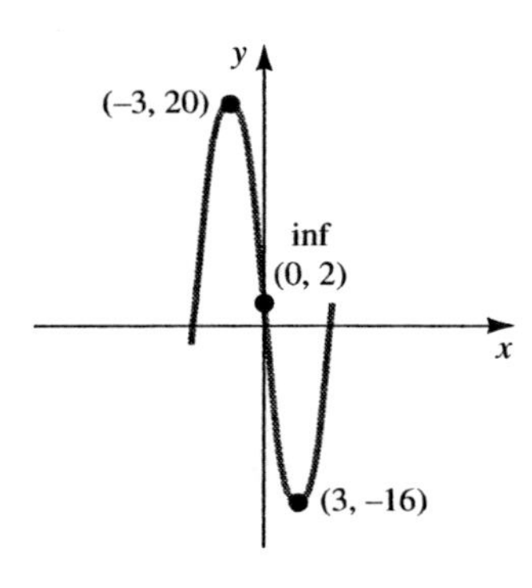

15.

$$f(x) = x^4 - 4x^3 + 10$$
$$f'(x) = 4x^3 - 12x^2 = 4x^2(x - 3)$$
$$f'(x) = 0 \text{ when } x = 0, 3$$
$$f''(x) = 12x^2 - 24x = 12x(x - 2)$$
$$f''(x) = 0 \text{ when } x = 0, 2$$

When $x < 0$, $f'(x) < 0$ so f is decreasing
$f''(x) > 0$ so f is concave up
$0 < x < 2$, $f'(x) < 0$ so f is decreasing
$f''(x) < 0$ so f is concave down
$2 < x < 3$, $f'(x) < 0$ so f is decreasing
$f''(x) > 0$ so f is concave up
$x > 3$, $f'(x) > 0$ so f is increasing
$f''(x) > 0$ so f is concave up.

Overall, f is increasing for $x > 3$; decreasing $x < 3$; concave up for $x < 0$ and $x > 2$; and concave down for $0 < x < 2$.
The critical value $x = 0$ corresponds to the point $(0, 10)$, which is not a relative extremum. However, the concavity charges at $x = 0$, so $(0, 10)$ is an inflection point. The concavity changes again at $x = 2$, so the point $(2, -6)$ is also an inflection point. The critical value $x = 3$ corresponds to the point $(3, -17)$, which is a relative minimum.

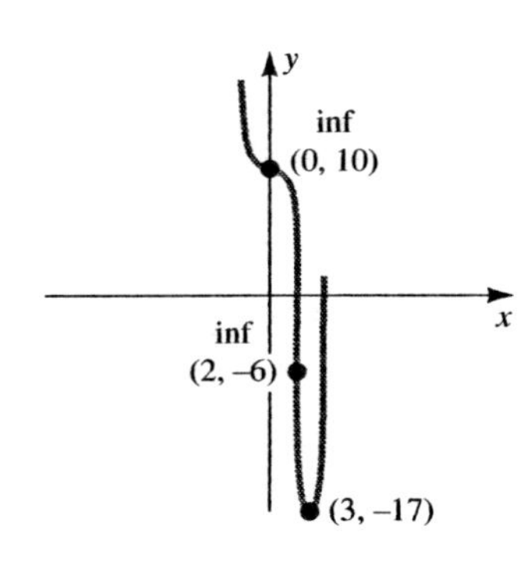

17.

$$f(x) = (x - 2)^3$$
$$f'(x) = 3(x - 2)^2(1)$$
$$f'(x) = 0 \text{ when } x = 2$$
$$f''(x) = 6(x - 2)$$
$$f''(x) = 0 \text{ when } x = 2$$

When $x < 2$, $f'(x) > 0$ so f is increasing
$f''(x) < 0$ so f is concave down
$x > 2$, $f'(x) > 0$ so f is increasing
$f''(x) > 0$ so f is concave up.

Overall, f is increasing for all values of x; concave up for $x > 2$; and concave down for $x < 2$.
The critical value $x = 2$ corresponds to the point $(2, 0)$ which is not a relative extremum. However, the concavity changes at $x = 2$, so $(2, 0)$ is an inflection point.

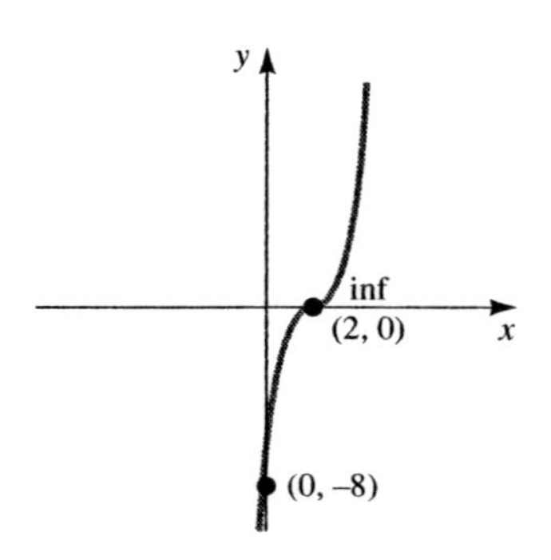

19. $f(x) = (x^2 - 5)^3$

$f'(x) = 3(x^2 - 5)^2(2x) = 6x(x^2 - 5)^2$

$f'(x) = 0$ when $x = -\sqrt{5}, 0, \sqrt{5}$

$$f''(x) = (6x)\left[2(x^2 - 5)(2x)\right] + (x^2 - 5)^2(6)$$
$$= 6(x^2 - 5)\left[4x^2 + x^2 - 5\right]$$
$$= 6(x^2 - 5)(5x^2 - 5)$$
$$= 30(x^2 - 5)(x + 1)(x - 1)$$

$f''(x) = 0$ when $x = -\sqrt{5}, -1, 1, \sqrt{5}$

When $x < -\sqrt{5}$, $f'(x) < 0$ so f is decreasing
$f''(x) > 0$ so f is concave up

$-\sqrt{5} < x < -1$, $f'(x) < 0$ so f is decreasing
$f''(x) < 0$ so f is concave down

$-1 < x < 0$, $f'(x) < 0$ so f is decreasing
$f''(x) > 0$ so f is concave up

$0 < x < 1$, $f'(x) > 0$ so f is increasing
$f''(x) > 0$ so f is concave up

$1 < x < \sqrt{5}$, $f'(x) > 0$ so f is increasing
$f''(x) < 0$ so f is concave down

$x > \sqrt{5}$, $f'(x) > 0$ so f is increasing
$f''(x) > 0$ so f is concave up.

Overall, f is increasing for $x > 0$; decreasing for $x < 0$; concave up for $x < -\sqrt{5}$, $-1 < x < 1$, and $x > \sqrt{5}$; and concave down for $-\sqrt{5} < x < -1$ and $1 < x < \sqrt{5}$.
The critical value $x = -\sqrt{5}$ corresponds to the point $(-\sqrt{5}, 0)$, which is not a relative extremum. However, the concavity changes at $x = -\sqrt{5}$, so $(-\sqrt{5}, 0)$ is an inflection point. The concavity changes again at $x = -1$, so the point $(-1, -64)$ is also an inflection point. The critical value $x = 0$ corresponds to the point $(0, -125)$, which is a relative minimum. The concavity next changes at $x = 1$, so the point $(1, -64)$ is an inflection point. The critical value $x = \sqrt{5}$ corresponds to the point $(\sqrt{5}, 0)$, which is not a relative extremum. However, the concavity changes at $x = \sqrt{5}$, so $(\sqrt{5}, 0)$ is an inflection point.

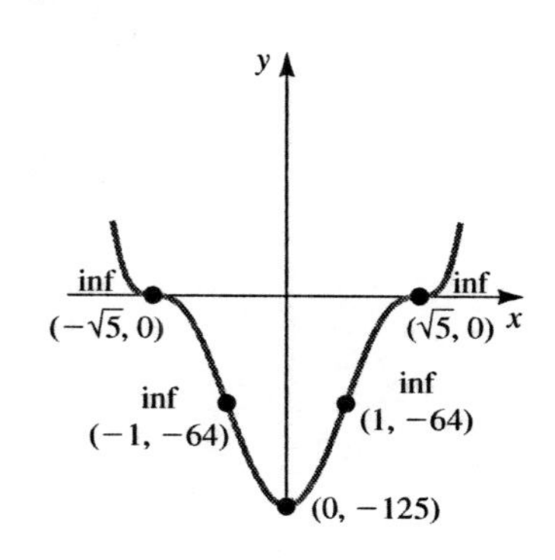

21. $f(s) = 2s(s + 4)^3$

$$f'(s) = (2s)\left[3(s + 4)^2(1)\right] + (s + 4)^3(2)$$
$$= 2(s + 4)^2[3s + s + 4]$$
$$= 8(s + 4)^2(s + 1)$$

$f'(s) = 0$ when $s = -4, -1$

$$f''(s) = 8\left[(s + 4)^2(1) + (s + 1)\,(2(s + 4)(1))\right]$$
$$= 8(s + 4)\,[s + 4 + 2(s + 1)]$$
$$= 24(s + 4)(s + 2)$$

$f''(s) = 0$ when $s = -4, -2$

When $s < -4$, $f'(s) < 0$ so f is decreasing
$f''(s) > 0$ so f is concave up

$-4 < s < -2$, $f'(s) < 0$ so f is decreasing
$f''(s) < 0$ so f is concave down

$-2 < s < -1$, $f'(s) < 0$ so f is decreasing
$f''(s) > 0$ so f is concave up

$s > -1$, $f'(s) > 0$ so f is increasing
$f''(s) > 0$ so f is concave up.

Overall, f is increasing for $s > -1$; decreasing for $s < -1$; concave up for $s < -4$ and $s > -2$; and concave down for $-4 < s < -2$.
The critical value $s = -4$ corresponds to the point $(-4, 0)$, which is not a relative extremum. However, the concavity changes at $s = -4$, so $(-4, 0)$ is an inflection point. The concavity changes again at

$s = -2$, so the point $(-2, -32)$ is also an inflection point. The critical value $s = -1$ corresponds to the point $(-1, -54)$, which is a relative minimum.

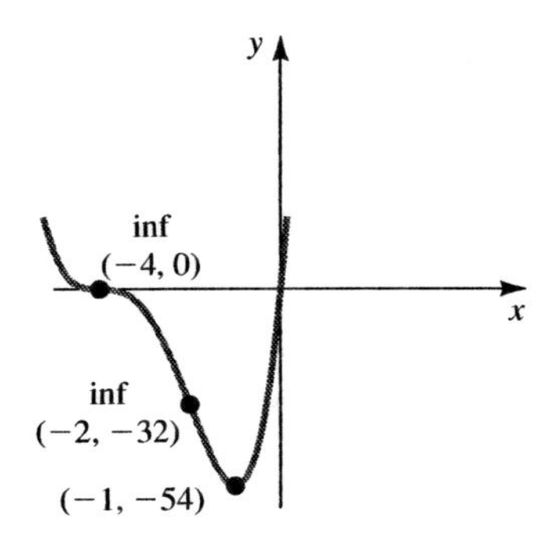

23. $g(x) = \sqrt{x^2+1} = (x^2+1)^{1/2}$

$$g'(x) = \frac{1}{2}(x^2+1)^{-1/2}(2x) = \frac{x}{\sqrt{x^2+1}}$$

$g'(x) = 0$ when $x = 0$

$$g''(x) = \frac{(x^2+1)^{1/2}(1) - (x)\left[\frac{1}{2}(x^2+1)^{-1/2}(2x)\right]}{\left(\sqrt{x^2+1}\right)^2}$$

$$g''(x) = \frac{(x^2+1)^{1/2} - \dfrac{x^2}{(x^2+1)^{1/2}}}{x^2+1} \cdot \frac{(x^2+1)^{1/2}}{(x^2+1)^{1/2}}$$

$$= \frac{x^2+1-x^2}{(x^2+1)^{3/2}} = \frac{1}{(x^2+1)^{3/2}}$$

When $x < 0$, $g'(x) < 0$ so g is decreasing
$g''(x) > 0$ so g is concave up
$x > 0$, $g'(x) > 0$ so g is increasing
$g''(x) > 0$ so g is concave up.

Overall, g is increasing for $x > 0$; decreasing for $x < 0$; and concave up for all values of x.
The critical value $x = 0$ corresponds to the point $(0, 1)$, which is a relative minimum.

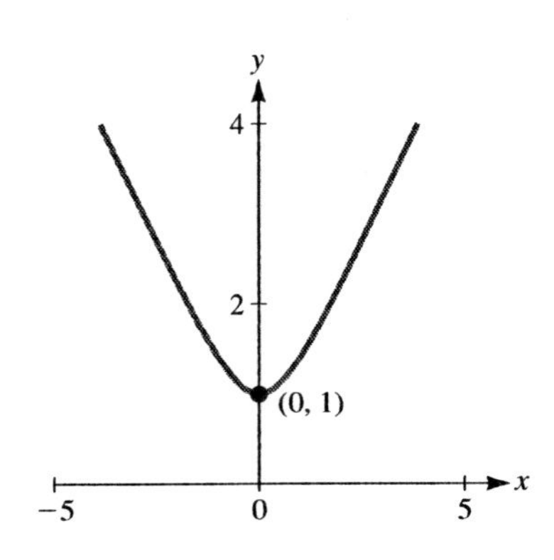

25. $f(x) = \dfrac{1}{x^2+x+1} = (x^2+x+1)^{-1}$

$$f'(x) = -(x^2+x+1)^{-2}(2x+1) = \frac{-(2x+1)}{(x^2+x+1)^2}$$

$f'(x) = 0$ when $x = -\dfrac{1}{2}$

$$f''(x) = \frac{1}{\left[(x^2+x+1)^2\right]^2}\left[(x^2+x+1)^2(-2) + (2x+1)(2)(x^2+x+1)(2x+1)\right]$$

$$= \frac{2(x^2+x+1)\left[-(x^2+x+1) + (2x+1)^2\right]}{(x^2+x+1)^4}$$

$$= \frac{6x(x+1)}{(x^2+x+1)^3}$$

$f''(x) = 0$ when $x = -1, 0$

When $x < -1$, $f'(x) > 0$ so f is increasing
$f''(x) > 0$ so f is concave up
$-1 < x < -\frac{1}{2}$, $f'(x) > 0$ so f is increasing
$f''(x) < 0$ so f is concave down
$-\frac{1}{2} < x < 0$, $f'(x) < 0$ so f is decreasing
$f''(x) < 0$ so f is concave down
$x > 0$, $f'(x) < 0$ so f is decreasing
$f''(x) > 0$ so f is concave up.

Overall, f is increasing for $x < -\frac{1}{2}$; decreasing for $x > -\frac{1}{2}$; concave up for $x < -1$ and $x > 0$; and concave down for $-1 < x < 0$.
At $x = -1$, the concavity changes, so the point $(-1, 1)$ is an inflection point. The critical value $x = -\frac{1}{2}$ corresponds to the point $\left(-\frac{1}{2}, \frac{4}{3}\right)$, which is relative maximum. The concavity changes again at $x = 0$, so the point $(0, 1)$ is an inflection point.

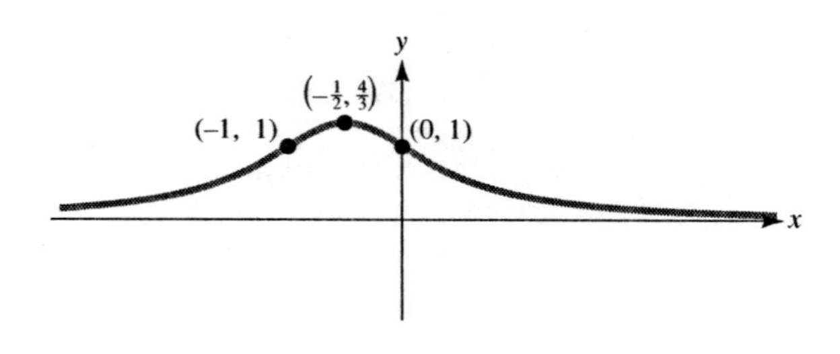

27.
$$f(x) = x^3 + 3x^2 + 1$$
$$f'(x) = 3x^2 + 6x = 3x(x+2)$$
$$f'(x) = 0 \text{ when } x = -2, 0$$
$$f''(x) = 6(x+1)$$
$$f''(0) = 6 > 0 \text{ and } f''(-2) = -6 < 0,$$
$$\text{and } f(-2) = 5 \text{ and } f(0) = 1.$$

So $(0, 1)$ is a relative minimum, and $(-2, 5)$ is a relative maximum.

29.
$$f(x) = (x^2 - 9)^2$$
$$f'(x) = 2(x^2 - 9)(2x) = 4x(x-3)(x+3)$$
$$f'(x) = 0 \text{ when } x = -3, 0, 3$$
$$f''(x) = 12(x^2 - 3)$$

$f''(-3) = 72 > 0$, $f''(0) = -36 < 0$, and $f''(3) = 72 > 0$; $f(\pm 3) = 0$ and $f(0) = 81$.
So $(0, 81)$ is a relative maximum, and $(-3, 0)$, $(3, 0)$ are relative minima.

31.
$$f(x) = 2x + 1 + \frac{18}{x}$$
$$f'(x) = 2 - \frac{18}{x^2} = \frac{2(x-3)(x+3)}{x^2}$$
$$f'(x) = 0 \text{ when } x = -3, 3$$
$$f''(x) = \frac{36}{x^3}$$

$f''(-3) = -\frac{4}{3} < 0$ and $f''(3) = \frac{4}{3} > 0$;
$f(-3) = -11$, $f(3) = 13$.
So $(-3, -11)$ is a relative maximum, $(3, 13)$ is a relative minimum.

33.
$$f(x) = x^2(x-5)^2 = x^4 - 10x^3 + 25x^2$$
$$f'(x) = 4x^3 - 30x^2 + 50x = 2x(x-5)(2x-5)$$
$$f'(x) = 0 \text{ when } x = 0, 2.5, 5$$
$$f''(x) = 12x^2 - 60x + 50$$

$f''(0) = 50 > 0$, $f''(2.5) = -25 < 0$, and $f''(5) = 50 > 0$; $f(0) = 0$, $f(2.5) = 39.0625$ and $f(5) = 0$. So $(0, 0)$ and $(5, 0)$ are relative minima and $(2.5, 39.065)$ is a relative maximum.

35.
$$h(t) = \frac{2}{1+t^2} = 2(1+t^2)^{-1}$$
$$h'(t) = -2(1+t^2)^{-2}(2t) = \frac{-4t}{(1+t^2)^2}$$
$$h'(t) = 0 \text{ when } t = 0$$
$$h''(t) = \frac{-4(1+t^2)^2 - (-4t)(2)(1+t^2)(2t)}{(1+t^2)^4} = \frac{4(1+t)\left[-(1+t^2) + 4t^2\right]}{(1+t^2)^4} = \frac{4(3t^2 - 1)}{(1+t^2)^3}$$

$h''(0) = -4 < 0$ and $h(0) = 2$. So, $(0, 2)$ is a relative maximum.

37.
$$f(x)=\frac{(x-2)^3}{x^2}$$
$$f'(x)=\frac{x^2\left[3(x-2)^2(1)\right]-(x-2)^3(2x)}{x^4}$$
$$=\frac{x(x-2)^2\left[3x-2(x-2)\right]}{x^4}$$
$$=\frac{(x-2)^2(x+4)}{x^3}$$
$f'(x)=0$ when $x=-4, 2$
$$f''(x)=\frac{1}{x^6}\left(x^3\left[(x-2)^2(1)+(x+4)(2)(x-2)\right]-\left[(x-2)^2(x+4)(3x^2)\right]\right)$$
$$=\frac{x^2(x-2)\left(x\left[(x-2)+2(x+4)\right]-3(x-2)(x+4)\right)}{x^6}$$
$$=\frac{24(x-2)}{x^4}$$
$f''(-4)=-\dfrac{9}{16}<0$ and $f(-4)=-13.5$. So, $(-4, -13.5)$ is a relative maximum. $f''(2)=0$, so the test fails.

39. $f''(x)=x^2(x-3)(x-1)$
$f''(x)=0$ when $x=0, 1, 3$

When $x<0$, $f''(x)>0$ so f is concave up
$0<x<1$, $f''(x)>0$ so f is concave up
$1<x<3$, $f''(x)<0$ so f is concave down
$x>3$, $f''(x)>0$ so f is concave up.

Overall, f is concave up for $x<0$, $0<x<1$, and $x>3$; concave down for $1<x<3$. There are inflection points at $x=1$ and $x=3$, as the concavity changes at those values.

41. $f''(x)=(x-1)^{1/3}$
$f''(x)=0$ when $x=1$

When $x<1$, $f''(x)<0$ so f is concave down
$x>1$, $f''(x)>0$ so f is concave up.

There is an inflection point at $x=1$, as the concavity changes at that value.

43. $f'(x)=x^2-4x$

(a) $f'(x)=x(x-4)$
$f'(x)=0$ when $x=0, 4$

When $x<0$, $f'(x)>0$ so f is increasing
$0<x<4$, $f'(x)<0$ so f is decreasing
$x>4$, $f'(x)>0$ so f is increasing.

(b) $f''(x)=2x-4=2(x-2)$
$f''(x)=0$ when $x=2$

When $x<2$, $f''(x)<0$ so f is concave down
$x>2$, $f''(x)>0$ so f is concave up.

(c) at $x=0$, there is a relative maximum;
at $x=4$, there is a relative minimum;
at $x=2$, there is an inflection point.

(d)

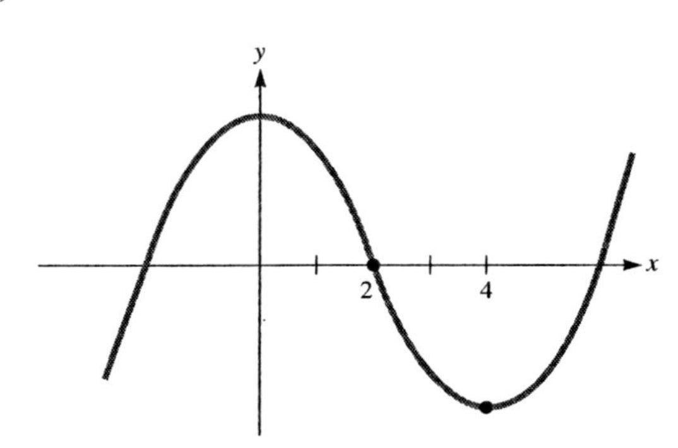

45. $f'(x)=5-x^2$
$f'(x)=0$ when $-\sqrt{5}, \sqrt{5}$

(a) When $x<-\sqrt{5}$, $f'(x)<0$ so f is decreasing
$-\sqrt{5}<x<\sqrt{5}$, $f'(x)>0$ so f is increasing
$x>\sqrt{5}$, $f'(x)<0$ so f is decreasing.

(b) $f''(x)=-2x$
$f''(x)=0$ when $x=0$

When $x<0$, $f''(x)>0$ so f is concave up
$x>0$, $f''(x)<0$ so f is concave down.

(c) at $x=-\sqrt{5}$, there is a relative minimum;
at $x=\sqrt{5}$, there is a relative maximum;

at $x = 0$, there is an inflection point.

(d)

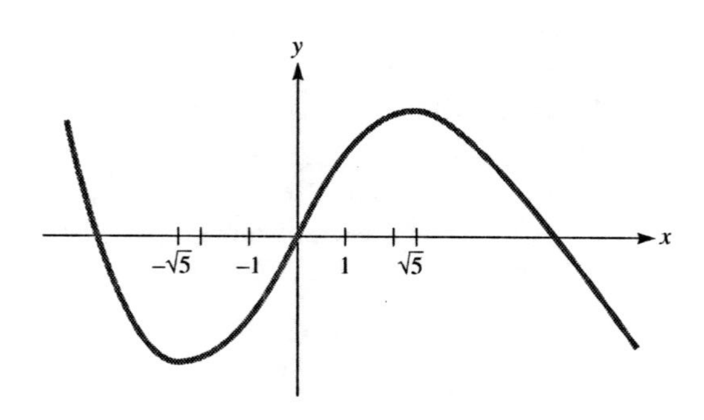

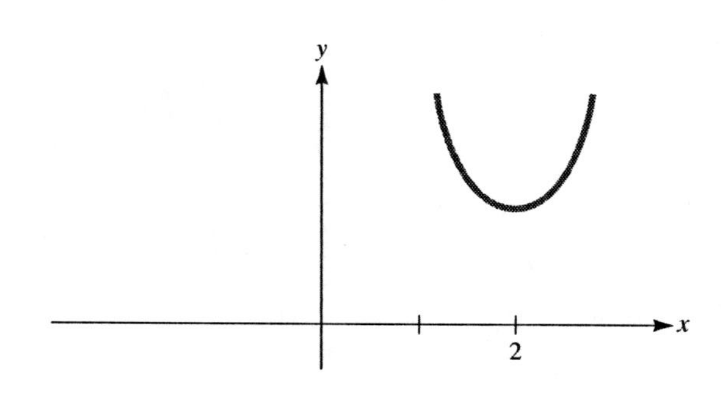

47. (a) The curve rises for $x < -1$ and $x > 3$.

(b) It falls when $-1 < x < 3$.

(c) The curve is concave down for $x < 2$.

(d) The curve is concave up for $x > 2$.
Here is a possible graph.

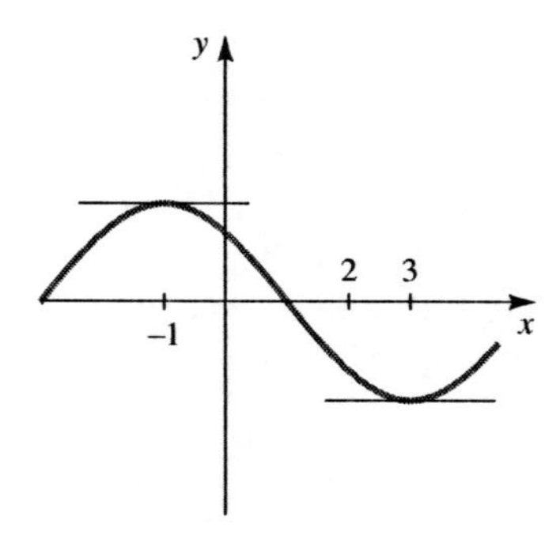

49. When $x < 2$, $f'(x) < 0$ so f is decreasing

$x = 2$, $f'(x) = 0$

and there is a relative minimum

$x > 2$, $f'(x) > 0$ so f is increasing.

Since f' is increasing for all values of x, its rate of change $f''(x) > 0$ for all x, and f is concave up for all x.

51. When $x < -3$, $f'(x) < 0$ so f is decreasing

$x = -3$, $f'(x) = 0$ so f levels

but there is not a relative extremum

$-3 < x < 2$, $f'(x) < 0$ so f is decreasing

$x = 2$, $f'(x) = 0$

and there is a relative minimum

$x > 2$, $f'(x) > 0$ so f is increasing

Since $f'(x)$ is increasing for $x < -3$ and for $x > -1$, $f''(x) > 0$ on these intervals and f is concave up. Since $f'(x)$ is decreasing for $-3 < x < -1$, $f''(x) < 0$ on that interval and f is concave down. Since the concavity changes at $x = -3$ and $x = -1$, there are inflection points at these values.

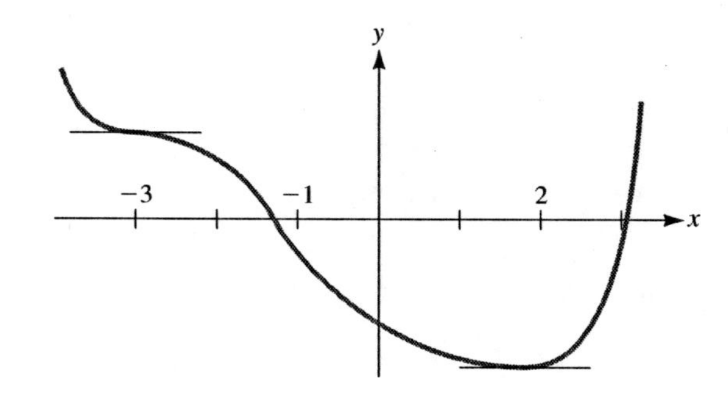

53. (a)

$$C(x) = 0.3x^3 - 5x^2 + 28x + 200$$

$$M(x) = C'(x) = 0.9x^2 - 10x + 28$$

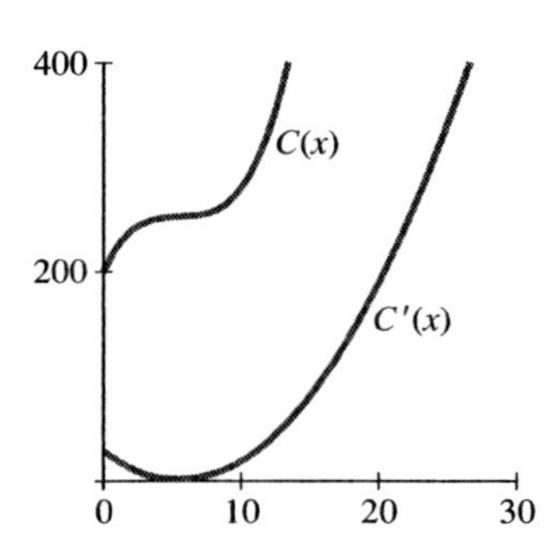

(b)
$$M'(x) = C''(x) = 1.8x - 10$$
$$C''(x) = 0 \text{ when } x \approx 5.56$$

Critical values of C'' are x values of possible extrema of C', which is the marginal cost function. $x = 5.56$ corresponds to a minimum on the graph of C'.

55. Need to maximize the rate of output on the interval $0 \leq t \leq 4$. Since the output is

$$Q(t) = -t^3 + \frac{9}{2}t^2 + 15t$$

the rate of change of the output is

$$R(t) = Q'(t) = -3t^2 + 9t + 15$$
$$R'(t) = Q''(t) = -3(2t - 3)$$
$$R'(t) = 0 \text{ when } t = 1.5$$

Usisng the interval endpoints and this critical value, $R(0) = 15$, $R(1.5) = 21.75$, and $R(4) = 3$.
So, an absolute maximum occurs at $t = 1.5$ and an absolute minimum when $t = 4$.

(a) The worker is performing most efficiently when $t = 1.5$, at 9:30 a.m.

(b) and least efficiently when $t = 4$, at 12:00 noon.

57. Need to optimize the rate of population growth on the interval $0 \leq t \leq 5$. Since the population is

$$P(t) = -t^3 + 9t^2 + 48t + 50$$

the rate of growth is

$$R(t) = P'(t) = -3t^2 + 18t + 48$$
$$R'(t) = P''(t) = -6t + 18$$
$$R'(t) = 0 \text{ when } t = 3$$

Using the interval endpoints and this critical value, $R(0) = 48$, $R(3) = 75$, and $R(5) = 63$.

(a) The rate of growth is greatest when $t = 3$, or 3 years from now.

(b) It is smallest when $t = 0$, or now.

(c) The rate the population growth changes most rapidly is when $R'(t)$ is a maximum. Since $R'(t) = -6t + 18$, is most rapid when $t = 0$, or now.

59. $N(t) = \dfrac{St}{12 + t^2}$

(a)
$$N'(t) = \frac{(12 + t^2)(5) - (5t)(2t)}{(12 + t^2)^2} = \frac{60 - 5t^2}{(12 + t^2)^2}$$

$$N''(t) = \frac{(12 + t^2)^2(-10t) - (60 - 5t^2) \cdot 2(12 + t^2)(2t)}{(12 + t^2)^4}$$
$$= \frac{2t(12 + t^2)\left[-5(12 + t^2) - 2(60 - 5t^2)\right]}{(12 + t^2)^4}$$
$$= \frac{2t(5t^2 - 180)}{(12 + t^2)^4} = \frac{10t(t^2 - 36)}{(12 + t^2)^4}$$

(b) $N'(t) = 0$ when $t = \sqrt{12} \approx 3.46$. Since $N''(3.46) < 0$, the maximum number of reported cases occurs after 3.46 weeks. The corresponding maximum number of new cases is
$N(3.46) = \dfrac{5\sqrt{12}}{12 + 12} \approx 0.7217$
or 722 new cases.

(c) $N''(t) = 0$ when $t = 0{,}6$

When $0 < t < 6$, $N''(t) < 0$ so N' is decreasing
$t > 0$, $N''(t) > 0$ so N' is increasing

So, the rate of reported cases N' is minimized after 6 weeks. The corresponding minimum number of new cases is
$N(6) = \dfrac{5(6)}{12 + 36} = 0.625$
or approximately 63 new cases.

61. $M(r)\dfrac{1+0.02r}{1+0.009r^2}$

(a) $M'r = \dfrac{(1+0.009r^2)(0.02)-(1+0.02r)(0.018r)}{(1+0.009r^2)^2}$

$= \dfrac{0.02-0.018r-0.00018r^2}{(1+0.009r^2)^2}$

$M''(r) = \Big[(1+0.009r^2)^2(-0.018-0.00036r)$

$-(0.02-0.018r-0.00018r^2)\cdot 2(1+0.09r^2)(0.018r)\Big] \Big/$

$(1+0.009r^2)^4$

$= \Big[0.018(1+0.009r^2)\Big[(1+0.009r^2)(-1-0.02r)$

$-2r(0.02-0.018r-0.00018r^2)\Big]\Big] \Big/$

$(1+0.009r^2)^4$

$= 0.018\Big[-1-0.06r+0.0027r^2+0.00018r^3\Big] \Big/$

$(1+0.009r^2)^3$

(b) Press [y=] and input $(1+0.02x) \div (1+0.009x^2)$ for $y_1 =$.
Use window dimensions [0, 20]0.05 by [0, 2]0.25.
Press [graph].

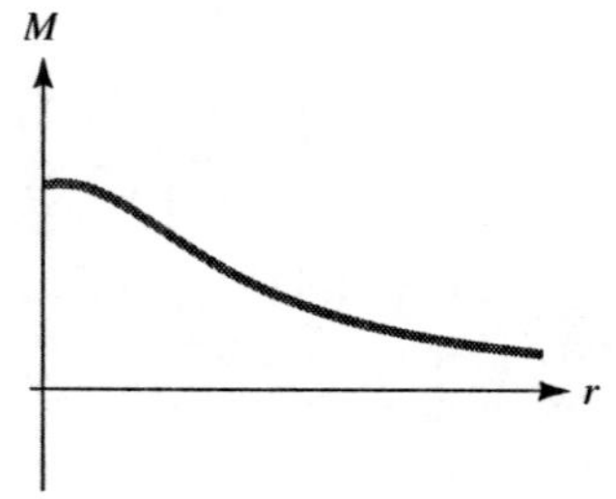

(c) To find the rate of interest at which the rate of construction of new houses is minimized, we must find r for which $M''(r)=0$. $M'(r)$ gives us the rate of construction and thus to minimize this, we take $M''(r)$ and set it equal to zero.
Press [y=] and input M'' for $y_1 =$.
Use window dimensions [−10, 10]1 by[−0.005, 0.005]0.001.
Press [graph]. Since we are only concerned with the positve zero, use the zero function under the calc menu with a close value to the positive x-intercept for the left bound, right bound and guess. We find the interest rate to be $r = 7.10\%$.

63.

$$100\frac{P'(t)}{P(t)} = A - BP(t)$$

$$P'(t) = \frac{P(t)[A-BP(t)]}{100}$$

$$P''(t) = \frac{1}{100}\left[P(t)\left(-BP'(t)\right) + (A-BP(t))\,P'(t)\right]$$

$$= \frac{1}{100}P'(t)\left[-BP(t) + A - BP(t)\right]$$

Substituting for $P'(t)$ from above,

$$= \frac{1}{100}\,\frac{P(t)\,[A-BP(t)]}{100}\,(A-2BP(t))$$

$P''(t) = 0$ when $P(t) = \dfrac{A}{B}$ or $P(t) = \dfrac{A}{2B}$.

When $P(t) = \dfrac{A}{B}$, the population is at the maximum possible and there is no change ($P'(t) = 0$).
When $P(t) = \dfrac{A}{2B}$, the population is changing most rapidly.

65. $\dfrac{dA}{dt} = k\sqrt{A(t)}\,[M - A(t)]$, $k > 0$

(a) $R(t) = \dfrac{dA}{dt} = k\,[A(t)]^{1/2}\,[M - A(t)]$

$$R'(t) = k\left[[A(t)]^{1/2}\left(-\frac{dA}{dt}\right) + (M - A(t))\left(\frac{1}{2}[A(t)]^{-1/2}\frac{dA}{dt}\right)\right]$$

$$= k\frac{dA}{dt}\left[-[A(t)]^{1/2}\cdot\frac{2\,[A(t)]^{1/2}}{2\,[A(t)]^{1/2}} + \frac{M-A(t)}{2\,[A(t)]^{1/2}}\right]$$

$$= k\frac{dA}{dt}\left[\frac{-2A(t)+M-A(t)}{2\,[A(t)]^{1/2}}\right]$$

$$= k\frac{dA}{dt}\left[\frac{M-3A(t)}{2[A(t)]^{1/2}}\right]$$

$R'(t) = 0$ when $M - 3A(t) = 0$, or $A(t) = \dfrac{M}{3}$.

(b) When $A(t) < \frac{M}{3}$, $M - 3A(t) > 0$ and $R'(t) > 0$, so R is increasing.
When $A(t) > \frac{M}{3}$, $M - 3A(t) < 0$ and $R'(t) < 0$, so R is decreasing.
So, when $A(t) = \frac{M}{3}$, the rate is the greatest.

(c) $R'(t) = A''(t)$, so graph of A has an inflection point when $A(t) = \frac{M}{3}$.

67.

$$f(x) = x^4 + x$$
$$f'(x) = 4x^3 + 1$$
$$f'(x) = 0 \text{ when } x = \sqrt[3]{-\frac{1}{4}} \approx -0.63$$
$$f''(x) = 12x^2$$
$$f''(x) = 0 \text{ when } x = 0$$

When $x < -0.63$, $f'(x) < 0$ so f is decreasing
$f''(x) > 0$ so f is concave up
$-0.63 < x < 0$, $f'(x) > 0$ so f is increasing
$f''(x) > 0$ so f is concave up
$x > 0$, $f'(x) > 0$ so f is increasing
$f''(x) > 0$ so f is concave up.

When $x = -0.63$, f has a relative minimum. When $x = 0$, f does not have a relative extremum, nor does f have an inflection point, as the concavity does not switch.
$f(-0.63) \approx -0.47$; $f(0) = 0$

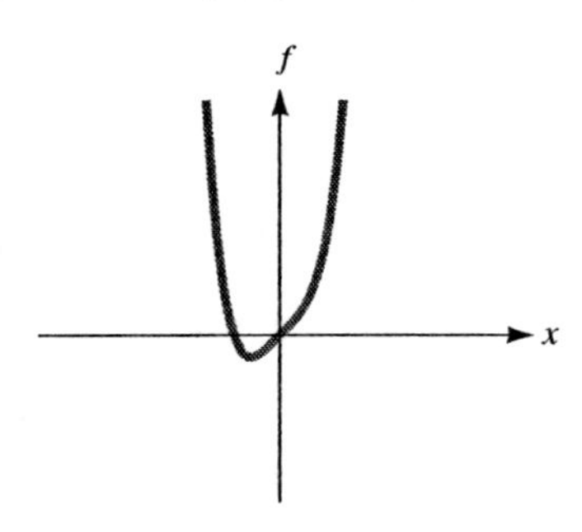

69. $f(x) = 2x^3 + 3x^2 - 12x - 7$

(a) To graph,

Press [y=] and input f for $y_1 =$
Use zstandard function of zoom for viewing window.
Press [Graph]
Change window dimensions to [−10, 10]1 by [−20, 20]2
Press [Graph]

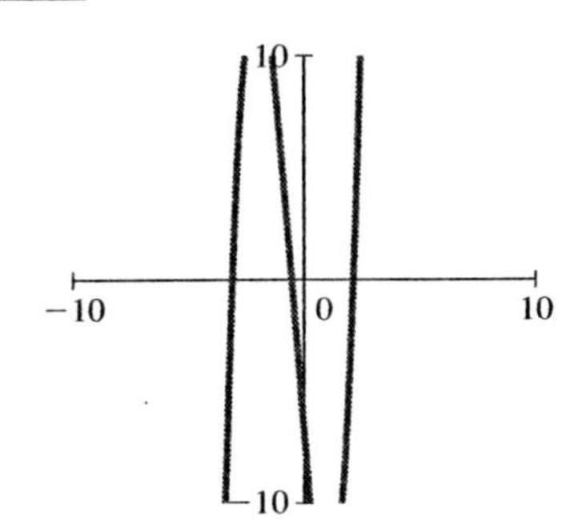

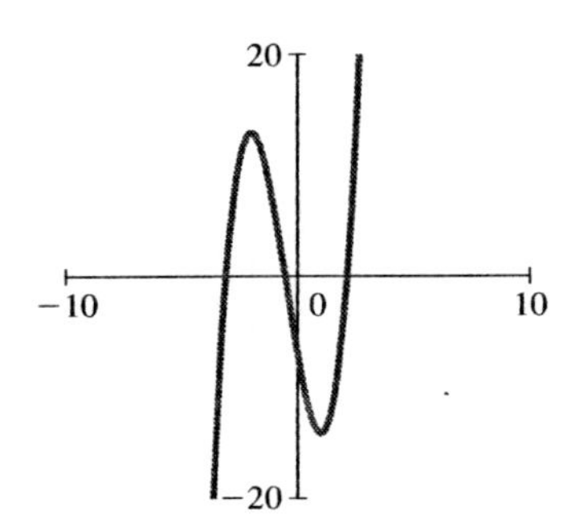

(b)

$$f'(x) = 6x^2 + 6x - 12$$
$$f''(x) = 12x + 6$$

To use the TI-84 to find these values, input f for $y_1 =$, f' for $y_2 =$, and f'' for $y_3 =$. De-select $y_2 =$ and $y_3 =$ so only $y_1 =$ is activated.
Use the value function in the calc menu. For $f(-4)$, input $x = -4$ and press [enter]. The display shows $y = -39$. Repeat process for $x = -2, -1, 0, 1$, and 2. For the $f'(x)$ values, de-select $y_1 =$ and activate $y_2 =$ and repeat process.
For the $f''(x)$ values, de-select $y_2 =$ and activate $y_3 =$ and repeat process.

x	−4	−2	−1	0	1	2

$f(x)$	-39	13	6	-7	-14	-3
$f'(x)$	60	0	-12	-12	0	24
$f''(x)$	-42	-18	-6	6	18	30

(c) To approximate the x-intercepts and y-intercept, Use zstandard function, activate $y_1 =$ and press [graph].
You may use [trace] and zoom-in to estimate x-intercepts to be $x_1 \approx -3.08$, $x_2 \approx -0.54$, and $x_3 \approx 2.11$.
An alternative is to use the zero function under the calc menu. Press [2nd] [calc] and enter zero function. The graph is displayed with left bound? For the left-most x-intercept, trace the graph to a value close to the intercept, but to the left of it and press [enter]. For the right bound? enter a value close to the x-intercept but to the right of it.
To guess a value, enter an x-value in between the bounds and press [enter].
The display shows the zero value of $x_1 \approx -3.08$. Repeat this process for the other two x-intercepts.
For the y-intercept, use zstandard and read the y-intercept as $y = -7$ (also given from the table in part (b)).

(d) To find the relative maximum and relative minimum points,
Use zstandard function with $y_1 =$ activated. Press [graph]. Trace graph left until off the screen (near the relative maximum) and press [enter]. This will move the viewing window to the relative maximum. Use trace and zoom functions to estimate the maximum point to be $(-2, 13)$.
As an alternative, use the maximum function under the calc menu. Enter a left bound, right bound, and guess. For the relative minimum, use z-standard to view the original graph and move cross-hair so relative minimum is in window. Use the minimum function under the calc menu to find the relative minimum to be $(1, -14)$.

(e) Using the graph and the information from part (d), f is increasing on $x < -2$ and $x > 1$.

(f) Using the graph and the information from part (d), f is decreasing on $-2 < x < 1$.

(g) There is an inflection value on $-2 < x < 1$, since the concavity changes from downward to upward.
On the TI-84, de-select $y_1 =$ and activate y_3. Press [graph]. Use the zero function under the calc menu to find the zero of f'' to be $x = -0.5$. Activate $y_1 =$ and use the value function under the calc menu to find $f(-0.5) = -0.5$.
The inflection point is $(-0.5, -0.5)$

(h) Using the graph of f and the information from the previous parts, f is concave upward on $x > -\frac{1}{2}$.

(i) Using the graph of f and the information from the previous parts, f is concave downward on $x < -\frac{1}{2}$.

(j) To verify f changes from concave downward to concave upward, use the value function under the calc menu to show

$$f''(-0.6) = -1.2$$
$$f''(-0.4) = 1.2$$

(Make sure that you have $y_3 = 12x + 6$ activated.)

(k) Relative minimum point: $(1, -14)$
Relative maximum point: $(-2, 13)$
Both of the x-values are within the specified interval. Check the endpoints of the interval. From part (a), $f(-4) = -39$ and $f(2) = -3$.
Absolute maximum value $= 13$
Absolute minimum value $= -39$.

3.3 Curve Sketching

1. $\lim_{x\to\infty} f(x) = +\infty$, so $x = 0$ is a vertical asymptote

$\lim_{x\to\pm\infty} f(x) = 0$, so $y = 0$ is a horizontal asymptote

3. There are no vertical asymptotes.

$\lim_{x\to-\infty} f(x) = 0$, so $y = 0$ is a horizontal asymptote

5. $\lim_{x\to -2} f(x) = +\infty$, so $x = -2$ is a vertical asymptote

$\lim_{x\to 2^-} f(x) = -\infty$ and $\lim_{x\to 2^+} f(x) = +\infty$,

so $x = 2$ is a vertical asymptote

$\lim_{x\to -\infty} f(x) = 0$, so $y = 0$ is a horizontal asymptote

$\lim_{x\to +\infty} f(x) = 2$, so $y = 2$ is a horizontal asymptote.

7. $\lim_{x\to 2^+} f(x) = +\infty$, so $x = 2$ is a vertical asymptote

$\lim_{x\to -\pm\infty} f(x) = 0$, so $y = 0$ is a horizontal asymptote

9. Since the denominator is zero when $x = -2$, $x = -2$ is a vertical asymptote.

$$\lim_{x\to\pm\infty} \frac{3x-1}{x+2} = \lim_{x\to\pm\infty} \frac{3 - \frac{1}{x}}{1 + \frac{2}{x}} = 3,$$

so $y = 3$ is a horizontal asymptote.

11. Since the denominator cannot be zero for any value of x, there are no vertical asymptotes.

$$\lim_{x\to\pm\infty} \frac{x^2+2}{x^2+1} = \lim_{x\to\pm\infty} \frac{1 + \frac{2}{x^2}}{1 + \frac{1}{x^2}} = 1,$$

so $y = 1$ is a horizontal asymptote.

13. $$f(t) = \frac{t^2+3t-5}{t^2-5t+6} = \frac{t^2+3t-5}{(t-2)(t-3)}$$

Since the denominator is zero when $t = 2, 3$, the vertical asymptotes are $t = 2$ and $t = 3$.

$$\lim_{x\to\pm\infty} \frac{t^2+3t-5}{t^2-5t+6} = \lim_{x\to\pm\infty} \frac{1 + \frac{3}{t} - \frac{5}{t^2}}{1 - \frac{5}{t} + \frac{6}{t^2}} = 1,$$

so $y = 1$ is a horizontal asymptote.

15. $$h(x) = \frac{1}{x} - \frac{1}{x-1} = \frac{-1}{x(x-1)} = \frac{-1}{x^2-x}$$

Since the denominator is zero when $x = 0, 1$, the vertical asymptotes are $x = 0$ and $x = 1$.

$$\lim_{x\to\pm\infty} \frac{-1}{x^2-x} = 0,$$

so $y = 0$ is a horizontal asymptote.

17. $f(x) = x^3 + 3x^2 - 2$
domain: all real numbers
intercepts
when $x = 0$, $f(0) = -2$; point $(0, -2)$
$f(x) = 0$ is too difficult to solve
asymptotes: no vertical or horizontal asymptotes

$$f'(x) = 3x^2 + 6x = 3x(x+2)$$
$$f'(x) = 0 \text{ when } x = -2, 0$$
$$f''(x) = 6x + 6 = 6(x+1)$$
$$f''(x) = 0 \text{ when } x = -1$$

When $x < -2$, $f'(x) > 0$ so f is increasing
$f''(x) < 0$ so f is concave down

$-2 < x < -1$, $f'(x) < 0$ so f is decreasing
$f''(x) < 0$ so f is concave down

$-1 < x < 0$, $f'(x) < 0$ so f is decreasing
$f''(x) > 0$ so f is concave up

$x > 0$, $f'(x) > 0$ so f is increasing
$f''(x) > 0$ so f is concave up.

$(-2, 2)$ is a relative maximum, $(-1, 0)$ is an inflection point, and $(0, -2)$ is a relative minimum.

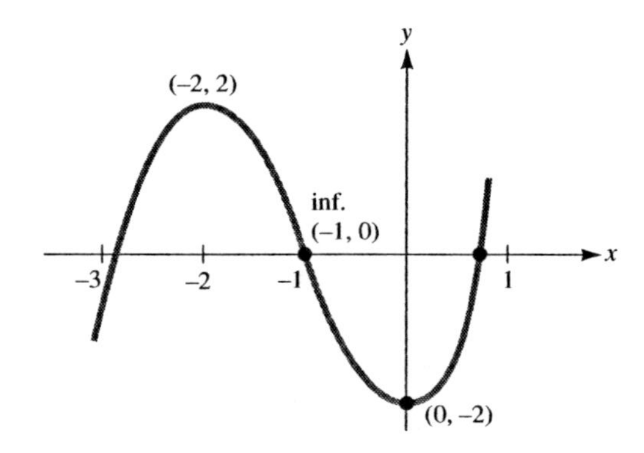

19. $f(x) = x^4 + 4x^3 + 4x^2 = x^2(x+2)^2$
domain: all real numbers

intercepts:

when $x = 0$, $f(0) = 0$; point (0, 0)

$f(x) = 0$, $x = 0, -2$; point $(-2, 0)$

asymptotes: no vertical or horizontal asymptotes.

$$f'(x) = 4x^3 + 12x^2 + 8x = 4x(x+1)(x+2)$$

$$f'(x) = 0 \text{ when } x = -2, -1, 0$$

$$f''(x) = 12x^2 + 24x + 8 = 4(3x^2 + 6x + 2)$$

$$f''(x) = 0 \text{ when } x = -1.6, -0.4$$

When $x < -2$, $f'(x) < 0$ so f is decreasing
$f''(x) > 0$ so f is concave up

$-2 < x < -1.6$, $f'(x) > 0$ so f is increasing
$f''(x) > 0$ so f is concave up

$-1.6 < x < -1$, $f'(x) > 0$ so f is increasing
$f''(x) < 0$ so f is concave down

$-1 < x < -0.4$, $f'(x) < 0$ so f is decreasing
$f''(x) < 0$ so f is concave down

$-0.4 < x < 0$, $f'(x) < 0$ so f is decreasing
$f''(x) > 0$ so f is concave up

$x > 0$, $f'(x) > 0$ so f is increasing
$f''(x) > 0$ so f is concave up.

$(-2, 0)$ is a relative minimum, $(-1.6, 0.4)$ is an inflection point, $(-1, 1)$ is a relative maximum, $(-0.4, 0.4)$ is an inflection point, and $(0, 0)$ is a relative minimum.

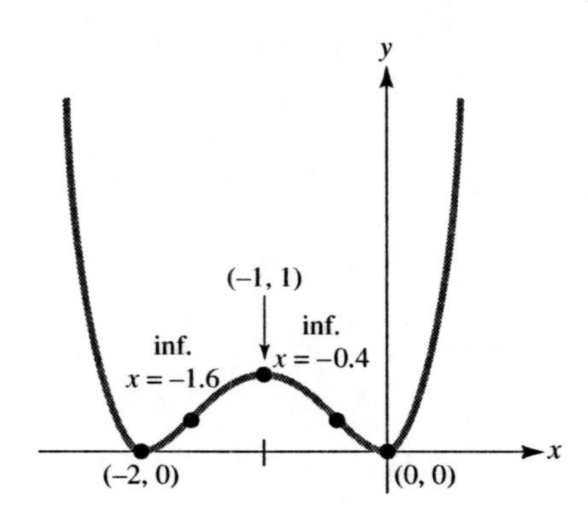

21. $f(x) = (2x-1)^2(x^2-9)$
$= (2x-1)^2(x+3)(x-3)$

domain: all real numbers

intercepts:

when $x = 0$, $f(0) = -9$; point $(0, -9)$

$f(x) = 0$, $x = \frac{1}{2}, -3, 3$; points

$\left(\frac{1}{2}, 0\right)$, $(-3, 0)$, $(3, 0)$

asymptotes: no vertical or horizontal asymptotes.

$$f'(x) = (2x-1)^2(2x) + (x^2-9)[2(2x-1)(2)]$$
$$= 2(2x-1)(4x^2 - x - 18)$$
$$= 2(2x-1)(4x-9)(x+2)$$

$$f'(x) = 0 \text{ when } x = -2, \frac{1}{2}, \frac{9}{4}$$

$$f''(x) = \left[(2x-1)(8x-1) + (4x^2 - x - 18)(2)\right]$$
$$= 2(24x^2 - 12x - 35)$$

$$f''(x) = 0 \text{ when } x = -0.98, 1.5$$

When $x < -2$, $f'(x) < 0$ so f is decreasing
$f''(x) > 0$ so f is concave up

$-2 < x < -0.98$, $f'(x) > 0$ so f is increasing
$f''(x) > 0$ so f is concave up

$-0.98 < x < 0.5$, $f'(x) > 0$ so f is increasing
$f''(x) < 0$ so f is concave down

$0.5 < x < 1.5$, $f'(x) < 0$ so f is decreasing
$f''(x) < 0$ so f is concave down

$1.5 < x < 2.25$, $f'(x) < 0$ so f is decreasing
$f''(x) > 0$ so f is concave up

$x > 2.25$, $f'(x) > 0$ so f is increasing
$f''(x) > 0$ so f is concave up.

$(-2, -125)$ is a relative minimum, $(-0.98, -70.4)$ is an inflection point, $(0.5, 0)$ is a relative maximum, $(1.5, -26.2)$ is an inflection point, and $(2.25, -48.2)$ is a relative minimum.

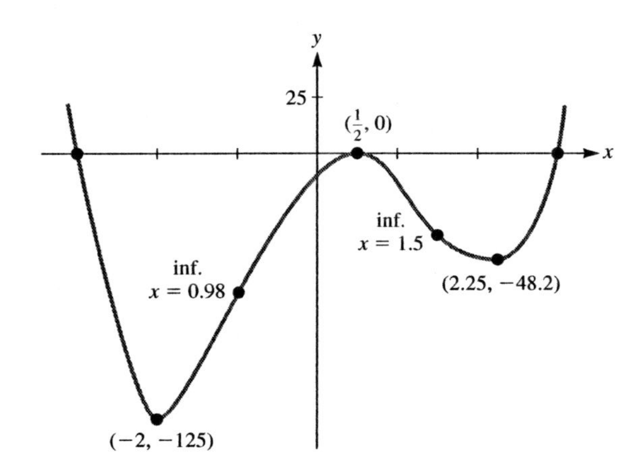

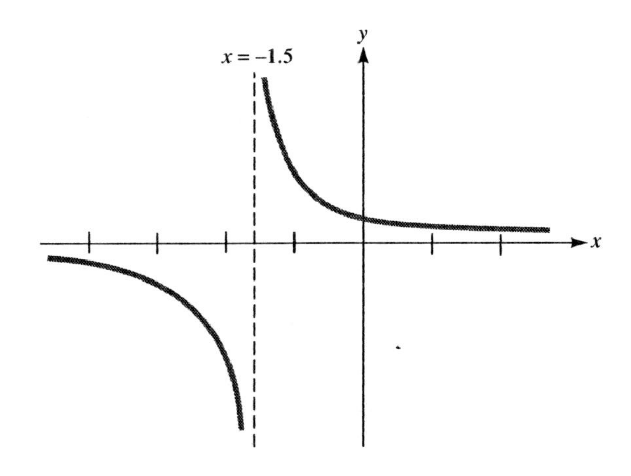

23. $f(x) = \dfrac{1}{2x+3}$

domain: $x \neq -\frac{3}{2}$

intercepts:

when $x = 0$, $f(0) = \frac{1}{3}$; point $\left(0, \frac{1}{3}\right)$

$f(x) \neq 0$ for any value of x

asymptotes: $x = -\frac{3}{2}$ is a vertical asymptote

$y = 0$ is a horizontal asymptote

$$f'(x) = -(2x+3)^{-2}(2) = \frac{-2}{(2x+3)^2}$$

note that $f'(x) < 0$ for all x in domain

$$f''(x) = -2\left[-2(2x+3)^{-3}(2)\right] = \frac{8}{(2x+3)^3}$$

When $x < -1.5$, $f'(x) < 0$ so f is decreasing

$f''(x) < 0$ so f is concave down

$x > -1.5$, $f'(x) < 0$ so f is decreasing

$f''(x) > 0$ so f is concave up.

f is undefined for $x = -1.5$, so there are no relative extrema or inflection points.

25. $f(x) = x - \dfrac{1}{x} = \dfrac{x^2-1}{x} = \dfrac{(x+1)(x-1)}{x}$

domain: $x \neq 0$

intercepts:

when $x = 0$, $f(0)$ undefined

$f(x) = 0$, $x = -1, 1$; points $(-1, 0)$, $(1, 0)$

asymptotes $x = 0$ is a vertical asymptote

no horizontal asymptote

(Note: $\displaystyle\lim_{x\to\pm\infty} \frac{x^2-1}{x} = \lim_{x\to\pm\infty} \frac{x - \frac{1}{x}}{1} = x$,

so $y = x$ is an oblique asymptote)

$$f'(x) = 1 + \frac{1}{x^2} = \frac{x^2+1}{x}$$

$$f''(x) = -\frac{2}{x^3}$$

When $x < 0$, $f'(x) > 0$ so f is increasing

$f''(x) > 0$ so f is concave up

$x > 0$, $f'(x) > 0$ so f is increasing

$f''(x) < 0$ so f is concave down.

f is undefined for $x = 0$, so there are no relative extrema or inflection points.

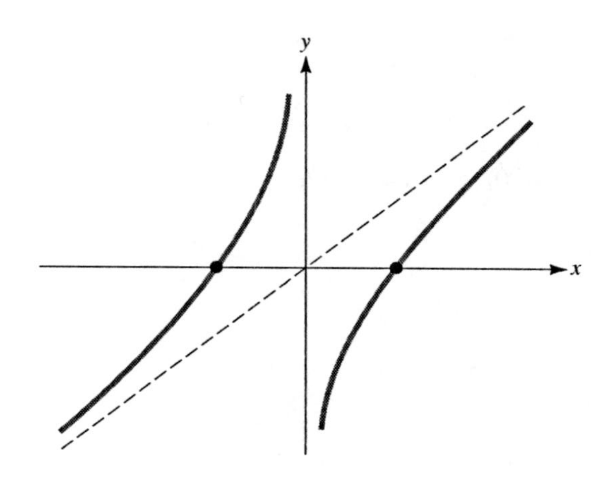

27. $$f(x) = \frac{1}{x^2 - 9} = \frac{1}{(x+3)(x-3)}$$

domain: $x \neq -3, 3$
intercepts:
when $x = 0$, $f(0) = -\frac{1}{9}$; point $\left(0, -\frac{1}{9}\right)$
$f(x) \neq 0$ for any value of x
asymptotes: $x = -3$ and $x = 3$ are vertical asymptotes
$y = 0$ is a horizontal asymptote

$$f'(x) = -(x^2 - 9)^2(2x) = \frac{-2x}{(x^2 - 9)^2}$$

$f'(x) = 0$ when $x = 0$

$$f''(x) = \frac{(x^2 - 9)^2(-2) - (-2x)(2(x^2 - 9)(2x))}{(x^2 - 9)^4}$$

$$= \frac{6(x^2 + 3)}{(x^2 - 9)^3}$$

When $x < -3$, $f'(x) > 0$ so f is increasing
$f''(x) > 0$ so f is concave up
$-3 < x < 0$, $f'(x) > 0$ so f is increasing
$f''(x) < 0$ so f is concave down
$0 < x < 3$, $f'(x) < 0$ so f is decreasing
$f''(x) < 0$ so f is concave down
$x > 3$, $f'(x) < 0$ so f is decreasing
$f''(x) > 0$ so f is concave up.

$\left(0, -\frac{1}{9}\right)$ is a relative maximum. Since f is undefined for $x = -3, 3$, there are no other relative extrema or inflection points.

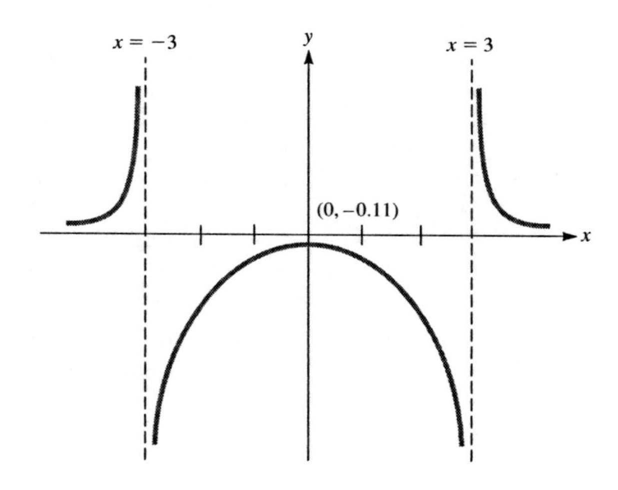

29. $$f(x) = \frac{x^2 - 9}{x^2 + 1} = \frac{(x+3)(x-3)}{x^2 + 1}$$

domain: all real numbers
intercepts:
when $x = 0$, $f(0) = -9$; point $(0, -9)$
$f(x) = 0$, $x = -3, 3$; points $(-3, 0)$, $(3, 0)$
asymptotes: no vertical asymptotes
$y = 1$ is a horizontal asymptote

$$f'(x) = \frac{(x^2 + 1)(2x) - (x^2 - 9)(2x)}{(x^2 + 1)^2} = \frac{20x}{(x^2 + 1)^2}$$

$f'(x) = 0$ when $x = 0$

$$f''(x) = \frac{(x^2 + 1)^2(20) - (20x)[2(x^2 + 1)(2x)]}{(x^2 + 1)^4}$$

$$= \frac{20(-3x^2 + 1)}{(x^2 + 1)^3}$$

$f''(x) = 0$ when $x = -\frac{1}{\sqrt{3}}, \frac{1}{\sqrt{3}}$

When $x < -\frac{1}{\sqrt{3}}$, $f'(x) < 0$ so f is decreasing

$f''(x) < 0$ so f is concave down

$-\frac{1}{\sqrt{3}} < x < 0$, $f'(x) < 0$ so f is decreasing

$f''(x) > 0$ so f is concave up

$0 < x < \frac{1}{\sqrt{3}}$, $f'(x) > 0$ so f is increasing

$f''(x) > 0$ so f is concave up

$x > \frac{1}{\sqrt{3}}$, $f'(x) > 0$ so f is increasing

$f''(x) < 0$ so f is concave down.

$(-0.58, -6.48)$ is an inflection point, $(0, -9)$ is a relative minimum, and $(0.58, -6.48)$ is an inflection point.

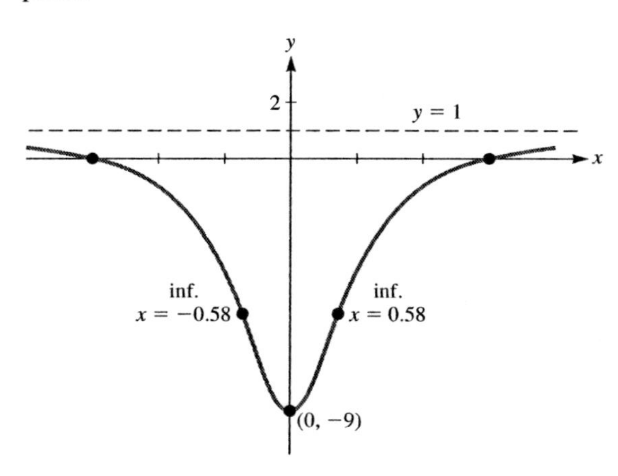

31. $f(x) = x^{3/2} = \sqrt{x^3}$
domain: $x \geq 0$
intercepts:
when $x = 0$, $f(0) = 0$; point $(0, 0)$
$f(x) = 0$, $x = 0$
asymptotes: no vertical or horizontal asymptotes

$$f'(x) = \frac{3}{2}x^{1/2} = \frac{3}{2}\sqrt{x}$$

$$f'(x) = 0 \text{ when } x = 0$$

$$f''(x) = \frac{3}{4}x^{-1/2} = \frac{3}{4\sqrt{x}}$$

When $x < 0$, f is undefined

$x > 0$, $f'(x) > 0$ so f is increasing

$f''(x) > 0$ so f is concave up.

$(0, 0)$ is a relative minimum.

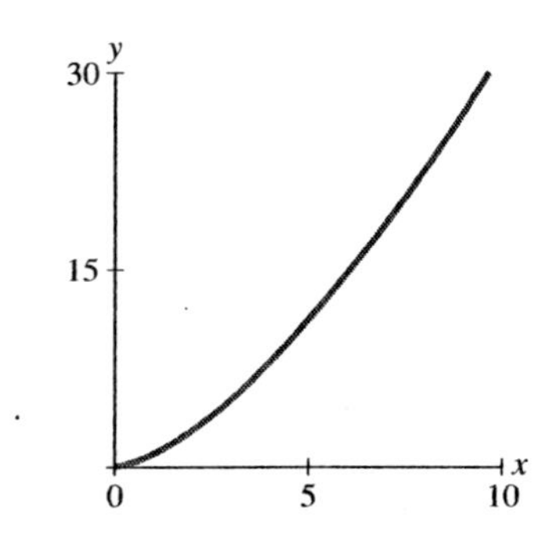

33. Answers will vary.

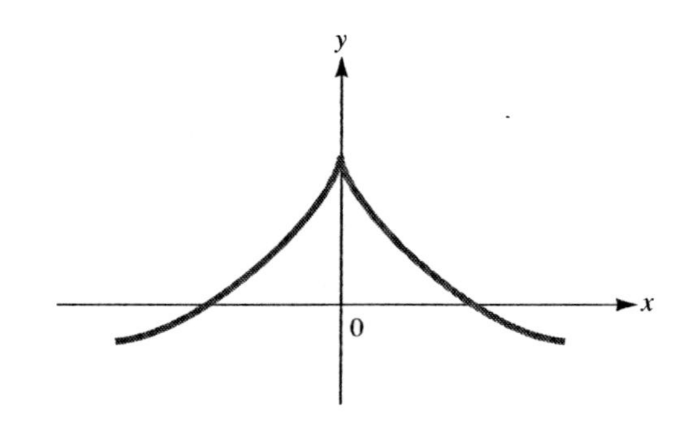

35. Answers will vary.

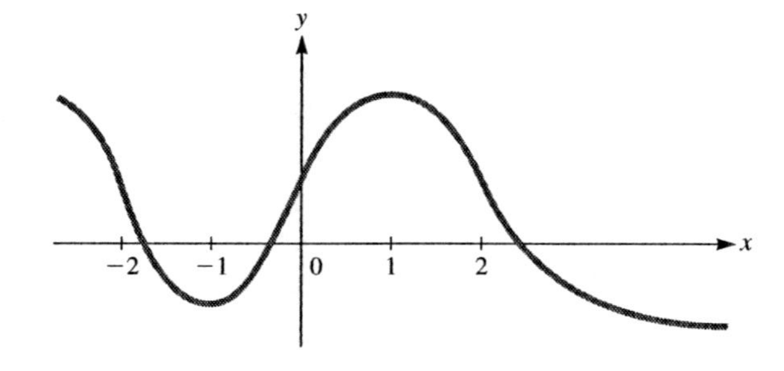

37. Answers will vary.

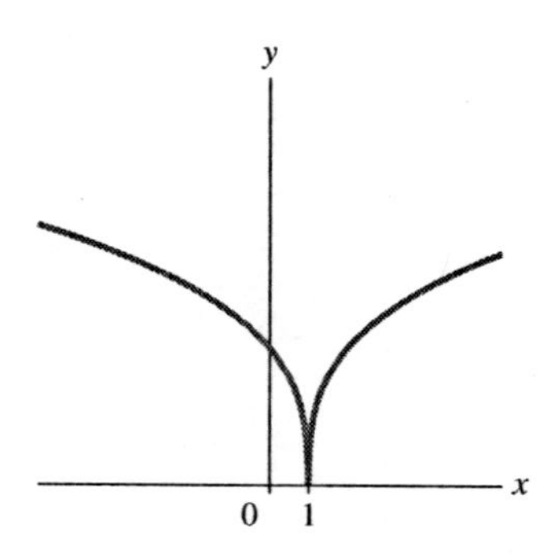

39. $f'(x) = x^3(x-2)^2$

(a) $f'(x) = 0$ when $x = 0, 2$

When $x < 0$, $f'(x) < 0$ so f is decreasing
$0 < x < 2$, $f'(x) > 0$ so f is increasing
$x > 2$, $f'(x) > 0$ so f is increasing.

(b) At $x = 0$, there is a relative minimum but there is no relative extrema at $x = 2$.

(c)
$$f''(x) = (x^3)[2(x-2)(1)] + (x-2)^2(3x^2) = x^2(x-2)(5x-6)$$

$f''(x) = 0$ when $x = 0, \frac{6}{5}, 2$

When $x < 0$, $f''(x) > 0$ so f is concave up
$0 < x < 1.2$, $f''(x) > 0$ so f is concave up
$1.2 < x < 2$, $f''(x) < 0$ so f is concave down
$x > 2$, $f''(x) > 0$ so f is concave up.

Overall, f is concave up when $x < 0$, $0 < x < 1.2$, and when $x > 2$; f is concave down when $1.2 < x < 2$.

(d) At $x = 1.2$ and $x = 2$, there are inflection points.

41. $f'(x) = \dfrac{x+3}{(x-2)^2}$

(a) $f'(x) = 0$ when $x = -3$
$f'(x)$ is undefined when $x = 2$

When $x < -3$, $f'(x) < 0$ so f is decreasing
$-3 < x < 2$, $f'(x) > 0$ so f is increasing
$x > 2$, $f'(x) > 0$ so f is increasing.

(b) When $x = -3$, there is a relative minimum but there is no relative extrema at $x = 2$.

(c)
$$f''(x) = \frac{(x-2)^2(1) - (x+3)\cdot 2(x-2)(1)}{(x-2)^4} = \frac{(x-2)^2\,[(x-2) - 2(x+3)]}{(x-2)^4} = \frac{-x-8}{(x-2)^3} = \frac{-(x+8)}{(x-2)^3}$$

$f''(x) = 0$ when $x = -8$
$f''(x)$ is undefined when $x = 2$

When $x < -8$, $f''(x) < 0$ so f is concave down
$-8 < x < 2$, $f''(x) > 0$ so f is concave up
$x > 2$, $f''(x) < 0$ so f is concave down.

(d) Since the concavity switches when $x = -8$, there is an inflection point when $x = -8$. The concavity switches when $x = 2$ as well, however f is undefined when $x = 2$.

43. To have a vertical asymptote of $x = 2$, the denominator must be zero for $x = 2$, so

$$5 + B(2) = 0 \rightarrow B = -\frac{5}{2}$$

To have a horizontal asymptote of $y = 4$

$$\lim_{x\to\pm\infty} \frac{Ax-3}{5-\frac{5}{2}x} = \lim_{x\to\pm\infty} \frac{A-\frac{3}{x}}{\frac{5}{x}-\frac{5}{2}} = \frac{A}{-\frac{5}{2}} = -\frac{2}{5}A = 4,$$

so $A = -10$.

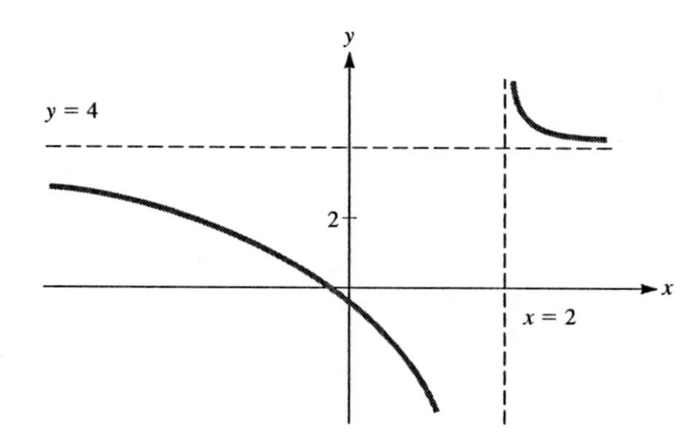

45. Answers will vary.

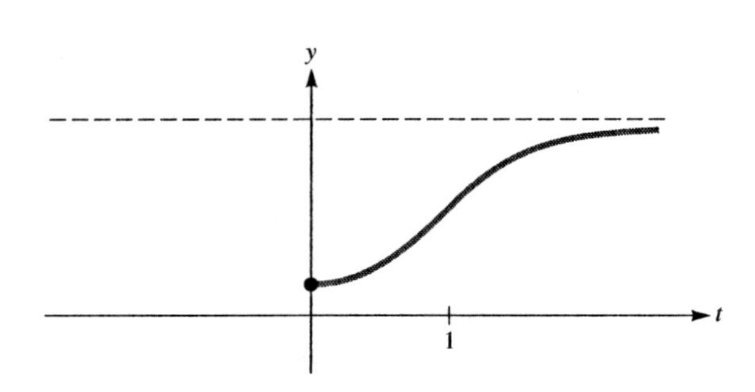

47. Answers will vary.

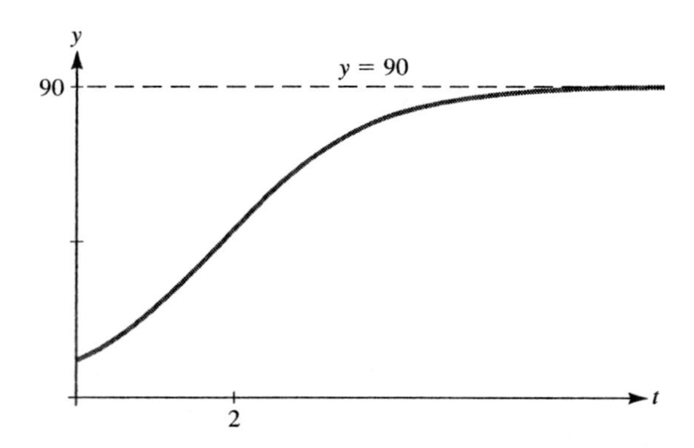

49.

$$C(x) = 3x^2 + x + 48$$

$$A(x) = 3x + 1 + \frac{48}{x} = \frac{3x^2 + x + 48}{x}$$

(a) $x = 0$ is a vertical asymptote; there are no horizontal asymptotes.

(b) As $x \to \infty$, the graph of A approaches the line $y = 3x + 1$ asymptotically.

(c)

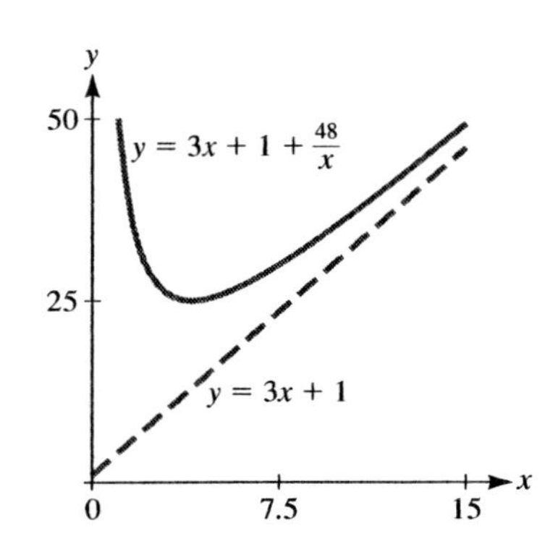

51. $W(x) = \dfrac{200x}{100 - x}$

(a) domain: $0 \le x < 100$
intercepts:
when $x = 0$, $W(0) = 0$; point $(0, 0)$
$W(x) = 0$, $x = 0$
asymptotes: $x = 100$ is a vertical asymptote
since $x \ge 0$, the is no horizontal asymptote

$$W'(x) = \frac{20{,}000}{(100 - x)^2}$$

$$W''(x) = \frac{40{,}000}{(100 - x)^3}$$

When $0 \le x < 100$, $W'(x) > 0$ so W is increasing
$W''(x) > 0$ so W is concave up.

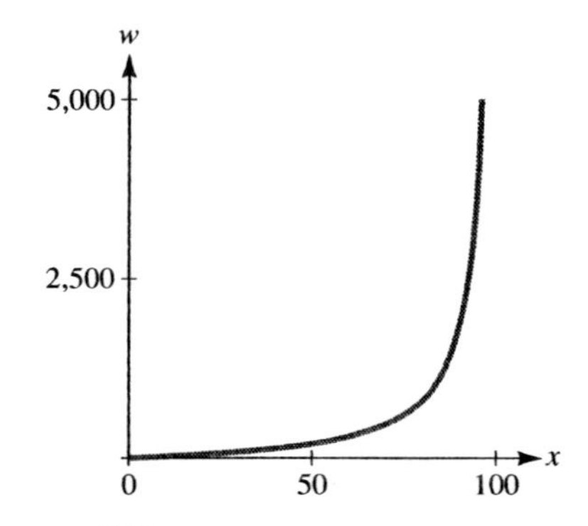

(b) $1500 = \dfrac{200x}{100 - x}$; $150{,}000 - 1500x = 200x$;
$x \approx 88.2\%$ will receive a new book, so
$100 - 88.2 = 11.8\%$ will not receive a new book.

53. $Q(x) = \dfrac{7x}{27 + x^2}$

(a) domain: $x \geq 0$
intercepts:
when $x = 0$, $Q(0) = 0$; point (0, 0)
$Q(x) = 0$, $x = 0$
asymptotes: the denominator is never zero, so there are no vertical asymptotes

$$\lim_{x\to\infty} Q(x) = \lim_{x\to\infty} \frac{\frac{7}{x}}{\frac{27}{x^2} + 1} = 0$$

so $y = 0$ is a horizontal asymptote

$$Q'(x) = \frac{(27 + x^2)(7) - (7x)(2x)}{(27 + x^2)^2} = \frac{189 - 7x^2}{(27 + x^2)^2}$$

$$Q''(x) = \frac{(27 + x^2)^2(-14x) - (189 - 7x^2) \cdot 2(27 + x^2)(2x)}{(27 + x^2)^4}$$

$$= \frac{2(27 + x^2)\left[-7x(27 + x^2) - 2x(189 - 7x^2)\right]}{(27 + x^2)^4}$$

$$= \frac{2(-567x + 7x^3)}{(27 + x^2)^3}$$

$Q'(x) = 0$ when $x = \sqrt{27} \approx 5.2$
$Q''(x) = 0$ when $x = 0,9$

When $0 < x < 5.2$, $Q'(x) > 0$ so Q is increasing
$Q''(x) < 0$ so Q is concave down.
$5.2 < x < 9$, $Q'(x) < 0$ so Q is decreasing
$Q''(x) < 0$ so Q is concave down.
$x > 9$, $Q'(x) < 0$ so Q is decreasing
$Q''(x) > 0$ so Q is concave up.

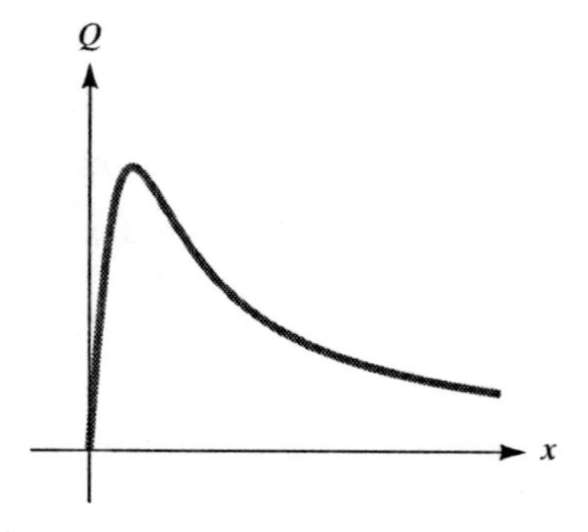

(b) Sales are maximized when $x = \sqrt{27}$, or a marketing expenditure of \$5,196. The corresponding maximum sales is
$Q(5.196) = \dfrac{7(5.196)}{27 + 27} \approx 0.6736$
or approximately 674 units.

55. $S(t) = \dfrac{100(t^2 - 3t + 25)}{t^2 + 7t + 25}$

(a) domain: $0 \leq t \leq 10$
intercepts:
when $t = 0$, $S(0) = 100$; point (0, 100)
$f(x) \neq 0$ for any x
asymptotes: there are no vertical asymptotes
$y = 100$ is a horizontal asymptote

$$S'(t) = \frac{100}{(t^2 + 7t + 25)^2}\Big((t^2 + 7t + 25)(2t - 3) - (t^2 - 3t + 25)(2t + 7)\Big)$$

$$= \frac{1000(t + 5)(t - 5)}{(t^2 + 7t + 25)^2} = \frac{1000(t^2 - 25)}{(t^2 + 7t + 25)^2}$$

$S'(t) = 0$ when $t = 5$

When $0 \leq t < 5$, $S'(t) < 0$ so S is decreasing
$5 < t \leq 10$, $S'(t) > 0$ so S is increasing.

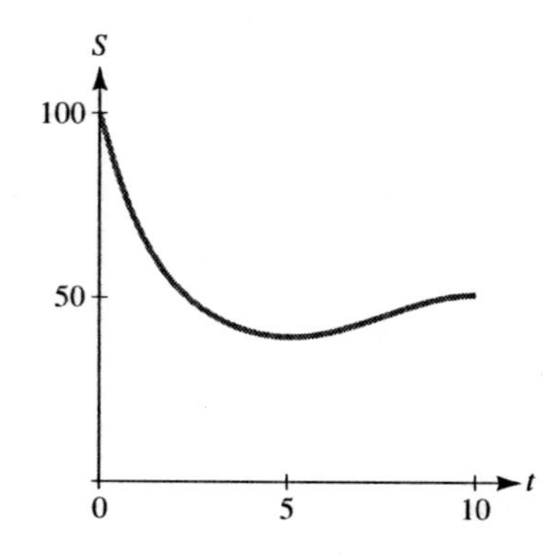

(b) When $t = 5$, there is a relative minimum, so her support is lowest when $t = 5$ and her minimum support level is $S(5) \approx 41.2\%$.

(c) When $t > 5$, $S'(t) > 0$ so $S'(10)$ is positive.

$$S''(t) = \frac{1000}{(t^2+7t+25)^4}\left[(t^2+7t+25)^2(2t) - (t^2-25)(2(t^2+7t+25)(2t+7))\right]$$

When $t = 10$, $S''(10) < 0$ so S', or her approval rate, is decreasing.

57. $G(x) = \dfrac{1}{2{,}000}\left(\dfrac{800}{x} + 5x\right)$

(a)

$$\begin{aligned}
\text{total cost} &= \text{cost driver} + \text{cost gas} \\
\text{cost driver} &= (\text{\#hrs})(\text{pay/hr}) \\
&= \left(\frac{\text{\#mi}}{\text{mi/hr}}\right)(\text{pay/hr}) \\
\text{cost gas} &= (\text{\#mi})(\text{gal/mi})(\text{cost/gal}) \\
C(x) &= \left(\frac{400}{x}\right)(18) \\
&\quad + (400)\left[\frac{1}{2{,}000}\left(\frac{800}{x} + 5x\right)\right](2.95) \\
&= \frac{7{,}672}{x} + 2.95x
\end{aligned}$$

domain: $30 \le x \le 65$
intercepts: none in domain
asymptotes: none in domain

$$C'(x) = -\frac{-7{,}672}{x^2} + 2.95$$

$C'(x) = 0$ when $x \approx 51$

When $30 \le x < 51$, $C'(x) < 0$ so C is decreasing
$51 < x \le 65$ $C'(t) > 0$ so C is increasing.

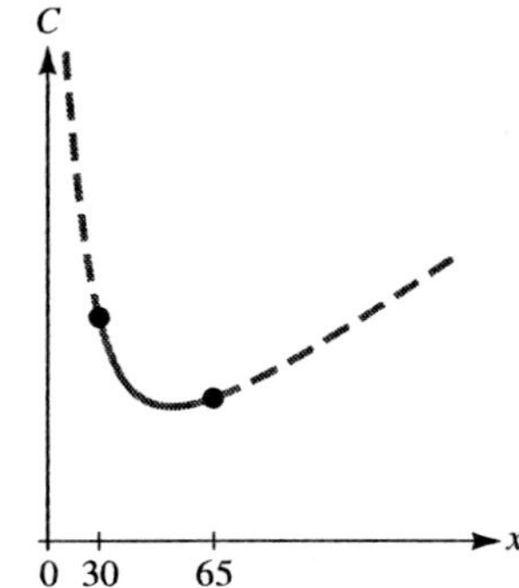

(b) When $x = 51$, there is a relative minimum, so cost is minimized when driver travels at 51 mph. The minimum cost is $C(51) \approx \$300.88$.

59. $f(x) = x^{2/3}(2x - 5)$

(a)

$$\begin{aligned}
f'(x) &= (x^{2/3})(2) + (2x-5)\left(\frac{2}{3}x^{-1/3}\right) \\
&= 2x^{2/3} + \frac{2(2x-5)}{3x^{1/3}} = \frac{10(x-1)}{3x^{1/3}} \\
&= \frac{10}{3}x^{2/3} - \frac{10}{3}x^{-1/3}
\end{aligned}$$

$f'(x) = 0$ when $x = 1$

When $x < 0$, $f'(x) > 0$ so f is increasing
$0 < x < 1$, $f'(x) < 0$ so f is decreasing
$x > 1$, $f'(x) > 0$ so f is increasing.

$(0, 0)$ is a relative maximum and $(1, -3)$ is a relative minimum.
Since f is defined but f' is undefined for $x = 0$, there is a vertical tangent at $x = 0$.

(b)

$$\begin{aligned}
f''(x) &= \frac{20}{9}x^{-1/3} + \frac{10}{9}x^{-4/3} \\
&= \frac{10(2x+1)}{9x^{4/3}}
\end{aligned}$$

$$f''(x) = 0 \text{ when } x = -\frac{1}{2}$$

When
$x < -0.5$, $f''(x) < 0$ so f is concave down
$-0.5 < x < 0$, $f''(x) > 0$ so f is concave up
$x > 0$, $f''(x) > 0$ so f is concave up.
$(-0.5, -3.8)$ is an inflection point.

(c) When $x = 0$, $f(0) = 0$, so y-intercept is 0. When $f(x) = 0$, $x = 0, 2.5$; so, x-intercepts are 0 and 2.5. There are no vertical or horizontal asymptotes.

(d)

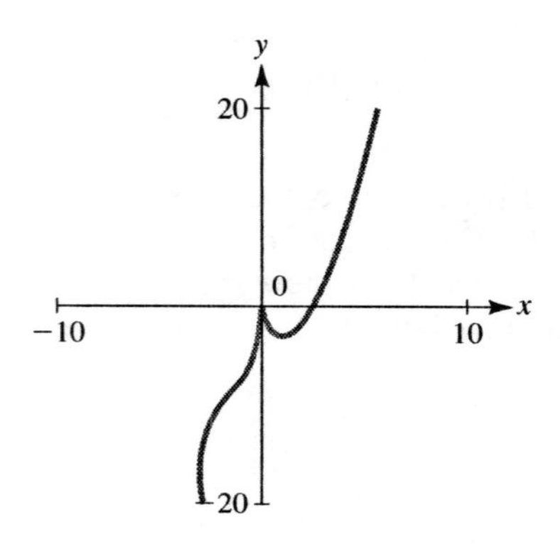

61. Let $f(x) = \dfrac{x-1}{x^2-1}$ and let $g(x) = \dfrac{x-1.01}{x^2-1}$.

(a) To use a graphing utility to sketch the graph of f,
Press [y=] and input $(x-1)/(x \wedge 2-1)$ and press [Graph].
At first appearance, the graph appears to be continuous at $x = 1$.
Use [2nd] [calc] and 1: value to evaluate $f(1)$. We see no y-value is displayed for $x = 1$ which means $f(1)$ is undefined. From algebra, $f(x) = \dfrac{x-1}{(x+1)(x-1)}$. We can cancel the common factor $x-1$, which leaves a "hole" in the graph of f at $x = 1$.

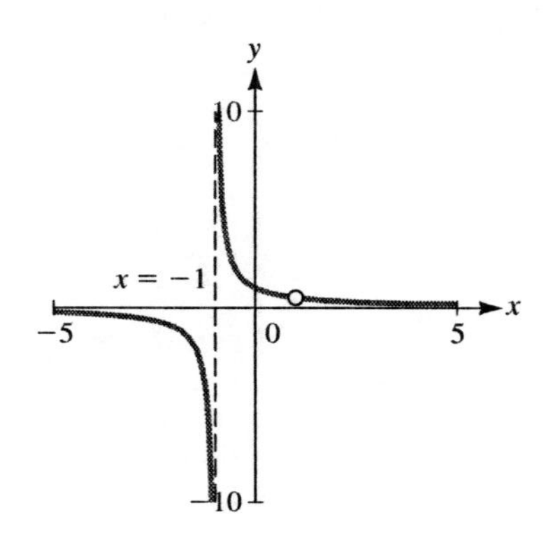

(b) To sketch a graph of g,
Press [y=] and input $(x-1.01)/(x \wedge 2-1)$ and press [Graph].

The graph of g appears to be the same as the graph for f. However, by tracing and zooming in at $x = 1$, we see the vertical asymptote appears at $x = 1$. In addition, using [2nd] [calc] and 1: value to evaluate $g(1)$ also produces an undefined y-value. The reason for this is not due to a "hole" in the graph for g but rather the vertical asymptote $x = 1$.

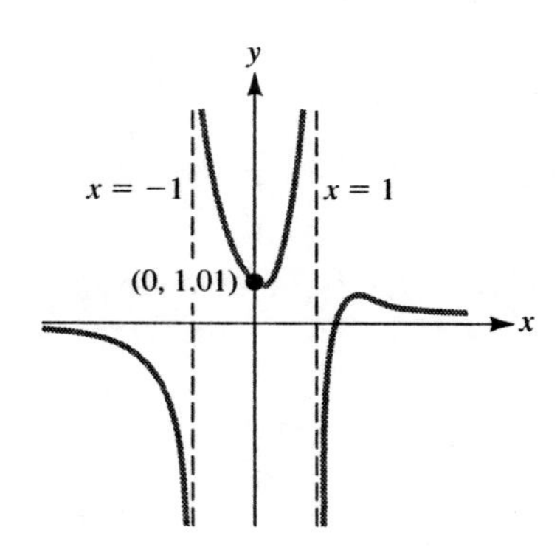

3.4 Optimization

1.
$$f(x) = x^2 + 4x + 5, \ -3 \le x \le 1$$
$$f'(x) = 2x + 4 = 2(x+2)$$

$f'(x) = 0$ when $x = -2$, which is in the interval

$$f(-2) = 1, \ f(-3) = 2 \text{ and } f(1) = 10$$

So, $f(1) = 10$ is the absolute maximum and $f(-2) = 1$ is the absolute minimum.

3.
$$f(x) = \frac{1}{3}x^3 - 9x + 2, \ 0 \le x \le 2$$
$$f'(x) = x^2 - 9 = (x+3)(x-3)$$

$f'(x) = 0$ when $x = -3$ and $x = 3$, which are not in the interval.
$f(0) = 2$, which is the absolute maximum and $f(2) = -\dfrac{40}{3}$, which is the absolute minimum.

5.
$$f(t) = 3t^5 - 5t^3, \ -2 \le t \le 0$$
$$f'(t) = 15t^4 - 15t^2 = 15t^2(t+1)(t-1)$$

$f'(t) = 0$ when $t = -1$, $t = 0$ and $t = 1$, of which $t = -1$ and $t = 0$ are in the interval.
$f(-1) = 2$, $f(0) = 0$, $f(-2) = -56$
So, $f(-1) = 2$ is the absolute maximum and $f(-2) = -56$ is the absolute minimum.

7. $f(x) = (x^2 - 4)^5$, $-3 \le x \le 2$

$$f'(x) = 5(x^2 - 4)^4(2x) = 10x(x + 2)^4(x - 2)^4$$

$f'(x) = 0$ when $x = -2$, $x = 0$, and $x = 2$, all of which are in the interval.

$$f(-2) = 0,\ f(0) = -1{,}024,\ f(2) = 0$$
$$\text{and } f(-3) = 3{,}125$$

So, $f(-3) = 3{,}125$ is the absolute maximum and $f(0) = -1{,}024$ is the absolute minimum.

9. $g(x) = x + \frac{1}{x}$, $\frac{1}{2} \le x \le 3$

$$g'(x) = 1 - \frac{1}{x^2} = \frac{x^2 - 1}{x^2} = \frac{(x + 1)(x - 1)}{x^2}$$

$g'(x) = 0$ when $x = -1$ and $x = 1$, of which $x = 1$ is in the interval.
$g'(x)$ is undefined at $x = 0$, however, $x = 0$ is not in the interval

$$g(1) = 2,\ g\left(\frac{1}{2}\right) = \frac{5}{2},\ g(3) = \frac{10}{3}$$

So, $g(3) = \frac{10}{3}$ is the absolute maximum and $g(1) = 2$ is the absolute minimum.

11. $f(u) = u + \frac{1}{u}$, $u > 0$

$$f'(u) = 1 - \frac{1}{u^2} = \frac{u^2 - 1}{u^2} = \frac{(u + 1)(u - 1)}{u^2}$$

$f'(u) = 0$ when $u = -1$ and $u = 1$, of which $u = 1$ is in the interval.
$f'(u)$ is undefined when $u = 0$, which is not in the interval

When $0 < x < 1$, $f'(x) < 0$ so f is decreasing
$x > 1$, $f'(x) > 0$ so f is increasing.

Since there are no endpoints, $f(1) = 2$ is the absolute minimum and there is no absolute maximum.

13. $f(x) = \frac{1}{x}$, $x > 0$

$$f'(x) = -\frac{1}{x^2}$$

$f'(x)$ is never zero and $f'(x)$ is undefined when $x = 0$, which is not in the domain. Also, there are no endpoints. So, there is no absolute maximum or absolute minimum.

15. $f(x) = \frac{1}{x + 1}$, $x \ge 0$

$$f'(x) = -(x + 1)^{-2}(1) = -\frac{1}{(x + 1)^2}$$

$f'(x)$ is never zero and $f'(x)$ is undefined when $x = -1$, which is not in the domain.
When $x > 0$, $f'(x) < 0$ so f is decreasing. So, $f(0) = 1$ is the absolute maximum and there is no absolute minimum.

17. $p(q) = 49 - q$ and $C(q) = \frac{1}{8}q^2 + 4q + 200$

(a)
$$R(q) = qp(q) = 49q - q^2$$
$$R'(q) = 49 - 2q$$
$$C'(q) = \frac{1}{4}q + 4$$

The profit function is

$$P(q) = R(q) - C(q) = -\frac{9}{8}q^2 + 45q - 200$$
$$P'(q) = -\frac{9}{4}q + 45$$

$P'(q) = 0$ when $q = 20$, so profit is maximized when 20 units are produced.

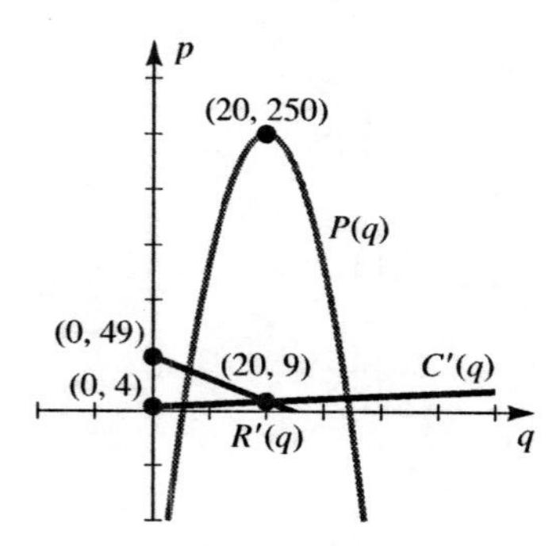

(b)

$$A(q) = \frac{C(q)}{q} = \frac{1}{8}q + 4 + \frac{200}{q}$$

$$A'(q) = \frac{1}{8} - \frac{200}{q^2}$$

$A'(q) = 0$ when $q = 40$, so the average cost is minimized when 40 units are produced.

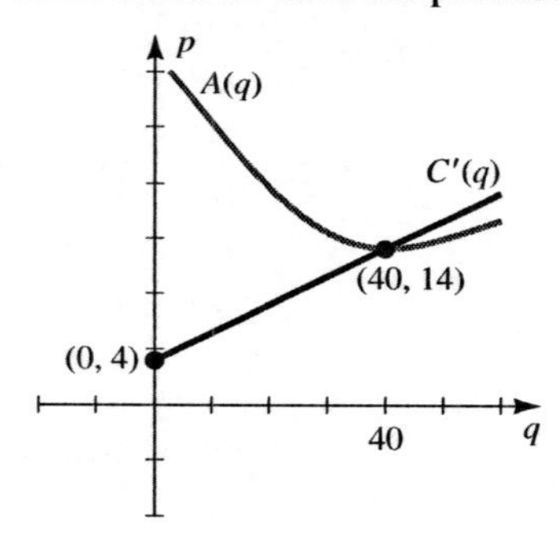

19. $p(q) = 180 - 2q$ and $C(q) = q^3 + 5q + 162$

(a)

$$R(q) = qp(q) = 180q - 2q^2$$
$$R'(q) = 180 - 4q$$
$$C'(q) = 3q^2 + 5$$

The profit function is $P(q) = R(q) - C(q)$

$$= -q^3 - 2q^2 + 175q - 162$$

$P'(q) = -3q^2 - 4q + 175$
$P'(q) = 0$ when $q = 7$ (rejecting negative solution), so profit is maximized when 7 units are produced.

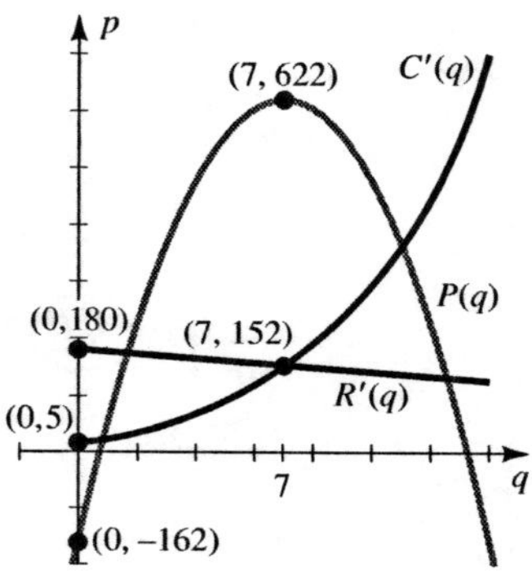

(b)

$$A(q) = \frac{C(q)}{q} = q^2 + 5 + \frac{162}{q}$$

$$A'(q) = 2q - \frac{162}{q^2}$$

$A'(q) = 0$ when $q \approx 4.327$, so the average cost is minimized when 4.327 units are produced.

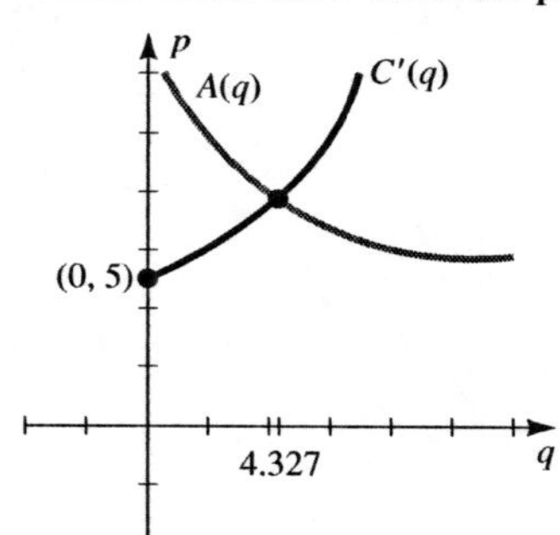

21. $p(q) = 1.0625 - 0.0025q$ and $C(q) = \dfrac{q^2 + 1}{q + 3}$

(a)

$$R(q) = qp(q) = 1.0625q - 0.0025q^2$$
$$R'(q) = 1.0625 - 0.005q$$
$$C'(q) = \frac{q^2 + 6q - 1}{(q + 3)^2}$$

The profit function is $P(q) = R(q) - C(q)$

$$= 1.0625q - 0.0025q^2 - \frac{q^2 + 1}{q + 3}$$

$$= \frac{1}{q + 3}[-0.0025q^3 + 0.055q^2 + 3.1875q - 1]$$

$$P'(q) = \frac{1}{(q+3)^2}\Big[(q+3)$$
$$(-0.0075q^2 + 0.11q + 3.1875)$$
$$+0.0025q^3 - 0.055q^2 - 3.1875q + 1\Big]$$
$$= \frac{1}{(q+3)^2}\Big[-0.005q^3$$
$$+0.0325q^2 + 0.33q + 10.5625\Big]$$

Press [y=] and input P, R', and C' for $y_1 =$, $y_2 =$, and $y_3 =$, respectively.
Use window dimensions [0, 45]5 by [0, 3]0.5
Press [graph]
Use the maximum function under the calc menu to find the relative maximum of P occurs at $x = 17.3361$.

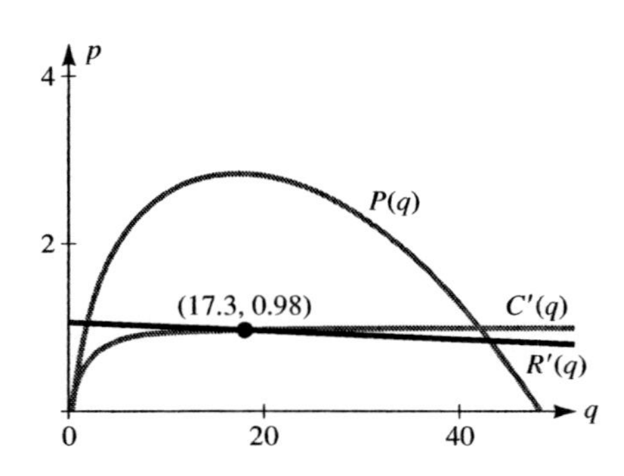

(b)

$$A(q) = \frac{C(q)}{q} = \frac{q^2+1}{q(q+3)}$$
$$A'(q) = \frac{1}{(q^2+3q)^2}\Big[2q(q^2+3q)$$
$$-(q^2+1)(2q+3)\Big]$$
$$= \frac{3q^2 - 2q - 3}{(q^2+3q)^2}$$

Press [y=] and input A and C' for $y_1 =$ and $y_2 =$, respectively. Use window dimensions of [0, 6] 0.5 by [0, 1.5] 0.2.
Press [graph]
Use the minimum function under the calc menu to find the relative minimum occurs at $q = 1.3874$.

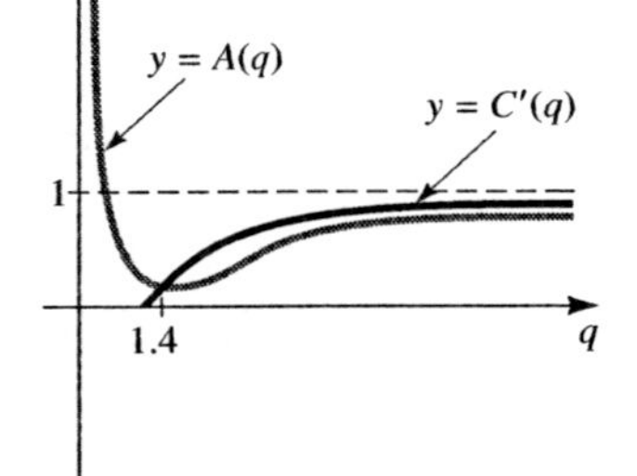

23.

$$D(p) = -1.3p + 10$$
$$E(p) = \frac{p}{D(p)}\frac{dD}{dp}$$
$$= \frac{p}{-1.3p+10}(-1.3)$$
$$E(4) = \frac{-1.3(4)}{-1.3(4)+10} = -\frac{13}{12}$$

$|E(4)| > 1$, so the demand is elastic.

25.

$$D(p) = 200 - p^2$$
$$E(p) = \frac{p}{D(p)}\frac{dD}{dp}$$
$$E(p) = \frac{p}{200-p^2}(-2p)$$
$$E(10) = \frac{-2(10)^2}{200-(10)^2} = -2$$

$|E(10)| > 1$, so the demand is elastic.

27.
$$D(p) = \frac{3{,}000}{p} - 100$$
$$E(p) = \frac{p}{D(p)}\frac{dD}{dp}$$
$$E(p) = \frac{p}{\frac{3{,}000}{p} - 100}\left(-\frac{3{,}000}{p^2}\right)$$
$$= \frac{p}{\frac{3{,}000 - 100p}{p}}\left(-\frac{3{,}000}{p^2}\right)$$
$$= \frac{p^2}{100(30-p)}\left(-\frac{3{,}000}{p^2}\right)$$
$$= -\frac{30}{30-p}$$
$$E(10) = -\frac{30}{30-10} = -\frac{3}{2}$$

$|E(10)| > 1$, so the demand is elastic.

29. (a) $R(q) = -2q^2 + 68q - 128$,
So average revenue per unit is

$$A(q) = \frac{R(q)}{q} = -2q + 68 - \frac{128}{q}$$

and marginal revenue is

$$R'(q) = -4q + 68$$

$R'(q) = A(q)$ when

$$-4q + 68 = -2q + 68 - \frac{128}{q}$$
$$2q = \frac{128}{q},\ q^2 = 64,\ \text{or } q = 8.$$

(b) $A'(q) = -2 + \dfrac{128}{q^2}$

$$A'(q) = 0 \text{ when } 0 = -2 + \frac{128}{q^2}$$
$$2 = \frac{128}{q2}$$
$$q^2 = 64,\ \text{or } q = -8 \text{ and } q = 8$$

Noting that the domain is $q \geq 0$,

when $0 < q < 8$, $A'(q) > 0$, so A is increasing;
when $q > 8$, $A'(q) < 0$, so A is decreasing.

(c)

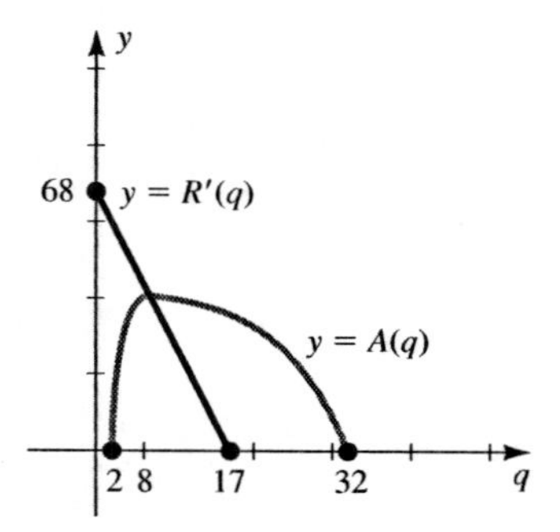

31. Need to find the maximum absolute value of the slope of the graph. The slope is

$$f'(x) = 4x - x^2$$

To maximize $|f'|$ on the interval $-1 \leq x \leq 4$,

$$f''(x) = 4 - 2x = 2(2-x)$$
$$f''(x) = 0 \text{ when } x = 2$$

Now,

$$|f'(2)| = |4| = 4$$
$$|f'(-1)| = |-5| = 5$$
$$|f'(4)| = |0| = 0$$

So, slope is steepest when $x = -1$, and its value is $f'(-1) = -5$.

33. (a) The membership of the association x years after 1992 is given by the function

$$f(x) = 100(2x^3 - 45x^2 + 264x)$$

The period of time from 1992 to 2006 corresponds to the interval $0 \leq x \leq 14$.

$$f'(x) = 100(6x^2 - 90x + 264)$$
$$= 600(x^2 - 15x + 44)$$
$$= 600(x-4)(x-11)$$

$f'(x) = 0$ when $x = 4$ and $x = 11$
$f(4) = 46{,}400$, $f(11) = 12{,}100$, and
$f(14) = 36{,}400$

$f(4) = 46{,}400$ is the absolute maximum.
So, the membership was greatest in 1996, four years after the founding of the association, when there were 46,400 members.

(b) During this period, the absolute minimum is $f(0) = 0$, or the membership was smallest at the time the club was founded, when there were no members.

35. $F(p) = p^n(1-p)^{m-n}$

$$\begin{aligned} F'(p) &= p^n(m-n)(1-p)^{m-n-1}(-1) \\ &\quad + (1-p)^{m-n}(n)(p^{n-1}) \\ &= -p^n(m-n)(1-p)^{m-n-1} \\ &\quad + p^{n-1}(n)(1-p)^{m-n} \\ &= p^{n-1}(1-p)^{m-n-1} \\ &\quad \left[-(m-n)p + n(1-p)\right] \\ &= p^{n-1}(1-p)^{m-n-1} \\ &\quad \left[-mp + np + n - np\right] \end{aligned}$$

$F'(p) = 0$ when $p = 0$, 1, and $\frac{n}{m}$
$F(0) = 0$, $F(1) = 0$
Since n, m are positive and $m > n$, $\frac{n}{m}$ is in interval.
$F\left(\frac{n}{m}\right) = \left(\frac{n}{m}\right)^n\left(1 - \frac{n}{m}\right)^{m-n}$, and $F\left(\frac{n}{m}\right) > 0$,
so, $p = \frac{n}{m}$ gives the absolute maximum.

37. $S(r) = c(R^2 - r^2)$, where c is a positive constant.
The relevant interval is $0 \le r \le R$.

$$S'(r) = -2cr$$
$$S'(r) = 0 \text{ when } r = 0$$

(the left-hand endpoint of the interval)
With $S(0) = cR^2$ and $S(r) = 0$, the speed of the blood is greatest when $r = 0$, that is, at the central axis.

39. $q^2 + 3pq = 22$

(a) Using implicit differentiation,

$$2q\frac{dq}{dp} + (3p)\frac{dq}{dp} + (q)(3) = 0$$
$$\frac{dq}{dp} = \frac{-3q}{2q+3p}$$
$$E(p) = \frac{p}{q}\frac{dq}{dp}$$
$$D(p) = q$$

So,

$$E(p) = \frac{p}{q}\left(\frac{-3q}{2q+3p}\right) = \frac{-3p}{2q+3p}$$

(b) When $p = 3$, $q^2 + 9q = 22$, or $q = 2$ (rejecting negative root).

$$|E(p)| = \left|\frac{-3\cdot 3}{2\cdot 2 + 3\cdot 3}\right| = \left|-\frac{9}{13}\right| = \frac{9}{13}$$

Since $\frac{9}{13} < 1$, demand is inelastic.

41. **(a)** When $q = 50$, $50 = 500 - 2p$, or $p = 225$.
Further, when $q = 0$, $0 = 500 - 2p$, or $p = 250$.
So, the range for price is

$$225 \le p \le 250$$

(b)
$$E(p) = \frac{p}{q}\frac{dq}{dp} = \frac{p}{500-2p}(-2) = -\frac{p}{250-p}$$

$$\left|-\frac{p}{250-p}\right| = 1 \text{ when } -\frac{p}{250-p} = \pm 1$$

or, when $p = 125$ and demand is of unit elasticity.
When $p < 125$, $|E_n| < 1$ and demand is inelastic.
When $p > 125$, $|E_n| > 1$ and demand is elastic.

(c) When the price is less than \$125, total revenue is increasing as price increases; when the price is \$125, total revenue is unaffected by a small change in price, when the price is more than \$125, total revenue is decreasing as price increases.

(d) If an unlimited number of prints is available, should charge \$125 each; if only 50 prints are available, should charge \$225, the value in the price interval which is closest to \$125.

43. $E(v) = \frac{1}{v}[0.074(v-35)^2 + 22]$

(a)

$$E'(v) = \left(\frac{1}{v}\right)[0.148(v-35)(1)] + \left[0.074(v-35)^2 + 22\right]\left(\frac{-1}{v^2}\right)$$

$$= \frac{1}{v}\left[0.148v - 5.18 - \frac{1}{v}(0.074v^2 - 5.18v + 112.65)\right]$$

$$= \frac{1}{v}\left(0.148v - 5.18 - 0.074v + 5.18 - \frac{112.65}{v}\right)$$

$$= \frac{1}{v}\left(0.074v - \frac{112.65}{v}\right)$$

So, $E'(v) = 0$ when

$$0.074v - \frac{112.65}{v} = 0$$

$$0.074v = \frac{112.65}{v}$$

$$v^2 \approx 1522.3$$

$$v \approx 39$$

$$E''(v) = \left(\frac{1}{v}\right)\left(0.074 + \frac{112.65}{v^2}\right) + \left(0.074v - \frac{112.65}{v}\right)\left(-\frac{1}{v^2}\right)$$

$E''(39) > 0$, so there is an absolute minimum when $v = 39$.

(b) Writing Exercise—Answers will vary.

45. $$R(D) = D^2\left(\frac{C}{2} - \frac{D}{3}\right) = \frac{C}{2}D^2 - \frac{1}{3}D^3$$

(a) To maximize $R'(D)$,

$$R'(D) = CD - D^2$$

$$R''(D) = C - 2D$$

$$R''(D) = 0 \text{ when } D = \frac{C}{2}$$

$$R'''(D) = -2$$

Since $R'''\left(\frac{C}{2}\right)$ is negative, $D = \frac{C}{2}$ is a maximum for sensitivity. The sensitivity when $D = \frac{C}{2}$ is

$$R'\left(\frac{C}{2}\right) = C\left(\frac{C}{2}\right) - \left(\frac{C}{2}\right)^2 = \frac{C^2}{4}$$

(b) The reaction when $D = \frac{C}{2}$ is

$$R\left(\frac{C}{2}\right) = \left(\frac{C}{2}\right)^2\left[\frac{C}{2} - \frac{C/2}{3}\right] = \frac{C^3}{12}$$

47. $$I = \frac{E}{r+R}, \quad P(r) = I^2R = \frac{E^2R}{(r+R)^2}$$

$$P'(R) = \frac{(r+R)^2(E^2) - (E^2R)[2(r+R)]}{(r+R)^4}$$

$$= \frac{E^2(r+R)[(r+R) - 2R]}{(r+R)^4}$$

$$= \frac{E^2(r-R)}{(r+R)^3}$$

$P'(R) = 0$ when $R = r$

When $R = 0$, $P(0) = 0$

$0 < R < r$, $P'(R) > 0$ so P is increasing

$R > r$, $P'(R) < 0$ so P is decreasing.

So, $R = r$ results in maximum power.

49. (a) $$P(x) = \frac{Ax}{B + x^m}$$

$$R(x) = P'(x) = A\frac{(B + x^m) - mxx^{m-1}}{(B + x^m)^2}$$

$$= \frac{A[B + (1-m)x^m]}{(B + x^m)^2}$$

$$R(x) = 0 \text{ when } x = \left(\frac{B}{m-1}\right)^{1/m}$$

(b) $$R'(x) = \frac{A}{(B+x^m)^4}\Big[(B+x^m)^2[m(1-m)x^{m-1}] - [B+(1-m)x^m][2(B+x^m)(mx^{m-1})]\Big]$$
$$= \frac{A(B+x^m)nx^{m-1}}{(B+x^m)^3}\Big[(B+x^m)(1-m) - 2(B+(1-m)x^m)\Big]$$
$$= \frac{-Amx^{m-1}\left[B(1+m)+x^m(1-m)\right]}{(B+x^m)^3}$$
$$R'(x) = 0 \text{ when } x = \left[\frac{B(m+1)}{m-1}\right]^{1/m}$$

(c) Assuming $m > 1$,

when $0 < x < \left[\frac{B(m+1)}{m-1}\right]^{1/m}$,

$R'(x) > 0$ so R is increasing;

when $x > \left[\frac{B(m+1)}{m-1}\right]^{1/m}$,

$R'(x) < 0$ so R is decreasing.

So there is a relative maximum when

$$x = \left[\frac{B(m+1)}{m-1}\right]^{1/m}$$

51. **(a)** $S(r) = ar^2(r_0 - r)$

$F(r) = \pi r^2 S(r) = a\pi(r_0 r^4 - r^5),\ 0 \le r \le r_0$

(b) $$F'(r) = a\pi(4r_0r^3 - 5r^4) = a\pi r^3(4r_0 - 5r)$$
$$F'(r) = 0 \text{ when } r = \frac{4}{5}r_0$$
$$F(0) = F(r_0) = 0, \text{ and } F\left(\frac{4r_0}{5}\right) > 0,$$

so $F(r)$ is maximized for $r = \frac{4r_0}{5}$.

53. $q = b - ap$

(a) $$E(p) = \frac{p}{q}\frac{dq}{dp} = \frac{p}{b-ap}(-a) = \frac{ap}{ap-b}$$

(b) $|E(p)| = 1$ when $\left|\frac{ap}{ap-b}\right| = 1$,

or when $\frac{ap}{ap-b} = \pm 1$, or $p = \frac{b}{2a}$

(c) $|E(p)| < 1$ when $p < \frac{b}{2a}$, so demand is inelastic when $0 \le p < \frac{b}{2a}$

$|E(p)| > 1$ when $p > \frac{b}{2a}$, so demand is elastic when $\frac{b}{2a} < p \le \frac{b}{a}$

55. $q = \frac{a}{p^m} = ap^{-m}$

The elasticity of demand is

$$E(p) = \frac{p}{q}\frac{dq}{dp} = \frac{p}{a/p^m}(-amp^{-m-1}) = \frac{p^{m+1}}{a}\left(-\frac{am}{p^{m+1}}\right) = -m$$

When

$0 < m < 1$, $|E(p)| < 1$ and demand is inelastic

$m = 1$ $|E(p)| = 1$ and demand is of unit elasticity

$m > 1$, $|E(p)| > 1$ and demand is elastic.

3.5 Additional Applied Optimization

1. Let x denote the number that exceeds its square, x^2, by the largest amount. Then,

$$f(x) = x - x^2$$

is the function to be maximized.

$$f'(x) = 1 - 2x$$
$$f'(x) = 0 \text{ when } x = \frac{1}{2}$$
$$f''(x) = -2, \text{ so } f''\left(\frac{1}{2}\right) < 0$$

and there is a relative maximum when $x = \frac{1}{2}$. Further, since $f''(x) < 0$ for all x, it is the absolute maximum. So, $x = \frac{1}{2}$ is the desired number.

3. Let x be the first number and y be the second. Then,

$$P = xy, \text{ or since } y = 50 - x,$$
$$P(x) = x(50 - x) = 50x - x^2$$

which is the function to be maximized.

$$P'(x) = 50 - 2x$$
$$P'(x) = 0 \text{ when } x = 25$$
$$P''(x) = -2, \text{ so } P''(25) < 0$$

and there is a relative maximum when $x = 25$. Further, since $P''(x) < 0$ for all x in the domain $0 < x < 50$, it is the absolute maximum. So, $x = 25$ and $y = 50 - 25 = 25$ are the desired numbers.

5. Let x be the \$1.00 price increments above \$40.00. Then $40 + x$ will be the price per computer game, $50 - 3x$ will be the number of units sold per month, and the profit will be

$$P(x) = (50 - 3x)[(40 + x) - 25] = 750 + 5x - 3x^2$$

which is the function to be maximized.

$$P'(x) = 5 - 6x$$
$$P'(x) = 0 \text{ when } x = \frac{5}{6}$$
$$P''(x) = -6, \text{ so } P''\left(\frac{5}{6}\right) < 0$$

and there is a relative maximum when $x = \frac{5}{6}$. Further, since $P''(x) < 0$ for all x in the domain $x \geq 0$, it is the absolute maximum. So, the selling price for maximum profit is $40 + \frac{5}{6} \approx \$41$.

7. Let x be the number of additional trees planted per acre. The number of oranges per tree will be $400 - 4x$ and the number of trees per acre $60 + x$. The yield per acre is

$$y(x) = \left(\frac{\text{\# of oranges}}{\text{tree}}\right)\left(\frac{\text{\# of trees}}{\text{acre}}\right)$$
$$= (400 - 4x)(60 + x)$$
$$= 24{,}000 + 160x - 4x^2$$

$$y'(x) = 160 - 8x$$
$$y'(x) = 0 \text{ when } x = 20$$
$$y''(x) = -8, \text{ so } y''(20) < 0$$

and there is a relative maximum when $x = 20$. Further, since $y''(x) < 0$ for all x in the domain $x \geq 0$, it is the absolute maximum. So, the yield is maximized when there are $60 + 20 = 80$ trees per acre.

9. Let x be the length of the field and y be the width. The amount of fencing is the perimeter of the field, or

$$P = 2x + 2y$$

Since the area is 3,600,

$$A = xy$$
$$3{,}600 = xy, \text{ or } y = \frac{3{,}600}{x}$$

and

$$P(x) = 2x + 2\left(\frac{3{,}600}{x}\right) = 2x + \frac{7{,}200}{x}$$

which is the function to be minimized.

$$P'(x) = 2 - \frac{7{,}200}{x^2}$$
$$P'(x) = 0 \text{ when } x = 60$$
$$P''(x) = \frac{14{,}400}{x^3}, \text{ so } P''(60) > 0$$

and there is a relative minimum when $x = 60$. Further, since $P''(x) > 0$ for all x in the domain $x > 0$, it is the absolute maximum. So, the field should have a length of 60 meters and a width of 60 meters.

11. Let x be the length of the rectangle and y be the width. The area is

$$A = xy$$

Since the perimeter is fixed, let C represent its fixed value. Then,

$$P = 2x + 2y$$
$$C = 2x + 2y, \text{ so } y = \frac{C - 2x}{2}$$

and

$$A(x) = x\left(\frac{C-2x}{2}\right) = \frac{C}{2}x - x^2$$

which is the function to be maximized.

$$A'(x) = \frac{C}{2} - 2x$$

$$A'(x) = 0 \text{ when } x = \frac{C}{4}$$

$$A''(x) = -2, \text{ so } A''\left(\frac{C}{4}\right) < 0$$

and there is a relative maximum when $x = \frac{C}{4}$. Further, since $A''(x) < 0$ for all x in the domain $0 < x < \frac{C}{2}$, it is the absolute maximum. When $x = \frac{C}{4}$, $y = \frac{C}{4}$. So for any given perimeter, a square is the rectangle having the maximum area.

13. Let x be the length of the rectangle and let y be the vertical distance above the rectangle along the side of length 5. Then, $5 - y$ is the width of the rectangle. The area of the rectangle is

$$A = x(5 - y)$$

By similar triangles,

$$\frac{12}{5} = \frac{x}{y}, \text{ or } y = \frac{5}{12}x$$

and

$$A(x) = x\left(5 - \frac{5}{12}x\right) = 5x - \frac{5}{12}x^2$$

which is the function to be maximized.

$$A'(x) = 5 - \frac{5}{6}x$$

$$A'(x) = 0 \text{ when } x = 6$$

$$A''(x) = -\frac{5}{6}, \text{ so } A''(6) < 0$$

and there is a relative maximum when $x = 6$. Further, since $A''(x) < 0$ for all x in the domain $0 < x < 12$, it is the absolute maximum. The dimensions of the rectangle having the maximum area are $x = 6$ and $y = \frac{5}{12}(6) = \frac{5}{2}$.

15. Let x be the length of the side of the square base and y be the height of the box. The volume of the box is

$$V = x^2 y$$

The cost of the four sides is

$$4 \text{ (cost per unit area)(area)}$$

$$= 4(3)(xy) = 12xy$$

The cost of the bottom of the box is

$$\text{(cost per unit area)(area)}$$

$$= 4(x^2)$$

Since there is 48 dollars available to build the box,

$$48 = 12xy + 4x^2, \text{ or}$$

$$y = \frac{48 - 4x^2}{12x} = \frac{4}{x} - \frac{x}{3}$$

and

$$V(x) = x^2\left(\frac{4}{x} - \frac{x}{3}\right) = 4x - \frac{1}{3}x^3$$

which is the function to be maximized.

$$V'(x) = 4 - x^2$$

$$V'(x) = 0 \text{ when } x = 2 \text{ (rejecting the negative solution)}$$

$$V''(x) = -2x, \text{ so } V''(2) < 0$$

and there is a relative maximum when $x = 2$. Further, since $V''(x) < 0$ for all x in the domain $x > 0$, it is the absolute maximum. So the box has a maximum volume when its dimensions are 2 meters by 2 meters by $y = \frac{4}{2} - \frac{2}{3} = \frac{4}{3}$ meters.

17. Let x be the number of hours since the truck and the car were originally 300 miles apart. The distance the truck travels in x hours is $30x$, so the distance from the truck, after x hours to the car's original location is $300 - 30x$. This distance and the distance the car travels, which is $60x$, form the legs of a right triangle, the hypotenuse of which is the distance between the vehicles. Since the distance between the vehicles and the square of this distance both increase and decrease simultaneously, the minimum of the distance occurs when the minimum of its square occurs. Let D be the square of the distance. Then

$$D(x) = (60x)^2 + (300 - 30x)^2$$
$$= 4{,}500x^2 - 18{,}000x + 90{,}000$$
$$D'(x) = 9{,}000x - 18{,}000$$
$$D'(x) = 0 \text{ when } x = 2$$
$$D''(x) = 9{,}000, \text{ so } D''(2) > 0$$

and there is a relative minimum when $x = 2$. Further, since $D''(x) > 0$ for all x in the domain $0 \le x \le 10$, it is the absolute minimum. So the distance is minimized after 2 hours.

19. The amount of material is the amount for the circular top and bottom, and the amount for the curved side.

$$m = 2\pi r^2 + 2\pi r h$$

Since the volume is 6.89π,

$$V = \pi r^2 h$$
$$6.89\pi = \pi r^2 h, \text{ or } h = \frac{6.89}{r^2}$$
$$\text{and } m(r) = 2\pi r^2 + 2\pi r\left(\frac{6.89}{r^2}\right)$$
$$= 2\pi r^2 + \frac{13.78\pi}{r}$$

which is the function to be minimized.

$$m'(r) = 4\pi r - \frac{13.78\pi}{r^2}$$
$$m'(r) = 0 \text{ when } r \approx 1.51$$
$$m''(r) = 4\pi + \frac{27.56}{r^3}, \text{ so } m''(1.51) > 0$$

and there is a relative minimum when $r = 1.51$. Further, since $m''(r) > 0$ for all r in the domain $r > 0$, it is an absolute minimum. So, the minimum material is when the can's radius is approximately 1.51 inches and its height is approximately $\frac{6.89}{(1.51)^2} \approx 3.02$ inches. (These dimensions are not used due to packaging and handling concerns.)

21. Let x be the distance along the riverbank where cable is not run underground. Then, $2{,}000 - x$ is the distance the cable is run underground, and its cost is $20(2{,}000 - x)$. The length of the cable run underwater is the hypotenuse of a right triangle having legs of length 1,200 and x, or $\sqrt{(1{,}200)^2 + x^2}$. It's cost is $25\sqrt{(1{,}200)^2 + x^2}$. The total cost is

$$C(x) = 20(2{,}000 - x) + 25\sqrt{(1{,}200)^2 + x^2}$$

which is the function to be minimized.

$$C'(x) = -20 + 25\left[\frac{1}{2}\left(1{,}440{,}000 + x^2\right)^{-1/2}(2x)\right]$$
$$= -20 + \frac{25x}{\sqrt{1{,}440{,}000 + x^2}}$$
$$C'(x) = 0 \text{ when } 20 = \frac{25x}{\sqrt{1{,}440{,}000 + x^2}}$$
$$\sqrt{1{,}440{,}000 + x^2} = \frac{5x}{4}$$
$$1{,}440{,}000 + x^2 = \frac{25x^2}{16}$$
$$\text{or, } x = 1{,}600$$

When $0 < x < 1{,}600$, $C'(x) < 0$ so C is decreasing
$x > 1{,}600$, $C'(x) > 0$ so C is increasing.

So, there is a relative minimum when $x = 1{,}600$. The domain of C is $0 \le x \le 2{,}000$, and $C(1{,}600) = 58{,}000$, $C(0) = 70{,}000$, $C(2{,}000) \approx 58{,}310$.
So, the cost is minimized when the cable reaches the riverbank $2000 - 1{,}600 = 400$ feet from the factory.

23. The volume of a can is

$$V = \pi r^2 h$$

The amount of metal for the can is 27π, so

$$m = \pi r^2 + 2\pi r h$$
$$27\pi = \pi r^2 + 2\pi r h, \text{ or } h = \frac{27 - r^2}{2r}$$

and

$$V(r) = \pi r^2\left(\frac{27 - r^2}{2r}\right) = \frac{27\pi}{2}r - \frac{\pi}{2}r^3$$

which is the function to be maximized.

$$V'(r) = \frac{27\pi}{2} - \frac{3\pi}{2}r^2$$
$$V'(r) = 0 \text{ when } r = 3$$
$$V''(r) = -3\pi r, \text{ so } V''(3) < 0$$

and there is a relative maximum when $r = 3$. Further, since $V''(x) < 0$ for all r in the domain $r > 0$, it is an absolute maximum. So, when the can's radius is 3 inches, it has its maximum volume of $\frac{27\pi}{2}(3) - \frac{\pi}{2}(3)^3 = 27\pi$ cubic inches.

25. The cost of the material is the cost of the circular bottom and the cost of the curved side.

$$C = 3(\pi r^2) + 2(2\pi r h)$$

Since the volme is to be fixed, let K represent this fixed value.

$$V = \pi r^2 h$$
$$K\pi r^2 h, \text{ or } h = \frac{K}{\pi r^2}$$

and

$$C(r) = 3\pi r^2 + 4\pi r\left(\frac{K}{\pi r^2}\right) = 3\pi r^2 + \frac{4K}{r}$$

which is the function to be minimized.

$$C'(r) = 6\pi r - \frac{4K}{r^2}$$
$$C'(r) = 0 \text{ when } \frac{3\pi}{2}r^3 = K$$
$$\frac{3\pi}{2}r^3 = \pi r^2 h$$
$$\text{or, } r = \frac{2}{3}h$$
$$C''(r) = 6\pi + \frac{8K}{r^3}, \text{ so } C''\left(\frac{2}{3}h\right) > 0$$

and there is a relative minimum when $r = \frac{2}{3}h$. Further, since $C''(r) > 0$ for all r in the domain $r > 0$, it is an absolute minimum. So, a can with a fixed volume has its cost minimized whenever $r = \frac{2}{3}h$.

27. **(a)** Let x denote the number of machines used and $C(x)$ the corresponding total cost. Then

$$C(x) = \text{set up cost} + \text{operating cost}$$
$$= 20\,(\text{number of machines}) + 15\,(\text{number of hours}).$$

Since each machine produces 30 kickboards per hour, x machines produce $30x$ kickboards per hour and the number of hours required to produce 8,000 kickboards is $\frac{8{,}000}{30x}$. So,

$$C(x) = 20x + 15\left(\frac{8{,}000}{30x}\right) = 20x + \frac{4{,}000}{x}$$
$$C'(x) = 20 - \frac{4{,}000}{x^2}$$
$$C'(x) = 0 \text{ when } x \approx 14$$

Since the company owns 10 machines, the domain of C is $1 \le x \le 10$. Further, $C(1) = 4{,}020$ and $C(10) = 600$, so cost is minimized when 10 machines are used.

(b) When 10 machines are used, the number of hours to produce the kickboards is $\frac{8{,}000}{30(10)}$ and the supervisor would be paid $15\left(\frac{8{,}000}{300}\right) = \$400$.

(c) The cost of setting up 10 machines is $20(10) = \$200$.

29. **(a)** Let x be the number of bottles in each shipment. The costs include:

$$\text{purchase cost} = (800)(20) = 16{,}000$$
$$\text{ordering cost} = \left(\frac{800}{x}\right)(10)$$
$$\text{ordering cost} = \left(\frac{x}{2}\right)(0.4)$$

So, the total cost is

$$C(x) = 16{,}000 + \frac{8{,}000}{x} + 0.2x$$

which is the function to be minimized

$$C'(x) = -\frac{8{,}000}{x^2} + 0.2$$

$$C'(x) = 0 \text{ when } x = 200$$

$$C''(x) = \frac{16{,}000}{x^3}, \text{ so } C''(200) > 0$$

and there is a relative minimum when $x = 200$. $C(200) = 16{,}080$, $C(1) = 17{,}000.20$, $C(800) = 16{,}170$. So, cost is minimized when 200 bottles are ordered in each shipment.

(b) The number of shipments is $\frac{800}{200} = 4$ times a year, so the store orders every 3 months.

31. (a) Since there are N withdrawals per year, the transaction fees for the year are $8N$. For the loss of interest, since $I = prt$, need to find the amount of each withdrawal and the length of time for which interest is lost. Since there is a total of \$10,000 withdrawn in N draws, the amount of each withdrawal is $\frac{10{,}000}{N}$ dollars. Assuming the first withdrawal occurs at the beginning of the year, the loss of interest is

$$0.04\left(\frac{10{,}000}{N}\right)1$$

When the second withdrawal occurs, $\frac{12}{N}$ months have passed, so the loss of interest is (converting to years)

$$0.04\left(\frac{10{,}000}{N}\right)\frac{\left(12 - \frac{12}{N}\right)}{12}$$

$$= 0.04\left(\frac{10{,}000}{N}\right)\left(\frac{N-1}{N}\right)$$

When the third withdrawal occurs, $2\left(\frac{12}{N}\right)$ months have passed, so the loss of interest is

$$0.004\left(\frac{10{,}000}{N}\right)\left(\frac{12 - \frac{24}{N}}{12}\right)$$

$$= 0.04\left(\frac{10{,}000}{N}\right)\left(\frac{N-2}{N}\right)$$

The pattern continues, with the last withdrawal having $(N-1)\left(\frac{12}{N}\right)$ months passed, so it occurs a loss of interest of

$$0.04\left(\frac{10{,}000}{N}\right)\left(\frac{12 - (N-1)\left(\frac{12}{N}\right)}{12}\right)$$

$$= 0.04\left(\frac{10{,}000}{N}\right)\left(\frac{1}{N}\right)$$

The total loss of interest is

$$0.04\left(\frac{10{,}000}{N}\right)\left[1 + \frac{N-1}{N} + \frac{N-2}{N} + \frac{N-3}{N} + \cdots + \frac{1}{N}\right]$$

$$= \frac{400}{N^2}[N + N - 1 + N - 2 + N - 3 + \cdots + 2 + 1]$$

Since $1 + 2 + 3 + \cdots + N = \frac{N(N+1)}{2}$,

$$= \frac{400}{N^2}\left[\frac{N(N+1)}{2}\right] = \frac{200(N+1)}{N}$$

The cost function is

$$C(N) = 8N + \frac{200(N+1)}{N}$$

(b)

$$C(N) = 8N + \frac{200N + 200}{N}$$

$$C'(N) = 8 + \frac{(N)(200) - (200N + 200)(1)}{N^2}$$

$$= 8 - \frac{200}{N^2}$$

$$C'(N) = 0 \text{ when}$$

$$0 = 8 - \frac{200}{N^2}$$

$$0 = 8N^2 - 200$$

$$N = 5C''(N) = \frac{400}{N^3}$$

When $N = 5$, $C''(5) > 0$, so the minimum occurs when the number of withdrawals is 5.

33. Let $P(x)$ be the profit from the sale of the wine at time x in years.
profit = value − purchase cost − storage cost
Let $V(x)$ be the value of the wine at time x, and let

C be the purchase cost of the wine. Since the storage cost is $3x$,

$$P(x) = V(x) - C - 3x$$

which is the function to maximize and

$$P'(x) = V'(x) - 3$$

Since the rate of change of value is $53 - 10x$,

$$P'(x) = 50 - 10x$$
$$P'(x) = 0 \text{ when } x = 5$$
$$P''(x) = -10, \text{ so } P''(5) < 0$$

and there is a relative maximum when $x = 5$. Further, since $P''(x) < 0$ for all x in the domain $x \geq 0$, it is the absolute maximum. So, the wine should be sold 5 years from the time of purchase to maximize profit.

35. The volume of the parcel is

$$V = x^2 y$$

The restriction given is

$$4x + y = 108(\text{max}), \text{ or } y = 108 - 4x$$

and

$$V(x) = x^2(108 - 4x) = 108x^2 - 4x^3$$

which is the function to be maximized.

$$V'(x) = 216x - 12x^2 = 12x(18 - x)$$
$$V'(x) = 0 \text{ when } x = 18 \text{ (rejecting } x = 0)$$
$$V''(x) = 216 - 24x, \text{ so } V''(18) < 0$$

and there is a relative minimum when $x = 18$.

When $0 < x < 18$, $V'(x) > 0$ so V is increasing

$x > 18$, $V'(x) < 0$ so V is decreasing.

So, the relative maximum is the absolute maximum. The maximum volume is $108(18)^2 - 4(18)^3 = 11{,}664$ cubic inches.

37. Let x be the number of units and $C(x)$ be the cost of producing those units. Then,

$$C(x) = 1{,}200 + 1.2x + \frac{100}{x^2}$$

which is the function to be minimized.

$$C'(x) = 1.2 - \frac{200}{x^3}$$
$$C'(x) = 0 \text{ when } x \approx 5.503 \approx 6$$
$$C''(x) = \frac{600}{x^4}, \text{ so } C''(6) > 0$$

and there is a relative minimum when $x = 6$. Further, since $C''(x) > 0$ for all x in the domain $0 < x < 100$, it is the absolute minimum. So, producing 6 units daily minimizes the cost.

39. Let x be the distance along the shoreline from A to P. Then, the distance from B to P is the hypotenuse of a right triangle,

$$d(B, P) = \sqrt{25 + x^2}$$

The total distance along the shoreline from A to L is the leg of a right triangle,

$$d(A, L) = \sqrt{(13)^2 - (5)^2} = 12$$

So, the distance from P to L is

$$d(P, L) = 12 - x$$

The path of the bird is from B to P, and then from P to L. If e is the energy per mile to fly over land (a constant), then the energy to fly this path is

$$E(x) = 2e\sqrt{25 + x^2} + e(12 - x)$$

which is the function to be minimized.

$$E'(x) = e(25 + x^2)^{-1/2}(2x) - e$$
$$= \frac{2ex}{(25 + x^2)^{1/2}} - e$$
$$E'(x) = 0 \text{ when } \frac{2ex}{(25 + x^2)^{1/2}} = e$$
$$\frac{2x}{(25 + x^2)^{1/2}} = 1$$
$$2x = (25 + x^2)^{1/2}$$
$$4x^2 = 25 + x^2$$
$$\text{or, } x = \sqrt{\frac{25}{3}} = \frac{5\sqrt{3}}{3}$$

Since $0 \le x \le 12$,

$$E\left(\frac{5\sqrt{3}}{3}\right) \approx 20.7e;\ E(0) = 22e;\ E(12) = 26e$$

So, to minimize energy expended, the bird should fly to point P which is $\sqrt{\frac{25}{3}} \approx 2.9$ miles from point A.

41. Let S be the stiffness of the beam. Then,

$$S = kwh^3,$$

where k is a constant of proportionality. Since $w^2 + h^2 = 225$, or $h = \sqrt{225 - w^2}$, S can be expressed as a function of w,

$$S(w) = kw(225 - w^2)^{3/2}$$

which is the function to be maximized.

$$S'(w) = k\left[w \cdot \frac{3}{2}(225 - w^2)^{1/2}(-2w) + (225 - w^2)^{3/2}(1)\right]$$
$$= k(225 - w^2)^{1/2}\left[-3w^2 + 225 - w^2\right]$$
$$= k(225 - w^2)^{1/2}(225 - 4w^2)$$

$S'(w) = 0$ when $w = \frac{15}{2}$ (rejecting the solution $w = 15$, which is not possible given the diameter)

When $0 < w < \frac{15}{2}$, $S'(w) > 0$ so C is increasing

$\frac{15}{2} < w < 15$, $S'(x)) < 0$ so S is decreasing.

So, the dimensions for maximum stiffness are $w = \frac{15}{2}$ inches and $h = \sqrt{225 - \left(\frac{15}{2}\right)^2} \approx 13.0$ inches.

43. Let x be the number of miles from the house to plant A. Then, $18 - x$ is its distance from plant B, and $1 \le x \le 16$. Let $P(x)$ be the concentration of particulate matter at the house. Then,

$$P(x) = \frac{80}{x} + \frac{720}{18 - x}$$

which is the function to minimize.

$$P'(x) = -\frac{80}{x^2} + \frac{0 - (720)(-1)}{(18 - x)^2}$$

$$P'(x) = 0 \text{ when } \frac{80}{x^2} = \frac{720}{(18 - x)^2}$$

$$2x^2 + 9x - 81 = 0$$

or, $x = \frac{9}{2}$ (rejecting negative solution)

$P(4.5) = 0$, $P(1) \approx 122.4$, $P(16) = 365$;

So, the total pollution is minimized when the house is 4.5 miles from plant A.

45. Let $C(N)$ be the total cost of using N machines. Now, the setup cost of N machines is aN and the operating cost of N machines is $\frac{b}{N}$. So,

$$C(N) = aN + \frac{b}{N}$$

which is the function to minimize.

$$C'(N) = a = \frac{b}{N^2}$$

$$C'(N) = 0 \text{ when } a = \frac{b}{N^2},$$

or when $aN = \frac{b}{N}$ (setup cost = operating cost)

$$C''(N) = \frac{2b}{N^3},$$

which is positive for all N in the domain $N \ge 1$, so there is an absolute minimum when setup cost equals operating cost.

47. Frank is right. In the cost function,

$$C(x) = 5\sqrt{(900)^2 + x^2} + 4(3{,}000 - x)$$

note where the distance downstream appears. Since it is only part of the constant term in $C(x)$, it drops out when finding $C'(x)$. So, the critical value is always $x = 1{,}200$ (as long as the distance downstream is at least 1,200 meters).

When $0 \le x < 1{,}200$, $C'(x) < 0$ so C is decreasing

$x > 1{,}200$, $C'(x)) > 0$ so C is increasing

So, the absolute minimum cost is always when the cable reaches the bank 1,200 meters downstream.

49. (a) Let x be the number of machines and let t be the number of hours required to produce q units. The set up cost is xs and the operating cost is pt. Since each machine produces n units per hour, then $q = xnt$, or $t = \frac{q}{nx}$. The total cost is

$$C(x) = xs + p\frac{q}{nx}$$

which is the function to be minimized.

$$C'(x) = s - \frac{pq}{nx^2}$$

$$C'(x) = 0 \text{ when } s = \frac{pq}{nx^2}$$

$$\text{or, } x = \left(\frac{pq}{ns}\right)^{1/2}$$

$$C''(x) = \frac{2pq}{nx^3}, \text{ so } C''\left[\left(\frac{pq}{ns}\right)^{1/2}\right] > 0$$

and there is a relative minimum when $x = \left(\frac{pq}{ns}\right)^{1/2}$. Further, since $C''(x) > 0$ for all values of x in the domain $x \geq 1$, it is the absolute minimum.

(b) The setup cost xs, at this minimum, becomes

$$xs = s\sqrt{\frac{pq}{ns}} = \sqrt{\frac{pqs}{n}}$$

and the operating cost pt, at this minimum, becomes

$$P\frac{q}{n\sqrt{\frac{pq}{ns}}} = \frac{pq}{\sqrt{\frac{pqn}{s}}}$$

$$= pq\sqrt{\frac{s}{pqn}} = \sqrt{\frac{pqs}{n}}$$

So, the setup cost equals the operating cost when the total cost is minimized.

51. (a) Let x be the number of units produced, $p(x)$ the price per unit, t the tax per unit, and $C(x)$ the total cost.

$$C(x) = \frac{7x^2}{8} + 5x + 100$$

Since $p(x) = 15 - \frac{3x}{8}$, the revenue is

$$R(x) = xp(x) = 15x - \frac{3x^2}{8}$$

Now profit is

$$P(x) = \text{revenue–taxation–cost}$$

$$P(x) = 15x - \frac{3x^2}{8} - tx - \frac{7x^2}{8} - 5x - 100$$

which is the function to be maximized.

$$P'(x) = 15 - \frac{3x}{4} - t - \frac{7x}{4} - 5$$

$$= -\frac{5}{2}x + 10 - t$$

$$P'(x) = 0 \text{ when } x = \frac{2(10-t)}{5}$$

$$P''(x) = -\frac{5}{2}, \text{ so } P''\left(\frac{2(10-t)}{5}\right) < 0$$

and there is a relative maximum when $x = \frac{2}{5}(10-t)$. Further, since $P''(x) < 0$ for all x in the domain $x > 0$, it is the absolute maximum.

(b) The government share is

$$G(x) = tx = \left(\frac{2}{5}\right)(10t - t^2)$$

which is the function to be maximized.

$$G'(t) = \left(\frac{2}{5}\right)(10 - 2t)$$

$$G'(t) = 0 \text{ when } t = 5$$

$$G''(t) = -\frac{4}{5}, \text{ so } G''(5) < 0$$

and there is a relative maximum when $t = 5$. Further, since $G''(t) < 0$ for all t in the domain $t > 0$, it is the absolute maximum.

(c) From part (a), with $t = 0$,

$$x = \frac{2(10-0)}{5} = 4, \text{ and with } t = 5,$$

$$x = \frac{2(10-5)}{5} = 2.$$

The price per unit for the two quantities produced is, respectively,

$$p(4) = 15 - \frac{3(4)}{8} = \$13.50 \text{ and}$$

$$p(2) = 15 - \frac{3(2)}{8} = \$14.25$$

The difference between the two unit prices is 14.25 − 13.50 or 75 cents, which represents the amount of tax passed on to the consumer. The monopolist will absorb $4.25 of the tax.

(d) Writing Exercise— Answers will vary.

Checkup for Chapter 3

1. Graph (a) is the graph of f, while graph (b) is the gtaph of f'; possible explanations include:

(i) the degree of (a) is one larger than the degree of (b)

(ii) the x-intercepts of (b) correspond to the relative extrema of (a)

2. (a)

$$f(x) = -x^4 + 4x^3 + 5$$

$$f'(x) = -4x^3 + 12x^2 = -4x^2(x-3)$$

$$f'(x) = 0 \text{ when } x = 0, 3$$

When $x < 0$, $f'(x) > 0$ so f is increasing

$0 < x < 3$, $f'(x) > 0$ so f is increasing

$x > 3$, $f'(x)) < 0$ so f is decreasing.

There is no relative extrema when $x = 0$, but when $x = 3$, f has a relative maximum.

(b)

$$f(t) = 2t^3 - 9t^2 + 12t + 5$$

$$f'(t) = 6t^2 - 18t + 12 = 6(t-1)(t-2)$$

$$f'(t) = 0 \text{ when } t = 0, 3$$

When $t < 1$, $f'(t) > 0$ so f is increasing

$1 < t < 2$, $f'(t) < 0$ so f is decreasing

$t > 2$, $f'(t) > 0$ so f is increasing.

When $t = 1$, f has a relative maximum, and when $t = 2$, f has a relative minimum.

(c)

$$g(t) = \frac{t}{t^2 + 9}$$

$$g'(t) = \frac{(t^2+9)(1) - (t)(2t)}{(t^2+9)^2} = \frac{(3+t)(3-t)}{(t^2+9)^2}$$

$$g'(t) = 0 \text{ when } t = -3, 3$$

When $t < -3$, $g'(t) < 0$ so g is decreasing

$-3 < t < 3$, $g'(t) > 0$ so g is increasing

$t > 3$, $g'(t) < 0$ so g is decreasing.

When $t = -3$, g has a relative minimum, and when $t = 3$, g has a relative maximum.

(d)

$$g(x) = \frac{4-x}{x^2+9}$$

$$g'(x) = \frac{(x^2+9)(-1) - (4-x)(2x)}{(x^2+9)^2}$$

$$= \frac{(x+1)(x-9)}{(x^2+9)^2}$$

$$g'(x) = 0 \text{ when } x = -1, 9$$

When $x < -1$, $g'(x) > 0$ so g is increasing

$-1 < x < 9$, $g'(x) < 0$ so g is decreasing

$x > 9$, $g'(x) > 0$ so g is increasing.

When $x = -1$, g has a relative maximum, and when $x = 9$, g has a relative minimum.

3. (a)

$$f(x) = 3x^5 - 10x^4 + 2x - 5$$

$$f'(x) = 15x^4 - 40x^3 + 2$$

$$f''(x) = 60x^3 - 120x^2 = 60x^2(x-2)$$

$$f''(x) = 0 \text{ when } x = 0, 2$$

When $x < 0$, $f''(x) < 0$ so f is concave down

$0 < x < 2$, $f''(x) < 0$ so f is concave down

$x > 2$, $f''(x) > 0$ so f is concave up.

There is an inflection point when $x = 2$.

(b)
$$f(x) = 3x^5 + 20x^4 - 50x^3$$
$$f'(x) = 15x^4 + 80x^3 - 150x^2$$
$$f''(x) = 60x^3 + 240x^2 - 300x = 60x(x+5)(x-1)$$
$$f''(x) = 0 \text{ when } x = -5,\ 0,\ 1$$

When $x < -5$, $f''(x) < 0$ so f is concave down
$-5 < x < 0$, $f''(x) > 0$ so f is concave up
$0 < x < 1$, $f''(x)) < 0$ so f is concave down
$x > 1$, $f''(x) > 0$ so f is concave up.

There are inflection points when $x = -5, 0, 1$.

(c)
$$f(t) = \frac{t^2}{t-1}$$
$$f'(t) = \frac{(t-1)(2t) - (t^2)(1)}{(t-1)^2} = \frac{t^2 - 2t}{(t-1)^2}$$
$$f''(t) = \frac{(t-1)^2(2t-2) - (t^2-2t)(2(t-1)(1))}{(t-1)^4}$$
$$f''(t) = \frac{2(t-1)^3 - 2t(t-2)(t-1)}{(t-1)^4}$$
$$f''(t) = \frac{2(t-1)\left[(t-1)^2 - t(t-2)\right]}{(t-1)^4}$$
$$f''(t) = \frac{2}{(t-1)^3}$$

$f''(t)$ is never zero, so there are no inflection points; $f''(t)$ is undefined for $t = 1$.

When $t < 1$, $f''(t) < 0$ so f is concave down
$t > 1$, $f''(t) > 0$ so f is concave up.

(d)
$$g(t) = \frac{3t^2 + 5}{t^2 + 3}$$
$$g'(t) = \frac{(t^2+3)(6t) - (3t^2+5)(2t)}{(t^2+3)^2} = \frac{8t}{(t^2+3)^2}$$
$$g''(t) = \frac{(t^2+3)^2(8) - (8t)(2(t^2+3)(2t))}{(t^2+3)^4}$$
$$= \frac{8(t^2+3)\left[(t^2+3) - 4t^2\right]}{(t^2+3)^4}$$
$$= \frac{24(1+t)(1-t)}{(t^2+3)^3}$$
$$g''(t) = 0 \text{ when } t = -1,\ 1$$

When $t < -1$, $g''(t) < 0$ so g is concave down
$-1 < t < 1$, $g''(t) > 0$ so f is concave up
$t > 1$, $g''(t) < 0$ so g is concave down.

There are inflection points when $t = -1, 1$.

4. **(a)** $f(x) = \dfrac{2x-1}{x+3}$

$x + 3 = 0$ when $x = -3$, so there is a vertical asymptote of $x = -3$.

$$\lim_{x\to\pm\infty} \frac{2x-1}{x+3} = \lim_{x\to\pm\infty} \frac{2 - \frac{1}{x}}{1 + \frac{3}{x}} = \frac{2}{1} = 2$$

so there is a horizontal asymptote of $y = 2$.

(b) $f(x) = \dfrac{x}{x^2-1}$

$x^2 - 1 = (x+1)(x-1) = 0$ when $x = -1, 1$; so, there are vertical asymptotes of $x = -1$ and $x = 1$.

$$\lim_{x\to\infty} \frac{x}{x^2-1} = \lim_{x\to\infty} \frac{\frac{1}{x}}{1 - \frac{1}{x^2}} = \frac{0}{1} = 0;$$

so, there is a horizontal asymptote of $y = 0$.

(c) $f(x) = \dfrac{x^2 + x - 1}{2x^2 + x - 3}$

$2x^2 + x - 3 = (2x+3)(x-1) = 0$ when $x = -\dfrac{3}{2}$, 1; so, there are vertical asymptotes of $x = -\dfrac{3}{2}$ and $x = 1$.

$$\lim_{x\to\pm\infty} \frac{x^2+x-1}{2x^2+x-3} = \lim_{x\to\pm\infty} \frac{1+\frac{1}{x}-\frac{1}{x^2}}{1+\frac{1}{x}-\frac{3}{x^2}} = \frac{1}{2};$$

so there is a horizontal asymptote of $y = \frac{1}{2}$.

(d)

$$f(x) = \frac{1}{x} - \frac{1}{\sqrt{x}} = \frac{\sqrt{x}-x}{x\sqrt{x}} = \frac{x^{1/2}-x}{x^{3/2}}$$

$x^{3/2} = 0$ when $x = 0$; so, there is a vertical asymptote of $x = 0$.

$$\lim_{x\to\pm\infty} \frac{x^{1/2}-x}{x^{3/2}} = \lim_{x\to\pm\infty} \frac{\frac{1}{x^{1/2}}-1}{x^{1/2}} = 0;$$

so, there is a horizontal asymptote of $y = 0$.

5. (a) $f(x) = 3x^4 - 4x^3$
When $x = 0$, $f(0) = 0$ so $(0, 0)$ is an intercept.

When $f(x) = 0$, $3x^4 - 4x^3 = x^3(3x-4) = 0$

so $f(x) = 0$ when $x = 0, \frac{4}{3}$, and $\left(\frac{4}{3}, 0\right)$ is an intercept.
There are no asymptotes.

$$f'(x) = 12x^3 - 12x^2 = 12x^2(x-1)$$
$$f'(x) = 0 \text{ when } x = 0, 1$$
$$f''(x) = 36x^2 - 24x = 12x(3x-2)$$
$$f''(x) = 0 \text{ when } x = 0, \frac{2}{3}$$

When $x < 0$, $f'(x) < 0$ so f is decreasing
$f''(x) > 0$ so f is concave up

$0 < x < \frac{2}{3}$, $f'(x) < 0$ so f is decreasing
$f''(x) < 0$ so f is concave down

$\frac{2}{3} < x < 1$, $f'(x) < 0$ so f is decreasing
$f''(x) < 0$ so f is concave up

$x > 1$, $f'(x) > 0$ so f is increasing
$f''(x) > 0$ so f is concave up.

There is a relative minimum when $x = 1$, or $(1, -1)$. There are inflection points when $x = 0, \frac{2}{3}$, or $(0, 0)$ and $\left(\frac{2}{3}, -\frac{16}{27}\right)$.

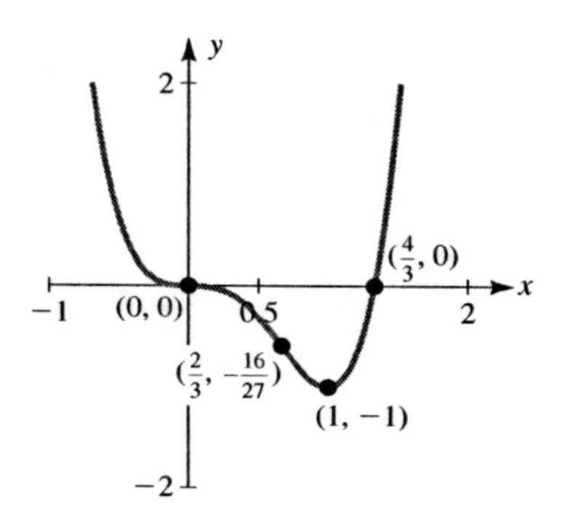

(b) $f(x) = x^4 - 3x^3 + 3x^2 + 1$
When $x = 0$, $f(0) = 1$, so $(0, 1)$ is an intercept.
$f(x) = 0$ is too difficult to solve.
There are no asymptotes.

$$f'(x) = 4x^3 - 9x^2 + 6x = x(4x^2 - 9x + 6)$$
$$f'(x) = 0 \text{ when } x = 0$$
$$f''(x) = 12x^2 - 18x + 6 = 6(2x-1)(x-1)$$
$$f''(x) = 0 \text{ when } x = \frac{1}{2}, 1$$

When $x < 0$, $f'(x) < 0$ so f is decreasing
$f''(x) > 0$ so f is concave up

$0 < x < \frac{1}{2}$, $f'(x) > 0$ so f is increasing
$f''(x) > 0$ so f is concave up

$\frac{1}{2} < x < 1$, $f'(x) > 0$ so f is increasing
$f''(x) < 0$ so f is concave down

$x > 1$, $f'(x) > 0$ so f is increasing
$f''(x) > 0$ so f is concave up.

There is a relative minimum when $x = 0$, or $(0, 1)$. There are inflection points when $x = \frac{1}{2}, 1$, or $\left(\frac{1}{2}, \frac{23}{16}\right)$ and $(1, 2)$.

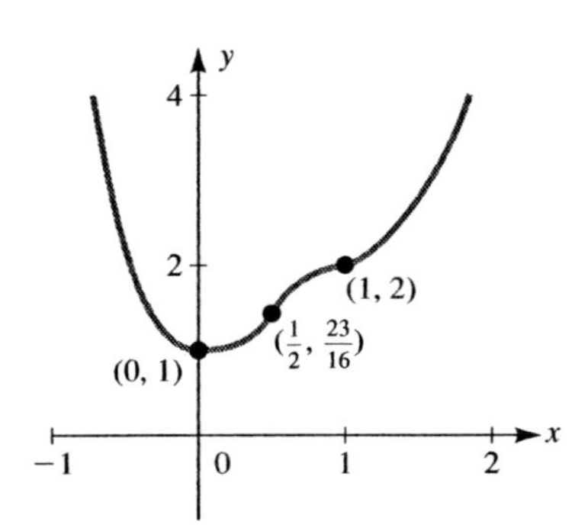

(c) $f(x) = \dfrac{x^2+2x+1}{x^2}$

When $x=0$, $f(0)$ is undefined.
When $f(x)=0$, $x^2+2x+1=(x+1)^2=0$, so $f(x)=0$ when $x=-1$, and $(-1, 0)$ is an intercept.
$x^2=0$ when $x=0$, so there is a vertical asymptote of $x=0$.

$$\lim_{x\to\pm\infty}\frac{x^2+2x+1}{x^2}=\lim_{x\to\pm\infty}\frac{1+\frac{2}{x}+\frac{1}{x^2}}{1}=\frac{1}{1},$$

so there is a horizontal aymptote of $y=1$
Note: $\dfrac{x^2+2x+1}{x^2}=1$ when $x^2+2x+1=x^2$, $2x+1=0$, or $x=-\frac{1}{2}$, so the graph will cross this asymptote at $\left(-\frac{1}{2}, 1\right)$.

$$f'(x)=\frac{(x^2)(2x+2)-(x^2+2x+1)(2x)}{x^4}$$
$$=\frac{-2x^2-2x}{x^4}=\frac{-2x(x+1)}{x^4}=\frac{-2(x+1)}{x^3}$$

$f'(x)=0$ when $x=-1$ and $f'(x)$ is undefined when $x=0$.

$$f''(x)=\frac{(x^3)(-2)-(-2(x+1)(3x^2))}{x^6}$$
$$=\frac{2x^2\left[-x+3(x+1)\right]}{x^6}=\frac{2(2x+3)}{x^4}$$

$f''(x)=0$ when $x=-\dfrac{3}{2}$ and $f''(x)$ is undefined when $x=0$.

When $x<-\dfrac{3}{2}$, $f'(x)<0$ so f is decreasing
$f''(x)<0$ so f is concave down
$-\dfrac{3}{2}<x<-1$, $f'(x)<0$ so f is decreasing
$f''(x)>0$ so f is concave up
$-1<x<0$, $f'(x)>0$ so f is increasing
$f''(x)>0$ so f is concave up
$x>0$, $f'(x)<0$ so f is decreasing
$f''(x)>0$ so f is concave up.

There is a relative minimum when $x=-1$, or $(-1, 0)$. There is an inflection point when $x=-\frac{3}{2}$, or $\left(-\frac{3}{2}, \frac{1}{9}\right)$.

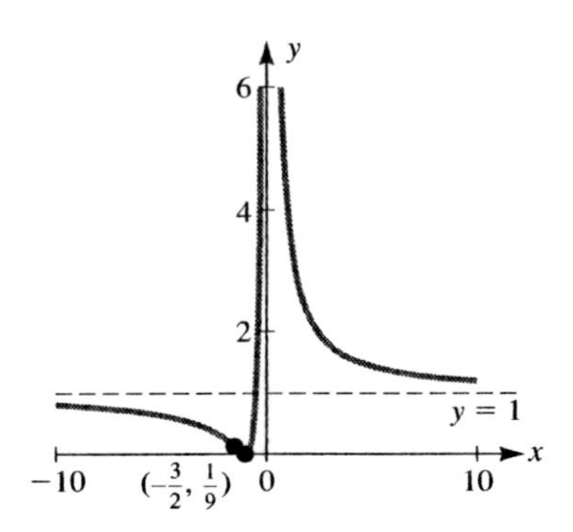

(d) $f(x)=\dfrac{1-2x}{(x-1)^2}$

When $x=0$, $f(0)=1$ so $(0, 1)$ is an intercept.
When $f(x)=0$, $1-2x=0$, or $x=\dfrac{1}{2}$ so $\left(\dfrac{1}{2}, 0\right)$ is an intercept.
$(x-1)^2=0$ when $x=1$, so there is a vertical asymptote of $x=1$.

$$\lim_{x\to\pm\infty}\frac{1-2x}{x^2-2x+1}=\lim_{x\to\pm\infty}\frac{\frac{1}{x}-2}{x-2+\frac{1}{x}}=0,$$

so there is a horizontal asymptote of $y=0$.

$$f'(x) = \frac{(x-1)^2(-2) - (1-2x)(2(x-1)(1))}{(x-1)^4}$$

$$= \frac{-2(x-1)\,[(x-1)+(1-2x)]}{(x-1)^4} = \frac{2x}{(x-1)^3}$$

$f'(x) = 0$ when $x = 0$ and $f'(x)$ is undefined when $x = 1$.

$$f''(x) = \frac{(x-1)^3(2) - (2x)(3(x-1)^2(1))}{(x-1)^6}$$

$$= \frac{2(x-1)^2[(x-1)-3x]}{(x-1)^6} = \frac{-2(1+2x)}{(x-1)^4}$$

$f''(x) = 0$ when $x = -\dfrac{1}{2}$ and $f''(x)$ is undefined when $x = 1$.

When $x < -\dfrac{1}{2}$, $f'(x) > 0$ so f is increasing

$f''(x) > 0$ so f is concave up

$-\dfrac{1}{2} < x < 0$, $f'(x) > 0$ so f is increasing

$f''(x) < 0$ so f is concave down

$0 < x < 1$, $f'(x) < 0$ so f is decreasing

$f''(x) < 0$ so f is concave down

$x > 1$, $f'(x) > 0$ so f is increasing

$f''(x) < 0$ so f is concave down.

There is a relative maximum when $x = 0$, or $(0, 1)$. There is an inflection point when $x = -\frac{1}{2}$, or $\left(-\frac{1}{2}, \frac{8}{9}\right)$.

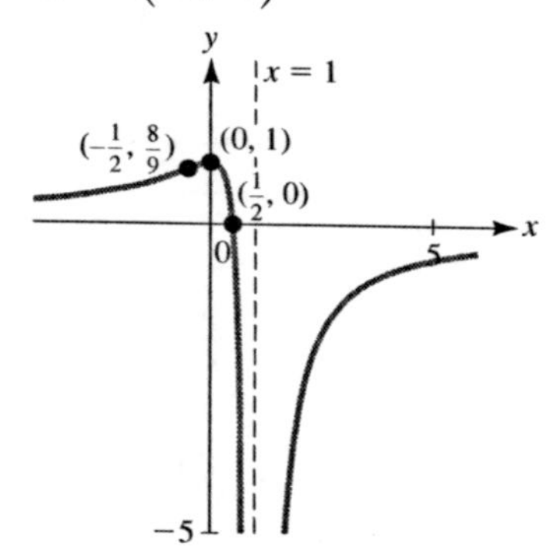

6. (a) Graph of f is increasing when $x < 0$ and $0 < x < 2$.
(b) Graph of f is decreasing when $x > 2$.
(c) Graph of f levels when $x = 0$ and $x = 2$; from parts (a) and (b), $x = 0$ is not a relative extremum and $x = 2$ is a relative maximum.
(d) Graph of f is concave down when $x < 0$ and $x > 1$.
(e) Graph of f is concave up when $0 < x < 1$; from parts (d) and (e), there are inflection points when $x = 0$ and $x = 1$.
(f) Graph of f goes through points $(-1, 0)$, $(4, 0)$, $(0, 1)$, $(1, 2)$ and $(2, 3)$.

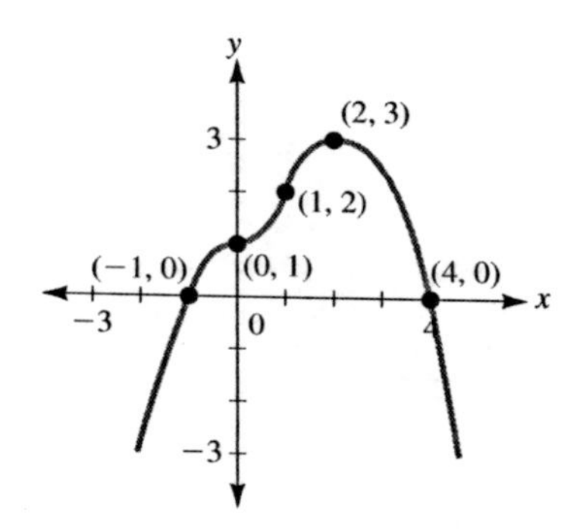

7. $f(t) = -t^3 + 7t^2 + 200t$ is the number of letters the clerk can sort in t hours. The clerk's rate of output is

$$R(t) = f'(t) = -3t^2 + 14t + 200$$

letters per hour. The relevant interval is $0 \le t \le 4$.

$$R'(t) = f''(t) = -6t + 14$$

$$R'(t) = 0 \text{ when } t = \frac{7}{3}$$

$$R\left(\frac{7}{3}\right) = 216.33, \; R(0) = 200, \text{ and } R(4) = 208$$

So, the rate of output is greatest when $t = \frac{7}{3}$ hours; that is, after 2 hours and 20 minutes, at 8:20 a.m.

8. Profit $=$ revenue–cost

$= $ (#sold)(selling price)–(#sold)(cost per unit)

$$P(x) = (120 - x)x - (120 - x)20$$
$$= (120 - x)(x - 20)$$

and the relevant domain is $x \geq 20$.

$$P'(x) = (120 - x)(1) + (x - 20)(-1)$$
$$= 140 - 2x$$
$$P'(x) = 0 \text{ when } x = 70$$

When $x = 20$, $P(20) = 0$.

When $20 < x < 70$, $P'(x) > 0$ so P is increasing

$x > 70$, $P'(x) < 0$ so P is decreasing.

So, when the selling price is \$70, the profit is maximized.

9. $C(t) = \dfrac{0.05t}{t^2 + 27}$

(a) The relevant domain of the function is $t \geq 0$.
When $t = 0$, $C(0) = 0$ so $(0, 0)$ is an intercept.
When $C(t) = 0$, $t = 0$.
$t^2 + 27$ is never zero, so there are no vertical asymptotes.

$$\lim_{t \to \pm\infty} \frac{0.05t}{t^2 + 27} = \lim_{t \to \pm\infty} \frac{0.05}{t + \dfrac{27}{t}} = 0,$$

so there is a horizontal asymptote of $y = 0$.

$$C'(t) = \frac{(t^2 + 27)(0.05) - (0.05t)(2t)}{(t^2 + 27)^2}$$
$$= \frac{1.35 - 0.05t^2}{(t^2 + 27)^2}$$

$C'(t) = 0$ when $t = \sqrt{27}$

$$C''(t) = \frac{1}{(t^2 + 27)^4}\Big[(t^2 + 27)^2(-0.1t) - (1.35 - 0.05t^2)(2(t^2 + 27)(2t))\Big]$$
$$= \frac{1}{(t^2 + 27)^4}\Big(t(t^2 + 27)\big[-0.1(t^2 + 27) - 4(1.35 - 0.05t^2)\big]\Big)$$
$$= \frac{t(0.1t^2 - 8.1)}{(t^2 + 27)^3}$$

$C''(t) = 0$ when $t = 9$

When

$0 < t < \sqrt{27}$, $C'(t) > 0$ so C is increasing
$C''(t) < 0$ so C is concave down

$\sqrt{27} < t < 9$, $C'(t) < 0$ so C is decreasing
$C''(t) < 0$ so C is concave down

$t > 9$, $C'(t) < 0$ so C is decreasing
$C''(t) > 0$ so C is concave up.

There is an absolute maximum when $t = \sqrt{27}$, or approximately (5.20, 0.005). There is an inflection point when $t = 9$, or approximately (9,0.004).

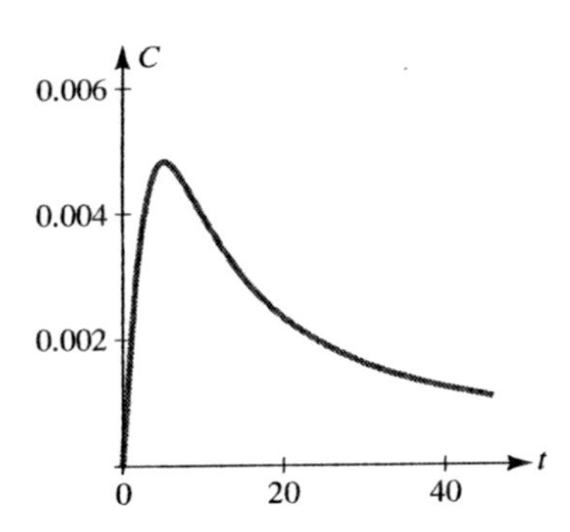

(b) $C'(t) < 0$ when $t > \sqrt{27}$, so C is decreasing when $t > \sqrt{27}$. The rate of decrease is maximized when $C''(t) = 0$ for $t > \sqrt{27}$, or when $t = 9$.

(c) $\displaystyle\lim_{t \to +\infty} \frac{0.05t}{t^2 + 27} = 0$, so the concentration tends to zero in the long run.

10. $P(t) = \dfrac{15t^2 + 10}{t^3 + 6}$
The relevant domain is $t \geq 0$.

(a) When $t = 0$, $P(0) = \dfrac{10}{6}$ or 1.667 million bacteria.

(b)
$$P'(t) = \frac{(t^3 + 6)(30t) - (15t^2 + 10)(3t^2)}{(t^3 + 6)^2} = \frac{-15t(t^3 + 2t - 12)}{(t^3 + 6)^2}$$

$P(t) = 0$ when $t = 0, 2$
When $t = 0$, $P(0) = \dfrac{10}{6}$.

When $0 < t < 2$, $P'(t) > 0$ so P is increasing
$t > 2$, $P'(t) < 0$ so P is decreasing.

So, when $t = 2$, the bacteria population is maximized and the maximum population is 5 million.

(c)
$$\lim_{t \to \infty} \frac{15t^2 + 10}{t^3 + 6} = \lim_{t \to \infty} \frac{15 + \frac{10}{t^2}}{t + \frac{6}{t^2}} = 0,$$

so in the long run, the bacteria population dies out.
Use all of the above information to graph P, noting also that $P(t)$ is never zero, so there are no other intercepts. $t^3 + 6$ is never zero, so there are no vertical asymptotes. $y = 0$ is the horizontal asymptote, and $P(2) = 5$.

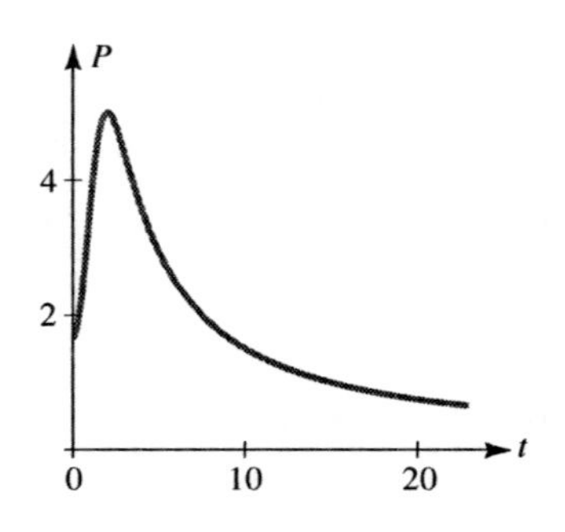

Review Problems

1. $f(x) = -2x^3 + 3x^2 + 12x - 5$
When $x = 0$, $f(0) = -5$ so $(0, -5)$ is an intercept.
$f(x) = 0$ is too difficult to solve.
There are no asymptotes.

$$f'(x) = -6x^2 + 6x + 12 = -6(x + 1)(x - 2)$$

$f'(x) = 0$ when $x = -1, 2$

$$f''(x) = -12x + 6 = -6(2x - 1)$$

$f''(x) = 0$ when $x = \dfrac{1}{2}$

When $x < -1$, $f'(x) < 0$ so f is decreasing
$f''(x) > 0$ so f is concave up
$-1 < x < \dfrac{1}{2}$, $f'(x) > 0$ so f is increasing
$f''(x) > 0$ so f is concave up
$\dfrac{1}{2} < x < 2$, $f'(x) > 0$ so f is increasing
$f''(x) < 0$ so f is concave down
$x > 2$, $f'(x) < 0$ so f is decreasing
$f''(x) < 0$ so f is concave down.

Overall, f is decreasing when $x < -1$ and $x > 2$
f is increasing when $-1 < x < 2$
f is concave down when $x > \dfrac{1}{2}$
f is concave up when $x < \dfrac{1}{2}$.

There is a relative minimum when $x = -1$, or $(-1, -12)$, and a relative maximum when $x = 2$, or $(2, 15)$. There is an inflection point when $x = \frac{1}{2}$, or $\left(\frac{1}{2}, \frac{3}{2}\right)$.

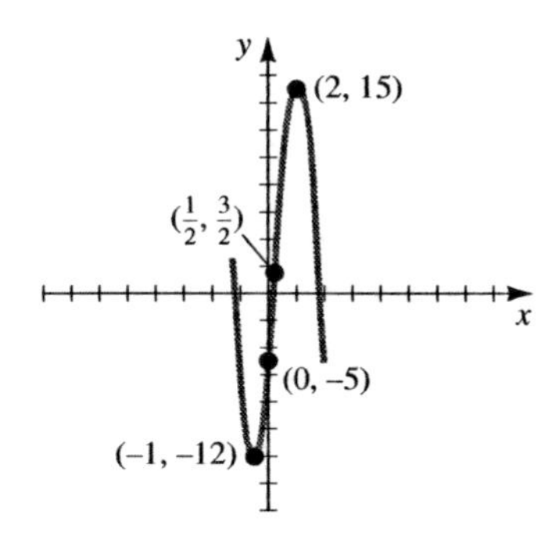

3. $f(x) = 3x^3 - 4x^2 - 12x + 17$
When $x = 0$, $f(0) = 17$ so $(0, 17)$ is an intercept
$f(x) = 0$ is too difficult to solve.
There are no asymptotes.

$$f'(x) = 9x^2 - 8x - 12$$

$f'(x) = 0$ when $x \approx -0.79, 1.68$

$$f''(x) = 18x - 8 = 2(9x - 4)$$

$f''(x) = 0$ when $x = \dfrac{4}{9}$

When $x < -0.79$, $f'(x) > 0$ so f is increasing
$f''(x) < 0$ so f is concave down

$-0.79 < x < \dfrac{4}{9}$, $f'(x) < 0$ so f is decreasing
$f''(x) < 0$ so f is concave down

$\dfrac{4}{9} < x < 1.68$, $f'(x) < 0$ so f is decreasing
$f''(x) > 0$ so f is concave up

$x > 1.68$, $f'(x) > 0$ so f is increasing
$f''(x) > 0$ so f is concave up.

Overall, f is decreasing when $-0.79 < x < 1.68$
f is increasing when $x < -0.79$ and $x > 1.68$

f is concave down when $x < \dfrac{4}{9}$

f is concave up when $x < \dfrac{4}{9}$.

There is a relative maximum when $x = -0.79$, or $(-0.79, 22.51)$, and a relative minimum when $x = 1.68$, or $(1.68, -0.23)$. There is an inflection point when $x = \frac{4}{9}$, or $(0.44, 11.1)$.

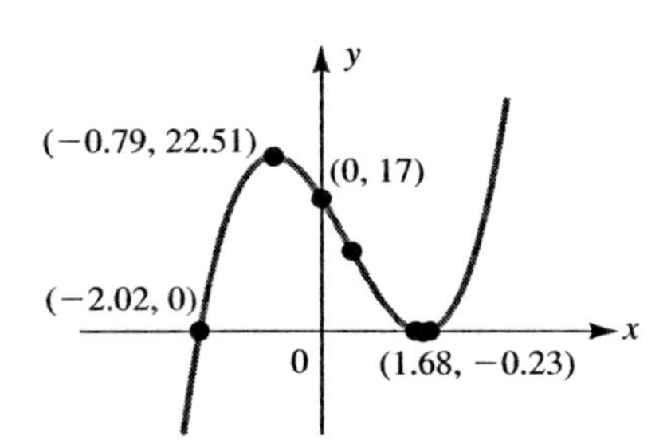

5. $f(t) = 3t^5 - 20t^3$
When $t = 0$, $f(0) = 0$ so $(0, 0)$ is an intercept.
When $f(t) = 3t^5 - 20t^3 = t^3(3t^2 - 20) = 0$ so

$$t = 0, \pm\sqrt{\frac{20}{3}} \text{ and } \left(\pm\sqrt{\frac{20}{3}}, 0\right)$$

are intercepts.
There are no asymptotes.

$$f'(t) = 15t^4 - 60t^2 = 15t^2(t + 2)(t - 2)$$

$f'(t) = 0$ when $t = -2, 0, 2$

$$f''(t) = 60t^3 - 120t = 60t(t^2 - 2)$$

$f''(t) = 0$ when $t = -\sqrt{2}, 0, \sqrt{2}$

When $t < -2$, $f'(t) > 0$ so f is increasing
$f''(t) < 0$ so f is concave down

$-2 < t < -\sqrt{2}$, $f'(t) < 0$ so f is decreasing
$f''(t) < 0$ so f is concave down

$-\sqrt{2} < t < 0$, $f'(t) < 0$ so f is decreasing
$f''(t) > 0$ so f is concave up

$0 < t < \sqrt{2}$, $f'(t) < 0$ so f is decreasing
$f''(t) < 0$ so f is concave down

$\sqrt{2} < t < 2$, $f'(t) < 0$ so f is decreasing
$f''(t) > 0$ so f is concave up

$t > 2$, $f'(t) > 0$ so f is increasing
$f''(t) > 0$ so f is concave up.

Overall,

f is decreasing when $-2 < t < 2$

f is increasing when $t < -2$ and $t > 2$

f is concave down when $t < -\sqrt{2}$ and $0 < t < \sqrt{2}$

f is concave up when $-\sqrt{2} < t < 0$ and $t > \sqrt{2}$.

There is a relative maximum when $t = -2$, or $(-2, 64)$, and a relative minimum when $t = 2$, or $(2, 64)$. There are inflection points when $t = -\sqrt{2}, \sqrt{2}$, or $(-1.4, 39.6)$ and $(1.4, -39.6)$.

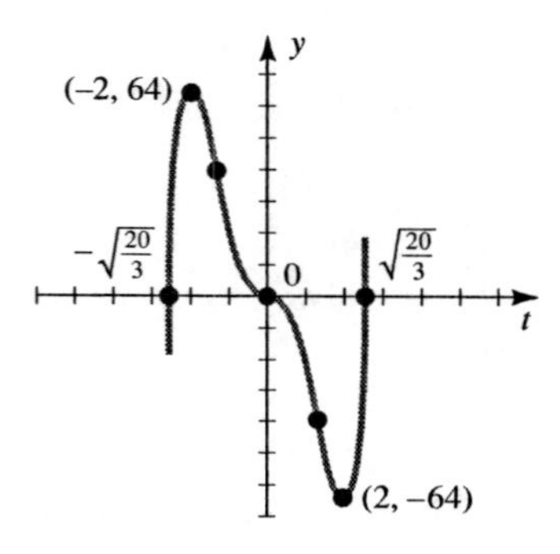

7. $g(t) = \dfrac{t^2}{t+1}$

When $t = 0$, $g(0) = 0$ so $(0, 0)$ is an intercept.
When $g(t) = 0$, $t = 0$.
$t + 1 = 0$ when $t = -1$, so there is a vertical asymptote of $t = -1$.

$$\lim_{t\to\pm\infty} \frac{t^2}{t+1} = \lim_{t\to\pm\infty} \frac{t}{1+\frac{1}{t}} = \pm\infty,$$

so there are no horizontal asymptotes.
Note: $y = t - 1$ is an oblique asymptote.

$$g'(t) = \frac{(t+1)(2t) - (t^2)(1)}{(t+1)^2}$$
$$= \frac{t^2 + 2t}{(t+1)^2} = \frac{t(t+2)}{(t+1)^2}$$

$g'(t) = 0$ when $t = -2, 0$ and $g'(t)$ is undefined when $t = -1$.

$$g''(t) = \frac{(t+1)^2(2t+2) - (t^2+2t)(2(t+1)(1))}{(t+1)^4}$$
$$= \frac{2(t+1)\left[(t+1)^2 - (t^2+2t)\right]}{(t+1)^4}$$
$$= \frac{2}{(t+1)^3}$$

$g''(t)$ is never zero and $g''(t)$ is undefined when $t = -1$.

When $t < -2$, $g'(t) > 0$ so g is increasing
$g''(t) < 0$ so g is concave down
$-2 < t < -1$, $g'(t) < 0$ so g is decreasing
$g''(t) < 0$ so g is concave down
$-1 < t < 0$, $g'(t) < 0$ so g is decreasing
$g''(t) > 0$ so f is concave up
$t > 0$, $g'(t) > 0$ so g is increasing
$g''(t) > 0$ so g is concave up.

Overall,

g is decreasing when $-2 < t < -1$ and $-1 < t < 0$

g is increasing when $t < -2$ and $t > 0$

g is concave down when $t < -1$

g is concave up when $t > -1$.

There is a relative maximum when $t = -2$, or $(-2, -4)$, and a relative minimum when $t = 0$, or $(0, 0)$. There are no inflection points.

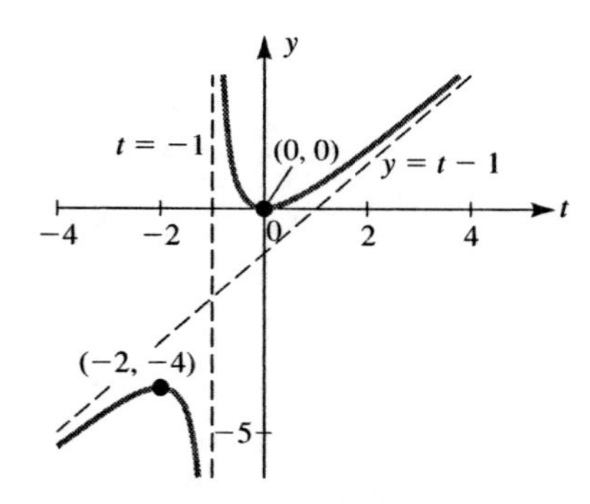

9. $F(x) = 2x + \dfrac{8}{x} + 2 = \dfrac{2x^2 + 2x + 8}{x}$

When $x = 0$, $F(0)$ is undefined.
$F(x) = 0$, $2(x^2 + x + 4) = 0$, which has no solution.

Denominator is zero 0 when $x = 0$, so there is a vertical asymptote of $x = 0$.

$$\lim_{x\to\pm\infty} \frac{2x^2 + 2x + 8}{x} = \lim_{x\to\pm\infty} \frac{2x + 2 + \frac{8}{x}}{1} = \pm\infty,$$

so there are no horizontal asymptotes.
Note: $y = 2x + 2$ is an oblique asymptote.

$$F'(x) = 2 - \frac{8}{x^2}$$

$F'(x) = 0$ when $x = -2$, 2 and $F'(x)$ is undefined when $x = 0$.

$$F''(x) = \frac{16}{x^3}$$

$F''(x)$ is never zero and $F''(x)$ is undefined when $x = 0$.

When $x < -2$, $F'(x) > 0$ so F is increasing
$F''(x) < 0$ so F is concave down
$-2 < x < 0$, $F'(x) < 0$ so F is decreasing
$F''(x) < 0$ so F is concave down
$0 < x < 2$, $F'(x) < 0$ so F is decreasing
$F''(x) > 0$ so F is concave up
$x > 2$, $F'(x) > 0$ so F is increasing
$F''(x) > 0$ so F is concave up.

Overall,
F is decreasing when $-2 < x < 0$ and $0 < x < 2$
F is increasing when $x < -2$ and $x > 2$
F is concave down when $x < 0$
F is concave up when $x > 0$.

There is a relative maximum when $x = -2$, or $(-2, -6)$, and a relative minimum when $x = 2$, or $(2, 10)$. There are no inflection points.

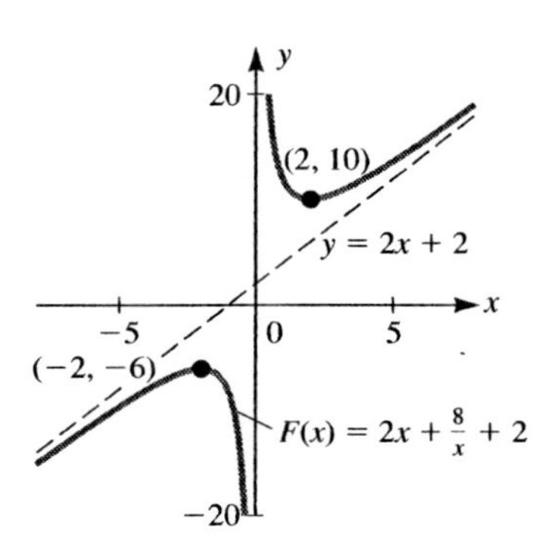

11. Graph (b) is the graph of f, and graph (a) is the graph of f'. Possible reasons include:

(i) The degree of graph (b) is one greater than the degree of graph (a).
(ii) Graph (a) is always positive, and graph (b) is always increasing.

13. $f'(x) = x^3(2x - 3)^2(x + 1)^5(x - 7)$
$f'(x) = 0$ when $x = -1, 0, \frac{3}{2}, 7$

When $x < -1$, $f'(x) < 0$ so f is decreasing
$-1 < x < 0$, $f'(x) > 0$ so f is increasing
$0 < x < \frac{3}{2}$, $f'(x) < 0$ so f is decreasing
$\frac{3}{2} < x < 7$, $f'(x) < 0$ so f is decreasing
$x > 7$, $f'(x) > 0$ so f is increasing.

There is a relative minimum when $x = -1$ and $x = 7$. There is a relative maximum when $x = 0$. There is no relative extremum when $x = \frac{3}{2}$.

15. $F'(x) = \dfrac{x(x - 2)^2}{x^4 + 1}$
$f'(x) = 0$, when $x = 0, 2$

When $x < 0$, $f'(x) < 0$ so f is decreasing
$0 < x < 2$, $f'(x) > 0$ so f is increasing
$0 > 2$, $f'(x) < 0$ so f is increasing.

There is a relative minimum when $x = 0$, but there is no relative extrema when $x = 2$.

17. **(a)** $f'(x) > 0$ so f is increasing when $x < 0$ and $x > 5$.

(b) $f'(x) < 0$ so f is decreasing when $0 < x < 5$.

(c) $f''(x) > 0$ so f is concave up when $-6 < x < -3$ and $x > 2$.

(d) $f''(x) < 0$ so f is concave down when $x < -6$ and $-3 < x < 2$.

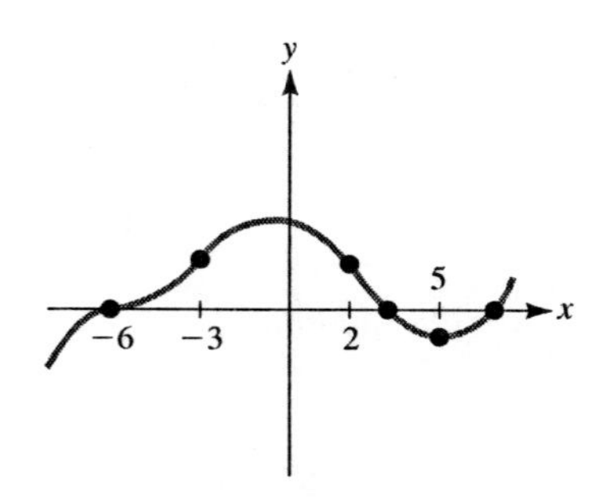

Note: since there are no points given, graphs can shift in y-direction, although not in x-direction.

19. **(a)** $f'(x) > 0$ so f is increasing when $1 < x < 2$

(b) $f'(x) < 0$ so f is decreasing when $x < 1$ and $x > 2$

(c) $f''(x) > 0$ so f is concave up when $x < 2$ and $x > 2$

(d) $f'(1) = 0$, so graph levels when $x = 0$
$f'(2)$ is undefined, so graph has a vertical asymptote, hole or vertical tangent when $x = 2$.

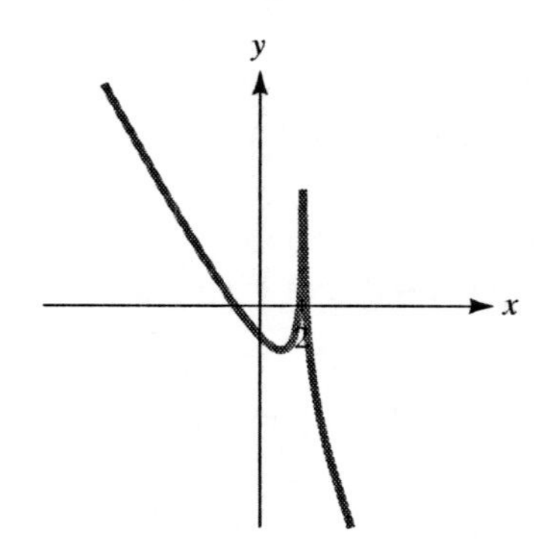

Note: since there are no points given, graphs can shift in y-direction, although not in x-direction.

21.

$$f(x) = -2x^3 + 3x^2 + 12x - 5$$
$$f'(x) = -6x^2 + 6x + 12 = -6(x+1)(x-2)$$

$f'(x) = 0$ when $x = -1, 2$
$f''(x) = -12x + 6$
$f''(-1) = 18 > 0$, so there is a relative minimum when $x = -1$, or $(-1, -12)$; $f''(2) = -18 < 0$, so there is a relative maximum when $x = 2$, or $(2, 15)$.

23. $f(x) = \dfrac{x^2}{x+1}$

$$f'(x) = \frac{(x+1)(2x) - (x^2)(1)}{(x+1)^2} = \frac{x(x+2)}{(x+1)^2}$$

$f'(x) = 0$ when $x = -2, 0$

$$f''(x) = \frac{(x+1)^2(2x+2) - (x^2+2x)(2(x+1)(1))}{(x+1)^2}$$

$f''(-2) = -2 < 0$, so there is a relative maximum when $x = -2$, or $(-2, -4)$; $f''(0) = 2 > 0$, so there is a relative minimum when $x = 0$, or $(0, 0)$.

25.

$$f(x) = -2x^3 + 3x^2 + 12x - 5$$
$$f(x) = -6x^2 + 6x + 12$$
$$f'(x) = -6(x+1)(x-2)$$

$f'(x) = 0$ when $x = -1, 2$, both of which are in the interval $-3 \le x \le 3$.
$f(-1) = -12$, $f(2) = 15$, $f(-3) = 40$, $f(3) = 4$.
So, $f(-3) = 40$ is the absolute maximum and $f(-1) = -12$ the absolute minimum.

27.

$$g(s) = \frac{s^2}{s+1}$$
$$g'(s) = \frac{(s+1)(2s) - (s^2)(1)}{(s+1)^2}$$
$$g'(s) = \frac{s(s+2)}{(s+1)^2}$$

$g'(s) = 0$ when $s = -2, 0$, of which only $s = 0$ is in the interval $-\dfrac{1}{2} \le s \le 1$.

$$g\left(-\frac{1}{2}\right) = \frac{1}{2},\ g(0) = 0,\ \text{and } g(1) = \frac{1}{2}$$

So, $g\left(-\frac{1}{2}\right) = g(1) = \frac{1}{2}$ is the absolute maximum and $g(0) = 0$ the absolute minimum.

29. $f'(x) = x(x-1)^2$

(a) $f'(x) = 0$ when $x = 0, 1$

When $x < 0$, $f'(x) < 0$ so f is decreasing
$0 < x < 1$, $f'(x) > 0$ so f is increasing
$x > 1$, $f'(x)) > 0$ so f is increasing.

(b)
$$f''(x) = x[2(x-1)(1)] + (x-1)^2(1) = (3x-1)(x-1)$$

$f''(x) = 0$ when $x = \frac{1}{3}, 1$

When $x < \frac{1}{3}$, $f''(x) > 0$ so f is concave up
$\frac{1}{3} < x < 1$, $f''(x) < 0$ so f is concave down
$x > 1$, $f''(x) > 0$ so f is concave up

(c) There is a relative minimum when $x = 0$ and there are inflection points when $x = \frac{1}{3}$ and $x = 1$.

(d)

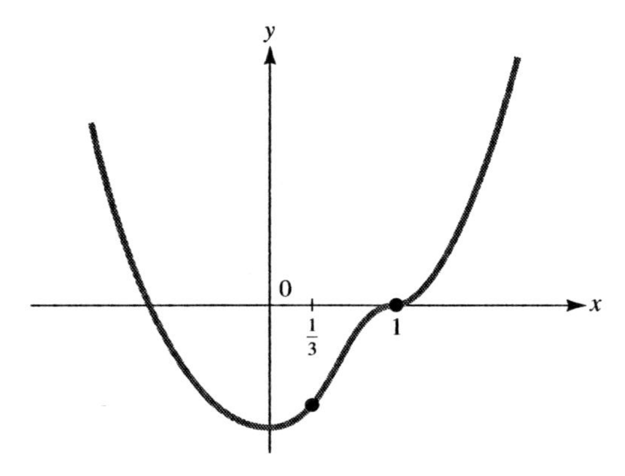

Note: since there are no points given, graph can shift in y-direction, although not in x-direction.

31. Profit = revenue–cost
= (#sold)(selling price)–(#sold)(cost per unit)

$$P(x) = (20-x)x - (20-x)5 = (20-x)(x-5)$$

The relevant domain is $x \geq 5$.

$$P'(x) = (20-x)(1) + (x-5)(-1) = 25 - 2x$$

$P'(x) = 0$ when $x = \frac{25}{2}$

When $5 \leq x < \frac{25}{2}$, $P'(x) > 0$ so P is increasing
$x > \frac{25}{2}$, $P'(x) < 0$ so P is decreasing

So, the profit is maximized when the selling price is \$12.50.

33. Let R denote the rate at which the rumor is spreading, N the number of people who have heard the rumor, and P the total population of the community. Then $R(N) = kN(P-N)$ where k is a positive constant of proportionality.

$$R'(N) = kN(-1) + (P-N)(k) = Pk - 2kN$$

$R'(N) = 0$ when $2kN = Pk$, or $N = \frac{P}{2}$, which is in the interval $0 \leq N \leq P$.

$$R''(N) = -2k$$
$$R''\left(\frac{P}{2}\right) < 0,$$

so the absolute maximum is reached when half the people have heard the rumor.

35. Let r denote the radius, h the height, C the (fixed) cost (in cents), and V the volume of the container.

$$V = \pi r^2 h$$
$$C = \text{cost of bottom} + \text{cost of side} = 3(\text{area of bottom}) + 2(\text{area of side})$$
$$\text{or } C = 3\pi r^2 + 4\pi rh$$

Solving for h,

$$h = \frac{C - 3\pi r^2}{4\pi r}$$

and

$$V(r) = \pi r^2 \left(\frac{C - 3\pi r^2}{4\pi r}\right)$$
$$= \frac{rC}{4} - \frac{3\pi r^3}{4}$$
$$V'(r) = \frac{C}{4} - \frac{9\pi r^2}{4}$$

$V'(r) = 0$ when $\frac{C}{4} = \frac{9\pi r^2}{4}$, or $C = 9\pi r^2$. So, $h = \frac{9\pi r^2 - 3\pi r^2}{4\pi r}$, or $h = \frac{3r}{2}$.

$$V''(r) = -\frac{9\pi r}{2}$$

$V''\left(\frac{3}{2}r\right) < 0$, so there is a relative maximum when $h = 1.5r$. Further, $V''(r) < 0$ for all r, so the volume is maximized when the height is 1.5 times the radius of the cylindrical container.

37. Profit = (#sold)(profit per card)
Let x be the number of 25 cent reductions in price. The profit per card will be
(selling price)−(cost to obtain)
$= (10 - 0.25x) - 5 = 5(1 - 0.05x)$
while the number of cards sold will be

$$25 + 5x = 5(5 + x)$$

The total profit will be

$$P(x) = 25(5 + x)(1 - 0.05x)$$
$$= 25(5 + 0.75x - 0.05x^2)$$
$$P'(x) = 18.75 - 2.5x$$
$$P'(x) = 0 \text{ when } x = 7.5$$

Since the number of 25 cent reductions must be an integer, and since $P(7) = P(8) = 195$, the store should lower the price by 8 reductions. That is, sell the cards for $10 - .25(8) = \$8$ per card. (Seven reductions yields the same profit, but using 8 reductions instead of 7 is good for the store's image.)

39. Let n denote the number of floors and $A(n)$ the corresponding average cost. Since the total cost is

$$C(n) = 2n^2 + 500n + 600 \text{ thousand dollars}$$
$$A(n) = \frac{C(n)}{n} = 2n + 500 + \frac{600}{n}$$

The relevant interval is $n > 0$.

$$A'(n) = 2 - \frac{600}{n^2} = \frac{2(n^2 - 300)}{n^2}$$

$A'(n) = 0$ when $n = \sqrt{300} \approx 17.32$

When $0 < n < 17.32$, $A'(n) < 0$ so A is decreasing
$n > 17.32$, $A'(n) > 0$ so A is increasing.

Since the number of floors must be an integer and $A(17) \approx 569.29$ and $A(18) \approx 569.33$, the average cost per floor is minimized when 17 floors are built.

41. Let Q be the point on the opposite bank straight across from the starting point. With $QP = x$, the distance walked along the bank is $1 - x$. The distance across the water is given by the pythagorean theorem to be $\sqrt{1 + x^2}$. The time t is

$$t = \text{time in the water} + \text{time on the land}$$
$$= \frac{\text{distance in the water}}{\text{speed in the water}} + \frac{\text{distance on the land}}{\text{speed on the land}}$$
$$= \frac{1}{4}(1 + x^2)^{1/2} + \frac{1}{5}(1 - x)$$

The relevant interval is $0 \leq x \leq 1$ and

$$t'(x) = \frac{x}{4\sqrt{1 + x^2}} - \frac{1}{5}$$

$t'(x) = 0$ when

$$\frac{x}{4\sqrt{1 + x^2}} = \frac{1}{5}$$
$$5x = 4\sqrt{1 + x^2}$$
$$25x^2 = 16 + 16x^2, \text{ or } x = \pm\frac{4}{3}$$

Neither of these critical values is in the interval $0 \leq x \leq 1$. So, the absolute minimum must occur at an endpoint.

$$t(0) = 0.45; \; t(1) = \frac{\sqrt{2}}{4} \approx 0.354$$

The minimum time is when $x = 1$. That is, when you row all the way to town.

43. Let x denote the number of machines used and $C(x)$ the corresponding cost of producing the 400,000 medals. Then

$$\begin{aligned} C(x) &= \text{set-up cost} + \text{operating cost} \\ &= 80 \text{ (number of machines)} \\ &\quad + 5.76 \text{ (number of hours)} \end{aligned}$$

Each machine can produce 200 medals per hour, so x machines can produce $200x$ medals per hour, and it will take $\frac{400{,}000}{200x}$ hours to produce the 400,000 medals. So,

$$\begin{aligned} C(x) &= 80x + 5.76\left(\frac{400{,}000}{200x}\right) \\ &= 80x + \frac{11{,}520}{x} \\ C'(x) &= 80 - \frac{11{,}520}{x^2} \\ &= \frac{80(x-12)(x+12)}{x^2} \end{aligned}$$

$C'(x) = 0$ when $x = 12$

When $0 < x < 12$, $C'(x) < 0$ so C is decreasing
$x > 12$, $C'(x) > 0$ so C is increasing.

So, the cost is minimized when 12 machines are used.

45. **(a)**

$$\begin{aligned} E(p) &= \frac{p}{q} \cdot \frac{dq}{dp} \\ &= \frac{p}{200 - 2p^2}(-4p) = -\frac{2p^2}{100 - p^2} \end{aligned}$$

(b) $E(6) = -\dfrac{2(6)^2}{100 - (6)^2} = -1.125$
A 1% increase in price will produce a decrease in demand of 1.125%.

(c) $-1 = \dfrac{-2p^2}{100 - p^2}$ or $p = \$5.77$

47. **(a)**

$$\begin{aligned} E(p) &= \frac{p}{q}\frac{dq}{dp} \\ &= \frac{p}{300 - 0.7p^2} \cdot -1.4p \\ &= \frac{-1.4p^2}{300 - 0.7p^2} \end{aligned}$$

(b) $E(8) = \dfrac{-1.4(8)^2}{300 - 0.7(8)^2} \approx -0.351$
Since $|E(8)| = 0.35 < 1$, revenue increases as the price increases. So, the cruise line should raise the price.

49. Let A be the amount of light per square foot transmitted through stained glass. Then $2A$ is the amount transmitted through the clear glass. The total light transmitted is

$$\begin{aligned} \text{total light} &= \text{(area rectangle)(2A)} \\ &\quad + \text{(area triangle)}(A) \end{aligned}$$

Let x be the dimension of one side of the triangle. Then the length of the rectangle is also x. Let y be the dimension of the width of the rectangle.

$$\text{area rectangle} = xy$$

but the total perimeter is 20, so

$$3x + 2y = 30, \text{ or } y = \frac{20 - 3x}{2}$$

$$\begin{aligned} \text{area triangle} &= \frac{1}{2}bh \\ &= \frac{1}{2}xh \end{aligned}$$

Using half of the triangle, h is the leg of a right triangle, with $\frac{x}{2}$ as its base, so

$$h = \sqrt{x^2 - \left(\frac{1}{2}x\right)^2} = \frac{\sqrt{3}}{2}x$$

The total light function, $L(x)$, is

$$L(x) = x\left(\frac{20-3x}{2}\right)(2A) + \frac{1}{2}(x)\left(\frac{\sqrt{3}}{2}x\right)(A)$$

$$= A\left(20x - 3x^2 + \frac{\sqrt{3}}{4}x^2\right)$$

$$L'(x) = A\left(20 - 6x + \frac{\sqrt{3}}{2}x\right)$$

$L'(x) = 0$ when

$$0 = 20 + \left(\frac{\sqrt{3}}{2} - 6\right)x, \text{ or}$$

$$x = \frac{20}{6 - \frac{\sqrt{3}}{2}} \approx 3.8956$$

When $0 < x < 3.896$, $L'(x) > 0$ so L is increasing

$x > 3.896$, $L'(x) < 0$ so L is decreasing

So, the light transmitted is maximized when the sides of the triangle and length of the rectangle are 3.896 feet, and the width of the rectangle is $\frac{20 - 3(3.8956)}{2} = 4.1566$ feet.

51. The relationship between the number of Moppsy dolls and Floppsy dolls is given by

$$y = \frac{82 - 10x}{10 - x}$$

with the relevant interval $0 \le x \le 8$.
Let C be the amount received from the sale of Floppsy doll. Then, $2C$ is the amount received from the sale of each Moppsy doll. The total revenue from the sale of both dolls is

$$R(x) = Cx + \frac{2C(82 - 10x)}{10 - x}$$

$$= C\left(\frac{164 - 10x - x^2}{10 - x}\right)$$

$$R'(x) = \frac{C}{(10 - x^2)}\Big[(10 - x)(-10 - 2x) - (164 - 10x - x^2)(-1)\Big]$$

$$= C\left(\frac{x^2 - 20x + 64}{(10 - x)^2}\right)$$

$$= C\frac{(x - 16)(x - 4)}{(10 - x)^2}$$

$R'(x) = 0$ when $x = 4$ ($x = 16$ is not in the interval)
Since $R(4) = 18C$, $R(0) = 16.4C$, and $R(8) = 10C$, revenue is maximized when 400 Floppsy and $\frac{82 - 10(4)}{10 - 4}$, or 700 Moppsy dolls are produced.

53. Let x denote the number of maps per batch and $C(x)$ the corresponding cost. Then,

$$C(x) = \text{(storage cost)} + \text{(production cost)} + \text{(set-up cost)}$$

The relevant interval is $0 < x \le 16{,}000$.

$$\text{storage cost} = \begin{pmatrix}\text{average}\\ \text{\#maps}\end{pmatrix}\begin{pmatrix}\text{storage cost}\\ \text{per map}\end{pmatrix} = \left(\frac{x}{2}\right)(0.20) = 0.1x$$

$$\text{production cost} = \begin{pmatrix}\text{total}\\ \text{\#maps}\end{pmatrix}\begin{pmatrix}\text{cost per}\\ \text{map}\end{pmatrix} = (16{,}000)(0.06) = 960$$

$$\text{set-up cost} = (\text{\#batches})\begin{pmatrix}\text{setup cost}\\ \text{per batch}\end{pmatrix} = \left(\frac{16{,}000}{x}\right)(100) = \frac{1{,}600{,}000}{x}$$

So,

$$C(x) = 0.1x + 960 + \frac{1{,}600{,}000}{x}$$

$$C'(x) = 0.1 - \frac{1{,}600{,}000}{x^2}$$

$C'(x) = 0$ when

$$0.1x^2 = 1{,}600{,}000$$

$$x^2 = 16{,}000{,}000, \text{ or}$$

$$x = 4{,}000$$

Using the second derivative test, since

$$C''(x) = \frac{3{,}200{,}000}{x^3}$$

$$C'(4{,}000) > 0$$

So cost is minimized when there are 4,000 maps in each batch.

55. Let x be the number of units ordered and k_1, k_2 constants of proportionality. Since the storage cost is $C_s = k_1 x$ and the ordering cost $C_0 = \frac{k_2}{x}$, the total cost is

$$C(x) = k_1 x + \frac{k_2}{x}$$

$$C'(x) = k_1 - \frac{k_2}{x^2}$$

$C'(x) = 0$ when $x = \sqrt{\frac{k_2}{k_1}}$

Using the second derivative test, since

$$C''(x) = \frac{2k_2}{x^3}$$

$$C''\left(\sqrt{\frac{k_2}{k_1}}\right) > 0$$

So cost is minimized when

$$C_s = k_1\sqrt{\frac{k_2}{k_1}} = \sqrt{k_1 k_2}$$

$$C_0 = \frac{k_2}{\sqrt{\frac{k_2}{k_1}}} = k_2\sqrt{\frac{k_1}{k_2}} = \sqrt{k_1 k_2}$$

That is, when the storage cost equals the ordering cost.

57. $f(x) = \frac{K(1+c^2x^3)}{(1+x)^3}$

(a)

$$f'(x) = \frac{K}{(1+x)^6}\Big[(1+x)^3(3c^2x^2) - (1+c^2x^3)(3(1+x)^2(1))\Big]$$

$$= \frac{3K(c^2x^2-1)}{(1+x)^4}$$

$f'(x) = 0$ when $c^2x^2 - 1 = 0$, or $x = \frac{1}{c}$

$$f''(x) = \frac{3K}{(1+x)^8}\Big[(1+x)^4(2c^2x) - (c^2x^2-1)(4(1+x)^3(1))\Big]$$

$$= 3K\frac{2(1+x)^3(2+c^2x-c^2x^2)}{(1+x)^8}$$

$$= 6K\frac{(2+c^2x-c^2x^2)}{(1+x)^5}$$

$$f''\left(\frac{1}{c}\right) = 6K\frac{(2+c-1)}{\left(1+\frac{1}{c}\right)^5}$$

$$= 6K\frac{(1+c)}{\left(1+\frac{1}{c}\right)^5} > 0$$

So, there is a relative minimum when $x = \frac{1}{c}$.

(b) With $c = 1$, $f'(x) = 0$ when $x = 1$

$$f(1) = \frac{2\pi}{3}\frac{(1+1)}{(1+1)^3} = \frac{\pi}{6} \approx 0.524$$

$$f\left(\sqrt{2}-1\right) = \frac{2\pi}{3}\frac{\left[1+\left(\sqrt{2}-1\right)^3\right]}{\left[1+\left(\sqrt{2}-1\right)\right]^3} \approx 0.793$$

So the minimum is 0.524 and the maximum is 0.793.

(c) With $C = \sqrt{2}$, $f'(x) = 0$ when $x = \frac{1}{\sqrt{2}}$

$$f\left(\frac{1}{\sqrt{2}}\right) = \frac{\sqrt{3}\pi}{16} \frac{\left[1+2\left(\frac{1}{\sqrt{2}}\right)^3\right]}{\left[1+\left(\frac{1}{\sqrt{2}}\right)\right]^3} \approx 0.117$$

$$f(0) = \frac{\sqrt{3}\pi}{16} \frac{(1+0)}{(1+0)^3} \approx 0.340$$

$$f(1) = \frac{\sqrt{3}\pi}{16} \frac{(1+2)}{(1+1)^3} \approx 0.128$$

So, the minimum is 0.117 and the maximum is 0.340.

(d) $$f(x) = \frac{K + Kc^2x^3}{1+3x+3x^2+x^3}$$

$$\lim_{x\to\infty} f(x) = \lim_{x\to\infty} \frac{\frac{K}{x^3} + Kc^2}{\frac{1}{x^3} + \frac{3}{x^2} + \frac{3}{x} + 1} = Kc^2$$

So, when r is much larger than R, the packing fraction depends only on the cell numbers c and K.

(e) Writing Exercise—Answers will vary.

59. $$R(S) = \frac{cS}{a+S+bS^2}$$

(a) domain: using the quadratic formula, the denominator is never zero, so the practical domain is $[0, \infty)$
intercepts: when $S = 0$, $R(0) = 0$; pt $(0, 0)$
when $R(S) = 0$, $S = 0$
asymptotes: no vertical asymptotes (since denominator is never zero)

$$\lim_{S\to\infty} \frac{\frac{C}{S}}{\frac{a}{S^2} + \frac{1}{S} + b} = 0,$$

so $y = 0$ is a horizontal asymptote

$$R'(S) = c\left[\frac{(a+S+bS^2)(1) - (S)(1+2bS)}{(a+S+bS^2)^2}\right]$$

$$= c\frac{a-bS^2}{(a+S+bS^2)^2}$$

$R'(S) = 0$ when $a - bS^2 = 0$
or $S = \sqrt{\frac{a}{b}}$ (rejecting negative answer)

When $0 \le S < \sqrt{\frac{a}{b}}$, $R'(S) > 0$, so R is increasing

$S > \sqrt{\frac{a}{b}}$, $R'(S) < 0$, so R is decreasing.

So, there is a relative maximum (which is also the absolute maximum) when $S = \sqrt{\frac{a}{b}}$.
(Note: the second derivative is too complex to use in sketching the graph.)

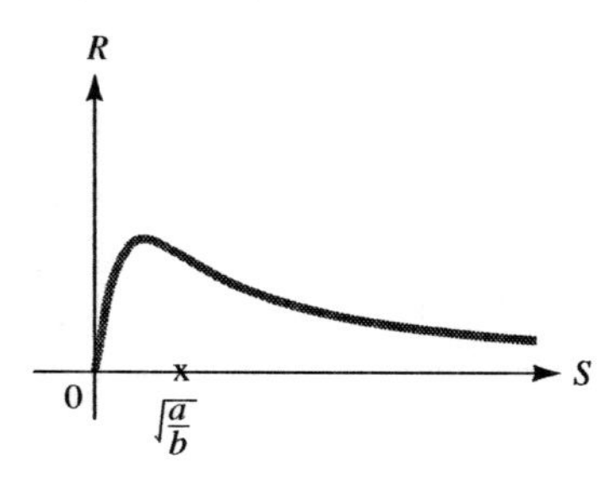

The lowest point is (0, 0). Since the graph starts concave down but then approaches the S axis asymptotically, there must be an inflection point. Since R approaches zero as S gets larger and larger, the growth rate, R', must also approach zero.

(b) Writing exercise—Answers will vary.

61. $N_1 = 3{,}351p^{0.287}$; $N_2 = 207.8p^{0.349}$

(a) $$E_1(p) = \frac{P}{N_1} \cdot \frac{dN_1}{dp}$$

$$= \frac{p}{3351p^{0.287}}(3351)(0.287)p^{-0.713}$$

$$= 0.287$$

So, the percentage increase for each 1% increase in price is 0.287%. When the price increases 2%, property crimes increase by 0.574%. When the price increases 5%, property crimes increase by 1.435%.

(b) $$E_2(p) = \frac{P}{N_2} \cdot \frac{dN_2}{dp}$$
$$= \frac{p}{207.8p^{0.349}}(207.8)(0.349)p^{-0.651}$$
$$= 0.349$$

When prices increase by 2%, personal crimes increase by 0.698%. When prices increase by 5%, personal crimes increase by 1.745%.

(c) $N = N_1 + N_2 = 3351p^{0.287} + 207.8p^{0.349}$

$$E(p) = \frac{p}{N} \cdot \frac{dN}{dp}$$
$$= \frac{p}{3351p^{0.287} + 207.8p^{0.349}}\left(961.737p^{-0.713} + 72.5222p^{-0.651}\right)$$
$$= \frac{p}{3351p^{0.287} + 207.8p^{0.349}}\left(\frac{961.737}{p^{0.713}} + \frac{72.5222}{p^{0.651}}\right)$$

When $p = \$75$, $E(75) \approx 0.292$ and a 5% increase in price results in a 1.46% increase in total crimes.

(d) From part c,

$$E(p) = \frac{p}{3351p^{0.287} + 207.8p^{0.349}}\left(\frac{961.737}{p^{0.713}} + \frac{72.5222}{p^{0.651}}\right)$$
$$E(p) = \frac{p}{3351p^{0.287} + 207.8p^{0.349}}\left(\frac{961.737p^{0.651} + 72.5222p^{0.713}}{p^{0.713}p^{0.651}}\right)$$
$$E(p) = \frac{961.737p^{1.651} + 72.5222p^{1.713}}{3351p^{1.651} + 207.8p^{1.713}}$$

Since we wish to find the price p that a 17% increase in price results in a 5% increase in total crime, we must solve
$$E(p) = \frac{5}{17}$$
Press [y=] and input

$$\frac{961.737x^{1.651} + 72.5222x^{1.713}}{3351x^{1.651} + 207.8x^{1.713}} - \frac{5}{17} \text{ for } y_1 = .$$

Note we are using x in place of p on the calculator.

Press [math]. Use the solver option under the math menu and set $y_1 = 0$. You can find y_1 under the y-vars menu with the [vars] options. Enter function and choose y_1. This gives the answer of $x = 148{,}000$.

(e) Writing exercise—Answers will vary.

Chapter 4

Exponential and Logarithmic Functions

4.1 Exponential Functions

1. Using the TI-84 Plus, find e_2 by pressing [2nd] e^x, then 2, right parenthesis, and [enter] to get $e^2 \approx 7.389$ Similarly $e^{-2} \approx 0.135$, $e^{0.05} \approx 1.051$, $e^{-0.05} \approx 0.951$, $e^0 = 1$, $e \approx 2.718$, $\sqrt{e} \approx 1.649$, $\frac{1}{\sqrt{e}} \approx 0.607$.

3.

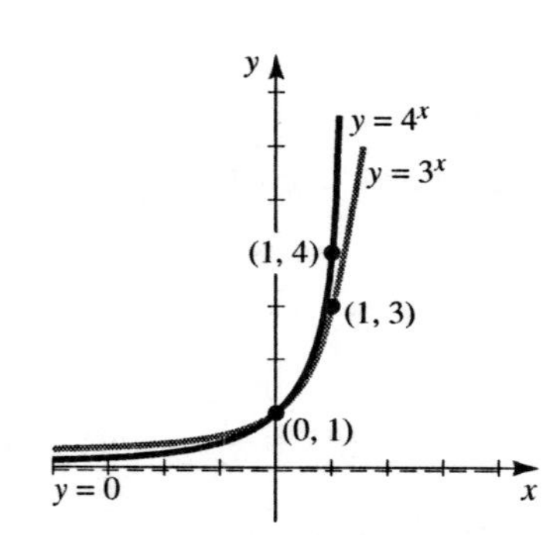

5. **(a)** $27^{2/3} = (27^{1/3})^2 = 3^2 = 9$

(b) $\left(\frac{1}{9}\right)^{3/2} = \frac{1}{(9^{1/2})^3} = \frac{1}{3^3} = \frac{1}{27}$

7. **(a)** $8^{2/3} + 16^{3/4} = \left(8^{1/3}\right)^2 + \left(16^{1/4}\right)^3$
$= 2^2 + 2^3 = 4 + 8 = 12$

(b) $\left(\frac{27+36}{121}\right)^{3/2} = \frac{\left(63^{1/2}\right)^3}{\left(121^{1/2}\right)^3} = \frac{\left(3\sqrt{7}\right)^3}{11^3} = \frac{189\sqrt{7}}{1331}$

9. **(a)** $(3^3)(3^{-2}) = 3^{3+-2} = 3$

(b) $(4^{2/3})(2^{2/3}) = (2^2)^{2/3}(2^{2/3})$
$= (2^{4/3})(2^{2/3}) = 2^{6/3} = 4$

11. **(a)** $(3^2)^{5/2} = 3^{2(5/2)} = 3^5 = 243$

(b) $(e^2 e^{3/2})^{4/3} = (e^{2+\frac{3}{2}})^{4/3} = (e^{7/2})^{4/3}$
$= e^{(7/2)(4/3)} = e^{14/3}$

13. **(a)** $(27x^6)^{2/3} = 27^{2/3} \cdot x^{6(2/3)}$
$= (27^{1/3})^2 x^4 = (3)^2 x^4 = 9x^4$

(b) $(8x^2y^3)^{1/3} = 8^{1/3}x^{2(\frac{1}{3})}y^{3(\frac{1}{3})} = 2x^{2/3}y$

15. **(a)** $\frac{(x+y)^0}{(x^2y^3)^{1/6}} = \frac{1}{x^{2(1/6)}y^{3(1/6)}} = \frac{1}{x^{1/3}y^{1/2}}$

(b) $(x^{1.1}y^2)(x^2+y^3)^0 = x^{1.1}y^2 \cdot 1 = x^{1.1}y^2$

17. **(a)** $(t^{5/6})^{-6/5} = t^{(5/6)(-6/5)} = t^{-1} = \frac{1}{t}$

(b) $(t^{-3/2})^{-2/3} = t^{(-3/2)(-2/3)} = t^1 = t$

19.

$$\begin{aligned} 4^{2x-1} &= 16 \\ (2^2)^{2x-1} &= (2)^4 \\ 2^{2(2x-1)} &= 2^4 \\ 2^{4x-2} &= 2^4 \end{aligned}$$

By the equality rule of exponential functions, $4x - 2 = 4$, or $x = \frac{3}{2}$.

21.

$$\begin{aligned} 2^{3-x} &= 4^x \\ 2^{3-x} &= (2^2)^x \\ 2^{3-x} &= 2^{2x} \end{aligned}$$

By the equality rule of exponential functions, $3 - x = 2x$, or $x = 1$.

23. $(2.14)^{x-1} = (2.14)^{1-x}$
By the equality rule of exponential functions, $x - 1 = 1 - x$, or $x = 1$.

25.
$$y = Cb^x$$
$$12 = Cb^2 \text{ and } 24 = Cb^3$$

Solving the first equation for C, $C = \frac{12}{b^2}$
Substituting into second equation,

$$24 = \left(\frac{12}{b^2}\right) b^3$$
$$24 = 12b$$
$$2 = b$$
$$C = \frac{12}{(2)^2} = 3$$

27. The balance after t years, if P dollars is invested at an annual interest rate r and interest is compounded k times per year, will be

$$B(t) = P\left(1 + \frac{r}{k}\right)^{kt} \text{ dollars}$$

continuously, will be $B(t) = Pe^{rt}$ dollars.
When $P = \$1{,}000$, $r = 0.07$, $t = 10$, and

(a) $k = 1$

$$B(10) = 1,000\left(1 + \frac{0.07}{1}\right)^{1 \cdot 10} \approx \$1,967.15$$

(b) $k = 4$

$$B(10) = 1,000\left(1 + \frac{0.07}{4}\right)^{4 \cdot 10} \approx \$2,001.60$$

(c) $k = 12$

$$B(10) = 1,000\left(1 + \frac{0.07}{12}\right)^{12 \cdot 10} \approx \$2,009.66$$

(d) compounded continuously

$$B(10) = 1,000e^{0.07(10)} \approx \$2,013.75$$

29. If $B(t) = \$5{,}000$, $r = 0.07$, $k = 4$, and $t = 5$,

$$5000 = P\left(1 + \frac{0.07}{4}\right)^{4 \cdot 5}$$

Solving for P,

$$P = 5000\left(1 + \frac{0.07}{4}\right)^{-20}$$
$$P \approx \$3,534.12$$

31. If $B(t) = \$9{,}000$, $r = 0.07$, $t = 5$, and

(a) $k = 4$

$$9{,}000 = P\left(1 + \frac{0.07}{4}\right)^{4 \cdot 5}$$
$$P = 9,000\left(1 + \frac{0.07}{4}\right)^{-20}$$
$$P \approx \$6{,}361.42$$

(b) Compounded continuously

$$9{,}000 = Pe^{0.07(5)}$$
$$P = 9000e^{-0.35}$$
$$P \approx \$6,342.19$$

33. $P(t) = 50e^{0.02t}$

(a) For the current population, $t = 0$ so

$$P(0) = 50e^0 = 50$$

so the current population is 50 million.

(b) When $t = 30$,

$$P(30) = 50e^{0.02(30)} \approx 91.11$$

so the population will be approx. 91.11 million

35. $f(x) = e^{kx}$

$$f(1) = e^{k \cdot 1} = 20$$
$$f(2) = e^{2k} = (e^k)^2 = (20)^2 = 400$$

37. $f(x) = A \cdot 2^{kx}$

$$f(0) = A \cdot 2^0 = A = 20, \text{ so}$$
$$f(x) = 20(2^{kx})$$
$$f(2) = 20(2^{2k}) = 40$$
$$2^{2k} = 2$$
$$f(8) = 20(2^{8k}) = 20[(2^{2k})^4] = 20(2)^4 = 320$$

39. Since the money doubles in 15 years,

$$B(15) = P\left(1+\frac{r}{4}\right)^{4\cdot 15}$$
$$= P\left(1+\frac{r}{4}\right)^{60} = 2P$$

Now,

$$B(30) = P\left(1+\frac{r}{4}\right)^{4\cdot 30}$$
$$= \left[P\left(1+\frac{r}{4}\right)^{60}\right]^2 = (2P)^2 = 4P$$

So the money quadruples in 30 years.

41. Since the pattern of daily growth is

$$P_0 \to 1.031P_0$$
$$\to 1.031(1.031P_0)$$
$$\to 1.031\left[1.031(1.031P_0)\right]$$

it can be modeled by the function

$$P(t) = P_0(1.031)^t$$

Since the initial population is $P_0 = 10,000$, the population after 10 days is

$$P(10) = 10,000(1.031)^{10}$$
$$\approx 13,570 \text{ bacteria}$$

43. Since the growth is exponential, $P(t) = P_0e^{kt}$ where the initial number of bacteria is $P_0 = 5{,}000$ Also, $P(10) = 5{,}000e^{10k} = 8{,}000$, so $e^{10k} = \frac{8}{5}$

Now, $P(30) = 5000(e^{30k})$

$$= 5000(e^{10k})^3$$
$$= 5000\left(\frac{8}{5}\right)^3$$
$$= 20,480 \text{ bacteria}$$

45. Since the decay is exponential,

$$S(t) = S_0e^{-kt}$$

as of the time publicity is discontinued. So, the initial number of sales is $S_0 = 25{,}000$. Also,

$$S(1) = 25{,}000e^{-k\cdot 1} = 10{,}000,$$

so $e^{-k} = \frac{2}{5}$.

Now, $S(2) = 25{,}000e^{-k\cdot 2}$

$$= 25{,}000(e^{-k})^2$$
$$= 25{,}000\left(\frac{2}{5}\right)^2$$
$$= 4{,}000 \text{ copies.}$$

47. $D(x) = 12e^{-0.07x}$

(a) At the center of the city, the density is $D(0) = 12$, or 12,000 people per square mile.

(b) Ten miles from the center, the density is

$$D(10) = 12e^{-0.07(10)} = 12e^{-0.7} \approx 5.959,$$

or 5,959 people per square mile.

49. Since the decay is exponential and 500 grams were present initially,

$$Q(t) = 500e^{-kt}$$

Also, $Q(50) = 500e^{-50k} = 400$, so $e^{-50k} = \frac{4}{5}$

Now, $Q(200) = 500e^{-200k}$

$$= 500(e^{-50k})^4$$
$$= 500\left(\frac{4}{5}\right)^4$$
$$= 204.8 \text{ grams}$$

51. $f(t) = e^{-0.2t}$

(a) The fraction of toasters still working after 3 years is

$$f(3) = e^{-0.2(3)} \approx 0.5488.$$

(b) The fraction which fail during the first year is

$$f(0) - f(1)$$
$$= e^0 - e^{-0.2(1)} \approx 1 - 0.8187 = 0.1813$$

(c) The fraction which fail during the third year is

$$\begin{aligned} &f(2) - f(3) \\ &= e^{-0.2(2)} - e^{-0.2(3)} \\ &\approx 0.6703 - 0.5488 = 0.1215 \end{aligned}$$

53. $I = I_0 e^{-kx}$
When $x = 3$ meters, $I = 0.1 I_0$. So

$$\begin{aligned} 0.1 I_0 &= I_0 e^{-k \cdot 3}, \text{ or} \\ 0.1 &= e^{-3k} \end{aligned}$$

When $x = 1$ meter,

$$\begin{aligned} I &= I_0 e^{-k \cdot 1} \\ &= I_0 (e^{-3k})^{1/3} \\ &= I_0 (0.1)^{1/3} \\ &\approx 0.46 I_0 \end{aligned}$$

55. When interest is compounded quarterly, the effective rate is

$$\left(1 + \frac{0.06}{4}\right)^4 - 1 \approx .06136,$$

or 6.136%.
When interest is compounded continuously, the effective rate is

$$e^{0.06} - 1 \approx .06184,$$

or 6.184%.
So, continuously compounding exceeds quarterly compounding by approx. 0.05%.

57. $N(t) = N_0 e^{-0.217t}$

(a) Let $t = 0$ in the year 200 B.C. In the year 2010, 2210 years have passed, so $t = 2.21$. Since $N_0 = 500$,

$$\begin{aligned} N(2.21) &= 500 e^{-0.217(2.21)} \\ &\approx 309.5, \text{ or } 310 \text{ words} \end{aligned}$$

(b) Using $t = 0$ in the year 950 A.D., $t = 1$ in the year 1950; since $N_0 = 210$, the formula predicts

$$\begin{aligned} N(1) &= 210 e^{-0.217(1)} \\ &\approx 169.04, \text{ or } 169 \text{ words} \end{aligned}$$

Although not exact, this is a very good estimate of the actual word count. Recall that the formula is based on experiments and, therefore, can only be used as an estimate.

(c) Writing Exercise—Answers will vary.

59. $M = \dfrac{Ai}{1 - (1+i)^{-n}}$
When $A = 150{,}000$, $i = \dfrac{0.09}{12} = 0.0075$, $n = 360$,

$$M = \frac{150,000(0.0075)}{1 - (1.0075)^{-360}} = \$1,206.93$$

for the monthly payment

61. (a) The potential buyer is offering to pay you,

$$1000 + 160(36) = \$6{,}760$$

Using the amortization formula, monthly payments would be

$$M = \frac{5{,}000\left(\frac{0.12}{12}\right)}{1 - \left(1 + \frac{0.12}{12}\right)^{-12 \cdot 3}} \approx \$166.07$$

This way, you would receive

$$\$1{,}000 + (166.07)(36) = \$6{,}978.52$$

This is $\$6{,}978.52 - \$6{,}760 = \$218.52$ more than the potential buyer is offering.

(b) Writing Exercise—Answers will vary.

63. $f(x) = \dfrac{1}{2}\left(\dfrac{1}{4}\right)^x$

x	−2.2	−1.5	0	1.5	2.3
$f(x)$	10.5561	4	0.5	0.0625	0.02062

Press [y=] and input .5(.25 ∧ x) for $y_1 =$.
Press [2nd] [TBLSET] and enter ask independent with auto dependent. Press [2nd] [table] and enter $x = -2.2$, -1.5, 0, 1.5, and 2.3. The output values are displayed automatically.

65. To use a calculator to evaluate $\left(1 + \dfrac{1}{n}\right)^n$ for $n = -1{,}000, -2{,}000 \ldots - 50{,}000$, press [y=] and input (1 + (1 ÷ x)) ∧ x for $y_1 =$.

Press [2nd] [TBLSET] and input TblStart = − 1,000, Δ Tbl = − 1,000 and auto independent with auto dependent. Press [2nd] [table].
Following are some values from this table:

x	y_1
−1,000	2.7196
−2,000	2.719
−3,000	2.7187
−4,000	2,7186
−5,000	2.7186
⋮	⋮
−48,000	2.7183
−49,000	2.7183
−50,000	2.7183

As n decreases without bound, $\left(1+\frac{1}{n}\right)^n$ approaches $e \approx 2.71828$.

67. To use a calculator to estimate $\lim_{x\to\infty}\left(2-\frac{5}{2n}\right)^{n/3}$,
Press [y =] and input
(2 − (5 ÷ (2x))) ∧ (x ÷ 3) for $y_1 =$.
Press [2nd] [TBLSET] and input Tblstart = 10 and Δ Tbl = 10. Use auto independent with auto dependent. Press [2nd][table]. The following are a few values from the table:

x	y
10	6.4584
20	66.071
⋮	⋮
100	7.12×10^9

These values suggest that

$$\lim_{n\to+\infty}\left(2-\frac{5}{2n}\right)^{n/3} = +\infty.$$

4.2 Logarithmic Functions

1. Using the TI-84 Plus, press LN, the number, a right parenthesis, and then ENTER. When combined with powers of e, press LN, 2nd, e^x, the power of e, right parenthesis, and then ENTER. So, $\ln 1 = 0$, $\ln 2 \approx 0.693$, $\ln e = 1$, $\ln 5 \approx 1.609$, $\ln(1/5) \approx -1.609$, $\ln e^2 = 2$. Since $\ln x$ has a domain of $x > 0$, ln 0 and ln(−2) yield ERR: DOMAIN.

3. Since $\ln x$ and e^x are inverse operations, $\ln e^3 = 3$.

5. Since e^x and $\ln x$ are inverse operations, $e^{\ln 5} = 5$.

7.
$$\begin{aligned} e^{3\ln 2-2\ln 5} &= e^{\ln 2^3-\ln 5^2} \\ &= e^{\ln 8-\ln 25} \\ &= e^{\ln\left(\frac{8}{25}\right)} \\ &= \frac{8}{25}. \end{aligned}$$

9. $\log_3 270 = \log_3 27 + \log_3 10$
$= \log_3 27 + \log_3 5 + \log_3 2$
$= \log_3 3^3 + \log_3 5 + \log_3 2$
Since $\log_3 x$ and 3^x are inverse operations,
$= 3 + \log_3 5 + \log_3 2$.

11.
$$\begin{aligned} \log_3 100 &= \log_3(10)^2 \\ &= 2\log_3 10 \\ &= 2\log_3(2\cdot 5) \\ &= 2(\log_3 2 + \log_3 5) \\ &= 2\log_3 2 + 2\log_3 5. \end{aligned}$$

13. $\log_2(x^4y^3) = \log_2 x^4 + \log_2 y^3$
$= 4\log_2 x + 3\log_2 y$.

15.
$$\begin{aligned} \ln\sqrt[3]{x^2-x} &= \ln(x^2-x)^{1/3} \\ &= \frac{1}{3}\ln(x^2-x) \\ &= \frac{1}{3}\ln[x(x-1)] \\ &= \frac{1}{3}[\ln x + \ln(x-1)] \\ &= \frac{1}{3}\ln x + \frac{1}{3}\ln(x-1). \end{aligned}$$

17.
$$\begin{aligned}&\ln\left[\frac{x^2(3-x)^{2/3}}{\sqrt{x^2+x+1}}\right]\\&=\ln\left[x^2(3-x)^{2/3}\right]-\ln\sqrt{x^2+x+1}\\&=\ln x^2+\ln(3-x)^{2/3}-\ln(x^2+x+1)^{1/2}\\&=2\ln x+\frac{2}{3}\ln(3-x)-\frac{1}{2}\ln(x^2+x+1).\end{aligned}$$

19.
$$\begin{aligned}\ln(x^3e^{-x^2})&=\ln x^3+\ln e^{-x^2}\\&=3\ln x-x^2.\end{aligned}$$

21. $4^x=53$
Taking the natural log of both sides gives
$\ln 4^x=\ln 53$.
Using a rule of logarithms gives
$x\ln 4=\ln 53$
$x=\dfrac{\ln 53}{\ln 4}\approx 2.864$

23. $\log_3(2x-1)=2$
Rwriting in exponential form gives
$2x-1=3^2$
or $x=5$

25. $2=e^{0.06x}$
Taking the natural log of both sides gives
$$\ln 2=0.06x,\text{ or}$$
$$x=\frac{\ln 2}{0.06}\approx 11.552$$

27. $3=2+5e^{-4x}$
$$1=5e^{-4x}$$
$$\frac{1}{5}=e^{-4x}$$
Taking the natural log of both sides gives
$$\ln\frac{1}{5}=-4x,\text{ or}$$
$$x=\frac{\ln(1/5)}{-4}$$
Since $\ln\dfrac{1}{5}=\ln 1-\ln 5=0-\ln 5=-\ln 5$,
$x=\dfrac{-\ln 5}{-4}=\dfrac{\ln 5}{4}\approx 0.402.$

29.
$$\begin{aligned}-\ln x&=\frac{t}{50}+C\\\ln x&=\frac{-t}{50}-C\\e^{\ln x}&=e^{(-t/50)-C},\text{ or}\\x&=e^{(-t/50)-C}\end{aligned}$$

31.
$$\begin{aligned}\ln x&=\frac{1}{3}(\ln 16+2\ln 2)\\&=\frac{1}{3}(\ln 16+\ln 4)\\&=\frac{1}{3}\ln(16\cdot 4)\\&=\ln 64^{1/3}\end{aligned}$$
So, $\ln x=\ln 4$
$e^{\ln x}=e^{\ln 4}$
or $x=4$.

33. $3^x=e^2$
Taking the natural log of both sides gives
$$\ln 3^x=\ln e^2$$
So, $x\ln 3=2$
$$x=\frac{2}{\ln 3}\approx 1.820.$$

35.
$$\begin{aligned}\frac{25e^{0.1x}}{e^{0.1x}+3}&=10\\25e^{0.1x}&=10(e^{0.1x}+3)\\25e^{0.1x}&=10e^{0.1x}+30\\15e^{0.1x}&=30\\e^{0.1x}&=2\\\ln e^{0.1x}&=\ln 2\\0.1x&=\ln 2\\x&=\frac{\ln 2}{0.1}=10\ln 2\approx 6.9315\end{aligned}$$

37. $\log_2 x=5$
Rewriting in exponential form,
$$x=2^5$$
$\ln x=\ln 2^5$
$\ln x=5\ln 2\approx 3.4657$

39. $\log_5(2x) = 7$
Rewriting in exponential form,

$$\begin{aligned} 2x &= 5^7 \\ x &= \frac{5^7}{2} \\ \ln x &= \ln\left(\frac{5^7}{2}\right) \\ &= \ln 5^7 - \ln 2 \\ &= 7\ln 5 - \ln 2 \approx 10.5729. \end{aligned}$$

41.
$$\begin{aligned} \ln\frac{1}{\sqrt{ab^3}} &= \ln 1 - \ln\left(\sqrt{ab^3}\right) \\ &= 0 - \ln(ab^3)^{1/2} \\ &= -\frac{1}{2}\ln(ab^3) \\ &= -\frac{1}{2}[\ln a + \ln b^3] \\ &= -\frac{1}{2}\ln a - \frac{1}{2}\ln b^3 \\ &= -\frac{1}{2}\ln a - \frac{3}{2}\ln b \end{aligned}$$

Since $\ln a = 2$ and $\ln b = 3$,

$$= -\frac{1}{2}(2) - \frac{3}{2}(3) = -\frac{11}{2}.$$

43. $B(t) = Pe^{rt}$
After a certain time, the investment will have grown to $B(t) = 2P$ at the interest rate of 0.06. So,

$$\begin{aligned} 2P &= Pe^{0.06t} \\ 2 &= e^{0.06t} \\ \ln 2 &= \ln e^{0.06t} \\ \ln 2 &= 0.06t \end{aligned}$$

and $t = \dfrac{\ln 2}{0.06} = 11.55$ years.

45. $B(t) = Pe^{rt}$
Since money doubles in 13 years,

$$\begin{aligned} 2P &= B(13) = Pe^{13r} \\ 2 &= e^{13r} \\ \ln 2 &= \ln e^{13r} \\ \ln 2 &= 13r \end{aligned}$$

and $r = \dfrac{\ln 2}{13} = 0.0533$. The annual interest rate is 5.33%.

47. $B(t) = Pe^{rt}$
Since money doubles in 12 years,
$2P = B(12) = Pe^{12r}$
$2 = e^{12r}$
$\ln 2 = 12r$
and $r = \dfrac{\ln 2}{12} \approx 0.05776.$
To find t when money triples,

$$\begin{aligned} 3P &= B(t) = Pe^{0.05776t} \\ 3 &= e^{0.5776t} \\ \ln 3 &= 0.05776t \\ t &= \frac{\ln 3}{0.05776} \approx 19.02 \text{ years} \end{aligned}$$

49. $C(t) = 0.4(2 - 0.13e^{-0.02t})$

(a) After 20 seconds, the drug concentration is
$C(20) = 0.4(2 - 0.13e^{-0.02(20)})$
$\approx 0.765 \text{ g/cm}^3$
After 60 seconds, it is
$C(60) = 0.4(2 - 0.13e^{-0.02(60)})$
$\approx 0.784 \text{ g/cm}^3$

(b) To find the time for the given concentration,

$$\begin{aligned} 0.75 &= 0.4(2 - 0.13e^{-0.02t}) \\ 1.875 &= 2 - 0.13e^{-0.02t} \\ 0.13e^{-0.02t} &= 0.125 \\ e^{-0.02t} &\approx 0.9615 \\ -0.02t &\approx \ln 0.9615 \\ t &\approx \frac{\ln 0.9615}{-0.02} \approx 1.96 \text{ seconds} \end{aligned}$$

51. The decay function is of the form

$$Q(t) = Q_0e^{-kt}$$

Since the half-life is 1,690 years, $Q(1690) = \frac{1}{2}Q_0$

$$\text{and}\frac{1}{2}Q_0 = Q_0 e^{-k(1,690)}$$
$$\frac{1}{2} = e^{-1,690k}$$
$$\ln\frac{1}{2} = \ln e^{-1,690k}$$
$$\ln\frac{1}{2} = -1,690k$$
$$\frac{\ln\frac{1}{2}}{-1,690} = k$$

or $k = \dfrac{\ln 2}{1,690}$

The initial amount, $Q_0 = 50$ grams, will reduce to 5 grams when

$$5 = 50e^{-kt}$$
$$\frac{1}{10} = e^{-kt}$$
$$\ln\frac{1}{10} = \ln e^{-kt},$$
$$\ln\frac{1}{10} = -kt$$
$$\frac{\ln\frac{1}{10}}{-k} = t$$

or $t = \dfrac{\ln 10}{k}$

Substituting k from above,

$$t = \frac{\ln 10}{\left(\frac{\ln 2}{1,690}\right)} = \frac{1,690\ln 10}{\ln 2} \approx 5,614 \text{ years.}$$

53. $Q(t) = Q_0 e^{kt}$

Since initial amount, $Q_0 = 6{,}000$,

$$Q(t) = 6{,}000e^{kt}$$

When $t = 20$ minutes, $Q(20) = 9{,}000$.

So,
$$9{,}000 = 6{,}000e^{k(20)}$$
$$\frac{3}{2} = e^{20k}$$
$$\ln\frac{3}{2} = \ln e^{20k}$$
$$\ln\frac{3}{2} = 20k$$
$$\frac{\ln\frac{3}{2}}{20} = k$$

In general, $Q(t) = 6,000e^{\frac{\ln 1.5}{20}t}$

When $t = 60$ minutes,

$$Q(60) = 6,000e^{\frac{\ln 1.5}{20}(60)}$$
$$= 20{,}250 \text{ bacteria.}$$

55. $Q(t) = 500 - Ae^{-kt}$

When $t = 0$, $Q(0) = 300$ and

$$300 = 500 - Ae^{-k(0)}$$
$$300 = 500 - A, \text{ or } A = 200$$

So, $\quad Q(t) = 500 - 200e^{-kt}$

When $t = 6$ months, $Q(6) = 410$ and

$$410 = 500 - 200e^{-k(6)}$$
$$200e^{-6k} = 90$$
$$e^{-6k} = \frac{9}{20}$$
$$\ln e^{-6k} = \ln\frac{9}{20}$$
$$-6k = \ln\frac{9}{20}$$
$$k = \frac{\ln\frac{9}{20}}{-6}$$

So, $\quad Q(t) = 500 - 200e^{\frac{\ln 0.45}{6}t}$

When $t = 12$ months,

$$Q(12) = 500 - 200e^{\frac{\ln 0.45}{6}(12)}$$
$$= 459.5 \text{ units}$$

57. The decay function is of the form $R(t) = R_0 e^{-kt}$.

e((ln 0.45 ÷ 6)(12)) × 200 = 40.5

From the text page 306, the half-life of ^{14}C is 5,730 years, so

$$\frac{1}{2}R_0 = R_0e^{-k(5,730)}$$
$$\frac{1}{2} = e^{-5,730k}$$
$$\ln\frac{1}{2} = \ln e^{-5,730k}$$
$$\ln\frac{1}{2} = -5,730k$$
$$\frac{\ln\frac{1}{2}}{-5,730} = k$$
$$\text{or } k = \frac{\ln 2}{5,730}$$

When 28% of the original amount remains,

$$0.28R_0 = R_0e^{-kt}$$
$$0.28 = e^{-kt}$$
$$\ln 0.28 = \ln e^{-kt}$$
$$\ln 0.28 = -kt$$
$$\frac{\ln 0.28}{-k} = t$$

Substituting k from above,

$$t = \frac{\ln 0.28}{-\left(\frac{\ln 2}{5,730}\right)} = \frac{-5,730 \ln 0.28}{\ln 2} \approx 10,523 \text{ years.}$$

59. The decay function is of the form $R(t) = R_0e^{-kt}$. From the text page 306, the half-life of ^{14}C is 5,730 years, so

$$\frac{1}{2}R_0 = R_0e^{-k(5,730)}$$
$$\frac{1}{2} = e^{-5,730k}$$
$$\ln\frac{1}{2} = \ln e^{-5,730k}$$
$$\ln\frac{1}{2} = -5,730k$$
$$\frac{\ln\frac{1}{2}}{-5,730} = k$$

$$\text{or } k = \frac{\ln 2}{5,730}$$

When 99.7% of the original amount remains,

$$0.997R_0 = R_0e^{-kt}$$
$$0.997 = e^{-kt}$$
$$\ln 0.997 = \ln e^{-kt}$$
$$\ln 0.997 = -kt$$
$$\frac{\ln 0.997}{-k} = t$$

Substituting k from above,

$$t = \frac{\ln 0.997}{-\left(\frac{\ln 2}{5,730}\right)} = \frac{-5,730 \ln 0.997}{\ln 2} \approx 24.8 \text{ years.}$$

So, the painting in question was painted only 24.8 years ago. If the painting was actually $2003 - 1640 = 363$ years old, and p represents the percentage of ^{14}C currently present,

$$pR_0 = R_0e^{-\left(\frac{\ln 2}{5,730}\right)(363)}$$
$$p = e^{-\left(\frac{\ln 2}{5,730}\right)(363)}$$
$$p \approx 0.957, \quad \text{or} \quad 95.7\%$$

61. $f(t) = 70 - Ae^{-kt}$
When $t = 0$, $f(0) = 212$

So,
$$212 = 70 - Ae^{-k(0)}$$
$$212 = 70 - A, \text{ or}$$
$$A = -142$$
and
$$f(t) = 70 + 142e^{-kt}$$

Now, let t_i be the ideal drinking temperature. Then,
$t_i + 15 = f(2) = 70 + 142e^{-k(2)}$
or, $t_i = 55 + 142e^{-2k}$
Also, $t_i = f(4) = 70 + 142e^{-k(4)}$
so, $70 + 142e^{-4k} = 55 + 142e^{-2k}$
$142e^{-4k} - 142e^{-2k} + 15 = 0$
Letting $u = e^{-2k}$,
$142u^2 - 142u + 15 = 0$
Using the quadratic formula,

$$u = \frac{142 \pm \sqrt{(-142)^2 - (4)(142)(15)}}{2(142)}$$

so, $u \approx 0.1200445$

or, $e^{-2k} \approx 0.1200445$ and

$$\begin{aligned} t_i &= 55 + 142(0.1200445) \\ &\approx 72.05°\text{F}. \end{aligned}$$

63. $$T = T_a + (98.6 - T_a)(0.97)^t$$

When $T = 40°$F and $T_a = 10°$F,

$$\begin{aligned} 40 &= 10 + (98.6 - 10)(0.97)^t, \\ \frac{30}{88.6} &= (0.97)^t \\ \ln \frac{30}{88.6} &= \ln(0.97)^t \\ \ln \frac{30}{88.6} &= t \ln(0.97) \end{aligned}$$

so, $$t = \frac{\ln \frac{30}{88.6}}{\ln(0.97)} \approx 35.55 \text{ hrs}$$

This means the murder occurred around 1:27 a.m. on Wednesday. Blohardt was in jail at this time, so Scélérat must have committed the murder.

65. $$R = \frac{2}{3}\log_{10}\left(\frac{E}{10^{4.4}}\right)$$

(a) When $E = 5.96 \times 10^{16}$,

$$\begin{aligned} R &= \frac{2}{3}\log_{10}\left(\frac{5.96 \times 10^{16}}{10^{4.4}}\right) \\ &\approx 8.25 \end{aligned}$$

(b) When $R = 6.4$,

$$\begin{aligned} 6.4 &= \frac{2}{3}\log_{10}\left(\frac{E}{10^{4.4}}\right) \\ 9.6 &= \log_{10}\left(\frac{E}{10^{4.4}}\right) \\ 10^{9.6} &= 10^{\log_{10}\left(\frac{E}{10^{4.4}}\right)} \\ 10^{9.6} &= \frac{E}{10^{4.4}} \\ E &= (10^{9.6})(10^{4.4}) = 10^{14} \text{ joules} \end{aligned}$$

67. $R = \dfrac{\ln I}{\ln 10}$

(a) When $R = 8.3$,

$$\begin{aligned} 8.3 \ln 10 &= \ln I \\ \ln 10^{8.3} &= \ln I \\ e^{\ln 10^{8.3}} &= e^{\ln I} \\ 10^{8.3} &= I \end{aligned}$$

So, $I \approx 1.9953 \times 10^8$.

(b) When $R = 7.1$

$$\begin{aligned} 7.1 \ln 10 &= \ln I \\ \ln 10^{7.1} &= \ln I \\ 10^{7.1} &= I \\ I &\approx 1.2589 \times 10^7 \\ \frac{I_s}{I_k} &\approx \frac{1.9953 \times 10^8}{1.2589 \times 10^7} = \frac{1.9953}{0.12589} \\ &\approx 15.85 \text{ times more intense.} \end{aligned}$$

69. Intensity function is of the form $I(t) = I_0 e^{-kt}$
When $t = 20.9$ hours,
$I(20.9) = \frac{1}{2} I_0$

So, $$\begin{aligned} \frac{1}{2} I_0 &= I_0 e^{-k(20.9)} \\ \frac{1}{2} &= e^{-20.9k} \\ \ln \frac{1}{2} &= \ln e^{-20.9k} \\ \ln \frac{1}{2} &= -20.9k \\ \frac{\ln \frac{1}{2}}{-20.9} &= k \end{aligned}$$

or $k = \dfrac{\ln 2}{20.9}$

(a) When $t = 24$ hours,

$$\begin{aligned} I(24) &= I_0 e^{-\left(\frac{\ln 2}{20.9}\right)(24)} \\ &\approx I_0 \cdot 0.451 \end{aligned}$$

So approximately 45.1% of the original amount should be detected.

(b)

$$I(25) = I_0 e^{\left(\frac{\ln 0.5}{20.9}\right)(25)}$$
$$\approx I_0 \cdot 0.436$$

A total of 43.6% should remain in the entire body, and 43.6% − 41.3% = 2.3% remains outside of the thyroid gland.

71. Population function is of the form

$$P(t) = P_0 e^{kt}$$

Let $t = 0$ in 1960; then, $P_0 = 3$. In 1975, $t = 15$,

so,
$$4 = 3e^{k(15)}$$
$$\frac{4}{3} = e^{15k}$$
$$\ln(4/3) = \ln e^{15k}$$
$$\ln(4/3) = 15k$$
$$\frac{\ln(4/3)}{15} = k$$

So,
$$P(t) = 3e^{\left(\frac{\ln 4/3}{15}\right)t}.$$

The population will be 40 billion when

$$40 = 3e^{kt}$$
$$\frac{40}{3} = e^{kt}$$
$$\ln(40/3) = \ln e^{kt}$$
$$\ln(40/3) = kt$$
$$\frac{\ln(40/3)}{k} = t$$

Substituting k from above,

$$t = \frac{\ln(40/3)}{\frac{\ln(4/3)}{15}} = (\ln 40/3)\left(\frac{15}{\ln 4/3}\right) \approx 135.06 \text{ years}$$

Since 1960 + 135.06 = 2095.06, the maximum population will be reached in the year 2095.

73. (a)

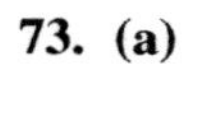

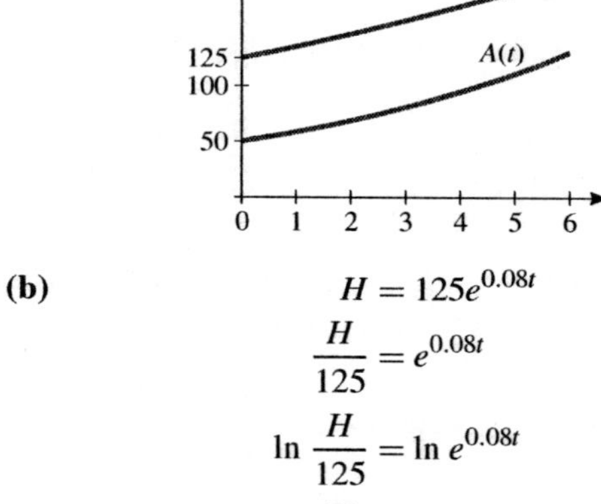

(b)

$$H = 125e^{0.08t}$$
$$\frac{H}{125} = e^{0.08t}$$
$$\ln \frac{H}{125} = \ln e^{0.08t}$$
$$\ln \frac{H}{125} = 0.08t$$
$$t = \frac{\ln \frac{H}{125}}{0.08}$$

Now, $A(t) = 50e^{0.16t}$

So,
$$A(H) = 50^{0.16(\ln(H/125)/0.08)}$$
$$= 50e^{2\ln(H/125)}$$
$$= 50e^{\ln(H/125)^2}$$
$$= 50\frac{H^2}{15{,}625}$$
$$= \frac{2H^2}{625}.$$

75. The midpoint of the segment joining the points (a, b) and (b, a) is

$$\left(\frac{a+b}{2}, \frac{b+a}{2}\right)$$

This point is on the line $y = x$; the slope of the line joining the points is

$$\frac{a-b}{b-a} = \frac{-(b-a)}{b-a} = -1$$

So the line is perpendicular to the line $y = x$, which has slope $= 1$. Now, using the midpoint found above, the distance from (a, b) to the line $y = x$ is

$$\sqrt{\left(a - \frac{a+b}{2}\right)^2 + \left(b - \frac{a+b}{2}\right)^2}$$

Similarly, the distance from (b, a) to the line $y = x$ is

$$\sqrt{\left(b - \frac{a+b}{2}\right)^2 + \left(a - \frac{a+b}{2}\right)^2}$$

which is the same distance. So, the reflection of the point (a, b) in the line $y = x$ is (b, a).

77.
$$\begin{aligned} y &= Cx^k \\ \ln y &= \ln(Cx^k) \\ &= \ln C + \ln x^k \\ &= \ln C + k \ln x \\ \ln y &= k \ln x + \ln c \\ \text{is of form} \quad Y &= mX + b \end{aligned}$$

So, $\ln y$ is a linear function of $\ln x$.

79.

$$x = \ln(3.42 \times 10^{-8.1})$$

Input $\ln(3.42 * 10 \wedge -8.1)$ and see that the output is approximately -17.4213. So $x \approx -17.4213$.
Note: Do not input $\ln(3.42$ [2nd] [EE] $- 8.1)$ as this results in an error.

81. $e^{0.113x} + 4.72 = 7.031 - x$
$x + e^{0.113x} - 2.311 = 0$
Press [y=] and input $x + e \wedge (0.113x) - 2.311$ for $y_1 =$
Press [graph]
Press [2nd] [calc] and use the zero function to find $x \approx 1.1697$.

83. (a) $(\log_a b)(\log_b a)$

$$= \left(\frac{\ln b}{\ln a}\right)\left(\frac{\ln a}{\ln b}\right) = 1$$

(b)
$$\begin{aligned} \frac{\log_b x}{\log_b a} &= \frac{\frac{\ln x}{\ln b}}{\frac{\ln a}{\ln b}} \\ &= \frac{\ln x}{\ln b} \cdot \frac{\ln b}{\ln a} \\ &= \frac{\ln x}{\ln a} = \log_a x \end{aligned}$$

4.3 Differentiation of Logarithmic and Exponential Functions

1. $f(x) = e^{5x}$

$$f'(x) = e^{5x} \frac{d}{dx}(5x) = 5e^{5x}$$

3.
$$\begin{aligned} f(x) &= xe^x \\ f'(x) &= x\frac{d}{dx}e^x + e^x\frac{d}{dx}x \\ &= x\left[e^x \frac{d}{dx}x\right] + e^x \cdot 1 \\ &= xe^x + e^x = e^x(x+1) \end{aligned}$$

5. $f(x) = 30 + 10e^{-0.05x}$

$$f'(x) = 0 + 10e^{-0.05x}\frac{d}{dx}(-0.05x) = -0.5e^{-0.05x}$$

7. $f(x) = (x^2 + 3x + 5)e^{6x}$

$$\begin{aligned} f'(x) &= (x^2+3x+5)\frac{d}{dx}e^{6x} + e^{6x}\frac{d}{dx}(x^2+3x+5) \\ &= (x^2+3x+5)\left[e^{6x}\frac{d}{dx}6x\right] + e^{6x}(2x+3) \\ &= 6(x^2+3x+5)(e^{6x}) + (2x+3)e^{6x} \\ &= e^{6x}\left[(6x^2+18x+30) + (2x+3)\right] \\ &= (6x^2+20x+33)e^{6x} \end{aligned}$$

9. $f(x) = (1 - 3e^x)^2$

$$f'(x) = 2(1-3e^x)\frac{d}{dx}(1-3e^x)$$
$$= 2(1-3e^x)(0-3e^x)$$
$$= -6e^x(1-3e^x)$$

11. $f(x) = e^{\sqrt{3x}}$

$$f'(x) = e^{\sqrt{3x}}\frac{d}{dx}\left(\sqrt{3x}\right)$$
$$= e^{\sqrt{3x}}\left(\sqrt{3}\frac{d}{dx}\sqrt{x}\right)$$
$$= \sqrt{3}e^{\sqrt{3x}}\left(\frac{1}{2}x^{-1/2}\right)$$
$$= \frac{\sqrt{3}}{2\sqrt{x}}e^{\sqrt{3x}} = \frac{3}{2\sqrt{3x}}e^{\sqrt{3x}}$$

13. $f(x) = \ln x^3 = 3\ln x$

$$f'(x) = 3\left(\frac{1}{x}\right) = \frac{3}{x}$$

15.

$$f(x) = x^2 \ln x$$
$$f'(x) = x^2\frac{d}{dx}(\ln x) + \ln x\frac{d}{dx}(x^2)$$
$$= x^2 \cdot \frac{1}{x} + 2x\ln x$$
$$= x + 2x\ln x$$
$$= x(1+2\ln x)$$

17. $f(x) = \sqrt[3]{e^{2x}} = e^{2x/3}$

$$f'(x) = e^{2x/3}\frac{d}{dx}\left(\frac{2x}{3}\right)$$
$$= \frac{2}{3}e^{2x/3}$$

19. $f(x) = \ln\left(\frac{x+1}{x-1}\right)$

$$f'(x) = \frac{1}{\left(\frac{x+1}{x-1}\right)}\frac{d}{dx}\left(\frac{x+1}{x-1}\right)$$
$$= \frac{x-1}{x+1}\left[\frac{(x-1)(1)-(x+1)(1)}{(x-1)^2}\right]$$
$$= \frac{-2}{(x+1)(x-1)}$$

21.

$$f(x) = e^{-2x} + x^3$$
$$f'(x) = e^{-2x}\frac{d}{dx}(-2x) + 3x^2$$
$$= -2e^{-2x} + 3x^2$$

23. $g(s) = (e^s + s + 1)(2e^{-s} + s)$

$$g'(s) = (e^s+s+1)\left(2e^{-s}\frac{d}{ds}(-s)+1\right)$$
$$+ (2e^{-s}+s)(e^s+1)$$
$$= (e^s+s+1)(-2e^{-s}+1) + (2e^{-s}+s)(e^s+1)$$
$$= -2e^0 - 2se^{-s} - 2e^{-s} + e^s + s + 1$$
$$+ 2e^0 + se^s + 2e^{-s} + s$$
$$= 1 + 2s + e^s + se^s - 2se^{-s}$$

25.

$$h(t) = \frac{e^t + t}{\ln t}$$
$$h'(t) = \frac{(\ln t)\frac{d}{dt}(e^t+t) - (e^t+t)\frac{d}{dt}(\ln t)}{(\ln t)^2}$$
$$= \frac{(\ln t)(e^t+1) - (e^t+t)\left(\frac{1}{t}\right)}{(\ln t)^2}$$
$$= \frac{t(\ln t)(e^t+1) - e^t - t}{t(\ln t)^2}$$

27.

$$f(x) = \frac{e^x + e^{-x}}{2} = \frac{1}{2}(e^x + e^{-x})$$
$$f'(x) = \frac{1}{2}(e^x - e^{-x})$$

29.

$$f(t) = \sqrt{\ln t + t} = (\ln t + t)^{1/2}$$
$$f'(t) = \frac{1}{2}(\ln t + t)^{-1/2}\frac{d}{dt}(\ln t + t)$$
$$= \frac{\frac{1}{t}+1}{2(\ln t + t)^{1/2}} = \frac{1+t}{2t\sqrt{\ln t + t}}$$

31.
$$f(x) = \ln(e^{-x} + x)$$
$$f'(x) = \frac{1}{e^{-x} + x}\frac{d}{dx}(e^{-x} + x)$$
$$= \frac{-e^{-x} + 1}{e^{-x} + x}$$

33. $g(u) = \ln\left(u + \sqrt{u^2 + 1}\right) = \ln\left(u + (u^2 + 1)^{1/2}\right)$

$$g'(u) = \frac{1}{u + (u^2 + 1)^{1/2}}\frac{d}{du}\left(u + (u^2 + 1)^{1/2}\right)$$
$$= \frac{1 + \frac{1}{2}(u^2 + 1)^{-1/2}(2u)}{u + (u^2 + 1)^{1/2}}$$
$$= \frac{1 + u(u^2 + 1)^{-1/2}}{u + (u^2 + 1)^{1/2}} \cdot \frac{u - (u^2 + 1)^{1/2}}{u - (u^2 + 1)^{1/2}}$$
$$= \frac{u + u^2(u^2 + 1)^{-1/2} - (u^2 + 1)^{1/2} - u}{u^2 - (u^2 + 1)}$$
$$= \frac{\frac{u^2}{(u^2+1)^{1/2}} - (u^2 + 1)^{1/2}}{-1}$$
$$= \frac{-u^2}{(u^2 + 1)^{1/2}} + (u^2 + 1)^{1/2}\frac{(u^2 + 1)^{1/2}}{(u^2 + 1)^{1/2}}$$
$$= \frac{-u^2 + u^2 + 1}{(u^2 + 1)^{1/2}} = \frac{1}{(u^2 + 1)^{1/2}}$$

35.
$$f(x) = \frac{\ln(x + 1)}{x + 1}, \quad 0 \le x \le 2$$
$$f'(x) = \frac{(x + 1) \cdot \frac{1}{x+1} - \ln(x + 1) \cdot 1}{(x + 1)^2}$$
$$= \frac{1 - \ln(x + 1)}{(x + 1)^2}$$

So, $f'(x) = 0$ when

$$1 - \ln(x + 1) = 0$$
$$1 = \ln(x + 1)$$
$$e^1 = e^{\ln(x+1)}$$
$$e = x + 1, \text{ or}$$
$$x = e - 1$$

$$f(e - 1) = \frac{\ln(e - 1 + 1)}{(e - 1) + 1} = \frac{1}{e} \approx 0.3679$$
$$f(0) = \frac{\ln(0 + 1)}{(0 + 1)^2} = 0$$
$$f(2) = \frac{\ln(2 + 1)}{2 + 1} \approx 0.3662$$

abs. max . $= \frac{1}{e}$; abs. min . $= 0$

37.
$$g(t) = t^{3/2}e^{-2t}; \quad 0 \le t \le 1$$
$$g'(t) = (t^{3/2})(e^{-2t} \cdot -2) + (e^{-2t})\left(\frac{3}{2}t^{1/2}\right)$$
$$= t^{1/2}e^{-2t}\left[-2t + \frac{3}{2}\right]$$

So, $g'(t) = 0$ when

$$t^{1/2} = 0 \to t = 0$$
$$e^{-2t} = 0 \to \text{no solution}$$
$$-2t + \frac{3}{2} = 0 \to t = \frac{3}{4}$$

$$g\left(\frac{3}{4}\right) = \left(\frac{3}{4}\right)^{3/2}\left(e^{-2\left(\frac{3}{4}\right)}\right) = \frac{3\sqrt{3}}{8}e^{-3/2} \approx 0.1449$$

$$g(0) = 0; \quad g(1) = e^{-2} \approx 0.1353$$

abs. max . $= \frac{3\sqrt{3}}{8}e^{-3/2}$; abs. min . $= 0$

39. $f(x) = xe^{-x}; \ x = 0$

$$f'(x) = (x)(e^{-x} \cdot -1) + (e^{-x})(1)$$
$$= e^{-x}(1 - x)$$

So, $\quad m = f'(0) = e^0(1 - 0) = 1$

Also, $f(0) = 0$, so point $(0, 0)$ is on tangent line and

$$y - 0 = 1(x - 0), \quad \text{or} \quad y = x.$$

41. $f(x) = \dfrac{e^{2x}}{x^2}, \quad x = 1$

$$f'(x) = \frac{(x^2)(e^{2x} \cdot 2) - (e^{2x})(2x)}{x^4}$$
$$= \frac{2xe^{2x}(x - 1)}{x^4} = \frac{2e^{2x}(x - 1)}{x^3}$$

so, $m = f'(1) = \dfrac{2e^2(1-1)}{1^3} = 0$

Since the slope of the line tangent is zero, the tangent line is horizontal and of the form $y = b$.
Since $f(1) = e^2$, the tangent line is $y = e^2$.

43.
$$f(x) = x^2 \ln \sqrt{x};\quad x = 1$$
$$f(x) = x^2 \ln x^{1/2} = \frac{1}{2}x^2 \ln x$$
$$f'(x) = \left(\frac{1}{2}x^2\right)\left(\frac{1}{x}\right) + (\ln x)(x)$$
$$= \frac{x}{2} + x \ln x$$

So, $m = f'(1) = \frac{1}{2} + \ln 1 = \frac{1}{2}$. Also, $f(1) = 1 \ln 1 = 0$, so the point $(1, 0)$ is on tangent line and

$$y - 0 = \frac{1}{2}(x-1),\ \text{ or } y = \frac{1}{2}x - \frac{1}{2}.$$

45.
$$f(x) = e^{2x} + 2e^{-x}$$
$$f'(x) = e^{2x} \cdot 2 + 2e^{-x} \cdot -1$$
$$= 2e^{2x} - 2e^{-x}$$
$$f''(x) = 2e^{2x} \cdot 2 - 2e^{-x} \cdot -1 = 4e^{2x} + 2e^{-x}$$

47.
$$f(t) = t^2 \ln t$$
$$f'(t) = (t^2)\left(\frac{1}{t}\right) + (\ln t)(2t)$$
$$= t(1 + 2\ln t)$$
$$f''(t) = (t)\left(2 \cdot \frac{1}{t}\right) + (1 + 2\ln t)(1)$$
$$= 2 + 1 + 2\ln t = 3 + 2\ln t$$

49.
$$f(x) = (2x+3)^2(x - 5x^2)^{1/2}$$
$$\ln f(x) = \ln\left[(2x+3)^2(x-5x^2)^{1/2}\right]$$
$$= \ln(2x+3)^2 + \ln(x-5x^2)^{1/2}$$
$$= 2\ln(2x+3) + \frac{1}{2}\ln(x-5x^2)$$

Differentiating,

$$\frac{f'(x)}{f(x)} = 2\left(\frac{1}{2x+3}\right)(2) + \frac{1}{2}\left(\frac{1}{x-5x^2}\right)(1-10x)$$
$$= \frac{4}{2x+3} + \frac{1-10x}{2(x-5x^2)}$$

Multiplying both sides by $f(x)$,

$$f'(x) = (2x+3)^2(x-5x^2)^{1/2}\left[\frac{4}{2x+3} + \frac{1-10x}{2(x-5x^2)}\right]$$

51. $f(x) = \dfrac{(x+2)^5}{\sqrt[6]{3x-5}}$.

$$\ln f(x) = \ln\left[\frac{(x+2)^5}{(3x-5)^{1/6}}\right]$$
$$= \ln(x+2)^5 - \ln(3x-5)^{1/6}$$
$$= 5\ln(x+2) - \frac{1}{6}\ln(3x-5)$$

Differentiating,

$$\frac{f'(x)}{f(x)} = \frac{5}{x+2} - \frac{3}{6(3x-5)}$$

Multiplying both sides by $f(x)$

$$f'(x) = \frac{(x+2)^5}{(3x-5)^{1/6}}\left[\frac{5}{x+2} - \frac{1}{2(3x-5)}\right]$$

53. $f(x) = (x+1)^3(6-x)^2\sqrt[3]{2x+1}$

$$\ln f(x) = \ln[(x+1)^3(6-x)^2(2x+1)^{1/3}]$$
$$= \ln(x+1)^3 + \ln(6-x)^2 + \ln(2x+1)^{1/3}$$
$$= 3\ln(x+1) + 2\ln(6-x) + \frac{1}{3}\ln(2x+1)$$

Differentiating,

$$\frac{f'(x)}{f(x)} = \frac{3}{x+1} + \frac{-2}{6-x} + \frac{2}{3(2x+1)}$$

Multiplying both sides by $f(x)$,

$$f'(x) = (x+1)^3(6-x)^2(2x+1)^{1/3} \cdot \left[\frac{3}{x+1} - \frac{2}{6-x} + \frac{2}{3(2x+1)}\right]$$

55.
$$f(x) = 5^{x^2}$$
$$\ln f(x) = \ln 5^{x^2}$$
$$= x^2 \ln 5$$

Differentiating,

$$\frac{f'(x)}{f(x)} = (\ln 5)2x$$

Multiplying both sides by $f(x)$,

$$f'(x) = (2 \ln 5) \cdot x \cdot 5^{x^2}$$

57. $D(p) = 3,000e^{-0.04p}$

(a)
$$E(p) = \frac{p}{q} \cdot \frac{dq}{dp}$$
$$= \frac{p}{3,000e^{-0.04p}}(3,000e^{-0.04p} \cdot -0.04)$$
$$= -0.04p$$
$$|E(p)| = |-0.04p| = 0.04p$$

Demand is of unit elasticity when $0.04p = 1$, or $p = 25$. Demand is elastic when $0.04p > 1$, or $p > 25$. Demand is inelastic when $0.04p < 1$, or $p < 25$.

(b) $E(15) = -0.04(15) = -0.60$, so a 2% increase in price results in a $(-0.60)(2) = -1.2$, or 1.2% decrease in demand.

(c) $R(p) = p \cdot 3000e^{-0.04p}$

$$R'(p) = (p)(3000e^{-0.04p} \cdot -0.04) + (3000e^{-0.04p})(1)$$
$$= 3000e^{-0.04p}(-0.04p + 1)$$

So $R'(p) = 0$ when $p = 25$.

59. $D(p) = 5000(p + 11)e^{-0.1p}$

(a) $E(p) = \dfrac{p}{q} \cdot \dfrac{dq}{dp}$

$$\frac{dq}{dp} = \frac{dD}{dp} = 5000\left[(p + 11)(e^{-0.1p} \cdot -0.1) + (e^{-0.1p})(1)\right]$$
$$= 5000e^{0.1p}\left[-0.1(p + 11) + 1\right]$$
$$= 5000e^{-0.1p}(-0.1p - 0.1)$$
$$= -500e^{-0.1p}(p + 1)$$

So,

$$E(p) = \frac{p}{5000(p + 11)e^{-0.1p}} \cdot -500e^{-0.1p}(p + 1)$$
$$= \frac{-p(p + 1)}{10(p + 11)}$$
$$|E(p)| = \left|\frac{-p(p + 1)}{10(p + 11)}\right| = \frac{p(p + 1)}{10(p + 11)}$$

Demand is of unit elasticity when

$$\frac{p(p + 1)}{10(p + 11)} = 1$$
$$p^2 + p = 10p + 110$$
$$p^2 - 9p - 110 = 0$$
$$p = \frac{9 \pm \sqrt{(-9)^2 - (4)(1)(-110)}}{2(1)} \approx 15.91$$

(rejecting the negative price)

Demand is elastic when $\dfrac{p(p + 1)}{10(p + 11)} > 1$

or $p > 15.91$

Demand is inelastic when $\dfrac{p(p + 1)}{10(p + 11)} < 1$

or $p < 15.91$

(b) $E(15) = \dfrac{-15(15 + 1)}{10(15 + 11)} \approx -0.923$, so a 2% increase in price results in a $(-0.923)(2) \approx -1.85$, or 1.85% decrease in demand.

(c) $R(p) = p \cdot 5000(p + 11)e^{-0.1p}$

$$R'(p) = 5000\left[(p^2+11p)(e^{-0.1p} \cdot -0.1) + (e^{-0.1p})(2p+11)\right]$$
$$= 5000e^{-0.1p}\left[-0.1(p^2+11p)+(2p+11)\right]$$
$$= 5000e^{-0.1p}(-0.1p^2+0.9p+11)$$

$R'(p) = 0$ when

$$-0.1p^2+0.9p+11=0$$
$$p = \frac{-0.9 \pm \sqrt{(0.9)^2-(4)(0.1)(11)}}{2(-0.1)}$$
$$\approx 15.91 \text{ (rejecting the negative price)}$$

61. $C(x) = e^{0.2x}$

(a) $C'(x) = 0.2e^{0.2x}$

(b) $A(x) = \frac{e^{0.2x}}{x}$
Marginal cost equals average cost when

$$0.2e^{0.2x} = \frac{e^{0.2x}}{x}$$

$0.2 = \frac{1}{x}$, or $x = 5$ units.

63. $C(x) = 12x^{1/2}e^{x/10}$

(a)

$$C'(x) = 12\left[(x^{1/2})\left(e^{x/10}\cdot\frac{1}{10}\right) + (e^{x/10})\left(\frac{1}{2}x^{-1/2}\right)\right]$$
$$= 6x^{-1/2}e^{x/10}\left(\frac{1}{5}x+1\right)$$
$$\frac{6e^{x/10}\left(\frac{1}{5}x+1\right)}{\sqrt{x}}$$

(b) $A(x) = \frac{12x^{1/2}e^{x/10}}{x} = \frac{12e^{x/10}}{x^{1/2}}$
Marginal cost equals average cost when

$$\frac{6e^{x/10}\left(\frac{1}{5}x+1\right)}{x^{1/2}} = \frac{12e^{x/10}}{x^{1/2}}$$

$\frac{1}{5}x+1=2$
or $x = 5$ units.

65. (a) $P(t) = 50e^{0.02t}$
The rate of change of the population t years from now will be

$$P'(t) = 50e^{0.02t}(0.02) = e^{0.02t}$$

So, the rate of change 10 years from now will be

$$P'(10) = e^{0.2} = 1.22 \text{ million per year.}$$

(b) The percentage rate of change t years from now will be

$$100\left[\frac{P'(t)}{P(t)}\right] = 100\left(\frac{e^{0.02t}}{50e^{0.02t}}\right)$$
$$= \frac{100}{50} = 2\% \text{ per year,}$$

which is a constant, independent of time.

67. (a) $Q(t) = 20{,}000e^{-0.4t}$
The rate of depreciation after t years is

$$Q'(t) = 20{,}000e^{-0.4t}(-0.4) = -8{,}000e^{-0.4t}$$

So, the rate after 5 years is

$$Q'(5) = -8,000e^{-2} = -\$1,082.68 \text{ per year.}$$

(b) The percentage rate of change t years from now will be

$$100\left[\frac{Q'(t)}{Q(t)}\right] = 100\left(\frac{-8,000e^{-0.4t}}{20,000e^{-0.4t}}\right) = -40\% \text{ per year,}$$

which is a constant, independent of time.

69. $f(x) = 20 - 15e^{-0.2x}$

(a) When a change in x is made, the corresponding change in f can be approximated by

$$\Delta f \approx f'(x)\Delta x$$

Here, $x = 10$ thousand initially, $\Delta x = 1$ thousand, and $f'(x) = -15e^{-0.2x}(-0.2) = 3e^{-0.2x}$
So,

$$\Delta f \approx f'(10)\cdot 1$$
$$= 3e^{-0.2(10)} \approx 0.406$$

An increase of one thousand additional complimentary copies will increase sales by 0.406 thousand, or 406 copies.

(b) The actual change in sales is

$$\begin{aligned}\Delta f &= f(11) - f(10)\\ &= (20 - 15e^{-2.2}) - (20 - 15e^{-2})\\ &= 0.368, \text{ or 368 copies.}\end{aligned}$$

71. $F(t) = B + (1 - B)e^{-kt}$

(a)
$$\begin{aligned}F'(t) &= 0 + (1 - B)e^{-kt}(-k)\\ &= -k(1 - B)e^{-kt}\end{aligned}$$

$F'(t)$ represents the rate at which recall is changing. That is, the rate at which you are forgetting material.

(b)
$$\begin{aligned}F - B &= B + (1 - B)e^{-kt} - B\\ &= (1 - B)e^{-kt}\end{aligned}$$

since
$$\begin{aligned}F'(t) &= -k(1 - B)e^{-kt},\\ F'(t) &= -k(F - B)\end{aligned}$$

That is, $F'(t)$ is proportional to $F - B$. This means that the rate you forget material is proportional to the fraction remaining that will be forgotten.

(c)

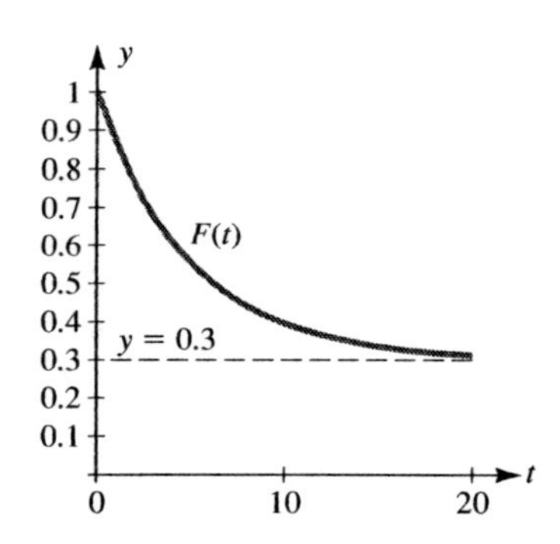

73. $D(p) = 3000e^{-0.01p}$

(a) $E(p) = p \cdot 3000e^{-0.01p}$

$$\begin{aligned}E'(p) &= (p)(3000e^{-0.01p} \cdot -0.01) + (3000e^{-0.01p})(1)\\ &= 3000e^{-0.01p}(-0.01p + 1)\end{aligned}$$

(b) $E'(p) = 0$ when $-0.01p + 1 = 0$, or $p = 100$

(c)
$$\begin{aligned}E''(p) &= (3000e^{-0.01p})(-0.01)\\ &\quad + (-0.01p + 1)\ (3000e^{-0.01p} \cdot -0.01)\\ &= -30e^{-0.01p}\left[1 + (-0.01p + 1)\right]\end{aligned}$$

$E''(p) = 0$ when $2 - 0.01p = 0$, or $p = 200$. When $0 < p < 200$, $E''(p) < 0$, so $E''(p)$ is decreasing. When $p = 200$, $E'(p) > 0$, so $E'(p)$ increasing. When $p > 200$, the rate of expenditure is neither increasing nor decreasing, but the rate then begins to increase.

75. $N(t) = \dfrac{600}{1 + 3e^{-0.02t}}$

(a)
$$\begin{aligned}N'(t) &= \frac{0 - (600)(3e^{-0.02t} \cdot -0.02)}{(1 + 3e^{-0.02t})^2}\\ &= \frac{36e^{-0.02t}}{(1 + 3e^{-0.02t})^2} \text{ individuals per year.}\end{aligned}$$

$N'(t)$ is never zero. $N'(t) > 0$ for all values of t, so the population is always increasing.

(b) Using logarithmic differentiation,

$$\begin{aligned}\ln N'(t) &= \ln\left[\frac{36e^{-0.02t}}{(1 + 3e^{-0.02t})^2}\right]\\ &= \ln 36 + \ln e^{0.02t} - \ln(1 + 3e^{-0.02t})^2\\ &= \ln 36 - 0.02t - 2\ln(1 + 3e^{-0.02t})\end{aligned}$$

$$\frac{N''(t)}{N'(t)} = -0.02 - 2\frac{-0.06e^{-0.02t}}{1 + 3e^{-0.02t}}$$

$$N''(t) = \left[\frac{-0.02(1 + 3e^{-0.02t}) + 0.12e^{-0.02t}}{1 + 3e^{-0.02t}}\right] N'(t)$$

$$N''(t) = \left[\frac{-0.02 + 0.06e^{-0.02t}}{1 + 3e^{-0.02t}}\right]\left[\frac{36e^{-0.02t}}{(1 + 3e^{-0.02t})^2}\right]$$

$$N''(t) = (-0.02 + 0.06e^{-0.02t})\left[\frac{36e^{-0.02t}}{1 + 3e^{-0.02t})^3}\right]$$

So $N''(t) = 0$ when

$$-0.02 + 0.06e^{-0.02t} = 0$$
$$e^{-0.02t} = \frac{1}{3}$$
$$\ln e^{-0.02t} = \ln \frac{1}{3}$$
$$-0.02t = \ln \frac{1}{3}$$
$$t = \frac{\ln \frac{1}{3}}{-0.02} = \frac{-\ln \frac{1}{3}}{0.02} = 50 \ln 3$$

When

$0 < t < 50 \ln 3$, $N''(t) > 0$, so $N'(t)$ is increasing

$t > 50 \ln 3$, $N''(t) < 0$, so $N'(t)$ is decreasing.

(c) As $t \to \infty$, $e^{-0.02t} \to 0$, so $\lim_{t\to\infty} \frac{600}{1+3e^{-0.02t}} =$ 600

So, in the long run, the number of individuals approaches 600.

77. $P_1(t) = \frac{21}{1+25e^{-0.3t}}$, $P_2(t) = \frac{20}{1+17e^{-0.6t}}$

(a) $P_1'(t) = \frac{0-(21)(25e^{-0.3t} \cdot -0.3)}{(1+25e^{-0.3t})^2}$

$$P_1'(10) = \frac{157.5e^{-3}}{(1+25e^{-3})^2} \approx 1.556 \text{ cm per day}$$
$$P_2'(t) = \frac{0-(20)(17e^{-0.6t} \cdot -0.6)}{(1+17e^{-0.6t})^2} = \frac{204e^{-0.6t}}{(1+17e^{-0.6t})^2}$$

Using logarithmic differentiation,

$$\ln P_2'(t) = \ln \left[\frac{204e^{-0.6t}}{(1+17e^{-0.6t})^2}\right]$$
$$= \ln 204 + \ln e^{0.6t} - \ln(1+17e^{-0.6t})^2$$
$$= \ln 204 - 0.6t - 2\ln(1+17e^{-0.6t})$$

$$\frac{P_2''(t)}{P_2'(t)} = -0.6 - 2\left(\frac{-10.2e^{-0.6t}}{1+17e^{-0.6t}}\right)$$
$$P_2''(t) = \left[\frac{-0.6(1+17e^{-0.6t})+20.4e^{-0.6t}}{1+17e^{-0.6t}}\right]P_2'(t)$$
$$= \left[\frac{-0.6+10.2e^{-0.6t}}{1+17e^{-0.6t}}\right]\left[\frac{204e^{-0.6t}}{(1+17e^{-0.6t})^2}\right]$$
$$= (-0.6+10.2e^{-0.6t})\left[\frac{204e^{-0.6t}}{(1+17e^{-0.6t})^3}\right]$$

Since $-0.6+10.2e^{-0.6(10)} < 0$, $P_2''(10) < 0$ so P_2' is decreasing. In other words, the rate of growth of the second plant is decreasing.

(b) $\frac{21}{1+25e^{-0.3t}} = \frac{20}{1+17e^{-0.6t}}$

$\frac{21}{1+25e^{-0.3t}} - \frac{20}{1+17e^{-0.6t}} = 0$

Press [y=] and input

21/(1 + 25e ∧ (−.3x)) − 20/(1 + 17e ∧ (−.6x)) for [y_1=].

Use window dimensions [0,30] 5 by [-10,10]1

Press [graph].

Use the zero function under the calc menu to find that the plants have the same height at approximately 20.71 days

$P_1(20.71) \approx 20$ cm

$P_1'(20.71) \approx 0.286$ cm/day and

$P_2'(20.71) \approx 0.000818$ cm/day, so P_1 is growing at a faster rate when they have the same height.

79. $R = E + T$

When $t = t_0$, $R = 11 + 8 = 9$

$$\text{Now, } \frac{E'(t_0)}{E(t_0)} = 0.09$$
$$\frac{T'(t_0)}{T(t_0)} = -0.02$$

or, $E'(t_0) = 0.09E(t_0)$ and $T'(t_0) = -0.02T(t_0)$.

Using logarithmic differentiation,

$$\ln R = \ln(E+T)$$
$$\frac{R'}{R} = \frac{E'+T'}{E+T}$$
$$\frac{R'(t_0)}{R(t_0)} = \frac{0.09(11)-0.02(8)}{11+8}$$
$$= 0.0437$$

So, 4.37% is the relative rate of growth of revenue when $t = t_0$.

81. $P(x) = 5,000\sqrt{x^2 + 4x + 19}$
Need to find

$$100\frac{P'(3)}{P(3)}$$

$$P'(t) = 5,000\left[\frac{1}{2}(x^2 + 4x + 19)^{-1/2}(2x + 4)\right]$$

$$P'(3) \approx 3952.85$$

$$P(3) \approx 31,622.78$$

So, $100\frac{3,952.85}{31,622.78} \approx 12.5\%$ per year

83. $f(x) = \frac{2^x}{x}$

$$f'(x) = \frac{x\frac{d}{dx}(2^x) - (2^x)(1)}{x^2} = \frac{x(\ln 2)2^x - 2^x}{x^2}$$

$$\frac{2^x(x \ln 2 - 1)}{x^2}$$

85. $f(x) = x \log_{10} x$

$$f'(x) = (x)\frac{d}{dx}(\log_{10} x) + (\log_{10} x)(1)$$

$$= x \cdot \frac{1}{\ln 10} \cdot \frac{1}{x} + \log_{10} x$$

$$= \frac{1}{\ln 10} + \log_{10} x$$

$$= \frac{1}{\ln 10} + \frac{\ln x}{\ln 10} = \frac{1 + \ln x}{\ln 10}$$

87. To use a numerical differentiation utility to find $f'(c)$, where $c = 0.65$ and

$$f(x) = \ln\left[\frac{\sqrt[3]{x+1}}{(1+3x)^4}\right]$$

Press [y=] and input f for $y_1 =$.
Press [2nd] [calc] and use dy/dx option.
Enter $x = 0.65$ and display shows $dy/dx = -3.866$

So, $f'(0.65) = -3.866$.
The slope of the line tangent at $x = 0.65$ is $m = -3.866$.
The point on the tangent line is $f(0.65) = -4.16$.
$y + 4.16 = -3.866(x - 0.65)$
$y = -3.866x - 1.6475$
Press [y=] and enter this line for $y_2 =$.
Use window dimensions [0,3]1 by [−10,2]1.
Press [graph].
An easier method is to
Press [y=] and input f for $y_1 =$.
Use window dimensions [−5, 5].5 by [−5, 5].5.
Press [graph].
Use the tangent function under the draw menu and enter $x = 0.65$.
The tangent line is drawn and the equation is displayed at the bottom of the screen.

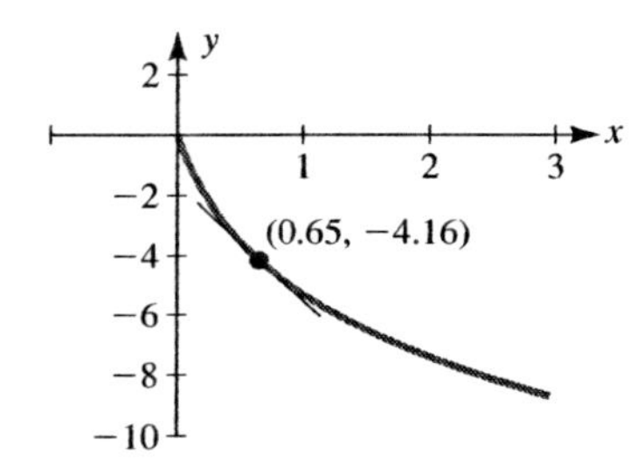

4.4 Additional Exponential Models

1. When $x = 1$, $f(1) = 0$ which eliminates f_1, f_3, and f_4. As $x \to \infty$, $f(x) \to 0$. Now, $\lim_{x\to\infty} x \ln x^5 = \infty$

$$\lim_{x\to\infty} \frac{5 \ln x}{x} = \lim_{x\to\infty} \frac{5/x}{1} = 0$$

$$\lim_{x\to\infty} (x-1)e^{-2x} = \lim_{x\to\infty} \frac{x-1}{e^{2x}} = \lim_{x\to\infty} \frac{1}{2e^{2x}} = 0$$

which eliminates f_2.

$$f_5'(x) = \frac{(x)\left(5 \cdot \frac{1}{x}\right) - (5 \ln x)(1)}{x^2} = \frac{5 - 5 \ln x}{x^2}$$

$f_5'(x) = 0$ when $5 - 5 \ln x = 0$

$$5 = 5 \ln x$$
$$1 = \ln x$$
$$e^1 = e^{\ln x}, \text{ or } x = e$$

$$f_6'(x) = (x-1)(-2e^{-2x}) + (e^{-2x})(1)$$
$$= e^{-2x}[-2(x-1)+1] = e^{-2x}(3-2x)$$

$f_6'(x) = 0$ when $3 - 2x = 0$, or $x = \frac{3}{2}$
which eliminates f_6. So, this is the graph of f_5.

3. As $x \to \infty$, $f(x) \to 2$.

$$\lim_{x\to\infty} 2 - e^{-2x} = 2$$
$$\lim_{x\to\infty} x \ln x^5 = \infty$$
$$\lim_{x\to\infty} \frac{2}{1-e^{-x}} = \frac{2}{1-0} = 2$$
$$\lim_{x\to\infty} \frac{2}{1+e^{-x}} = \frac{2}{1+0} = 2$$
$$\lim_{x\to\infty} \frac{\ln x^5}{x} = 0$$
$$\lim_{x\to\infty} (x-1)e^{-2x} = 0$$

which eliminates f_2, f_5, and f_6.
As $x \to 0^+$, $f(x) \to -\infty$.

$$\lim_{x\to 0^+} 2 - e^{-2x} = 2 - 1 = 1$$
$$\lim_{x\to 0^+} \frac{2}{1+e^{-x}} = \frac{2}{1+1} = 2$$

which eliminates f_1 and f_4. So, this is the graph of f_3.

5. $f(t) = 2 + e^t$
When $x = 0$, $f(0) = 3$ so $(0, 3)$ is an intercept. When $f(t) = 0$, $2 + e^t = 0$, $e^t = -2$ has no solution.
$\lim_{t\to-\infty} 2 + e^t = 2$, so y =2 is a horizontal asymptote.
$\lim_{t\to+\infty} 2 + e^t = +\infty$
$f'(t) = e^t$, so there are no critical values. $f'(t) > 0$ for all values of t, so f is always increasing. Since $f''(t) = e^t$ as well, $f''(t) > 0$ and f is always concave up.

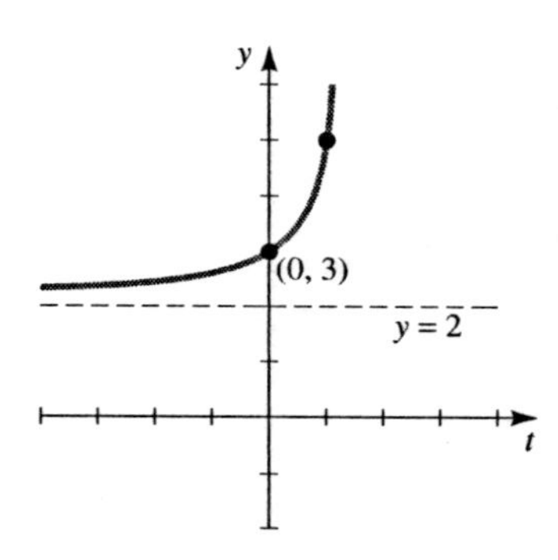

7. $g(x) = 2 - 3e^x$
When $x = 0$, $g(0) = -1$ so $(0, -1)$ is an intercept.
When $f(x) = 0$, $0 = 2 - 3e^x$, $3e^x = 2$, $e^x = \frac{2}{3}$
$\ln e^x = \ln \frac{2}{3}$, or $x = \ln \frac{2}{3}$.
So $\left(\ln \frac{2}{3}, 0\right)$ is an intercept.
$\lim_{x\to-\infty} 2 - 3e^x = 2$, so $y = 2$ is a horizontal asymptote.
$\lim_{x\to+\infty} 2 - 3e^x = -\infty$. $g'(x) = -3e^x$ so there are no critical values. $g'(x) < 0$ for all values of x, so g is always decreasing. Since $g''(x) = -3e^x$ as well, $g''(x) < 0$ and g is always concave down.

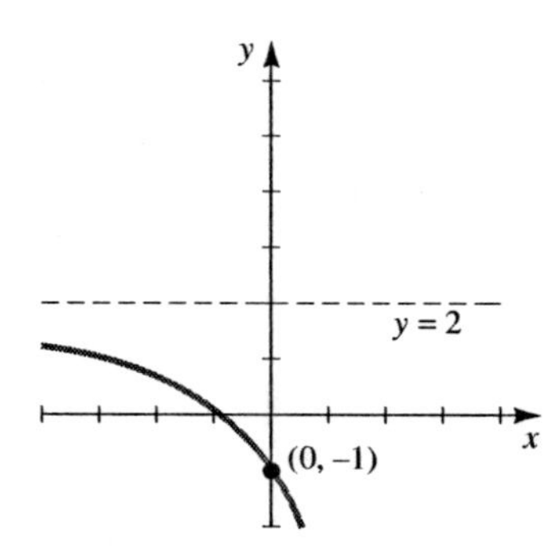

9. $f(x) = \dfrac{2}{1+3e^{-2x}}$
When $x = 0$, $f(0) = \frac{1}{2}$, so $\left(0, \frac{1}{2}\right)$ is an intercept.
$f(x) = 0$ has no solution.

$\lim_{x\to-\infty} \dfrac{2}{1+3e^{-2x}} = 0$, so $y = 0$

is a horizontal asymptote.

$\lim_{x \to +\infty} \dfrac{2}{1 + 3e^{-2x}} = 2$, so $y = 2$

is a horizontal asymptote.

$$f'(x) = \frac{0 - (2)(-6e^{-2x})}{(1 + 3e^{-2x})^2}, \text{ so } f'(x) = 0$$

when $12e^{-2x} = 0$.

Since $12e^{-2x}$ is never zero, there are no critical values. $f'(x) > 0$ for all values of x, so f is always increasing. Using logarithmic differentiation,

$$\ln f'(x) = \ln\left[\frac{12e^{-2x}}{(1 + 3e^{-2x})^2}\right]$$

$$= \ln 12 + \ln e^{-2x} - \ln(1 + 3e^{-2x})^2$$

$$= \ln 12 - 2x - 2\ln(1 + 3e^{-2x})$$

$$\frac{f''(x)}{f'(x)} = -2 - 2 \cdot \frac{-6e^{-2x}}{1 + 3e^{-2x}}$$

$$f''(x) = \left[\frac{-2(1 + 3e^{-2x}) + 12e^{-2x}}{1 + 3e^{-2x}}\right] f'(x)$$

$$= \left[\frac{-2 + 6e^{-2x}}{1 + 3e^{-2x}}\right]\left[\frac{12e^{-2x}}{(1 + 3e^{-2x})^2}\right]$$

$$= (-2 + 6e^{-2x})\left[\frac{12e^{-2x}}{(1 + 3e^{-2x})^3}\right]$$

So $f''(x) = 0$ when $-2 + 6e^{-2x} = 0$;

$$e^{-2x} = \frac{1}{3}$$

$$\ln e^{-2x} = \ln\frac{1}{3}$$

$$-2x = \ln\frac{1}{3}$$

$$x = \frac{\ln\frac{1}{3}}{-2},$$

or $x = \dfrac{\ln 3}{2} \approx 0.549$.

When $0 < x < 0.549$, $f''(x) > 0$, so f is concave up $x > 0.549$, $f''(x) < 0$, so f is concave down.

Since the concavity changes at $x = 0.549$, the point $(0.549, 1)$ is an inflection point.

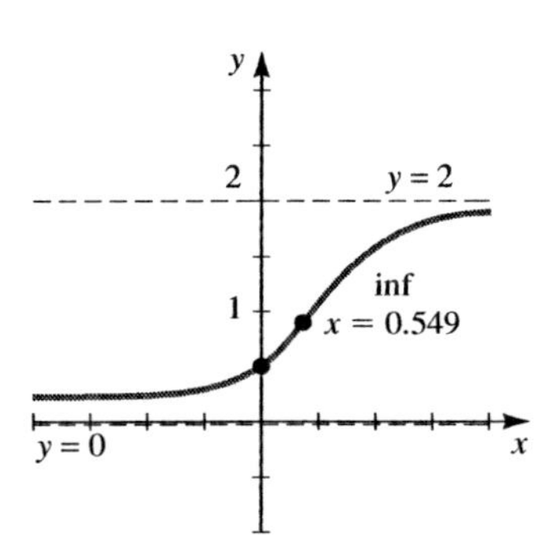

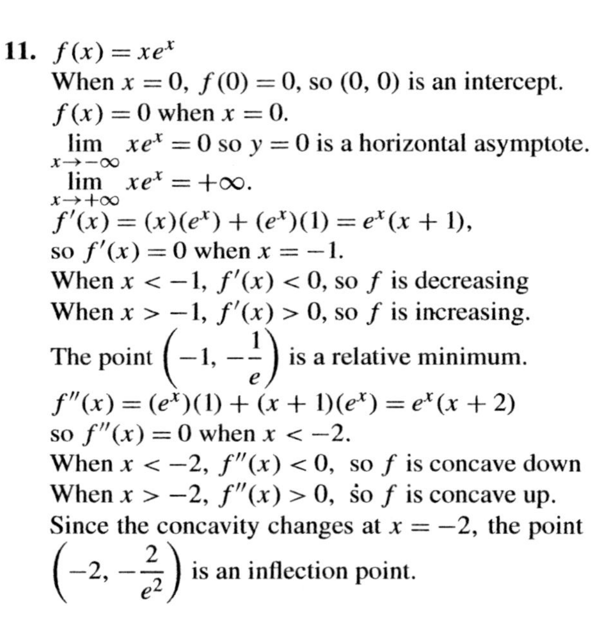

11. $f(x) = xe^x$

When $x = 0$, $f(0) = 0$, so $(0, 0)$ is an intercept.

$f(x) = 0$ when $x = 0$.

$\lim_{x \to -\infty} xe^x = 0$ so $y = 0$ is a horizontal asymptote.

$\lim_{x \to +\infty} xe^x = +\infty$.

$f'(x) = (x)(e^x) + (e^x)(1) = e^x(x + 1)$,

so $f'(x) = 0$ when $x = -1$.

When $x < -1$, $f'(x) < 0$, so f is decreasing

When $x > -1$, $f'(x) > 0$, so f is increasing.

The point $\left(-1, -\dfrac{1}{e}\right)$ is a relative minimum.

$f''(x) = (e^x)(1) + (x + 1)(e^x) = e^x(x + 2)$

so $f''(x) = 0$ when $x < -2$.

When $x < -2$, $f''(x) < 0$, so f is concave down

When $x > -2$, $f''(x) > 0$, so f is concave up.

Since the concavity changes at $x = -2$, the point $\left(-2, -\dfrac{2}{e^2}\right)$ is an inflection point.

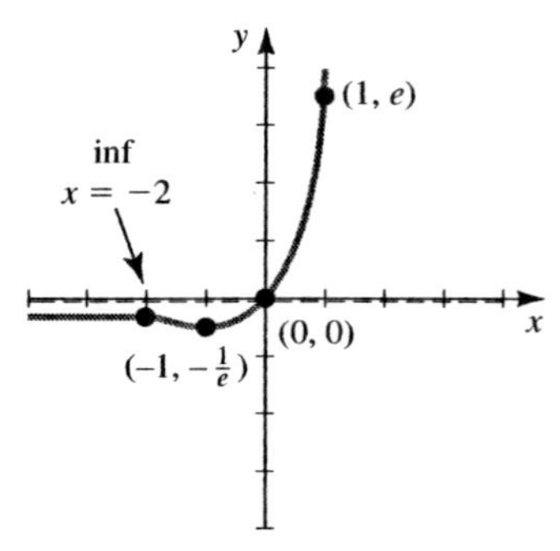

13. $f(x) = xe^{2-x}$

When $x = 0$, $f(0) = 0$, so (0,0) is an intercept.

$f(x) = 0$ when $x = 0$.

$\lim_{x \to -\infty} xe^{2-x} = -\infty$

$\lim_{x \to +\infty} xe^{2-x} = \lim_{x \to +\infty} \frac{x}{e^{x-2}} = \lim_{x \to +\infty} \frac{1}{e^x} = 0,$

so $y = 0$ is a horizontal asymptote.

$f'(x) = (x)(-e^{2-x}) + (e^{2-x})(1) = e^{2-x}(1 - x),$

so $f'(x) = 0$ when $x = 1$.

When $x < 1$, $f'(x) > 0$, so f is increasing

$x > 1$, $f'(x) < 0$, so f is decreasing.

The point $(1, e)$ is a relative maximum.

$f''(x) = (e^{2-x})(-1) + (1 - x)(-e^{2-x})$

$= e^{2-x}(x - 2)$

So $f''(x) = 0$ when $x = 2$.

When $x < 2$, $f''(x) < 0$ so f is concave down

$x > 2$, $f''(x) > 0$ so f is concave up.

Since the concavity changes at $x = 2$, the point (2, 2) is an inflection point.

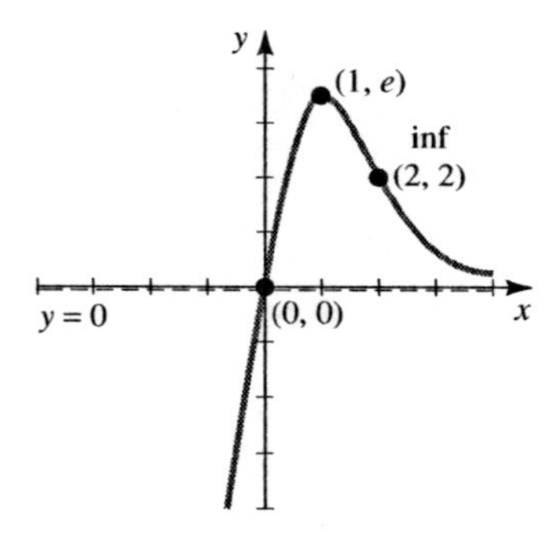

15. $f(x) = x^2 e^{-x}$

When $x = 0$, $f(0) = 0$ so (0, 0) is an intercept.

$f(x) = 0$ when $x = 0$.

$\lim_{x \to -\infty} x^2 e^{-x} = +\infty.$

$\lim_{x \to +\infty} x^2 e^{-x} = \lim_{x \to +\infty} \frac{x^2}{e^x} = \lim_{x \to +\infty} \frac{2x}{e^x}$

$= \lim_{x \to +\infty} \frac{2}{e^x} = 0$, so $y = 0$ is a horizontal asymptote.

$f'(x) = (x^2)(-e^{-x}) + (e^{-x})(2x)$

$= e^{-x}(2x - x^2) = e^{-x}(2 - x)x$. So, $f'(x) = 0$ when $x = 0, 2$.

When $x < 0$, $f'(x) < 0$, so f is decreasing.

$0 < x < 2$, $f'(x) > 0$, so f is increasing.

$x > 2$, $f'(x) < 0$, so f is decreasing.

The point (0, 0) is a relative minimum and the point $\left(2, \frac{4}{e^2}\right)$ is a relative maximum.

$$\begin{aligned} f''(x) &= (e^{-x})(2 - 2x) + (2x - x^2)(-e^{-x}) \\ &= e^{-x}[(2 - 2x) - (2x - x^2)] \\ &= e^{-x}(x^2 - 4x + 2). \end{aligned}$$

So $f''(x) = 0$ when $x = 2 \pm \sqrt{2}$.

When $x < 2 - \sqrt{2}$, $f''(x) > 0$ so f is concave up.

$2 - \sqrt{2} < x < 2 + \sqrt{2}$, $f''(x) < 0$ so f is concave down.

$x > 2 + \sqrt{2}$, $f''(x) > 0$, so f is concave up.

The points (0.59, 0.19) and (3.41, 0.38) are inflection points.

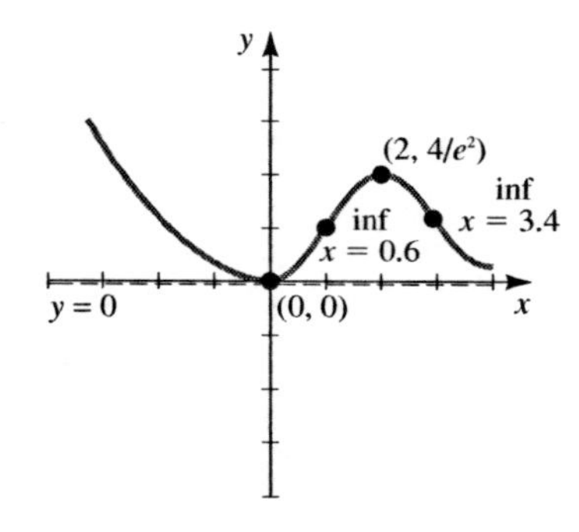

17. $f(x) = \dfrac{6}{1 + e^{-x}}$

When $x = 0$, $f(0) = 3$, so (0,3) is an intercept.

$f(x) = 0$ has no solution.

$\lim_{x \to -\infty} \dfrac{6}{1 + e^{-x}} = 0$, so $y = 0$ is a horizontal asymptote

$\lim_{x \to +\infty} \dfrac{6}{1 + e^{-x}} = 6$, so $y = 6$ is a horizontal asymptote.

$f'(x) = \dfrac{0 - (6)(-e^{-x})}{(1 + e^{-x})^2}$ so $f'(x) = 0$ when $6e^{-x} = 0$

Since $6e^{-x}$ is never zero, there are no critical values. $f'(x) > 0$ for all values of x, so f is always increasing. Using logarithmic differentiation,

$$\ln f'(x) = \ln\left[\frac{6e^{-x}}{(1+e^{-x})^2}\right]$$
$$= \ln 6 + \ln e^{-x} - \ln(1+e^{-x})^2$$
$$= \ln 6 - x - 2\ln(1+e^{-x})$$
$$\frac{f''(x)}{f'(x)} = -1 - 2\cdot\frac{-e^{-x}}{1+e^{-x}}$$

$$f''(x) = \left[\frac{-(1+e^{-x})+2e^{-x}}{1+e^{-x}}\right] f'(x)$$
$$= \left[\frac{-1+e^{-x}}{1+e^{-x}}\right]\left[\frac{6e^{-x}}{(1+e^{-x})^2}\right]$$
$$= (-1+e^{-x})\left[\frac{6e^{-x}}{(1+e^{-x})^3}\right]$$

So $f''(x) = 0$ when $-1+e^{-x} = 0$, or when $x = 0$.

When $x < 0$, $f''(x) > 0$, so f is concave up.
$x > 0$, $f''(x) < 0$, so f is concave down.

The point (0,3) is an inflection point.

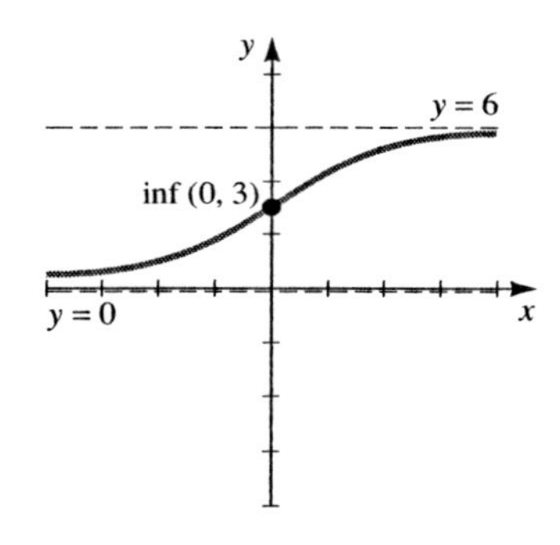

19. $f(x) = (\ln x)^2$, $x > 0$
When $f(x) = 0$, $(\ln x)^2 = 0$; $\ln x = 0$, or $x = 1$.
So (1,0) is an intercept. $\lim_{x\to 0^+} (\ln x)^2 = \infty$ so $x = 0$ is a vertical asymptote.
$\lim_{x\to+\infty} (\ln x)^2 = +\infty$.
$f'(x) = 2(\ln x)\left(\frac{1}{x}\right)$ so $f'(x) = 0$ when $x = 1$.

When $0 < x < 1$, $f'(x) < 0$, so f is decreasing.
$x > 1$, $f'(x) > 0$, so f is increasing.

The point (1, 0) is a relative minimum.

$$f''(x) = \left(\frac{2}{x}\right)\left(\frac{1}{x}\right) + (\ln x)\left(-\frac{2}{x^2}\right) = \frac{2}{x^2}(1-\ln x)$$

So $f''(x) = 0$ when $1 - \ln x = 0$
$1 = \ln x$
$e^1 = e^{\ln x}$, or $x = e$.

When $0 < x < e$, $f''(x) > 0$, so f is concave up
$x > e$, $f''(x) < 0$, so f is concave down.

The point $(e, 1)$ is an inflection point.

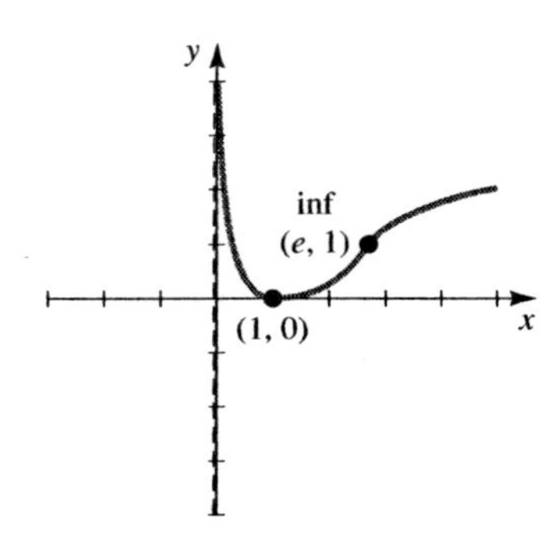

21. $f(t) = 1 - e^{-0.03t}$

(a) When $t = 0$, $f(0) = 0$, and $f(t) = 0$, when $t = 0$.

$$\lim_{t\to\infty} 1 - e^{-0.03t} = 1,$$
so $y = 1$ is a horizontal asymptote.
$f'(t) = 0.03e^{-0.03t}$
$f'(t) > 0$ for all values of t,
so f is always increasing.
$f''(t) = -0.09e^{-0.03t}$
$f''(t) < 0$ for all values of t,
so f is always concave down.

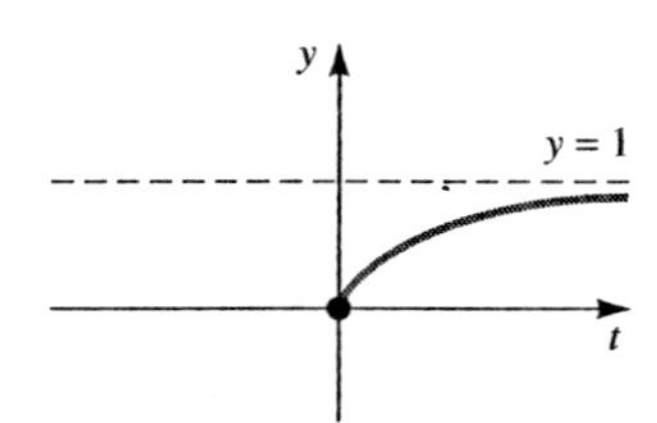

(b) The fraction of tankers that sink in fewer than 10 days is $f(10) = 1 - e^{-0.3}$.
The fraction of tankers that remain afloat for at least 10 days is therefore $1 - f(10) = e^{-0.3} = 0.7408$.

(c) The fraction of tankers that can be expected to sink between the 15th and 20th days is

$$\begin{aligned} f(20) - f(15) &= (1 - e^{-0.6}) - (1 - e^{-0.45}) \\ &= -e^{-0.6} + e^{-0.45} \\ &= -0.5488 + 0.6373 = 0.0888. \end{aligned}$$

23. $T(t) = -5 + Ae^{-kt}$

(a) When $t = 0$, $T(0) = 80$, so $80 = -5 + Ae^0$, or $A = 85$.

When $t = 20$, $T(20) = 25$, so

$$\begin{aligned} 25 &= -5 + 85e^{-k \cdot 20} \\ 30 &= 85e^{-20k} \\ \frac{6}{17} &= e^{-20k} \\ \ln\frac{6}{17} &= \ln e^{-20k} \\ \ln\frac{6}{17} &= -20k, \text{ or} \end{aligned}$$

$$k = \frac{\ln\frac{6}{17}}{-20} = \frac{-\ln\frac{6}{17}}{20} = \frac{\ln\frac{17}{6}}{20}$$

(b) $T(t) = -5 + 85e^{-0.052t}$
When $t = 0$, $T(0) = 80$, so $(0, 80)$ is an intercept.

$$\begin{aligned} \text{When } T(t) = 0, \quad 0 &= -5 + 85e^{-0.052t} \\ 5 &= 85e^{-0.052t} \\ \frac{1}{17} &= e^{-0.052t} \\ \ln\frac{1}{17} &= \ln e^{-0.052t} \end{aligned}$$

$$\ln\frac{1}{17} = -0.052t, \text{ so}$$

$$t = \frac{\ln\frac{1}{17}}{-0.052} = \frac{-\ln\frac{1}{17}}{0.052} = \frac{\ln 17}{0.052} \approx 54.5$$

So (54.5,0) is an intercept
$\lim_{t\to+\infty} -5 + 85e^{-0.052t} = -5$, so $y = -5$ is a horizontal asymptote. $T'(t) = -4.42e^{-0.052t}$
$T'(t) < 0$ for all values of t, so T is always decreasing.
$T''(t) = 0.23e^{-0.052t}$
$T''(t) > 0$ for all values of t, so T is always concave up.
As $t \to +\infty$, the temperature approaches -5°C.

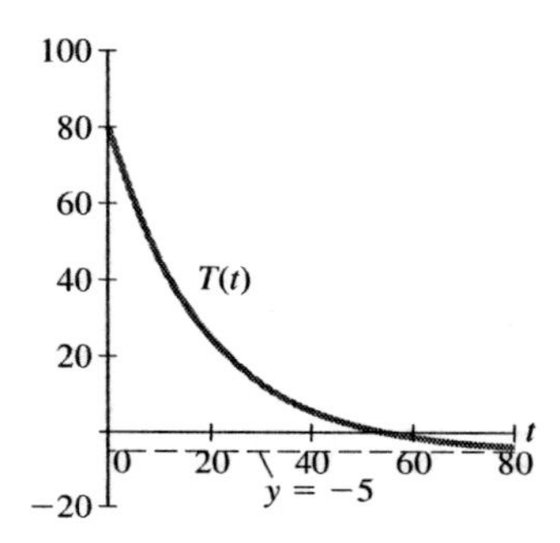

(c) $T(30) = -5 + 85e^{-0.052(30)} \approx 12.8$°C

(d) The temperature will be 0°C after approximately 54.5 minutes (see part a).

25. $f(t) = \dfrac{2}{1 + 3e^{-0.8t}}$

(a) When $t = 0$, $f(0) = \frac{1}{2}$, so $\left(0, \frac{1}{2}\right)$ is an intercept. $f(t) = 0$ has no solution.

$$\lim_{t\to\infty} \frac{2}{1 + 3e^{-0.8t}} = 2, \text{ so } y = 2$$

is a horizontal asymptote.

$$f'(t) = \frac{0 - (2)(-2.4e^{-0.8t})}{(1 + 3e^{-0.8t})^2} = \frac{4.8e^{-0.8t}}{(1 + 3e^{-0.8t})^2}$$

$f'(t) > 0$ for all values of t, so f is always increasing. Using logarithmic differentiation,

$$\begin{aligned} \ln f'(t) &= \ln\left[\frac{4.8e^{-0.8t}}{(1 + 3e^{-0.8t})^2}\right] \\ &= \ln 4.8 + \ln e^{-0.8t} - \ln(1 + 3e^{-0.8t})^2 \\ &= \ln 4.8 - 0.8t - 2\ln(1 + 3e^{-0.8t}) \end{aligned}$$

$$\frac{f''(t)}{f'(t)} = -0.8 - 2 \cdot \frac{-2.4e^{-0.8t}}{1+3e^{-0.8t}}$$

$$f''(t) = \left[\frac{-0.8(1+3e^{-0.8t}) + 4.8e^{-0.8t}}{1+3e^{-0.8t}}\right] f'(t)$$

$$= \left[\frac{-0.8+2.4e^{-0.8t}}{1+3e^{-0.8t}}\right]\left[\frac{4.8e^{-0.8t}}{(1+3e^{-0.8t})^2}\right]$$

$$= (-0.8+2.4e^{-0.8t})\left[\frac{4.8e^{-0.8t}}{(1+3e^{-0.8t})^3}\right]$$

$f''(t) = 0$ when $-0.8 + 2.4e^{-0.8t} = 0$

$$e^{-0.8t} = \frac{1}{3}$$

$$\ln e^{-0.8t} = \ln \frac{1}{3}$$

$$-0.8t = \ln \frac{1}{3}, \text{ or}$$

$$t = \frac{\ln \frac{1}{3}}{-0.8} = \frac{-\ln \frac{1}{3}}{0.8} = \frac{\ln 3}{0.8} = \frac{5 \ln 3}{4}$$

When $0 < t < \frac{5 \ln 3}{4}$, $f''(t) > 0$, so f is concave up

$t > \frac{5 \ln 3}{4}$, $f''(t) < 0$, so f is concave down.

The point (1.37,1) is an inflection point.

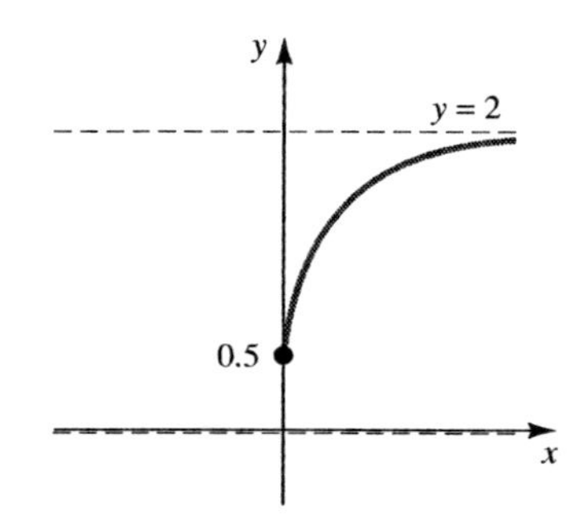

(b) $f(0) = 0.5$ thousand people, or 500 people.

(c) $f(3) = \dfrac{2}{1+3\ (0.907)} = 1.572$, so 1,572 people have caught the disease.

(d) $\lim_{t\to\infty} \dfrac{2}{1+3e^{-0.8t}} = 2$, so in the long run, approximately 2,000 people will contract the disease.

27. $Q(t) = 40 - Ae^{-kt}$
When $t = 0$, $Q(0) = 20$, so
$20 = 40 - Ae^{-k\cdot 0}$, or $A = 20$.
Now, $Q(t) = 40 - 20e^{-kt}$.
When $t = 1$, $Q(1) = 30$, so
$30 = 40 - 20e^{-k\cdot 1}$
$20e^{-k} = 10$; $e^{-k} = \frac{1}{2}$; $-k = \ln \frac{1}{2}$
$k = -\ln \frac{1}{2} = \ln 2$
Now, $Q(t) = 40 - 20e^{-(\ln 2)t}$.
When $t = 3$,
$Q(3) = 40 - 20e^{-(\ln 2)(3)} = 37.5$ units per day.

29. $f(x) = 15 - 20e^{-0.3x}$

(a) $\Delta f \approx f'(x)$, where x is the current number of complimentary copies. $f'(x) = 6e^{-0.3x}$
$\Delta f \approx f'(9) = 6e^{-0.3(9)} \approx 0.403$
So, approximately 403 additional copies will be sold.

(b)
$$\Delta f = f(x_2) - f(x_1) = f(10) - f(9)$$
$$f(10) = 15 - 20e^{-0.3(10)} \approx 14.004$$
$$f(9) = 15 - 20e^{-0.3(9)} \approx 13.656$$
$\Delta f = 0.348$, or 348 additional copies.

The approximation is off by 55 copies, or $\frac{55}{348} \approx 16\%$.

31. $L(t) = \dfrac{\ln(t+1)}{t+1}$

(a)
$$L'(t) = \frac{(t+1)\left(\frac{1}{t+1}\cdot 1\right) - \ln(t+1)(1)}{(t+1)^2} = \frac{1-\ln(t+1)}{(t+1)^2}$$

So, $L'(t) = 0$ when $1 - \ln(t+1) = 0$
$\ln(t+1) = 1$
$e^{\ln(t+1)} = e^1$
$t + 1 = e$, or $t = e - 1$.

When $0 \le t < e-1$, $L'(t) > 0$, so L is increasing. When $e-1 < t \le 5$, $L'(t) < 0$, so L is decreasing.

$$L(e-1) = \frac{\ln(e-1+1)}{e-1+1} = \frac{\ln e}{e} = \frac{1}{e} \approx 0.368$$

$$L(0) = \frac{\ln(1)}{1} = 0$$

$$L(1) = \frac{\ln 2}{2} \approx 0.347$$

So, at the age $e-1 \approx 1.7$ years of age, a child's learning capacity is the greatest.

(b) Need to maximize the rate of learning, or maximize the first derivative.

$$L''(t) = \frac{(t+1)^2\left(\frac{-1}{t+1}\right) - [1-\ln(t+1)]\,[2(t+1)(1)]}{(t+1)^4}$$

$$= \frac{-(t+1) - 2(t+1)\,[1-\ln(t+1)]}{(t+1)^4}$$

$$= \frac{(t+1)\,[-1-2(1-\ln(t+1))]}{(t+1)^4}$$

$$= \frac{-1-2\,[1-\ln(t+1)]}{(t+1)^3}$$

So $L''(t) = 0$ when
$-1-2\,[1-\ln(t+1)] = 0$, or

$$-2\,[1-\ln(t+1)] = 1$$

$$1-\ln(t+1) = -\frac{1}{2}$$

$$1+\frac{1}{2} = \ln(t+1)$$

$$e^{1.5} = e^{\ln(t+1)}$$

$$e^{1.5} = t+1, \text{ or}$$

$$t = e^{1.5}-1$$

When $0 \le t < e^{1.5}-1$, $L''(t) < 0$, so $L'(t)$ is decreasing.
$e^{1.5}-1 < t \le 5$, $L''(t) > 0$, so $L'(t)$ is increasing.

$$L'(e^{1.5}-1) = \frac{1-\ln(e^{1.5}-1+1)}{(e^{1.5}-1+1)^2} = \frac{1-\ln e^{1.5}}{(e^{1.5})^2}$$

$$= \frac{1-1.5}{e^3} \approx -0.025$$

$$L'(0) = \frac{1-\ln(0+1)}{(0+1)^2} = 0$$

$$L'(5) = \frac{1-\ln(5+1)}{(5+1)^2} \approx -0.022$$

So, a child's learning capability is increasing most rapidly at birth.

33. $p(t) = \dfrac{Ce^{kt}}{1+Ce^{kt}}$

Since

$$p(0) = \frac{1}{200},$$

$$\frac{1}{200} = \frac{Ce^{k\cdot 0}}{1+Ce^{k\cdot 0}}$$

$$\frac{1}{200} = \frac{C}{1+C}$$

$$1+C = 200C, \text{ or } C = \frac{1}{199}$$

$$\text{So, } p(t) = \frac{\frac{1}{199}e^{kt}}{1+\frac{1}{199}e^{kt}} = \frac{e^{kt}}{199+e^{kt}}$$

Since $\quad p(4) = \dfrac{1}{100},$

$$\frac{1}{100} = \frac{e^{k\cdot 4}}{199+e^{k\cdot 4}}$$

$$199+e^{4k} = 100e^{4k}$$

$$\frac{199}{99} = e^{4k}$$

$$\ln\frac{199}{99} = 4k$$

$$k = \frac{\ln\frac{199}{99}}{4} \approx 0.1745$$

$$\text{So, } p(t) = \frac{e^{0.1745t}}{199+e^{0.1745t}}.$$

Using logarithmic differentiation to find the rate of change,

$$\ln p(t) = \ln\left[\frac{e^{0.1745t}}{199 + e^{0.1745t}}\right]$$
$$= 0.1745t - \ln(199 + e^{0.1745t})$$
$$\frac{p'(t)}{p(t)} = 0.1745 - \frac{0.1745e^{0.1745t}}{199 + e^{0.1745t}}$$
$$= \frac{(0.1745)(199)}{199 + e^{0.1745t}} = \frac{34.7255}{199 + e^{0.1745t}}$$

So,

$$p'(t) = \frac{34.7255}{199 + e^{0.1745t}}\left(\frac{e^{0.1745t}}{199 + e^{0.1745t}}\right)$$
$$= \frac{34.7255e^{0.1745t}}{(199 + e^{0.1745t})^2}.$$

To maximize this rate, use logarithmic differentiation again.

$$\ln p'(t) = \ln\left[\frac{34.7255e^{0.1745t}}{(199 + e^{0.1745t})^2}\right]$$
$$= \ln 34.7255 + 0.1745t - 2\ln(199 + e^{0.1745t})$$
$$\frac{p''(t)}{p'(t)} = 0 + 0.1745 - 2\left(\frac{0.1745e^{0.1745t}}{199 + e^{0.1745t}}\right)$$
$$= \frac{(0.1745)(199) - 0.1745e^{0.1745t}}{199 + e^{0.1745t}}$$
$$= \frac{34.7255 - 0.1745e^{0.1745t}}{199 + e^{0.1745t}}$$

and,

$$p''(t) = \left[\frac{34.7255 - 0.1745e^{0.1745t}}{199 + e^{0.1745t}}\right]\left[\frac{34.7255e^{0.1745t}}{(199 + e^{0.1745t})^2}\right]$$

$p''(t) = 0$ when $34.7255 - 0.1745e^{0.1745t} = 0$

$\frac{34.7255}{0.1745} = e^{0.1745t}$

$\ln 199 = 0.1745t$,

or $t = \frac{\ln 199}{0.1745} \approx 30.33$ weeks

$p(30.33) \approx 0.5$, so roughly half of the trading volume is due to day trading.

35. (a) Profit = (# sold) (profit on each)

$$P(x) = (1,000e^{-0.02x})(x - 125)$$

When $x = 0$, $P(0) = -125{,}000$ which is not in the practical domain. When $P(x) = 0$, $x = 125$ so (125,0) is an intercept.
$\lim_{x\to\infty} (1000e^{-0.02x})(x - 125) = 0$, so $y = 0$ is a horizontal asymptote.

$$P'(x) = 1000\left[(e^{-0.02x})(1) + (x - 125)(-0.02^{-0.02x})\right]$$
$$= 1000e^{-0.02x}\left[1 - 0.02(x - 125)\right]$$
$$= 1000e^{-0.02x}(3.5 - 0.02x)$$

So, $P'(x) = 0$ when $3.5 - 0.02x = 0$
$3.5 = 0.02x$, or $x = 175$.
When $125 < x < 175$, $P'(x) > 0$, so P is increasing. When $x > 175$, $P'(x) < 0$, so P is decreasing.
The point (175,1510) is a relative maximum.
Using logarithmic differentiation,

$$\ln P'(x) = \ln\left[1{,}000e^{-0.02x}(3.5 - 0.02x)\right]$$
$$= \ln 1{,}000 + \ln e^{-0.02x} + \ln(3.5 - 0.02x)$$
$$= \ln 1{,}000 - 0.02x + \ln(3.5 - 0.02x)$$
$$\frac{P''(x)}{P'(x)} = -0.02 + \frac{-0.02}{3.5 - 0.02x}$$
$$P''(x) = \left(\frac{-0.02(3.5 - 0.02x) - 0.02}{3.5 - 0.02x}\right)P'(x)$$
$$= \left[\frac{0.0004x - 0.09}{3.5 - 0.02x}\right]\left[1{,}000e^{-0.02x}(3.5 - 0.02x)\right]$$
$$= (0.4x - 90)e^{-0.02x}$$

So $P''(x) = 0$ when $0.4x - 90 = 0$, or $x = 225$.
When $125 < x < 225$, $P''(x) < 0$, so P is concave down
$x > 225$, $P''(x) > 0$, so P is concave up.
The point (225, 1111) is an inflection point.

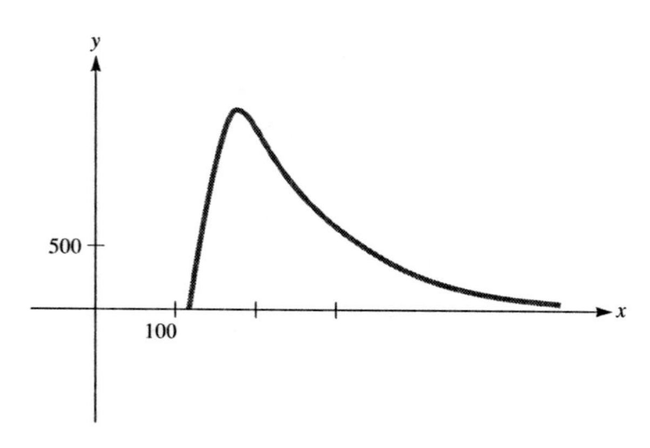

(b) Since $P''(175) < 0$, the relative maximum is the absolute maximum, so the selling price should be \$175.

37. $V(t) = 8,000e^{\sqrt{t}}$
The prevailing interest rate of 6% is the same as the percentage rate of change of V, so
$6 = 100\frac{V'(t)}{V(t)}$
Now, $V'(t) = 8,000e^{\sqrt{t}}\left(\frac{1}{2}t^{-1/2}\right)$

$$\text{So, } 100\frac{V'(t)}{V(t)} = 100\frac{8,000e^{\sqrt{t}}\left(\frac{1}{2}t^{-1/2}\right)}{8,000e^{\sqrt{t}}} = \frac{50}{t^{1/2}}$$

and

$$6 = \frac{50}{t^{1/2}}, \quad t^{1/2} = \frac{25}{3}, \quad \text{or}$$
$$t = \left(\frac{25}{3}\right)^2 \approx 69.44$$

When $0 < t < \frac{625}{9}$, the percentage rate of growth $100\frac{V'(t)}{V(t)} > 6\%$. When $t > \frac{625}{9}$, the percentage rate of growth is $100\frac{V'(t)}{V(t)} < 6\%$. So, the land should be sold approximately 69.44 years from now.

39. $p(x) = e^{-0.2x}$; $f(x) = 5x^{0.9}$
The per capita rate of increase function is

$$R(x) = \frac{\ln\left[e^{-0.2x}(5x^{0.9})\right]}{x}$$
$$= -0.2 + [\ln 5 + 0.9\ln x]\frac{1}{x}$$

So,

$$R'(x) = 0 + [\ln 5 + 0.9\ln x]\left(\frac{-1}{x^2}\right) + \left(\frac{1}{x}\right)\left(0 + \frac{0.9}{x}\right)$$
$$= \frac{-\ln 5 - 0.9\ln x + 0.9}{x^2}$$

So, $R'(x) = 0$ when

$$0 = -\ln 5 - 0.9\ln x + 0.9$$
$$0.9\ln x = -\ln 5 + 0.9$$
$$\ln x = \frac{-\ln 5 + 0.9}{0.9}$$
$$e^{\ln x} = e^{(-\ln 5 + 0.9)/0.9}$$
$$x = e^{(-\ln 5 + 0.9)/0.9} \approx 0.45$$

Since

$$R''(x) = \frac{(x^2)\left(\frac{-0.9}{x}\right) + [-\ln 5 - 0.9\ln x + 0.9](2x)}{x^4}$$
$$= \frac{x\,[-0.9 + 2(-\ln 5 - 0.9\ln x + 0.9)]}{x^4}$$
$$= \frac{-2\ln 5 - 1.8\ln x + 0.9}{x^3}$$

and $R''(0.45) < 0$, so $x = 0.45$ corresponds to the absolute maximum. The ideal reproductive age is 0.45 years.

41. $E(C) = C\left(aR + \frac{b}{C}\right)^2$

(a)
$$E'(C) = (C)\left[2\left(aR + \frac{b}{C}\right)\left(\frac{-b}{C^2}\right)\right] + \left(aR + \frac{b}{C}\right)^2(1)$$
$$= \frac{-2b}{C}\left(aR + \frac{b}{C}\right) + \left(aR + \frac{b}{C}\right)^2$$
$$= \left(aR + \frac{b}{C}\right)\left[\frac{-2b}{C} + aR + \frac{b}{C}\right]$$
$$= \left(aR + \frac{b}{C}\right)\left(aR - \frac{b}{C}\right)$$

So $E'(C)=0$ when

$$aR-\frac{b}{C}=0 \text{ (rejecting the negative solution)}$$

$$\text{aR} = \frac{\text{b}}{\text{C}}$$

$$C=\frac{b}{aR}$$

$$E''(C)=\left(aR+\frac{b}{C}\right)\left(\frac{b}{C^2}\right)+\left(aR-\frac{b}{C}\right)\left(\frac{-b}{C^2}\right)$$

$$=\frac{b}{C^2}\left[\left(aR+\frac{b}{C}\right)-\left(aR-\frac{b}{C}\right)\right]$$

$$=\frac{b}{C^2}\left(\frac{2b}{C}\right)=\frac{2b^2}{C^3}$$

When $C=\dfrac{b}{aR}$,

$$E''\left(\frac{b}{aR}\right)=\frac{2b^2}{\left(\frac{b}{aR}\right)^3}$$

Since a, b, R are all positive,

$$E''\left(\frac{b}{aR}\right)>0$$

So, the absolute minimum occurs when

$$C=\frac{b}{aR}.$$

(b) $E(C)=mCe^{k/C}$
Using logarithmic differentiation,

$$\ln E(C)=\ln mCe^{k/C}$$

$$\ln E(C)=\ln m+\ln C+\ln e^{k/C}$$

$$\ln E(C)=\ln m+\ln C+k/C$$

$$\frac{E'(C)}{E(C)}=\frac{1}{C}-\frac{k}{C^2}$$

or,

$$E'(C)=\frac{C-k}{C^2}E(C)$$

$$=\frac{C-k}{C^2}(mCe^{k/C})$$

$$=\frac{(C-k)m}{C}e^{k/C}$$

So, $E'(C)=0$ when $C-k=0$, or $C=k$. We want the same value of C for a minimum in both models, so

$$k=\frac{b}{aR}$$

From the first model, the minimum value is

$$E\left(\frac{b}{aR}\right)=\frac{b}{aR}\left(aR+b\cdot\frac{aR}{b}\right)^2$$

$$=\frac{b}{aR}(2aR)^2$$

$$=4abR.$$

For the second model to have this same minimum,

$$E(k)=mke^{k/k}$$

$$4abR=m\left(\frac{b}{aR}\right)e,\quad \text{so } m=4a^2R^2e^{-1}.$$

43. $w(t)=\dfrac{10}{1+15e^{-0.05t}}$; $p(t)=e^{-0.01t}$

(a) Total weight
=(weight per fish) (number of fish)
=(weight per fish) [(beginning number fish) (proportion remaining)]

$$E(t)=\left(\frac{10}{1+15e^{-0.05t}}\right)(1,000e^{-0.01t})$$

$$E(t)=10{,}000\frac{e^{-0.01t}}{1+15e^{-0.05t}}$$

(b) Using logarithmic differentiation,

$$\ln E(t) = \ln\left[10{,}000\frac{e^{-0.01t}}{1+15e^{-0.05t}}\right]$$
$$= \ln 10{,}000 + \ln e^{-0.01t} - \ln(1+15e^{-0.05t})$$
$$= \ln 10{,}000 - 0.01t - \ln(1+15e^{-0.05t})$$
$$\frac{E'(t)}{E(t)} = -0.01 - \frac{-0.75e^{-0.05t}}{1+15e^{-0.05t}}$$
$$E'(t) = \left[\frac{-0.01(1+15e^{-0.05t}) + 0.75e^{-0.05t}}{1+15e^{-0.05t}}\right]E(t)$$
$$= \left[\frac{-0.01+0.6e^{-0.05t}}{1+15e^{-0.05t}}\right]\left[10,000\frac{e^{-0.01t}}{1+15e^{-0.05t}}\right]$$
$$= \left[-0.01+0.6e^{-0.05t}\right]\left[10,000\frac{e^{-0.01t}}{(1+15e^{-0.05t})^2}\right]$$

So, $E((t) = 0$ when $-0.01 + 0.6e^{-0.05t} = 0$

$$0.6e^{-0.05t} = 0.01$$
$$e^{-0.05t} = \frac{1}{60}$$
$$\ln e^{-0.05t} = \ln\frac{1}{60}$$
$$-0.05t = \ln\frac{1}{60}, \text{ or}$$
$$t = \frac{\ln\frac{1}{60}}{-0.05} = \frac{-\ln\frac{1}{60}}{0.05} = \frac{\ln 60}{0.05} \approx 81.9$$

For the domain $t \geq 0$,

when $0 \leq t < \frac{\ln 60}{0.05}$, $E'(t) > 0$, so E is increasing

$t > \frac{\ln 60}{0.05}$, $E'(t) < 0$, so E is decreasing.

So, the relative maximum is also the absolute maximum.
When $t \approx 81.9$ days, the yield is the maximum, namely

$$E(81.9) = 10,000\frac{e^{-0.01(81.9)}}{1+15e^{-0.05(81.9)}}$$
$$\approx 3,527 \text{ pounds}$$

(c)

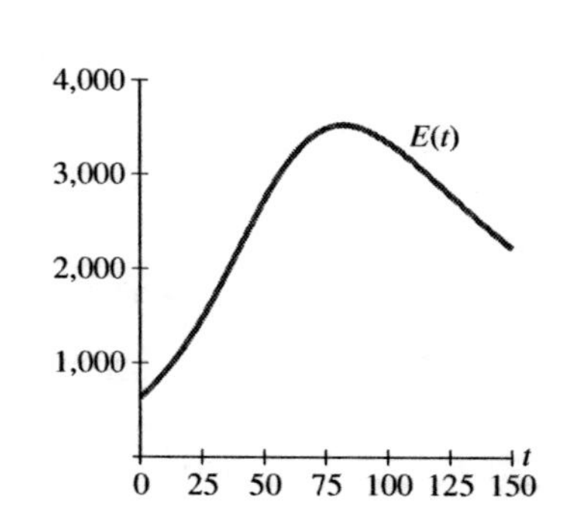

45.

$$N(t) = \frac{B}{1+Ce^{-kt}}$$

(a) When $t = 0$, $N(0) = 0.1B$ so

$$0.1B = \frac{B}{1+Ce^{-k\cdot 0}}$$
$$0.1 = \frac{1}{1+C}$$
$$1+C = \frac{1}{0.1}$$
$$C = 9$$

When $t = 2$, $N(2) = 0.25B$, so

$$0.25B = \frac{B}{1+9e^{-k(2)}}$$
$$0.25 = \frac{1}{1+9e^{-2k}}$$
$$1+9e^{-2k} = \frac{1}{0.25}$$
$$9e^{-2k} = 3$$
$$e^{-2k} = \frac{1}{3}$$
$$\ln e^{-2k} = \ln\frac{1}{3}$$
$$-2k = \ln\frac{1}{3}, \text{ or}$$
$$k = \frac{\ln\frac{1}{3}}{-2} = \frac{-\ln\frac{1}{3}}{2} = \frac{\ln 3}{2}$$

(b)

$$N(t) = \frac{B}{1 + 9e^{-\left(\frac{\ln 3}{2}\right)t}}$$

$$0.5B = \frac{B}{1 + 9e^{(-t/2)(\ln 3)}}$$

$$1 + 9e^{(-t/2)(\ln 3)} = \frac{1}{0.5}$$

$$9e^{(-t/2)(\ln 3)} = 1$$

$$e^{\ln 3^{-t/2}} = \frac{1}{9}$$

$$3^{-t/2} = \frac{1}{9}$$

$$\ln 3^{-t/2} = \ln \frac{1}{9}$$

$$-\frac{t}{2} \ln 3^{-t/2} = \ln \frac{1}{9}$$

$$-\frac{t}{2} = \frac{\ln \frac{1}{9}}{\ln 3}, \text{ or}$$

$$t = \frac{-2 \ln \frac{1}{9}}{\ln 3} = \frac{2 \ln 9}{\ln 3} = 4 \text{ hours}$$

(c) Need to maximize the rate at which news is spreading (maximize the first derivative).

$$N(t) = \frac{B}{1 + 9e^{-(\ln 3/2)t}}$$

To use the result from page 331, consider

$$\frac{\ln 3}{2} = B\left(\frac{\ln 3}{2B}\right) = Bk$$

Then, $N''(t) = 0$ when

$$t = \frac{\ln 9}{\ln \frac{3}{2}} = \frac{2 \ln 9}{\ln 3}$$

$$= \frac{\ln 81}{\ln 3} = \log_3 81 = 4$$

So, the news is spreading most rapidly after 4 hours.

47. $N(t) = 500(0.03)^{(0.4)^t}$

(a) When $t = 0$, $N(0) = 500(0.03)^{(0.4)^0} = 15$ employees.

When $t = 5$, $N(5) = 500(0.03)^{(0.4)^5} \approx 482$ employees.

$$300 = 500(0.03)^{(0.4)^t}$$

$$\frac{3}{5} = (0.03)^{(0.4)^t}$$

$$\ln 0.6 = \ln(0.03)^{(0.4)^t}$$

$$\ln 0.6 = (0.4)^t \ln(0.03)$$

$$\frac{\ln 0.6}{\ln 0.03} = (0.4)^t$$

$$0.145677 \approx (0.4)^t$$

$$\ln 0.145677 \approx \ln(0.4)^t$$

$$\ln 0.145677 \approx t \ln(0.4), \text{ so}$$

$$t \approx \frac{\ln 0.145677}{\ln 0.4} \approx 2.10 \text{ years}$$

Since $\lim_{t\to+\infty} (0.4)^t = 0$,

$\lim_{t\to+\infty} 500(0.03)^{(0.4)^t} = 500$ employees.

(b) To sketch the graphs of N and $F(t) = 500(0.03)^{-(0.4)^{-t}}$ on the same graph,
Press [y=] and input N for $Y_1 =$.
Use window dimensions of [−6, 6]2 by [0,1000]100
Press [graph].
Press [y=] and input F for $Y_2 =$.
Press [graph].

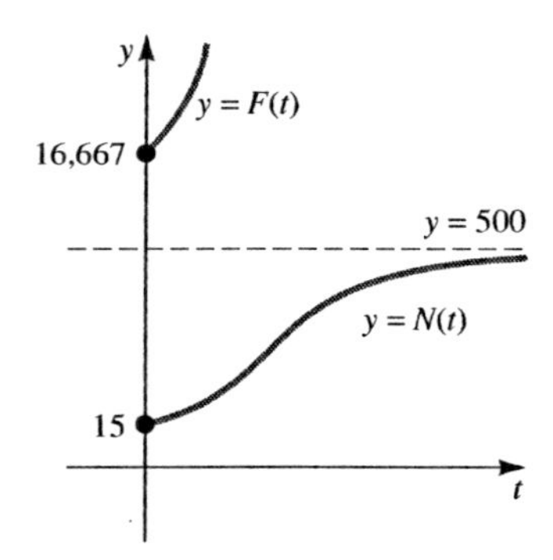

Writing Exercise—Answers will vary.

49. $C(t) = Ate^{-kt}$

(a)
$$C'(t) = A\left[(t)(-ke^{-kt}) + (e^{-kt})(1)\right]$$
$$= Ae^{-kt}(1-kt)$$

So $C'(t) = 0$ when $1 - kt = 0$, or $t = \frac{1}{k}$.

When $0 \le t < \frac{1}{k}$, $C'(t) > 0$, so C is increasing

$t > \frac{1}{k}$, $C'(t) < 0$, so C is decreasing.

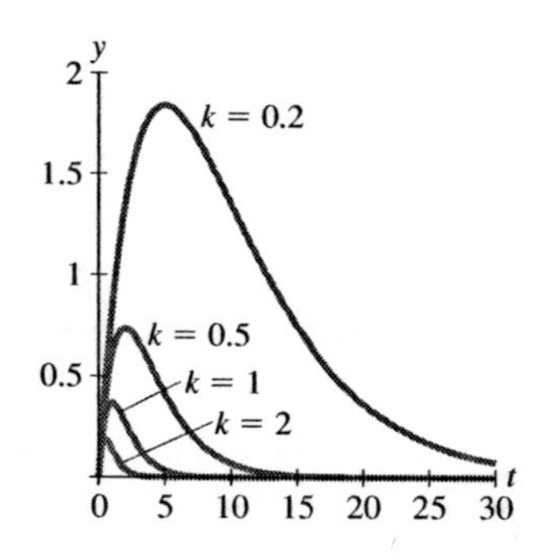

$$C''(t) = A\left[(e^{-kt})(-k) + (1-kt)(-ke^{-kt})\right]$$
$$= Ake^{-kt}(kt-2)$$

$C''\left(\frac{1}{k}\right) < 0$, so the absolute maximum concentration occurs when $t = \frac{1}{k}$ and has a value of

$$C\left(\frac{1}{k}\right) = A\left(\frac{1}{k}\right)e^{-k(1/k)} = \frac{A}{ke}.$$

(b) From above, $C''(t) = Ake^{-kt}(kt-2)$, so $C''(t) = 0$ when $kt - 2 = 0$, or $t = \frac{2}{k}$.

When $0 \le t < \frac{2}{k}$, $C''(t) < 0$, so C is concave down;

$t > \frac{2}{k}$, $C''(t) > 0$, so C is concave up.

The point $\left(\frac{2}{k}, \frac{2A}{ke^2}\right)$ is an inflection point.
The zeros of the second derivative are relative extrema of the first derivative, or in this case, the rate of change of drug concentration.

When $0 < t < \frac{2}{k}$, $C''(t) < 0$, so C' is decreasing;

$t > \frac{2}{k}$, $C''(t) > 0$, so C' is increasing.

So the inflection point corresponds to the minimum rate of change of drug concentration.

(c) The maximum point shifts to the left and the height of the curve decreases.

51. (a) Assuming continuous growth, the situation can be modeled by a function of the form

$$Q(t) = Q_0e^{kt}$$

Let $t = 0$ be the year 1947. Since $r = 0.06$ and $Q_0 = 1{,}139$,

$$Q(t) = 1{,}139e^{0.06t}$$

In the year 1954, $t = 7$ and

$$Q(7) = 1{,}139e^{0.06(7)}$$
$$\approx 1{,}733 \text{ staff members}$$

(b) Let the original size of the staff be Q_0 and double the staff be $2Q_0$. Then,

$$2Q_0 = Q_0e^{0.06t}$$
$$2 = e^{0.06t}$$
$$\ln 2 = \ln e^{0.06t}$$
$$\ln 2 = 0.06t, \text{ or } t = \frac{\ln 2}{0.06} = 11.55$$

So, any size staff doubles in approximately 11.55 years.

(c) Writing Exercise—Answers will vary.

53. $p(x) = Ax^sc^{-sx/r}$

(a)
$$p'(x) = A\left[(x^s)\left(\frac{-s}{r}e^{-sx/r}\right) + \left(e^{-sx/r}\right)\left(sx^{s-1}\right)\right]$$
$$= sAx^{s-1}e^{-sx/r}\left[\frac{-x}{r} + 1\right]$$

So $p'(x) = 0$ when $\frac{-x}{r} + 1 = 0$, or $x = r$.

When $0 \le x < r$, $p'(x) > 0$, so p is increasing
$x > r$, $p'(x) < 0$ so p is decreasing.

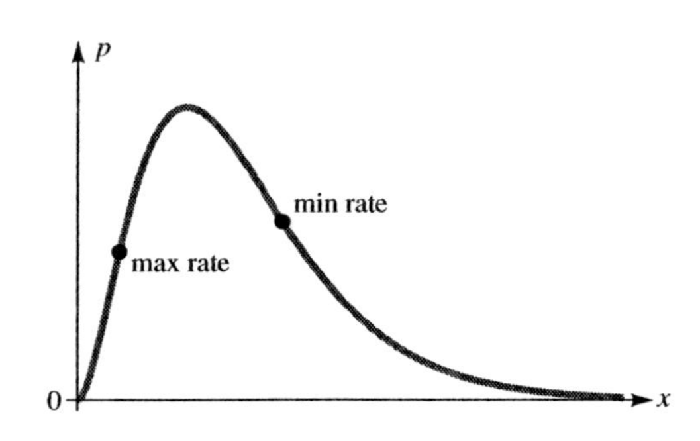

Since the domain of p is $p \ge 0$, this means the absolute maximum occurs when $x = r$.

(b) Rewrite $p'(x)$ as

$$p'(x) = sAe^{-sx/r}\left(\frac{-x^s}{r} + x^{s-1}\right)$$

Then $p''(x) =$

$$sA\left[\left(e^{-sx/r}\right)\left(\frac{-sx^{s-1}}{r} + (s-1)x^{s-2}\right) + \left(\frac{-x^s}{r} + x^{s-1}\right)\left(\frac{-s}{r}e^{-sx/r}\right)\right]$$

$$= sAx^{s-2}e^{-sx/r}\left[\frac{-sx}{r} + s - 1 + \left(\frac{-x^2}{r} + x\right)\frac{-s}{r}\right]$$

$$= sAx^{s-2}e^{-sx/r}\left(\frac{-s}{r}x + s - 1 + \frac{s}{r^2}x^2 - \frac{s}{r}x\right)$$

$$= sAx^{s-2}e^{-sx/r}\left[\frac{s}{r^2}x^2 - \frac{2s}{r}x + (s-1)\right]$$

$$= r^2sAe^{s-2}e^{-sx/r}\left[sx^2 - 2rsx + r^2(s-1)\right]$$

Using the quadratic formula,

$$x = \frac{2rs \pm \sqrt{(2rs)^2 - (4)(s)r^2(s-1)}}{2(s)}$$

$$x = \frac{2rs \pm 2r\sqrt{s^2 - s(s-1)}}{2s}$$

$$x = \frac{rs \pm r\sqrt{s}}{s} = \frac{r}{s}\left(s \pm \sqrt{s}\right)$$

So, there are two possible inflection points. (Checking with $p''(x)$ shows that they both are inflection points.)

(c) When $0 < s < 1$, $s - \sqrt{s} < 0$, so $x < 0$. Since the practical domain is $x > 0$, this value is rejected and there is only one inflection point.

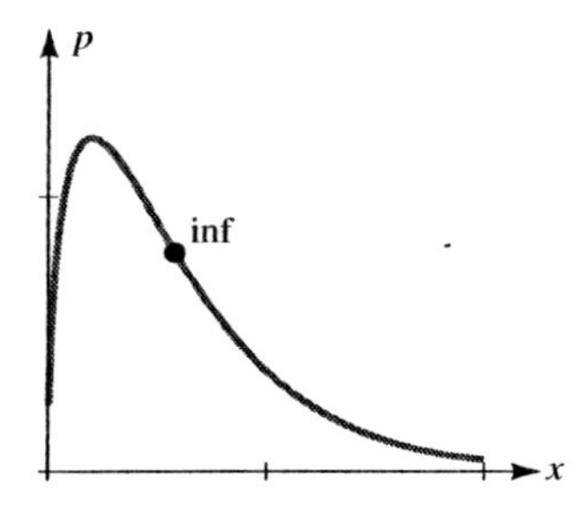

55. $f(t) = \dfrac{A}{1 + Ce^{-kt}}$

The epidemic is spreading most rapidly when the rate of change, or derivative, is maximized

$$f'(t) = \frac{0 - (A)(-kCe^{-kt})}{(1 + Ce^{-kt})^2}$$

$$= \frac{kACe^{-kt}}{(1 + Ce^{-kt})^2} = kAC\frac{e^{-kt}}{(1 + Ce^{-kt})^2}$$

The possible min/max of f' are the zeros of f''.

$$f''(t) = kAC\left[\frac{(1 + Ce^{-kt})^2(-ke^{-kt})}{(1 + Ce^{-kt})^4} - \frac{(e^{-kt})\left[2(1 + Ce^{-kt})(-kCe^{-kt})\right]}{(1 + Ce^{-kt})^4}\right]$$

$$= -k^2ACe^{-kt}(1 + Ce^{-kt})\left[\frac{1 + Ce^{-kt} - 2Ce^{-kt}}{(1 + Ce^{-kt})^4}\right]$$

So, $f''(t) = 0$ when

$$1 - Ce^{-kt} = 0$$
$$1 = Ce^{-kt}$$
$$\frac{1}{C} = e^{-kt}$$
$$\ln \frac{1}{C} = \ln e^{-kt}$$
$$\ln \frac{1}{C} = -kt, \text{ or}$$
$$t = \frac{\ln \frac{1}{C}}{-k} = \frac{-\ln \frac{1}{C}}{k} = \frac{\ln C}{k}$$

Checking with f''' shows this value of t corresponds to the absolute maximum. The absolute maximum is

$$f\left(\frac{\ln C}{k}\right) = \frac{A}{1 + Ce^{-k(\ln C/k)}}$$
$$= \frac{A}{1 + Ce^{-\ln C}}$$
$$= \frac{A}{1 + Ce^{\ln(1/C)}}$$
$$= \frac{A}{1 + C \cdot \frac{1}{C}} = \frac{A}{2}$$

So the epidemic is spreading most rapidly when half of those susceptible are infected.

57. $N(t) = 2(1 - e^{-.037t})$

To graph this function and see what happens as $t \to \infty$, press [y=] and input N for $y_1 =$.

Use window dimensions of [0, 200]10 by [0, 3]1.

Press [graph].

The value of N approaches the maximum of 2 million viewers.

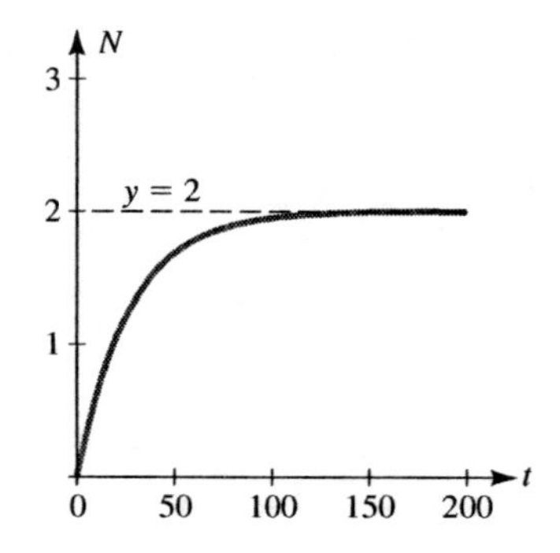

59. $y(t) = \frac{c}{b - a}(e^{-at} - e^{-bt})$

(a) $y'(t) = \frac{c}{b - a}(-ae^{-at} + be^{-bt})$

So, $y'(t) = 0$ when

$$-ae^{-at} + be^{-bt} = 0$$
$$be^{-bt} = ae^{-at}$$
$$\frac{e^{-bt}}{e^{-at}} = \frac{a}{b}$$
$$e^{-bt+at} = \frac{a}{b}$$
$$\ln e^{-bt+at} = \ln \frac{a}{b}$$
$$(a - b)t = \ln \frac{a}{b}$$
$$t = \frac{\ln \frac{a}{b}}{a - b} = \frac{\ln \frac{b}{a}}{b - a}$$

$$y''(t) = \frac{c}{b - a}(a^2 e^{-at} - b^2 e^{-bt})$$

$y''\left(\frac{\ln \frac{a}{b}}{a - b}\right) < 0$, so the maximum occurs

when $t = \frac{\ln \frac{a}{b}}{a - b}$.

In the long run,

$$\lim_{t \to +\infty} \frac{c}{b - a}(e^{-at} - e^{-kt}) = \frac{c}{b - a}(0 - 0) = 0.$$

So, the concentration approaches zero.

(b)

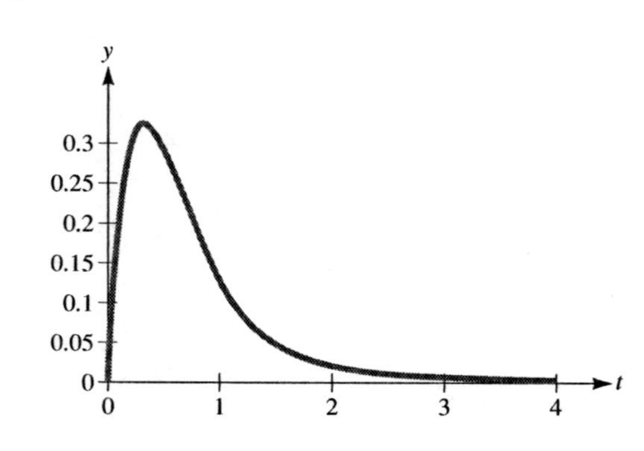

(c) Writing Exercise—Answers will vary.

61. $V(t) = V_0\left(1 - \frac{2}{L}\right)^t$

(a) When $L = 8$, $V(t) = 875\left(1 - \frac{2}{8}\right)^t = 875(0.75)^t$.
When $t = 5$, $V(5) = 875(0.75)^5 \approx \$207.64$
The annual rate of depreciation is the derivative, and logarithmic differentiation must be used.

$$\begin{aligned}\ln V &= \ln[8.75(0.75)^t] \\ &= \ln 8.75 + \ln(0.75)^t \\ &= \ln 8.75 + t\ln 0.75\end{aligned}$$

Differentiating,

$$\frac{V'(t)}{V(t)} = 0 + \ln 0.75$$

$$V'(t) = (\ln 0.75)V(t) = (\ln 0.75)(875)(0.75^t)$$

(b) In general, the percentage rate of change is

$$\begin{aligned}100\frac{V'(t)}{V(t)} &= 100\frac{\ln\left(1 - \frac{2}{L}\right)V(t)}{V(t)} \\ &= 100\ln\left(1 - \frac{2}{L}\right).\end{aligned}$$

Checkup for Chapter 4

1. (a)

$$\frac{(3^{-2})(9^2)}{(27)^{2/3}} = \frac{\left(\frac{1}{3^2}\right)(9^2)}{\left(\sqrt[3]{27}\right)^2} = \frac{\left(\frac{1}{9}\right)(81)}{(3)^2} = 1$$

(b)

$$\begin{aligned}\sqrt[3]{(25)^{1.5}\left(\frac{8}{27}\right)} &= \sqrt[3]{(25)^{1.5}}\sqrt[3]{\frac{8}{27}} \\ &= \left[(25)^{1.5}\right]^{1/3}\frac{\sqrt[3]{8}}{\sqrt[3]{27}} = (25)^{0.5}\left(\frac{2}{3}\right) \\ &= \sqrt{25}\left(\frac{2}{3}\right) = \frac{10}{3}\end{aligned}$$

(c) $\log_2 4 + \log_4 16^{-1}$

$\log_2 4 = a$ if and only if $2^a = 4$, so $a = 2$

$\log_4 16^{-1} = \log_4\left(\frac{1}{16}\right)$. Now,

$\log_4\left(\frac{1}{16}\right) = b$ if and only if $4^b = \frac{1}{16}$, so $b = -2$

$\log_2 4 + \log_4 16^{-1} = 2 - 2 = 0$

(d)

$$\begin{aligned}\left(\frac{8}{27}\right)^{-2/3}\left(\frac{16}{81}\right)^{3/2} &= \left(\frac{27}{8}\right)^{2/3}\left(\frac{16}{81}\right)^{3/2} \\ &= \left(\sqrt[3]{\frac{27}{8}}\right)^2\left(\sqrt{\frac{16}{81}}\right)^3 = \left(\frac{3}{2}\right)^2\left(\frac{4}{9}\right)^3 \\ &= \left(\frac{9}{4}\right)\left(\frac{64}{729}\right) = \frac{16}{81}\end{aligned}$$

2. (a)

$$\begin{aligned}(9x^4y^2)^{3/2} &= 9^{3/2}(x^4)^{3/2}(y^2)^{3/2} \\ &= \left(\sqrt{9}\right)^3(x^6)(y^3) \\ &= 27x^6y^3\end{aligned}$$

(b)

$$\begin{aligned}(3x^2y^{4/3})^{-1/2} &= \left(\frac{1}{3x^2y^{4/3}}\right)^{1/2} \\ &= \frac{(1)^{1/2}}{(3)^{1/2}(x^2)^{1/2}(y^{4/3})^{1/2}} = \frac{\sqrt{1}}{\left(\sqrt{3}\right)(x)(y^{2/3})} \\ &= \frac{1}{\sqrt{3}xy^{2/3}}\end{aligned}$$

(c)

$$\begin{aligned}\left(\frac{y}{x}\right)^{3/2}\left(\frac{x^{2/3}}{y^{1/6}}\right)^2 &= \left(\frac{y^{3/2}}{x^{3/2}}\right)\left(\frac{x^{4/3}}{y^{1/3}}\right) \\ &= \left(x^{4/3-3/2}\right)\left(y^{3/2-1/3}\right) \\ &= x^{-1/6}y^{7/6} = \frac{y^{7/6}}{x^{1/6}}\end{aligned}$$

(d)

$$\left(\frac{x^{0.2}y^{-1.2}}{x^{1.5}y^{0.4}}\right)^5 = \left[\left(x^{0.2-1.5}\right)\left(y^{-1.2-0.4}\right)\right]^5$$

$$= \left(x^{-1.3}y^{-1.6}\right)^5 = (x^{-1.3})^5(y^{-1.6})^5$$
$$= x^{-6.5}y^{-8} = \frac{1}{x^{6.5}y^8}$$

3. **(a)** $4^{2x-x^2} = \frac{1}{64}$

$4^{2x-x^2} = 4^{-3}$

So, $2x - x^2 = -3$

$$0 = x^2 - 2x + 3$$
$$0 = (x-3)(x+1)$$
$$x = 3, -1$$

(b)
$$e^{1/x} = 4$$
$$\ln e^{1/x} = \ln 4$$
$$\frac{1}{x} = \ln 4$$
$$x = \frac{1}{\ln 4}$$

(c) $\log_4 x^2 = 2$ if and only if $4^2 = x^2$, so, $x = \pm 4$.

(d) $\frac{25}{1 + 2e^{-0.5t}} = 3$

$$\frac{25}{3} = 1 + 2e^{-0.5t}$$
$$\frac{22}{3} = 2e^{-0.5t}$$
$$\frac{11}{3} = e^{-0.5t}$$
$$\ln\frac{11}{3} = \ln e^{-0.5t}$$
$$\ln\frac{11}{3} = -0.5t, \text{ or}$$
$$t = \frac{\ln\frac{11}{3}}{-0.5} = -2\ln\frac{11}{3} = 2\ln\frac{3}{11}$$

4. **(a)** (a) $y = \frac{e^x}{x^2 - 3x}$

$$\frac{dy}{dx} = \frac{(x^2-3x)(e^x \cdot 1) - (e^x)(2x-3)}{(x^2-3x)^2}$$
$$= \frac{e^x\left[(x^2-3x) - (2x-3)\right]}{(x^2-3x)^2}$$
$$= \frac{e^x(x^2-5x+3)}{(x^2-3x)^2}$$

(b) $y = \ln(x^3 + 2x^2 - 3x)$

$$\frac{dy}{dx} = \frac{1}{x^3+2x^2-3x}(3x^2+4x-3)$$
$$= \frac{3x^2+4x-3}{x^3+2x^2-3x}$$

(c) $y = x^3 \ln x$

$$\frac{dy}{dx} = (x^3)\left(\frac{1}{x}\cdot 1\right) + (\ln x)(3x^2)$$
$$= x^2 + 3x^2\ln x$$
$$= x^2(1 + 3\ln x)$$

(d) $y = \frac{e^{-2x}(2x-1)^3}{1-x^2}$

Using logarithmic differentiation,

$$\ln y = \ln\left[\frac{e^{-2x}(2x-1)^3}{1-x^2}\right]$$
$$= \ln e^{-2x} + \ln(2x-1)^3 - \ln(1-x^2)$$
$$= -2x + 3\ln(2x-1) - \ln(1-x^2)$$

$$\frac{y'}{y} = -2 + 3\cdot\frac{2}{2x-1} - \frac{-2x}{1-x^2}$$
$$y' = \left(-2 + \frac{6}{2x-1} + \frac{2x}{1-x^2}\right)y$$
$$= \left(-2 + \frac{6}{2x-1} + \frac{2x}{1-x^2}\right)\left[\frac{e^{-2x}(2x-1)^3}{1-x^2}\right]$$
$$= \left(-1 + \frac{3}{2x-1} + \frac{x}{1-x^2}\right)\left[\frac{2e^{-2x}(2x-1)^3}{1-x^2}\right]$$

5. **(a)** $y = x^2e^{-x}$

When $x = 0$, $y = 0$ so $(0, 0)$ is an intercept.

When $y = 0$, $x = 0$.

Also, $\lim_{x\to-\infty} x^2e^{-x} = +\infty$

$\lim_{x\to+\infty} x^2e^{-x} = \lim_{x\to+\infty} \frac{x^2}{e^x} = \lim_{x\to+\infty} \frac{2x}{e^x} =$

$\lim_{x\to+\infty} \frac{2}{e^x} = 0$ so $y = 0$ is a horizontal asymptote. $y' = (x^2)(-e^{-x}) + (e^{-x})(2x)$
$= xe^{-x}(2-x)$
so $y' = 0$ when $x = 0, 2$.
Rewriting, $y' = e^{-x}(2x - x^2)$, so

$$\begin{aligned} y'' &= (e^{-x})(2-2x) + (2x - x^2)(-e^{-x}) \\ &= e^{-x}\left[(2-2x) - (2x - x^2)\right] \\ &= e^{-x}(2 - 4x + x^2) \end{aligned}$$

So, $y'' = 0$ when $2 - 4x + x^2 = 0$.
Using the quadratic formula,

$$x = 2 \pm \sqrt{2}$$

When $x < 0$, $y' < 0$, so y is decreasing, $y'' > 0$, so y is concave up;
$0 < x < 2 - \sqrt{2}$, $y' > 0$, so y is increasing, $y'' > 0$, so y is concave up;
$2 - \sqrt{2} < x < 2$, $y' > 0$, so y is increasing, $y'' < 0$, so y is concave down;
$2 < x < 2 + \sqrt{2}$, $y' < 0$, so y is decreasing, $y'' < 0$, so y is concave down;
$x > 2 + \sqrt{2}$, $y' < 0$, so y is decreasing, $y'' > 0$, so y is concave up.

Overall, y is increasing when $0 < x < 2$;
y is decreasing when $x < 0$ and $x > 2$;
y is concave up when $x < 2 - \sqrt{2}$ and $x > 2 + \sqrt{2}$;
y is concave down when $2 - \sqrt{2} < x < 2 + \sqrt{2}$.
The point $(0, 0)$ is a relative minimum, the point $\left(2, \frac{4}{e^2}\right)$ is a relative maximum, and the points $(0.59, 0.19)$, $(3.41, 0.38)$ are inflection points.

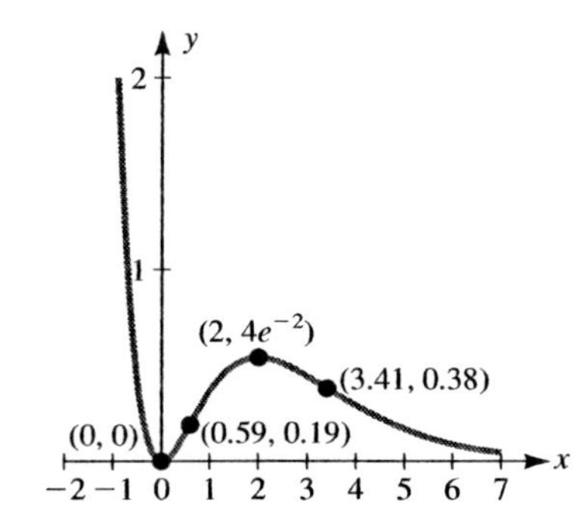

(b) (b) $y = \frac{\ln\sqrt{x}}{x^2} = \frac{\ln x^{1/2}}{x^2} = \frac{\frac{1}{2}\ln x}{x^2} = \frac{\ln x}{2x^2}$

Note that the domain of y is $x > 0$, and $x = 0$ is a vertical asymptote.
When $y = 0$, $x = 1$ so $(1, 0)$ is an intercept.

$$\lim_{x\to\infty} \frac{\ln x}{2x^2} = \lim_{x\to\infty} \frac{\frac{1}{x}}{4x} = \lim_{x\to\infty} \frac{1}{4x^2} = 0$$

so $y = 0$ is a horizontal asymptote.

$$\begin{aligned} y' &= \frac{(2x^2)\left(\frac{1}{x}\right) - (\ln x)(4x)}{(2x^2)^2} \\ &= \frac{2x(1 - 2\ln x)}{4x^4} = \frac{1 - 2\ln x}{2x^3} \end{aligned}$$

So $y' = 0$ when

$$\begin{aligned} 1 - 2\ln x &= 0 \\ 1 &= 2\ln x \\ \frac{1}{2} &= \ln x \\ e^{1/2} &= e^{\ln x}, \text{ or} \\ x &= e^{1/2} \end{aligned}$$

$$\begin{aligned} y'' &= \frac{(2x^3)\left(-2\cdot\frac{1}{x}\right) - (1 - 2\ln x)(6x^2)}{(2x^3)^2} \\ &= \frac{2x^2\,[(-2 - 3(1 - 2\ln x)]}{4x^6} \\ &= \frac{-2 - 3 + 6\ln x}{4x^6} = \frac{-5 + 6\ln x}{4x^6} \end{aligned}$$

So, $y'' = 0$ when

$$-5+6\ln x=0$$
$$6\ln x=5$$
$$\ln x=\frac{5}{6}$$
$$e^{\ln x}=e^{5/6}$$
$$x=e^{5/6}$$

When $0<x<e^{1/2}$, $y'>0$, so y is increasing,
$y''<0$, so y is concave down;
$e^{1/2}<x<e^{5/6}$, $y'<0$, so y is decreasing,
$y''<0$, so y is concave down;
$x>e^{5/6}$, $y'<0$, so y is decreasing,
$y''>0$, so y is concave up.

Overall, y is increasing when $0<x<e^{1/2}$;
y is decreasing when $x>e^{1/2}$;
y is concave up when $x>e^{5/6}$;
y is concave down when $0<x<e^{5/6}$.
The point $\left(e^{1/2}, \frac{1}{4e}\right)$ is a relative maximum and the point $\left(e^{5/6}, \frac{5}{12e^{5/3}}\right)$ is an inflection point.

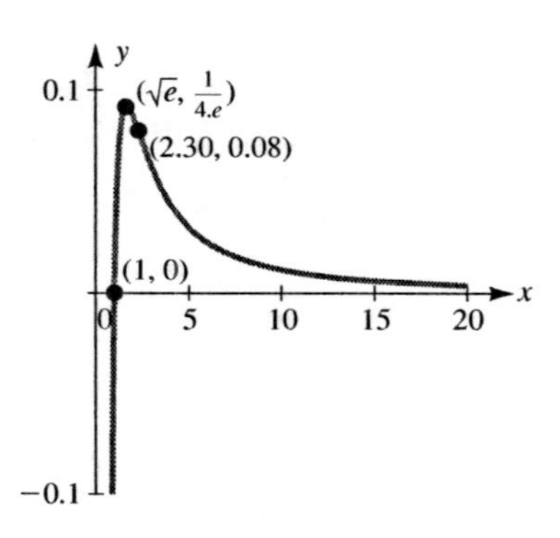

(c) (c) $y=\ln\left(\sqrt{x}-x\right)^2=2\ln(x^{1/2}-x)$
Note that the domain of y is $x>0$ and $x\neq 1$.
When $y=0$, $\ln(x^{1/2}-x)^2=0$
$x^{1/2}-x=\pm 1$
$0=x-x^{1/2}+1$ has no solution.
$0=x-x^{1/2}-1$ is solved by letting $u=x^{1/2}$, so $0=u^2-u-1$

$$u=\frac{1\pm\sqrt{1+4}}{2}\approx 1.62$$

(rejecting the negative solution)
$x^{1/2}\approx 1.62$ so $x\approx 2.6$
So, (2.6, 0) is an intercept. Since y is undefined when $x=1$, there is a vertical asymptote at $x=1$. Similarly, there is a vertical asymptote at $x=0$.

$$\text{Since } \lim_{x\to+\infty}\left(\sqrt{x}-x\right)^2=+\infty,$$
$$\lim_{x\to+\infty}\ln\left(\sqrt{x}-x\right)^2=+\infty$$

$$y'=2\cdot\frac{1}{x^{1/2}-x}\left(\frac{1}{2}x^{-1/2}-1\right)$$
$$=\frac{\frac{1}{x^{1/2}}-2}{x^{1/2}-x}\cdot\frac{x^{1/2}}{x^{1/2}}$$
$$=\frac{1-2x^{1/2}}{x-x^{3/2}}$$

So $y'=0$ when $1-2x^{1/2}=0$

$$1=2x^{1/2}$$
$$\frac{1}{2}=x^{1/2}, \text{ or}$$
$$x=\frac{1}{4}$$

$$y''=\frac{(x-x^{3/2})(-x^{-1/2})-(1-2x^{1/2})\left(1-\frac{3}{2}x^{1/2}\right)}{(x-x^{3/2})^2}$$
$$=\frac{-x^{1/2}+x-\left(1-2x^{1/2}-\frac{3}{2}x^{1/2}+3x\right)}{(x-x^{3/2})^2}$$
$$=\frac{-2x+\frac{5}{2}x^{1/2}-1}{(x-x^{3/2})^2}$$

So $y''=0$ when $-2x+\frac{5}{2}x^{1/2}-1=0$.
To solve, let $u=x^{1/2}$, so $-2u^2+\frac{5}{2}u-1=0$.
Using the quadratic formula, there are no solutions.

When $0 < x < \frac{1}{4}$, $y' > 0$, so y is increasing,

$y'' < 0$, so y is concave down;

$\frac{1}{4} < x < 1$, $y' < 0$, so y is decreasing,

$y'' < 0$, so y is concave down;

$x > 1$, $y' > 0$, so y is increasing,

$y'' < 0$, so y is concave down.

Overall, y is increasing when $0 < x < \frac{1}{4}$ and $x > 1$;
y is decreasing when $\frac{1}{4} < x < 1$;
y is concave down when $0 < x < 1$ and $x > 1$.
The point $\left(\frac{1}{4}, \ln \frac{1}{16}\right)$ is a relative maximum and there are no inflection points.

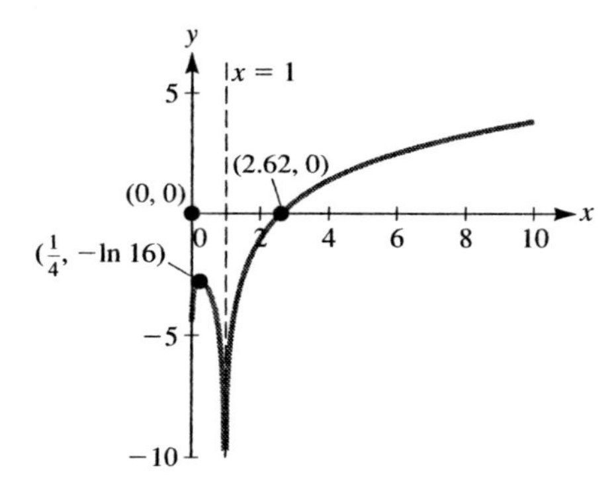

(d) $y = \dfrac{4}{1+e^{-x}}$
When $x = 0$, $y = 2$ so $(0, 2)$ is an intercept.
$y = 0$, has no solution.
$\lim\limits_{x\to-\infty} \dfrac{4}{1+e^{-x}} = 0$ so $y = 0$ is a horizontal asymptote.
$\lim\limits_{x\to+\infty} \dfrac{4}{1+e^{-x}} = 4$ so $y = 4$ is a horizontal asymptote.

$$y' = \frac{0-(4)(-e^{-x})}{(1+e^{-x})^2} = \frac{4e^{-x}}{(1+e^{-x})^2}$$

So y' is never zero. Further, $y' > 0$ for all values of x, so y is always increasing.
Using logarithmic differentiation,

$$\ln y' = \ln\left[\frac{4e^{-x}}{(1+e^{-x})^2}\right]$$
$$= \ln 4 + \ln e^{-x} - \ln(1+e^{-x})^2$$
$$= \ln 4 - x - 2\ln(1+e^{-x})$$

$$\frac{y''}{y'} = -1 - 2\cdot\frac{-e^{-x}}{1+e^{-x}}$$
$$= -1 + \frac{2e^{-x}}{1+e^{-x}}$$

$$y'' = \left[\frac{-(1+e^{-x})+2e^{-x}}{1+e^{-x}}\right]y'$$
$$= \left[\frac{-1+e^{-x}}{1+e^{-x}}\right]\left[\frac{4e^{-x}}{(1+e^{-x})^2}\right]$$
$$= (-1+e^{-x})\left[\frac{4e^{-x}}{(1+e^{-x})^3}\right]$$

So, $y'' = 0$ when

$$-1+e^{-x} = 0$$
$$e^{-x} = 1$$
$$-x = \ln 1$$
$$\text{or } x = 0$$

When $x < 0$, $y'' > 0$ so y is concave up

$x > 0$, $y'' < 0$ so y is concave down.

The point $(0, 2)$ is an inflection point.

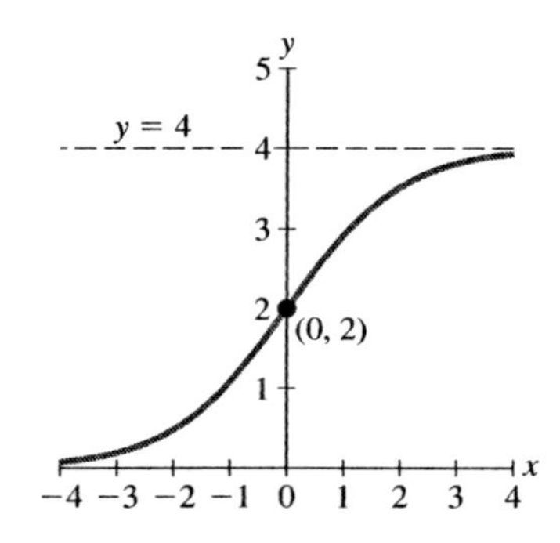

6. In general, $B(t) = Pe^{rt}$.
Here, $B(t) = 2000e^{0.05(t)}$.
When

$$t = 3,\quad B(3) = 2000e^{0.05(3)}$$
$$= 2000e^{0.15}$$
$$\approx \$2,323.67$$

For a balance of \$3,000,

$$3000 = 2000e^{0.05t}$$
$$\frac{3}{2} = e^{0.05t}$$
$$\ln\frac{3}{2} = \ln e^{0.05t}$$
$$\ln\frac{3}{2} = 0.05t, \text{ or}$$
$$t = \frac{\ln(3/2)}{0.05} \approx 8.1 \text{ years}$$

7. $p = \dfrac{\ln(t+1)}{t+1} + 5$

Note that the domain is $t > -1$.

(a)
$$p'(t) = \frac{(t+1)\left(\frac{1}{t+1}\cdot 1\right) - [\ln(t+1)](1)}{(t+1)^2} + 0$$
$$= \frac{1 - \ln(t+1)}{(t+1)^2}$$

So, $p'(t) = 0$ when

$$1 - \ln(t+1) = 0$$
$$1 = \ln(t+1)$$
$$e^1 = e^{\ln(t+1)}$$
$$e = t+1, \text{ or}$$
$$t = e - 1$$

When $-1 < t < e-1$, $p' > 0$ so p is increasing

$t > e-1$, $p' < 0$ so p is decreasing.

(b) The price is decreasing most rapidly when the first derivative is maximized.

$$p'' = \frac{(t+1)^2\left(\frac{-1}{t+1}\cdot 1\right) - [1 - \ln(t+1)][2(t+1)(1)]}{(t+1)^4}$$
$$= \frac{(t+1)[-1 - 2(1 - \ln(t+1))]}{(t+1)^4}$$
$$= \frac{-3 + 2\ln(t+1)}{(t+1)^3}$$

So, $p'' = 0$ when

$$-3 + 2\ln(t+1) = 0$$
$$2\ln(t+1) = 3$$
$$\ln(t+1) = \frac{3}{2}$$
$$e^{\ln(t+1)} = e^{3/2}$$
$$t + 1 = e^{3/2}, \text{ or}$$
$$t = e^{3/2} - 1$$

$$p''' = \frac{(t+1)^3\left(\frac{2}{t+1}\right) - [-3 + 2\ln(t+1)]\left[3(t+1)^2\right]}{(t+1)^6}$$
$$= \frac{(t+1)^2[2 - 3(-3 + 2\ln(t+1))]}{(t+1)^6}$$
$$= \frac{11 - 6\ln(t+1)}{(t+1)^6}$$

So, when $t = e^{3/2} - 1$,

$$p''' = \frac{11 - 6\ln(e^{3/2} - 1 + 1)}{(e^{3/2} - 1 + 1)^6}$$
$$= \frac{11 - 6e^{3/2}}{e^9}$$

Since $p''' < 0$, $t = e^{3/2} - 1$ is a maximum.

(c)
$$\lim_{t\to\infty} \frac{\ln(t+1)}{t+1} + 5$$
$$= \lim_{t\to\infty} \frac{\ln(t+1)}{t+1} + \lim_{t\to\infty} 5$$
$$= \lim_{t\to\infty} \frac{\frac{1}{t+1}\cdot 1}{1} + 5$$
$$= \lim_{t\to\infty} \frac{1}{t+1} + 5 = 0 + 5 = 5$$

So, in the long run, the price approaches \$500.

8. $D = q(p) = 1{,}000(p+2)e^{-p}$

(a)
$$q'(p) = 1{,}000\left[(p+2)(-e^{-p}) + (e^{-p})(1)\right]$$
$$= -1{,}000e^{-p}\left[(p+2) - 1\right]$$
$$= -1{,}000e^{-p}(p+1)$$

So, $q'(p) = 0$ when $p = -1$.

For the practical domain $p \geq 0$, $q'(p) < 0$ so q decreases.

(b) $R = pq = 1{,}000p(p+2)e^{-p}$
Rewriting R as $1{,}000(p^2+2p)e^{-p}$,

$$R'(p) = 1{,}000\left[(p^2+2p)(-e^{-p}) + (e^{-p})(2p+2)\right]$$
$$= -1{,}000e^{-p}\left[(p^2+2p) - (2p+2)\right]$$
$$= -1{,}000e^{-p}(p^2-2)$$

So $R'(p) = 0$ when $p = \sqrt{2}$.

When $0 \leq x < \sqrt{2}$, $R'(q) > 0$, so R is increasing
$x > \sqrt{2}$, $R'(q) < 0$ so R is decreasing.

$$R''(p) = -1{,}000\left[(e^{-p})(2p) + (p^2-2)(-e^{-p})\right]$$
$$= 1{,}000e^{-p}[p^2 - 2p - 2]$$

So $R''\left(\sqrt{2}\right) < 0$ and the maximum revenue occurs when the price is approximately \$141.42. The maximum revenue is

$$R\left(\sqrt{2}\right) = 1{,}000\left(\sqrt{2}\right)\left(\sqrt{2}+2\right)e^{-\sqrt{2}}$$
$$\approx 1,173.8714 \text{ hundred or } \$117,387.14$$

9. $R(t) = R_0 e^{-kt}$
Since the half-life is 5,730 years,

$$\frac{1}{2}R_0 = R_0 e^{-k(5{,}730)}$$
$$\ln\frac{1}{2} = \ln e^{-5{,}730k}$$
$$\ln\frac{1}{2} = -5,730k, \text{ so}$$
$$k = \frac{\ln(1/2)}{-5{,}730} = \frac{-\ln(1/2)}{5{,}730} = \frac{\ln 2}{5{,}730}$$

So, $R(t) = R_0 e^{-(\ln 2/5{,}730)t}$
When 45% remains,

$$0.45R_0 = R_0 e^{-(\ln 2/5{,}730)t}$$
$$\ln 0.45 = \ln e^{-(\ln 2/5{,}730)t}$$
$$\ln 0.45 = -(\ln 2/5{,}730)t$$
$$t = \frac{-5{,}730\ln 0.45}{\ln 2} \approx 6{,}601 \text{ years old}$$

10. $N(T) = 10{,}000(8+t)e^{-0.1t}$

(a) When $t = 0$, $N(0) = 10{,}000(8)e^0 = 80{,}000$ bacteria

(b) $N'(t) = 10{,}000\left[(8+t)(-0.1e^{-0.1t}) + (e^{-0.1t})(1)\right]$

$$= 10{,}000e^{-0.1t}[-0.1(8+t)+1]$$
$$= 10{,}000e^{-0.1t}(0.2-0.1t)$$

So, $N'(t) = 0$ when

$$0.2 - 0.1t = 0$$
$$0.2 = 0.1t, \text{ or}$$
$$t = 2$$

$$N''(t) = 10{,}000\left[\begin{array}{l}(e^{-0.1t})(-0.1)\\ +(0.2-0.1t)(-0.1e^{-0.1t})\end{array}\right]$$
$$= 10{,}000e^{-0.1t}[-0.1 - 0.1(0.2-0.1t)]$$
$$= 10{,}000e^{-0.1t}(-0.12+0.01t)$$

When $t = 2$, $N''(2) < 0$, so the maximum occurs when $t = 2$ and is

$$N(2) = 10{,}000(8+2)e^{-0.1(2)}$$
$$\approx 81,873 \text{ bacteria}$$

(c) $\lim_{t\to\infty} 10{,}000(8+t)e^{-0.1t} = 10{,}000 \lim_{t\to\infty} \frac{8+t}{e^{0.1t}}$

$$= 10{,}000 \lim_{t\to\infty} \frac{1}{0.1e^{0.1t}} = 10{,}000(0) = 0$$

So, the bacterial colony dies off in the long run.

Review Problems

1. $f(x) = 5^x$
When $x = 0$, $f(0) = 1$, so (0, 1) is an intercept.
$f(x) = 0$ has no solution.

$\lim_{x\to-\infty} 5^x = \lim_{x\to-\infty} \frac{1}{5^{-x}} = 0$ so $y = 0$ is a horizontal asymptote.
$\lim_{x\to+\infty} 5^x = +\infty$ so $f(x)$ increases without bound as x increases.

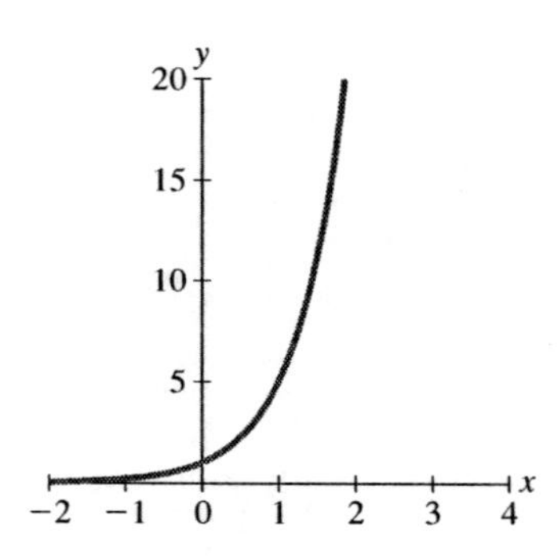

3. $f(x) = \ln x^2$
Note that the domain of f is $x =\neq 0$, so $x = 0$ is a vertical asymptote.
When $f(x) = 0$, $x = \pm 1$ so $(-1, 0)$ and $(1, 0)$ are intercepts.
$\lim_{x\to-\infty} \ln x^2 = \lim_{x\to\infty} \ln x^2 = +\infty$, so f increases without bound as x decreases and as x increases.

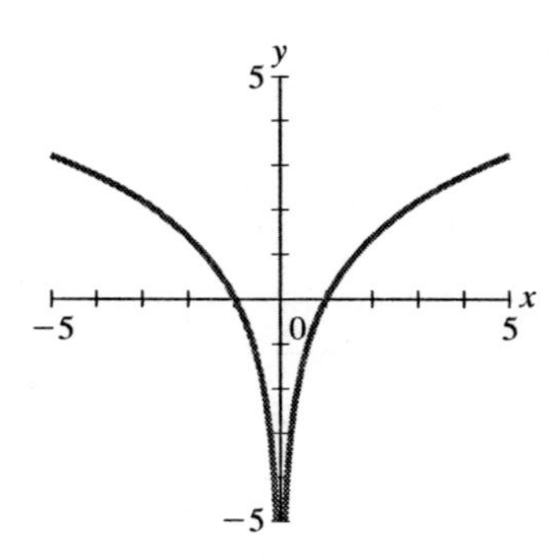

5. **(a)** $f(x) = Ae^{-kx}$
Since $f(0) = 10$, $10 = Ae^0$, or $A = 10$ and $f(x) = 10e^{-kx}$.

Since $f(1) = 25$,

$$25 = 10e^{-k(1)}$$
$$\frac{5}{2} = e^{-k}$$
$$\ln\frac{5}{2} = \ln e^{-k}$$
$$k = -\ln\frac{5}{2}$$

$$\text{So, } f(x) = 10e^{-(\ln\frac{5}{2})t}$$
$$= 10e^{(\ln\frac{5}{2})t}$$
$$f(4) = 10e^{-4\ln(5/2)}$$
$$= 10e^{\ln(5/20)^4}$$
$$= 10\left(\frac{5}{2}\right)^4 = \frac{3125}{8}$$

(b) $f(x) = Ae^{kx}$
Since $f(1) = 3$,
$3 = Ae^{k(1)}$, or $A = 3e^{-k}$.
Since $f(2) = 10$,
$10 = Ae^{k(2)}$, or $A = 10e^{-2k}$.
So, $3e^{-k} = 10e^{-2k}$
$\frac{3}{10} = e^{-k}$
and $A = 3\left(\frac{3}{10}\right) = \frac{9}{10}$.

$$f(3) = \frac{9}{10}e^{k(3)}$$
$$= \frac{9}{10}(e^{-k})^{-3} = \frac{9}{10}\left(\frac{3}{10}\right)^{-3}$$
$$= \frac{9}{10}\left(\frac{1000}{27}\right) = \frac{100}{3}$$

(c) $f(x) = 30 + Ae^{-kx}$
Since $f(0) = 50$,
$50 = 30 + Ae^0$, or $A = 20$ and
$f(x) = 30 + 20e^{-kx}$
Since $f(3) = 40$,
$40 = 30 + 20e^{-k(3)}$
$10 = 20e^{-3k}$
$\frac{1}{2} = e^{-3k}$

$$f(9) = 30 + 20e^{-k(9)}$$
$$= 30 + 20(e^{-3k})^3$$
$$= 30 + 20\left(\frac{1}{2}\right)^3$$
$$= 30 + \frac{5}{2} = \frac{65}{2}$$

(d) $f(t) = \dfrac{6}{1 + Ae^{-kt}}$

Since $f(0) = 3$,

$$3 = \frac{6}{1 + Ae^0}, \text{ or } A = 1.$$

Now, $f(t) = \dfrac{6}{1 + e^{-kt}}$.

Since $f(5) = 2$,

$2 = \dfrac{6}{1 + e^{-k(5)}}$,

$1 + e^{-5k} = 3$,

$e^{-5k} = 2$.

So,

$$f(10) = \frac{6}{1 + e^{-k(10)}}$$
$$= \frac{6}{1 + (e^{-5k})^2}$$
$$= \frac{6}{1 + (2)^2} = \frac{6}{5}.$$

7.
$$8 = 2e^{0.04x},$$
$$e^{0.04x} = 4,$$
$$0.04x = \ln 4$$
$$x \approx 25 \ln 4.$$

9. $4 \ln x = 8$, $\ln x = 2$, or $x = e^2 \approx 7.389$.

11. $\log_9(4x - 1) = 2$ if and only if

$$4x - 1 = 9^2$$
$$4x = 82, \text{ or } x = \frac{41}{2}$$

13. $e^{2x} + e^x - 2 = 0$

Letting $u = e^x$, $u^2 + u - 2 = 0$

$(u + 2)(u - 1) = 0$

or, $u = -2, 1$.

If $u = -2$, $e^x = -2$ and there is no solution.

If $u = 1$, $e^x = 1$, so $x = 0$.

15.
$$y = x^2 e^{-x}$$
$$\frac{dy}{dx} = (x^2)(-e^{-x}) + (e^{-x})(2x)$$
$$= xe^{-x}(-x + 2)$$

17.
$$y = x \ln x^2 = 2x \ln x$$
$$\frac{dy}{dx} = (2x)\left(\frac{1}{x}\right) + (\ln x)(2)$$
$$= 2(1 + \ln x)$$

19.
$$y = \log_3(x^2) = \frac{\ln(x^2)}{\ln 3}$$
$$= \frac{2}{\ln 3} \ln x$$
$$\frac{dy}{dx} = \frac{2}{\ln 3} \cdot \frac{1}{x} = \frac{2}{x \ln 3}$$

21. $y = \dfrac{e^{-x} + e^x}{1 + e^{-2x}}$

$$\frac{dy}{dx} = \frac{(1 + e^{-2x})(-e^{-x} + e^x) - (e^{-x} + e^x)(-2e^{-2x})}{(1 + e^{-2x})^2}$$
$$= \frac{-e^{-x} - e^{-3x} + e^x + e^{-x} + 2e^{-3x} + 2e^{-x}}{(1 + e^{-2x})^2}$$
$$= \frac{e^{-3x} + 2e^{-x} + e^x}{(1 + e^{-2x})^2}$$
$$= \frac{(e^{-2x} + 1)(e^{-x} + e^x)}{(1 + e^{-2x})^2}$$
$$= \frac{e^{-x} + e^x}{1 + e^{-2x}} = \frac{e^{-x} + e^x}{1 + e^{-2x}} \cdot \frac{e^{-x}}{e^{-x}} = e^x$$

23. $y = \ln(e^{-2x} + e^{-x})$

$$\frac{dy}{dx} = \frac{1}{e^{-2x}+e^{-x}}(-2e^{-2x}-e^{-x})$$
$$= \frac{-e^{-x}(2e^{-x}+1)}{e^{-x}(e^{-x}+1)}$$
$$= -\frac{2e^{-x}+1}{e^{-x}+1}$$

25. $y = \dfrac{e^{-x}}{x+\ln x}$

$$\frac{dy}{dx} = \frac{(x+\ln x)(-e^{-x})-(e^{-x})\left(1+\frac{1}{x}\right)}{(x+\ln x)^2}$$
$$= \frac{-xe^{-x}-e^{-x}\ln x-e^{-x}-\frac{e^{-x}}{x}}{(x+\ln x)^2}\cdot\frac{x}{x}$$
$$= \frac{-x^2e^{-x}-xe^{-x}\ln x-xe^{-x}-e^{-x}}{x(x+\ln x)^2}$$
$$= \frac{-e^{-x}(x^2+x\ln x+x+1)}{x(x+\ln x)^2}$$

27. $ye^{x-x^2} = x+y$

$$(y)\left[\left(e^{x-x^2}\right)(1-2x)\right]+\left(e^{x-x^2}\right)\frac{dy}{dx} = 1+\frac{dy}{dx}$$
$$\left(e^{x-x^2}\right)\frac{dy}{dx}-\frac{dy}{dx} = 1-y\left(e^{x-x^2}\right)(1-2x)$$
$$\left(e^{x-x^2}-1\right)\frac{dy}{dx} = 1-y\left(e^{x-x^2}\right)(1-2x)$$
$$\frac{dy}{dx} = \frac{1-y\left(e^{x-x^2}\right)(1-2x)}{e^{x-x^2}-1}$$
$$= \frac{1+y\left(e^{x-x^2}\right)(2x-1)}{e^{x-x^2}-1}$$

29. $y = \dfrac{(x^2+e^{2x})^3e^{-2x}}{(1+x-x^2)^{2/3}}$

Using logarithmic differentiation,

$$\ln y = \ln\left[\frac{(x^2+e^{2x})^3e^{-2x}}{(1+x-x^2)^{2/3}}\right]$$
$$= \ln(x^2+e^{2x})^3+\ln e^{-2x}-\ln(1+x-x^2)^{2/3}$$
$$= 3\ln(x^2+e^{2x})-2x-\frac{2}{3}\ln(1+x-x^2)$$

$$\frac{y'}{y} = 3\cdot\frac{2x+2e^{2x}}{x^2+e^{2x}} - 2 - \frac{2}{3}\cdot\frac{1-2x}{1+x-x^2}$$
$$y' = \left[\frac{6(x+e^{2x})}{x^2+e^{2x}} - 2 - \frac{2(1-2x)}{3(1+x-x^2)}\right]\left[\frac{(x^2+e^{2x})^3e^{-2x}}{(1+x-x^2)^{2/3}}\right]$$
$$= \left[\frac{3(x+e^{2x})}{x^2+e^{2x}} - 1 - \frac{(1-2x)}{3(1+x-x^2)}\right]\left[\frac{2(x^2+e^{2x})^3e^{-2x}}{(1+x-x^2)^{2/3}}\right]$$

31. $f(x) = e^x - e^{-x}$

When $x=0$, $f(0)=0$ so $(0, 0)$ is an intercept.

When $f(x)=0$, $x=0$.

$\lim\limits_{x\to-\infty} e^x - e^{-x} = -\infty$ so f decreases without bound as x decreases.

$\lim\limits_{x\to+\infty} e^x - e^{-x} = +\infty$ so f increases without bound as x increases.

$f'(x) = e^x + e^{-x}$

$f'(x)$ is never zero; further, $f'(x) > 0$ for all values of x, so f is always increasing.

$f''(x) = e^x - e^{-x}$

So, $f''(x) = 0$ when $x = 0$.

When $x < 0$, $f''(x) < 0$ so f is concave down

$x > 0$, $f''(x) > 0$ so f is concave up.

The point $(0, 0)$ is an inflection point.

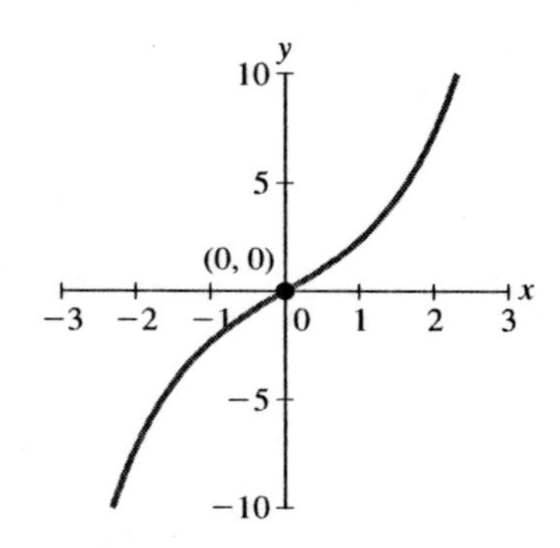

33. $f(t) = t + e^{-t}$

When $t=0$, $f(0)=1$ so $(0, 1)$ is an intercept.

$f(t)=0$ has no solution.

$\lim\limits_{t\to-\infty} t+e^{-t}=+\infty$ (since e^{-t} increases more rapidly than t decreases).
$\lim\limits_{t\to+\infty} t+e^{-t}=t$, so $y=t$ is an oblique asymptote.
$f'(t)=1-e^{-t}$
So $f'(t)=0$ when $1-e^{-t}=0$

$$1=e^{-t}$$
$$\ln 1=t, \text{ or } t=0.$$

When $t<0$, $f'(t)<0$ so f is decreasing
$t>0$, $f'(t)>0$ so f is increasing.

The point (0, 1) is a relative minimum.
$f''(t)=e^{-t}$
So, $f''(t)$ is never zero; further $f''(t)>0$ for all values of t, so f is always concave up.

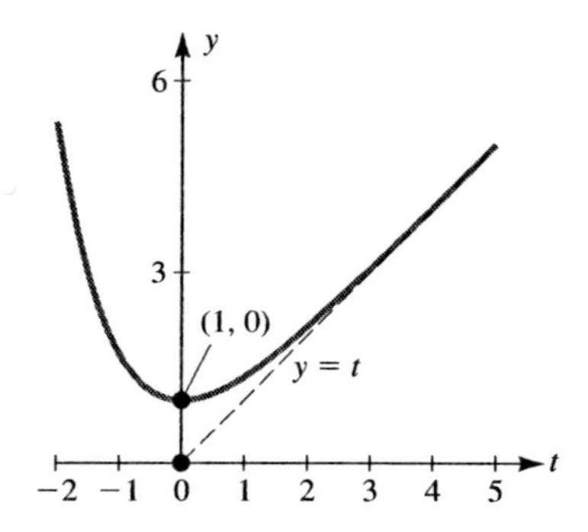

35. $F(u)=u^2+2\ln(u+2)$
Note that the domain is $u>-2$, so $u=-2$ is a vertical asymptote.
When $u=0$, $F(0)=2\ln 2$ so $(0, 2\ln 2)$ is an intercept.
$F(u)=0$ is too difficult to solve.
$\lim\limits_{u\to+\infty} u^2+2\ln(u+2)=+\infty$ so F increases without bound as u increases.

$$F'(u)=2u+2\cdot\frac{1}{u+2}\cdot 1=2\left(u+\frac{1}{u+2}\right)$$
$$=2\frac{u^2+2u+1}{u+2}=2\frac{(u+1)^2}{u+2}$$

So, $F'(u)=0$ when $u=-1$.

When $-2<u<-1$, $F'(u)>0$ so F increases
$u>-1$, $F'(u)>0$ so F increases.

$$F''(u)=2\left[\frac{(u+2)(2u+2)-(u^2+2u+1)(1)}{(u+2)^2}\right]$$
$$=2(u+1)\left[\frac{2(u+2)-(u+1)}{(u+2)^2}\right]$$
$$=2(u+1)\left[\frac{u+3}{(u+2)^2}\right]$$

So, $F''(u)=0$ when u $=-1$ (rejecting $u=-3$ since it is not in the domain of F).
When

$-2<u<-1$, $F''(u)<0$ so F is concave down
$u>-1$, $F''(u)>0$ so F is concave up.

The point $(-1, 1)$ is an inflection point.

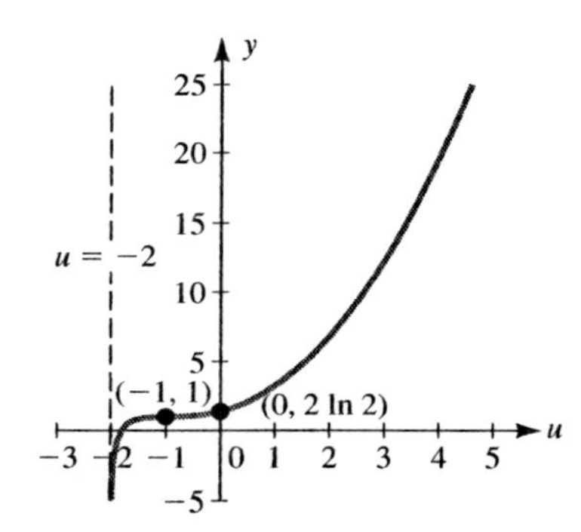

37. $G(x)=\ln(e^{-2x}+e^{-x})$
When $x=0$, $G(0)=\ln 2$ so $(0, \ln 2)$ is an intercept.
When $G(x)=0$, $\ln(e^{-2x}+e^{-x})=0$;
$e^{-2x}+e^{-x}=1$;
$e^{-2x}+e^{-x}-1=0$
Letting $u=e^{-x}$,

$$u^2+u-1=0$$
$$u=\frac{-1\pm\sqrt{1+(4)(1)(1)}}{2(1)}=\frac{-1\pm\sqrt{5}}{2}$$

So,

$$e^{-x} = \frac{-1 \pm \sqrt{5}}{2}$$

$$\ln e^{-x} = \ln\left(\frac{-1+\sqrt{5}}{2}\right) \quad \text{(rejecting negative value)}$$

$$-x = \ln\left(\frac{-1+\sqrt{5}}{2}\right), \quad \text{or}$$

$$x = -\ln\left(\frac{-1+\sqrt{5}}{2}\right) \approx 0.48$$

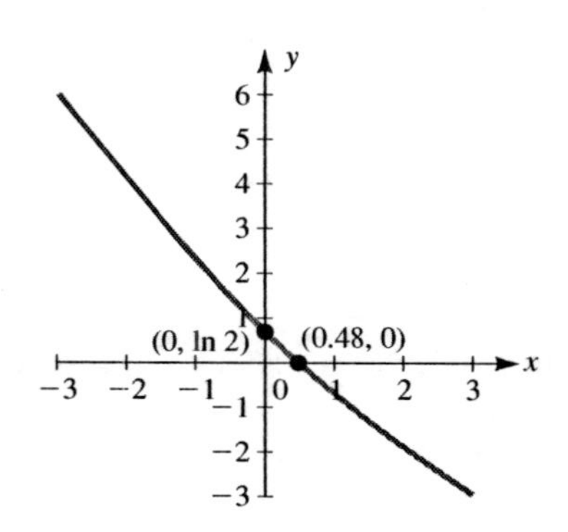

So, (0.48, 0) is an intercept.
$\lim_{x\to-\infty} \ln(e^{-2x}+e^{-x}) = +\infty$ so G increases without bound as x decreases.
$\lim_{x\to+\infty} \ln(e^{-2x}+e^{-x}) = \lim_{x\to 0^+} \ln x = -\infty$ so G decreases without bound as x increases.
$G'(x) = \frac{1}{e^{-2x}+e^{-x}}(-2e^{-2x}-e^{-x})$
So, $G'(x) = 0$ when

$$-2e^{-2x} - e^{-x} = 0$$

$$-e^{-x}(2e^{-x}+1) = 0$$

$$2e^{-x}+1 = 0 \quad \text{(since } e^{-x} \text{ is never zero)}$$

$$e^{-x} = -\frac{1}{2} \text{ has no solution.}$$

$G'(x)$ is never zero; further, $G'(x) < 0$ for all x so G is always decreasing.

$$G''(x) = \left[\frac{(e^{-2x}+e^{-x})(4e^{-2x}+e^{-x})}{(e^{-2x}+e^{-x})^2} - \frac{(-2e^{-2x}-e^{-x})(-2e^{-2x}-e^{-x})}{(e^{-2x}+e^{-x})^2}\right]$$

$$= \frac{4e^{-4x}+5e^{-3x}+e^{-2x}-(4e^{-4x}+4e^{-3x}+e^{-2x})}{(e^{-2x}+e^{-x})^2}$$

$$= \frac{e^{-3x}}{(e^{-2x}+e^{-x})^2}$$

Since e^{-3x} is never zero, $G''(x)$ is never zero; further $G''(x) > 0$ for all x so G is always concave up.

39.

$$f(x) = \ln(4x - x^2), \quad 1 \le x \le 3$$

$$f'(x) = \frac{4-2x}{4x-x^2}$$

So, $f'(x) = 0$ when $4 - 2x = 0$, or $x = 2$.
$f(2) = \ln 4$; $f(1) = \ln 3$; $f(3) = \ln 3$
The function's largest value is ln 4 and its smallest value is ln 3.

41.

$$h(t) = (e^{-t}+e^{t})^5, \ -1 \le t \le 1$$

$$h'(t) = 5(e^{-t}+e^{t})^4(-e^{-t}+e^{t})$$

So, $h'(t) = 0$ when

$$-e^{-t}+e^{t} = 0 \text{ (since } e^{-t}+e^{t} \text{ is never zero)}$$

$$e^{-t}(-1+e^{2t}) = 0$$

$$e^{2t} = 1, \quad \text{or } t = 0$$

$$h(0) = 32; h(-1) = \left(e+\frac{1}{e}\right)^5 \approx 280,\ h(1) = \left(e+\frac{1}{e}\right)^5$$

So, the function's largest value is $\left(e+\frac{1}{e}\right)^5$ and its smallest value is 32.

43. $y = \ln x^2$, $x = 1$
When $x = 1$, $y = \ln 1 = 0$ so point (1, 0) is on the tangent line.

$$y' = (x)\left(\frac{2x}{x^2}\right) + (\ln x^2) = 2 + \ln x^2$$

$$\text{slope } = y' = 2 + \ln(1)^2 = 2$$

So, the equation of the tangent line is

$$y - 0 = 2(x - 1), \text{ or}$$
$$y = 2x - 2.$$

45. $y = x^3e^{2-x}$, $x = 2$
When $x = 2$, $y = 8$ so point (2, 8) is on the tangent line.
$y' = (x^3)(-e^{2-x}) + (e^{2-x})(3x^2)$
slope $= y' = (2)^3(-e^0) + (e^0)(3 \cdot 4) = 4$
So, the equation of the tangent line is

$$y - 8 = 4(x - 2), \text{ or}$$
$$y = 4x.$$

47. $Q(x) = 50 - 40e^{-0.1x}$

(a) When $x = 0$, $Q(0) = 10$ so (0, 10) is an intercept.

$$Q(x) = 0 \text{ when } 50 - 40e^{-0.1x} = 0$$
$$50 = 40e^{-0.1x}$$
$$\frac{5}{4} = e^{-0.1x}$$
$$\ln\frac{5}{4} = -0.1x, \text{ or}$$
$$x = \frac{\ln\frac{5}{4}}{-0.1}$$

Since the relevant domain is $x \geq 0$, this intercept will not be on graph.
$\lim_{x\to\infty} 50 - 40e^{-0.1x} = 50$, so $y = 50$ is a horizontal asymptote.
$Q'(x) = 4e^{-0.1x}$
Now, $Q'(x)$ is never zero. Further, $Q'(x) > 0$ for all x so Q is always increasing.
$Q''(x) = -0.4e^{-0.1x}$
which is never zero. Further, $Q''(x) < 0$ for all x so Q is always concave down.

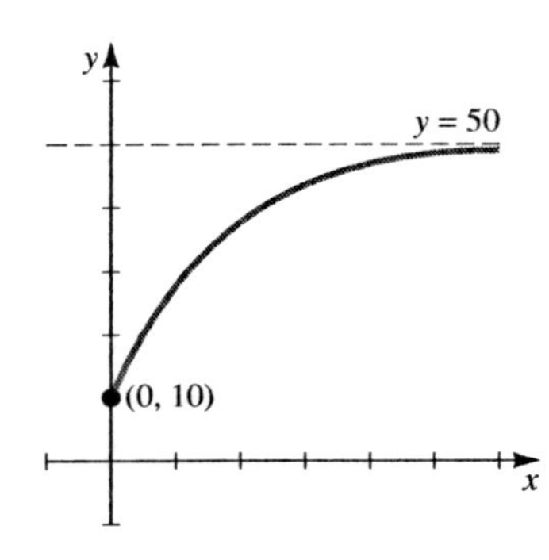

(b) When no money is spent on advertising,
$Q(0) = 50 - 40e^0 = 10$
So, 10,000 units will be sold.

(c) If 8 thousand dollars are spent on advertising,
$Q(8) = 50 - 40e^{-0.1(8)} \approx 32.027$
So, approximately 32,027 units will be sold.

(d) For sales of 35 thousand units,

$$35 = 50 - 40e^{-0.1x}$$
$$40e^{-0.1x} = 15$$
$$e^{-0.1x} = \frac{3}{8}$$
$$-0.1x = \ln\frac{3}{8}$$
$$x = \frac{\ln\frac{3}{8}}{-0.1} = 10\ln\frac{8}{3} \approx 9.81$$

So, approximately $9,810 dollars must be spent on advertising.

(e) $\lim_{x\to\infty} 50 - 40e^{-0.1x} = 50$
So, approximately (just less than) 50,000 units is the optimal sales projection.

49. $P(t) = \dfrac{30}{1 + 2e^{-0.05t}}$

(a) When $t = 0$, $P(0) = 10$ so (0, 10) is an intercept.
$P(t) = 0$ has no solution.
$\lim_{x\to\infty} \dfrac{30}{1 + 2e^{-0.05t}} = 30$, so $y = 30$ is a horizontal asymptote.

$$P'(t) = \frac{0 - (30)(-0.1e^{-0.05t})}{(1 + 2e^{-0.05t})^2} = \frac{3e^{-0.05t}}{(1 + 2e^{-0.05t})^2}$$

Since $3e^{-0.05t}$ is never zero, $P'(t)$ is never zero. Further, $P'(t) > 0$ for all t, so P is always increasing. Using logarithmic differentiation,

$$\ln P'(t) = \ln\left[\frac{3e^{-0.05t}}{(1+2e^{-0.05t})^2}\right]$$
$$= \ln 3 + \ln e^{-0.05t} - \ln(1+2e^{-0.05t})^2$$
$$= \ln 3 - 0.05t - 2\ln(1+2e^{-0.05t})$$

$$\frac{P''(t)}{P(t)} = -0.05 - 2\cdot\frac{-0.1e^{-0.05t}}{1+2e^{-0.05t}}$$
$$P''(t) = \left[\frac{-0.05(1+2e^{-0.05t})+0.2e^{-0.05t}}{1+2e^{-0.05t}}\right]P'(t)$$
$$= \left[\frac{-0.05+0.1e^{-0.05t}}{1+2e^{-0.05t}}\right]\left[\frac{3e^{-0.05t}}{(1+2e^{-0.05t})^2}\right]$$
$$= (-0.05+0.1e^{-0.05t})\left[\frac{3e^{-0.05t}}{(1+2e^{-0.05t})^3}\right]$$

So $P''(t) = 0$ when

$$-0.05 + 0.1e^{-0.05t} = 0$$
$$e^{-0.05t} = 0.5$$
$$-0.05t = 0.5, \text{ or}$$
$$t = \frac{\ln 0.5}{-0.05} = 20\ln 2 \approx 13.9$$

When $0 < t < 13.9$, $P''(t) > 0$, so P is concave up. When $t > 13.9$, $P''(t) < 0$ so P is concave down. The point (13.9, 15.0) is an inflection point.

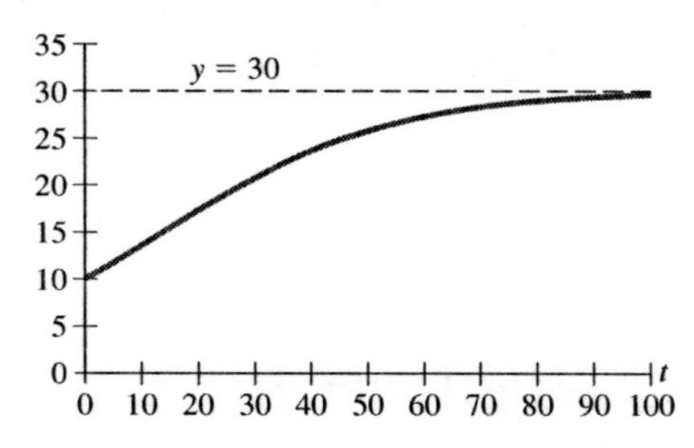

(b) The current population is
$P(0) = \dfrac{30}{1+2e^0} = 10$, or 10,000,000 people.

(c) The population in 20 years will be

$$P(20) = \frac{30}{1+2e^{-0.05(20)}}$$
$$= \frac{30}{1+2e^{-1}} = 17.2835$$

or 17,283,500 people.

(d) $\lim\limits_{x\to\infty} \dfrac{30}{1+2e^{-0.05t}} = 30$
So, the population approaches 30,000,000 in the long run.

51. When interest is compounded quarterly, the effective rate is
$\left(1+\dfrac{.0825}{4}\right)^4 - 1 \approx 0.08509$, or 8.51%.
When interest is compounded continuously, the effective rate is
$e^{.082} - 1 \approx 0.08546$, or 8.55%.
So, 8.20% compounded continuously has the greater effective interest rate.

53. (a)
$$8{,}000 = (P)\left(1+\frac{0.0625}{2}\right)^{2(10)}$$
$$P = \frac{8{,}000}{(1.03125)^{20}} = \$4{,}323.25$$

(b)
$$8{,}000 = Pe^{0.0625(10)}$$
$$P = 8{,}000e^{-0.625} = \$4{,}282.09$$

55. At 6% compounded annually, the effective interest rate is

$$\left(1+\frac{r}{k}\right)^k - 1 = \left(1+\frac{0.06}{1}\right)^1 - 1$$
$$= 0.06.$$

At r% compounded continuously, the effective interest rate is $e^r - 1$. Setting the two effective rates equal to each other yields

$$e^r - 1 = 0.06,$$
$$e^r = 1.06,\ r = \ln 1.06 = 0.0583 \text{ or } 5.83\%.$$

57. (a) The rate of change of the carbon monoxide level t years from now is $Q'(t) = 0.12e^{0.03t}$. The rate two years from now is $Q'(2) = 0.12e^{0.03(2)} =$ 0.13 parts per million per year.

(b) The percentage rate of change of the carbon monoxide level t years from now is $100\left[\frac{Q'(t)}{Q(t)}\right] = 100\left(\frac{0.12e^{0.03t}}{4e^{0.03t}}\right) = 3\%$ per year, which is a constant (independent of time).

59. $V(t) = 2{,}000e^{\sqrt{2t}}$

The percentage rate of change of the value of the asset is

$$100\frac{V'(t)}{V(t)} = 100\frac{2000e^{\sqrt{2t}}\left[\frac{1}{2}(2t)^{-1/2}(2)\right]}{2{,}000e^{\sqrt{2t}}}$$

$$= 1000\frac{1}{\sqrt{2t}}$$

Which will equal the prevailing interest rate when

$$\frac{1}{\sqrt{2t}} = 0.05$$

$$\sqrt{2t} = \frac{1}{0.05} = 20$$

$$2t = 400$$

$$t = 200 \text{ years}$$

When $0 < t < 200$, the percentage rate is more than the prevailing rate. When $t > 200$, the prevailing rate is greater, so, it's best to sell the asset after 200 years.

61. $Q(t) = Q_0e^{-kt}$

(a) When $t = \lambda$, $Q(\lambda) = \frac{1}{2}Q_0$, so

$$\frac{1}{2}Q_0 = Q_0e^{-k(\lambda)}$$

$$\frac{1}{2} = e^{-k(\lambda)}$$

$$\ln\frac{1}{2} = -k\lambda, \text{ or}$$

$$k = \frac{\ln\frac{1}{2}}{-\lambda} = \frac{\ln 2}{\lambda}$$

So, $Q(t) = Q_0e^{-\left(\frac{\ln 2}{\lambda}\right)t}$

(b)

$$Q_0e^{-\left(\frac{\ln 2}{\lambda}\right)t} = Q_0(0.5)^{kt}$$

$$e^{-\left(\frac{\ln 2}{\lambda}\right)t} = (0.5)^{kt}$$

$$-\frac{\ln 2}{\lambda}t = kt\ln 0.5$$

$$k = \frac{-\ln 2}{\lambda\ln 0.5} = \frac{\ln\frac{1}{2}}{\lambda\ln\frac{1}{2}} = \frac{1}{\lambda}$$

63. $R(t) = R_0e^{-\left(\frac{\ln 2}{5{,}730}\right)t}$

Since the Bronze age began about 5,000 years ago, the maximum percentage is

$$\frac{R(5{,}000)}{R_0} = \frac{R_0e^{-(\ln 2/5{,}730)(5{,}000)}}{R_0}$$

$$\approx 0.5462, \text{ or } 54.62\%.$$

65. $T(t) = 35e^{-0.32t}$

$27 = 35e^{-0.32t}$ or $t = 0.811$ min .

Rescuers have about 49 seconds before the girl looses consciousness.

$$\frac{dT}{dt} = -35(0.32)e^{-0.32t}$$

So, when $t = 0.811$,

$$\frac{dT}{dt} = (-35)(0.32)(e^{-0.32(0.811)}) \approx -8.64$$

So, the girl's temperature is dropping at a rate of 8.64 °C per minute.

67. $C(t) = Ate^{-kt}$

(a)

$$C'(t) = A\left[(t)(-ke^{-kt}) + (e^{-kt})(1)\right]$$

$$= Ae^{-kt}(-kt+1)$$

So, $C'(t) = 0$ when $-kt + 1 = 0$, or $t = \frac{1}{k}$

When $0 < t < \frac{1}{k}$, $C'(t) > 0$, so C is increasing

$t > \frac{1}{k}$, $C'(t) < 0$, so C is decreasing.

So, the maximum occurs when $t = \frac{1}{k}$. Since the maximum occurs after 2 hours,
$2 = \frac{1}{k}$, or $k = \frac{1}{2}$
The maximum is 10, so

$$10 = A(2)e^{-\frac{1}{2}(2)}, \text{ or}$$
$$A = 5e$$

(b) To find when the concentration falls to 1 microgram / ml,

$$C(t) = 5ete^{-0.5t}$$
$$5ete^{-0.5t} = 1$$
$$5ete^{-0.5t} - 1 = 0$$

Press [y=] and input
$5e \wedge (1)xe \wedge (-.5x) - 1$ for $y_1 =$.
Use window dimensions of
$[-5, 20]2$ by $[-10, 10]1$
Press [graph].
Press [2nd] [calc] and use the zero function to find $t \approx 9.78$ hours.

69. $Q(t) = Q_0 e^{-0.0015t}$

(a) The percentage rate is

$$100\frac{Q'(t)}{Q(t)}$$
$$= 100\frac{-0.0015Q_0e^{-0.0015t}}{Q_0e^{-0.0015t}} = -0.15\% \text{ per year}$$

(b) When 10% is depleted, 90% remains, so

$$0.9Q_0 = Q_0e^{-0.0015t}$$
$$0.9 = e^{-0.0015t}$$
$$\ln 0.9 = -0.0015t, \text{ or}$$
$$t = \frac{\ln 0.9}{-0.0015} \approx 70.24 \text{ years}$$

The percentage rate of change is constant, so the rate at this time is 0.15%.

71. $pH = -\log_{10}[H_3O^+]$
For milk and lime,
$pHm = 3pH_l$.
For lime and orange,
$pH_l = \frac{1}{2}pH_0$.
If $pH_0 = 3.2$,
$pH_l = \frac{1}{2}(3.2) = 1.6$
Then,

$$1.6 = -\log_{10}[H_3O^+]_l$$
$$-1.6 = \log_{10}[H_3O^+]_l$$
$$10^{-1.6} = 10^{\log_{10}[H_3O^+]_l} \text{ or}$$
$$[H_3O^+]_l = 10^{-1.6} \approx 0.0251$$

73. (a) $D(t) = (D_0 - 0.00046)e^{-0.162t} + 0.00046$
With

$$D_0 = 0.008,$$
$$D(10) = (0.008 - 0.00046)e^{-0.162(10)} + 0.00046$$
$$= 0.00195, \text{ or } 1.95 \text{ deaths per 1,000 women.}$$

$D(25) = 0.000590$, or 0.59 deaths per 1,000 women.

(b) When $t = 0$, $D(0) = 0.008$ so $(0, 0.008)$ is an intercept.
When $D(t) = 0$, $0.00754e^{-0.162t} + 0.00046 = 0$
$e^{-0.162t} = -0.061008$, which has no solution.
$D'(t) = -0.00122e^{-0.162t}$
So $D'(t)$ is never zero. Further, $D'(t) < 0$ for all t, so D is always decreasing.
$D''(t) = 0.000198e^{-0.162t}$
$D''(t)$ is never zero. Further, $D''(t) > 0$ for all t, so D is always concave up.

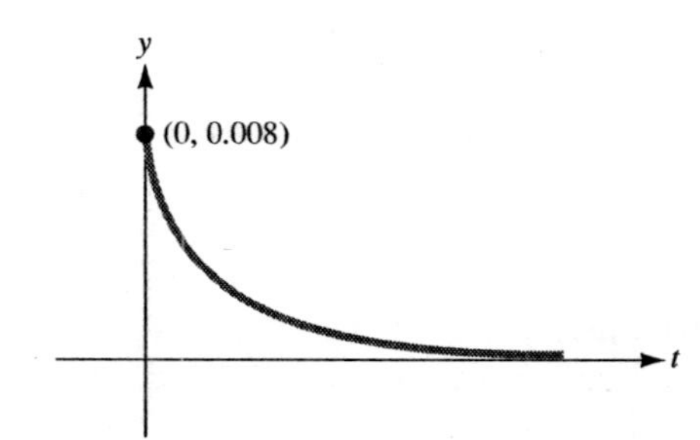

75. $R(t) = R_0e^{-\left(\frac{\ln 2}{5,730}\right)t}$

(a)
$$R(3.8 \times 10^6) = R_0e^{-\left(\frac{\ln 2}{5,730}\right)(3.8\times10^6)}$$
$$= R_0e^{-459.7}$$

Note: different calculators evaluate $e^{-459.7}$ differently; as a result, you may get 0 or you may get $\frac{1}{(2.3)^{200}}$. In either case, $R(3.8 \times 10^6) \approx 0$
Since $\lim_{t\to+\infty} e^{-t} = 0$, we can't distinguish ages for large values of t.

(b) Writing Exercise—Answers will vary.

77. $P(t) = \dfrac{202.31}{1 + e^{3.938-0.314t}}$

(a) To use this formula to compute the population of US for the years 1790, 1800, 1830, 1860, 1880, 1900, 1920, 1940, 1960, 1980, 1990, and 2000,
Press [y =] and input $P(t)$ for $y_1 =$.
Press [2nd] [tblset] and use Tblstart $= 0$, ΔTbl $= 1$, auto independent and auto dependent.
Press [2nd] [table].
Given below are the parts of the table corresponding to the years above.

Year	t	Population (in millions)
1790	0	3.8671
1800	1	5.2566
1830	4	12.957
1860	7	30.207
1880	9	50.071
1900	11	77.142
1920	13	108.43
1940	15	138.37
1960	17	162.29
1980	19	178.78
1990	20	184.57
2000	21	189.03

(b) Press [y =] and input $P(t)$ for $y_1 =$.
Use window dimensions [0,28]4 by [0,200]25
Press [graph].
The rate the population is growing is given by
$P'(t) = \dfrac{63.52534e^{3.938-0.314t}}{(1 + e^{3.938-0.314t})^2}$
Press [y =] and input $P'(t)$ for $y_2 =$.
Deselect $y_1 =$ so only $P'(t)$ is active.
Use window dimensions [0, 28]4 by [0, 20]2
Use the maximum function under the calc menu to find that the maximum of $P'(t)$ occurs at $x \approx 12.5$. So, the population is growing most rapidly when $t = 12.5$ or in 1915.

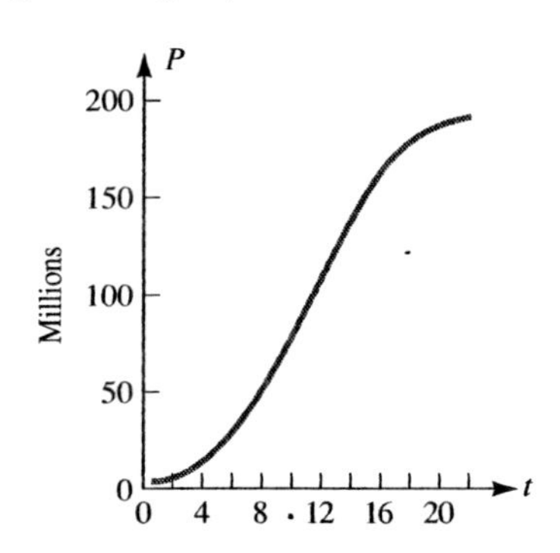

(c) Writing Exercise—Answers will vary.

79. To draw graphs of $y = \sqrt{3^x}$ $y = \sqrt{3^{-x}}$ and $y = 3^{-x}$ on the same axes,
Press [y =] and input $\sqrt{\ }(3 \wedge x)$ for $y_1 =$,
$\sqrt{\ }(3 \wedge (-x))$ for $y_2 =$, and
$3 \wedge (-x)$ for $y_3 =$.
Use window dimensions [−3, 3]1 by [−3, 3]1.
Press [graph].
The graph of $y = \sqrt{3^{-x}}$ is a reflection of the graph of $y = \sqrt{3^x}$ across the y-axis. The graph of $y = \sqrt{3^{-x}}$ is the graph of $y = 3^{-x}$ vertically compressed. Similarly, the graph of $y = \sqrt{3^x}$ is vertically compressed in addition to being reflected across the y-axis.

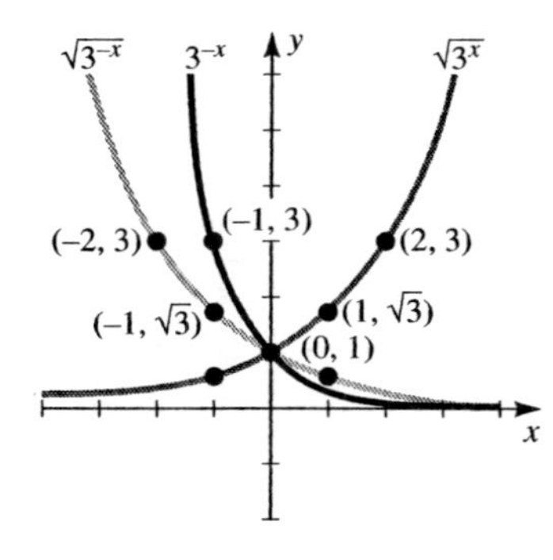

81. Using the conversion formula for logarithms, we will change all logarithms to natural logarithms:

$$\log_5(x+5) - \log_2 x - \log_{10}(x^2+2x)^2 = 0$$

$$\frac{\ln(x+5)}{\ln 5} - \frac{\ln(x)}{\ln 2} - \frac{\ln(x^2+2x)^2}{\ln 10} = 0$$

Press [y=] and input
$\ln(x+5)/\ln(5) - \ln(x)/\ln(2) - \ln\left((x^2+2x)^2\right)/\ln(10)$ for $y_1 =$.
Press [graph].
Use the zero function under the calc menu to find that $x \approx 1.066$ is a root. There is no other real root because x^2 increases much more rapidly than any other argument, making $y_1 =$ monotonically decreasing.

83. To make a table for $(\sqrt{n})^{\sqrt{n+1}}$ and $\left(\sqrt{n+1}\right)^{\sqrt{n}}$ with $n = 8, 9, 12, 20, 25, 31, 37, 38, 43, 50, 100$, and 1,000, press [y=] and input $\sqrt{n} \wedge \sqrt{n+1}$ for $y_1 =$ and $\sqrt{n+1} \wedge \sqrt{n}$ for $y_2 =$.
Press [2nd] [tblset] and use ask independent and auto dependent.
Press [2nd] [table] and input each value of n given.

n	$(\sqrt{n})^{\sqrt{n+1}}$	$\left(\sqrt{n+2}\right)^{\sqrt{n}}$
8	22.63	22.36
9	32.27	31.62
12	88.21	85.00
20	957.27	904.84
25	3,665	3,447
31	16,528	15,494
37	68,159	63,786
38	85,679	80,166
43	261,578	244,579
50	1.17×10^6	1.09×10^6
1000	1.1×10^{10}	1.1×10^{10}
1000	2.9×10^{47}	2.8×10^{47}

$$(\sqrt{n})^{\sqrt{n+1}} \geq \left(\sqrt{n+1}\right)^{\sqrt{n}}$$

This inequality holds for all $n \geq 8$. To confirm, since $(n+1)^{\sqrt{n}} \leq (n+1)^{\sqrt{n+1}}$

$$\lim_{n\to\infty} \frac{(n+1)^{\sqrt{n}}}{n^{\sqrt{n+1}}} \leq \lim_{n\to\infty} \frac{(n+1)^{\sqrt{n+1}}}{n^{\sqrt{n+1}}}$$

$$\leq \lim_{n\to\infty} \left(\frac{n+1}{n}\right)^{\sqrt{n+1}}$$

$$\leq \lim_{n\to\infty} e^{\ln\left(\frac{n+1}{n}\right)^{\sqrt{n+1}}}$$

$$\leq \lim_{n\to\infty} e^{\sqrt{n+1}\ln\left(\frac{n+1}{n}\right)}$$

$$\leq \lim_{n\to\infty} e^{[(\ln(n+1)/n)/(n+1)^{-1/2}]}$$

$$\leq e^{\lim_{n\to\infty} \frac{\ln\left(\frac{n+1}{n}\right)}{(n+1)^{-1/2}}}$$

Using l'Hopital's rule,

$$\leq e^{\lim_{n\to\infty} \frac{\left(\frac{n}{n+1}\right)\frac{n(1)-(n+1)(1)}{n^2}}{-\frac{1}{2}(n+1)^{-3/2}(1)}}$$

$$\leq e^{\lim_{n\to\infty} \frac{-\frac{1}{n}}{-\frac{1}{2}(n+1)^{-3/2}(n+1)}}$$

$$\leq e^{\lim_{n\to\infty} \frac{2(n+1)^{1/2}}{n}}$$

Using l'Hopital's rule again,

$$\leq e^{\lim_{n\to\infty} \frac{1}{(n+1)^{1/2}}}$$

$$\leq e^0$$

$$\leq 1$$

Since the ratio of $(n+1)^{\sqrt{n}}$ to $n^{\sqrt{n+1}}$ is less than or equal to one,

$$(n+1)^{\sqrt{n}} \leq n^{\sqrt{n+1}}.$$

Chapter 5

Integration

5.1 Antidifferentiation; the Indefinite Integral

1.
$$I = \int -3\,dx = -3x + C.$$

3.
$$I = \int x^5\,dx = \frac{x^6}{6} + C.$$

5.
$$\begin{aligned} I &= \int \frac{1}{x^2}\,dx = \int x^{-2}\,dx \\ &= -x^{-1} + C = -\frac{1}{x} + C. \end{aligned}$$

7.
$$\begin{aligned} I &= \int \frac{2}{\sqrt{t}}\,dt = 2\int t^{-1/2}\,dt \\ &= 2\frac{t^{1/2}}{1/2} + C \\ &= 4t^{\frac{1}{2}} + C = 4\sqrt{t} + C. \end{aligned}$$

9.
$$\begin{aligned} I &= \int u^{-2/5}\,du \\ &= \frac{u^{3/5}}{3/5} + C = \frac{5}{3}u^{3/5} + C. \end{aligned}$$

11.
$$\begin{aligned} I &= \int (3t^2 - \sqrt{5t} + 2)\,dt \\ &= 3\int t^2\,dt - \sqrt{5}\int t^{1/2}\,dt + 2\int dt \\ &= 3\left(\frac{t^3}{3}\right) - \sqrt{5}\left(\frac{t^{3/2}}{\frac{3}{2}}\right) + 2t + C \\ &= t^3 - \frac{2\sqrt{5}}{3}t^{3/2} + 2t + C. \end{aligned}$$

13.
$$\begin{aligned} I &= \int \left(3\sqrt{y} - 2y^{-3}\right)\,dy \\ &= 3\int y^{1/2}\,dy - 2\int y^{-3}\,dy \\ &= 3\frac{y^{3/2}}{3/2} - 2\frac{y^{-}2}{-2} + C \\ &= 2y^{3/2} + y^{-2} + C \\ &= 2y^{3/2} + \frac{1}{y^2} + C. \end{aligned}$$

15.
$$\begin{aligned} I &= \int \left(\frac{e^x}{2} + x\sqrt{x}\right)\,dx \\ &= \frac{1}{2}\int e^x\,dx + \int x^{3/2}\,dx \\ &= \frac{1}{2}e^x + \frac{x^{5/2}}{5/2} + C \\ &= \frac{e^x}{2} + \frac{2}{5}x^{5/2} + C. \end{aligned}$$

17.
$$\begin{aligned} I &= \int u^{1.1}\left(\frac{1}{3u} - 1\right)\,du \\ &= \int \left(\frac{u^{1.1}}{3u} - u^{1.1}\right)\,du \\ &= \int \left(\frac{u^{0.1}}{3} - u^{1.1}\right)\,du \\ &= \frac{1}{3}\int u^{0.1}\,du - \int u^{1.1}\,du \\ &= \frac{1}{3}\cdot\frac{u^{1.1}}{1.1} - \frac{u^{2.1}}{2.1} + C \\ &= \frac{u^{1.1}}{3.3} - \frac{u^{2.1}}{2.1} + C. \end{aligned}$$

19.
$$I=\int \frac{x^2+2x+1}{x^2}\,dx$$
$$=\int\left(1+\frac{2}{x}+\frac{1}{x^2}\right)\,dx$$
$$=\int dx+2\int \frac{1}{x}\,dx+\int x^{-2}\,dx$$
$$=x+2\ln|x|+\frac{x^{-1}}{-1}+C$$
$$=x+2\ln|x|-\frac{1}{x}+C$$
$$=x+\ln x^2-\frac{1}{x}+C.$$

21.
$$I=\int (x^3-2x^2)\left(\frac{1}{x}-5\right)\,dx$$
$$=\int (x^2-2x-5x^3+10x^2)\,dx$$
$$=\int\left(-5x^3+11x^2-2x\right)\,dx$$
$$=-5\int x^3\,dx+11\int x^2\,dx-2\int x\,dx$$
$$=-\frac{5x^4}{4}+\frac{11x^3}{3}-\frac{2x^2}{2}+C$$
$$=-\frac{5}{4}x^4+\frac{11}{3}x^3-x^2+C.$$

23.
$$I=\int \sqrt{t}(t^2-1)\,dt$$
$$=\int (t^{5/2}-t^{1/2})\,dt$$
$$=\int t^{5/2}\,dt-\int t^{1/2}\,dt$$
$$=\frac{2t^{7/2}}{7}-\frac{2t^{3/2}}{3}+C$$
$$=\frac{2}{7}t^{7/2}-\frac{2}{3}t^{3/2}+C.$$

25.
$$I=\int (e^t+1)^2\,dt$$
$$=\int (e^{2t}+2e^t+1)\,dt$$
$$=\int e^{2t}\,dt+2\int e^t\,dt+\int dt$$
$$=\frac{1}{2}e^{2t}+2e^t+t+C.$$

27.
$$I=\int\left(\frac{1}{3y}-\frac{5}{\sqrt{y}}+e^{-y/2}\right)\,dy$$
$$=\frac{1}{3}\int \frac{1}{y}\,dy-5\int y^{-1/2}\,dy+\int e^{-\frac{1}{2}y}\,dy$$
$$=\frac{1}{3}\ln|y|-5\frac{y^{1/2}}{1/2}+\frac{1}{-1/2}e^{-\frac{1}{2}y}+C$$
$$=\frac{1}{3}\ln|y|-10\sqrt{y}-2e^{-y/2}+C.$$

29.
$$I=\int t^{-1/2}(t^2-t+2)\,dt$$
$$=\int (t^{3/2}-t^{1/2}+2t^{-1/2})\,dt$$
$$=\int t^{3/2}\,dt-\int t^{1/2}\,dt+2\int t^{-1/2}\,dt$$
$$=\frac{t^{5/2}}{5/2}-\frac{t^{3/2}}{3/2}+2\frac{t^{1/2}}{1/2}+C$$
$$=\frac{2}{5}t^{5/2}-\frac{2}{3}t^{3/2}+4t^{1/2}+C.$$

31.
$$\frac{dy}{dx}=3x-2$$
$$\int \frac{dy}{dx}\,dx=\int (3x-2)\,dx$$
$$\int \frac{dy}{dx}\,dx=3\int x\,dx-2\int dx$$
$$y=3\frac{x^2}{2}-2x+C$$
$$y=\frac{3}{2}x^2-2x+C$$

Since $y=2$ when $x=-1$,
$$2=\frac{3}{2}(-1)^2-2(-1)+C$$
$$2=\frac{3}{2}+2+C,\ \text{or}$$

$$C = -\frac{3}{2}$$

So, $y = \frac{3}{2}x^2 - 2x - \frac{3}{2}$.

33.
$$\frac{dy}{dx} = \frac{2}{x} - \frac{1}{x^2}$$
$$\int \frac{dy}{dx}\,dx = \int \left(\frac{2}{x} - \frac{1}{x^2}\right) dx$$
$$\int \frac{dy}{dx}\,dx = 2\int \frac{1}{x}\,dx - \int x^{-2}\,dx$$
$$y = 2\ln|x| - \frac{x^{-1}}{-1} + C$$
$$= \ln x^2 + \frac{1}{x} + C$$

Since $y = -1$ when $x = 1$,

$$-1 = \ln 1 + \frac{1}{1} + C$$
$$-1 = 0 + 1 + C, \text{ or}$$
$$C = -2$$

So, $y = \ln x^2 + \frac{1}{x} - 2$.

35.
$$f'(x) = 4x + 1$$
$$\int f'(x)\,dx = \int (4x + 1)\,dx$$
$$\int f'(x)\,dx = 4\int x\,dx + \int dx$$
$$f(x) = 4\frac{x^2}{2} + x + C$$
$$= 2x^2 + x + C$$

Since the function goes through the point (1, 2),

$$2 = 2(1)^2 + 1 + C, \text{ or}$$
$$C = -1$$

So, $f(x) = 2x^2 + x - 1$.

37.
$$f'(x) = x^3 - \frac{2}{x^2} + 2$$
$$\int f'(x)\,dx = \int (x^3 - \frac{2}{x^2} + 2)\,dx$$
$$\int f'(x)\,dx = \int x^3\,dx - 2\int x^{-2}\,dx + 2\int dx$$
$$f(x) = \frac{x^4}{4} - 2\frac{x^{-1}}{-1} + 2x + C$$
$$= \frac{1}{4}x^4 + \frac{2}{x} + 2x + C$$

Since the function goes through the point (1, 3),

$$3 = \frac{1}{4}(1)^4 + \frac{2}{1} + 2(1) + C, \text{ or}$$
$$C = -\frac{5}{4}$$

So, $f(x) = \frac{1}{4}x^4 + \frac{2}{x} + 2x - \frac{5}{4}$.

39.
$$f'(x) = e^{-x} + x^2$$
$$\int f'(x)\,dx = \int (e^{-x} + x^2)\,dx$$
$$\int f'(x)\,dx = \int e^{-x}\,dx + \int x^2\,dx$$
$$f(x) = \frac{1}{-1}e^{-x} + \frac{x^3}{3} + C$$
$$= -e^{-x} + \frac{1}{3}x^3 + C$$

Since the function goes through the point (0, 4),

$$4 = -e^0 + \frac{1}{3}(0) + C, \text{ or}$$
$$C = 5$$

So, $f(x) = -e^{-x} + \frac{1}{3}x^3 + 5$.

41. Let $P(t)$ be the population of the town t months from now. Since

$$\frac{dP}{dt} = 4 + 5t^{2/3},$$

then, $P(t) = \int \frac{dP}{dt}\, dt$

$$\begin{aligned} &= \int (4 + 5t^{2/3})\, dt \\ &= 4\int dt + 5\int t^{2/3}\, dt \\ &= 4t + 5\frac{t^{5/3}}{5/3} + C \\ &= 4t + 3t^{5/3} + C \end{aligned}$$

Since the population is 10,000 when $t = 0$,

$$10{,}000 = 4(0) + 3(0) + C, \text{ or}$$
$$C = 10{,}000$$

So, $P(t) = 4t + 3t^{5/3} + 10{,}000$.

When $t = 8$,

$$\begin{aligned} P(8) &= 4(8) + 3(8)^{5/3} + 10{,}000 \\ &= 10{,}128 \text{ people.} \end{aligned}$$

43.

$$\begin{aligned} M(t) &= \int M'(t)\, dt \\ &= \int 0.5e^{0.2t}\, dt \\ &= 0.5\int e^{0.2t}\, dt \\ &= 0.5 \cdot \frac{1}{0.2}e^{0.2t} + C \\ &= 2.5e^{0.2t} + C \end{aligned}$$

Growth during the second hour is

$$\begin{aligned} &M(2) - M(1) \\ &= (2.5e^{0.2(2)} + C) = (2.5e^{0.2(1)} + C) \\ &= 2.5(e^{0.4} - e^{0.2}) \text{ grams} \end{aligned}$$

45.

$$\begin{aligned} C(q) &= \int C'(q)\, dq \\ &= \int (3q^2 - 24q + 48)\, dq \\ &= 3\int q^2\, dq - 24\int q\, dq + 48\int dq \\ &= 3\frac{q^3}{3} - 24\frac{q^2}{2} + 48q + C \\ &= q^3 - 12q^2 + 48q + C \end{aligned}$$

Since the cost is $5,000 for producing 10 units,

$$5000 = (10)^3 - 12(10)^2 + 48(10) + C, \text{ or}$$
$$C = 4720$$

So, $C(q) = q^3 - 12q^2 + 48q + 4720$.

When 30 units are produced, the cost is

$$\begin{aligned} C(30) &= (30)^3 - 12(30)^2 + 48(30) + 4720 \\ &= \$22{,}360. \end{aligned}$$

47. $M'(t) = 0.4t - 0.005t^2$

(a)

$$\begin{aligned} M(t) &= \int M'(t)\, dt \\ &= \int (0.4t - 0.005t^2)\, dt \\ &= 0.4\int t\, dt - 0.005\int t^2\, dt \\ &= 0.4\frac{t^2}{2} - 0.005\frac{t^3}{3} + C \\ &= 0.2t^2 - \frac{0.005}{3}t^3 + C \end{aligned}$$

Since $M(t) = 0$ when $t = 0$, $C = 0$.

So, $M(t) = 0.2t^2 - \frac{0.005}{3}t^3$.

In ten minutes, Bob can memorize

$$\begin{aligned} M(10) &= 0.2(10)^2 - \frac{0.005}{3}(10)^3 \\ &= 18\frac{1}{3} \text{ items.} \end{aligned}$$

(b)

$$\begin{aligned} &M(20) - M(10) \\ &= \left[0.2(20)^2 - \frac{0.005}{3}(20)^3\right] - 18\frac{1}{3} \\ &= 66\frac{2}{3} - 18\frac{1}{3} = 48\frac{1}{3} \text{ items.} \end{aligned}$$

49.

$$N(t) = \int N'(t)\,dt$$
$$= \int (154t^{2/3} + 37)\,dt$$
$$= 154\int t^{2/3}\,dt + 37\int dt$$
$$= 154\frac{t^{5/3}}{5/3} + 37t + C$$
$$= \frac{462}{5}t^{5/3} + 37t + C$$

Since there are no subscribers when $t = 0$,

$$C = 0$$

So, $N(t) = \frac{462}{5}t^{5/3} + 37t$.

Eight months from now, the number of subscribers will be

$$N(8) = \frac{462}{5}(8)^{5/3} + 37(8)$$
$$\approx 3{,}253 \text{ subscribers.}$$

51.

$$T'(t) = 7e^{-0.35t}$$

(a)

$$T(t) = \int T'(t)\,dt$$
$$= \int 7e^{-0.35t}\,dt$$
$$= 7\int e^{-0.35t}\,dt$$
$$= 7\cdot\frac{1}{-0.35}e^{-0.35t} + C$$
$$= -20e^{-0.35t} + C$$

Since the temperature was −4°C when $t = 0$,

$$-4 = -20e^0 + C, \text{ or}$$
$$C = 16$$

So, $T(t) = -20e^{-0.35t} + 16$.

(b) After two hours,

$$T(2) = -20e^{-0.35(2)} + 16$$
$$\approx 6.07°C.$$

(c) For the temperature to reach 10°C,

$$10 = -20e^{-0.35t} + 16$$
$$6 = 20e^{-0.35t}$$
$$\frac{3}{10} = e^{-0.35t}$$
$$\ln\frac{3}{10} = \ln e^{-0.35t}$$
$$\ln\frac{3}{10} = -0.35t, \text{ or}$$
$$t - \frac{\ln\frac{3}{10}}{-0.35} = \frac{-20}{7}\ln\frac{3}{10}$$
$$= \frac{20}{7}\ln\frac{10}{3} \approx 3.44 \text{ hours.}$$

53.

$$R'(q) = 100 - 2q$$

(a) Since $P'(q) = R'(q)$,

$$P(q) = \int R'(q)\,dq$$
$$= \int (100 - 2q)\,dq$$
$$= 100\int dq - 2\int q\,dq$$
$$= 100q - 2\frac{q^2}{2} + C$$
$$= 100q - q^2 + C$$

Since the profit is \$700 when 10 units are produced,

$$700 = 100(10) - (10)^2 + C, \text{ or}$$
$$C = -200$$

So, $P(q) = 100q - q^2 - 200$.

(b) Since $R'(q) = P'(q)$, to maximize P,

$$R'(q) = 0 \text{ when } 100 - 2q = 0, \text{ or } q = 50$$

Further, $R''(q) = -2$, so $R''(50) < 0$ and the maximum profit occurs when $q = 50$. The maximum profit is

$$P(50) = 100(50) - (50)^2 - 200$$
$$= \$2{,}300.$$

55.

$$\begin{aligned} c(x) &= \int c'(x)\,dx \\ &= \int (0.9 + 0.3\sqrt{x})\,dx \\ &= 0.9\int dx + 0.3\int x^{1/2}\,dx \\ &= 0.9x + 0.3\frac{x^{3/2}}{3/2} + C \\ &= 0.9x + 0.2x^{3/2} + C \end{aligned}$$

Since the consumption is 10 billion when $x = 0$,

$$10 = 0.9(0) + 0.2(0) + C, \text{ or}$$

$$C = 10$$

So, $c(x) = 0.9x + 0.2x^{3/2} + 10$.

57.

$$f'(x) = 0.1(10 + 12x - 0.6x^2)$$

(a) To maximize the rate of learning,

$$f''(x) = 0.1(12 - 1.2x)$$

So $f''(x) = 0$ when $12 - 1.2x = 0$, or

$$x = 10$$

Further, $f''(x) = 0.1(-1.2) = -0.12$ so $f'''(10) < 0$ and the absolute maximum occurs when $x = 10$. The maximum rate is

$$\begin{aligned} f'(10) &= 0.1[10 + 12(10) - 0.6(10)^2] \\ &= 7 \text{ items per minute.} \end{aligned}$$

(b)

$$\begin{aligned} f(x) &= \int f'(x)\,dx \\ &= \int [0.1(10 + 12x - 0.6x^2)]\,dx \\ &= \int (1 + 1.2x - 0.06x^2)\,dx \\ &= \int dx + 1.2\int x\,dx - 0.06\int x^2\,dx \\ &= x + 1.2\frac{x^2}{2} - 0.06\frac{x^3}{3} + C \\ &= x + 0.6x^2 - 0.02x^3 + C \end{aligned}$$

Since no items are memorized when $t = 0$,

$$C = 0$$

So, $f(x) = x + 0.6x^2 - 0.02x^3$.

(c)

$$\begin{aligned} f'(x) &= 0.1(10 + 12x - 0.6x^2) \\ &= 1 + 1.2x - 0.06x^2 \end{aligned}$$

So, $f'(x) = 0$ when

$$x = \frac{-1.2 \pm \sqrt{(1.2)^2 - 4(-0.06)(1)}}{2(-0.06)}$$

or, $x \approx 20.8$ (rejecting the negative solution)

$f''(20.8) < 0$, so the absolute maximum is

$$\begin{aligned} f(20.8) &= (20.8) + 0.6(20.8)^2 - 0.02(20.8)^3 \\ &\approx 100 \text{ items} \end{aligned}$$

59. Since profit = revenue − cost,
marginal profit = marginal revenue − marginal cost.

$$\begin{aligned} P'(q) &= R'(q) - C'(q) \\ &= 200q^{-1/2} - 0.4q \end{aligned}$$

$$\begin{aligned} P(q) &= \int P'(q)\,dq \\ &= \int (200q^{-1/2} - 0.4q)\,dq \\ &= 200\int q^{-1/2}\,dq - 0.4\int q\,dq \\ &= 200\frac{q^{1/2}}{1/2} - 0.4\frac{q^2}{2} + C \\ &= 400q^{1/2} - 0.2q^2 + C \end{aligned}$$

Since the profit is \$2,000 when 25 units are produced,

$$2000 = 400(25)^{1/2} - 0.2(25)^2 + C, \text{ or}$$

$$C = 125$$

So, $P(q) = 400q^{1/2} - 0.2q^2 + 125$

and the profit when 36 units are produced is

$$\begin{aligned} P(36) &= 400(36)^{1/2} - 0.2(36)^2 + 125 \\ &= \$2{,}265.80 \end{aligned}$$

61.
$$v'(r) = -ar$$
$$v(r) = \int v'(r)\,dr$$
$$= \int -ar\,dr = -a\int r\,dr$$
$$= -a\frac{r^2}{2} + C = -\frac{a}{2}r^2 + C$$

Since $v(R) = 0$,

$$0 = -\frac{a}{2}(R)^2 + C, \text{ or}$$
$$C = \frac{aR^2}{2}$$

So, $v(r) = -\frac{a}{2}r^2 + \frac{aR^2}{2} = \frac{a}{2}(R^2 - r^2)$.

63.
$$v(t) = 3 + 2t + 6t^2$$

Since velocity is the derivative of distance,

$$s(t) = \int v(t)\,dt$$
$$= \int (3 + 2t + 6t^2)\,dt$$
$$= 3\int dt + 2\int t\,dt + 6\int t^2\,dt$$
$$= 3t + 2\frac{t^2}{2} + 6\frac{t^3}{3} + C$$
$$= 3t + t^2 + 2t^3 + C$$

The distance traveled during the second minute is

$$s(2) - s(1)$$
$$= [3(2) + (2)^2 + 2(2)^3 + C] - [3(1) + (1)^2 + 2(1)^3 + C]$$
$$= 20 \text{ meters.}$$

65.
$$\int b^x\,dx = \int e^{x\ln b}\,dx = \int e^{(\ln b)x}\,dx$$
$$= \frac{1}{\ln b}e^{x\ln b} + C = \frac{1}{\ln b}b^x + C$$

67.
$$a(t) = -23$$

(a) Since acceleration is the derivative of velocity,

$$v(t) = \int -23\,dt$$
$$= -23t + C$$

The velocity when the brakes are applied is 67 ft/sec, so

$$67 = -23(0) + C, \text{ or } C = 67$$
$$\text{and } v(t) = -23t + 67$$

Since velocity is the derivative of distance,

$$s(t) = \int v(t)\,dt$$
$$= \int (-23t + 67)\,dt$$
$$= -23\int t\,dt + 67\int dt$$
$$= -23\frac{t^2}{2} + 67t + C$$
$$= -\frac{23}{2}t^2 + 67t + C$$

Since the distance is to be measured from the point the brakes are applied, $s(0) = 0$ and

$$0 = -\frac{23}{2}(0) + 67(0) + C,$$
$$\text{or } C = 0$$

So, $s(t) = -\frac{23}{2}t^2 + 67t$.

(b) To use the graphing utility to sketch graphs of $v(t)$ and $s(t)$ on same screen,
Press [y=] and input $v(t)$ for $y_1 =$ and input $s(t)$ for $y_2 =$.
Use window dimensions [0, 5]1 by [0, 200]10.
Press [graph].

(c) The car comes to a complete stop when $v(t) = 0$.
Press [trace] and verify that the cross-hairs are on the line $y_1 = -23t + 67$.
Move along line until it appears to be at the t-intercept.

Use the zoom-in function under the zoom menu to find that the velocity $= 0$ when $t \approx 2.9$ seconds.

To find how far the car travels in 2.9 seconds, go back to the original graphing screen. Use the value function under the calc menu and input 2.9 for x and press enter. Use the ↑ arrow to verify that $y_2 = -\frac{23}{2}t^2 + 67t$ is displayed. The car travels 97.6 feet in 2.9 seconds.

To find how fast the car travels when $s = 45$ feet, trace along the parabola $s(t)$ and use the zoom-in function to find that it takes approximately 0.77 seconds and 5.05 seconds to travel 45 feet. Next, go back to the original graphing screen and use the value fucntion under the calc menu. Input $x = 0.77$ and verify $y_1 = -23t + 67$ is displayed. The car is traveling 49.2 feet/sec when it has traveled 45 feet. Repeat this process with $x = 5.05$ to find the velocity at 5.05 is 49.15 (decelerating).

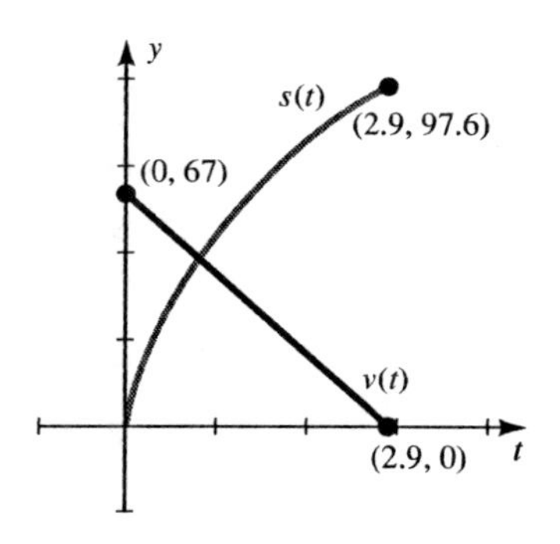

5.2 Integration by Substitution

1. **(a)** $u = 3x + 4$
(b) $u = 3 - x$
(c) $u = 2 - t^2$
(d) $u = 2 + t^2$

3. Let $u = 2x + 6$. Then $du = 2\,dx$ or $dx = \frac{du}{2}$.

$$\text{So} \int (2x+6)^5\,dx = \frac{1}{2}\int u^5\,du = \frac{(2x+6)^6}{12} + C.$$

5. Let $u = 4x - 1$. Then $du = 4\,dx$ or $dx = \frac{du}{4}$.

$$\text{So} \int \sqrt{4x-1}\,dx = \frac{1}{4}\int u^{1/2}\,du = \frac{1}{4}\frac{2u^{3/2}}{3} + C = \frac{(4x-1)^{3/2}}{6} + C.$$

7. Let $u = 1 - x$. Then $du = -\,dx$ or $dx = -\,du$.

$$\text{So} \int e^{1-x}\,dx = -\int e^u\,du = -e^{1-x} + C.$$

9. Let $u = x^2$. Then $\frac{du}{dx} = 2x$ or $\frac{1}{2}\,du = x\,dx$.

$$\int xe^{x^2}\,dx = \int e^{x^2}\cdot x\,dx = \int e^u \cdot \frac{1}{2}\,du = \frac{1}{2}\int e^u\,du = \frac{1}{2}e^{x^2} + C.$$

11. Let $u = t^2 + 1$. Then $\frac{du}{dt} = 2t$ or $\frac{1}{2}\,du = t\,dt$.

$$\int t(t^2+1)^5\,dt = \int (t^2+1)^5 t\,dt = \int u^5 \cdot \frac{1}{2}\,du = \frac{1}{2}\int u^5\,du = \frac{(t^2+1)^6}{12} + C.$$

13. Let $u = x^3 + 1$. Then $\frac{du}{dx} = 3x^2$ or $\frac{1}{3}\,du = x^2\,dx$.

$$\int x^2(x^3+1)^{3/4}\,dx = \int (x^3+1)^{3/4}x^2\,dx = \int u^{3/4}\cdot\frac{1}{3}\,du = \frac{1}{3}\int u^{3/4}\,du$$

$$= \frac{4(x^3+1)^{7/4}}{21} + C.$$

15. Let $u = y^5 + 1$. Then $\dfrac{du}{dy} = 5y^4$, or $\dfrac{1}{5}\,du = y^4\,dy$.

$$\int \frac{2y^4}{y^5+1}\,dy = 2\int \frac{1}{y^5+1}y^4\,dy = 2\int \frac{1}{u}\cdot\frac{1}{5}\,du$$
$$= \frac{2}{5}\int \frac{1}{u}\,du = \frac{2}{5}\ln|y^5+1| + C.$$

17. Let $u = x^2 + 2x + 5$. Then $\dfrac{du}{dx} = 2x + 2$ $= 2(x+1)$, or $\dfrac{1}{2}\,du = (x+1)\,dx$.

$$\int (x+1)(x^2+2x+5)^{12}\,dx$$
$$= \int (x^2+2x+5)^{12}(x+1)\,dx$$
$$= \int u^{12}\cdot\frac{1}{2}\,du = \frac{1}{2}\int u^{12}\,du$$
$$= \frac{(x^2+2x+5)^{13}}{26} + C.$$

19. Let $u = x^5 + 5x^4 + 10x + 12$. Then $\dfrac{du}{dx} = 5x^4 + 20x^3 + 10 = 5(x^4 + 4x^3 + 2)$, or $\dfrac{1}{5}\,du = (x^4 + 4x^3 + 2)\,dx$.

$$\int \frac{3x^4+12x^3+6}{x^5+4x^3+2}\,dx = \int \frac{3(x^4+4x^3+2)}{x^5+4x^3+2}\,dx$$
$$= 3\int \frac{1}{x^5+4x^3+2}(x^4+4x^3+2)\,dx$$
$$= 3\int \frac{1}{u}\cdot\frac{1}{5}\,du = \frac{3}{5}\int \frac{1}{u}\,du$$
$$= \frac{3}{5}\ln|x^5+5x^4+10x+12| + C.$$

21. Let $t = u^2 - 2u + 6$. Then $\dfrac{dt}{du} = 2u - 2 = 2(u-1)$, or $\dfrac{1}{2}\,dt = (u-1)\,du$.

$$\int \frac{3u-3}{(u^2-2u+6)^2}\,du = \int \frac{3(u-1)}{(u^2-2u+6)^2}\,du$$
$$= 3\int \frac{1}{(u^2-2u+6)^2}(u-1)\,du$$
$$= 3\int \frac{1}{t^2}\cdot\frac{1}{2}\,dt = \frac{3}{2}\int t^{-2}\,dt$$
$$= \frac{-3}{2(u^2-2u+6)} + C.$$

23. Let $u = \ln 5x$. Then $\dfrac{du}{dx} = \dfrac{1}{5x}\cdot 5 = \dfrac{1}{x}$, or $du = \dfrac{1}{x}\,dx$.

$$\int \frac{\ln 5x}{x}\,dx = \int \ln 5x\cdot\frac{1}{x}\,dx$$
$$= \int u\,du = \frac{(\ln 5x)^2}{2} + C.$$

25. Let $u = \ln x$. Then $\dfrac{du}{dx} = \dfrac{1}{x}$, or $du = \dfrac{1}{x}\,dx$.

$$\int \frac{1}{x(\ln x)^2}\,dx = \int \frac{1}{(\ln x)^2}\cdot\frac{1}{x}\,dx$$
$$= \int \frac{1}{u^2}\,du = -\frac{1}{\ln x} + C.$$

27. Let $u = x^2 + 1$. Then $\dfrac{du}{dx} = 2x$, or $\dfrac{1}{2}\,du = x\,dx$.

$$\int \frac{2x\ln(x^2+1)}{x^2+1}\,dx = 2\int \frac{\ln(x^2+1)}{x^2+1}\cdot x\,dx$$
$$= 2\int \frac{\ln u}{u}\cdot\frac{1}{2}\,du = \int \frac{\ln u}{u}\,du$$

Substitution must be used a second time. Let $t = \ln u$. Then $\dfrac{dt}{du} = \dfrac{1}{u}$, or $dt = \dfrac{1}{u}\,du$.

$$\int \frac{\ln u}{u}du = \int \ln u\cdot\frac{1}{u}\,du$$
$$= \int t\,dt = \frac{t^2}{2} + C$$
$$= \frac{(\ln u)^2}{2} + C = \frac{[\ln(x^2+1)]^2}{2} + C.$$

29. Let $u = e^x - e^{-x}$. Then $\dfrac{du}{dx} = e^x + e^{-x}$, or $du = (e^x + e^{-x})\,dx$.

$$\int \frac{e^x + e^{-x}}{e^x - e^{-x}}\,dx = \int \frac{1}{e^x - e^{-x}}(e^x + e^{-x})\,dx$$
$$= \int \frac{1}{u}\,du = \ln|e^x - e^{-x}| + C.$$

31. Let $u = 2x + 1$. Then $\frac{du}{dx} = 2$, or $\frac{1}{2}\,du = dx$.

Further, $x = \frac{u-1}{2}$.

$$\int \frac{x}{2x+1}\,dx = \frac{1}{4}\int \frac{u-1}{u}\,du$$
$$= \frac{1}{4}\int \left(1 - \frac{1}{u}\right)du = \frac{1}{4}\int du - \frac{1}{4}\int \frac{1}{u}\,du$$
$$= \frac{1}{4}u - \frac{1}{4}\ln|u| + C = \frac{1}{4}(2x+1) - \frac{1}{4}\ln|2x+1| + C.$$

This can also be written as

$$= \frac{1}{2}x + \frac{1}{4} - \frac{1}{4}\ln|2x+1| + C$$
$$= \frac{1}{2}x - \frac{1}{4}\ln|2x+1| + C,$$

where the $\frac{1}{4}$ has been added to the constant C. (In mathematics, the same C is often used for the original constant and for the constant after it is changed.)

33. Let $u = 2x + 1$. Then $\frac{du}{dx} = 2$, or $\frac{1}{2}\,du = dx$.

Further, $x = \frac{u-1}{2}$.

$$\int x\sqrt{2x-1}\,dx = \frac{1}{4}\int (u-1)u^{1/2}\,du$$
$$= \frac{1}{4}\int (u^{3/2} - u^{1/2})\,du$$
$$= \frac{1}{4}\left(\frac{2}{5}(2x+1)^{5/2} - \frac{2}{3}(2x+1)^{3/2}\right) + C$$
$$= \frac{1}{10}(2x+1)^{5/2} - \frac{1}{6}(2x+1)^{3/2} + C.$$

35. Let $u = \sqrt{x} + 1$. Then $\frac{du}{dx} = \frac{1}{2}x^{-1/2} = \frac{1}{2x^{1/2}}$, or

$2\,du = \frac{1}{\sqrt{x}}\,dx$.

$$\int \frac{1}{\sqrt{x}(\sqrt{x}+1)}\,dx = \int \frac{1}{\sqrt{x}+1}\cdot\frac{1}{\sqrt{x}}\,dx$$
$$= 2\int \frac{1}{u}\,du = 2\ln|\sqrt{x}+1| + C$$
$$= 2\ln(\sqrt{x}+1) + C.$$

37.
$$y = \int \frac{dy}{dx}\,dx = \int \frac{1}{x+1}\,dx$$

Let $u = x + 1$. Then $\frac{du}{dx} = 1$, or $du = dx$.

$$\int \frac{1}{x+1}\,dx = \int \frac{1}{u}\,du$$
$$= \ln|x+1| + C$$

Since $y = 1$ when $x = 0$,

$$1 = \ln|0+1| + C, \text{ or}$$
$$C = 1$$

So, $y = \ln|x+1| + 1$.

39.
$$y = \int \frac{dy}{dx}\,dx = \int \frac{x+2}{x^2+4x+5}\,dx$$

Let $u = x^2 + 4x + 5$. Then $\frac{du}{dx} = 2x + 4$ $= 2(x+2)$, or $\frac{1}{2}\,du = (x+2)\,dx$.

$$\int \frac{x+2}{x^2+4x+5}\,dx = \int \frac{1}{x^2+4x+5}(x+2)\,dx$$
$$= \frac{1}{2}\int \frac{1}{u}\,du = \frac{1}{2}\ln|x^2+4x+5| + C$$

Since $y = 3$ when $x = -1$,

$$3 = \frac{1}{2}\ln|(-1)^2 + 4(-1) + 5| + C$$
$$\text{or, } C = 3 - \frac{1}{2}\ln 2$$

So, $y = \frac{1}{2}\ln|x^2+4x+5| + 3 - \frac{1}{2}\ln 2$.

41.
$$f(x) = \int f'(x)\,dx = \int (1-2x)^{3/2}\,dx$$

Let $u = 1 - 2x$. Then $\frac{du}{dx} = -2$, or $-\frac{1}{2}\,du = dx$.

$$\int (1-2x)^{3/2}\,dx = -\frac{1}{2}\int u^{3/2}\,du$$
$$= -\frac{1}{2}\left[\frac{2}{5}(1-2x)^{5/2}\right] + C$$
$$= -\frac{1}{5}(1-2x)^{5/2} + C$$

Since the function goes through the point (0, 0),

$$0 = -\frac{1}{5}[1-2(0)]^{5/2} + C, \text{ or}$$
$$C = \frac{1}{5}$$

So, $f(x) = -\frac{1}{5}(1-2x)^{5/2} + \frac{1}{5}$.

43. $$f(x) = \int f'(x)\,dx = \int xe^{4-x^2}\,dx$$

Let $u = 4 - x^2$. Then $\frac{du}{dx} = -2x\,dx$, or $-\frac{1}{2}\,du = x\,dx$.

$$\int xe^{4-x^2}\,dx = \int e^{4-x^2}\cdot x\,dx$$
$$= -\frac{1}{2}\int e^u\,du = -\frac{1}{2}e^{4-x^2} + C$$

Since $y = 1$ when $x = -2$,

$$1 = -\frac{1}{2}e^{4-(-2)^2} + C, \text{ or}$$
$$C = \frac{3}{2}$$

So, $f(x) = -\frac{1}{2}e^{4-x^2} + \frac{3}{2}$.

45. (a) $$x(t) = \int x'(t)\,dt$$
$$= \int -2(3t+1)^{1/2}\,dt$$

Let $u = 3t + 1$. Then $\frac{du}{dt} = 3$, or $\frac{1}{3}\,du = dt$.

$$= -\frac{2}{3}\int u^{1/2}\,dt = -\frac{4}{9}(3t+1)^{3/2} + C$$

When $t = 0$, $x(0) = 4$, so

$$4 = -\frac{4}{9}[3(0)+1]^{3/2} + C, \text{ or}$$
$$C = \frac{40}{9}$$

So, $x(t) = -\frac{4}{9}(3t+1)^{3/2} + \frac{40}{9}$.

(b) $$x(4) = -\frac{4}{9}[3(4)+1]^{3/2} + \frac{40}{9}$$
$$\approx -16.4$$

(c) $$3 = -\frac{4}{9}(3t+1)^{3/2} + \frac{40}{9}$$
$$\frac{13}{4} = (3t+1)^{3/2}$$
$$t = \frac{\left(\frac{13}{4}\right)^{2/3} - 1}{3} \approx 0.4$$

47. (a) $$x(t) = \int x'(t)\,dt$$
$$= \int \frac{1}{\sqrt{2t+1}}\,dt$$

Let $u = 2t + 1$. Then, $\frac{1}{2}\,du = dt$.

$$= \frac{1}{2}\int u^{-1/2}\,dt$$
$$= \frac{1}{2}(2u^{1/2}) + C$$
$$= (2t+1)^{1/2} + C$$

When $t = 0$, $x(0) = 0$ so $C = -1$ and $x(t) = (2t+1)^{1/2} - 1$.

(b) When $t = 4$,

$$x(4) = [2(4)+1]^{1/2} - 1 = 2$$

(c) $$3 = (2t+1)^{1/2} - 1$$
$$16 = 2t + 1,$$
$$\text{or } t = \frac{15}{2}.$$

49. (a)

$$C(q) = \int C'(q)\,dq$$
$$= \int 3(q-4)^2\,dq$$

Let $u = q - 4$. Then $\frac{du}{dq} = 1$, or $du = dq$.

$$= 3\int u^2\,du = (q-4)^3 + C$$

Let C_0 represent the overhead. Then

$$C_0 = C(0) = (0-4)^3 + C,$$
$$\text{or } C = C_0 + 64$$

So, $C(q) = (q-4)^3 + 64 + C_0$.

(b) When $C_0 = 436$,

$$C(q) = (q-4)^3 + 500$$
$$\text{and } C(14) = (14-4)^3 + 500$$
$$= \$1{,}500$$

51. Let $G(x)$ represent the height in meters of the tree in x years.

$$G(x) = \int G'(x)\,dx$$
$$= \int \left(1 + \frac{1}{(x+1)^2}\right)dx$$
$$= \int dx + \int \frac{1}{(x+1)^2}\,dx$$

Let $u = x + 1$. Then $\frac{du}{dx} = 1$, or $du = dx$.

$$= \int dx + \int \frac{1}{u^2}\,du$$
$$= x - \frac{1}{x+1} + C.$$

Since the height was 5 meters after 2 years,

$$5 = 2 - \frac{1}{2+1} + C, \text{ or}$$
$$C = \frac{10}{3}$$

So, $G(x) = x - \frac{1}{x+1} + \frac{10}{3}$ and

$$G(0) = 0 - \frac{1}{0+1} + \frac{10}{3}$$
$$= \frac{7}{3} \text{ meters tall.}$$

53. (a)

$$R(x) = \int R'(x)\,dx$$
$$= \int (50 + 3.5xe^{-0.01x^2})\,dx$$
$$= 50\int dx + 3.5\int xe^{-0.01x^2}\,dx$$

Let $u = -0.01x^2$. Then $\frac{du}{dx} = -0.02x$, or $-50\,du = x\,dx$.

$$= 50\int dx - 3.5\int e^{-0.01x^2}x\,dx$$
$$= 50\int dx - 175\int e^u\,du$$
$$= 50x - 175e^{-0.01x^2} + C$$

Since $R(0) = 0$,

$$0 = 50(0) - 175e^0 + C, \text{ or}$$
$$C = 175$$

So, $R(x) = 50x - 175e^{-0.01x^2} + 175$.

(b) $R(1000) = 50(1000) - 175e^{-0.01(1000)} + 175$
$\approx \$50,175$

55. (a)

$$C(t) = \int C'(t)\,dt$$
$$= \int \frac{-0.01e^{0.01t}}{(e^{0.01t}+1)^2}\,dt$$

Let $u = e^{0.01t} + 1$. Then $\frac{du}{dt} = 0.01e^{0.01t}$, or

$$100\ du = e^{0.01t}\,dt.$$
$$= -0.01\int \frac{1}{(e^{0.01t}+1)^2}e^{0.01t}\,dt$$
$$= -\int \frac{1}{u^2}\,du = \frac{1}{e^{0.01t}+1} + C$$

When the shot is initially administered, $t = 0$ and

$$0.5 = \frac{1}{e^0 + 1} + C, \quad \text{or } C = 0$$

So, $C(t) = \dfrac{1}{e^{0.01t} + 1}$.

(b) After one hour, when $t = 60$ minutes, the concentration is

$$C(60) = \frac{1}{e^{0.01(60)} + 1} \approx 0.3543 \text{ mg/cm}^3$$

After three hours, when $t = 180$ minutes, the concentration is

$$C(180) = \frac{1}{e^{0.01(180)} + 1} \approx 0.1419 \text{ mg/cm}^3$$

(c) To determine how much time passes before next injection is given,
Press [y=] and input $C(t) = 1/(e \wedge (0.01t) + 1)$ for $y_1 =$.
Use window dimensions [0, 500]50 by [0, 1]0.02.
Press [trace] and move along the curve until $y \approx 0.05$. Use the zoom-in function under the zoom menu to get a more accurate reading.
A new injection is given after approximately 294 minutes.

57. (a)

$$L(t) = \int L'(t)\,dt$$

$$= \int \frac{0.24 - 0.03t}{\sqrt{36 + 16t - t^2}}\,dt$$

Let $u = 36 + 16t - t^2$. Then $\dfrac{du}{dt} = 16 - 2t = 2(8 - t)$, or $\dfrac{1}{2}\,du = (8 - t)\,dt$.

$$= \int \frac{0.03(8 - t)}{(36 + 16t - t^2)^{1/2}}\,dt = \frac{0.03}{2}\int u^{-1/2}\,du$$

$$= 0.03(36 + 16t - t^2)^{1/2} + C$$

At 7:00 a.m., $t = 0$ and $L(0) = 0.25$, so

$$0.25 = 0.03\sqrt{36 + 16(0) - (0)} + C, \quad \text{or}$$

$$C = 0.07$$

So, $L(t) = 0.03\sqrt{36 + 16t - t^2} + 0.07$. To find the peak level,

$L'(t) = 0$ when $0.24 - 0.03t = 0$,

or when $t = 8$

Further, when $0 \le t < 8$, $L'(t) > 0$

so L is increasing;

when $t > 8$, $L'(t) < 0$ so L is decreasing

so, the absolute maximum occurs when $t = 8$, or 3:00 p.m. The maximum is

$$L(8) = 0.03\sqrt{36 + 16(8) - (8)^2} + 0.07$$
$$= 0.37 \text{ parts per million}$$

(b) To use graphing utility to graph $L(t)$ and answer the questions in part (a),
press [y=] and input $L(t) = 0.03\sqrt{(-t^2 + 16t + 36)} + 0.07$ for $y_1 =$.
Use window dimensions [0, 16]2 by [0.24, −4]0.04.
Press [graph].
Press [trace] and move along curve to the maximum point and use zoom-in if necessary.
We find the maximum point occurs when $t = 8$ (at 3:00 p.m.). The ozone level is 0.37 ppm at this time.
At 11:00 a.m., $t = 4$. Use the value function under the calc menu to find the ozone level is 0.34 ppm at 11:00 a.m. Trace along the curve to find when the y-value is 0.34 ppm. We find that the ozone level is 0.34 ppm again when $t = 12$, or at 7:00 p.m.

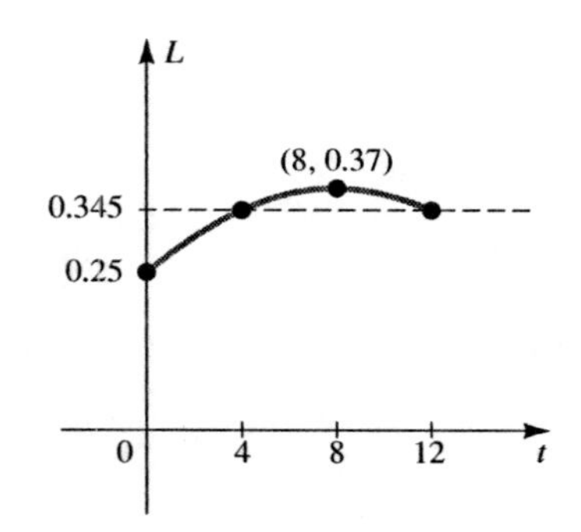

59. (a)

$$p(x) = \int p'(x)\,dx$$
$$= \int \frac{-300x}{(x^2+9)^{3/2}}\,dx$$

Let $u = x^2 + 9$. Then $\frac{du}{dx} = 2x$, or $\frac{1}{2}\,du = x\,dx$.

$$= -300 \int \frac{1}{(x^2+9)^{3/2}}\,x\,dx$$
$$= -150 \int u^{-3/2}\,du$$
$$= \frac{300}{\sqrt{x^2+9}} + C$$

When the price is $75, 4 hundred pair are demanded, so

$$75 = \frac{300}{\sqrt{(4)^2+9}} + C$$
$$\text{or, } C = 15$$

So, $p(x) = \frac{300}{\sqrt{x^2+9}} + 15$.

(b) When $x = 5$ hundred,

$$p(5) = \frac{300}{\sqrt{(5)^2+9}} + 15$$
$$= \$66.45 \text{ per pair}$$
$$p(0) = \frac{300}{\sqrt{0+9}} + 15 = \$115 \text{ per pair}$$

(c)

$$90 = \frac{300}{\sqrt{x^2+9}} + 15$$

$\sqrt{x^2+9} = 4$, or $x \approx 2.65$, or 265 pairs.

61. Since profit = revenue − costs, marginal profit = marginal revenue − marginal costs.

$$P'(x) = R'(x) - C'(x)$$
$$= \frac{11-x}{\sqrt{14-x}} - (2 + x + x^2)$$

$$P(x) = \int P'(x)\,dx$$
$$= \int \left(\frac{11-x}{\sqrt{14-x}} - 2 - x - x^2\right)dx$$
$$= \int \frac{11-x}{\sqrt{14-x}}\,dx - 2\int dx - \int x\,dx - \int x^2\,dx$$

Let $u = 14 - x$. Then $\frac{du}{dx} = -1$, or $-du = dx$. Further, $x = 14 - u$.

$$= -\int \frac{-3+u}{u^{1/2}}\,du - 2\int dx - \int x\,dx - \int x^2\,dx$$
$$= -\left[\int -3u^{-1/2}\,du + \int u^{1/2}\,du\right] - 2\int dx - \int x\,dx - \int x^2\,dx$$
$$= 3\int u^{-1/2}\,du - \int u^{1/2}\,du - 2\int dx - \int x\,dx - \int x^2$$
$$= 6\sqrt{14-x} - \frac{2}{3}(14-x)^{3/2} - 2x - \frac{x^2}{2} - \frac{x^3}{3} + C$$

When production is raised from 5 to 9, $P(9) - P(5)$

$$= \left[6\sqrt{14-9} - \frac{2}{3}(14-9)^{3/2} - 2(9) - \frac{(9)^2}{2} - \frac{(9)^3}{3} + C\right]$$
$$- \left[6\sqrt{14-5} - \frac{2}{3}(14-5)^{3/2} - 2(5) - \frac{(5)^2}{2} - \frac{(5)^3}{3} + C\right]$$
$$\approx -231.37, \text{ or a loss of } \$231.37.$$

63. Let $u = x^{2/3} + 1$. Then $\frac{du}{dx} = \frac{2}{3}x^{-1/3}$, or $\frac{3}{2}\,du = x^{-1/3}\,dx$. Further, $x^{2/3} = u - 1$.

$$\int x^{1/3}(x^{2/3}+1)^{3/2}\,dx$$
$$= \int (x^{2/3}+1)^{3/2}x^{2/3}x^{-1/3}\,dx$$

$$= \frac{3}{2}\int u^{3/2}(u-1)\,du = \frac{3}{2}\int u^{5/2} - u^{3/2}\,du$$
$$= \frac{3}{2}\left[\frac{2}{7}(x^{2/3}+1)^{7/2} - \frac{2}{5}(x^{2/3}+1)^{5/2}\right] + C$$
$$= \frac{3}{7}(x^{2/3}+1)^{7/2} - \frac{3}{5}(x^{2/3}+1)^{5/2} + C.$$

65. Let $u = 1 + e^x$. Then $\dfrac{du}{dx} = e^x$, or $du = e^x\,dx$. Further $e^x = u - 1$.

$$\int \frac{e^{2x}}{1+e^x}\,dx = \int \frac{e^x}{1+e^x}e^x\,dx$$
$$= \int \frac{u-1}{u}\,du = \int \left(1 - \frac{1}{u}\right)du$$
$$= 1 + e^x - \ln|1+e^x| + C$$
$$= 1 + e^x - \ln(1+e^x) + C.$$

5.3 The Definite Integral and the Fundamental Theorem of Calculus

1. $$\int_{-1}^{2} 5\,dx = 5x\Big|_{-1}^{2} = 5(2) - 5(-1) = 15$$

3. $$\int_0^5 (3x+2)\,dx = \left(\frac{3x^2}{2} + 2x\right)\Big|_0^5$$
$$= \left[\frac{3(5)^2}{2} + 2(5)\right] - 0 = \frac{95}{2}$$

5. $$\int_{-1}^{1} 3t^4\,dt = \frac{3t^5}{5}\Big|_{-1}^{1} = \frac{3(1)^5}{5} - \frac{3(-1)^5}{5} = \frac{6}{5}$$

7. $$\int_{-1}^{1}(2u^{1/3} - u^{2/3})\,du = \left(\frac{3}{2}u^{4/3} - \frac{3}{5}u^{5/3}\right)\Big|_{-1}^{1}$$
$$= \left[\frac{3}{2}(1)^{4/3} - \frac{3}{5}(1)^{5/3}\right]$$
$$- \left[\frac{3}{2}(-1)^{4/3} - \frac{3}{5}(-1)^{5/3}\right] = -\frac{6}{5}$$

9. $$\int_0^1 e^{-x}(4 - e^x)\,dx = \int_0^1 (4e^{-x} - e^0)\,dx$$
$$= (-4e^{-x} - x)\Big|_0^1 = (-4e^{-1} - 1) - (-4e^0 - 0)$$
$$= 3 - \frac{4}{e}$$

11. $$\int_0^1 (x^4 + 3x^3 + 1)\,dx = \left(\frac{x^5}{5} + \frac{3x^4}{4} + x\right)\Big|_0^1$$
$$= \left[\frac{(1)^5}{5} + \frac{3(1)^4}{4} + 1\right] - 0 = \frac{39}{20} = 1.95$$

13. $$\int_2^5 (2 + 2t + 3t^2)\,dt = (2t + t^2 + t^3)\Big|_2^5$$
$$= \left[2(5) + (5)^2 + (5)^3\right] - \left[2(2) + (2)^2 + (2)^3\right] = 144$$

15. $$\int_1^3 \left(1 + \frac{1}{x} + \frac{1}{x^2}\right)dx = \left(x + \ln|x| - \frac{1}{x}\right)\Big|_1^3$$
$$= \left(3 + \ln 3 - \frac{1}{3}\right) - (1 + \ln 1 - 1) = \frac{8}{3} + \ln 3.$$

17. $$\int_{-3}^{-1} \frac{t+1}{t^3}\,dt = \int_{-3}^{-1}\left(\frac{1}{t^2} + \frac{1}{t^3}\right)dt$$
$$= \left(-\frac{1}{t} - \frac{1}{2t^2}\right)\Big|_{-3}^{-1} = \left[\frac{-1}{-1} - \frac{1}{2(-1)^2}\right]$$
$$- \left[\frac{-1}{-3} - \frac{1}{2(-3)^2}\right] = \frac{2}{9}$$

19. $$\int_1^2 (2x-4)^4\,dx$$

Let $u = 2x - 4$. Then $\frac{1}{2}\,du = dx$, and the limits of integration become $2(1) - 4 = -2$ and $2(2) - 4 = 0$.

$$= \frac{1}{2}\int_{-2}^{0} u^4\,du = \frac{1}{2}\left(\frac{u^5}{5}\right)\Big|_{-2}^{0}$$
$$= \frac{1}{10}(u^5)\Big|_{-2}^{0} = \frac{1}{10}\left[0 - (-2)^5\right] = 3.2$$

21. $$\int_0^4 \frac{1}{\sqrt{6t+1}}\,dt$$

Let $u = 6t + 1$. Then, $\frac{1}{6}\,du = dt$, and the limits of integration become $6(0) + 1 = 1$ and $6(4) + 1 = 25$.

$$= \frac{1}{6}\int_1^{25} u^{-1/2}\,du = \frac{1}{6}(2\sqrt{u})\Big|_1^{25}$$
$$= \frac{1}{3}(\sqrt{u})\Big|_1^{25} = \frac{1}{3}(\sqrt{25}-\sqrt{1}) = \frac{4}{3}$$

23. $$\int_0^1 (x^3+x)\sqrt{x^4+2x^2+1}\,dx$$

Let $u = x^4+2x^2+1$. Then $\frac{1}{4}\,du = (x^3+x)\,dx$, and the limits of integration become $(0)+2(0)+1=1$ and $(1)^4+2(1)^2+1=4$.

$$= \frac{1}{4}\int_1^4 u^{1/2}\,du = \frac{1}{4}\left(\frac{2}{3}u^{3/2}\right)\Big|_1^4 = \frac{1}{6}(u^{3/2})\Big|_1^4$$
$$= \frac{1}{6}\left[(4)^{3/2}-(1)^{3/2}\right] = \frac{7}{6}$$

25. $$\int_1^{e+1} \frac{x}{x-1}\,dx$$

Let $u = x-1$. Then $du = dx$ and $x = u+1$. Further, the limits of integration become $2-1=1$ and $(e+1)-1=e$.

$$= \int_1^e \frac{u+1}{u}\,du = \int_1^e \left(1+\frac{1}{u}\right)\,du$$
$$= (u+\ln|u|)\Big|_1^e$$
$$= (e+\ln e)-(1+\ln 1) = e$$

27. $$\int_1^{e^2} \frac{(\ln x)^2}{x}\,dx$$

Let $u = \ln x$. Then $du = \frac{1}{x}\,dx$, and the limits of integration become $\ln 1 = 0$ and $\ln(e)^2 = 2$.

$$\int_0^2 u^2\,du = \frac{1}{3}(u^3)\Big|_0^2$$
$$= \frac{1}{3}\left[(2)^3-(0)\right] = \frac{8}{3}$$

29. $$\int_{1/3}^{1/2} \frac{e^{1/x}}{x^2}\,dx$$

Let $u = \frac{1}{x}$. Then $-du = \frac{1}{x^2}\,dx$, and the limits of integration become $\frac{1}{1/3}=3$ and $\frac{1}{1/2}=2$.

$$= -\int_3^2 e^u\,du = \int_2^3 e^u\,du$$
$$= (e^u)\Big|_2^3 = e^3-e^2$$

31. $$\int_{-3}^2 \left[-2f(x)+5g(x)\right]\,dx$$

$$= -2\int_{-3}^2 f(x)\,dx + 5\int_{-3}^2 g(x)\,dx$$
$$= -2(5)+5(-2) = -20$$

33. $$\int_4^4 g(x)\,dx = G(4)-G(4) = 0,$$

where $G(x)$ is the antiderivative of $g(x)$.

35. $$\int_1^2 \left[3f(x)+2g(x)\right]\,dx$$

$$= 3\int_1^2 f(x)\,dx + 2\int_1^2 g(x)\,dx$$
$$= 3\left[\int_{-3}^2 f(x)\,dx - \int_{-3}^1 f(x)\,dx\right]$$
$$+2\left[\int_{-3}^2 g(x)\,dx - \int_{-3}^1 g(x)\,dx\right]$$
$$= 3(5-0)+2(-2-4) = 3$$

37. $$\int_{-1}^2 x^4\,dx = \frac{1}{5}(x^5)\Big|_{-1}^2$$
$$= \frac{1}{5}\left[(2)^5-(-1)^5\right] = \frac{33}{5}$$

39. $$\int_0^{\ln 3} e^{2x}\,dx = \frac{1}{2}(e^{2x})\Big|_0^{\ln 3}$$
$$= \frac{1}{2}(e^{2\ln 3}-e^0) = \frac{1}{2}(e^{\ln 3^2}-1) = 4$$

41. $$\int_0^4 (3x+4)^{1/2}\,dx$$

Let $u = 3x+4$. Then $\frac{1}{3}\,du = dx$, and the limits of integration become $3(0)+4=4$ and $3(4)+4=16$.

$$= \frac{1}{3}\int_4^{16} u^{1/2}\,du = \frac{1}{3}\left(\frac{2}{3}u^{3/2}\right)\Big|_4^{16}$$
$$= \frac{2}{9}(u^{3/2})\Big|_4^{16} = \frac{2}{9}\left[(16)^{3/2} - (4)^{3/2}\right] = \frac{112}{9}$$

43. $$\int_0^5 V'(t)\,dt = V(5) - V(0)$$

45. The number of pounds of soybeans stored per week x weeks from now is $12{,}000 - 300x$, a function that decreases linearly from 12,000 to 0 in 40 weeks. The weekly cost rate will be $0.2(12{,}000 - 300x)$ cents per week. The cost over the next 40 weeks

$$= \int_0^{40} 0.2(12{,}000 - 300x)\,dx$$
$$= 0.2(12{,}000x - 150x^2)\Big|_0^{40}$$
$$= 48{,}000 \text{ cents, or } \$480$$

47. $$L(3) - L(0) = \int_0^3 (0.1t + 0.1)\,dt$$
$$= (0.05t^2 + 0.1t)\Big|_0^3$$
$$= \left[0.05(3)^2 + 0.1(3)\right] - 0 = 0.75 \text{ ppm}$$

49. $$P(8) - P(0) = \int_0^8 (5 + 3t^{2/3})\,dt$$
$$= \left(5t - \frac{9}{5}t^{5/3}\right)\Big|_0^8 = \left[5(8) + \frac{9}{5}(8)^{5/3}\right] - 0$$
$$= \frac{488}{5} \approx 98 \text{ people}$$

51. Let $V(t)$ be the value of the crop, in dollars, after t days. Then

$$\frac{dV}{dt} = 3(0.3t^2 + 0.6t + 1)$$

The change in value will be

$$V(5) - V(0) = \int_0^5 3(0.3t^2 + 0.6t + 1)\,dt$$
$$= 3(0.1t^3 + 0.3t^2 + t)\Big|_0^5$$
$$= 3\left[(0.1(5)^3 + 0.3(5)^2 + 5) - 0\right] = \$75$$

53. $$P(3) - P(2) = \int_2^3 1500\left(2 - \frac{t}{2t+5}\right)dt$$
$$= 3000\int_2^3 dt - 1500\int_2^3 \frac{t}{2t+5}\,dt$$

Let $u = 2t + 5$. Then $\frac{1}{2}\,du = dt$, and $t = \frac{u-5}{2}$. Further, the limits of integration beome $2(2) + 5 = 9$ and $2(3) + 5 = 11$.

$$= 3000\int_2^3 dt - 750\int_9^{11} \frac{u-5}{2u}\,du$$
$$= 3000\int_2^3 dt - 375\int_9^{11}\left(1 - \frac{5}{u}\right)du$$
$$= 3000(t)\Big|_2^3 - 375(u - 5\ln|u|)\Big|_9^{11}$$
$$= 3000(3-2) - 375\left[(11 - 5\ln 11) - (9 - 5\ln 9)\right]$$
$$= 3000 - 375(2 - 5\ln 11 + 5\ln 9)$$
$$\approx 2{,}626 \text{ telephones}$$

55. $$C(4) - C(0) = \int_0^4 \frac{-0.33t}{\sqrt{0.02t^2 + 10}}\,dt$$
$$= -0.33\int_0^4 \frac{t}{\sqrt{0.02t^2 + 10}}\,dt$$

Let $u = 0.02t^2 + 10$. Then $25\,du = t\,dt$, and the limits of integration become $0.02(0) + 10 = 10$ and $0.02(4)^2 + 10 = 10.32$.

$$= -8.25\int_{10}^{10.32} u^{-1/2}du - -8.25(2u^{1/2})\Big|_{10}^{10.32}$$
$$= -16.5(\sqrt{u})\Big|_{10}^{10.32} = -16.5(\sqrt{10.32} - \sqrt{10})$$
$$\approx -0.8283,$$

or the concentration decreases by approximately 0.8283 mg/cm^3.

57. Let $V(x)$ be the value of the machine, in dollars, after t years. Then,

$$\frac{dV}{dt} = 220(x - 10)$$

$$V(2) - V(1) = \int_1^2 220(x - 10)\, dx$$

$$= \int_1^2 (220x - 2200)\, dx$$

$$= (110x^2 - 2200x)\Big|_1^2$$

$$= \left[110(2)^2 - 2200(2)\right] - [110(1) - 2200(1)]$$

$$= -1870,$$

or the machine depreciates by \$1,870.

59.

$$P(2) - P(5) = -[P(5) - P(2)]$$

$$= -\int_2^5 -\frac{2}{t+1}\, dt$$

Let $u = t + 1$. Then $du = dt$, and the limits of integration become $2 + 1 = 3$ and $5 + 1 = 6$.

$$= 2\int_3^6 \frac{1}{u}\, du = 2(\ln|u|)\Big|_3^6$$

$$= 2(\ln 6 - \ln 3) = 2(\ln \frac{6}{3})$$

$$= 2 \ln 2 \approx 1.386 \text{ grams.}$$

61.

$$L(10) - L(5) = \int_5^{10} \frac{4}{\sqrt{t+1}}\, dt$$

Let $u = t + 1$. Then $du = dt$, and the limits of integration become $5 + 1 = 6$ and $10 + 1 = 11$.

$$= 4\int_6^{11} u^{-1/2}\, du = 4(2u^{1/2})\Big|_6^{11} = 8(\sqrt{u})\Big|_6^{11}$$

$$= 8(\sqrt{11} - \sqrt{6}) \approx 7 \text{ facts.}$$

63. Let $s(t)$ be the distance traveled, in feet, after t seconds. Since velocity is the derivative of distance,

$$s(3) - s(0) = \int_0^3 (-32t + 80)\, dt$$

$$= (-16t^2 + 80t)\Big|_0^3 = \left[-16(3)^2 + 80(3)\right] - 0$$

$$= 96 \text{ feet.}$$

65. **(a)** $\int_0^1 \sqrt{1 - x^2}\, dx$ represents the area under the curve $\sqrt{1 - x^2}$, above the x-axis, from $x = 0$ to $x = 1$. But the graph of $y = \sqrt{1 - x^2}$ is a semi-circle, having radius 1 and center (0, 0) since

$$y = \sqrt{1 - x^2}$$

$$y^2 = 1 - x^2$$

$$x^2 + y^2 = 1$$

The area from $x = 0$ to $x = 1$ corresponds to a quarter of the circle's area.

$$= \frac{1}{4}\left[\pi(1)^2\right] = \frac{\pi}{4}$$

(b) Similarly, the graph of $\sqrt{2x - x^2}$ is the same semicircle, shifted one unit to the right since

$$y = \sqrt{2x - x^2}$$

$$y^2 = 2x - x^2$$

$$x^2 - 2x + y^2 = 0$$

$$(x^2 - 2x + 1) + y^2 = 1$$

$$(x - 1)^2 + y^2 = 1$$

So, the area from $x = 1$ to $x = 2$ still corresponds to a quarter of the circle $= \dfrac{\pi}{4}$.

5.4 Applying Definite Integration: Area Between Curves and Average Value

1. The limits of integration are

$$x^3 - \sqrt{x};\ x^3 - x^{1/2} = 0;$$

$$x^{1/2}(x^{5/2} - 1) = 0$$

so $x = 0$ and $x = 1$.
The shaded area is

$$\int_0^1 (\sqrt{x} - x^3)\,dx$$
$$= \left(\frac{2}{3}x^{3/2} - \frac{x^4}{4}\right)\Bigg|_0^1$$
$$= \frac{5}{12}$$

3. The limits of integration are $x = 0$ and

$$x = \frac{2}{x+1}, \quad x^2 + x = 2$$
$$x^2 + x - 2 = 0, \quad (x+2)(x-1) = 0,$$

$x = 1$ (rejecting $x = -2$ since shaded area starts at $x = 0$).
The shaded area is

$$\int_0^1 \left(\frac{2}{x+1} - x\right) dx$$
$$= \left(2\ln|x+1| - \frac{x^2}{2}\right)\Bigg|_0^1$$
$$= 2\ln 2 - \frac{1}{2}$$

5. The shaded area is

$$\int_0^1 [x - (-x)]\,dx$$
$$= (x^2)\Big|_0^1 = 1$$

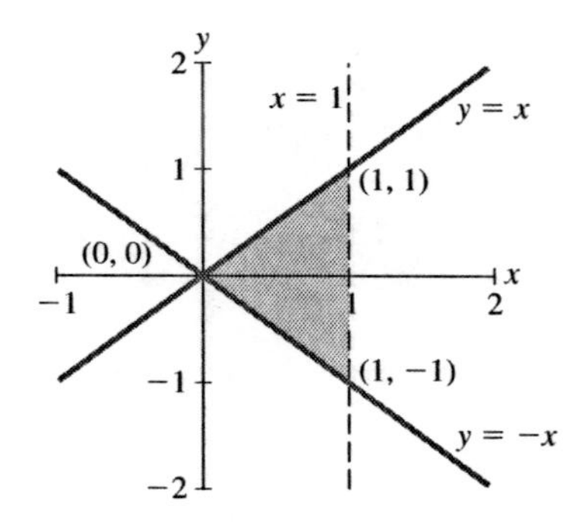

7. The shaded area is

$$\int_1^3 \left[(-x^2 + 4x - 3) - 0\right] dx$$
$$= \left(-\frac{x^3}{3} + 2x^2 - 3x\right)\Bigg|_1^3$$
$$= \frac{4}{3}$$

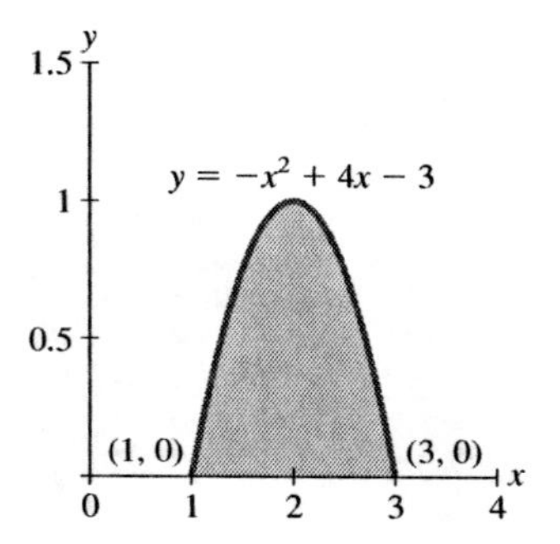

9. The shaded area is

$$\int_0^2 \left[0 - (x^2 - 2x)\right] dx$$
$$= \left(-\frac{x^3}{3} + x^2\right)\Bigg|_0^2$$
$$= \frac{4}{3}$$

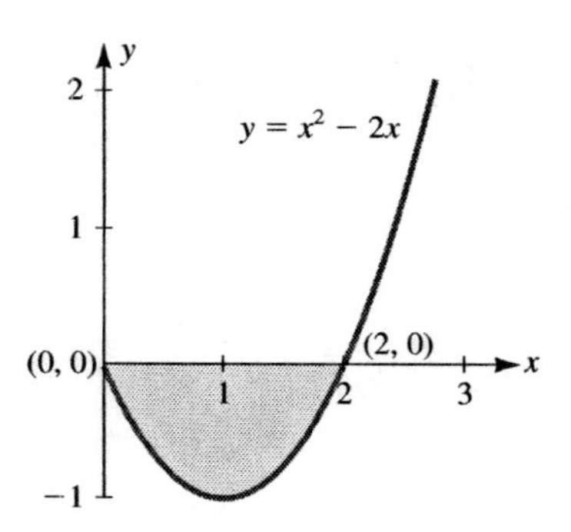

11. The limits of integration are

$$x^2 - 2x = -x^2 + 4$$
$$2x^2 - 2x + 4 = 0$$
$$2(x-2)(x+1) = 0$$

$x = -1$ and $x = 2$.

The shaded area is

$$\int_{-1}^{2}\left[(-x^2+4)-(x^2-2x)\right]\,dx$$

$$\left(-\frac{2x^3}{3}+x^2+4x\right)\Big|_{-1}^{2}=9$$

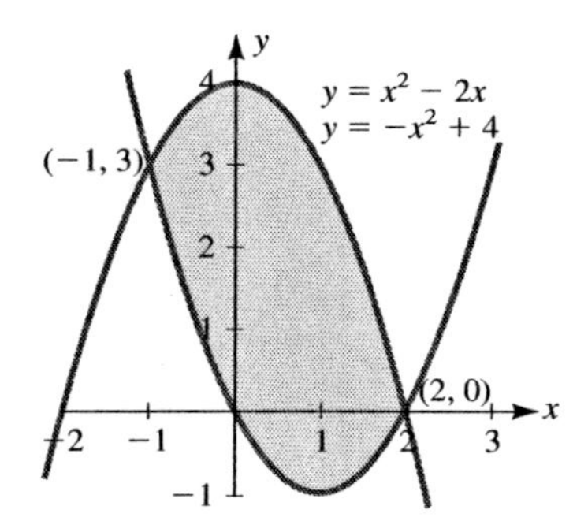

13. The points of intersection are

$$x^3-3x^2=x^2+5x$$
$$x^3-4x^2-5x=0$$
$$x(x-5)(x+1)=0.$$

There are two shaded areas

$$\int_{-1}^{0}\left[(x^3-3x^2)-(x^2+5x)\right]\,dx$$

$$+\int_{0}^{5}\left[(x^2+5x)-(x^3-3x^2)\right]\,dx$$

$$=\left(\frac{x^4}{4}-\frac{4x^3}{3}-\frac{5x^2}{2}\right)\Big|_{-1}^{0}+\left(-\frac{x^4}{4}+\frac{4x^3}{3}+\frac{5x^2}{2}\right)\Big|_{0}^{5}$$

$$=\frac{11}{12}+\frac{825}{12}=\frac{443}{6}.$$

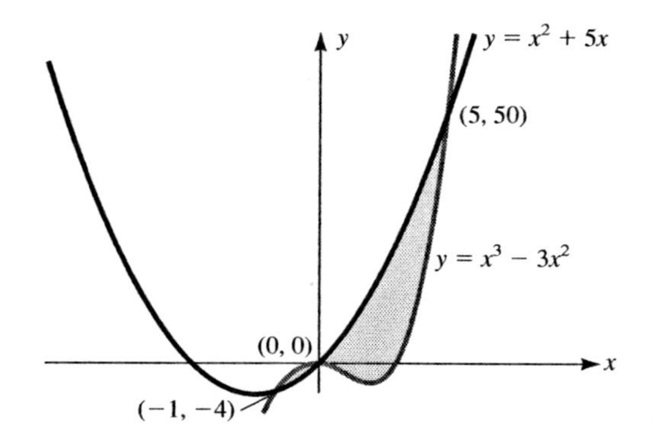

15. The equation of the top curve is the equation of the line through the points $(-4, 0)$ and $(2, 6)$.

$$m=\frac{6}{6}=1,\ \text{so } y=x+4.$$

The shaded area is

$$\int_{-4}^{2}[(x+4)-0]\,dx=\left(\frac{x^2}{2}+4x\right)\Big|_{-4}^{2}=18.$$

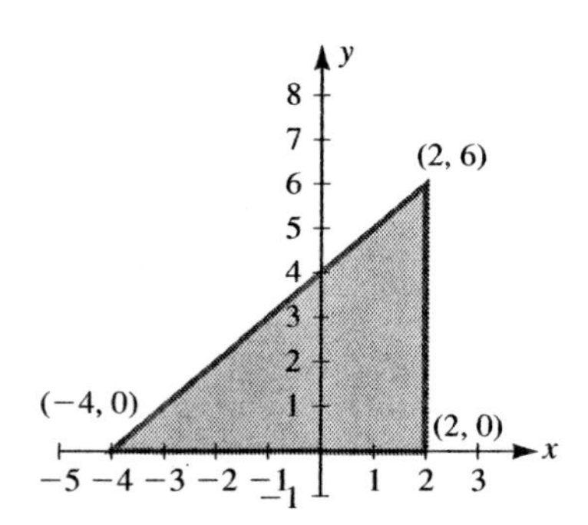

17. The equation of the top curve is the equation of the line through the points $(0, 6)$ and $(2, 8)$.

$$m=\frac{8-6}{2-0}=1,\ \text{so } y=x+6$$

The shaded area is

$$\int_{0}^{2}[(x+6)-0]\,dx=\left(\frac{x^2}{2}+6x\right)\Big|_{0}^{2}=14$$

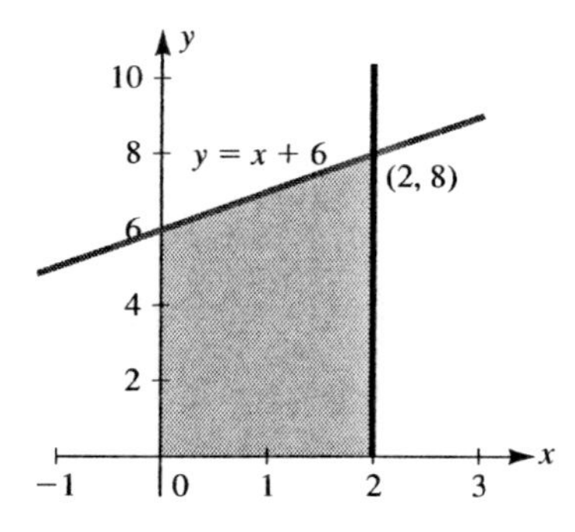

19.

$$f_{av}=\frac{1}{3-(-3)}\int_{-3}^{3}(1-x^2)\,dx$$

$$=\frac{1}{6}\left(x-\frac{x^3}{3}\right)\Big|_{-3}^{3}=-2$$

21.
$$f_{av} = \frac{1}{1-(-1)} \int_{-1}^{1} \left[e^{-x}(4 - e^{2x}) \right] dx$$
$$= \frac{1}{2} \int_{-1}^{1} (4e^{-x} - e^{x})\, dx$$
$$= \frac{1}{2}(-4e^{-x} - e^{x}) \Big|_{-1}^{1}$$
$$= \frac{1}{2}\left(\frac{-3}{e} + 3e\right) = \frac{3}{2}\left(e - \frac{1}{e}\right)$$

23.
$$f_{av} = \frac{1}{\ln 3 - 0} \int_{0}^{\ln 3} \left(\frac{e^{x} - e^{-x}}{e^{x} + e^{-x}} \right) dx$$

Using substitution with $u = e^{x} + e^{-x}$,

$$= \frac{1}{\ln 3} \int_{2}^{10/3} \frac{1}{u}\, du = \frac{1}{\ln 3}(\ln u)\Big|_{2}^{10/3}$$
$$\frac{1}{\ln 3}\left(\ln \frac{10}{3} - \ln 2\right) = \frac{1}{\ln 3}(\ln 10 - \ln 3 - \ln 2)$$
$$= \frac{1}{\ln 3}(\ln 5 - \ln 3)$$

25. $$f_{av} = \frac{1}{2-0} \int_{0}^{2} (2x - x^{2})\, dx = \frac{1}{2}\left(x^{2} - \frac{x^{3}}{3}\right)\Big|_{0}^{2} = \frac{2}{3}$$

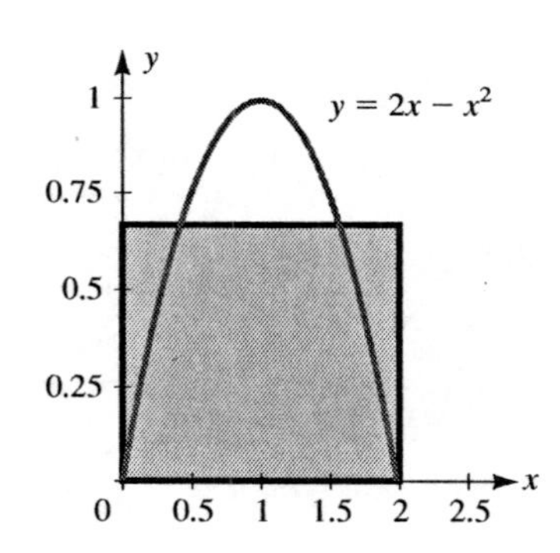

27. $$f_{av} = \frac{1}{4-2} \int_{2}^{4} \frac{1}{u}\, du = \frac{1}{2}\,[\ln |u|]\Big|_{0}^{2} = \frac{1}{2} \ln 2$$

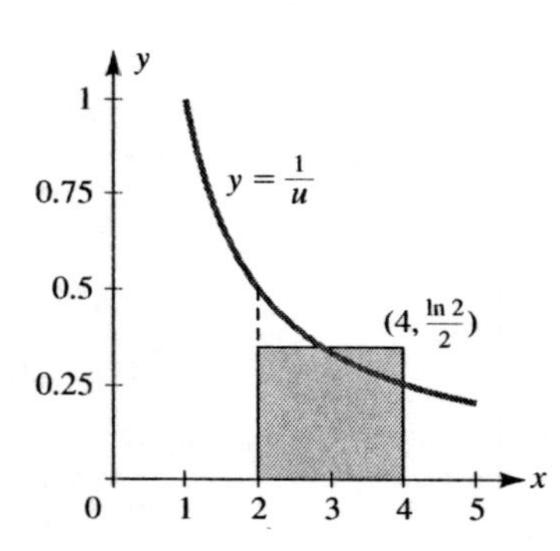

29. $$GI = 2\int_{0}^{1} (x - x^{3})\, dx = \left(x^{2} - \frac{x^{4}}{2}\right)\Big|_{0}^{1} = \frac{1}{2}$$

31.
$$GI = 2\int_{0}^{1} (x - 0.55x^{2} - 0.45x)\, dx$$
$$= 2\left(\frac{0.55x^{2}}{2} - \frac{0.55x^{3}}{3}\right)\Big|_{0}^{1}$$
$$= 0.183$$

33.
$$GI = 2\int_{0}^{1} \left(x - \frac{2}{3}x^{3.7} - \frac{1}{3}x\right) dx$$
$$= 2\left(\frac{x^{2}}{3} - \frac{2x^{4.7}}{3(4.7)}\right)\Big|_{0}^{1}$$
$$= 0.383$$

35.
$$P_{av} = \frac{1}{3-0} \int_{0}^{3} (0.09t^{2} - 0.2t + 1.6)\, dt$$
$$= \frac{1}{3}(0.03t^{3} - 0.1t^{2} + 1.6t)\Big|_{0}^{3}$$
$$= \$1.57 \text{ pound}$$

37.
$$Q_{av} = \frac{1}{5-0} \int_{0}^{5} 2{,}000e^{0.05t} dt$$
$$= \frac{400}{0.05} e^{0.05t}\Big|_{0}^{5} = 2{,}272 \text{ bacteria}$$

39. The equation of the function is the equation of the line joining (0, 60,000) and (1, 0).

$$m = \frac{60{,}000}{-1}, \text{ so } y = -60{,}000(t - 1)$$
$$y_{av} = \frac{1}{1-0} \int_{0}^{1} -60{,}000(t - 1)\, dt$$
$$= -60{,}000\left(\frac{t^{2}}{2} - t\right)\Big|_{0}^{1} = 30{,}000 \text{ kilograms}$$

41. **(a)** Testing a couple of values shows that P_2' is initially more profitable. It will stay more profitable until $P_2'(t) = P_1'(t)$.

$$306 + 5t = 130 + t^2$$
$$0 = t^2 - 5t - 176,$$
$$0 = (t - 16)(t + 11)$$

or $t = 16$ years (rejecting the negative solution).

(b)
$$\text{Excess} = \int_0^{16} \left[(306 + 5t) - (130 + t^2)\right] dt$$
$$= \left(176t + \frac{5t^2}{2} - \frac{t^3}{3}\right)\Big|_0^{16}$$
$$= 2{,}090.67, \text{ or } \$209{,}067.$$

(c)

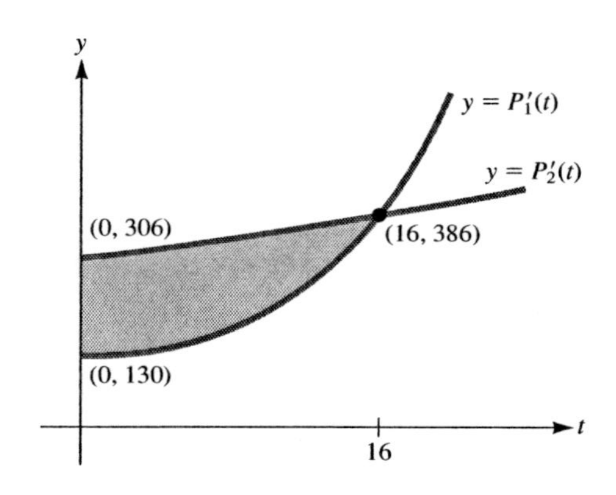

43. **(a)** Testing a couple of values shows that P_2' is initially more profitable. It will stay more profitable until $P_2'(t) = P_1'(t)$.

$$140e^{0.07t} = 90e^{0.1t}$$
$$\frac{14}{9}e^{0.07t} = e^{0.1t}$$
$$\ln\left(\frac{14}{9}e^{0.07t}\right) = \ln e^{0.1t}$$
$$\ln\frac{14}{9} + \ln e^{0.07t} = 0.1t$$
$$\ln\frac{14}{9} + 0.07t = 0.1t,$$

or $t \approx 14.7$ years.

(b)
$$\text{Excess} = \int_0^{14.7} (140e^{0.07t} - 90e^{0.1t})\, dt$$
$$= (2{,}000e^{0.07} - 900e^{0.1t})\Big|_0^{14.7}$$
$$\approx 582.22, \text{ or } \$582{,}220$$

(c)

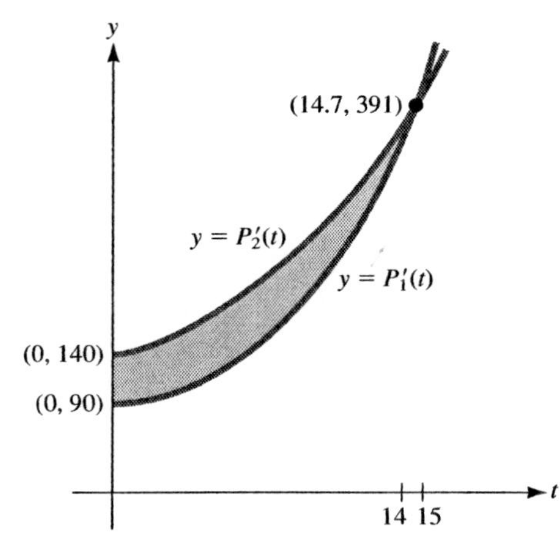

45.
$$P_{av} = \frac{1}{10 - 0}\int_0^{10} \frac{e^{0.2t}}{4 + e^{0.2t}}\, dt$$

Using substitution with $u = 4 + e^{0.2t}$,

$$= \frac{1}{2}\int_5^{4+e^2} \frac{1}{u}\, du$$
$$= \frac{1}{2}(\ln|u|)\Big|_5^{4+e^2}$$
$$= \frac{1}{2}\left[\ln(4 + e^2) - \ln 5\right] \approx 0.411607$$

or 411,607 people.

47.
$$C_{av} = \frac{1}{8 - 0}\int_0^{8} \frac{3t}{(t^2 + 36)^{3/2}}\, dt$$

Using substitution with $u = t^2 + 36$,

$$= \frac{3}{16}\int_{36}^{100} u^{-3/2}\, du = \frac{3}{16}\left(-\frac{2}{\sqrt{u}}\right)\Big|_{36}^{100}$$
$$= \frac{1}{40}\text{ mg/cm}^3.$$

49. **(a)** $T(t) = 3 - \frac{1}{3}(t-5)^2$ Since $t = 2$ at 8:00a.m., $a = 2$. Since $t = 11$ at 5:00 p.m., $b = 11$. So, the average temperature is

$$T_{av} = \frac{1}{11-2}\int_2^{11} 3 - \frac{1}{3}(t-5)^2 dt$$

Using substitution for the second term with $u = t - 5$, $du = dt$, $u_1 = -3$ and $u_2 = 6$,

$$= \frac{1}{9}\left[3t\Big|_2^{11} - \frac{1}{9}u^3\Big|_{-3}^{6}\right]$$

$$= \frac{1}{9}\left[(33-6) - \frac{1}{9}(216+27)\right]$$

$$= 0°C$$

(b) Need to find t when $T(t) = 0$, so

$$0 = -\frac{1}{3}(t-5)^2$$

$$\frac{1}{3}(t-5)^2 = 3$$

$$(t-5)^2 = 9$$

$$t - 5 = \pm 3$$

$$t = 2.8$$

When $t = 2$, the time is 8:00 a.m. and when $t = 8$, the time is 2:00 p.m.

51. **(a)**

$$S_{av} = \frac{1}{6-1}\int_1^6 (t^3 - 10.5t^2 + 30t + 20)\, dt$$

$$= \frac{1}{5}\left(\frac{t^4}{4} - 3.5t^3 + 15t^2 + 20t\right)\Big|_1^6$$

$$= 39.25 \text{ mph.}$$

(b) Need to find t when $S(t) = 39.25$, so
$39.25 = t^3 - 10.5t^2 + 30t + 20$
$0 = t^3 - 10.5t^2 + 30t - 19.25$
To solve $t^3 - 10.5t^2 + 30t - 19.25 = 0$, press [y=] and enter $x \wedge 3 - 10.5x \wedge 2 + 30x - 19.25$ for $y_1 =$. Use zstandard under the zoom menu and the graph of $y_1 =$ is displayed. The graph has 3 x-intercepts. To find the first, use the zero function under the calc menu. Enter a left bound close to the first x-intercept, a right bound, and a guess. The first x-intercept is approximately $x \approx 0.902$. Repeat this process to find the other two x-intercepts are $x = 3.5$ and $x \approx 6.10$. The only intercept corresponding to a time between 1:00 and 6:00 p.m. is $x = 3.5$, which is 3:30 p.m.

53. **(a)**

$$M_{av} = \frac{1}{12-0}\int_0^{12} (M_0 + 50te^{-0.1t^2})\, dt$$

$$= \frac{1}{12}\left[\int_0^{12} M_0\, dt + 50\int_0^{12} (te^{-0.1t^2})\, dt\right]$$

Using substitution with $u = -0.1t^2$,

$$= \frac{1}{12}\left[\int_0^{12} M_0\, dt - 250\int_0^{-14.4} e^u\, du\right]$$

$$= \frac{1}{12}\left[\int_0^{12} M_0\, dt + 250\int_{-14.4}^{0} e^u\, du\right]$$

$$= \frac{1}{12}\left[M_0 t\Big|_0^{12} + 250(e^u)\Big|_{-14.4}^{0}\right]$$

$$= M_0 + 20.83 \text{ kilo-Joules per hour.}$$

(b) When $t = 0$, $M(0) = M_0$ so $(0, M_0)$ is an intercept.
$\lim_{t\to+\infty} (M_0 + 50te^{-0.1t^2}) = M_0$, so $y = M_0$ is a horizontal asymptote.

$$M'(t) = 50\left[(t)(e^{-0.1t^2} \cdot -0.2t) + (e^{-0.1t^2})(1)\right]$$

$$= 50e^{-0.1t^2}(-0.2t^2 + 1).$$

So $M'(t) = 0$ when $-0.2t^2 + 1 = 0$, or $t = \sqrt{5}$. The peak metabolic rate is

$$M\left(\sqrt{5}\right) = M_0 + 50\sqrt{5}e^{-0.5}$$

$$= M_0 + 50\sqrt{\frac{5}{e}}$$

$$M''(t) = 50\left[(e^{-0.1t^2})(-0.4) + (-0.2t^2 + 1)(e^{-0.1t^2} \cdot -0.2t)\right]$$

$$= -10e^{-0.1t^2}\left[2 + (-0.2t^2 + 1)\right]$$

So $M''(t) = 0$ when $3 - 0.2t^2 = 0$, or $t = \sqrt{15}$.

When $0 < t < \sqrt{5}$, $M'(t) > 0$ so m is increasing
$M''(t) < 0$ so m is concave down
$\sqrt{5} < t < \sqrt{15}$, $M'(t) < 0$ so m is decreasing
$M''(t) < 0$ so m is concave down
$t > \sqrt{15}$, $M'(t) < 0$ so m is decreasing
$M''(t) > 0$ so m is concave up.

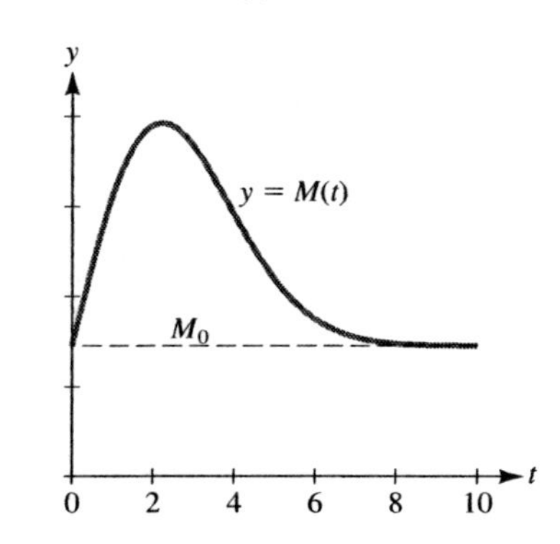

55.
$$GI_1 = 2\int_0^1 \left(x - \frac{2}{3}x^3 - \frac{1}{3}x\right) dx$$
$$= 2\left(\frac{x^2}{3} - \frac{x^4}{6}\right)\Big|_0^1 = \frac{1}{3} \approx 0.33$$
$$GI_2 = 2\int_0^1 \left(x - \frac{5}{6}x^2 - \frac{1}{6}x\right) dx$$
$$= 2\left(\frac{5}{12}x^2 - \frac{5}{18}x^3\right)\Big|_0^1 = \frac{5}{18} \approx 0.28$$
$$GI_3 = 2\int_0^1 \left(x - \frac{3}{5}x^4 - \frac{2}{5}x\right) dx$$
$$= 2\left(\frac{3}{10}x^2 - \frac{3}{25}x^5\right)\Big|_0^1 = \frac{9}{25} = 0.36$$
So, football is the most equitable, basketball is the least equitable.

57.
$$\text{Excess} = \int_0^{10} \left(10e^{0.02t} - \frac{20e^{0.02t}}{1 + e^{0.02t}}\right) dt$$
$$= 10\int_0^{10} e^{0.02t}\, dt - 20\int_0^{10} \frac{e^{0.02t}}{1 + e^{0.02t}}\, dt$$
Using substitution with $u = 1 + e^{0.02t}$,
$$= 10\int_0^{10} e^{0.02t}\, dt - 1{,}000\int_2^{1+e^{0.2}} \frac{1}{u}\, du$$
$$= 500(e^{0.02t})\Big|_0^{10} - 1{,}000(\ln|u|)\Big|_2^{1+e^{0.2}}$$
≈ 5.710, or 5,710 people.

59. Total cost = cost of cabin + cost of land
cost of cabin = (area of cabin)(price per sq. yard)
= (64)(2,000) = \$128,000
cost of land = (area of land)(price per sq. yard)
area of land = area under curve − area of cabin
$$= \int_0^{15} 10e^{0.04x}\, dx - 64$$
$$= 250(e^{0.04x})\Big|_0^{15} - 64 \approx 141.53$$
cost of land = (141.53)(800) = \$113,224

So, the total cost is \$241,224.

61. (a) $S = F'(M) = \frac{1}{3}(2kM - 3M^2)$

We need to maximize S.
$$F''(M) = \frac{1}{3}(2k - 6M)$$
So $F''(M) = 0$ when $2k - 6M = 0$, or $M = \dfrac{k}{3}$.
$F'''(M) = -2$, so $F'''\left(\dfrac{k}{3}\right) < 0$, so the absolute maximum occurs when $M = \dfrac{k}{3}$.

(b)
$$F_{av} = \frac{1}{k/3 - 0}\int_0^{k/3} \frac{1}{3}(kM^2 - M^3)\, dM$$
$$= \frac{1}{k}\left(\frac{kM^3}{3} - \frac{M^4}{4}\right)\Big|_0^{k/3} = \frac{k^3}{108}$$

63. Press [y=] and input $\sqrt{\frac{2}{5}x^2 - 2}$ for $y_1 =$,
input $-\sqrt{\frac{2}{5}x^2 - 2}$ for $y_2 =$,
and input $x \wedge 3 - 8.9x^2 + 26.7x - 27$ for $y_3 =$.

Use window dimensions [−5, 5]1 by [−4, 4]0.5
Press [graph].
Use trace and zoom-in to find the points of intersection are (4.2, 2.25) and (2.34, −0.44).
An alternative to using trace and zoom is to use the intersect function under the calc menu. To find the first point, use ↑ and ↓ arrows to verify $y_1 = \sqrt{\frac{2}{5}x^2 - 2}$ is displayed. Enter and value close to the point of intersection.
Then, verify $y_3 = x^3 - 8.9x^2 + 26.7x - 27$ is displayed and enter a value close and finally, enter a guess. This gives the point (4.2, 2.25)
Repeat this process using $y_2 = -\sqrt{\frac{2}{5}x^2 - 2}$ and $y_3 = x^3 - 8.9x^2 + 26.7x - 27$ to find the second point (2.34, −0.44).
To find the area bounded by the curves, we also find the positive x-intercept of $\frac{x^2}{5} - \frac{y^2}{2} = 1$ to be $x = 2.236$
The area is given by

$$\int_{2.236}^{2.34} y_1 - y_2 + \int_{2.34}^{4.2} y_1 - y_3 = \int_{2.236}^{2.34} y_1 - \int_{2.236}^{2.34} y_2$$
$$+ \int_{2.34}^{4.2} y_1 - \int_{2.34}^{4.2} y_3$$

Use the $\int f(x)\,dx$ function under the calc menu making sure the correct y equation is displayed in the upper left corner for each integral to find the area is $0.03008441 - (-0.0300844) + 2.7254917 - 0.68880636 \approx 2.097$
An easier alternative to evaluating each separate integral is to use the f_nInt function. From the home screen, select f_nInt from the math menu and enter $f_n\text{Int}(y_1 - y_2, x, 2.236, 2.34) + f_n\text{Int}(y_1 - y_3, x, 2.34, 4.2)$ to find the area. You input the y equations by pressing [vars] and selecting which y equation you want from the function window under y-vars.

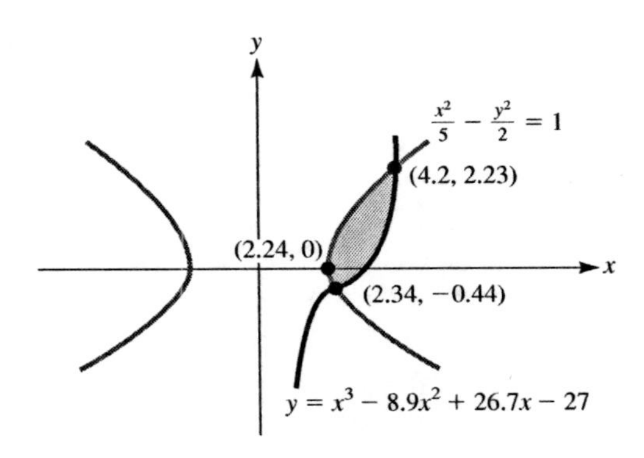

65. Let t_1 represent the starting time of an arbitrary time interval and let t_2 represent the ending time. Also, let $S(t)$ represent the distance function. Then, the average value of the velocity is

$$\frac{S(t_2) - S(t_1)}{t_2 - t_1}$$

The averagae velocity is

$$\frac{1}{t_2 - t_1}\int_{t_1}^{t_2} v(t)dt$$

Since distance is the integral of velocity,

$$= \frac{1}{t_2 - t_1}\left[S(t)\Big|_{t_1}^{t_2}\right]$$
$$= \frac{1}{t_2 - t_1}\left[S(t_2) - S(t_1)\right]$$
$$= \frac{S(t_2) - S(t_1)}{t_2 - t_1}$$

5.5 Additional Applications to Business and Economics

1. (a)

$$D(q) = 2(64 - q^2)$$
$$A(6) = 2\int_0^6 (64 - q^2)\,dq$$
$$= 2\left(64q - \frac{q^3}{3}\right)\Big|_0^6 = \$624$$

(b) The consumer's willingness to spend in part (a) is the area under the demand curve from $q = 0$ to $q = 6$.

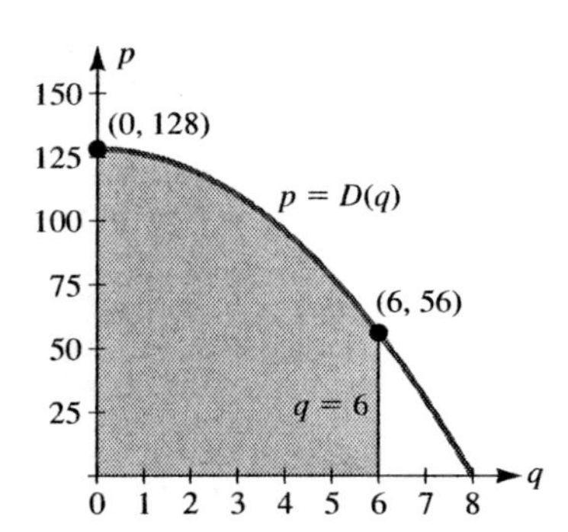

(b) The consumer's willingness to spend in part (a) is the area under the demand curve from $q = 0$ to $q = 10$.

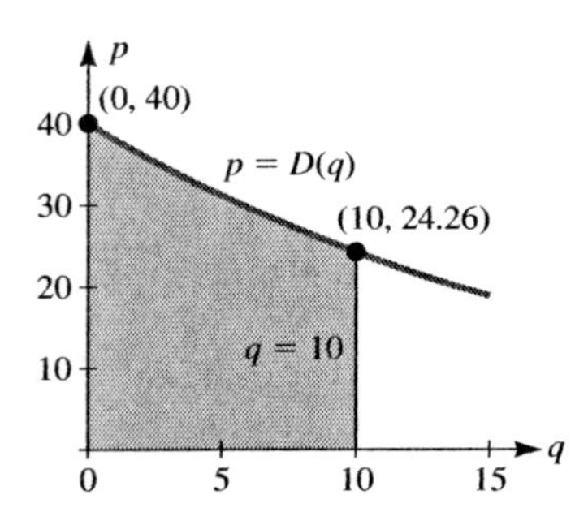

3. (a)

$$D(q) = \frac{400}{0.5q + 2}$$

$$A(12) = 2\int_0^{12} \frac{400}{0.5q + 2}\,dq$$

$$= 800 \ln |0.5q + 2|\Big|_0^{12}$$

$$= 800 \ln 4 = \$1,109.04$$

(b) The consumer's willingness to spend in part (a) is the area under the demand curve from $q = 0$ to $q = 12$.

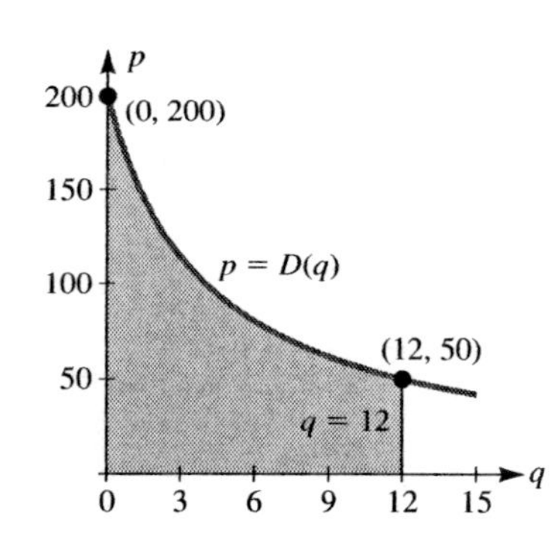

5. (a)

$$D(q) = 40e^{-0.05q}$$

$$A(10) = 40\int_0^{10} e^{-0.05q}\,dq$$

$$= -800e^{-0.05q}\Big|_0^{10} = \$314.78$$

7. $D(q) = p_0$ if $110 = 2(64 - q^2)$ or $q = 3$. The consumer's surplus is

$$CS = \int_0^3 2(64 - q^2)\,dq - 3(110)$$

$$= 2\left(64q - \frac{q^3}{3}\right)\Big|_0^3 - 330 = \$36$$

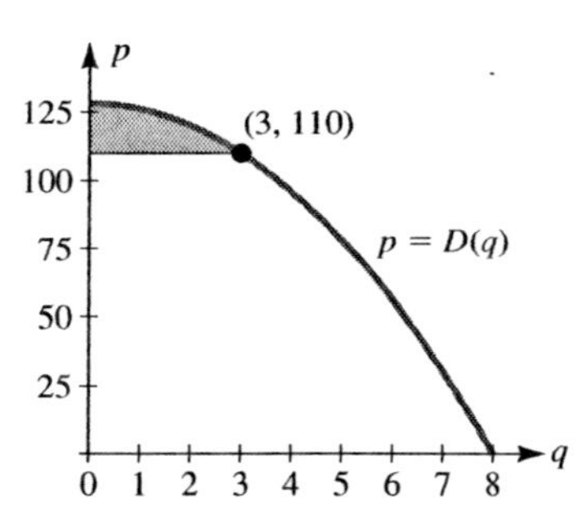

9. $D(q) = p_0$ if $31.15 = 40e^{-0.25}$ or $q = 5$. The consumer's surplus is

$$CS = \int_0^5 (40e^{-0.05q})\,dq - 5(31.15)$$

$$= -800e^{-0.05q}\Big|_0^5 - 93.45$$

$$= \$21.20$$

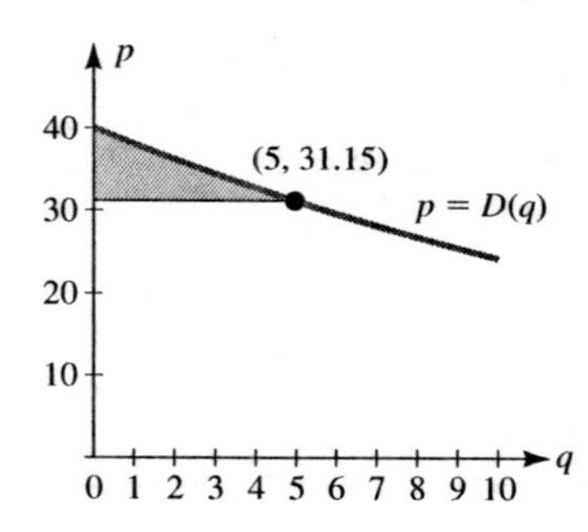

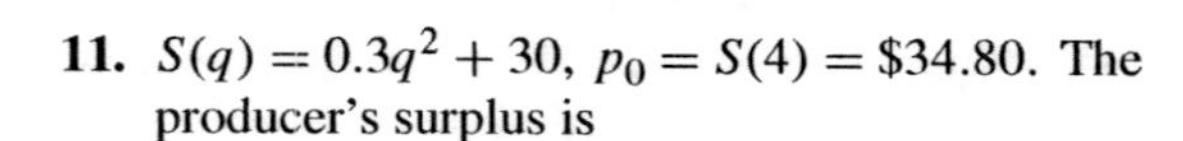

11. $S(q) = 0.3q^2 + 30$, $p_0 = S(4) = \$34.80$. The producer's surplus is

$$PS = 4(34.80) - \int_0^4 (0.3q^2 + 30)\, dq$$

$$= 139.20 - (0.1q^3 + 30q)\Big|_0^4$$

$$= \$12.80$$

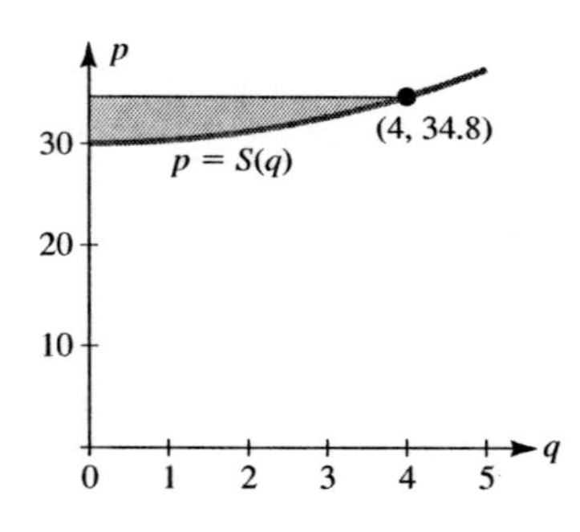

13. $S(q) = 10 + 15e^{0.03q}$, $p_0 = S(3) = \$26.41$. The producer's surplus is 3187

$$PS = 3(26.41) - \int_0^3 (10 + 15e^{0.3q})\, dq$$

$$= 79.23 - (10q + 500e^{0.03q})\Big|_0^3$$

$$= \$2.14$$

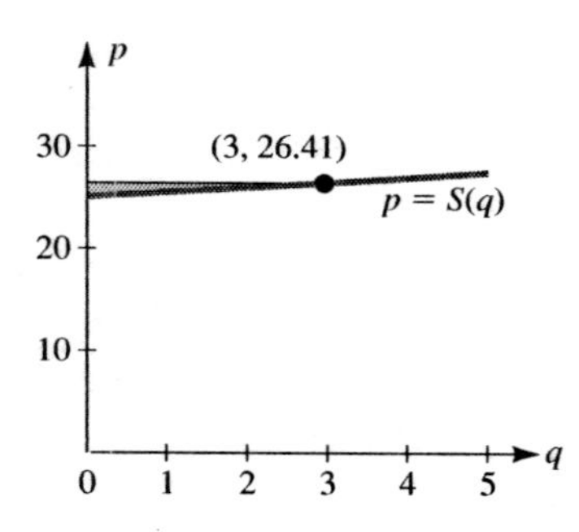

15. (a) The supply equals demand when

$$50 + \frac{2}{3}q^2 = 131 - \frac{1}{3}q^2$$

$$q^2 = 81, \text{ or } q = 9$$

So, the equilibrium price is

$$p_e = D(9) = 131 - \frac{1}{3}(9)^2 = \$104$$

(b) The corresponding consumer's surplus is

$$CS = \int_0^9 \left(131 - \frac{1}{3}q^2\right) dq - 9(104)$$

$$= \left(131q - \frac{1}{9}q^3\right)\Big|_0^9 - 936$$

$$= \$162$$

and the corresponding producer's surplus is

$$PS = (9)(104) - \int_0^9 \left(50 + \frac{2}{3}q^2\right) dq$$

$$= 936 - \left(50q + \frac{2}{9}q^3\right)\Big|_0^9$$

$$= \$324$$

17. (a) The supply equals demand when

$$-0.3q^2 + 70 = 0.1q^2 + q + 20$$

$$0 = 0.4q^2 + q - 50$$

$$q = \frac{-1 \pm \sqrt{1 + 4(0.4)(50)}}{2(0.4)} = 10$$

So, the equilibrium price is

$$p_e = D(10) = -0.3(10)^2 + 70 = \$40$$

(b) The corresponding consumer's surplus is

$$CS = \int_0^{10} (-0.3q^2 + 70)\, dq - 10(40)$$
$$= (-0.1q^3 + 70q)\Big|_0^{10} - 400$$
$$= \$200$$

and the corresponding producer's surplus is

$$PS = 10(40) - \int_0^{10} (0.1q^2 + q + 20)\, dq$$
$$= 400 - \left(\frac{0.1}{3}q^3 + \frac{q^2}{2} + 20q\right)\Big|_0^{10}$$
$$\approx \$116.67$$

19. (a) The supply equals demand when

$$\frac{1}{3}(q+1) = \frac{16}{q+2} - 3$$
$$\frac{(q+1)}{3} = \frac{10-3q}{q+2}$$
$$0 = q^2 + 12q - 28$$
$$q = \frac{-12 \pm \sqrt{(12)^2 + 4(1)(28)}}{2(1)}$$

or, $q = 2$

So, the equilibrium price is

$$p_e = D(2) = \frac{16}{2+2} - 3 = \$1$$

(b) The corresponding consumer's surplus is

$$\int_0^2 \left(\frac{16}{q+2} - 3\right) dq - 2(1)$$
$$= (16 \ln |q+2| - 3q)\Big|_0^2 - 2$$
$$= \$3.09$$

and the corresponding producer's surplus is

$$PS = 2(1) - \int_0^2 \frac{1}{3}(q+1)\, dq$$
$$= 2 - \frac{1}{3}\left(\frac{q^2}{2} + q\right)\Big|_0^2$$
$$= \$0.67$$

21. (a) The use of the machine will be profitable as long as the rate at which revenue is generated is greater than the rate at which costs accumulate. That is, until

$$R'(t) = C'(t)$$
$$7{,}250 - 18t^2 = 3{,}620 + 12t^2$$

or $t = 11$ years.

(b) The rate at which net earnings are generated by the machine is

$$R'(t) = C'(t)$$

So, the net earnings over the next 11 years is

$$\int_0^{11} \left[R'(t) - C'(t)\right] dt$$
$$= \int_0^{11} \left[(7{,}250 - 18t^2) - (3{,}620 + 12t^2)\right] dt$$
$$= \int_0^{11} (3{,}630 - 30t^2)\, dt$$
$$= (3{,}630t - 10t^3)\Big|_0^{11} = \$26{,}620$$

(c)

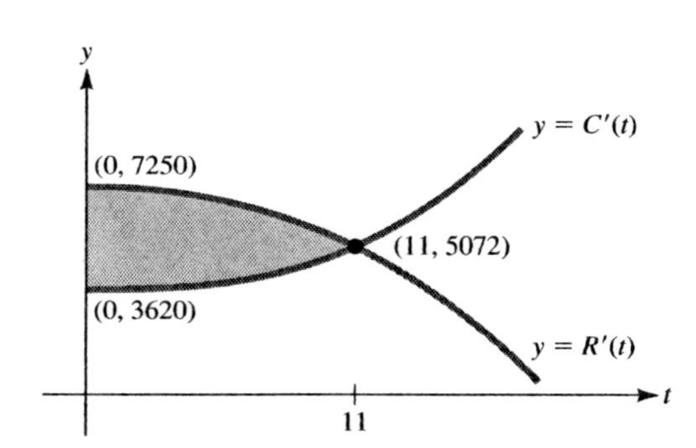

23. (a) The drive is profitable as long as rate of revenue exceeds weekly expenses.

$$e^{-0.3t} = \frac{593}{6{,}537}$$
$$= 0.090714,$$
$$-0.3t = \ln 0.090714,$$

or $t = 8$ weeks.

(b) The net earnings during the first 8 weeks are

$$N = \int_0^8 (6{,}537e^{-0.3t} - 593)\,dt$$
$$= \left(-\frac{6{,}537}{0.3}e^{-0.3t} - 593t\right)\Big|_0^8$$
$$= \frac{6{,}537}{0.3}(1 - 0.09072) - (593)(8)$$
$$= 19{,}813.26 - 4{,}744 = \$15{,}069.26.$$

(c) In geometric terms, the net earnings in part (b) is the area of the region between the curves

$$y = R'(t) \text{ and } y = E'(t)$$
$$\text{rewrite as: } y = R'(t) = 6{,}537e^{-0.3t}$$
$$\text{and } y = E'(t) = 593$$

25.
$$\text{amount} = \int_0^{10} 1000e^{0.1(10-t)}\,dt$$
$$= 1000e^1 \int_0^{10} e^{-0.1t}\,dt$$
$$= -10{,}000e(e^{-0.1t})\Big|_0^{10}$$
$$= -10{,}000e(e^{-1} - e^0)$$
$$\approx \$17{,}182.82$$

27. At age 60, Tom would have

$$\int_0^{35} 2500e^{0.05(35-t)}\,dt$$
$$= 2500e^{1.75} \int_0^{35} e^{-0.05t}\,dt$$
$$= -50{,}000e^{1.75}(e^{-0.05t})\Big|_0^{35}$$
$$= -50{,}000e^{1.75}(e^{-1.75} - e^0)$$
$$\approx \$237{,}730.13$$

At age 65, Tom would have

$$\int_0^{40} 2500e^{0.05(40-t)}\,dt$$
$$= 2500e^2 \int_0^{40} e^{-0.05t}\,dt$$
$$= -50{,}000e^2(e^{-0.05t})\Big|_0^{40}$$
$$= -50{,}000e^2(e^{-2} - e^0)$$
$$\approx \$319{,}452.80$$

29.
$$PV = \int_0^5 1200e^{-0.05t}\,dt$$
$$= -24{,}000(e^{-0.05t})\Big|_0^5$$
$$= -24{,}000(e^{-0.25} - e^0)$$
$$\approx \$5{,}308.78$$

31. The net income of the first investment is

$$\int_0^5 15{,}000e^{0.06(5-t)}\,dt - 50{,}000$$
$$= 15{,}000e^{0.3} \int_0^5 e^{-0.06t}\,dt - 50{,}000$$
$$= 87{,}464.70 - 50{,}000 = \$37{,}464.70$$

The net income of the second investment is

$$\int_0^5 9000e^{0.06(5-t)}\,dt - 30{,}000$$
$$= 9000e^{0.3} \int_0^5 e^{-0.06t}\,dt - 30{,}000$$
$$= 52{,}478.82 - 30{,}000 = \$22{,}478.82$$

So, the first investment will generate more income.

33. (a) The profit function is

$$P(q) = (110 - q)q - (q^3 - 25q^2 + 2q + 3{,}000)$$
$$= 110q - q^2 - q^3 + 25q^2 - 2q - 3{,}000$$
$$= -q^3 + 24q^2 + 108q - 3{,}000$$

(b)
$$P'(q) = -3q^2 + 48q + 108$$
$$= -3(q^2 - 16q - 36)$$

So, $P'(q) = 0$ when

$$q = \frac{24 \pm \sqrt{24^2 + 3(108)}}{3} = 18$$

$$P''(q) = -6q + 48$$

and $P''(18) < 0$, so $q = 18$ corresponds to the maximum profit.

(c) When $q = 18$, the price is

$$p = 110 - 18 = 92$$

and the corresponding consumer's surplus is

$$\begin{aligned} CS &= \int_0^{18} (110 - q)\,dq - 18(92) \\ &= \left(110q - \frac{q^2}{2}\right)\Big|_0^{18} - 1656 \\ &= \$162 \end{aligned}$$

35. (a)

$$\begin{aligned} P(t) &= \int P'(t)\,dt \\ &= \int 1.3e^{0.04t}\,dt \\ &= 1.3 \int e^{0.04t}\,dt \\ &= 32.5e^{0.04t} + C \end{aligned}$$

When $t = 0$, $P(0) = 0$ so $C = -32.5$ and $P(t) = 32.5e^{0.04t} - 32.5$.
When $t = 3$, $P(3) = 32.5e^{0.04(3)} - 32.5 \approx 4.14$ billion barrels.
Over the following three years, the amount pumped is $P(6) - P(3)$, or

$$\begin{aligned} &= (32.5e^{0.04(6)} - 32.5) - 4.14 \\ &\approx 4.68 \text{ billion barrels} \end{aligned}$$

(b) The field stops operating when it uses up the 20 billion barrels it holds, or when

$$\begin{aligned} 20 &= 32.5e^{0.04t} - 32.5 \\ \frac{21}{13} &= e^{0.04t} \\ \ln \frac{21}{13} &= \ln e^{0.04t}, \text{ or} \\ t &- \frac{\ln \frac{21}{13}}{0.04}, \text{ or approximately 12 years} \end{aligned}$$

(c)

$$\begin{aligned} PV &= \int V(t)e^{-rt}\,dt \\ &= \int 56P'(t)e^{-rt}\,dt \\ &= \int_0^{12} 56(1.3e^{0.04t})e^{-0.05t}\,dt \\ &= 72.8 \int_0^{12} e^{-0.01t}\,dt \\ &= -7{,}280(e^{-0.01t})\Big|_0^{12} \\ &\approx 823.22 \text{ billion dollars} \end{aligned}$$

(d) Writing exercise—Answers will vary.

37. (a)

$$\begin{aligned} P(t) &= \int P'(t)\,dt \\ &= \int 1.2e^{0.02t}\,dt \\ &= 1.2 \int e^{0.02t}\,dt \\ &= 60e^{0.02t} + C \end{aligned}$$

When $t = 0$, $P(0) = 0$ so $C = -60$ and $P(t) = 60e^{0.02t} - 60$.
When $t = 3$, $P(3) = 60e^{0.02(3)} - 60$
≈ 3.71 billion barrels
$P(6) - P(3) = (60e^{0.02(6)} - 60) - 3.71$
≈ 3.94 billion barrels

(b)

$$\begin{aligned} 12 &= 60e^{0.02t} - 60 \\ \frac{6}{5} &= e^{0.02t} \\ \ln \frac{6}{5} &= \ln e^{0.02t} \\ \ln \frac{6}{5} &= 0.02t, \text{ or} \\ t &= \frac{\ln \frac{6}{5}}{0.02}, \text{ or approximately 9.12 years} \end{aligned}$$

(c) Since the annual revenue is $A(t)P(t)$, the rate of annual revenue is, using the product rule,

$$A(t)P'(t) + A'(t)P(t)$$

$$= (56e^{0.015t})(1.2e^{0.02t}) + (60e^{0.02t} - 60)(0.39e^{0.015t})$$
$$= 67.2e^{0.035t} + 23.4e^{0.035t} - 23.4e^{0.015t}$$
$$= 90.6e^{0.035t} - 23.4e^{0.015t}$$

$$PV = \int_0^{9.12} (90.6e^{0.035t} - 23.4e^{0.015t})e^{-0.05t}\,dt$$
$$= \int_0^{9.12} \left(90.6e^{-0.015t} - 23.4e^{-0.035t}\right) dt$$
$$= \left[\frac{90.6}{-0.015}e^{-0.015t} + \frac{23.4}{0.035}e^{-0.035t}\right]_0^{9.12}$$
$$= \left[\frac{90.6}{-0.015}e^{-0.015(9.12)} + \frac{23.4}{0.035}e^{-0.035(9.12)}\right] - \left[\frac{90.6}{-0.015}e^{0} + \frac{23.4}{0.035}e^{0}\right]$$
$$\approx 589.55 \text{ billion years}$$

(d) Writing exercise—Answers will vary.

39. $PV = 10 \text{ million} = \int_0^6 Ae^{-0.05t}\,dt$

$$10 = -20A(e^{-0.05t})\Big|_0^6$$
$$10 = -20A(e^{-0.3} - e^0)$$
$$\frac{1}{-2(e^{-0.3} - 1)} = A \approx 1.929148 \text{ million, or}$$

$\$1,929,148$

41. $A(t) = 10e^{1-0.05t}$

(a)
$$FV = \int_0^5 10e^{1-0.05t} \cdot e^{1-0.05(5-t}\,dt$$
$$= 10\int_0^5 e^{(1-0.05t)+(0.25-0.05t)}\,dt$$
$$= 10\int_0^5 e^{1.25-0.1t}\,dt$$
$$= 10e^{1.25}\int_0^5 e^{-0.1t}\,dt$$
$$= 10e^{1.25}\left(-10e^{-0.1t}\Big|_0^5\right)$$
$$= -100e^{1.25}\left(e^{-0.1t}\Big|_0^5\right)$$
$$= -100e^{1.25}\left(e^{-0.5} - e^0\right)$$
$$= -100e^{1.25}\left(e^{0.5} - 1\right)$$
$$\approx 137.33429$$

or \$137,334.29

(b)
$$PV\int_1^3 10e^{1-0.05t} \cdot e^{-0.05t}\,dt$$
$$= 10\int_1^3 e^{(1-0.05t)-0.05t}\,dt$$
$$= 10\int_1^3 e^{1-0.1t}\,dt$$
$$= 10e^1\int_1^3 e^{-0.1t}\,dt$$
$$= 10e\left(-10e^{-0.1t}\Big|_1^3\right)$$
$$= -100e\left(e^{-0.1t}\Big|_1^3\right)$$
$$= -100e\left(e^{-0.3} - e^0\right)$$
$$= -100e\left(e^{-0.3} - 1\right)$$
$$\approx 70.45291$$

or \$70,452.91

43. (a)
$$R'(t) = 300(18 + 0.3t^{1/2})$$
$$FV = \int_0^{36} 300(18 + 0.3t^{1/2})\,dt$$
$$= 300\int_0^{36} (18 + 0.3t^{1/2})\,dt$$
$$= 300(18t + 0.2t^{3/2})\Big|_0^{36}$$
$$= 300(648 + 43.2) = \$207,360$$

(b) Writing exercise—Answers will vary.

5.6 Additional Applications to the Life and Social Sciences

1. After 5 months, the number of the original population surviving is

$$50{,}000e^{-0.1(5)}.$$

The number of new members surviving after 5 months is

$$\int_0^5 40e^{-0.1(5-t)}\,dt.$$

So, the total will be

$$= 50{,}000e^{-0.5} + 40e^{-0.5}\int_0^5 e^{0.1t}\,dt$$

$$= e^{-0.5}\left[50{,}000 + 400(e^{0.1t})\Big|_0^5\right]$$

$$\approx 30{,}484 \text{ members.}$$

3. After 3 years, the number of the original population surviving is

$$500{,}000e^{-0.011(3)}.$$

The number of new members surviving after 3 years is

$$\int_0^3 800e^{-0.011(3-t)}\,dt.$$

So, the total will be

$$500{,}000e^{-0.033} + 800e^{-0.033}\int_0^3 e^{0.011t}\,dt$$

$$= 800e^{-0.033}\left[625 + \frac{1}{0.011}(e^{0.011t})\Big|_0^3\right]$$

$$\approx 486{,}130 \text{ members}$$

5. After 8 years, the number of the original population surviving is $500{,}000e^{-0.013(8)}$
The number of new members surviving after 8 years is

$$\int_0^8 100e^{0.01t} - e^{-0.013(8-t)}dt$$

$$= 100\int_0^8 e^{0.01t-0.104+0.013t}dt$$

$$= 100e^{-0.104}\int_0^8 e^{0.023t}dt$$

So, the total will be

$$500{,}000e^{-0.104} + 100e^{-0.104}\int_0^8 e^{0.023t}dt$$

$$= 100e^{-0.104}\left[5000 + \frac{1}{0.023}e^{0.023t}\Big|_0^8\right]$$

$$= 100e^{-0.104}\left[5000 + \frac{1000}{23}\left(e^{0.023t}\Big|_0^8\right)\right]$$

$$= 100{,}000e^{-0.104}\left[5 + \frac{1}{23}\left(e^{0.184} - e^{0}\right)\right]$$

$$= 100{,}000e^{-0.104}\left[5 + \frac{1}{23}\left(e^{0.184} - 1\right)\right]$$

$$\approx 451{,}404 \text{ members}$$

7. Volum of $S = \pi\int_0^1 (3x+1)^2dx$

$$= \pi\int_0^1 (9x^2 + 6x + 1)dx$$

$$= \pi\left(3x^3 + 3x^2 + x\Big|_0^1\right)$$

$$= \pi\,[(3+3+1) - (0)] = 7\pi$$

9. Volume of $S = \pi\int_{-1}^3 (x^2+2)^2dx$

$$= \pi \int_{-1}^{3} (x^4 + 4x^2 + 4) dx$$
$$= \left(\frac{x^5}{5} + \frac{4x^3}{3} + 4x \Big|_{-1}^{3} \right)$$
$$= \pi \left[\left(\frac{243}{5} + \frac{108}{3} + 12 \right) - \left(-\frac{1}{5} - \frac{4}{3} - 4 \right) \right]$$
$$= \pi \left(\frac{729}{15} + \frac{540}{15} + \frac{180}{15} + \frac{3}{15} + \frac{20}{15} + \frac{60}{15} \right)$$
$$= \frac{1532}{15} \pi$$

11. Volume of $S = \pi \int_{-2}^{2} \left(\sqrt{4 - x^2} \right)^2 dx$

$$= \pi \int_{-2}^{2} (4 - x)\, dx = \pi \left(4x - \frac{x^3}{3} \Big|_{-2}^{2} \right)$$
$$= \left[\left(8 - \frac{8}{3} \right) - \left(-8 + \frac{8}{3} \right) \right]$$
$$= \pi \left(\frac{24}{3} - \frac{8}{3} + \frac{24}{3} - \frac{8}{3} \right)$$
$$= \frac{32}{3} \pi$$

13. Volume of $S = \pi \int_{1}^{e^2} \left(\frac{1}{\sqrt{x}} \right)^2 dx$

$$= \pi \int_{1}^{e^2} \frac{1}{x} = \pi \left(\ln x \Big|_{1}^{e^2} \right)$$
$$= \pi \left(\ln e^2 - \ln 1 \right)$$
$$= \pi (2 - 0) = 2\pi$$

15.
$$P(t) = \int P'(t)\, dt$$
$$= \int e^{0.02t}\, dt$$
$$= 50e^{0.02t} + C$$

When $t = 0$, $P(0) = 50$

$$50 = 50e^0 + C, \text{ or } C = 0.$$

So, $P(t) = 50e^{0.02t}$ and

$$P(10) = 50e^{0.02(10)} \approx 61.07 \text{ million,}$$

or 61,070,138 people.

17. After 8 months, the number of the original members remaining is

$$200e^{-0.2(8)}$$

The number of new members remaining is

$$\int_0^8 10e^{-0.2(8-t)}\, dt.$$

So, the total will be

$$200e^{-1.6} + 10e^{-1.6} \int_0^8 e^{0.2t}\, dt$$
$$= 10e^{-1.6}[20 + 5(e^{0.2t}) \Big|_0^8]$$
$$\approx 80 \text{ members}$$

19. After 30 days, the number of those originally infected who still have the disease is

$$5000e^{-0.02(30)}$$

The number of those since infected who still have the disease is

$$\int_0^{30} 60e^{-0.02(30-t)}\, dt.$$

So, the total still infected will be

$$20e^{-0.6} \left(250 + 3 \int_0^{30} e^{0.02t}\, dt \right)$$
$$= 20e^{-0.6} \left[250 + 150(e^{0.02t}) \Big|_0^{30} \right]$$
$$= 1{,}000e^{-0.6} \left[5 + 3(e^{0.02t}) \Big|_0^{30} \right]$$
$$\approx 4{,}098 \text{ people}$$

21.
$$\int_0^{10} 30e^{0.1t}\, dt = 300(e^{0.1t}) \Big|_0^{10}$$
$$\approx 515.48 \quad \text{billion barrels.}$$

23. After 10 months, the number of the origianl members remaining is

$$8{,}000e^{-10/10}.$$

The number of new members remaining is

$$\int_0^{10} 200e^{-(10-t)/10}\,dt$$

So, the total will be

$$200e^{-1}(40 + \int_0^{10} e^{t/10}\,dt)$$

$$= 200e^{-1}[40 + 10(e^{t/10})\Big|_0^{10}]$$

$$\approx 4{,}207 \text{ members}$$

25. (a)

$$\int_0^{24} (-0.028t^2 + 0.672t)\,dt$$

$$= \left(\frac{-0.028}{3}t^3 + 0.336t^2\right)\Big|_0^{24}$$

$$= 64.512.$$

So, the cardiac output is

$$R = \frac{5}{64.512} \approx 0.0775 \text{ liters/sec.}$$

(b) When $t = 0$, $C = 0$ so $(0, 0)$ is an intercept. $C(t) = 0$, $0 = -0.028t(t - 24)$, or $t = 24$, so $(24, 0)$ is an intercept.
The vertex is

$$\left(-\frac{b}{2a}, f\left(-\frac{b}{2a}\right)\right)$$

$$h = -\frac{b}{2a} = -\frac{0.672}{2(-0.028)} = 12$$

$$k = C(12) = -0.028(12)^2 + 0.672(12) \approx 4.03.$$

So, the vertex is $(12, 4.03)$.

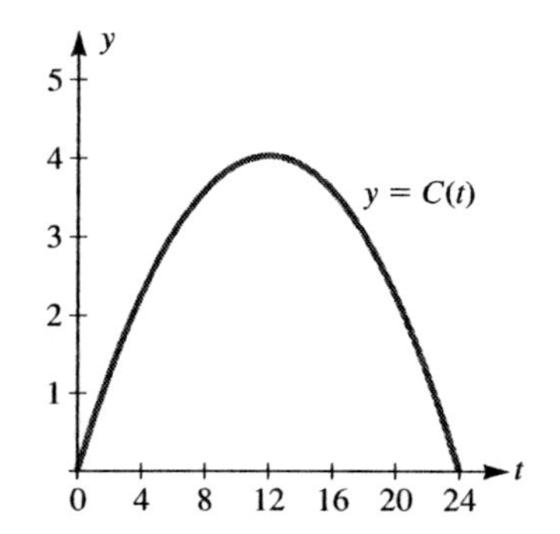

(c) Writing exercise—Answers will vary.

27. (a)

$$\int_0^{24} \frac{1}{12{,}312}\left(t^4 - 48t^3 + 378t^2 + 4{,}752t\right)dt$$

$$= \frac{1}{12{,}312}\left(\frac{t^5}{5} - 12t^4 + 126t^3 + 2{,}376t^2\right)\Big|_0^{24}$$

$$\approx 58.611.$$

So, the cardiac output is

$$R = \frac{5}{58.611} \approx 0.0853 \text{ liters/sec.}$$

(b) To sketch the graph of $C(t)$,
Press [y=] and input $C(t)$ for $y_1 =$.
Use window dimensions [0, 24]4 by [0, 5]1.

Writing exercise—Answers will vary.

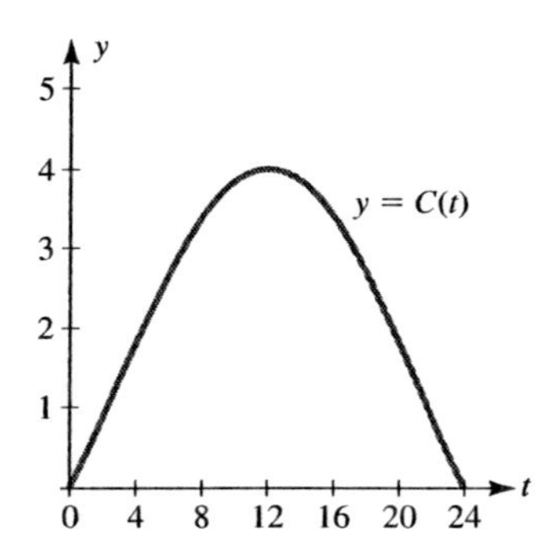

29.

$$\#\text{ people} = \int_1^2 2\pi r(25{,}000e^{-0.05r^2})\,dr$$

$$= 50{,}000\pi \int_1^2 re^{-0.05r^2}\,dr$$

Let $u = -0.05r^2$; then $-10\,du = r\,dr$ and the limits of integration become $-0.05(1)^2 = -0.05$ and $-0.05(2)^2 = -0.2$ So,

$$= 50{,}000\pi \int_{-0.05}^{-0.2} e^u \cdot -10\, du$$

$$= 500{,}000\pi \int_{-0.2}^{-0.05} e^u\, du$$

$$= 500{,}000\pi \left(e^u \Big|_{-0.2}^{-0.05} \right)$$

$$= 500{,}000\pi (e^{-0.05} - e^{-0.2})$$

$$\approx 208{,}128 \text{ people}$$

31. (a)

$$\int_0^3 0.3t(49 - t^2)^{0.4}\, dt$$

Using substitution with $u = 49 - t^2$,

$$= \frac{-0.3}{2} \int_{49}^{40} u^{0.4}\, du$$

$$= \frac{0.3}{2} \int_{40}^{49} u^{0.4}\, du$$

$$= \frac{0.3}{2.8} (u^{1.4}) \Big|_{40}^{49}$$

$$\approx 6.16,$$

so LDL decreases by approximately 6.16 units.

(b)

$$L(t) = \int L'(t)\, dt$$

$$= \int 0.3t(49 - t^2)^{0.4}\, dt$$

$$= \frac{3}{28}(49 - t^2)^{1.4} + C$$

When $t = 0$, $L(t) = 150$ so

$$150 = \frac{3}{28}(49)^{1.4} + C,$$

$$\text{or } C = 150 - \frac{3}{28}(49)^{1.4}$$

So,

$$L(t) = \frac{3}{28}(49 - t^2)^{1.4} + 150 - \frac{3}{28}(49)^{1.4}$$

$$= \frac{3}{28}(49 - t^2)^{1.4} + 150 - \frac{21}{4}(49)^{0.4}$$

(c) To find how many days it takes for patient's LDL level to be safe,

Press [y=]

Input $(3/28)(49 - x^2) \wedge (1.4) + 150-$ $((21/4)(49) \wedge 0.4)$ for $y_1 =$.

Use window dimensions [0, 10]1 by [0, 200]20.

Press [graph].

Use [trace] and zoom-in to find that $y = 130$ when $x \approx 5.8$

Therefore, it takes approximately 5.8 days for the LDL level to be safe.

33. For the first colony, the number of bacteria after 50 days will be

$$100{,}000e^{-0.011(50)} + \int_0^{50} 50e^{-0.011(50-t)}\, dt$$

$$= 100{,}000e^{-0.55} + 50e^{-0.55} \int_0^{50} e^{0.011t}\, dt$$

$$= 50e^{-0.55} \left[2{,}000 + \frac{1}{0.011}(e^{0.011t}) \Big|_0^{50} \right]$$

$$\approx 59{,}618$$

The number in the second colony will be

$$P(50) = \frac{5{,}000}{1 + 49e^{0.009(50)}}$$

$$\approx 64.228, \text{ or } 64{,}228$$

So, after 50 days, the population is larger in the second colony.

Similarly, after 100 days, the first colony's population will be

$$100{,}000e^{-0.011(100)} + \int_0^{100} 50e^{-0.011(100-t)}\, dt$$

$$\approx 36{,}320$$

and the second colony will be

$$P(100) = \frac{5{,}000}{1 + 49e^{0.009(100)}}$$

$$\approx 41.145, \text{ or } 41{,}145$$

So, the second colony is still larger after 100 days.

Similarly, after 300 days, the first will be

$$100{,}000e^{-0.011(300)} + \int_0^{300} 50e^{-0.011(300-t)}\, dt$$

$$\approx 8{,}066$$

and the second will be

$$P(300) = \frac{5{,}000}{1+49e^{0.009(300)}} \approx 6{,}848$$

So, after 300 days, the first colony is now larger.

35. Using the result of problem #24,

$$P(10) = 3{,}000e^{-0.07(10)} + \int_0^{10} 10e^{0.01t}e^{-0.07(10-t)}\,dt$$

$$= 3{,}000e^{-0.7} + 10e^{-0.7}\int_0^{10} e^{0.08t}\,dt$$

$$= 10e^{-0.7}\left[300 + \frac{1}{0.08}(e^{0.08t})\Big|_0^{10}\right]$$

$\approx 1{,}566$ members of the species.

37. Using the result of problem #24,

$$P(10) = 85{,}000\left(\frac{1}{10+1}\right) + \int_0^{10} 1{,}000\frac{1}{(10-t)+1}dt$$

$$= \frac{85{,}000}{11} + 1{,}000\int_0^{10}\frac{1}{11-t}\,dt$$

$$= \frac{85{,}000}{11} - 1{,}000\int_{11}^{1}\frac{1}{u}\,du$$

$$= \frac{85{,}000}{11} + 1{,}000\int_{1}^{11}\frac{1}{u}\,du$$

$$= \frac{85{,}000}{11} + 1{,}000\ln|u|\Big|_1^{11}$$

$\approx 10{,}125$ people.

39.

$$D(t) = \int D'(t)\,dt$$

$$= \int 0.12 + \frac{0.08}{t+1}\,dt$$

$$= 0.12t + 0.08\ln|t+1| + C$$

When $t = 0$, $D(0) = 0$ so $C = 0$ and

$$D(t) = 0.12t + 0.08\ln|t+1|$$

When $t = 12$ months (1 year),

$$D(12) = 0.12(12) + 0.08\ln|12+1| \approx 1.65,$$ or 165 infected people of those inoculated.

Of those not inoculated,

$$W(t) = \int W'(t)\,dt = \int \frac{0.8e^{0.13t}}{(1+e^{0.13t})^2}\,dt$$

Using substitution, with $u = 1 + e^{0.13t}$,

$$= 0.8\int \frac{1}{(1+e^{0.13t})^2}e^{0.13t}\,dt$$

$$= \frac{0.8}{0.13}\int u^{-2}\,du$$

$$= \frac{80}{13}\left[\frac{-1}{(1+e^{0.13t})}\right] + C$$

When $t = 0$, $W(0) = 0$, so

$$0 = \frac{80}{13}\left(\frac{-1}{2}\right) + C,$$

or $C = \frac{40}{13}$

and $W(t) = \frac{-80}{13(1+e^{0.13t})} + \frac{40}{13}$.

So, after 12 months,

$$W(12) = \frac{-80}{13(1+e^{0.13(12)})} + \frac{40}{13} \approx 2.01,$$ or approximately 201 people infected.

So, approximately $201 - 165 = 36$ people protected by the drug, or

$$\frac{W(12) - D(12)}{W(12)} \approx 18.1\%.$$

41. **(a)** At birth,

$$L(0) = \frac{110e^0}{1+e^0} = 55 \text{ years of age}$$

(b)

$$L_{av} = \frac{1}{70-10}\int_{10}^{70}\frac{110e^{0.015t}}{1+e^{0.015t}}\,dt$$

Using substitution, with $u = 1 + e^{0.015t}$,

$$= \frac{110}{60}\int_{10}^{70} \frac{1}{1+e^{0.015t}} e^{0.015t}\,dt$$

$$= \frac{11}{6(0.015)}\int_{1+e^{0.15}}^{1+e^{1.05}} \frac{1}{u}\,du$$

$$= \frac{11}{0.09}(\ln|u|)\Big|_{1+e^{0.15}}^{1+e^{1.05}}$$

≈ 70.78 years of age

(c) To find the age T such that $L(T) = T$, we must find T such that

$$\frac{110e^{0.015T}}{1+e^{0.015T}} = T$$

$$110e^{0.015T} - T(1+e^{0.015T}) = 0$$

Press [y=] and input $110e \wedge (0.015x)-$ $(x * (1+e \wedge (0.015x)))$ for $y_1 =$.
Use window dimensions [0, 100]10 by [−10, 120]20.
Press [graph].
Use the zero function under the calc menu to find that $T \approx 86.4$ years.
On the average, this is how long people in this country live.

(d) $$L_e = \frac{1}{86.4-0}\int_0^{86.4} \frac{110e^{0.015t}}{1+e^{0.015t}}dt$$

Using substitution as before,

$$= \frac{110}{(86.4)(0.015)}\int_2^{1+e^{1.296}} \frac{1}{u}du$$

$$= \frac{110}{1.296}\left[\ln(1+e^{1.296}) - \ln 2\right]$$

≈ 71.7 years of age

43. (a)

$$0 = -0.41t^2 + 0.97t$$
$$= t(0.97 - 0.41t)$$

so $R(t) = 0$ when $t = 0$ and when $t \approx 2.37$ sec.

(b) $$\text{Volume} = \int_0^{2.37} (-0.41t^2 + 0.97t)\,dt$$

$$= \left(\frac{-0.41}{3}t^3 + \frac{0.97}{2}t^2\right)\Big|_0^{2.37}$$

≈ 0.905 liters

(c) $$R_{av} = \frac{1}{2.37-0}\int_0^{2.37} (-0.41t^2 + 0.97t)\,dt$$

$$\approx \frac{0.905}{2.37} \approx 0.382 \text{ liters/sec.}$$

45. $T(r) = \dfrac{3}{2+r} = 3(2+r)^{-1}$

(a) domain: $[0, \infty)$

intercepts: when $r = 0$, $T(0) = \frac{3}{2}$; point $\left(0, \frac{3}{2}\right)$
when $T(r) = 0$, no solution

vertical asymptote outside of domain ($r = -2$)

horizontal aymptote

$$\lim_{r\to\infty} \frac{\frac{3}{r}}{\frac{2}{r}+1} = 0, \text{ or } y = 0$$

$$T'(r) = -\frac{3}{(2+r)^2} = -3(2+r)^{-2}$$

$$T''(r) = \frac{6}{(2+r)^3}$$

When $r \geq 0$, $T'(r) < 0$ so T is decreasing
$T''(r) > 0$ so T is concave up.

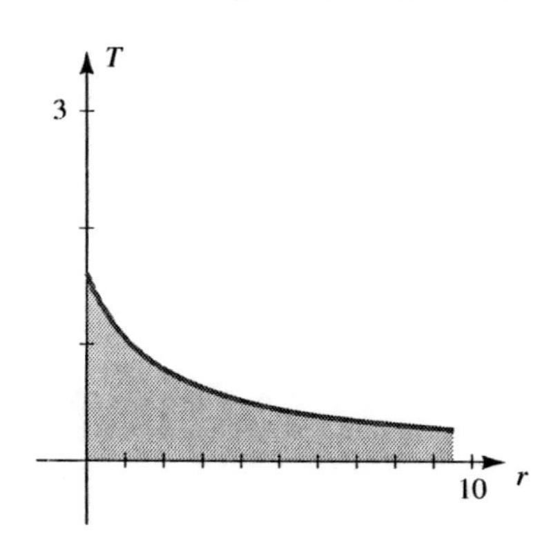

(b)

$$T(r) = \frac{3}{2+r}$$

$$2 + r = \frac{3}{T}$$

$$r(T) = \frac{3}{t} - 2$$

Graph is relection of graph in part(a) over the line $y = x$.

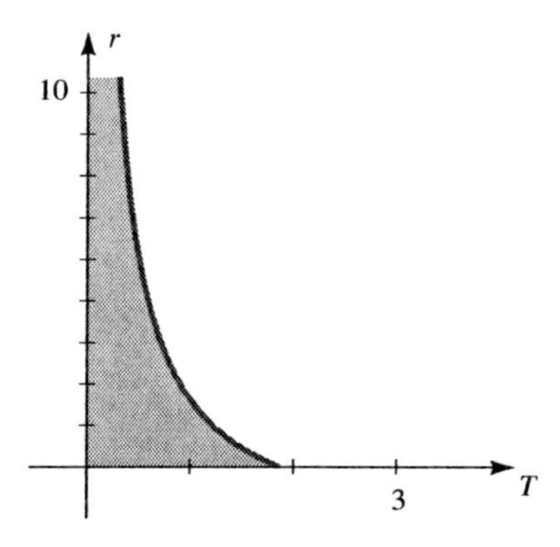

(c) When $r = 0$, $T = \frac{3}{2}$ and when $r = 7$, $T = \frac{1}{3}$.

Volume

$$= \pi \int_{1/3}^{3/2} \left(\frac{3}{T} - 2\right)^2 dT$$

$$= \pi \int_{1/3}^{3/2} \left(\frac{9}{T^2} - \frac{12}{T} + 4\right) dT$$

$$= \pi \left[-\frac{9}{T} - 12 \ln T + 4T \Big|_{1/3}^{3/2} \right]$$

$$= \pi \left[\left(-6 - 12 \ln \frac{3}{2} + 6\right) - \left(-27 - 12 \ln \frac{1}{3} + \frac{4}{3}\right) \right]$$

$$= \pi \left[-12 \ln \frac{3}{2} + \frac{81}{3} + 12 \ln \frac{1}{3} - \frac{4}{3} \right]$$

$$= \pi \left[12 \ln \frac{1}{3} - 12 \ln \frac{3}{2} + \frac{77}{3} \right] \approx 23.93\text{ft}^3$$

47. $p(r) = \dfrac{200}{5 + 2r^2}$

(a) Since the pollution is distributed in a circular fashion about the smoke stack,

$$\text{pollution} = 2\pi \int_0^3 r \left(\frac{200}{5 + 2r^2}\right) dr$$

$$= 400\pi \int_0^3 \frac{r}{5 + 2r^2} dr$$

Using substitution with $u = 5 + 2r^2$, $\frac{1}{4} du = r\,dr$ and limits of integration $u_1 = 5$ and $u_2 = 23$,

$$= 400\pi \int_5^{23} \frac{1}{u} \cdot \frac{1}{4} du$$

$$= 100\pi \int_5^{23} \frac{1}{u} du$$

$$= 100\pi \left(\ln u \Big|_5^{23} \right)$$

$$= 100\pi (\ln 23 - \ln 5)$$

$$= 100\pi \ln \frac{23}{5} \approx 479.42 \text{ units}$$

(b) $4 = \dfrac{200}{5 + 2r^2}$

$$L = r = \sqrt{\frac{45}{2}} = \frac{3\sqrt{10}}{2} \approx 4.74 \text{ miles}$$

$$\text{amt of pollution} = 2\pi \int_0^{\frac{3\sqrt{10}}{2}} r \left(\frac{200}{5 + 2r^2}\right) dr$$

$$= 100\pi \left(\ln u \Big|_5^{50} \right)$$

$$= 100\pi \; (\ln 50 - \ln 5)$$

$$= 100\pi \ln 10 \approx 723.38 \text{ units}$$

49. Volume $= \displaystyle\int_0^h \pi y^2 dx$

Since the hypotenuse of the triangle is along the line $y = \frac{r}{h} x$,

$$= \pi \int_0^h \left(\frac{r}{h} x\right)^2 dx$$

$$= \frac{\pi r^2}{h^2} \int_0^h x^2 dx$$

$$= \frac{\pi r^2}{h^2} \left(\frac{x^3}{3}\Big|_0^h\right)$$

$$= \frac{\pi r^2}{h^2} \left(\frac{h^3}{3} - 0\right)$$

$$= \frac{1}{3} \pi r^2 h$$

Checkup for Chapter 5

1. (a)

$$\int \left(x^3 - \sqrt{3x} + 5e^{-2x}\right) dx$$

$$= \int x^3\,dx - \sqrt{3}\int x^{1/2}\,dx + 5\int e^{-2x}\,dx$$

$$= \frac{x^4}{4} - \frac{2\sqrt{3}}{3}x^{3/2} - \frac{5}{2}e^{-2x} + C$$

(b)

$$\int \frac{x^2 - 2x + 4}{x}\,dx$$

$$= \int \left(x - 2x + \frac{4}{x}\right) dx$$

$$= \int x\,dx - 2\int dx + 4\int \frac{1}{x}\,dx$$

$$= \frac{x^2}{2} - 2x + 4\ln|x| + C$$

(c)

$$\int \sqrt{x}\left(x^2 - \frac{1}{x}\right) dx$$

$$= \int \left(x^{5/2} - x^{-1/2}\right) dx$$

$$= \frac{2}{7}x^{7/2} - 2x^{1/2} + C$$

(d) $\displaystyle\int \frac{x\,dx}{(3+2x^2)^{3/2}}$

Let $u = 3 + 2x^2$; then $\frac{1}{4}\,du = x\,dx$

$$= \frac{1}{4}\int u^{-3/2}\,du = \frac{1}{4}(-2u^{-1/2}) + C$$

$$= \frac{-1}{2\sqrt{3+2x^2}} + C$$

(e)

$$\int \frac{\ln\sqrt{x}}{x}\,dx$$

$$= \int \frac{\frac{1}{2}\ln x}{x}\,dx$$

let $u = \ln x$; then $du = \frac{1}{x}\,dx$

$$\frac{1}{2}\int (\ln x)\frac{1}{x}\,dx$$

$$= \frac{1}{2}\int u\,du = \frac{1}{4}(\ln x)^2 + C$$

(f) $\displaystyle\int xe^{1+x^2}\,dx$

Let $u = 1 + x^2$; then $\frac{1}{2}\,du = x\,dx$

$$= \int (e^{1+x^2})x\,dx$$

$$= \frac{1}{2}\int e^u\,du = \frac{1}{2}e^{1+x^2} + C$$

2. (a)

$$\int_1^4 \left(x^{3/2} + \frac{2}{x}\right) dx$$

$$= \int_1^4 x^{3/2}\,dx + 2\int_1^4 \frac{1}{x}\,dx$$

$$= \frac{2}{5}x^{5/2}\Big|_1^4 + 2(\ln|x|)\Big|_1^4$$

$$= \frac{2}{5}\left[(4)^{5/2} - (1)^{5/2}\right] + 2[\ln 4 - \ln 1]$$

$$= \frac{62}{5} + 2\ln 4$$

$$= \frac{62}{5} + 2\ln 2^2 = \frac{62}{5} + 4\ln 2$$

(b) $\displaystyle\int_0^3 e^{3-x}\,dx$

Let $u = 3 - x$; then $-\,du = dx$ and the limits of integration become $3 - 3 = 0$ and $3 - 0 = 3$

$$= -\int_3^0 e^u\,du$$

$$= \int_0^3 e^u\,du = e^3 - e^0 = e^3 - 1$$

(c) $\displaystyle\int_0^1 \frac{x}{x+1}\,dx$

Let $u = x + 1$; then $du = dx$ and $x = u - 1$. Further, the limits of integration become $0 + 1 = 1$ and $1 + 2 = 2$

$$= \int_1^2 \frac{u-1}{u}\,du$$
$$= \int_1^2 \left(1 - \frac{1}{u}\right) du$$
$$= (u - \ln|u|)\Big|_1^2$$
$$= (2 - \ln 2) - (1 - \ln 1) = 1 - \ln 2$$

(d) $\displaystyle\int_0^3 \frac{(x+3)\,dx}{\sqrt{x^2+6x+4}}$

Let $u = x^2 + 6x + 4$; then $du = (2x+6)\,dx$ or, $\frac{du}{2} = (x+3)\,dx$. Further, the limits of integration become $0 + 6(0) + 4 = 4$ and $(3)^2 + 6(3) + 4 = 31$

$$= \frac{1}{2}\int_4^{31} u^{-1/2}\,du = \frac{1}{2}(2u^{1/2})\Big|_4^{31}$$
$$= u^{1/2}\Big|_4^{31} = \sqrt{31} - 2$$

3. (a) $$\text{Area} = \int_1^4 \left[(x + \sqrt{x}) - 0\right] dx$$
$$= \int_1^4 x + x^{1/2}\,dx$$
$$= \left(\frac{x^2}{2} + \frac{2}{3}x^{3/2}\right)\Big|_1^4$$
$$= \left[\frac{(4)^2}{2} + \frac{2}{3}(4)^{3/2}\right] - \left[\frac{1}{2} + \frac{2}{3}(1)^{3/2}\right]$$
$$= \frac{73}{6} \text{ sq. units.}$$

(b) The limits of integration are

$$x^2 - 3x = x + 5$$
$$x^2 - 4x - 5 = 0$$
$$(x-5)(x+1) = 0$$
$$x = -1, 5$$

Further, from a sketch of the graphs, or by comparing function values between $-1 < x < 5$, $y = x + 5$ is the top curve

$$\text{area} = \int_{-1}^5 \left[(x+5) - (x^2 - 3x)\right] dx$$
$$= \int_{-1}^5 (4x + 5 - x^2)\,dx$$
$$= \left(2x^2 + 5x - \frac{x^3}{3}\right)\Big|_{-1}^5$$
$$= \left[2(5)^2 + 5(5) - \frac{(5)^3}{3}\right]$$
$$- \left[2(-1)^2 + 5(-1) - \frac{(-1)^3}{3}\right]$$
$$= 36 \text{ sq. units.}$$

4. $$f_{av} = \frac{1}{2-1}\int_1^2 \frac{x-2}{x}\,dx$$
$$= \int_1^2 \left(1 - \frac{2}{x}\right) dx$$
$$= (x - 2\ln|x|)\Big|_1^2$$
$$= (2 - 2\ln 2) - (1 - 2\ln 1)$$
$$= 1 - 2\ln 2$$

5. Net change $= \displaystyle\int_a^b R'(q)\,dq$

$$= \int_4^9 q(10 - q)\,dq$$
$$= \int_4^9 (10q - q^2)\,dq$$
$$= \left(5q^2 - \frac{q^3}{3}\right)\Big|_4^9$$
$$= \left[5(9)^2 - \frac{(9)^3}{3}\right] - \left[5(4)^2 - \frac{(4)^3}{3}\right]$$
$$= \frac{310}{3} \text{ hundred, or approximately \$10,333.33}$$

6. The rate the trade deficit is changing
= rate of change of imports–rate of change of exports.

$$D'(t) = I'(t) - E'(t)$$

So, the change over the next five years is

$$\int_0^5 [E'(t) - I'(t)]\,dt$$

$$= \int_0^5 [12.5e^{0.2t} - (1.7t + 3)]\,dt$$

$$= \left[12.5\left(\frac{1}{0.2}e^{0.2t}\right) - \frac{1.7}{2}t^2 - 3t\right]_0^5$$

$$= \left[62.5e^{0.2(5)} - \frac{1.7}{2}(5)^2 - 3(5)\right] - \left[62.5e^0 - 0 - 0\right]$$

≈ 71.14, or the trade deficit will increase by approximately 71.14 billion dollars.

7. When $q_0 = 4$, $p_0 = 25 - (4)^2 = 9$

$$CS = \int_0^4 (25 - q^2)\,dq - (4)(9)$$

$$= \left(25q - \frac{q^3}{3}\right)\Big|_0^4 - 36$$

$$= \left[25(4) - \frac{(4)^3}{3}\right] - 36$$

≈ 42.6667, or approximately \$4,266.67

8.

$$FV = \int_0^3 5{,}000e^{0.05(3-t)}\,dt$$

$$= 5{,}000e^{0.15}\int_0^3 e^{-0.05t}\,dt$$

$$= \frac{5{,}000e^{0.15}}{-0.05}\left(e^{0.05t}\right)\Big|_0^3$$

$$= \frac{5{,}000e^{0.15}}{-0.05}\left(e^{0.05(3)} - e^0\right)$$

$\approx \$16{,}183.42$

9. The number of the original 50,000 people remaining after 20 years is

$$50{,}000e^{-0.02(20)}$$

The number of new arrivals remaining after 20 years is

$$\int_0^{20} 700e^{-0.02(20-t)}\,dt$$

So, the total will be

$$50{,}000e^{-0.4} + 700e^{-0.4}\int_0^{20} e^{0.02t}\,dt$$

$$= 100e^{-0.4}\left[500 + 7\left(\frac{1}{0.02}e^{0.02t}\right)\Big|_0^{20}\right]$$

$$= 100e^{-0.4}\left[500 + 350\left(e^{0.02t}\right)\Big|_0^{20}\right]$$

$\approx 45{,}055$ people

10.

$$C_{av} = \frac{1}{3-0}\int_0^3 \frac{0.3t}{(t^2+16)^{1/2}}\,dt$$

Let $u = t^2 + 16$; then $\frac{1}{2}\,du = t\,dt$, and the limits of integration become $0 + 16 = 16$ and $(3)^2 + 16 = 25$

$$= \frac{0.3}{3}\int_0^3 \frac{1}{(t^2+16)^{1/2}}t\,dt$$

$$= \frac{0.1}{2}\int_{16}^{25} u^{-1/2}\,du$$

$$= 0.05(2u^{1/2})\Big|_{16}^{25}$$

$$= 0.1(u^{1/2})\Big|_{16}^{25}$$

$$= 0.1\left(\sqrt{25} - \sqrt{16}\right) = 0.1 \text{ mg/cm}^3.$$

Review Problems

1.

$$\int (x^3 + \sqrt{x} - 9)\,dx$$

$$= \int x^3\,dx + \int x^{1/2}\,dx - 9\int dx$$

$$= \frac{x^4}{4} + \frac{2}{3}x^{3/2} - 9x + C$$

3.

$$\int (x^4 - 5e^{-2x})\,dx$$

$$= \int x^4\,dx - 5\int e^{-2x}\,dx$$

$$= \frac{x^5}{5} + \frac{5}{2}e^{-2x} + C$$

5.

$$\int\left(\frac{5x^3-3}{x}\right)dx$$
$$=\int\left(5x^2-\frac{3}{x}\right)dx$$
$$=5\int x^2\,dx-3\int\frac{1}{x}\,dx$$
$$=\frac{5x^3}{3}-3\ln|x|+C$$

7.

$$\int\left(t^5-3t^2+\frac{1}{t^2}\right)dt$$
$$=\int t^5\,dt-3\int t^2\,dt+\int t^{-2}\,dt$$
$$=\frac{t^6}{6}-t^3-\frac{1}{t}+C$$

9.

$$\int\sqrt{3x+1}\,dx$$
$$=\int(3x+1)^{1/2}\,dx$$

Let $u=3x+1$; then $\frac{1}{3}\,du=dx$

$$=\frac{1}{3}\int u^{1/2}\,du$$
$$=\frac{2}{9}(3x+1)^{3/2}+C$$

11.

$$\int(x+2)(x^2+4x+2)^5\,dx$$

Let $u=x^2+4x+2$; then $du=(2x+4)\,dx$, or $\frac{1}{2}\,du=(x+2)\,dx$

$$=\int(x^2+4x+2)^5(x+2)\,dx$$
$$=\frac{1}{2}\int u^5\,du$$
$$=\frac{1}{12}(x^2+4x+2)^6+C$$

13.

$$\int\frac{3x+6}{(2x^2+8x+3)^2}\,dx$$

Let $u=2x^2+8x+3$; then, $du=(4x+8)\,dx$, or $\frac{1}{4}\,du=(x+2)\,dx$

$$=\int\frac{3(x+2)}{(2x^2+8x+3)^2}\,dx$$
$$=\frac{3}{4}\int u^{-2}\,du$$
$$=\frac{-3}{4(2x^2+8x+3)}+C$$

15. $\int v(v-5)^{12}\,dv$

Let $u=v-5$; then, $du=dv$ and $v=u+5$

$$=\int(u+5)u^{12}\,du$$
$$=\int(u^{13}+5u^{12})\,du$$
$$=\frac{(v-5)^{14}}{14}+\frac{5(v-5)^{13}}{13}+C$$

17. $\int 5xe^{-x^2}\,dx$

Let $u=-x^2$; then $-\frac{1}{2}\,du=x\,dx$

$$=5\int(e^{-x^2})x\,dx$$
$$=-\frac{5}{2}\int e^u\,du=-\frac{5}{2}e^{-x^2}+C$$

19. $\int\left(\frac{\sqrt{\ln x}}{x}\right)dx$

Let $u=\ln x$; then $du=\frac{1}{x}\,dx$

$$=\int(\ln x)^{1/2}\cdot\frac{1}{x}\,dx$$
$$=\int u^{1/2}\,du=\frac{2}{3}(\ln x)^{3/2}+C$$

21.

$$\int_0^1(5x^4-8x^3+1)\,dx$$
$$=(x^5-2x^4+x)\Big|_0^1=(1-2+1)-0=0$$

23. $\int_0^1 (e^{2x} + 4\sqrt[3]{x})\,dx$

$= \int_0^1 (e^{2x} + 4x^{1/3})\,dx$

$= \left(\frac{1}{2}e^{2x} + 3x^{4/3}\right)\Big|_0^1 = \left[\frac{1}{2}e^2 + 3(1)\right] - \left[\frac{1}{2}e^0 + 3(0)\right]$

$= \frac{1}{2}e^2 + \frac{5}{2}$

25. $\int_{-1}^{2} 30(5x-2)^2\,dx$

Let $u = 5x - 2$; then $\frac{1}{5}\,du = dx$, and the limits of integration become $5(-1) - 2 = -7$ and $5(2) - 2 = 8$

$$= \frac{30}{5}\int_{-7}^{8} u^2\,du = 6\left(\frac{u^3}{3}\right)\Big|_{-7}^{8}$$

$$= 6\left[\frac{(8)^3}{3} - \frac{(-7)^3}{3}\right] = 1{,}710$$

27. $\int_0^1 2te^{t^2-1}\,dt$

Let $u = t^2 - 1$; then $du = 2t\,dt$, and the limits of integration become $(0) - 1 = -1$ and $(1)^2 - 1 = 0$

$$= \int_{-1}^{0} e^u\,du = (e^u)\Big|_{-1}^{0} = e^0 - e^{-1}$$

$$= 1 - \frac{1}{e}$$

29. $\int_0^{e-1} \left(\frac{x}{x+1}\right)dx$

Let $u = x + 1$; then $du = dx$, $x = u - 1$, and the limits of integration become $0 + 1 = 1$ and $(e-1) + 1 = e$

$$= \int_1^e \left(\frac{u-1}{u}\right)du = \int_1^e \left(1 - \frac{1}{u}\right)du$$

$$= (u - \ln|u|)\Big|_1^e = (e - \ln e) - (1 - \ln 1)$$

$$= e - 2$$

31. 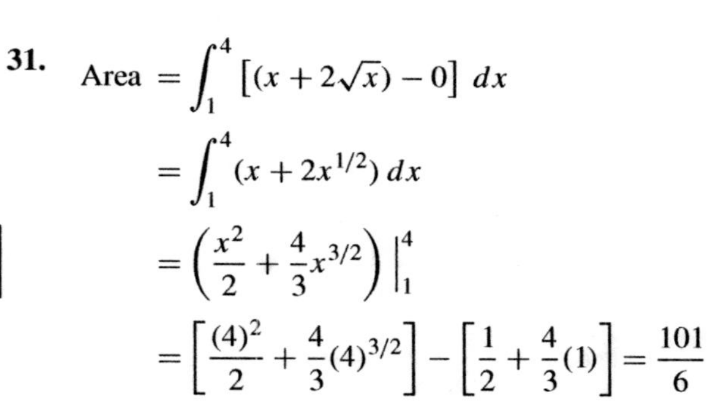

Area $= \int_1^4 \left[(x + 2\sqrt{x}) - 0\right]dx$

$= \int_1^4 (x + 2x^{1/2})\,dx$

$= \left(\frac{x^2}{2} + \frac{4}{3}x^{3/2}\right)\Big|_1^4$

$= \left[\frac{(4)^2}{2} + \frac{4}{3}(4)^{3/2}\right] - \left[\frac{1}{2} + \frac{4}{3}(1)\right] = \frac{101}{6}$

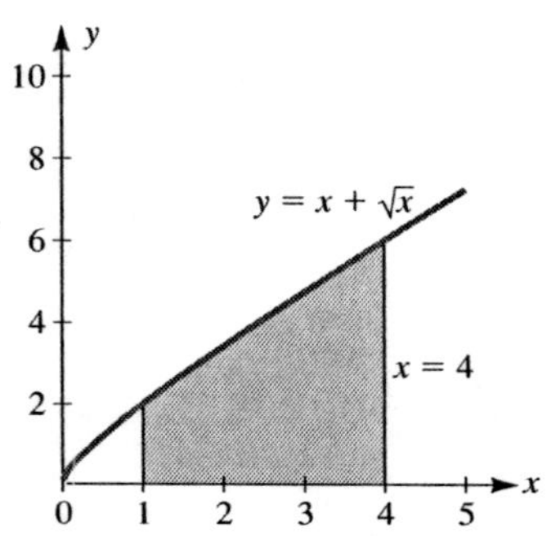

33. 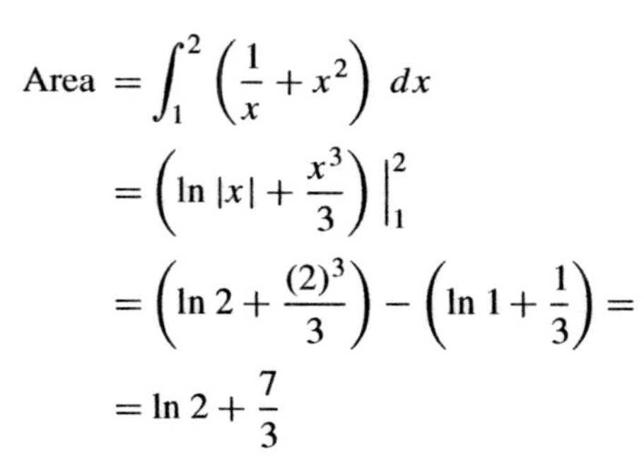

Area $= \int_1^2 \left(\frac{1}{x} + x^2\right)dx$

$= \left(\ln|x| + \frac{x^3}{3}\right)\Big|_1^2$

$= \left(\ln 2 + \frac{(2)^3}{3}\right) - \left(\ln 1 + \frac{1}{3}\right) =$

$= \ln 2 + \frac{7}{3}$

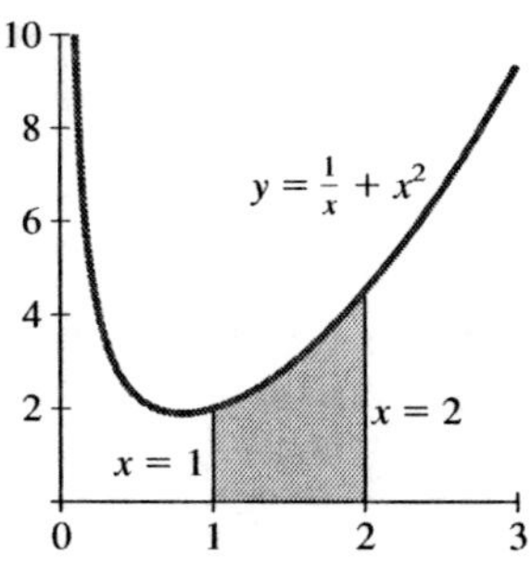

35. The limits of integration are

$$\frac{4}{x} = 5 - x$$

$$4 = 5x - x^2$$
$$x^2 - 5x + 4 = 0$$
$$(x-4)(x-1) = 0$$
$$x = 1, 4$$

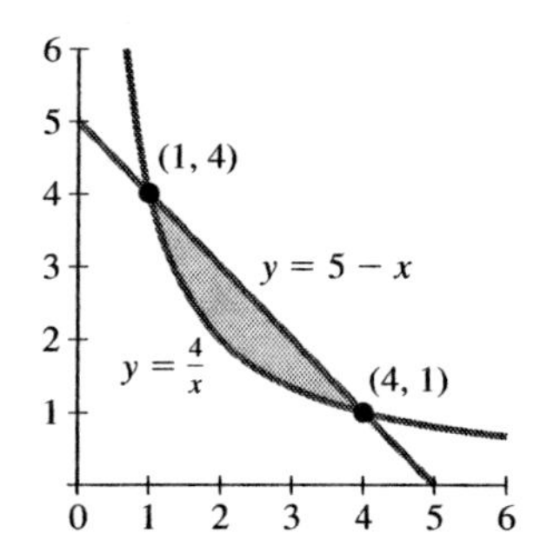

Noting that $y = 5 - x$ is the top curve,

$$\begin{aligned}
\text{Area} &= \int_1^4 \left[(5-x) - \left(\frac{4}{x}\right)\right] dx \\
&= \int_1^4 \left(5 - x - \frac{4}{x}\right) dx \\
&= \left(5x - \frac{x^2}{2} - 4 \ln |x|\right) \Big|_1^4
\end{aligned}$$

$$= \left[5(4) - \frac{(4)^2}{2} - 4 \ln 4\right] - \left[5(1) - \frac{1}{2} - 4 \ln 1\right]$$
$$= \frac{15}{2} - 4 \ln 4 = \frac{15}{2} - 4 \ln (2)^2$$
$$= \frac{15}{2} - 8 \ln 2$$

37. The graph of $y = 2 + x - x^2$ intersects $y = 0$ when $0 = 2 + x - x^2$

$$x^2 - x - 2 = 0$$
$$(x-2)(x+1) = 0$$
$$\text{or, } x = -1, 2$$

So, the limits of integration are $x = -1$ and $x = 2$

$$\begin{aligned}
\text{Area} &= \int_{-1}^2 \left[(2 + x - x^2) - 0\right] dx \\
&= \int_{-1}^2 (2 + x - x^2)\, dx \\
&= \left(2x + \frac{x}{2} - \frac{x^3}{3}\right) \Big|_{-1}^2
\end{aligned}$$

$$= \left[2(2) + \frac{2}{2} - \frac{(2)^3}{3}\right] - \left[2(-1) + \frac{-1}{2} - \frac{(-1)^3}{3}\right]$$
$$= \frac{9}{2}$$

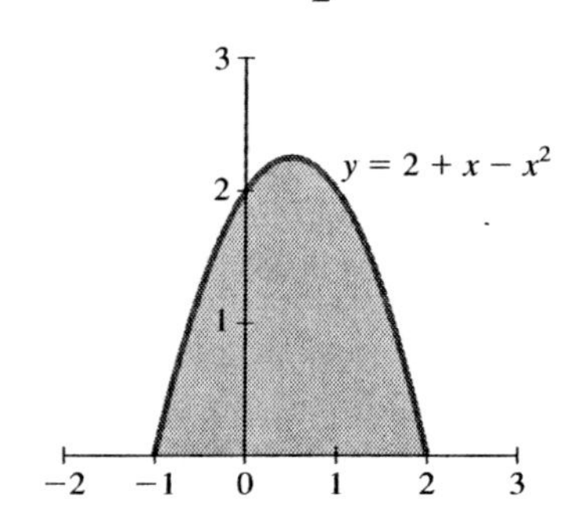

39.

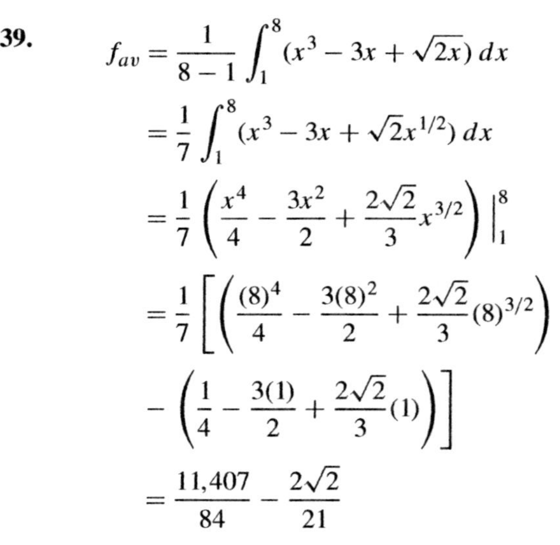

$$\begin{aligned}
f_{av} &= \frac{1}{8-1} \int_1^8 (x^3 - 3x + \sqrt{2x})\, dx \\
&= \frac{1}{7} \int_1^8 (x^3 - 3x + \sqrt{2}x^{1/2})\, dx \\
&= \frac{1}{7} \left(\frac{x^4}{4} - \frac{3x^2}{2} + \frac{2\sqrt{2}}{3} x^{3/2}\right) \Big|_1^8 \\
&= \frac{1}{7} \left[\left(\frac{(8)^4}{4} - \frac{3(8)^2}{2} + \frac{2\sqrt{2}}{3}(8)^{3/2}\right) \right. \\
&\quad \left. - \left(\frac{1}{4} - \frac{3(1)}{2} + \frac{2\sqrt{2}}{3}(1)\right)\right] \\
&= \frac{11{,}407}{84} - \frac{2\sqrt{2}}{21}
\end{aligned}$$

41. $g_{av} = \dfrac{1}{2-0} \displaystyle\int_0^2 ve^{-v^2} dv$

Let $u = -v^2$; then $du = -2v\,dv$, or $-\dfrac{1}{2}\, du = v\, dv$. Further, the limits of integration become 0 and

$-(2)^2 = -4$

$$= -\frac{1}{4}\int_0^{-4} e^u\,du$$
$$= \frac{1}{4}\int_{-4}^{0} e^u\,du$$
$$= \frac{1}{4}(e^u)\Big|_{-4}^{0} = \frac{1}{4}(e^0 - e^{-4})$$
$$= \frac{1}{4}\left(1 - \frac{1}{e^4}\right)$$

43. When $q_0 = 2$, $p_0 = 4[36 - (2)^2] = \$128$

$$CS = \int_0^2 4(36 - q^2)\,dq - 2(128)$$
$$= 4\left(36q - \frac{q^3}{3}\right)\Big|_0^2 - 256$$
$$= 4\left[\left(36(2) - \frac{(2)^3}{3}\right) - 0\right] - 256$$
$$= \frac{64}{3}, \text{ or approximately } \$21.33$$

45. When $q_0 = 4$, $p_0 = 10e^{-0.1(4)} \approx \$6.70$

$$CS = \int_0^4 10e^{-0.1q}\,dq - 4(6.70)$$
$$= 10\int_0^4 e^{-0.1q}\,dq - 26.80$$
$$= -100(e^{-0.1q})\Big|_0^4 - 26.80$$
$$= -100(e^{-0.1(4)} - e^0) - 26.80$$
$$\approx \$6.17$$

47.
$$GI = 2\int_0^1 (x - x^{3/2})\,dx$$
$$= 2\left(\frac{x^2}{2} - \frac{2}{5}x^{5/2}\right)\Big|_0^1$$
$$= 2\left[\left(\frac{1}{2} - \frac{2}{5}(1)\right) - 0\right] = \frac{1}{5}$$

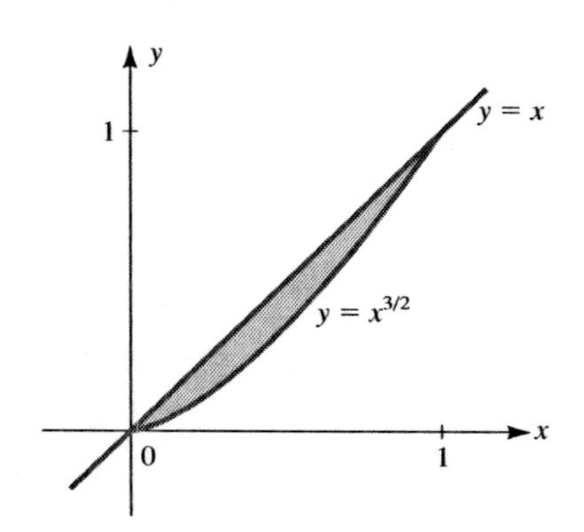

49.

$$GI = 2\int_0^1 \left[x - (0.3x^2 + 0.7x)\right]\,dx$$
$$= 2\int_0^1 (0.3x - 0.3x^2)\,dx$$
$$= 0.6\int_0^1 (x - x^2)\,dx$$
$$= 0.6\left(\frac{x^2}{2} - \frac{x^3}{3}\right)\Big|_0^1$$
$$= 0.6\left[\left(\frac{1}{2} - \frac{1}{3}\right) - 0\right] = 0.1$$

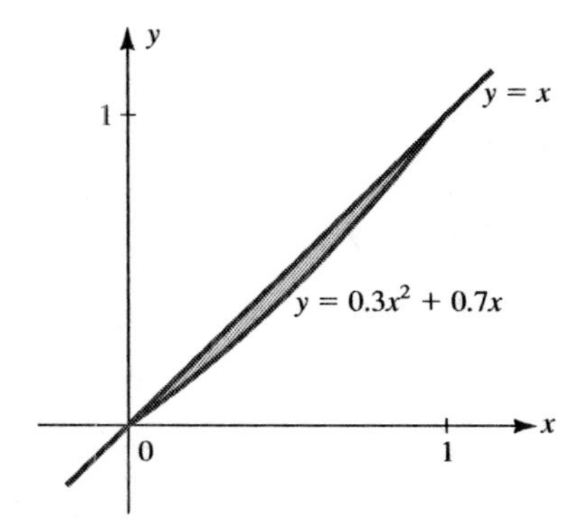

51. After 6 months, the number of the original population surviving is $75{,}000\,e^{-0.09(6)}$
The number of new members surviving is

$$\int_0^6 60e^{-0.09(6-t)}dt$$
$$= 60e^{-0.54}\int_0^6 e^{0.09t}dt$$

So, the total will be

$$75{,}000e^{-0.54} + 60e^{-0.54}\int_0^6 e^{0.09t}dt$$
$$= 60e^{-0.54}\left[1250 + \frac{1}{0.09}e^{0.09t}\Big|_0^6\right]$$
$$= 60e^{-0.54}\left[1250 + \frac{100}{9}\left(e^{0.09t}\Big|_0^6\right)\right]$$
$$= 60e^{-0.54}\left[1250 + \frac{100}{9}\left(e^{0.54} - 1\right)\right]$$
$$\approx 43{,}984 \text{ members}$$

53. After 10 years, the number of the original population surviving is 100,000 $e^{-0.2(10)}$
The number of new members surviving is

$$\int_0^{10}\left[90e^{0.1t}\right]\left[e^{-0.2(10-t)}\right]dt$$
$$= 90\int_0^{10} e^{0.1t-2+0.2t}dt$$
$$= 90e^{-2}\int_0^{10} e^{0.3t}dt$$

So, the total will be

$$100{,}000e^{-2} + 90e^{-2}\int_0^{10} e^{0.3t}dt$$
$$= 10e^{-2}\left[10{,}000 + 9\left(\frac{1}{0.3}e^{0.3t}\Big|_0^{10}\right)\right]$$
$$= 10e^{-2}\left[10{,}000 + 30\left(e^{0.3t}\Big|_0^{10}\right)\right]$$
$$= 100e^{-2}\left[1{,}000 + 3(e^3 - 1)\right]$$
$$\approx 14{,}308 \text{ members}$$

55. Volume of $S = \pi\int_{-1}^{2}(x^2+1)^2dx$

$$= \pi\int_{-1}^{2}(x^4 + 2x^2 + 1)dx$$
$$= \pi\left[\frac{x^5}{5} + \frac{2x^3}{3} + x\Big|_{-1}^{2}\right]$$
$$= \pi\left[\left(\frac{32}{5} + \frac{16}{3} + 2\right) - \left(-\frac{1}{5} - \frac{2}{3} - 1\right)\right]$$
$$= \pi\left[\frac{33}{5} + \frac{18}{3} + 3\right] = \frac{78}{5}\pi$$
$$= \pi \approx 49.01$$

57. Volume of $S = \pi\int_1^3\left(\frac{1}{\sqrt{x}}\right)^2dx$

$$= \pi\int_1^3 \frac{1}{x}dx$$
$$= \pi\left(\ln x\Big|_1^3\right)$$
$$= \pi\ (\ln 3 - \ln 1)$$
$$= \pi \ln 3 \approx 3.45$$

59.
$$y = \int \frac{dy}{dx}dx$$
$$= \int 2\,dx = 2x + C$$
$$4 = 2(-3) + C, \text{ or } C = 10$$
so, $y = 2x + 10$

61.
$$x = \int \frac{dx}{dt}dt$$
$$= \int e^{-2t}\,dt$$
$$= -\frac{1}{2}e^{-2t} + C$$
$$4 = -\frac{1}{2}e^0 + C, \text{ or } C = \frac{9}{2}$$
so, $x = \frac{1}{2}(9 - e^{-2t})$

63. Since slope $= \dfrac{dy}{dx}$,

$$y = \int x(x^2+1)^{-1}\,dx$$

let $u = x^2 + 1$; then $du = 2x\,dx$, or $\frac{1}{2}\,du = x\,dx$

$$= \frac{1}{2}\int \frac{1}{u}\,du$$

$$= \frac{1}{2}\ln\left|x^2+1\right| + C$$

Since the graph of y passes through the point (1,5)

$$5 = \frac{1}{2}\ln 2 + C, \text{ or } C = 5 - \frac{1}{2}\ln 2$$

$$\text{so, } y = \frac{1}{2}\ln(x^2+1) + 5 - \frac{1}{2}\ln 2$$

65. $V'(t) = 2[0.5t^2 + 4(t+1)^{-1}]$

$$\text{increase } = \int_0^6 \left[t^2 + \frac{8}{(t+1)}\right] dt$$

Let $u = t + 1$; then $du = dt$, and the limits of integration become $0 + 1 = 1$ and $6 + 1 = 7$

$$= \int_0^6 t^2\,dt + 8\int_1^7 \frac{1}{u}\,du$$

$$= \left(\frac{t^3}{3}\right)\Big|_0^6 + 8(\ln|u|)\Big|_1^7$$

$$= (72 - 0) + 8(\ln 7 - \ln 1)$$

$$= 72 + 8\ln 7 \approx \$87.57$$

67. Since $t = 1$ at 10:00 a.m., and $t = 3$ at noon, the number of people will be

$$\int_1^3 \left[-4(t+2)^3 + 54(t+2)^2\right] dt$$

Let $u = t + 2$; then $du = dt$, and the limits of integration become $1 + 2 = 3$ and $3 + 2 = 5$

$$= \int_3^5 (-4u^3 + 54u^2)\,du$$

$$= (-u^4 + 18u^3)\Big|_3^5$$

$$= \left[-(5)^4 + 18(5)^3\right] - \left[-(3)^4 + 18(3)^3\right]$$

$$= 1{,}220 \text{ people}$$

69.

$$C(x) = \int C'(x)\,dx$$

$$= \int (18x^2 + 500)\,dx$$

$$= 6x^3 + 500x + C$$

When $x = 0$, $C(0) = 8{,}000$ so $C = 8{,}000$, and

$$C(x) = 6x^3 + 500x + 8{,}000$$

So,

$$C(5) = 6(5)^3 + 500(5) + 8{,}000$$

$$= 11{,}250 \text{ commuters}$$

71.

$$D'(t) = \frac{1}{1+2t}$$

The amount of oil demanded during the year 2006 will be

$$D(t) = \int_1^2 \frac{1}{1+2t}\,dt$$

Using substitution with $u = 1 + 2t$, $\frac{1}{2}\,du = dt$ and the limits of integration become $1 + 2(1) = 3$ and $1 + 2(2) = 5$

$$= \frac{1}{2}\int_3^5 \frac{1}{u}\,du = \frac{1}{2}\left(\ln|u|\Big|_3^5\right)$$

$$= \frac{1}{2}(\ln 5 - \ln 4) \approx 0.2554 \text{ billion barrels}$$

Similarly, the amount of oil demanded during the year 2009 will be

$$= \int_4^5 \frac{1}{1+2t}\,dt$$

$$= \frac{1}{2}\int_9^{11} \frac{1}{u}\,du = \frac{1}{2}\left(\ln|u|\Big|_9^{11}\right)$$

$$= \frac{1}{2}(\ln 11 - \ln 9) \approx 0.1003 \text{ billion barrels}$$

So, more oil will be demanded in 2006.

73.

$$\begin{aligned} FV &= \int_0^5 1{,}200e^{0.08(5-t)}\,dt \\ &= 1{,}200e^{0.4}\int_0^5 e^{-0.08t}\,dt \\ &= \frac{1{,}200e^{0.4}}{-0.08}(e^{-0.08t})\Big|_0^5 \\ &= -15{,}000e^{0.4}(e^{-0.4}-e^0) \\ &\approx \$7{,}377.37 \end{aligned}$$

75. The number of the original houses still on the market after 10 weeks is

$$200e^{-0.2(10)}$$

The number of new listings which will still be on the market after 10 weeks is

$$\int_0^{10} 8e^{-0.2(10-t)}\,dt$$

So, the total will be

$$\begin{aligned} &200e^{-2} + 8e^{-2}\int_0^{10} e^{0.2t}\,dt \\ &= 8e^{-2}\left(25 + \int_0^{10} e^{0.2t}\,dt\right) \\ &= 8e^{-2}\left[25 + \frac{1}{0.2}(e^{0.2t})\Big|_0^{10}\right] \\ &= 8e^{-2}\left[25 + 5(e^2 - e^0)\right] \\ &\approx 62 \text{ houses.} \end{aligned}$$

77. The decay function is of the form

$$Q(t) = Q_0e^{-kt}$$

Since the half-life is 35 years,

$$\begin{aligned} \frac{Q_0}{2} &= Q_0e^{-k(35)} \\ \ln\frac{1}{2} &= \ln e^{-35k} \\ k &= \frac{\ln\frac{1}{2}}{-35} = \frac{-\ln\frac{1}{2}}{35} = \frac{\ln 2}{35} \\ &\approx 0.0198 \end{aligned}$$

The amount remaining

$$\begin{aligned} &= \int_0^{200} 300e^{-0.0198(200-t)}\,dt \\ &= 300e^{-3.96}\int_0^{200} e^{0.0198t}\,dt \\ &= 300^{-3.96}\left(\frac{1}{0.0198}e^{0.0198t}\right)\Big|_0^{200} \\ &= \frac{300e^{-3.96}}{0.0198}(e^{3.96} - e^0) \\ &\approx 14{,}863 \text{ pounds} \end{aligned}$$

79. Rate revenue changes = (#barrels) (rate selling price changes)

$$R'(t) = 900(16 + 0.08t)$$

Since time is measured in months,

$$\begin{aligned} \text{revenue} &= \int_0^{36} 900(16+0.08t)\,dt \\ &= 900(16t + 0.04t^2)\Big|_0^{36} \\ &= 900\left[(16(36) + 0.04(36)^2) - 0\right] \\ &\approx \$565{,}056 \end{aligned}$$

81.

$$\begin{aligned} P_{av} &= \frac{1}{6-0}\int_0^6 (0.06t^2 - 0.2t + 1.2)\,dt \\ &= \frac{1}{6}(0.02t^3 - 0.1t^2 + 1.2t)\Big|_0^6 \\ &= \frac{1}{6}\left[\left(0.02(6)^3 - 0.1(6)^2 + 1.2(6)\right) - 0\right] \\ &= \$1.32 \text{ per pound} \end{aligned}$$

83. At 8:00 a.m., $t = 8$ and at 8:00 p.m., $t = 20$ so the change in temperature will be

$$\int_8^{20} -0.02(t-7)(t-14)\,dt$$

$$= -0.02\int_8^{20}(t^2 - 21t + 98)\,dt$$
$$= -0.02\left(\frac{t^3}{3} - \frac{21t^2}{2} + 98t\right)\Big|_8^{20}$$
$$= -0.02\left[\left(\frac{(20)^3}{3} - \frac{21(20)^2}{2} + 98(20)\right) - \left(\frac{(8)^3}{3} - \frac{21(8)^2}{2} + 98(8)\right)\right]$$
≈ -2.88, or a decrease of approximately 2.88°C

85. (a)

$$p(x) = \int p'(x)\,dx = \int (0.2 + 0.003x^2)\,dx = 0.2x + 0.001x^3 + C$$

When $x = 0$, $p(0) = 250$ cents, so $C = 250$ and $p(x) = 0.2x + 0.001x^3 + 250$.
Press [y=] and input $p(x)$ for $y_1 =$.
Use window dimensions [0, 50]10 by [240, 340]20.
Press [graph].
Use the value function under the calc menu and input $x = 10$ to find the price of eggs 10 weeks from now is 253 cents or $2.53.

(b)

$$p(x) = \int (0.3 + 0.003x^2)\,dx = 0.3x + 0.001x^3 + C = 0.3x + 0.001x^3 + 250$$

Press [y=] and input $p_2(x)$ for $y_2 =$.
Use window dimensions [0, 50]10 by [240, 340]20.
Press [graph].
Use the value function under the calc menu and input $x = 10$. Verify that $p_2(x) = 0.3x + 0.001x^3 + 250$ is displayed on the upper left corner. $P_2(10) = 254$ cents or $2.54.

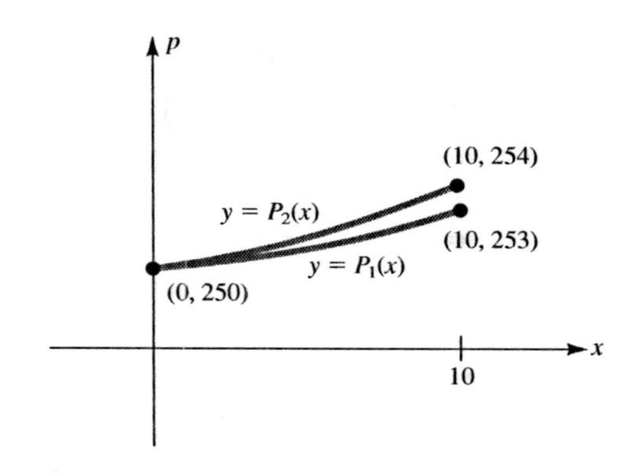

87. Let $s(t)$ be the distance traveled, in meters, after time t, in minutes. Then,

$$s(t) = \int_2^3 v(t)\,dt$$
$$\int_2^3 (1 + 4t + 3t^2)\,dt$$
$$= (t + 2t^2 + t^3)\Big|_2^3$$
$$= \left[3 + 2(3)^2 + (3)^3\right] - \left[2 + 2(2)^2 + (2)^3\right]$$
$$= 30 \text{ meters}$$

89.

$$GI_{sw} = 2\int_0^1 (x - x^{1.6})\,dx = 2\left(\frac{x^2}{2} - \frac{x^{2.6}}{2.6}\right)\Big|_0^1 = 2\left[\left(\frac{1}{2} - \frac{1}{2.6}\right) - 0\right] \approx 0.2308$$

$$GI_{PT} = 2\int_0^1 \left[x - (0.65x^2 + 0.35x)\right]dx$$
$$= 2\int_0^1 (0.65x - 0.65x^2)\,dx$$
$$= 1.3\int_0^1 (x - x^2)\,dx$$
$$= 1.3\left(\frac{x^2}{2} - \frac{x^3}{3}\right)\Big|_0^1$$
$$= 1.3\left[\left(\frac{1}{2} - \frac{1}{3}\right) - 0\right] \approx 0.2167$$

So, income is more equitably distributed for physical therapists.

91. $2x^2 + 3y^2 = 6$ The equation for the bottom half of the curve is

$$y = -\sqrt{\frac{6-2x^2}{3}}$$

The volume, in cubic miles, of the lake is *half* the volume of the solid generated by this curve. Since when $y = 0$, $x = \pm\sqrt{3}$, want

$$\frac{\pi}{2}\int_{-\sqrt{3}}^{\sqrt{3}} \left(-\sqrt{\frac{6-2x^2}{3}}\right)^2 dx$$

$$= \frac{\pi}{2}\int_{-\sqrt{3}}^{\sqrt{3}} \left(2 - \frac{2}{3}x^2\right) dx$$

$$= \frac{\pi}{2}\left[2x - \frac{2x^3}{9}\Big|_{-\sqrt{3}}^{\sqrt{3}}\right]$$

$$= \frac{\pi}{2}\left[\left(2\sqrt{3} - \frac{2(\sqrt{3})^3}{9}\right) - \left(-2\sqrt{3} + \frac{2(\sqrt{3})^3}{9}\right)\right]$$

$$= \frac{\pi}{2}\left[4\sqrt{3} - \frac{4(3^{3/2})}{9}\right] \approx 7.255$$

To have 1,000 trout per cubic mile, need $1{,}000(7.255) = 7{,}255$ trout.
So, need an additional 2,255 trout.

93. **(a)** $S_{av} = \dfrac{1}{N-0}\displaystyle\int_0^N S(t)\,dt$

(b) Since velocity is the derivative of distance,

$$D(t) = \int_0^N S(t)\,dt$$

(c) Average speed $= \dfrac{\text{distance traveled}}{\text{time elapsed}}$

95. Press [y=] and input $(x-2)/(x+1)$ for $y_1 =$ and input $\sqrt{}(25 - x^2)$ for $y_2 =$.
Use window dimensions $[-5, 5]1$ by $[-1, 6]1$.
Press [graph].

Use trace and zoom-in to find the points of intersection are $(-4.66, 1.82)$, $(-1.82, 4.66)$, and $(4.98, 0.498)$.
An alternative to using trace and zoom is to use the intersect function under the calc menu. Enter a value close to the point of intersection on $y_1 =$ and enter a value close to the same point of intersection on $y_2 =$ and finally, enter a guess for the point of intersection. Repeat this process for the other two points of intersections.
The curves are bounded by the points of intersection given by $x = -4.66$ and $x = -1.82$. To find the area bounded by the curves, we need to find

$$\int_{-4.66}^{-1.82} (y_2 - y_1)\,dx = \int_{-4.66}^{-1.82} y_2\,dx - \int_{-4.66}^{-1.82} y_1\,dx$$

For each separate integral, use the $\int f(x)\,dx$ function under the calc menu making sure that the correct y equation is displayed in the upper left corner. We find the area to be

$$10.326439 - 7.32277423 \approx 3$$

An alternative to finding each separate integral is to use the fnInt function from the home screen. Select fnInt function from the math menu and enter fnInt $(y_2 - y_1, x, -4.66, -1.82)$. You input the y equations by pressing [vars] and selecting which y equation you want from the function window under y-vars.

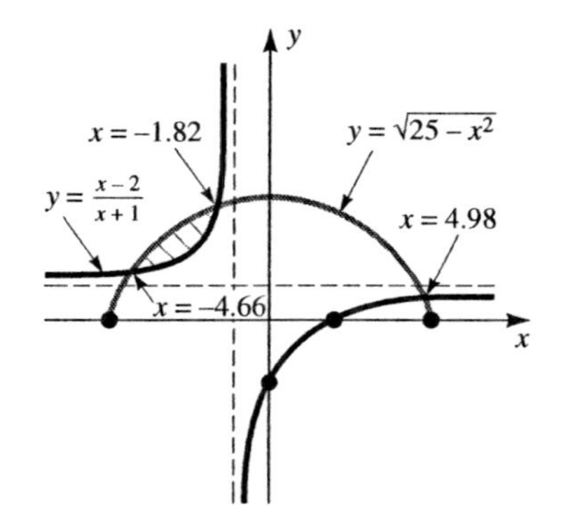

Chapter 6

Additional Topics in Integration

6.1 Integration by Parts; Integral Tables

1. Both terms are easy to integrate; however, the derivative of x becomes simpler while the derivative of e^{-x} does not. So,

$$u = x \quad \text{and} \quad dV = e^{-x}\,dx$$
$$du = dx \qquad V = -e^{-x}$$

and

$$\begin{aligned}\int xe^{-x}\,dx &= -xe^{-x} - \int -e^{-x}\,dx\\ &= -xe^{-x} + \int e^{-x}\,dx\\ &= -xe^{-x} - e^{-x} + C\\ &= -(x+1)e^{-x} + C\end{aligned}$$

3. Both terms are easy to integrate; however, the derivative of $1 - x$ becomes simpler while the derivative of e^x does not. So,

$$u = 1 - x \qquad dV = e^x\,dx$$
$$du = -dx \qquad V = e^x$$

and

$$\begin{aligned}\int (1-x)e^x\,dx &= (1-x)e^x - \int e^x - dx\\ &= (1-x)e^x + \int e^x\,dx\\ &= (1-x)e^x + e^x + C\\ &= [(1-x)+1]e^x + C\\ &= (2-x)e^x + C\end{aligned}$$

5. $\ln 2t$ cannot be easily integrated. So,

$$u = \ln 2t \quad \text{and} \quad dV = t\,dt$$
$$du = \frac{1}{2t}\cdot 2dt \qquad V = \frac{t^2}{2}$$
$$= \frac{1}{t}\,dt$$

and

$$\begin{aligned}\int t\ln 2t\,dt &= \frac{t^2}{2}\ln 2t - \int \frac{t^2}{2}\cdot\frac{1}{t}\,dt\\ &= \frac{t^2}{2}\ln 2t - \frac{1}{2}\int t\,dt\\ &= \frac{t^2}{2}\ln 2t - \frac{1}{4}t^2 + C\\ &= \frac{t^2}{2}\left(\ln 2t - \frac{1}{2}\right) + C\end{aligned}$$

7. Both terms are easy to integrage; however, the derivative of v becomes simpler while the derivative of $e^{-v/5}$ does not. So,

$$u = v \quad \text{and} \quad dV = e^{-v/5}dv$$
$$du = dv \qquad V = -5e^{-v/5}$$

and

$$\begin{aligned}\int ve^{-v/5}dv &= -5ve^{-v/5} - \int -5e^{-v/5}dv\\ &= -5ve^{-v/5} + 5\int e^{-v/5}dv\\ &= -5ve^{-v/5} - 25e^{-v/5} + C\\ &= -5(v+5)e^{-v/5} + C\end{aligned}$$

9. Both terms are easy to integrate; however, the derivative of x becomes simpler while the derivative of $\sqrt{x-6}$ does not. So,

$$u = x \quad \text{and} \quad dV = (x-6)^{1/2}\,dx$$
$$du = dx \quad V = \frac{2}{3}(x-6)^{3/2}$$

and

$$\begin{aligned}\int x\sqrt{x-6}\,dx &= \frac{2}{3}x(x-6)^{3/2} - \int \frac{2}{3}(x-6)^{3/2}\,dx \\ &= \frac{2}{3}x(x-6)^{3/2} - \frac{2}{3}\int (x-6)^{3/2}\,dx \\ &= \frac{2}{3}x(x-6)^{3/2} - \frac{4}{15}(x-6)^{5/2} + C\end{aligned}$$

11. Both terms are easy to integrate; however, the derivative of x becomes simpler while the derivative of $(x+1)^8$ does not. So,

$$u = x \quad \text{and} \quad dV = (x+1)^8\,dx$$
$$du = dx \quad V = \frac{1}{9}(x+1)^9$$

and

$$\begin{aligned}\int x(x+1)^8\,dx &= \frac{1}{9}x(x+1)^9 - \int \frac{1}{9}(x+1)^9\,dx \\ &= \frac{1}{9}x(x+1)^9 - \frac{1}{9}\int (x+1)^9\,dx \\ &= \frac{1}{9}x(x+1)^9 - \frac{1}{90}(x+1)^{10} + C\end{aligned}$$

13. Rewriting, $\int \frac{x}{\sqrt{x+2}}\,dx = \int x(x+2)^{-1/2}\,dx$, both terms are easy to integrate; however, the derivative of x becomes simpler while the derivative of $(x+2)^{-1/2}$ does not. So,

$$u = x \quad \text{and} \quad dV = (x+2)^{-1/2}\,dx$$
$$du = dx \quad V = 2(x+2)^{1/2}$$

and

$$\begin{aligned}\int \frac{x}{\sqrt{x+2}}\,dx &= 2x(x+2)^{1/2} - \int 2(x+2)^{1/2}\,dx \\ &= 2x(x+2)^{1/2} - 2\int (x+2)^{1/2}\,dx \\ &= 2x\sqrt{x+2} - \frac{4}{3}(x+2)^{3/2} + C\end{aligned}$$

15. Rewriting, $\int_{-1}^{4} \frac{x}{\sqrt{x+5}}\,dx = \int_{-1}^{4} x(x+5)^{-1/2}\,dx$, both terms are easy to integrate; however, the derivative of x becomes simpler while the derivative of $(x+5)^{-1/2}$ does not. So,

$$u = x \quad \text{and} \quad dV = (x+5)^{-1/2}\,dx$$
$$du = dx \quad V = 2(x+5)^{1/2}$$

and

$$\begin{aligned}\int_{-1}^{4} \frac{x}{\sqrt{x+5}}\,dx &= 2x(x+5)^{1/2}\Big|_{-1}^{4} - \int_{-1}^{4} 2(x+5)^{1/2}\,dx \\ &= 2x(x+5)^{1/2}\Big|_{-1}^{4} - 2\int_{-1}^{4} (x+5)^{1/2}\,dx \\ &= \left[2x\sqrt{x+5} - \frac{4}{3}(x+5)^{3/2}\right]\Big|_{-1}^{4} \\ &= \left[2(4)\sqrt{4+5} - \frac{4}{3}(4+5)^{3/2}\right] \\ &\quad - \left[2(-1)\sqrt{-1+5} - \frac{4}{3}(-1+5)^{3/2}\right] \\ &= \frac{8}{3}\end{aligned}$$

17. Rewriting, $\int_0^1 \frac{x}{e^{2x}}\,dx = \int_0^1 xe^{-2x}\,dx$, both terms are easy to integrate; however, the derivative of x becomes simpler while the derivative of e^{-2x} does not. So,

$$u = x \quad \text{and} \quad dV = e^{-2x}\,dx$$
$$du = dx \quad V = -\frac{1}{2}e^{-2x}$$

and

$$\int_0^1 \frac{x}{e^{2x}}\,dx = -\frac{x}{2}e^{-2x}\Big|_0^1 - \int_0^1 -\frac{1}{2}e^{-2x}\,dx$$
$$= -\frac{x}{2}e^{-2x}\Big|_0^1 + \frac{1}{2}\int_0^1 e^{-2x}dx$$
$$= \left[-\frac{x}{2}e^{-2x} - \frac{1}{4}e^{-2x}\right]\Big|_0^1$$
$$= \left[-\frac{1}{2}e^{-2} - \frac{1}{4}e^{-2}\right] - \left[0 - \frac{1}{4}e^{0}\right]$$
$$= -\frac{3}{4}e^{-2} + \frac{1}{4} = \frac{1}{4}(1 - 3e^{-2})$$

19. $\ln \sqrt[3]{x}$ cannot be easily integrated. So,

$$u = \ln \sqrt[3]{x} \quad \text{and} \quad dV = x\,dx$$
$$= \ln(x)^{1/3} \qquad V = \frac{x^2}{2}$$
$$= \frac{1}{3}\ln x$$
$$du = \frac{1}{3x}\,dx$$

and

$$\int_1^{e^2} x \ln \sqrt[3]{x}\,dx = \frac{x^2}{6}\ln x\Big|_1^{e^2} - \int_1^{e^2} \frac{x^2}{2}\cdot\frac{1}{3x}\,dx$$
$$= \frac{x^2}{6}\ln x\Big|_1^{e^2} - \frac{1}{6}\int_1^{e^2} x\,dx$$
$$= \left(\frac{x^2}{6}\ln x - \frac{x^2}{12}\right)\Big|_1^{e^2}$$
$$= \left[\frac{(e^2)^2}{6}\ln(e^2) - \frac{(e^2)^2}{12}\right] - \left[\frac{1}{6}\ln 1 - \frac{1}{12}\right]$$
$$= \frac{1}{12}\left(3e^4 + 1\right)$$

21. $\ln 2t$ cannot be easily integrated. So,

$$u = \ln 2t \quad \text{and} \quad dV = t\,dt$$
$$du = \frac{1}{2t}\cdot 2dt \qquad V = \frac{t^2}{2}$$
$$= \frac{1}{t}\,dt$$

and

$$\int_{1/2}^{e/2} t\ln 2t\,dt = \frac{t^2}{2}\ln 2t\Big|_{1/2}^{e/2} - \int_{1/2}^{e/2} \frac{t^2}{2}\cdot\frac{1}{t}\,dt$$
$$= \frac{t^2}{2}\ln 2t\Big|_{1/2}^{e/2} - \frac{1}{2}\int_{1/2}^{e/2} t\,dt$$
$$= \left(\frac{t^2}{2}\ln 2t - \frac{t^2}{4}\right)\Big|_{1/2}^{e/2}$$
$$= \left[\frac{\left(\frac{e}{2}\right)^2}{2}\ln 2\left(\frac{e}{2}\right) - \frac{\left(\frac{e}{2}\right)^2}{4}\right] - \left[\frac{\left(\frac{1}{2}\right)^2}{2}\ln 2\left(\frac{1}{2}\right) - \frac{\left(\frac{1}{2}\right)^2}{4}\right]$$
$$= \frac{1}{16}(e^2 + 1)$$

23. Rewriting, $\int \frac{\ln x}{x^2}\,dx = \int x^{-2}\ln x\,dx$, $\ln x$ cannot be easily integrated. So,

$$u = \ln x \quad \text{and} \quad dV = x^{-2}\,dx$$
$$du = \frac{1}{x}\,dx \qquad V = -\frac{1}{x}$$

and

$$\int \frac{\ln x}{x^2}\,dx = -\frac{1}{x}\ln x - \int -\frac{1}{x}\cdot\frac{1}{x}\,dx$$
$$= -\frac{1}{x}\ln x + \int x^{-2}\,dx$$
$$= -\frac{1}{x}\ln x - \frac{1}{x} + C$$
$$= -\frac{1}{x}(\ln x + 1) + C$$

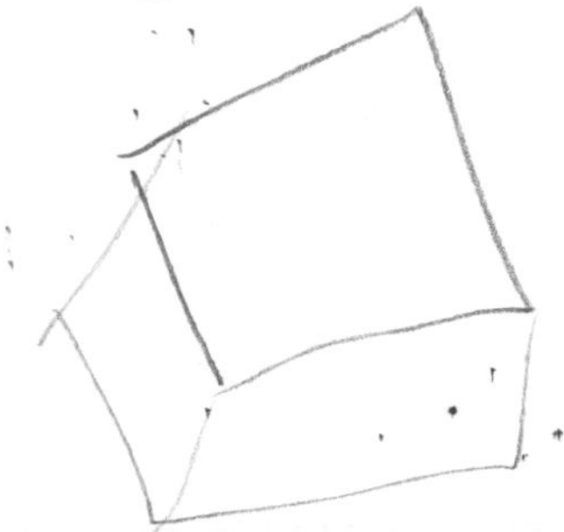

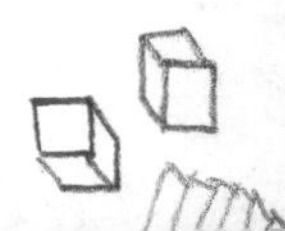

25. Using the hint,

$$u = x^2 \quad \text{and} \quad dV = xe^{x^2}\,dx$$
$$du = 2x\,dx \quad \text{let } u = x^2; \frac{1}{2}\,du = x\,dx$$
$$V = \frac{1}{2}e^{x^2}$$

and

$$\int x^3 e^{x^2} dx = \frac{x^2}{2}e^{x^2} - \int \frac{1}{2}e^{x^2} \cdot 2x\,dx$$
$$= \frac{x^2}{2}e^{x^2} - \frac{1}{2}e^{x^2} + C$$
$$= \frac{1}{2}e^{x^2}(x^2 - 1) + C$$

27. Rewriting, $\int \frac{x\,dx}{3 - 5x} = \int \frac{x\,dx}{3 + -5x}$ which is of the form $\int \frac{u\,du}{a + bu}$ (formula #1). Using $u = x, du = dx$, $a = 3$, and $b = -5$, the formula yields

$$\int \frac{x\,dx}{3 - 5x} = \frac{1}{(-5)^2}\,[3 + -5x - 3\ln|\,3 + -5x\,|] + C$$
$$= \frac{1}{25}\,(3 - 5x - 3\ln|\,3 - 5x\,|) + C$$

29. Rewriting, $\int \frac{\sqrt{4x^2 - 9}}{x^2}\,dx = \int \frac{\sqrt{(2x)^2 - (3)^2}}{x^2}\,dx$ most closely resembles $\int \frac{\sqrt{u^2 - a^2}}{u^2}\,du$ (formula #19). Now,

$$\int \frac{\sqrt{(2x)^2 - (3)^2}}{x^2}\,dx = \int \frac{4\sqrt{(2x)^2 - (3)^2}}{4x^2}\,dx$$
$$= 2\int \frac{\sqrt{(2x)^2 - (3)^2}}{(2x)^2}\,2\,dx$$

and formula #19 can be used with $u = 2x, du = 2\,dx$, and $a = 3$. So,

$$\int \frac{\sqrt{4x^2 - 9}}{x^2}\,dx$$
$$= 2\left[\frac{-\sqrt{4x^2 - 9}}{2x} + \ln\left|2x + \sqrt{4x^2 - 9}\right|\right] + C$$
$$= \frac{-\sqrt{4x^2 - 9}}{x} + 2\ln\left|2x + \sqrt{4x^2 - 9}\right| + C$$

31. As written, $\int \frac{dx}{x(2 + 3x)}$ is of the form $\int \frac{du}{u(a + bu)}$ (formula #6). Using $u = x,\ du = dx,\ a = 2$, and $b = 3$, the formula yields $\int \frac{dx}{x(2 + 3x)} = \frac{1}{2}\ln\left|\frac{x}{2 + 3x}\right| + C.$

33. Rewriting, $\int \frac{du}{16 - 3u^2} = \int \frac{du}{3\left(\frac{16}{3} - u^2\right)} = \frac{1}{3}\int \frac{du}{\frac{16}{3} - u^2} = \frac{1}{3}\int \frac{du}{\left(\frac{4}{\sqrt{3}}\right)^2 - u^2}$ which is of the form $\int \frac{du}{a^2 - u^2}$ (formula #16). Using $a = \frac{4}{\sqrt{3}}$, the formula yields

$$\int \frac{du}{16 - 3u^2} = \frac{1}{3}\left[\frac{1}{2\left(\frac{4}{\sqrt{3}}\right)}\ln\left|\frac{\frac{4}{\sqrt{3}} + u}{\frac{4}{\sqrt{3}} - u}\right|\right] + C$$
$$= \frac{\sqrt{3}}{24}\ln\left|\frac{\frac{4+\sqrt{3}u}{\sqrt{3}}}{\frac{4-\sqrt{3}u}{\sqrt{3}}}\right| + C$$
$$= \frac{\sqrt{3}}{24}\ln\left|\frac{4 + \sqrt{3}u}{4 - \sqrt{3}u}\right| + C$$

35. $\int (\ln x)^3\,dx$ is of the form $\int (\ln u)^n\,du$ (formula #27). Using $u = x$, the formula yields

$$\int (\ln x)^3\,dx = x(\ln x)^3 - 3\int (\ln x)^2\,dx$$

Applying the formula again to the last term

$$= x(\ln x)^3 - 3\left[x(\ln x)^2 - 2\int \ln x\,dx\right]$$
$$= x(\ln x)^3 - 3x(\ln x)^2 + 6\int \ln x\,dx$$

Applying the formula one more time (or using formula #23),

$$= x(\ln x)^3 - 3x(\ln x)^2 + 6[x\ln x - x] + C$$
$$= x(\ln x)^3 - 3x(\ln x)^2 + 6x\ln x - 6x + C$$

37. $\int \frac{dx}{x^2(5+2x)^2}$ is of the form $\int \frac{du}{u^2(a+bu)^2}$ (formula #8). Using $u = x$, $du = dx$, $a = 5$, and $b = 2$, the formula yields

$$\int \frac{dx}{x^2(5+2x)^2}$$
$$= -\frac{1}{25}\left[\frac{5+4x}{x(5+2x)} + \frac{4}{5}\ln\left|\frac{x}{5+2x}\right|\right] + C$$

39.

$$\text{Slope} = y' = (x+1)e^{-x}$$
$$y = \int y'\,dx$$
$$= \int (x+1)e^{-x}\,dx$$
$$u = x+1 \quad \text{and} \quad dV = e^{-x}\,dx$$
$$du = dx \qquad V = -e^{-x}$$
$$y = -(x+1)e^{-x} - \int -e^{-x}\,dx$$
$$= -(x+1)e^{-x} + \int e^{-x}\,dx$$
$$= -(x+1)e^{-x} - e^{-x} + C$$

Since the graph of y passes through the point (1,5), $5 = -(1+1)e^{-1} - e^{-1} + C$ or, $C = 5 + \frac{3}{e}$. So,

$$y = -(x+1)e^{-x} - e^{-x} + 5 + \frac{3}{e}$$
$$= 5 + \frac{3}{e} - \frac{x+2}{e^x}$$

41.

$$\text{Velocity} = \text{derivative of distance}$$
$$s(t) = \int v(t)\,dt$$
$$= \int te^{-t/2}\,dt$$
$$u = t \quad \text{and} \quad dV = e^{-t/2}\,dt$$
$$du = dt \qquad = -2e^{-t/2}$$
$$s(t) = -2te^{-t/2} - \int -2e^{-t/2}\,dt$$
$$= -2te^{-t/2} + 2\int e^{-t/2}\,dt$$
$$= -2te^{-t/2} - 4e^{-t/2} + C$$

Assuming $s(0) = 0$, $0 = 0 - 4e^0 + C$, or $C = 4$. So, $s(t) = -2e^{-t/2}(t+2) + 4$.

43.

$$Q(t) = \int Q'(t)\,dt$$
$$= \int_0^5 2000te^{-0.2t}\,dt$$
$$= 2000\int_0^5 te^{-0.2t}\,dt$$
$$u = t \quad \text{and} \quad dV = e^{-0.2t}\,dt$$
$$du = dt \qquad V = -5e^{-0.2t}$$
$$Q(t) = 2000\left[-5te^{-0.2t}\Big|_0^5 - \int_0^5 -5e^{-0.2t}\,dt\right]$$
$$= 2000\left[-5te^{-0.2t}\Big|_0^5 + 5\int_0^5 e^{-0.2t}\,dt\right]$$
$$= 2000\left[-5te^{-0.2t} - 25e^{-0.2t}\right]\Big|_0^5$$
$$= 2000\left[\left(-25e^{-1} - 25e^{-1}\right) - \left(0 - 25e^0\right)\right]$$
$$= 2000\left(\frac{-50}{e} + 25\right) \approx \$13{,}212.06$$

45.

$$P(t) = \int P'(t)\,dt$$
$$= \int t\ln\sqrt{t+1}\,dt$$

$$u = \ln\sqrt{t+1} \quad \text{and} \quad dV = t\,dt$$
$$= \ln(t+1)^{1/2} \quad \text{and} \quad dV = \frac{t^2}{2}$$
$$= \frac{1}{2}\ln(t+1)$$
$$du = \frac{1}{2(t+1)}\,dt$$
$$P(t) = \frac{t^2}{4}\ln(t+1) - \int \frac{t^2}{2}\cdot\frac{1}{2(t+1)}\,dt$$
$$= \frac{t^2}{4}\ln(t+1) - \frac{1}{4}\int \frac{t^2}{t+1}\,dt$$

Rewriting,

$$\int \frac{t^2}{t+1}\,dt = \int \frac{1+t^2-1}{t+1}\,dt$$
$$= \int \frac{1+(t+1)(t-1)}{t+1}\,dt$$
$$= \int \frac{1}{t+1}\,dt + \int (t-1)\,dt$$

So,

$$P(t) = \frac{t^2}{4}\ln(t+1) - \frac{1}{4}\left[\ln|t+1| + \frac{(t-1)^2}{2}\right] + C$$
$$= \frac{t^2}{4}\ln(t+1) - \frac{1}{4}\ln|t+1| - \frac{(t-1)^2}{8} + C$$

When $t = 0$, $P(0) = 2000$ thousand, so

$$2000 = 0 - \frac{1}{4}\ln 1 - \frac{1}{8} + C,$$
$$\text{or } C = 2000.125$$

So,

$$P(t) = \frac{t^2}{4}\ln(t+1) - \frac{1}{4}\ln|t+1| - \frac{(t-1)^2}{8} + 2000.125$$

and when $t = 5$,

$$P(5) = \frac{25}{4}\ln 6 - \frac{1}{4}\ln 6 - 2 + 2000.125$$
$$= 6\ln 6 + 1998.125 \approx 2{,}008.8756 \text{ thousand.}$$

The population will be approximately 2,008,876 people.

47.

$$C_{av} = \frac{1}{6-0}\int_0^6 4te^{(2-0.3t)}\,dt$$
$$= \frac{2}{3}e^2\int_0^6 te^{-0.3t}\,dt$$
$$u = t \quad \text{and} \quad dV = e^{-0.3t}\,dt$$
$$du = dt \qquad V = -\frac{10}{3}e^{-0.3t}$$

So,

$$C_{av} = \frac{2}{3}e^2\left[-\frac{10}{3}te^{-0.3t}\Big|_0^6 - \int_0^6 -\frac{10}{3}e^{-0.3t}\,dt\right]$$
$$= \frac{2}{3}e^2\left[-\frac{10}{3}te^{-0.3t}\Big|_0^6 + \frac{10}{3}\int_0^6 e^{-0.3t}\,dt\right]$$
$$= \frac{2}{3}e^2\left[-\frac{10}{3}te^{-0.3t} - \frac{100}{9}e^{-0.3t}\right]\Big|_0^6$$
$$= \frac{2}{3}e^2\left[\left(-\frac{10}{3}(6)e^{-1.8} - \frac{100}{9}e^{-1.8}\right) - \left(0 - \frac{100}{9}e^0\right)\right]$$
$$= \frac{2}{3}e^2\left(-\frac{280}{9}e^{-1.8} + \frac{100}{9}\right) \approx 29.4 \text{ mg/ml}$$

49.

$$FV = \int_0^{10} (3{,}000 + 5t)e^{0.05(10-t)}\,dt$$
$$= e^{0.5}\int_0^{10} (3{,}000 + 5t)e^{-0.05t}\,dt$$
$$u = 3{,}000 + 5t \quad \text{and} \quad dV = e^{-0.05t}\,dt$$
$$du = 5\,dt \qquad V = -20e^{-0.05t}$$

So,

$$FV = e^{0.5}\left[-20(3{,}000+5t)e^{-0.05t}\Big|_0^{10} - \int_0^{10} -20e^{-0.05t}\cdot 5\,dt\right]$$
$$= e^{0.5}\left[-20(3{,}000+5t)e^{-0.05t}\Big|_0^{10} + 100\int_0^{10} e^{-0.05t}\,dt\right]$$

$$= e^{0.5}\left[-20(3{,}000+5t)e^{-0.05t} - 2{,}000e^{-0.05t}\right]\Big|_0^{10}$$
$$= e^{0.5}\left(\left[-20(3{,}000+5(10))e^{-0.5t} - 2{,}000e^{-0.5}\right] - \left[-20(3{,}000+0)e^{0} - 2{,}000e^{0}\right]\right)$$
$$= e^{0.5}\left(-63{,}000e^{-0.5} + 62{,}000\right) \approx \$39{,}220.72$$

51.
$$PV = \int_0^5 (20+3t)e^{-0.07t}\,dt$$

$u = 20+3t$ and $dV = e^{-0.07t}\,dt$

$du = 3\,dt$ $\quad V = -\frac{100}{7}e^{-0.07t}$

So,

$$PV = -\frac{100}{7}(20+3t)e^{-0.07t}\Big|_0^5 - \int_0^5 -\frac{100}{7}e^{-0.07t}\cdot 3\,dt$$
$$= -\frac{100}{7}(20+3t)e^{-0.07t}\Big|_0^5 + \frac{300}{7}\int_0^5 e^{-0.07t}\,dt$$
$$= \left[-\frac{100}{7}(20+3t)e^{-0.07t} - \frac{30{,}000}{49}e^{-0.07t}\right]\Big|_0^5$$
$$= \left[-\frac{100}{7}(20+3(5))e^{-0.35} - \frac{30{,}000}{49}e^{-0.35}\right] - \left[-\frac{100}{7}(20+0)e^{0} - \frac{30{,}000}{49}e^{0}\right]$$
$$= \left(-500e^{-0.35} - \frac{30{,}000}{49}e^{-0.35}\right) - \left(-\frac{2{,}000}{7} - \frac{30{,}000}{49}\right)$$

≈ 114.17345 hundred, or \$11,417.35

53. From section 5.6, problem #24,

$$= P_0S(N) + \int_0^N R(t)S(N-t)\,dt$$

Here,

$$\#\text{ members} = 5{,}000e^{-0.02(9)} + \int_0^9 5te^{-0.02(9-t)}\,dt$$
$$= 5{,}000e^{-0.18} + 5e^{-0.18}\int_0^9 te^{0.02t}\,dt$$
$$= 5e^{-0.18}\left[1{,}000 + \int_0^9 te^{0.02t}\,dt\right]$$

$u = t$ and $dV = e^{0.02t}\,dt$

$du = dt$ $\quad V = 50e^{0.02t}$

$$= 5e^{-0.18}\left(1{,}000 + 50te^{0.02t}\Big|_0^9 - \int_0^9 50e^{0.02t}\,dt\right)$$
$$= 5e^{-0.18}\left(1{,}000 + 50te^{0.02t}\Big|_0^9 - 50\int_0^9 e^{0.02t}\,dt\right)$$
$$= 5e^{-0.18}\left[1{,}000 + \left(50te^{0.02t} - 2{,}500e^{0.02t}\right)\Big|_0^9\right]$$
$$= 5e^{-0.18}\left[1{,}000 + \left(50(9)e^{0.02(9)} - 2{,}500e^{0.02(9)}\right) - \left(0 - 2{,}500e^{0}\right)\right]$$

$\approx 4{,}367$ members

55. (a)

$p = D(q)$

$D(q) = 10 - qe^{0.02q}$

$D(5) = 10 - (5)e^{0.02(5)} = \$4.47$ each

(b)

$$CS = \int_0^5 \left(10 - qe^{0.02q}\right)\,dq - 5(4.47)$$
$$= \int_0^5 10\,dq - \int_0^5 qe^{0.02q}\,dq - 22.35$$

$u = q$ and $dV = e^{0.02q}\,dq$

$du = dq$ $\quad V = 50e^{0.02q}$

$$= 10q\Big|_0^5 - \left[50qe^{0.02q}\Big|_0^5 - \int_0^5 50e^{0.02q}\,dq\right] - 22.35$$

$$= 10q\Big|_0^5 - 50qe^{0.02q}\Big|_0^5 + 50\int_0^5 e^{0.02q}\,dq - 22.35$$

$$= \left(10q - 50qe^{0.02q} + 2{,}500e^{0.02q}\right)\Big|_0^5 - 22.35$$

$$= \left[10(5) - 50(5)e^{0.02(5)} + 2{,}500e^{0.02(5)}\right] - \left[0 - 0 + 2{,}500e^{0}\right] - 22.35$$

≈ 14.28456 thousand, or \$14,284.56

57.

$$GI = 2\int_0^1 \left(x - xe^{x-1}\right)\,dx$$

$$= 2\left[\int_0^1 x\,dx - \int_0^1 xe^{x-1}dx\right]$$

$$u = x \quad \text{and} \quad dV = e^{x-1}\,dx$$
$$du = dx \qquad V = e^{x-1}$$

$$= 2\left[\frac{x^2}{2}\Big|_0^1 - \left(xe^{x-1}\Big|_0^1 - \int_0^1 e^{x-1}\,dx\right)\right]$$

$$= 2\left(\frac{x^2}{2} - xe^{x-1} + e^{x-1}\right)\Big|_0^1$$

$$= 2\left[\left(\frac{1}{2} - 1e^0 + e^0\right) - \left(0 - 0 + e^{-1}\right)\right]$$

$$= 1 - \frac{2}{e} \approx 0.2642$$

59. From section 5.6, cardiac output is

$$R = \frac{D}{\int_0^{T_0} C(t)\,dt}$$

Here,

$$R = \frac{5}{\int_0^{20}\left(1.54te^{-0.12t} - 0.007t^2\right)\,dt}$$

where the denominator can be written as

$$1.54\int_0^{20} te^{-0.12t}\,dt - \int_0^{20} 0.007t^2\,dt$$

$$u = t \quad \text{and} \quad dV = e^{-0.12t}\,dt$$
$$du = dt \qquad V = -\frac{100}{12}e^{-0.12t}$$

$$= 1.54\left[-\frac{25}{3}te^{-0.12t}\Big|_0^{20} - \int_0^{20} -\frac{25}{3}e^{-0.12t}\,dt\right] - \frac{0.007}{3}t^3\Big|_0^{20}$$

$$= 1.54\left[-\frac{25}{3}te^{-0.12t}\Big|_0^{20} + \frac{25}{3}\int_0^{20} e^{-0.12t}\,dt\right] - \frac{0.007}{3}t^3\Big|_0^{20}$$

$$= 1.54\left(-\frac{25}{3}te^{-0.12t} - \frac{625}{9}e^{-0.12t}\right)\Big|_0^{20} - \frac{0.007}{3}t^3\Big|_0^{20}$$

$$= 1.54\left[\left(-\frac{25}{3}(20)e^{-0.12(20)} - \frac{625}{9}e^{-0.12(20)}\right) - \left(0 - \frac{625}{9}e^0\right)\right] - \left[\frac{0.007}{3}(20)^3 - 0\right]$$

$$\approx 55.2917$$

So, $R \approx \dfrac{5}{55.2917} \approx 0.0904$ bit/sec

61. $\int u^n e^{au}\,du$

Let

$$f = u^n \quad \text{and} \quad dV = e^{au}\,du$$
$$df = nu^{n-1}\,du \qquad V = \frac{1}{a}e^{au}$$

$$= \frac{1}{a}u^n e^{au} - \int \frac{1}{a}e^{au}\cdot nu^{n-1}\,du$$

$$= \frac{1}{a}u^n e^{au} - \frac{n}{a}\int u^{n-1}e^{au}\,du$$

63.

$$\text{area} = \int_0^{\ln 2} (2 - e^x)\, dx$$
$$= (2x - e^x)\Big|_0^{\ln 2}$$
$$= \left(2\ln 2 - e^{\ln 2}\right) - \left(0 - e^0\right)$$
$$= 2\ln 2 - 1 \approx 0.38629$$
$$\bar{x} = \frac{1}{0.38629}\int_0^{\ln 2} x\,(2 - e^x)\, dx$$
$$= 2.5887\left[\int_0^{\ln 2} 2x\, dx - \int_0^{\ln 2} xe^x\, dx\right]$$

Let $u = x$ and $dV = e^x\, dx$

$$= 2.5887\left[x^2\Big|_0^{\ln 2} - \left(xe^x\Big|_0^{\ln 2} - \int_0^{\ln 2} e^x\, dx\right)\right]$$
$$= 2.5887\left(x^2 - xe^x + e^x\right)\Big|_0^{\ln 2}$$
$$= 2.5887\left[\left((\ln 2)^2 - (\ln 2)(e^{\ln 2}) + e^{\ln 2}\right) - \left(0 - 0 + e^0\right)\right] \approx 0.244$$
$$\bar{y} = \frac{1}{2(0.38629)}\int_0^{\ln 2} (2 - e^x)^2\, dx$$
$$= 1.2944\int_0^{\ln 2}\left(4 - 4e^x + e^{2x}\right) dx$$
$$= 1.2944\left(4x - 4e^x + \frac{1}{2}e^{2x}\right)\Big|_0^{\ln 2}$$
$$= 1.2944\left[\left(4\ln 2 - 4e^{\ln 2} + \frac{1}{2}e^{2(\ln 2)}\right) - \left(0 - 4e^0 + \frac{1}{2}e^0\right)\right] \approx 0.353$$

So, the centroid is (0.244, 0.353).

65. (a) The kiosk should be located at the centroid. Using $y = \sqrt{2x^2 - 1}$,

$$\text{Area} = \int_1^5 \sqrt{2x^2 - 1}\, dx$$
$$= \int_1^5 \sqrt{(\sqrt{2}x)^2 - (1)^2}\, dx$$

which most closely resembles $\int \sqrt{u^2 - a^2}\, du$ (formula #18). Rewriting,

$$\int_1^5 \sqrt{(\sqrt{2}x)^2 - (1)^2}\, dx$$
$$= \frac{1}{\sqrt{2}}\int_1^5 \sqrt{(\sqrt{2}x)^2 - (1)^2}\cdot\sqrt{2}\, dx$$

The formula can be used with $u = \sqrt{2}x$, $du = \sqrt{2}\, dx$, and $a = 1$.

$$= \frac{1}{\sqrt{2}}\left[\frac{\sqrt{2}x}{2}\sqrt{2x^2 - 1} - \frac{1}{2}\ln\left|\sqrt{2}x + \sqrt{2x^2 - 1}\right|\right]_1^5$$
$$= \frac{1}{\sqrt{2}}\left[\left(\frac{\sqrt{2}(5)}{2}\sqrt{2(5)^2 - 1} - \frac{1}{2}\ln\left|\sqrt{2}(5) + \sqrt{2(5)^2 - 1}\right|\right) - \left(\frac{\sqrt{2}(1)}{2}\sqrt{2(1)^2 - 1} - \frac{1}{2}\ln\left|\sqrt{2}(1) + \sqrt{2(1)^2 - 1}\right|\right)\right]$$
$$= \frac{1}{\sqrt{2}}\left[\frac{35\sqrt{2}}{2} - \frac{1}{2}\ln\left|5\sqrt{2} + 7\right| - \frac{\sqrt{2}}{2} + \frac{1}{2}\ln\left(\sqrt{2} + 1\right)\right]$$
$$\approx 16.3768$$
$$\bar{x} = \frac{1}{16.3768}\int_1^5 x\sqrt{2x^2 - 1}\, dx$$

Using substitution with $u = 2x^2 - 1$, $\frac{1}{4}\, du = x\, dx$, and limits of integration of $2(1)^2 - 1 = 1$ and $2(5)^2 - 1 = 49$,

$$\bar{x} = 0.06106\left(\frac{1}{4}\int_1^{49} u^{1/2}\, du\right)$$
$$= 0.01527\left(\frac{2}{3}u^{3/2}\right)\Big|_1^{49}$$
$$= 0.010177\left[(49)^{3/2} - (1)^{3/2}\right] \approx 3.48$$

$$\bar{y} = \frac{1}{2(16.3768)} \int_1^5 \left(\sqrt{2x^2 - 1}\right)^2 dx$$
$$= 0.030531 \int_1^5 (2x^2 - 1)\, dx$$
$$= 0.030531 \left[\frac{2x^3}{3} - x\right]_1^5$$
$$= 0.030531 \left[\left(\frac{2(5)^3}{3} - 5\right) - \left(\frac{2(1)^3}{3} - 1\right)\right]$$
$$\approx 2.40$$

So, the kiosk should be located at the coordinates (3.48, 2.40).

(b) Writing Exercise—Answers will vary.

67. To use graphing utility to find where curves intersect and compute the area of region bounded by the curves,

Press [y=] and input $\sqrt{\left(\left(\frac{2}{5}\right)x^2 - 2\right)}$ for $y_1 =$.

Input $-\sqrt{\left(\left(\frac{2}{5}\right)x^2 - 2\right)}$ for $y_2 =$,

and input $x^\wedge 3 - 3.5x^2 + 2x$ for $y_3 =$.

Use window dimensions [−1, 4]1 by [−3, 5]1 for a good view of where the graphs intersect.
Use trace and zoom to find the points of intersection or use the intersect function under the calc menu to find (2.966, 1.232) and (2.608, −0.850) are the two points of intersection.
To find the area bounded by the curves, we must find that the x-intercept of the hyperbola is $x \approx 2.236$. Then we need

$$\int_{2.236}^{2.608} y_1 - y_2 + \int_{2.608}^{2.966} y_1 - y_3 = \int_{2.236}^{2.608} y_1$$
$$- \int_{2.236}^{2.608} y_2 + \int_{2.608}^{2.966} y_1 - \int_{2.608}^{2.966} y_3$$

Use the $\int f(x)\, dx$ function under the calc menu making sure the current equation is activated for each integral. The area is approximately 0.75834. Alternatively, you can use the *fnInt* function from the home screen under the math menu and enter:

$$fnInt(y_1 - y_2, x, 2.236, 2.608)$$
$$+ fnInt(y_1 - y_3, x, 2.608, 2.966)$$

You can insert y_1, y_2, y_3 by pressing [vars] and select Function under $Y - vars$ and then select which y function to insert.

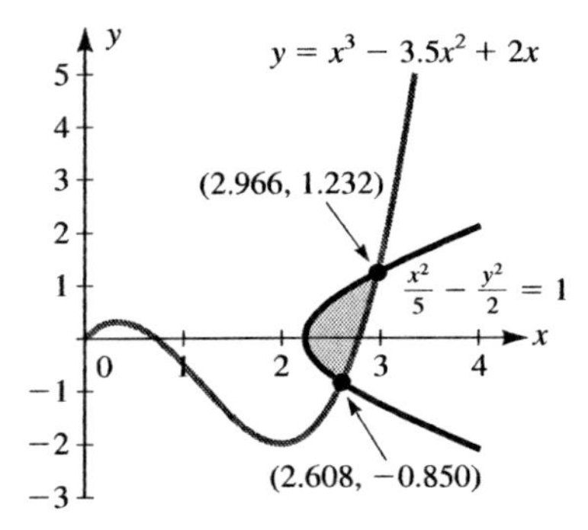

69. To use the numeric integration feature to evaluate the integral,

Press [y=] and input $\sqrt{(4x^2 - 7)}$ for $y_1 =$.
Use window dimensions [−1, 4]1 by [−3, 5]1.
Press [Graph].

Use the $\int f(x)\, dx$ function under the calc menu. Enter $x = 2$ for the lower limit and $x = 3$ for the upper limit. We see that $\int_2^3 \sqrt{4x^2 - 7}\, dx \simeq 4.227$. To verify, we use formula #18 on the table of integrals with

$$u = 2x$$
$$du = 2\, dx$$
$$dx = \frac{1}{2}\, du$$

When $x = 2$, $u = 4$;
when $x = 3$, $u = 6$.

So,

$$\int_2^3 \sqrt{4x^2-7}\,dx$$
$$=\frac{1}{2}\int_4^6 \sqrt{u^2-7}\,du$$
$$=\frac{1}{2}\left[\frac{u}{2}\sqrt{u^2-7}-\frac{7}{2}\ln\left|u+\sqrt{u^2-7}\right|\right]\Bigg|_4^6$$
$$=\frac{1}{2}\left[3\sqrt{29}-\frac{7}{2}\ln(6+\sqrt{29})-2(3)+\frac{7}{2}\ln 7\right]$$
$$=\frac{1}{2}(8.45309083)$$
$$\approx 4.227$$

71. To use the numeric integration feature to evaluate the integral,

Press [y=] and input $\sqrt{((x^2+2x))\,/\,((x+1)^2)}$ for $y_1 =$.
Use window dimensions [−1, 3]1 by [−1, 2]1.
Press [graph].

Use the $\int f(x)\,dx$ function under the calc menu with $x=0$ as the lower limit and $x=1$ the upper limit. We see that

$$\int_0^1 \frac{\sqrt{x^2+2x}}{(x+1)^2}\,dx \approx 0.4509$$

To verify, we use formula #19 on the table of integrals:

$$\int_0^1 \frac{\sqrt{x^2+2x}}{(x+1)^2}\,dx=\int_0^1 \frac{\sqrt{(x+1)^2-1}}{(x+1)^2}\,dx$$

Let

$$u=x+1$$
$$du=dx$$

When $x=0$, $u=1$;
when $x=1$, $u=2$.
So,

$$\int_1^2 \frac{\sqrt{u^2-1}}{u^2}\,du$$
$$=\left[-\frac{\sqrt{u^2-1}}{u}+\ln\left|u+\sqrt{u^2-1}\right|\right]\Bigg|_1^2$$
$$=-\frac{\sqrt{3}}{2}+\ln\left|2+\sqrt{3}\right|-\left(-\frac{\sqrt{0}}{1}+\ln\left|1+\sqrt{0}\right|\right)$$
$$=-\frac{\sqrt{3}}{2}+\ln(2+\sqrt{3})-\ln 1$$
$$\approx 0.4509$$

6.2 Improper Integrals

1.

$$\int_1^\infty \frac{1}{x^3}\,dx$$
$$=\lim_{N\to\infty}\int_1^N x^{-3}dx$$
$$=\lim_{N\to\infty}-\frac{1}{2}\left(\frac{1}{x^2}\right)\Bigg|_1^N$$
$$=-\frac{1}{2}\lim_{N\to\infty}\left(\frac{1}{N^2}-\frac{1}{1}\right)=-\frac{1}{2}(0-1)=\frac{1}{2}$$

3.

$$\int_1^\infty \frac{1}{\sqrt{x}}\,dx=\lim_{N\to\infty}\int_1^N x^{-1/2}dx$$
$$=\lim_{N\to\infty}2\left(x^{1/2}\right)\Bigg|_1^N$$
$$=2\lim_{N\to\infty}\left(x^{1/2}\right)\Bigg|_1^N$$
$$=2\left(\sqrt{N}-1\right)=\infty$$

So, the integral diverges.

5.
$$\int_3^\infty \frac{1}{2x-1}\,dx = \lim_{N\to\infty}\int_3^N \frac{1}{2x-1}\,dx$$
$$= \lim_{N\to\infty}\frac{1}{2}\ln|2x-1|\Big|_3^N$$
$$= \frac{1}{2}\lim_{N\to\infty}\ln|2x-1|\Big|_3^N$$
$$= \frac{1}{2}\lim_{N\to\infty}[\ln(2N-1)-\ln 7]$$
$$= \infty$$

So, the integral diverges.

7.
$$\int_3^\infty \frac{1}{(2x-1)^2}\,dx$$
$$= \lim_{N\to\infty}\int_3^N (2x-1)^{-2}\,dx$$
$$= \lim_{N\to\infty}\frac{1}{2}\left(-\frac{1}{2x-1}\right)\Big|_3^N$$
$$= \frac{1}{2}\lim_{N\to\infty}\left(-\frac{1}{2N-1}+\frac{1}{5}\right) = \frac{1}{2}\cdot\frac{1}{5} = \frac{1}{10}$$

9.
$$\int_0^\infty 5e^{-2x}\,dx$$
$$= \lim_{N\to\infty} 5\int_0^N e^{-2x}\,dx$$
$$= 5\lim_{N\to\infty} -\frac{1}{2}\left(e^{-2x}\right)\Big|_0^N$$
$$= -\frac{5}{2}\lim_{N\to\infty}\left(e^{-2N}-e^0\right)$$
$$= -\frac{5}{2}\cdot -1 = \frac{5}{2}$$

11.
$$\int_1^\infty \frac{x^2}{(x^3+2)^2}\,dx = \lim_{N\to\infty}\int_1^N \frac{x^2}{(x^3+2)^2}\,dx$$

Using substitution with $u = x^3+2$,

$$= \lim_{N\to\infty}\frac{1}{3}\int_3^{N^3+2} u^{-2}\,du$$
$$= \frac{1}{3}\lim_{N\to\infty}\left(-\frac{1}{u}\right)\Big|_3^{N^3+2}$$
$$= \frac{1}{3}\lim_{N\to\infty}\left(-\frac{1}{N^3+2}+\frac{1}{3}\right)$$
$$= \frac{1}{3}\cdot\frac{1}{3} = \frac{1}{9}$$

13.
$$\int_1^\infty \frac{x^2}{\sqrt{x^3+2}}\,dx = \lim_{N\to\infty}\int_1^N \frac{x^2}{(x^3+2)^{1/2}}\,dx$$

Using substitution with $u = x^3+2$,

$$= \lim_{N\to\infty}\frac{1}{3}\int_3^{N^3+2} u^{-1/2}\,du$$
$$= \frac{1}{3}\lim_{N\to\infty} 2\left(u^{1/2}\right)\Big|_3^{N^3+2}$$
$$= \frac{2}{3}\lim_{N\to\infty}\left(\sqrt{N^3+2}-\sqrt{3}\right) = \infty$$

So, the integral diverges.

15.
$$\int_1^\infty \frac{e^{-\sqrt{x}}}{\sqrt{x}}\,dx = \lim_{N\to\infty}\int_1^N \frac{e^{-\sqrt{x}}}{\sqrt{x}}\,dx$$

Using substitution with $u = -\sqrt{x}$,

$$= \lim_{N\to\infty} -2\int_{-1}^{-\sqrt{N}} e^u\,du = 2\int_{-\sqrt{N}}^{-1} e^u\,du$$
$$= 2\lim_{N\to\infty}\left(e^u\right)\Big|_{-\sqrt{N}}^{-1}$$
$$= 2\lim_{N\to\infty}\left(\frac{1}{e}-\frac{1}{e^{\sqrt{N}}}\right) = \frac{2}{e}$$

17.
$$\int_0^\infty 2xe^{-3x}\,dx = \lim_{N\to\infty} 2\int_0^N xe^{-3x}\,dx$$

Using integration by parts, with $u = x$ and $dV = e^{-3x}\,dx$,

$$= 2 \lim_{N\to\infty} \left[-\frac{x}{3}e^{-3x} \Big|_0^N - \int_0^N -\frac{1}{3}e^{-3x}\,dx \right]$$

$$= 2 \lim_{N\to\infty} \left[-\frac{x}{3}e^{-3x} \Big|_0^N + \frac{1}{3}\int_0^N e^{-3x}\,dx \right]$$

$$= 2 \lim_{N\to\infty} \left(-\frac{x}{3}e^{-3x} - \frac{1}{9}e^{-3x} \right) \Big|_0^N$$

$$= 2 \lim_{N\to\infty} \left[\left(-\frac{N}{3}e^{-3N} - \frac{1}{9}e^{-3N} \right) - \left(0 - \frac{1}{9}e^0 \right) \right]$$

$$= 2 \lim_{N\to\infty} \left(-\frac{N}{3}e^{-3N} - \frac{1}{9}e^{-3N} + \frac{1}{9} \right)$$

$$= 2 \cdot \frac{1}{9} = \frac{2}{9}$$

19. $$\int_1^\infty \frac{\ln x}{x}\,dx = \lim_{N\to\infty} \int_1^N \frac{\ln x}{x}\,dx$$

Using substitution with $u = \ln x$,

$$= \lim_{N\to\infty} \int_0^{\ln N} u\,du = \lim_{N\to\infty} \left(\frac{u^2}{2} \right) \Big|_0^{\ln N}$$

$$= \frac{1}{2} \lim_{N\to\infty} \left(u^2 \right) \Big|_0^{\ln N}$$

$$= \frac{1}{2} \lim_{N\to\infty} \left[(\ln N)^2 - 0 \right] = \infty$$

So, the integral diverges.

21. $$\int_2^\infty \frac{1}{x \ln x}\,dx = \lim_{N\to\infty} \int_2^N \left(\frac{1}{\ln x} \right) \frac{1}{x}\,dx$$

Using substitution with $u = \ln x$,

$$= \lim_{N\to\infty} \int_{\ln 2}^{\ln N} \frac{1}{u}\,du = \lim_{N\to\infty} (\ln |u|) \Big|_{\ln 2}^{\ln N}$$

$$= \lim_{N\to\infty} [\ln(\ln N) - \ln(\ln 2)] = \infty$$

So, the integral diverges.

23. $$\int_0^\infty x^2 e^{-x}\,dx = \lim_{N\to\infty} \int_0^N x^2 e^{-x}\,dx$$

Using integration by parts with $u = x^2$ and $dV = e^{-x}\,dx$,

$$= \lim_{N\to\infty} \left[-x^2 e^{-x} \Big|_0^N - \int_0^N -2xe^{-x}\,dx \right]$$

$$= \lim_{N\to\infty} \left[-x^2 e^{-x} \Big|_0^N + 2\int_0^N xe^{-x}\,dx \right]$$

Using integration by parts with $u = x$ and $dV = e^{-x}\,dx$,

$$= \lim_{N\to\infty} \left[-x^2 e^{-x} \Big|_0^N + 2\left(-xe^{-x} \Big|_0^N - \int_0^N -e^{-x}\,dx \right) \right]$$

$$= \lim_{N\to\infty} \left[-x^2 e^{-x} + 2\left(-xe^{-x} \Big|_0^N + \int_0^N e^{-x}\,dx \right) \right]$$

$$= \lim_{N\to\infty} \left[-x^2 e^{-x} - 2xe^{-x} - 2e^{-x} \right] \Big|_0^N$$

$$= \lim_{N\to\infty} \left[\left(-N^2 e^{-N} - 2Ne^{-N} - 2e^{-N} \right) - \left(0 - 0 - 2e^0 \right) \right] = 2$$

25. $$PV = \int_0^\infty 2{,}400 e^{-0.04t}\,dt$$

$$= \lim_{N\to\infty} 2{,}400 \int_0^N e^{-0.04t}\,dt$$

$$= 2{,}400 \lim_{N\to\infty} \left(-25e^{-0.04t} \right) \Big|_0^N$$

$$= -60{,}000 \lim_{N\to\infty} \left(e^{-0.04t} \right) \Big|_0^N$$

$$= -60{,}000 \lim_{N\to\infty} \left(e^{-0.04N} - e^0 \right)$$

$$= -60{,}000 \cdot -1 = \$60{,}000$$

27. $$PV = \int_0^\infty (12{,}000 + 900t)e^{-0.05t}\,dt$$

$$= \lim_{N\to\infty} \int_0^N (12{,}000 + 900t)e^{-0.05t}\,dt$$

Using integration by parts with $u = 12{,}000 + 900t$ and $dV = e^{-0.05t}\,dt$

$$= \lim_{N\to\infty} \left[-20(12{,}000 + 900t)e^{-0.05t} \Big|_0^N - \int_0^N -18{,}000e^{-0.05t}\, dt \right]$$

$$= \lim_{N\to\infty} \left[-20(12{,}000 + 900t)e^{-0.05t} \Big|_0^N + 18{,}000 \int_0^N e^{-0.05t}\, dt \right]$$

$$= -20 \lim_{N\to\infty} \left[(12{,}000 + 900t)\, e^{-0.05t} + 18{,}000e^{-0.05t} \right] \Big|_0^N$$

$$= -20 \lim_{N\to\infty} \left[\left(12{,}000 + 900N)e^{-0.05N} + 18{,}000e^{-0.05N}\right) - \left(12{,}000e^0 + 18{,}000e^0\right) \right]$$

$$= -20 \lim_{N\to\infty} \left[30{,}000e^{-0.05N} + 900Ne^{-0.05N} - 30{,}000 \right]$$

$$= -20(-30{,}000) = \$600{,}000$$

29. Number of patients

$$= \lim_{N\to\infty} \int_0^N 10e^{-(N-t)/20}\, dt$$

$$= \lim_{N\to\infty} 10e^{-N/20} \int_0^N e^{t/20}\, dt$$

$$= 10 \lim_{N\to\infty} e^{-N/20} \left(20e^{t/20}\right) \Big|_0^N$$

$$= 200 \lim_{N\to\infty} e^{-N/20} \left(e^{N/20} - e^0\right)$$

$$= 200 \lim_{N\to\infty} \left(e^0 - e^{-N/20}\right)$$

$$= 200 \cdot 1 = 200 \text{ patients.}$$

31.

$$\text{Amount of drug} = \lim_{N\to\infty} \int_0^N 5e^{-(N-t)/10}\, dt$$

$$= \lim_{N\to\infty} 5e^{-N/10} \int_0^N e^{t/10}\, dt$$

$$= 5 \lim_{N\to\infty} e^{-N/10} \left(10e^{t/10}\right) \Big|_0^N$$

$$= 50 \lim_{N\to\infty} e^{-N/10} \left(e^{N/10} - e^0\right)$$

$$= 50 \lim_{N\to\infty} \left(e^0 - e^{-N/10}\right)$$

$$= 50 \cdot 1 = 50 \text{ units}$$

33. The cost of maintaining
machine 1: $m_1(t) = 1{,}000(1 + 0.06t)$ dollars
machine 2: $m_2(t) = 1{,}100$ dollars

(a) The present value of maintaining machine 1 is

$$PV_1 = \int_0^\infty 1{,}000(1 + 0.06t)e^{-0.09t}\, dt$$

$$\lim_{N\to\infty} 1{,}000 \int_0^N (1 + 0.06t)e^{-0.09t}\, dt$$

Using integration by parts with $u = 1 + 0.06t$ and $dV = e^{-0.09t}dt$,

$$= 1{,}000 \lim_{N\to\infty} \left[(1 + 0.06t) \cdot \frac{-100}{9} e^{-0.09t} \Big|_0^N - \int_0^N 0.06 \cdot \frac{-100}{9} e^{-0.09}\, dt \right]$$

$$= 1{,}000 \lim_{N\to\infty} \left[(1 + 0.06t) \cdot \frac{-100}{9} e^{-0.09t} \Big|_0^N + \frac{2}{3} \int_0^N e^{-0.09t}\, dt \right]$$

$$= 1{,}000 \lim_{N\to\infty} \left[(1 + 0.06t) \cdot \frac{-100}{9} e^{-0.09t} \Big|_0^N + \frac{2}{3} \left(\frac{-100}{9} e^{-0.09t} \Big|_0^N \right) \right]$$

$$= 1{,}000 \lim_{N\to\infty} \left[(1 + 0.06t) \cdot \frac{-100}{9} e^{-0.09t} - \frac{200}{27} e^{-0.09t} \Big|_0^N \right]$$

$$= 1{,}000 \lim_{N\to\infty}\left[\left((1+0.06N)\cdot\frac{-100}{9}e^{-0.09N} - \frac{200}{27}e^{-0.09N}\right) - \left(1\cdot\frac{-100}{9}e^{0} - \frac{200}{27}e^{0}\right)\right]$$

$$\approx 1{,}000[18.519] = \$18{,}519$$

So, the capitalized cost of machine 1
$= \$10{,}000 + \$18{,}519 = \$28{,}519.$
The present value of maintaining machine 2 is

$$PV_2 = \int_0^\infty 1{,}100e^{-0.09t}\,dt$$
$$= \lim_{N\to\infty} 1{,}100\int_0^N e^{-0.09t}\,dt$$
$$= 1{,}100\lim_{N\to\infty}\left(-\frac{100}{9}e^{-0.09t}\Big|_0^N\right)$$
$$= -\frac{110{,}000}{9}\lim_{N\to\infty}\left(e^{-0.09N} - e^{0}\right)$$
$$= -\frac{110{,}000}{9}(-1) \approx \$12{,}222$$

So, the capitalized cost of machine 2
$= \$8{,}000 + \$12{,}222 = \$20{,}222$
The company should purchase machine 2.

(b) Writing Exercise—Answers will vary.

35. Using $P = B(t)e^{-rt}$,

$$\int_0^\infty Qe^{-rt}\,dt = \lim_{N\to\infty}\int_0^N Qe^{-rt}\,dt$$
$$= \lim_{N\to\infty} Q\int_0^N e^{-rt}\,dt$$
$$= Q\lim_{N\to\infty}\left(\frac{1}{-r}e^{-rt}\Big|_0^N\right)$$
$$= Q\lim_{N\to\infty}\left(-\frac{e^{-Nt}}{r} + \frac{e^{0}}{r}\right)$$
$$= Q\left(0 + \frac{1}{r}\right) = \frac{Q}{r}$$

6.3 Numerical Integration

1. For $\int_1^2 x^2\,dx$ with $n = 4$, $\Delta x = \frac{2-1}{4} = 0.25$, and $x_1 = 1$, $x_2 = 1.25$, $x_3 = 1.50$, $x_4 = 1.75$, $x_5 = 2$.

(a) By the trapezoidal rule, $\int_1^2 x^2\,dx$

$$= \frac{\Delta x}{2}\left[f(x_1) + 2f(x_2) + 2f(x_3) + 2f(x_4) + f(x_5)\right]$$
$$= \frac{0.25}{2}\left[1^2 + 2(1.25)^2 + 2(1.5)^2 + 2(1.75)^2 + 2^2\right]$$
$$\approx 2.3438.$$

(b) By Simpson's rule, $\int_1^2 x^2\,dx$

$$= \frac{\Delta x}{3}\left[f(x_1) + 4f(x_2) + 2f(x_3) + 4f(x_4) + f(x_5)\right]$$
$$= \frac{0.25}{3}\left[1^2 + 4(1.25)^2 + 2(1.5)^2 + 4(1.75)^2 + 2^2\right]$$
$$\approx 2.3333.$$

3. For $\int_0^1 \frac{1}{1+x^2}\,dx$ with $n = 4$, $\Delta x = \frac{1-0}{4} = 0.25$, and $x_1 = 0$, $x_2 = 0.25$, $x_3 = 0.50$, $x_4 = 0.75$, $x_5 = 1$.

(a) By the trapezoidal rule, $\int_0^1 \frac{1}{1+x^2}\,dx$

$$= \frac{\Delta x}{2}\left[f(x_1) + 2f(x_2) + 2f(x_3) + 2f(x_4) + f(x_5)\right]$$
$$= \frac{0.25}{2}\left[1 + \frac{2}{1+(0.25)^2} + \frac{2}{1+(0.5)^2} + \frac{2}{1+(0.75)^2} + \frac{1}{2}\right] \approx 0.7828.$$

(b) By Simpson's rule, $\int_0^1 \frac{1}{1+x^2}\,dx$

$$= \frac{\Delta x}{3}\left[f(x_1)+4f(x_2)+2f(x_3)+4f(x_4)+f(x_5)\right]$$
$$= \frac{0.25}{3}\left[1+\frac{4}{1+(0.25)^2}+\frac{2}{1+(0.5)^2}\right.$$
$$\left.+\frac{4}{1+(0.75)^2}+\frac{1}{2}\right] \approx 0.7854.$$

5. For $\int_{-1}^{0}\sqrt{1+x^2}dx$ with $n=4$, $\Delta x = \frac{0-(-1)}{4} = 0.25$, and $x_1=-1$, $x_2=-0.75$, $x_3=-0.5$, $x_4=-0.25$, $x_5=0$.

(a) By the trapezoidal rule, $\int_1^2 \sqrt{1+x^2}\,dx$

$$= \frac{\Delta x}{2}\left[f(x_1)+2f(x_2)+2f(x_3)+2f(x_4)+f(x_5)\right]$$
$$= \frac{0.25}{2}\left[\sqrt{1+(-1)^2}+2\sqrt{1+(-0.75)^2}\right.$$
$$\left.+2\sqrt{1+(-0.5)^2}+2\sqrt{1+(-0.25)^2}+\sqrt{1+(0)^2}\right]$$
$$\approx 1.1515.$$

(b) By Simpson's rule, $\int_1^2 \sqrt{1+x^2}\,dx$

$$= \frac{\Delta x}{3}\left[f(x_1)+4f(x_2)+2f(x_3)+4f(x_4)+f(x_5)\right]$$
$$= \frac{0.25}{3}\left[\sqrt{1+(-1)^2}+4\sqrt{1+(-0.75)^2}\right.$$
$$\left.+2\sqrt{1+(-0.5)^2}+4\sqrt{1+(-0.25)^2}+\sqrt{1+(0)^2}\right]$$
$$\approx 1.1478.$$

7. For $\int_0^1 e^{-x^2}\,dx$ with $n=4$, $\Delta x = \frac{1-0}{4} = 0.25$, and $x_1=0$, $x_2=0.25$, $x_3=0.50$, $x_4=0.75$, $x_5=1$.

(a) By the trapezoidal rule, $\int_1^2 e^{-x^2}\,dx$

$$= \frac{\Delta x}{2}\left[f(x_1)+2f(x_2)+2f(x_3)+2f(x_4)+f(x_5)\right]$$
$$= \frac{0.25}{2}\left[1+2e^{-(0.25)^2}+2e^{-(0.5)^2}+2e^{-(0.75)^2}+e^{-1}\right]$$
$$\approx 0.7430.$$

(b) By Simpson's rule, $\int_1^2 e^{-x^2}\,dx$

$$= \frac{\Delta x}{3}\left[f(x_1)+4f(x_2)+2f(x_3)+4f(x_4)+f(x_5)\right]$$
$$= \frac{0.25}{3}\left[1+4e^{-(0.25)^2}+2e^{-(0.5)^2}+4e^{-(0.75)^2}+e^{-1}\right]$$
$$\approx 0.7469.$$

9. For $\int_2^4 \frac{dx}{\ln x}$ with $n=6$, $\Delta x = \frac{4-2}{6} = \frac{1}{3}$ and $x_1=2$, $x_2=\frac{7}{3}$, $x_3=\frac{8}{3}$, $x_4=3$, $x_5=\frac{10}{3}$, $x_6=\frac{11}{3}$, $x_7=4$.

(a) By the trapezoidal rule, $\int_2^4 \frac{dx}{\ln x}$

$$\approx \frac{\Delta x}{2}\left[f(x_1)+2f(x_2)+2f(x_3)+2f(x_4)\right.$$
$$\left.+2f(x_5)+2f(x_6)+f(x_7)\right]$$
$$= \frac{\frac{1}{3}}{2}\left[\frac{1}{\ln 2}+\frac{2}{\ln\frac{7}{3}}+\frac{2}{\ln\frac{8}{3}}+\frac{2}{\ln 3}+\frac{2}{\ln\frac{10}{3}}+\frac{2}{\ln\frac{11}{3}}+\frac{1}{\ln 4}\right]$$
$$\approx 1.9308$$

(b) By Simpson's rule, $\int_2^4 \frac{dx}{\ln x}$

$$\approx \frac{\Delta x}{3}\left[f(x_1)+4f(x_2)+2f(x_3)+4f(x_4)\right.$$
$$\left.+2f(x_5)+4f(x_6)+f(x_7)\right]$$
$$= \frac{\frac{1}{3}}{3}\left[\frac{1}{\ln 2}+\frac{4}{\ln\frac{7}{3}}+\frac{2}{\ln\frac{8}{3}}+\frac{4}{\ln 3}+\frac{2}{\ln\frac{10}{3}}+\frac{4}{\ln\frac{11}{3}}+\frac{1}{\ln 4}\right]$$
$$\approx 1.9228$$

11. For $\int_0^1 \sqrt[3]{1+x^2}dx$ with $n=4$, $\Delta x = \frac{1-0}{4} = 0.25$ and $x_1=0$, $x_2=0.25$, $x_3=0.05$, $x_4=0.75$, $x_5=1$.

(a) By the trapezoidal rule, $\int_0^1 \sqrt[3]{1+x^2}dx$

$$\approx \frac{\Delta x}{2}\left[f(x_1)+2f(x_2)+2f(x_3)+2f(x_4)+f(x_5)\right]$$
$$= \frac{0.25}{2}\left[1+2\sqrt[3]{1.0625}+2\sqrt[3]{1.25}+2\sqrt[3]{1.5625}+\sqrt[3]{2}\right]$$
$$\approx 1.0970$$

(b) By Simpson's rule, $\int_0^1 \sqrt[3]{1+x^2}dx$

$$\approx \frac{\Delta x}{3}\left[f(x_1)+4f(x_2)+2f(x_3)+4f(x_4)+f(x_5)\right]$$
$$= \frac{0.25}{3}\left[1+4\sqrt[3]{1.0625}+2\sqrt[3]{1.25}+4\sqrt[3]{1.5625}+\sqrt[3]{2}\right]$$
$$\approx 1.0948$$

13. For $\int_0^2 e^{-\sqrt{x}}dx$ with $n=8$, $\Delta x = \frac{2-0}{8} = 0.25$ and $x_1=0$, $x_2=0.25$, $x_3=0.5$, $x_4=0.75$, $x_5=1$, $x_6=1.25$, $x_7=1.5$, $x_8=1.75$, $x_9=2$.

(a) By the trapezoidal rule, $\int_0^2 e^{-\sqrt{x}}dx$

$$\approx \frac{\Delta x}{2}\left[f(x_1)+2f(x_2)+2f(x_3)+2f(x_4)+2f(x_5)\right.$$
$$\left.+2f(x_6)+2f(x_7)+2f(x_8)+f(x_9)\right]$$
$$= \frac{0.25}{2}\left[1+2e^{-\sqrt{0.25}}+2e^{-\sqrt{0.5}}+2e^{-\sqrt{0.75}}+2e^{-1}\right.$$
$$\left.+2e^{-\sqrt{1.25}}+2e^{-\sqrt{1.5}}+2e^{-\sqrt{1.75}}+e^{-\sqrt{2}}\right]$$
$$\approx 0.8492$$

(b) By Simpson's rule, $\int_0^2 e^{-\sqrt{x}}dx$

$$\approx \frac{\Delta x}{3}\left[f(x_1)+4f(x_2)+2f(x_3)+4f(x_4)+2f(x_5)\right.$$
$$\left.+4f(x_6)+2f(x_7)+4f(x_8)+f(x_9)\right]$$
$$= \frac{0.25}{3}\left[1+4e^{-\sqrt{0.25}}+2e^{-\sqrt{0.5}}+4e^{-\sqrt{0.75}}+2e^{-1}\right.$$
$$\left.+4e^{-\sqrt{1.25}}+2e^{-\sqrt{1.5}}+4e^{-\sqrt{1.75}}+e^{-\sqrt{2}}\right]$$
$$\approx 0.8362$$

15. For $\int_1^2 \frac{1}{x^2}\,dx$ with $n=4$, $\Delta x = \frac{2-1}{4} = 0.25$, and $x_1=1$, $x_2=1.25$, $x_3=1.50$, $x_4=1.75$, $x_5=2$.

(a) By the trapezoidal rule, $\int_1^2 \frac{1}{x^2}\,dx$

$$= \frac{\Delta x}{2}\left[f(x_1)+2f(x_2)+2f(x_3)+2f(x_4)+f(x_5)\right]$$
$$= \frac{0.25}{2}\left[1+\frac{2}{(1.25)^2}+\frac{2}{(1.5)^2}+\frac{2}{(1.75)^2}+\frac{1}{2^2}\right]$$
$$\approx 0.5090.$$

The error estimate is $|E_n| \le \frac{M(b-a)^3}{12n^2}$. For $n=4$, $a=1$, and $b=2$, $|E_4| \le \frac{M(2-1)^2}{12(4^2)} = \frac{M}{192}$, where M is the maximum value of $|f''(x)|$ on $1 \le x \le 2$. Now $f(x)=x^{-2}$, $f'(x)=-2x^{-3}$, and $f''(x)=6x^{-4}$. For $1 \le x \le 2$, $|f''(x)| = \frac{6}{x^4} \le \frac{6}{1^4} = 6$. So, $|E_4| = \frac{6}{192} \approx 0.03125$.

(b) By Simpson's rule, $\int_1^2 \frac{1}{x^2}\,dx$

$$= \frac{\Delta x}{3}\left[f(x_1)+4f(x_2)+2f(x_3)+4f(x_4)+f(x_5)\right]$$
$$= \frac{0.25}{3}\left[1+\frac{4}{(1.25)^2}+\frac{2}{(1.5)^2}+\frac{4}{(1.75)^2}+\frac{1}{2^2}\right]$$
$$\approx 0.5004.$$

The error estimate is $|E_n| \le \frac{M(b-a)^5}{180n^4}$. For $n=4$, $a=1$, and $b=2$, $|E_4| \le \frac{M(2-1)^5}{180(4^4)} = \frac{M}{46{,}080}$ where M is the maximum value of $|f^{(4)}(x)|$ on $1 \le x \le 2$. Now $f''(x)=6x^{-4}$, $f^{(3)}(x)=-24x^{-5}$, and $f^{(4)}(x)=120x^{-6}$. For $1 \le x \le 2$, $|f^{(4)}(x)| = \frac{120}{x^6} \le \frac{120}{1^6} = 120$. So, $|E_4| \le \frac{120}{46{,}080} \approx 0.0026$.

17. For $\int_1^3 \sqrt{x}\,dx$ with $n=10$, $\Delta x = \frac{3-1}{10} = 0.2$, and $x_1=1$, $x_2=1.2$, $x_3=1.4$, $\ldots$, $x_{10}=2.8$, $x_{11}=3$.

(a) By the trapezoidal rule, $\int_1^3 \sqrt{x}\,dx$

$= \frac{\Delta x}{2}\Big[f(x_1) + 2f(x_2) + 2f(x_3) + \cdots$

$+ 2f(x_{10}) + f(x_{11})\Big]$

$= \frac{0.2}{2}\Big[1 + 2\sqrt{1.2} + 2\sqrt{1.4} + 2\sqrt{1.6} + 2\sqrt{1.8}$

$+ 2\sqrt{2}\,2\sqrt{2.2} + 2\sqrt{2.4} + 2\sqrt{2.6} + 2\sqrt{2.8} + \sqrt{3}\Big]$

$\approx 2.7967.$

The error estimate is $|\,E_n\,| \le \frac{M(b-a)^3}{12n^2}$. For $n = 10$, $a = 1$, and $b = 3$, $|\,E_{10}\,| \le \frac{M(3-1)^3}{12(10^2)} = \frac{8M}{1{,}200} = \frac{M}{150}$, where M is the maximum value of $|\,f''(x)\,|$ on $1 \le x \le 3$. Now, $f(x) = x^{1/2}$, $f'(x) = \frac{1}{2}x^{-1/2}$, and $f''(x) = -\frac{1}{4}x^{-3/2}$. For $1 \le x \le 3$, $|\,f''(x)\,| = \left|-\frac{1}{4}x^{-3/2}\right| \le \frac{1}{4}(1^{-3/2}) = \frac{1}{4}$. So, $|\,E_{10}\,| = \frac{1}{150}\left(\frac{1}{4}\right) \approx 0.0017.$

(b) By Simpson's rule, $\int_1^3 \sqrt{x}\,dx$

$= \frac{\Delta x}{3}\left[f(x_1) + 4f(x_2) + 2f(x_3) + \cdots + 4f(x_{10})\right.$

$\left.+ f(x_{11})\right]$

$= \frac{0.2}{3}\Big[1 + 4\sqrt{1.2} + 2\sqrt{1.4} + 4\sqrt{1.6} + 2\sqrt{1.8} + 4\sqrt{2}$

$+ 2\sqrt{2.2} + 4\sqrt{2.4} + 2\sqrt{2.6} + 4\sqrt{2.8} + \sqrt{3}\Big]$

$\approx 2.7974.$

The error estimate is $|\,E_n\,| \le \frac{M(b-a)^5}{180n^4}$. For $n = 10$, $a = 1$, and $b = 3$, $|\,E_{10}\,| \le \frac{M(3-1)^5}{180(10^4)} = \frac{32M}{180(10^4)}$, where M is the maximum value of $|\,f^{(4)}(x)\,|$ on $1 \le x \le 3$. Now $f''(x) = -\frac{1}{4}x^{-3/2}$, $f^{(3)}(x) = \frac{3}{8}x^{-5/2}$, and $f^{(4)}(x) = -\frac{15}{16}x^{-7/2}$. For $1 \le x \le 3$, $|\,f^{(4)}(x)\,| = \left|-\frac{15}{16}x^{-7/2}\right| \le \frac{15}{16}(1^{-7/2}) = \frac{15}{16}$. So, $|\,E_{10}\,| = \frac{32}{180(10{,}000)}\left(\frac{15}{16}\right) \approx 0.0000167.$

19. For $\int_0^1 e^{x^2}\,dx$ with $n = 4$, $\Delta x = \frac{1-0}{4} = 0.25$, and $x_1 = 0$, $x_2 = 0.25$, $x_3 = 0.50$, $x_4 = 0.75$, $x_5 = 1$.

(a) By the trapezoidal rule, $\int_1^2 e^{x^2}\,dx$

$= \frac{\Delta x}{2}\left[f(x_1) + 2f(x_2) + 2f(x_3) + 2f(x_4) + f(x_5)\right]$

$= \frac{0.25}{2}\left[1 + 2e^{(0.25)^2} + 2e^{(0.5)^2} + 2e^{(0.75)^2} + e^1\right] \approx 1.4907.$

The error estimate is $|\,E_n\,| \le \frac{M(b-a)^3}{12n^2}$. For $n = 4$, $a = 0$, and $b = 1$, $|\,E_4\,| \le \frac{M(1-0)^3}{12(4^2)} = \frac{M}{192}$, where M is the maximum value of $|\,f''(x)\,|$ on $0 \le x \le 1$. Now, $f(x) = e^{x^2}$, $f'(x) = -2xe^{-x^2}$, and $f''(x) = \left(4x^2 + 2\right)e^{x^2}$. For $0 \le x \le 1$, $|\,f''(x)\,| = \left[4\left(1^2\right) + 2\right]e^{1^2} = 6e$. So, $|\,E_4\,| = \frac{6e}{192} \approx 0.0849.$

(b) By Simpson's rule, $\int_1^2 e^{x^2}\,dx$

$= \frac{\Delta x}{3}\left[f(x_1) + 4f(x_2) + 2f(x_3) + 4f(x_4) + f(x_5)\right]$

$= \frac{0.25}{3}\left[1 + 4e^{(0.25)^2} + 2e^{(0.5)^2} + 4e^{(0.75)^2} + e^1\right]$

$\approx 1.4637.$

The error estimate is $|\,E_n\,| \le \frac{M(b-a)^5}{180n^4}$. For $n = 4$, $a = 0$, and $b = 1$, $|\,E_4\,| \le \frac{M(1-0)^5}{180(4^4)} = \frac{M}{46{,}080}$, where M is the maximum value of $|\,f^{(4)}(x)\,|$ on $0 \le x \le 1$. Now, $f''(x) = \left(4x^2 + 2\right)e^{x^2}$, $f^{(3)}(x) = \left(8x^3 + 12x\right)e^{x^2}$, and $f^{(4)}(x) =$

$(16x^4 + 48x^2 + 12)\, e^{x^2}$. For $0 \le x \le 1$, $|\, f^{(4)}(x) \,| = [16\,(1^4) + 48\,(1^2) + 12]\, e^{1^2} = 76e$. So, $|\, E_4 \,| \le \dfrac{76e}{46{,}080} \approx 0.0045$.

21. The integral to be approximated is $\int_1^3 \frac{1}{x}\, dx$. The derivatives of $f(x) = \frac{1}{x} = x^{-1}$ are $f'(x) = -x^{-2}$, $f''(x) = 2x^{-3}$, $f^{(3)}(x) = -6x^{-4}$, and $f^{(4)}(x) = 24x^{-5}$.

(a) For the trapezoidal rule, $|\, E_n \,| \le \dfrac{M(b-a)^3}{12n^2}$, where M is the maximum value of $|\, f''(x) \,|$ on $1 \le x \le 3$. Now $|\, f''(x) \,| = \dfrac{2}{x^3} \le \dfrac{2}{1^3} = 2$ on $1 \le x \le 3$. $|\, E_n \,| \le \dfrac{2(3-1)^3}{12n^2} = \dfrac{4}{3n^2}$, which is less than 0.00005 if $4 < 3(0.00005)n^2$ or $n > \sqrt{\dfrac{4}{3(0.00005)}} \approx 163.3$. So, 164 intervals should be used.

(b) For Simpson's rule, $|\, E_n \,| \le \dfrac{M(b-a)^5}{180n^4}$, where M is the maximum value of $|\, f^{(4)}(x) \,|$ on $1 \le x \le 3$. Now, $|\, f^{(4)}(x) \,| = \left|\dfrac{24}{x^5}\right| \le \dfrac{24}{1^5} = 24$ on $1 \le x \le 3$. $|\, E_n \,| \le \dfrac{24(3-1)^5}{180n^4} = \dfrac{768}{180n^4}$ which is less than 0.00005 if $768 < 180(0.00005)n^4$ or $n > \sqrt[4]{\dfrac{768}{180(0.00005)}} \approx 17.1$. So, 18 subintervals should be used.

23. The integral to be approximated is $\int_1^2 \frac{1}{\sqrt{x}}\, dx$. The derivatives of $f(x) = \frac{1}{\sqrt{x}} = x^{-1/2}$ are $f'(x) = -\frac{1}{2}x^{-3/2}$, $f''(x) = \frac{3}{4}x^{-5/2}$, $f^{(3)}(x) = -\frac{15}{8}x^{-7/2}$, and $f^{(4)}(x) = \frac{105}{16}x^{-9/2}$.

(a) For the trapezoidal rule, $|\, E_n \,| \le \dfrac{M(b-a)^3}{12n^2}$, where M is the maximum value of $|\, f''(x) \,|$ on $1 \le x \le 2$. Now $|\, f''(x) \,| = \dfrac{3}{4}x^{-5/2} \le \dfrac{3}{4}$ on $1 \le x \le 2$. $|\, E_n \,| \le \dfrac{3}{4}\dfrac{(2-1)^3}{12n^2} = \dfrac{1}{16n^2}$, which is less than 0.00005 if $1 < 16(0.00005)n^2$ or $n > \sqrt{\dfrac{1}{16(0.00005)}} \approx 35.4$. So, 36 intervals should be used.

(b) For Simpson's rule, $|\, E_n \,| \le \dfrac{M(b-a)^5}{180n^4}$, where M is the maximum value of $|\, f^{(4)}(x) \,|$ on $1 \le x \le 2$. Now $|\, f^{(4)}(x) \,| = \left|\dfrac{105}{16}x^{-9/2}\right| \le \dfrac{105}{16}$ on $1 \le x \le 2$. $|\, E_n \,|\ \dfrac{105(2-1)^5}{16(180)n^4} = \dfrac{7}{192n^4}$, which is less than 0.00005 if $7 < 192(0.00005)n^4$ or $n > \sqrt[4]{\dfrac{7}{192(0.00005)}} \approx 5.2$. So, 6 subintervals should be used.

25. The integral to be approximated is $\int_{1.2}^{2.4} e^x\, dx$.

(a) For the trapezoidal rule, $|\, E_n \,| \le \dfrac{M(b-a)^3}{12n^2}$, where M is the maximum value of $|\, f''(x) \,|$ on $1.2 \le x \le 2.4$. Now $|\, f''(x) \,| = |\, e^x \,| \le e^{2.4}$ on $1.2 \le x \le 2.4$. $|\, E_n \,| \le \dfrac{e^{2.4}(2.4-1.2)^3}{12n^2} = \dfrac{1.728e^{2.4}}{12n^2}$ which is less than 0.00005 if $1.728e^{2.4} < 12(0.00005)n^2$ or $n > \sqrt{\dfrac{1.728e^{2.4}}{12(0.00005)}} \approx 178.2$. So, 179 intervals should be used.

(b) For Simpson's rule, $|\, E_n \,| \le \dfrac{M(b-a)^5}{180n^4}$, where M is the maximum value of $|\, f^{(4)}(x) \,|$ on $1.2 \le x \le 2.4$. Now $|\, f^{(4)}(x) \,| = |\, e^x \,| \le e^{2.4}$ on $1.2 \le x \le 2.4$. $|\, E_n \,| \le \dfrac{e^{2.4}(2.4-1.2)^5}{180n^4}$ which is less than 0.00005 if $e^{2.4}(1.2)^5 < 180(0.00005)n^4$ or $n > \sqrt[4]{\dfrac{e^{2.4}(1.2)^5}{180(0.00005)}} \approx 7.4$. So, 8 subintervals should be used.

27. For $\int_0^1 \sqrt{1-x^2}\,dx$ with $n=8$, $\Delta x = \dfrac{1-0}{8} =$ 0.125, and $x_1 = 0$, $x_2 = 0.125$, $x_3 = 0.25, \ldots,$ $x_8 = 1.875$, $x_9 = 2$.

(a) By the trapezoidal rule, $\int_0^1 \sqrt{1-x^2}\,dx$

$$\begin{aligned}
&= \frac{\Delta x}{2}\left[f(x_1) + 2f(x_2) + 2f(x_3) + \cdots + 2f(x_8) + f(x_9)\right]\\
&= 0.0625\left[\sqrt{1-(0)^2} + 2\sqrt{1-(0.125)^2} + 2\sqrt{1-(0.25)^2} + 2\sqrt{1-(0.375)^2} + 2\sqrt{1-(0.5)^2} + 2\sqrt{1-(0.625)^2} + 2\sqrt{1-(0.75)^2} + 2\sqrt{1-(0.875)^2} + \sqrt{1-(1)^2}\right] \approx 0.7725
\end{aligned}$$

$(0.7725)(4) = 3.090$ as an approximation of π.

(b) By Simpson's rule, $\int_0^1 \sqrt{1-x^2}\,dx$

$$\begin{aligned}
&= \frac{\Delta x}{3}\left[f(x_1) + 4f(x_2) + 2f(x_3) + 4f(x_4) + \cdots + 4f(x_8) + f(x_9)\right]\\
&= \frac{1}{24}\left[\sqrt{1-(0)^2} + 4\sqrt{1-(0.125)^2} + 2\sqrt{1-(0.25)^2} + 4\sqrt{1-(0.375)^2} + 2\sqrt{1-(0.5)^2} + 4\sqrt{1-(0.625)^2} + 2\sqrt{1-(0.75)^2} + 4\sqrt{1-(0.875)^2} + \sqrt{1-(1)^2}\right] \approx 0.7803
\end{aligned}$$

$(0.7803)(4) = 3.121$ as an approximation of π.

29. For $\int_1^6 \dfrac{e^{-0.4x}}{x}\,dx$ with $n=10$, $\Delta x = \frac{6-1}{10} = 0.5$, and $x_1 = 1$, $x_2 = 1.5$, $x_3 = 2.0$, $x_4 = 2.5$, $x_5 = 3.0$, $x_6 = 3.5$, $x_7 = 4.0$, $x_8 = 4.5$, $x_9 = 5.0$, $x_{10} = 5.5$, $x_{11} = 6.0$.

By the trapezoidal rule, $\int_1^6 \dfrac{e^{-0.4x}}{x}\,dx$

$$\begin{aligned}
&= \frac{\Delta x}{2}\left[f(x_1) + 2f(x_2) + 2f(x_3) + 2f(x_4) + 2f(x_5) + 2f(x_6) + 2f(x_7) + 2f(x_8) + 2f(x_9) + 2f(x_{10}) + f(x_{11})\right]\\
&= 0.25\left[\left(\frac{e^{-0.4(1)}}{1}\right) + 2\left(\frac{e^{-0.4(1.5)}}{1.5}\right) + 2\left(\frac{e^{-0.4(2)}}{2}\right) + 2\left(\frac{e^{-0.4(2.5)}}{2.5}\right) + 2\left(\frac{e^{-0.4(3)}}{3}\right) + 2\left(\frac{e^{-0.4(3.5)}}{3.5}\right) + 2\left(\frac{e^{-0.4(4)}}{4}\right) + 2\left(\frac{e^{-0.4(4.5)}}{4.5}\right) + 2\left(\frac{e^{-0.4(5)}}{5}\right) + 2\left(\frac{e^{-0.4(5.5)}}{5.5}\right) + \left(\frac{e^{-0.4(6)}}{6}\right)\right] \approx 0.6929
\end{aligned}$$

So, the estimate of the average value is

$$\frac{1}{6-1}(0.6929) = 0.1386$$

31. Volumn of $S = \pi \int_0^1 \left(\dfrac{x}{1+x}\right)^2 dx$

Using the trapezoidal rule with $n=7$, $\Delta x = \dfrac{1-0}{7}$ and $x_1 = 0$, $x_2 = \dfrac{1}{7}$, $x_3 = \dfrac{2}{7}$, $x_4 = \dfrac{3}{7}$, $x_5 = \dfrac{4}{7}$, $x_6 = \dfrac{5}{7}$, $x_7 = \dfrac{6}{7}$, $x_8 = 1$.

$$\int_0^1 \left(\frac{x}{1+x}\right)^2 dx \approx \frac{\Delta x}{2}\left[f(x_1) + 2f(x_2) + 2f(x_3) + 2f(x_4) + 2f(x_5) + 2f(x_6) + 2f(x_7) + f(x_8)\right]$$

$$= \frac{\frac{1}{7}}{2}\left[0 + 2\left(\frac{1}{8}\right)^2 + 2\left(\frac{2}{9}\right)^2 + 2\left(\frac{3}{10}\right)^2 + 2\left(\frac{4}{11}\right)^2 + 2\left(\frac{5}{12}\right)^2 + 2\left(\frac{6}{13}\right)^2 + \left(\frac{1}{2}\right)^2\right]$$

$$\approx 0.114124$$

So, the volume is $\approx \pi(0.114124) \approx 0.3585$

33. $$FV = e^{rT}\int_0^T f(t)e^{-rt}dt$$
$$= e^{0.06(10)}\int_0^{10} \sqrt{t}e^{-0.06t}dt$$
$$= e^{0.6}\int_0^{10} \sqrt{t}e^{-0.06t}dt$$

Using the trapezoidel rule with $n = 5$, $\Delta t = \dfrac{10-0}{5}$ and $t_1 = 0$, $t_2 = 2$, $t_3 = 4$, $t_4 = 6$, $t_5 = 8$, $t_6 = 10$

$$\int_0^{10} \sqrt{t}e^{-0.06t}dt \approx \frac{\Delta t}{2}\left[f(t_1) + 2f(t_2) + 2f(t_3) + 2f(t_4) + 2f(t_5) + f(t_6)\right]$$
$$= \frac{2}{2}\left[0 + 2\sqrt{2}e^{-0.12} + 2\sqrt{4}e^{-0.24} + 2\sqrt{6}e^{-0.36} + 2\sqrt{8}e^{-0.48} + \sqrt{10}e^{-0.6}\right]$$
$$\approx 14.308884$$

So, $FV \approx e^{0.6}(14.308884) \approx 26.07249$ or \$26,072

35. $$P(T) = P_0S(T) + \int_0^T RS(T-t)dt$$
$$= 3000e^{-0.01(8)} + \int_0^8 50\sqrt{t}\cdot e^{-0.01(8-t)}dt$$
$$= 3000e^{-0.08} + 50e^{-0.08}\int_0^8 \sqrt{t}e^{0.01t}dt$$
$$= 50e^{-0.08}\left[60 + \int_0^8 \sqrt{t}e^{0.01t}dt\right]$$

Using Simpson's rule with $n = 8$, $\Delta t = \dfrac{8-0}{8}$ and $t_1 = 0$, $t_2 = 1$, $t_3 = 2$, …, $t_9 = 8$.

$$\int_0^8 \sqrt{t}e^{0.01t}dt \approx \frac{\Delta t}{3}\left[f(t_1) + 4f(t_2) + 2f(t_3) + 4f(t_4) + 2f(t_5) + 4f(t_6) + 2f(t_7) + 4f(t_8) + f(t_9)\right]$$
$$= \frac{1}{3}\left[0 + 4e^{0.01} + 2\sqrt{2}e^{0.02} + 4\sqrt{3}e^{0.03} + 2\sqrt{4}e^{0.04} + 4\sqrt{5}e^{0.05} + 2\sqrt{6}e^{0.06} + 4\sqrt{7}e^{0.07} + \sqrt{8}e^{0.08}\right]$$
$$\approx 15.749112$$

So, the number of people with the flu is $\approx 50e^{-0.08}[60 + 15.749112] \approx 3{,}496$ people.

37. Since distance is the integral of velocity, we need to approximate $\int_2^3 V(t)\,dt$ using the trapezoidal rule. Since the readings are every 5 minutes, $\Delta t = 5$ minutes $= \frac{1}{2}$ hour.

$$\text{Distance} \approx \frac{\frac{1}{12}}{2}[45 + 2(48) + 2(37) + 2(39) + 2(55) + 2(60) + 2(60) + 2(55) + 2(50) + 2(67) + 2(58) + 2(45) + 49]$$
$$\approx 51.75 \text{ miles}$$

39. We need to approximate

$$FV = \int_a^b (\text{rate income enters})e^{r(b-t)}\,dt$$

Since the readings are every 2 months, $\Delta t = 2$, $r = \dfrac{0.04}{12}$, $a = 0$, and $b = 12$.

Future value

$$\approx \frac{2}{3}\Big[(437)e^{(0.04/12)(12-0)} + 4(357)e^{(0.04/12)(12-2)} + 2(615)e^{(0.04/12)(12-4)} + 4(510)e^{(0.04/12)(12-6)} + 2(415)e^{(0.04/12)(12-8)} + 4(550)e^{(0.04/12)(12-10)} + (593)e^{(0.04/12)(12-12)}\Big]$$
$$\approx \$5949.70$$

41. We need to approximate $\int_a^b f(x) - g(x)\,dx$ using the trapzoidal rule. Since readings are made every 5 feet, $\Delta t = 5$.

$$\text{Area} \approx \frac{5}{2}\,[2 + 2(5) + 2(7) + 2(8) + 2(8) + 2(5) + 2(6) + 2(4) + 2(3) + 0]$$
$$\approx 235 \text{ square feet}$$

43. We need to approximate
$PS = p_0q_0 - \int_0^{q_0} S(q)dq$
using the trapezsoidal rule. Since data was collected in increments of 1 thousand units, $\Delta q = 1$;

$$\int_0^7 S(q)dq \approx \frac{1}{2}\,[1.21 + 2(3.19) + 2(3.97) + 2(5.31) + 2(6.72) + 2(8.16) + 2(9.54) + 11.03]$$
$$= 43.01$$

So, $PS \approx (11.03)(7) - 43.01 = 34.2$ or \$34,200.

45. We need to approximate $2\pi \int_0^{10} r \cdot D(r)dr$ using the trapezoidal rule. Since measurements were made every 2 miles, $\Delta r = 2$;

$$\int_0^{10} rD(r)dr = \frac{2}{2}[0 + 2(2)(2844) + 2(4)(2087) + 2(6)(1752) + 2(8)(1109) + (10)(879)]$$
$$= 75{,}630$$

So, the total population is $\approx 2\pi(75{,}630) \approx 475{,}197$ people.

47. **(a)** $N = \int_0^{11} D'(t)dt = D(11) - D(0)$
This is the difference in cummulative deaths from $t = 0$ (1990) to $t = 11$ (2001).

(b) Using the trapezoidal rule to estimate the integral with $\Delta t = 1$.

$$\approx \frac{1}{2}\,[31{,}120 + 2(36{,}175) + 2(40{,}587) + 2(45{,}850) + 2(50{,}842) + 2(51{,}670) + 2(38{,}296) + 2(22{,}245) + 2(18{,}823) + 2(18{,}249) + 2(16{,}672) + 15{,}603]$$
$$\approx 362{,}771 \text{ deaths}$$

(c) Writing exercise—Answers will vary.

Checkup for Chapter 6

1. **(a)** $\int \sqrt{2x}\ln x^2\,dx$

Let $u = \ln x^2 = 2\ln x$ and $dV = \sqrt{2}x^{1/2}\,dx$

$du = \frac{2}{x}\,dx \qquad V = \frac{2\sqrt{2}}{3}x^{3/2}$

$$= \frac{4\sqrt{2}}{3}x^{3/2}\ln x - \int \frac{2\sqrt{2}}{3}x^{3/2}\cdot\frac{2}{x}\,dx$$
$$= \frac{4\sqrt{2}}{3}x^{3/2}\ln x - \frac{4\sqrt{2}}{3}\int x^{1/2}\,dx$$
$$= \frac{4\sqrt{2}}{3}x^{3/2}\ln x - \frac{8\sqrt{2}}{9}x^{3/2} + C$$
$$= \frac{4\sqrt{2}}{9}x^{3/2}\,[3\ln|x| - 2] + C$$

(b) $\int_0^1 xe^{0.2x}\,dx$

Let $u = x$ and $dV = e^{0.2x}\,dx$

$du = dx \qquad = 5e^{0.2x}$

$$= 5xe^{0.2x}\Big|_0^1 - \int_0^1 5e^{0.2x}\,dx$$
$$= \left(5xe^{0.2x} - 25e^{0.2x}\right)\Big|_0^1$$
$$= \left[5(1)e^{0.2(1)} - 25e^{0.2(1)}\right] - \left[0 - 25e^0\right]$$
$$= 25 - 20e^{0.2}$$

(c) $\int_{-4}^0 x\sqrt{1-2x}\,dx$

Let $u = x$ and $dV = (1-2x)^{1/2}\,dx$

$du = dx \qquad = -\frac{1}{2}\cdot\frac{2}{3}(1-2x)^{3/2}$

$$= -\frac{x}{3}(1-2x)^{3/2}\Big|_{-4}^{0} - \int_{-4}^{0} -\frac{1}{3}(1-2x)^{3/2}\,dx$$

$$= -\frac{x}{3}(1-2x)^{3/2}\Big|_{-4}^{0} + \frac{1}{3}\int_{-4}^{0}(1-2x)^{3/2}\,dx$$

$$= \left[-\frac{x}{3}(1-2x)^{3/2} - \frac{1}{15}(1-2x)^{5/2}\right]\Big|_{-4}^{0}$$

$$= \left[0 - \frac{1}{15}(1)\right] - \left[\frac{4}{3}(9)^{3/2} - \frac{1}{15}(9)^{5/2}\right] = -\frac{298}{15}$$

(d) $\int \frac{x-1}{e^x}\,dx = \int (x-1)e^{-x}\,dx$

Let $u = x - 1$ and $dV = e^{-x}\,dx$

$du = dx$ $\quad V = -e^{-x}$

$$= -(x-1)e^{-x} - \int -e^{-x}\,dx$$

$$= -(x-1)e^{-x} + \int e^{-x}\,dx$$

$$= -(x-1)e^{-x} - e^{-x} + C$$

$$= [(-x+1) - 1]e^{-x} + C$$

$$= -xe^{-x} + C$$

2. **(a)** $\int_1^{\infty} \frac{1}{x^{1.1}}\,dx = \lim_{N\to\infty}\int_1^N x^{-1.1}\,dx$

$$= \lim_{N\to\infty}\left(-10x^{-0.1}\right)\Big|_1^N$$

$$= \lim_{N\to\infty}\left[-10N^{-0.1} + 10(1)^{-0.1}\right]$$

$$= 0 + 10 = 10$$

(b) $\int_1^{\infty} xe^{-2x}\,dx = \lim_{N\to\infty}\int_1^N xe^{-2x}\,dx$

Let $u = x$ and $dV = e^{-2x}$

$du = dx$ $\quad V = -\frac{1}{2}e^{-2x}$

$$= \lim_{N\to\infty}\left[-\frac{x}{2}e^{-2x}\Big|_1^N - \int_1^N -\frac{1}{2}e^{-2x}\,dx\right]$$

$$= \lim_{N\to\infty}\left[-\frac{x}{2}e^{-2x}\Big|_1^N + \frac{1}{2}\int_1^N e^{-2x}\,dx\right]$$

$$= \lim_{N\to\infty}\left[-\frac{x}{2}e^{-2x} - \frac{1}{4}e^{-2x}\right]\Big|_1^N$$

$$= \lim_{N\to\infty}\left[\left(-\frac{N}{2}e^{-2N} - \frac{1}{4}e^{-2N}\right) - \left(-\frac{1}{2}e^{-2(1)} - \frac{1}{4}e^{-2(1)}\right)\right]$$

$$= 0 + \frac{1}{2}e^{-2} + \frac{1}{4}e^{-2} = \frac{3}{4}e^{-2}$$

(c) $\int_1^{\infty} \frac{x}{(x+1)^2}\,dx = \lim_{N\to\infty}\int_1^N x(x+1)^{-2}\,dx$

Let $u = x$ and $dV = (x+1)^{-2}\,dx$

$du = dx$ $\quad V = -\frac{1}{(x+1)}$

$$= \lim_{N\to\infty}\left[-\frac{x}{x+1}\Big|_1^N - \int_1^N -\frac{1}{x+1}\,dx\right]$$

$$= \lim_{N\to\infty}\left[-\frac{x}{x+1}\Big|_1^N + \int_1^N \frac{1}{x+1}\,dx\right]$$

$$= \lim_{N\to\infty}\left[-\frac{x}{x+1} + \ln|x+1|\right]\Big|_1^N$$

$$= \lim_{N\to\infty}\left[\left(-\frac{N}{N+1} + \ln(N+1)\right) - \left(-\frac{1}{2} + \ln 2\right)\right]$$

Since $\lim_{N\to\infty} -\frac{N}{N+1} = \lim_{N\to\infty} -\frac{1}{1} = -1$, and $\lim_{N\to\infty}\ln(N+1) = \infty$,

$$= \lim_{N\to\infty}\left[-\frac{N}{N+1} + \ln(N+1) + \frac{1}{2} - \ln 2\right] = \infty$$

so, the integral diverges.

(d) $\int_0^{\infty} xe^{-x^2}\,dx = \lim_{N\to\infty}\int_0^N xe^{-x^2}\,dx$

Using substitution with $u = -x^2$ and $-\frac{1}{2}\,du = x\,dx$,

$$= \lim_{N\to\infty} -\frac{1}{2}\int_0^{-N^2} e^u\,du = \frac{1}{2}\lim_{N\to\infty}\int_{-N^2}^{0} e^u\,du$$
$$= \frac{1}{2}\lim_{N\to\infty}\left(e^u\Big|_{-N^2}^{0}\right)$$
$$= \frac{1}{2}\lim_{N\to\infty}(e^0 - e^{-N^2}) = \frac{1}{2}(1-0) = \frac{1}{2}$$

3. **(a)** $\int \left(\ln\sqrt{3x}\right)^2 dx$

$$= \int \ln(3x)^{1/2}\cdot\ln(3x)^{1/2}\,dx$$
$$= \int \frac{1}{2}\ln(3x)\cdot\frac{1}{2}\ln(3x)\,dx$$
$$= \frac{1}{4}\int(\ln 3x)^2\,dx$$

which most resembles $\int(\ln u)^n\,du$ (formula #27). Let $u = 3x$; then $du = 3\,dx$ or $\frac{1}{3}du = dx$,

$$= \frac{1}{4}\int(\ln u)^2\cdot\frac{1}{3}\,du = \frac{1}{12}\int(\ln u)^2\,du$$
$$= \frac{1}{12}\left[u(\ln u)^2 - 2\int\ln u\,du\right]$$

Using formula #23,

$$= \frac{1}{12}\left[u(\ln u)^2 - 2(u\ln|u| - u)\right] + C$$
$$= \frac{1}{12}\left[3x(\ln 3x)^2 - 2(3x)\ln|3x| + 3x\right] + C$$
$$= \frac{x}{4}(\ln 3x)^2 - \frac{x}{2}\ln 3x + \frac{x}{4} + C$$
$$= \frac{x}{4}\left[(\ln|3x|)^2 - 2\ln|3x| + 2\right] + C$$

(b) $\int\frac{dx}{x\sqrt{4+x^2}}$ is of the form $\int\frac{du}{u\sqrt{a^2+u^2}}$ (formula #11). Let $x = u$, $dx = du$, and $a = 2$,

$$= -\frac{1}{2}\ln\left|\frac{\sqrt{4+x^2}+2}{x}\right| + C$$

(c) $\int\frac{dx}{x^2\sqrt{x^2-9}}$ is of the form $\int\frac{du}{u^2\sqrt{u^2-a^2}}$ (formula #21). Let $x = u$, $dx = du$, and $a = 3$,

$$= \frac{\sqrt{x^2-9}}{9x} + C$$

(d) $\int\frac{dx}{3x^2-4x}$ can be written as $\int\frac{dx}{x(-4+3x)}$ so it is of the form $\int\frac{du}{u(a+bu)}$ (formula #6). Let $x = u$, $dx = du$, $a = -4$, and $b = 3$,

$$= -\frac{1}{4}\ln\left|\frac{x}{3x-4}\right| + C$$

4.

$$\lim_{N\to\infty}\int_0^N 300e^{0.001t}e^{-0.03(N-t)}\,dt$$
$$= \lim_{N\to\infty} 300e^{-0.03N}\int_0^N e^{0.031t}\,dt$$
$$= 300\lim_{N\to\infty} e^{-0.03N}\left(\frac{1000}{31}e^{0.031t}\Big|_0^N\right)$$
$$= \frac{300{,}000}{31}\lim_{N\to\infty} e^{-0.03N}\left(e^{0.031N} - e^0\right)$$
$$= \frac{300{,}000}{31}\lim_{N\to\infty}\left(e^{0.001N} - e^{-0.03N}\right) = \infty$$

5.

$$PV = \int_0^\infty (50+3t)e^{-0.06t}\,dt$$
$$\lim_{N\to\infty}\int_0^N (50+3t)e^{-0.06t}\,dt$$

Using integration by parts with $u = 50 + 3t$ and $dV = e^{-0.06t}\,dt$,

$$= \lim_{N\to\infty}\left[(50+3t) - \frac{50}{3}e^{-0.06t}\Big|_0^N - \int_0^N -\frac{50}{3}e^{-0.06t}3\,dt\right]$$

$$= \lim_{N\to\infty}\left[-\frac{50}{3}(50+3t)e^{-0.06t}\Big|_0^N + 50\int_0^N e^{-0.06t}\,dt\right]$$

$$= \lim_{N\to\infty} -\frac{50}{3}\left[(50+3t)e^{-0.06t} + 50e^{-0.06t}\right]\Big|_0^N$$

$$= -\frac{50}{3}\lim_{N\to\infty}\left[50e^{-0.06t} + 3te^{-0.06t} + 50e^{-0.06t}\right]\Big|_0^N$$

$$= -\frac{50}{3}\lim_{N\to\infty}\left(100e^{-0.06t} + 3te^{-0.06t}\right]\Big|_0^N$$

$$= -\frac{50}{3}\lim_{N\to\infty}\left[\left(100e^{-0.06N} + 3Ne^{-0.06N}\right) - \left(100e^{0}+0\right)\right]$$

$$= -\frac{50}{3}\cdot -100 = \frac{5{,}000}{3} \approx 1{,}666.6667 \text{ thousand,}$$

or approximately $1,666,666.67

6. Amount of drug $= \lim_{N\to\infty}\int_0^N 0.7e^{-0.2(N-t)}dt$

$$= \lim_{N\to\infty} 0.7e^{-0.2N}\int_0^N e^{0.2t}\,dt$$

$$= \lim_{N\to\infty} 0.7e^{-0.2N}\left(5e^{0.2t}\right)\Big|_0^N$$

$$= 3.5\lim_{N\to\infty}\left[e^{-0.2N}\left(e^{0.2N} - e^{0}\right)\right]$$

$$= 3.5\lim_{N\to\infty}\left(e^{0} - e^{-0.2N}\right)$$

$$= 3.5\cdot 1 = 3.5 \text{ mg}$$

7.

$$\int_0^\infty 800e^{-0.05t}\,dt$$

$$= \lim_{N\to\infty}\int_0^N 800e^{-0.05t}\,dt$$

$$= 800\lim_{N\to\infty}\int_0^N e^{-0.05t}\,dt$$

$$= 800\lim_{N\to\infty}\left(-20\cdot e^{-0.05t}\Big|_0^N\right)$$

$$= -16{,}000\lim_{N\to\infty}\left(e^{-0.05N} - e^{0}\right)$$

$$= -16{,}000(0-1) = 16{,}000 \text{ units}$$

8. To approximate $\int_3^4 \frac{\sqrt{25-x^2}}{x}\,dx$ using the trapezoidal rule with $n=8$, $\Delta x = \frac{4-3}{8} = 0.125$,

$$\approx \frac{0.125}{2}\left[\left(\frac{\sqrt{25-(3)^2}}{3}\right) + \left(\frac{\sqrt{25-(3.125)^2}}{3.125}\right)\right.$$

$$+2\left(\frac{\sqrt{25-(3.25)^2}}{3.25}\right) + 2\left(\frac{\sqrt{25-(3.375)^2}}{3.375}\right)$$

$$+2\left(\frac{\sqrt{25-(3.5)^2}}{3.5}\right) + 2\left(\frac{\sqrt{25-(3.625)^2}}{3.625}\right)$$

$$+2\left(\frac{\sqrt{25-(3.75)^2}}{3.75}\right) + 2\left(\frac{\sqrt{25-(3.875)^2}}{3.875}\right)$$

$$+\left(\frac{\sqrt{25-(4)^2}}{4}\right) \approx 1.027552$$

Using formula #17 with $x = u$, $dx = du$, and $a = 5$,

$$= \left[\sqrt{25-x^2} - 5\ln\left|\frac{5+\sqrt{25-x^2}}{x}\right|\right]\Big|_3^4$$

$$= \left[\sqrt{25-4^2} - 5\ln\left|\frac{5+\sqrt{25-4^2}}{4}\right|\right]$$

$$= -\left[\sqrt{25-3^2} - 5\ln\left|\frac{5+\sqrt{25-3^2}}{3}\right|\right]$$

$$= (3 - 5\ln 2) - (4 - 5\ln 3)$$

$$= -1 - \ln 2^5 + \ln 3^5$$

$$= -1 + \ln\left(\frac{3}{2}\right)^5 = -1 + 5\ln\left(\frac{3}{2}\right) \approx 1.027326$$

Review Problems

1.

$$\int te^{1-t}\,dt$$

Let $u = t$ and $dV = e^{1-t}\,dt$

$du = dt$ $\quad V = -e^{1-t}$

$$= -te^{1-t} - \int -e^{1-t}\, dt$$
$$= -te^{1-t} + \int e^{1-t}\, dt$$
$$= -te^{1-t} - e^{1-t} + C$$
$$= -e^{1-t}(t+1) + C$$

3. $$\int x(2x+3)^{1/2}\, dx$$

Let $u = x$ and $dV = (2x+3)^{1/2}$

$du = dx$

Using substitution with $u = 2x + 3$,

$$V = \frac{1}{2}\left[\frac{2}{3}(2x+3)^{3/2}\right]$$
$$= \frac{1}{3}(2x+3)^{3/2}$$

So, $\int x(2x+3)^{1/2}\, dx$

$$= \frac{x}{3}(2x+3)^{3/2} - \int \frac{1}{3}(2x+3)^{3/2}\, dx$$

Using substitution with $u = 2x + 3$,

$$= \frac{x}{3}(2x+3)^{3/2} - \frac{1}{3}\left(\frac{1}{2}\right)\left(\frac{2}{5}\right)(2x+3)^{5/2} + C$$
$$= \frac{x}{3}(2x+3)^{3/2} - \frac{1}{15}(2x+3)^{5/2} + C$$

5. $$\int_1^4 \frac{\ln\sqrt{S}}{\sqrt{S}}\, dS = \int_1^4 S^{-1/2} \ln S^{1/2}\, dS$$

Let $u = \ln S^{1/2} = \frac{1}{2}\ln S$ and $dV = S^{-1/2}\, dS$

$du = \frac{1}{2S}\, dS$ $\qquad V = 2S^{1/2}$

$$= S^{1/2}\ln S\Big|_1^4 - \int_1^4 2S^{1/2}\cdot\frac{1}{2S}\, dS$$
$$= S^{1/2}\ln S\Big|_1^4 - \int_1^4 S^{-1/2}\, dS$$
$$= \left(S^{1/2}\ln S - 2S^{1/2}\right)\Big|_1^4$$
$$= \left[\sqrt{4}\ln 4 - 2\sqrt{4}\right] - [1 \ln 1 - 2(1)]$$
$$= 2\ln 4 - 2 = 2\ln(2)^2 - 2 = 4\ln 2 - 2$$

7. $$\int_{-2}^{1} (2x+1)(x+3)^{3/2}\, dx$$

Let $u = 2x + 1$ and $dV = (x+3)^{3/2}\, dx$

$du = 2\, dx$ $\qquad V = \frac{2}{5}(x+3)^{5/2}$

$$= \frac{2}{5}(2x+1)(x+3)^{5/2}\Big|_{-2}^{1} - \int_{-2}^{1} \frac{2}{5}(x+3)^{5/2}\cdot 2\, dx$$
$$= \left[\frac{2}{5}(2x+1)(x+3)^{5/2} - \left(\frac{4}{5}\right)\left(\frac{2}{7}\right)(x+3)^{7/2}\right]\Big|_{-2}^{1}$$
$$= \left[\frac{2}{5}(2(1)+1)(1+3)^{5/2} - \frac{8}{35}(1+3)^{7/2}\right] - \left[\frac{2}{5}(2(-2)+1)(-2+3)^{5/2} - \frac{8}{35}(-2+3)^{7/2}\right]$$
$$= \frac{74}{7}$$

9. $$\int x^3\left(3x^2+2\right)^{1/2} dx = \int x^2 \cdot x\left(3x^2+2\right)^{1/2} dx$$

Let $u = x^2$ and $dV = x\left(3x^2+2\right)^{1/2} dx$

$du = 2x\, dx$

Using substitution with $u = 3x^2 + 2$,

$$V = \left(\frac{1}{6}\right)\left(\frac{2}{3}\right)\left(3x^2+2\right)^{3/2}$$
$$V = \frac{1}{9}\left(3x^2+2\right)^{3/2}$$

So, $\int x^2 \cdot x(3x^2+2)^{1/2}\,dx$

$$= \frac{x^2}{9}\left(3x^2+2\right)^{3/2} - \int \frac{1}{9}\left(3x^2+2\right)^{3/2} 2x\,dx$$
$$= \frac{x^2}{9}\left(3x^2+2\right)^{3/2} - \frac{2}{9}\int \left(3x^2+2\right)^{3/2} x\,dx$$
$$= \frac{x^2}{9}\left(3x^2+2\right)^{3/2}$$
$$-\left(\frac{2}{9}\right)\left(\frac{1}{6}\right)\left(\frac{2}{5}\right)\left(3x^2+2\right)^{5/2} + C$$
$$= \frac{x^2}{9}\left(3x^2+2\right)^{3/2} - \frac{2}{135}\left(3x^2+2\right)^{5/2} + C$$

11. $$\int \frac{5\,dx}{8-2x^2} = \int \frac{5\,dx}{2(4-x^2)} = \frac{5}{2}\int \frac{dx}{4-x^2}$$

which is of the form $\int \frac{du}{a^2-u^2}$ (formula #16). Let $x=u$, $dx=du$, and $a=2$,

$$= \frac{5}{2}\left[\frac{1}{2(2)}\ln\left|\frac{2+x}{2-x}\right| + C\right]$$
$$= \frac{5}{8}\ln\left|\frac{2+x}{2-x}\right| + C$$

13. $$\int w^2 e^{-w/3} dw = \int w^2 e^{-\frac{1}{3}w} dw$$

which is of the form $\int u^n e^{au}\,du$ (formula #26). Let $w=u$, $dw=du$, and $a=-\frac{1}{3}$,

$$= \frac{1}{-\frac{1}{3}} w^2 e^{-w/3} - \frac{2}{-\frac{1}{3}}\int w e^{-w/3}\,dw$$
$$= -3w^2 e^{-w/3} + 6\int w e^{-w/3}\,dw$$

Using formula #22,

$$= -3w^2 e^{-w/3} + 6\left[\frac{1}{(-\frac{1}{3})^2}\left(-\frac{1}{3}w - 1\right)e^{-w/3}\right] + C$$
$$= -3w^2 e^{-w/3} + 54\left(-\frac{1}{3}w - 1\right)e^{-w/3} + C$$
$$= -3w^2 e^{-w/3} - 18we^{-w/3} - 54e^{-w/3} + C$$

15. $$\int (\ln 2x)^3\,dx = \frac{1}{2}\int (\ln 2x)^3 \cdot 2\,dx$$

which is of the form $\int (\ln u)^n\,du$ (formula #27). Let $u=2x$, $du=2\,dx$, and $n=3$,

$$= \frac{1}{2}\left[2x(\ln 2x)^3 - 3\int (\ln 2x)^2 2\,dx\right]$$
$$= x(\ln 2x)^3 - \frac{3}{2}\left[2x(\ln 2x)^2 - 2\int (\ln 2x)2\,dx\right]$$
$$= x(\ln 2x)^3 - 3x(\ln 2x)^2 + 3\left[2x\ln|2x| - 2x\right] + C$$
$$= x(\ln 2x)^3 - 3x(\ln 2x)^2 + 6x\ln 2x - 6x + C$$
$$= x\left[(\ln 2x)^3 - 3(\ln 2x)^2 + 6(\ln 2x) - 6\right] + C$$

17. $$\int_0^\infty \frac{1}{\sqrt[3]{1+2x}}\,dx$$

Using substitution with $u = 1+2x$,

$$= \lim_{N\to\infty}\int_0^N (1+2x)^{-1/3}\,dx$$
$$= \lim_{N\to\infty}\frac{3}{4}(1+2x)^{2/3}\Big|_0^N = \infty.$$

So, the interval diverges.

19. $$\int_0^\infty \frac{3x}{x^2+1}\,dx$$

Using substitution with $u = x^2+1$,

$$= 3\lim_{N\to\infty}\int_0^N \frac{3x}{x^2+1}\,dx$$
$$= 3\lim_{N\to\infty}\frac{1}{2}\ln(x^2+1)\Big|_0^N = \infty.$$

So, the interval diverges.

21. $$\int_0^\infty xe^{-2x}\,dx$$

Using integration by parts with $u=x$ and $dV = e^{-2x}\,dx$,

$$= \lim_{N\to\infty} \int_0^N xe^{-2x}\,dx$$

$$= \lim_{N\to\infty} \left[\left(-\frac{1}{2}xe^{-2x} \right) \Big|_0^N + \frac{1}{2}\int_0^N e^{-2x}\,dx \right]$$

$$= \lim_{N\to\infty} \left(-\frac{1}{2}xe^{-2x} - \frac{1}{4}e^{-2x} \right) \Big|_0^N = \frac{1}{4}$$

23. $$\int_0^\infty x^2 e^{-2x}\,dx = \lim_{N\to\infty} \int_0^N x^2 e^{-2x}\,dx$$

Using integration by parts with $u = x^2$ and $dV = e^{-2x}\,dx$,

$$= -\lim_{N\to\infty} \frac{1}{2}x^2 e^{-2x}\Big|_0^N + \lim_{N\to\infty} \int_0^N xe^{-2x}\,dx$$

$$= -\lim_{N\to\infty} \frac{1}{2}x^2 e^{-2x}\Big|_0^N - \lim_{N\to\infty} \frac{1}{2}xe^{-2x}\Big|_0^N$$

$$- \lim_{N\to\infty} \frac{1}{4}xe^{-2x}\Big|_0^N = \frac{1}{4}$$

25. $$\int_1^\infty \frac{\ln x}{\sqrt{x}}\,dx = \lim_{N\to\infty} \int_1^N x^{-1/2}\ln x\,dx$$

Using integration by parts with $u = \ln x$ and $dV = x^{-1/2}\,dx$,

$$= \lim_{N\to\infty} \left[2x^{1/2}\ln x\Big|_1^N - \int_1^N 2x^{1/2}\cdot\frac{1}{x}\,dx \right]$$

$$= \lim_{N\to\infty} \left[2x^{1/2}\ln x\Big|_1^N - 2\int_1^N x^{-1/2}\,dx \right]$$

$$= \lim_{N\to\infty} 2\left[x^{1/2}\ln x - 2x^{1/2} \right]\Big|_1^N$$

$$= 2\lim_{N\to\infty} \left[\left(N^{1/2}\ln N - 2N^{1/2} \right) - (\ln 1 - 2) \right]$$

$= \infty$, so the integral diverges.

27. $$\lim_{N\to\infty} \int_0^N \left(300 - 200e^{-0.03t}\right)e^{-0.02(N-t)}\,dt$$

$$= \lim_{N\to\infty} \int_0^N 300e^{-0.02(N-t)} - 200e^{-0.02N-0.01t}\,dt$$

$$= \lim_{N\to\infty} 300e^{-0.02N} \int_0^N e^{0.02t}\,dt$$

$$- \lim_{N\to\infty} 200e^{-0.02N} \int_0^N e^{-0.01t}\,dt$$

$$= 300 \lim_{N\to\infty} e^{-0.02N} \int_0^N e^{0.02t}\,dt$$

$$- 200 \lim_{N\to\infty} e^{-0.02N} \int_0^N e^{-0.01t}\,dt$$

$$= 300 \lim_{N\to\infty} e^{-0.02N} \left(50e^{0.02t}\Big|_0^N \right)$$

$$- 200 \lim_{N\to\infty} e^{-0.02N} \left(-100e^{-0.01t}\Big|_0^N \right)$$

$$= 15{,}000 \lim_{N\to\infty} e^{-0.02N}\left(e^{0.02N} - e^0\right)$$

$$+ 20{,}000 \lim_{N\to\infty} e^{-0.02N}\left(e^{-0.01N} - e^0\right)$$

$$= 15{,}000 \lim_{N\to\infty} \left(e^0 - e^{-0.02N}\right)$$

$$+ 20{,}000 \lim_{N\to\infty} \left(e^{-0.03N} - e^{-0.02N}\right)$$

$= 15{,}000(1 - 0) + 20{,}000(0 - 0)$

$= 15{,}000$ pounds

29. In N years, the population of the city will be

$$P_0 f(N) + \int_0^N r(t) f(N-t)\,dt$$

where $P_0 = 100{,}000$ is the current population,

$$f(t) = e^{-t/20}$$

is the fraction of the residents remaining for at least t years, and

$$r(t) = 100t$$

is the rate of new arrivals. In the long run, the number of residents will be

$$\lim_{N\to\infty}\left[100{,}000e^{-N/20}+\int_0^N 100te^{-(N-t)/20}\,dt\right]$$
$$=0+\lim_{N\to\infty}\left[100e^{-N/20}\int_0^N te^{t/20}\,dt\right]$$
$$=\lim_{N\to\infty}100e^{-N/20}\left[20te^{t/20}-400e^{t/20}\,dt\right]\Big|_0^N$$
$$=\lim_{N\to\infty}100\left(20N-400+400e^{-N/20}\right)=\infty.$$

So, the population will increase without bound.

31. $PV=\lim_{N\to\infty}\int_0^N (8{,}000+400t)e^{-0.05t}dt$
Using integration by parts with $u=8{,}000+400t$ and $dV=e^{-0.05t}dt$

$$=\lim_{N\to\infty}\left[-20(8{,}000+400t)e^{-0.05t}\Big|_0^N-\int_0^N -8{,}000e^{-0.05t}dt\right]$$
$$=\lim_{N\to\infty}\left[-20(8{,}000+400t)e^{-0.05t}\Big|_0^N+8{,}000\int_0^N e^{-0.05t}\right]$$
$$=-20\lim_{N\to\infty}\left[(8{,}000+400t)e^{-0.05t}+8{,}000e^{-0.05t}\right]\Big|_0^N$$
$$=-20\lim_{N\to\infty}\left[\left((8{,}000+400N)e^{-0.05N}+8{,}000e^{-0.05N}\right)-(8{,}000e^0+8{,}000e^0)\right]$$
$$=-20\lim_{N\to\infty}\left[16{,}000e^{-0.05N}+400Ne^{-0.05N}-16{,}000\right]$$
$$=-20(-16{,}000)=\$320{,}000$$

33. For $\int_1^3 \frac{1}{x}\,dx$ with $n=10$, $\Delta x=\frac{3-1}{10}=0.2$, and $x_1=1, x_2=1.2, x_3=1.4, \ldots, x_{10}=2.8, x_{11}=3$.

(a) By the trapezoidal rule, $\int_1^3 \frac{1}{x}\,dx$

$$\approx\frac{\Delta x}{2}\left[f(x_1)+2f(x_2)+2f(x_3)+\cdots+2f(x_{10})+f(x_{11})\right]$$
$$=\frac{0.2}{2}\left[\frac{1}{1}+2\left(\frac{1}{1.2}\right)+2\left(\frac{1}{1.4}\right)+2\left(\frac{1}{1.6}\right)+2\left(\frac{1}{1.8}\right)+2\left(\frac{1}{2}\right)+2\left(\frac{1}{2.2}\right)+2\left(\frac{1}{2.4}\right)+2\left(\frac{1}{2.6}\right)+2\left(\frac{1}{2.8}\right)+\frac{1}{3}\right]$$
$$\approx 1.1016$$

The error estimate is $|E_{10}|\le\frac{m(3-1)^3}{12(10)^2}=\frac{m}{150}$ where m is the maximum value of $|f''(x)|$ on $1\le x\le 3$. Now, $f'(x)=\frac{-1}{x^2}$ and $f''(x)=\frac{2}{x^3}$. Since $f''(x)$ is always positive and decreasing on $1\le x\le 3$, $m=|f''(1)|=2$.

$$|E_{10}|\le\frac{2}{150}\approx 0.0133.$$

(b) By Simpson's rule,

$$\int_1^3\frac{1}{x}\,dx\approx\frac{\Delta x}{3}\left[f(x_1)+4f(x_2)+2f(x_3)+4f(x_4)+\cdots+4f(x_{10})+f(x_{11})\right]$$
$$=\frac{0.2}{3}\left[\frac{1}{1}+4\left(\frac{1}{1.2}\right)+2\left(\frac{1}{1.4}\right)+4\left(\frac{1}{1.6}\right)+2\left(\frac{1}{1.8}\right)+4\left(\frac{1}{2}\right)+2\left(\frac{1}{2.2}\right)+4\left(\frac{1}{2.4}\right)+2\left(\frac{1}{2.6}\right)+4\left(\frac{1}{2.8}\right)+\frac{1}{3}\right]$$
$$\approx 1.0987$$

For the error estimate,

$$|E_{10}| \le \frac{m(3-1)^5}{180(10)^4} = \frac{32m}{1{,}800{,}000} = \frac{m}{56{,}250}$$

where m is the maximum value of $|f^{(4)}(x)|$ on $1 \le x \le 3$.

$$f^{(3)}(x) = \frac{-6}{x^4} \text{ and } f^{(4)}(x) = \frac{24}{x^5}$$

Since $f^{(4)}(x)$ is always positive and decreasing on $1 \le x \le 3$, $m = |f^4(1)| = 24$.

$$|E_{10}| \le \frac{24}{56{,}250} \approx 0.000427.$$

35. For $\int_1^2 \frac{e^x}{x}\,dx$ with $n = 10$, $\Delta x = \frac{2-1}{10} = 0.1$, and $x_1 = 1, x_2 = 1.1, x_3 = 1.2, \ldots, x_{10} = 1.9, x_{11} = 2$.

(a) By the trapezoidal rule, $\int_1^2 \frac{e^x}{x}\,dx$

$$\approx \frac{\Delta x}{2}\left[f(x_1) + 2f(x_2) + 2f(x_3) + \cdots + 2f(x_{10}) + f(x_{11})\right]$$

$$= \frac{0.1}{2}\left[\frac{e^1}{1} + 2\left(\frac{e^{1.1}}{1.1}\right) + 2\left(\frac{e^{1.2}}{1.2}\right) + 2\left(\frac{e^{1.3}}{1.3}\right) + 2\left(\frac{e^{1.4}}{1.5}\right) + 2\left(\frac{e^{1.5}}{1.5}\right) + 2\left(\frac{e^{1.6}}{1.6}\right) + 2\left(\frac{e^{1.7}}{1.7}\right) + 2\left(\frac{e^{1.8}}{1.8}\right) + 2\left(\frac{e^{1.9}}{1.9}\right) + \frac{e^2}{2}\right]$$

$$\approx 3.0607$$

The error estimate is

$$|E_{10}| \le \frac{m(2-1)^2}{12(10)^2} = \frac{m}{1{,}200}$$

where m is the maximum value of $|f''(x)|$ on $1 \le x \le 2$. Rewriting $f(x) = x^{-1}e^x$,

$$f'(x) = (x^{-1})(e^x \cdot 1) + (e^x)(-x^{-2}) = e^x(x^{-1} - x^{-2})$$

$$f''(x) = (e^x)(-x^{-2} + 2x^{-3}) + (x^{-1} - x^{-2})(e^x \cdot 1) = e^x(-x^{-2} + 2x^{-3} + x^{-1} - x^{-2}) = e^x(2x^{-3} - 2x^{-2} + x^{-1}) = e^x\left(\frac{2}{x^3} - \frac{2}{x^2} + \frac{1}{x}\right)$$

The maximum value of $|f''(x)|$ on $1 \le x \le 2$ is $|f''(1)| = e$. (A graph is the easiest way to see this.)

$$|E_{10}| = \frac{e}{1{,}200} \approx 0.00227.$$

(b) By Simpson's rule,

$$\int_1^2 \frac{e^x}{x}\,dx \approx \frac{\Delta x}{3}\left[f(x_1) + 4f(x_2) + 2f(x_3) + 4f(x_4) + \cdots + 4f(x_{10}) + f(x_{11})\right]$$

$$= \frac{0.1}{3}\left[\frac{e^1}{1} + 4\frac{e^{1.1}}{1.1} + 2\frac{e^{1.2}}{1.2} + 4\frac{e^{1.3}}{1.3} + 2\frac{e^{1.4}}{1.4} + 4\frac{e^{1.5}}{1.5} + 2\frac{e^{1.6}}{1.6} + 4\frac{e^{1.7}}{1.7} + 2\frac{e^{1.8}}{1.8} + 4\frac{e^{1.9}}{1.9} + \frac{e^2}{2}\right]$$

$$\approx 3.0591$$

For the error estimate,

$$|E_{10}| \le \frac{m(2-1)^5}{180(10)^4} = \frac{m}{1{,}800{,}000}$$

where m is the maximum value of $|f^{(4)}(x)|$ on $1 \le x \le 2$.

$$f^{(3)}(x) = (e^x)(-6x^{-4} + 4x^{-3} - x^{-2}) + (2x^{-3} - 2x^{-2} + x^{-1})(e^x \cdot 1) = e^x(-6x^{-4} + 6x^{-3} - 3x^{-2} + x^{-1})$$

$$
\begin{aligned}
f^{(4)}(x) &= (e^x)(24x^{-5} - 18x^{-4} + 6x^{-3} - x^{-2}) \\
&\quad + (-6x^4 + 6x^{-3} - 3x^{-2} + x^{-1})(e^x \cdot 1) \\
&= e^x(24x^{-5} - 24x^{-4} + 12x^{-3} - 4x^{-2} + x^{-1} \\
&= e^x\left(\frac{24}{x^5} - \frac{24}{x^4} + \frac{12}{x^3} - \frac{4}{x^2} + \frac{1}{x}\right)
\end{aligned}
$$

The maximum value of $|f^{(4)}(x)|$ on $1 \le x \le 2$ is $|f^{(4)}(1)| + 9e$. (A graph is the easiest way to see this.)

$$|E_{10}| \le \frac{9e}{1{,}800{,}000} \approx 0.0000136.$$

37. (a)

$$
\begin{aligned}
|E_n| \le \frac{m(b-a)^3}{12n^2} &< 0.00005 \\
\frac{m(3-1)^3}{12n^2} &< 0.00005 \\
n^2 &> \frac{8m}{12(0.00005)} \\
n^2 &> 13{,}333.33m
\end{aligned}
$$

$$
\begin{aligned}
f(x) &= x^{\frac{1}{2}} \\
f'(x) &= \frac{1}{2}x^{-\frac{1}{2}} \\
f''(x) &= -\frac{1}{4}x^{-\frac{3}{2}} = -\frac{1}{4x^{\frac{3}{2}}}
\end{aligned}
$$

Since $f''(x)$ is always increasing for $1 \le x \le 3$, but less than zero, the maximum value m of $|f''(x)| = |f''(1)| = \frac{1}{4}$.

So, $n^2 > 3333.33$

$n > 57.735$, or $n = 58$.

(b)

$$
\begin{aligned}
|E_n| \le \frac{m(b-a)^5}{180n^4} &< 0.00005 \\
\frac{m(3-1)^5}{180n^4} &< 0.00005 \\
n^4 &> \frac{32m}{180(0.00005)} \\
n^4 &> 3{,}555.556m
\end{aligned}
$$

$$
\begin{aligned}
f^{(3)}(x) &= \frac{3}{8}x^{-\frac{5}{2}} \\
f^{(4)}(x) &= -\frac{15}{16}x^{-\frac{7}{2}} = -\frac{15}{16x^{\frac{7}{2}}}
\end{aligned}
$$

Since $f^{(4)}(x)$ is always increasing for $1 \le x \le 3$, but less than zero, the maximum value m of $|f^{(4)}(x)| = |f^{(4)}(1)| = \frac{15}{16}$.

So, $n^4 > 3333.33$

$n > 7.598$, or $n = 8$.

39. (a) The total cost of producing 8 units is

$$\int_0^8 \sqrt{q}e^{0.01q}\,dq$$

(b) Using $n = 8$, $\Delta x = \frac{8-0}{8} = 1$ and $x_1 = 0$, $x_2 = 1, x_3 = 2, \ldots, x_9 = 8$.

For the trapezoidal rule, $\int_0^8 \sqrt{q}e^{0.01q}\,dq$

$$
\begin{aligned}
&\approx \frac{\Delta x}{2}\left[f(x_1) + 2f(x_2) + 2f(x_3) + \cdots \right.\\
&\quad \left. + 2f(x_8) + f(x_9)\right] \\
&= \frac{1}{2}\left[0 + 2\sqrt{1}e^{0.01} + 2\sqrt{2}e^{0.02} + 2\sqrt{3}e^{0.03}\right.\\
&\quad + 2\sqrt{4}e^{0.04} + 2\sqrt{5}e^{0.05} + 2\sqrt{6}e^{0.06} \\
&\quad \left. + 2\sqrt{7}e^{0.07} + \sqrt{8}e^{0.08}\right] \\
&\approx \$15.64
\end{aligned}
$$

41. To use the graphing utility to find where the curves intersect, and then find the area region bounded by the curves,

Press [y=] and input $-x \wedge 3 - 2x^2 + 5x - 2$ for $y_1 =$ and input $x\ln(x)$ for $y_2 =$.

Use window dimensions [−4, 3]0.5 by [−0.8, 0.4]0.1

Press [graph].

Use trace and zoom to find the points of intersection or use the intersect function under the calc menu to find that (0.406, −0.37) and (1, 0) are the two points of intersection.

To find the area bounded by the curves, we must find

$$\int_{0.406}^{1} \left(-x^3 - 2x^2 + 5x - 2 \ - \ x \ln x\right) \, dx$$

Use the $\int f(x)\,dx$ function under the calc menu (making sure that y_1 is shown in the upper left corner) with $x = 0.406$ as the lower limit and $x = 1$ as the upper limit to find that $\int_{0.406}^{1} \left(-x^3 - 2x^2 + 5x - 2\right) \, dx = .03465167.$
Repeat this process with y_2 activated to find that $\int_{0.406}^{1} x \ln x \, dx \approx -.1344992.$
The area is $0.03465167 - (-0.1344992) \approx 0.1692.$ Alternatively, you can use fn Int function under the math menu:

$$\text{fn Int}(y_1 - y_2, x, 0.406, 1)$$

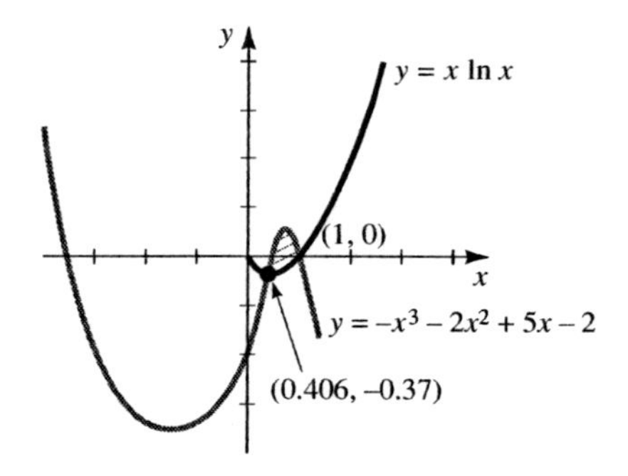

43. To use numeric integration feature to evaluate the integral,

Press [y=] and input $\dfrac{2}{(9 - x^2)}$ for $y_1 = .$

Use window dimensions [−5, 5]1 by [−5, 5]1.

Press [graph].

Use the $\int f(x)\,dx$ function under the calc menu with $x = -1$ as the lower limit and $x = 1$ as the upper limit to find $\int_{-1}^{1} \frac{2}{9 - x^2} \, dx \approx 0.4621.$

45. To use numeric integration feature to compute the integral,

Press [y=] and enter $\left(\dfrac{1}{\sqrt{\pi}}\right) * e^{\wedge(-x^2)}$ for $y_1 = .$

Use window dimensions [−50, 50]20 by [−3, 3]1.

Press [graph].

Use the $\int f(x)\,dx$ function under the calc menu with $x = 0$ as the lower limit and $x = 1$ as the upper limit to find

$$\int_0^1 \frac{1}{\sqrt{\pi}} e^{-x^2} \, dx = 0.4214$$

Repeat this process with $x = 10$ as the upper limit to find

$$\int_0^{10} \frac{1}{\sqrt{\pi}} e^{-x^2} \, dx = 0.5$$

Repeat this process with $x = 50$ as the upper limit to find

$$\int_0^{50} \frac{1}{\sqrt{\pi}} e^{-x^2} \, dx = 0.5$$

The improper integral $\int_0^{\infty} \frac{1}{\sqrt{\pi}} e^{-x^2}$ appears to converge to 0.5.

Chapter 7

Calculus of Several Variables

7.1 Functions of Several Variables

1. $f(x, y) = (x-1)^2 + 2xy^3.$

$$f(2, -1) = (2-1)^2 + 2(2)(-1)^3 = -3$$
$$f(1, 2) = (1-1)^2 + 2(1)(2)^3 = 16.$$

3. $g(x, y) = \sqrt{y^2 - x^2}$

$$g(4, 5) = \sqrt{5^2 - 4^2} = \sqrt{9} = 3$$
$$g(-1, 2) = \sqrt{2^2 - (-1)^2} = \sqrt{3} \approx 1.732$$

5. $f(r, s) = \dfrac{s}{\ln r}.$

$$f(e^3, 3) = \frac{3}{\ln e^2} = \frac{3}{2}$$
$$f(\ln 9, e^3) = \frac{e^3}{\ln(\ln 9)} \approx 25.515$$

7.
$$g(x, y) = \frac{y}{x} + \frac{x}{y}$$
$$g(1, 2) = \frac{2}{1} + \frac{1}{2} = \frac{5}{2}$$
$$g(2, -3) = -\frac{3}{2} + -\frac{2}{3} = -\frac{13}{6} \approx -2.167$$

9.
$$f(x, y, z) = xyz$$
$$f(1, 2, 3) = (1)(2)(3) = 6$$
$$f(3, 2, 1) = (3)(2)(1) = 6$$

11.
$$F(r, s, t) = \frac{\ln(r+t)}{r+s+t}$$
$$f(1, 1, 1) = \frac{\ln(2)}{3} \approx 0.2310$$
$$f(0, e^2, 3e^2) = \frac{\ln(3e^2)}{4e^2} = \frac{2 + \ln 3}{4e^2} \approx 0.1048$$

13. $f(x, y) = \dfrac{5x + 2y}{4x + 3y}$
The domain of f is the set of all real pairs (x, y) such that $4x + 3y \neq 0$, or $y \neq -\frac{4}{3}x$.

15. $f(x, y) = \sqrt{x^2 - y}$
The domain of f is the set of all real pairs (x, y) such that $x^2 - y \geq 0$, or $y \leq x^2$.

17. $f(x, y) = \ln(x + y - 4)$
The domain of f is the set of all real pairs (x, y) such that $x + y - 4 > 0$, or $y > 4 - x$.

19. $f(x, y) = x + 2y$
With $C = 1$, $C = 2$, and $C = -3$, the three sketched level curves have equations
$x + 2y = 1$, $x + 2y = 2$, and $x + 2y = -3$.

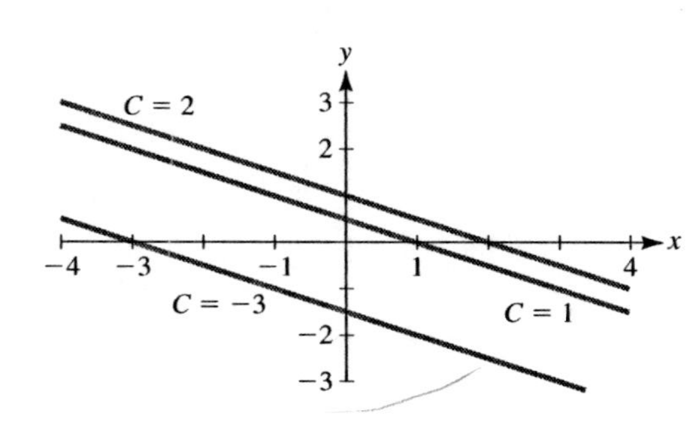

21. $f(x, y) = x^2 - 4x - y.$

With $C = -4$, and $C = 5$, the two sketched level curves have equations $x^2 - 4x - y = -4$ and $x^2 - 4x - y = 5$.

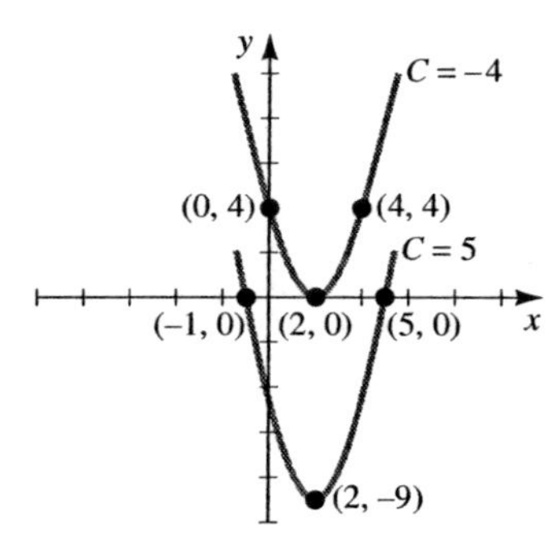

23. $f(x, y) = xy$.
With $C = 1$, $C = -1$, $C = 2$, and $C = -2$, the four sketched level curves have equations $xy = 1$, $xy = -1$, $xy = 2$, and $xy = -2$.

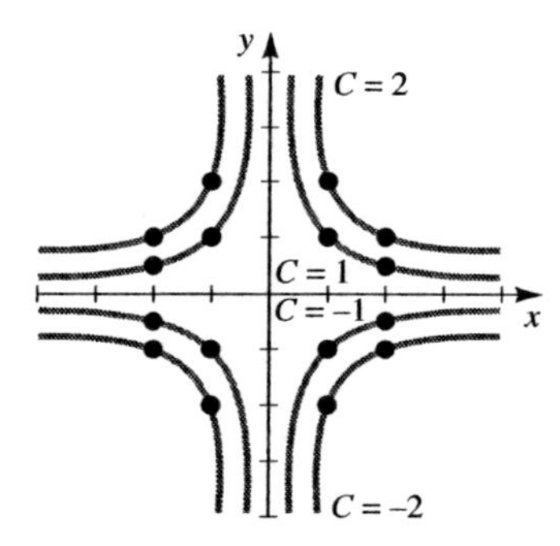

25. $f(x, y) = xe^y$.
With $C = 1$, and $C = e$, the two sketched level curves have equations $xe^y = 1$ and $xe^y = e$.

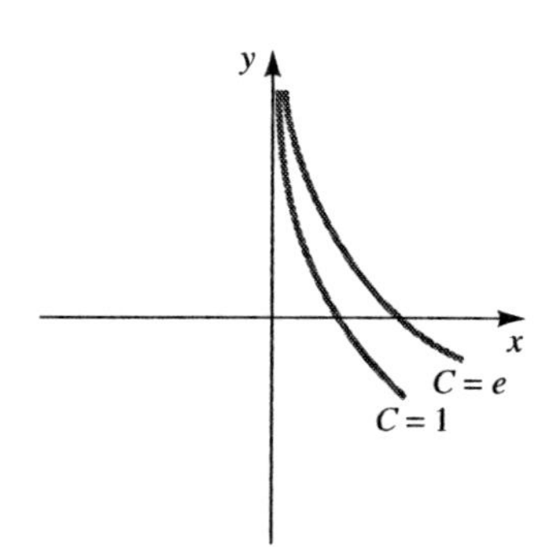

27. **(a)** $Q(x, y) = 10x^2y$ and $x = 20$, $y = 40$.
$Q(20, 40) = 10(20)^2(40) = 160{,}000$ units.

(b) With one more skilled worker, $x = 21$ and the additional output is

$$Q(21, 40) - Q(20, 40) = 16{,}400 \text{ units.}$$

(c) With one more unskilled worker, $y = 41$ and the additional output is

$$Q(20, 41) - Q(20, 40) = 4{,}000 \text{ units.}$$

(d) With one more skilled worker and one more unskilled worker, $x = 21$ and $y = 41$, so the additional output is

$$Q(21, 41) - Q(20, 40) = 20{,}810 \text{ units.}$$

29. **(a)** Let R denote the total monthly revenue. Then,

$$\begin{aligned} R &= (\text{revenue from the first brand}) \\ &\quad + (\text{revenue from the second brand}) \\ &= x_1D_1(x_1, x_2) + x_2D_2(x_1, x_2). \end{aligned}$$

So,

$$\begin{aligned} R(x_1, x_2) &= x_1(200 - 10x + 20x_2) \\ &\quad + x_2(100 + 5x_1 - 10x_2) \\ &= 200x_1 - 10x_1^2 + 25x_1x_2 \\ &\quad + 100x_2 - 10x_2^2. \end{aligned}$$

(b) If $x_1 = 6$ and $x_2 = 5$, then

$$\begin{aligned} R(6, 5) &= 200(6) - 10(6)^2 + 25(6)(5) \\ &\quad + 100(5) - 10(5)^2 \\ &= \$1{,}840. \end{aligned}$$

31. $f(x, y) = Ax^ay^b$.

$$\begin{aligned} f(2x, 2y) &= A(2x)^a(2y)^b = A(2)^ax^a(2)^by^b \\ &= (2^{a+b})Ax^ay^b. \end{aligned}$$

$x \geq 0$, $y \geq 0$, and $A > 0$.

(a) If $a + b > 1$, $2^{a+b} > 2$ and f more than doubles.

(b) If $a + b < 1$, $2^{a+b} < 2$ and f increases but does not double.

(c) If $a + b = 1$, $2^{a+b} = 2$ and f doubles (exactly).

33. Let R denote the manufacturer's revenue. Then R = (revenue from domestic sales) + (revenue from sales abroad)

$$R(x, y) = x\left(60 - \frac{x}{5} + \frac{y}{20}\right) + y\left(50 - \frac{y}{10} + \frac{x}{20}\right)$$
$$= 60x + 50y - \frac{x^2}{5} - \frac{y^2}{10} + \frac{xy}{10}.$$

35. (a) $S(15.83, 87.11)$
$= 0.0072(15.83 \wedge 0.425)(87.11 \wedge 0.725)$

Input into home screen to find $S(15.83, 87.11) \approx 0.5938$

To sketch several additional level curves of $S(W, H)$, we will use the list feature of the calculator.

In general, $0.0072W^{0.425}H^{0.725} = S$

$$H^{0.725} = \frac{S}{0.0072}W^{-0.425}$$

$$H = \left(\frac{S}{0.0072}W^{-0.425}\right)^{1/0.725}$$

We will use $S = 0.3$, 0.5938, and 1.5.

Press [y =].

Input $((L_1/0.0072) * x \wedge (-0.425)) \wedge (1/0.725)$ for $y_1 =$.

From the home screen, input {0.3, 0.5938, 1.5} [STO→] [2nd] L_1.

Use window dimensions [0, 400]50 by [0, 150]25.

Press [graph].

Different combinations of height and weight that result in the same surface area.

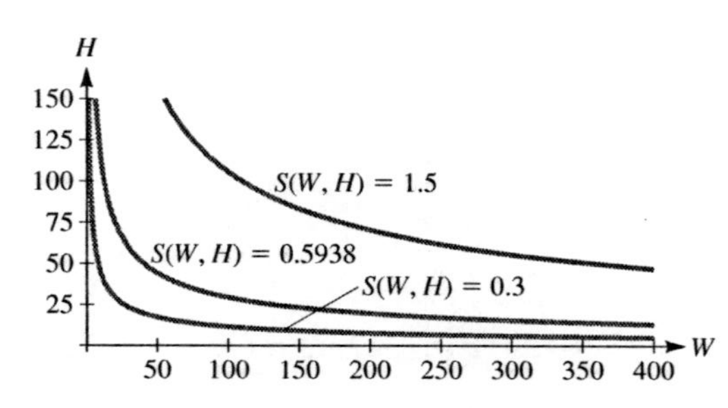

(b) $0.648 = 0.0072(18.37)^{0.425}H^{0.725}$,
$H^{0.725} = 26.121$, $H = 90.05$ cm.

(c) Let W_0, H_0 be Jenny's weight and height at birth. Then,

$$S(W_0, H_0) = 0.0072W_0^{0.425}H_0^{0.725}$$

When $W = 6W_0$ and $H = 2H_0$,

$$S(6W_0, 2H_0) = 0.0072(6W_0)^{0.425}(2H_0)^{0.725}$$
$$= 0.0072(6)^{0.425}W_0^{0.425}(2)^{0.725}H_0^{0.725}$$
$$\approx 3.53966S(W_0, H_0)$$

The % change in surface area is:

$$100\frac{3.53966S(W_0, H_0) - S(W_0, H_0)}{S(W_0, H_0)}$$
$$= 100\frac{2.53966S(W_0, H_0)}{S(W_0, H_0)} \approx 253.97\% \text{ increase.}$$

(d) Writing Exercise—Answers will vary.

37. (a) $Q(10, 20) = 30 + 40 = 70$ units

(b) $3x + 2y = 70$
or $y = -\frac{3}{2}x + 35$

(c)

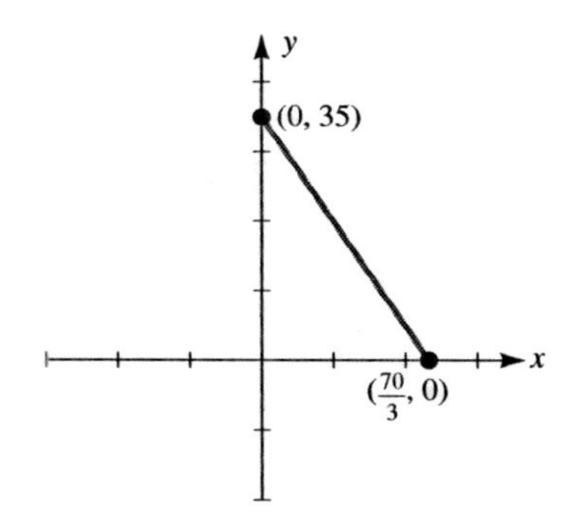

(d) $70 = 3 \cdot (12) + 2(20 + \Delta y)$

$$2\Delta y = 70 - 36 - 40$$
$$\Delta y = -\frac{6}{2} = -3, \text{ or decrease}$$

unskilled labor by 3 workers.

39. $U(25, 8) = (25 + 1)(8 + 2) = 260$

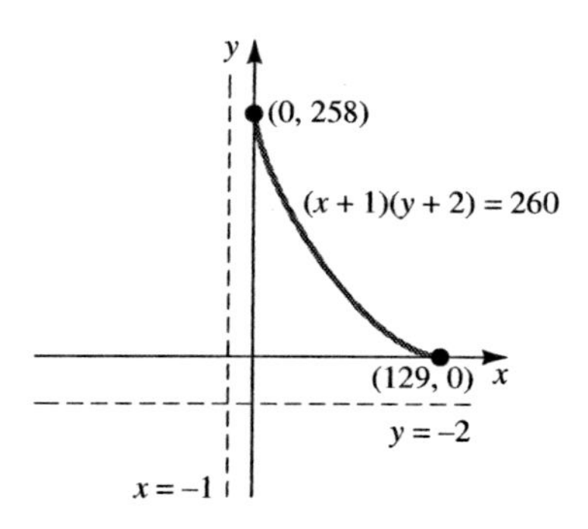

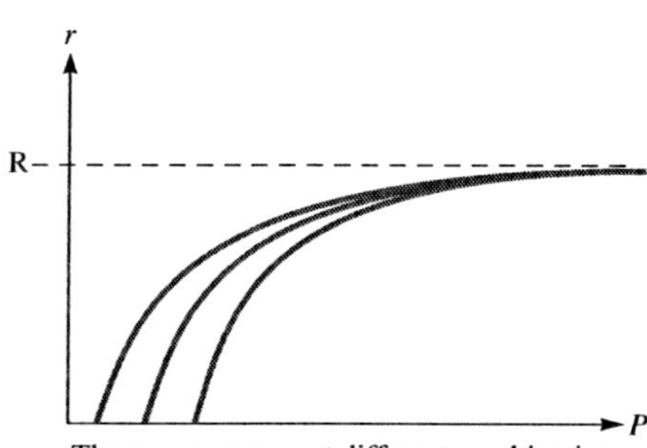

The curves represent different combinations of pressure and distance from the axis that result in the same speed.

41. (a)

$$V(3{,}875, 1.675, 0.004) = \frac{9.3(3,875)}{1.675}\left[(0.0075)^2 - (0.004)^2\right] \approx 0.866\text{cm/sec}$$

(b) For the fixed values of L and R,

$$V(P, r,) = \frac{9.3P}{1.675}\left[(0.0075)^2 - r^2\right] = 5.55P(0.0000563 - r^2)$$

To sketch several level curves of V, set $V(P, r) = C$ for several values of C and solve for P. We will use the list feature of the calculator with $C = 100$, 200 and 300. Setting $V(P, r) = C$

$$5.55(0.0000563 - r^2) = C$$

In general,

$$P = \frac{0.1802C}{0.0000563 - r^2}$$

Press [y =].
Input $(0.1802L_1)/(0.0000563 - x^2)$ for $y_1 =$.
From the home screen, enter
$\{100, 200, 300\}$ [STO→] [2nd] L_1.
Use z-standard function under the zoom menu for the standard window dimensions.
Press [graph].
Note that there are vertical asymptotes when $r = \pm\sqrt{0.0000563}$ but that the graph is defined in between these asymptotes as well.

43. (a) To sketch graphs of several level curves, for simplicity's sake, we will choose $a = b = 1$. We use the list feature of the calculator to sketch level curves for $T(P, V) = C$ for $C = -100, 0, 100$.
In general,

$$0.0122\left(P + \frac{1}{V^2}\right)(V - 1) - 273.15 = C$$

and $P = \dfrac{C + 273.15}{0.0122(V - 1)} - \dfrac{1}{V^2}$.
Press [y =].
Input $(L_1 + 273.15)/(0.0122(x - 1)) - \left(\frac{1}{x^2}\right)$ for $y_1 =$.
From the home screen, enter $\{-100,0,100\}$ [Sto→] [2nd] L_1.
Use window dimensions [0,35000]5,000 by [0,2.9]0.3.
Press [graph].

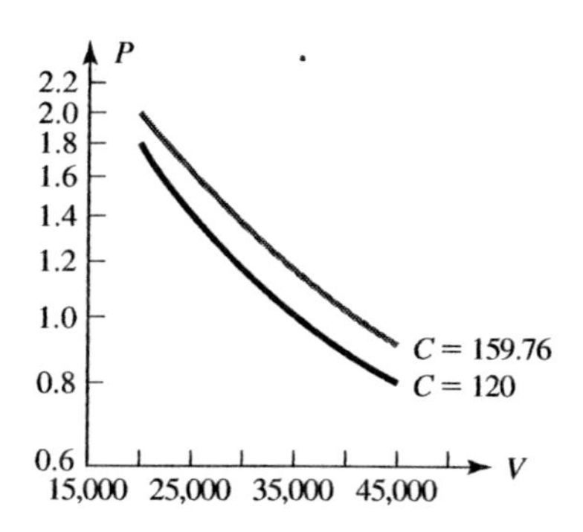

(b) To find $T(1.13, 31.275 \times 10^3)$,

From the home screen, enter 0.0122(1.13 + $(6.49 \times 10^6)/31{,}275^2)(31{,}275 - 56.2) - 273.15 \approx 159.76$.
Thus, the temperature is 159.76°C.

45. **(a)** $B_m(90, 190, 22) = 66.47 + 13.75(90) + 5.00(190) - 6.77(22) = 2{,}105.03$ kilo calories.

(b) $B_f(61, 170, 27) = 655.10 + 9.60(61) + 1.85(170) - 4.68(27) = 1{,}428.84$ kilo calories.

(c) $B_m(85, 193, A) = 66.47 + 13.75(85) + 5.00(193) - 6.77A$

$$2{,}018 = 2{,}200.22 - 6.77A$$
$$A \approx 26.9 \text{ years old.}$$

(d) $B_f(67, 173, A) = 655.10 + 9.60(67) + 1.85(173) - 4.68A$

$$1{,}504 = 1{,}618.35 - 4.68A$$
$$A \approx 24.4 \text{ years old.}$$

47. $m(A, n, i) = \dfrac{Ai}{1 - (1 + i)^{-12n}}$

(a) $$m\left(250000, 15, \frac{0.052}{12}\right) = \frac{250{,}000\left(\frac{0.052}{12}\right)}{1 - \left(1 + \frac{0.052}{12}\right)^{-12(15)}}$$

$\approx \$2{,}003.13$
The total amount paid is
$(2003.13)(12)(15) = \$360{,}563.40$
Since the original loan is for \$250000, the interest paid is $360{,}563.4 - 250{,}000 = \$110{,}563.40$

(b) $$m\left(250000, 30, \frac{0.056}{12}\right) = \frac{250{,}000\left(\frac{0.056}{12}\right)}{1 - \left(1 + \frac{0.056}{12}\right)^{-12(30)}}$$

$\approx \$1{,}435.20$
The total amount paid is
$(1435.20)(12)(30) = \$516{,}672$
Since the original loan is for \$250000, the interest paid is $516{,}672 - 250{,}000 = \$266{,}672$

49. $$P(2, 0.53, 23) = 0.075(2)(0.53)(273.15 + 23\ ) \approx 23.54 \text{ atmospheres.}$$

51.
$$Q(K, L) = AK^{\alpha}L^{1-\alpha}$$
$$Q(sK, sL) = A(sK)^{\alpha}(sL)^{1-\alpha} = As^{\alpha}K^{\alpha}s^{1-\alpha}L^{1-\alpha} = AsK^{\alpha}L^{1-\alpha} = sQ(K, L)$$

7.2 Partial Derivatives

1. $f(x, y) = 2xy^5 + 3x^2y + x^2$

$$f_x = 2y^5 + 6xy + 2x$$
$$f_y = 2x(5y^4) + 3x^2 = 10xy^4 + 3x^2$$

3.
$$z = (3x + 2y)^5$$
$$\frac{\partial z}{\partial x} = 5(3x + 2y)^4\frac{\partial}{\partial x}(3x + 2y) = 15(3x + 2y)^4$$
$$\frac{\partial z}{\partial y} = 5(3x + 2y)^4\frac{\partial}{\partial y}(3x + 2y) = 10(3x + 2y)^4$$

5.
$$f(s, t) = \frac{3t}{2s} = \frac{3}{2}s^{-1}t$$
$$f_s = \frac{3}{2}(-1)s^{-2}t = -\frac{3t}{2s^2}$$
$$f_t = \frac{3}{2}s^{-1} = \frac{3}{2s}$$

7.
$$z = xe^{xy}$$
$$\frac{\partial z}{\partial x} = x(ye^{xy}) + e^{xy}(1) = (xy + 1)e^{xy}$$
$$\frac{\partial z}{\partial y} = x(e^{xy})(x) = x^2e^{xy}$$

9.

$$f(x, y) = \frac{e^{2-x}}{y^2} = e^{2-x}y^{-2}$$

$$f_x = -e^{2-x}y^{-2} = -\frac{e^{2-x}}{y^2}$$

$$f_y = e^{2-x}(-2y^{-3}) = -\frac{2e^{2-x}}{y^3}$$

11.

$$f(x, y) = \frac{2x + 3y}{y - x}$$

$$f_x = \frac{(y - x)(2) - (2x + 3y)(-1)}{(y - x)^2} = \frac{5y}{(y - x)^2}$$

$$f_y = \frac{(y - x)(3) - (2x + 3y)(1)}{(y - x)^2} = -\frac{5x}{(y - x)^2}$$

13.

$$z = u \ln v$$

$$\frac{\partial z}{\partial u} = (1) \ln v = \ln v$$

$$\frac{\partial z}{\partial v} = u\left(\frac{1}{v}\right) = \frac{u}{v}$$

15.

$$f(x, y) = \frac{\ln(x + 2y)}{y^2}$$

$$f_x = \frac{(y^2)[1/(x + 2y)] - \ln(x + 2y)(0)}{y^4} = \frac{1}{y^2(x + 2y)}$$

$$f_y = \frac{(y^2)[2/(x + 2y)] - \ln(x + 2y)(2y)}{y^4} = \frac{(y)(2) - (x + 2y)\ln(x + 2y)(2)}{(x + 2y)y^3} = \frac{2[y - (x + 2y)\ln(x + 2y)]}{y^3(x + 2y)}$$

17.

$$f(x, y) = 3x^2 - 7xy + 5y^3 - 3(x + y) - 1$$

$$f_x = 6x - 7y - 3$$

$$f_y = -7x + 15y^2 - 3$$

$$f_x(-2, 1) = -12 - 7 - 3 = -22$$

$$f_y(-2, 1) = 14 + 15 - 3 = 26$$

19.

$$f(x, y) = xe^{-2y} + ye^{-x} + xy^2$$

$$f_x = e^{-2y} - ye^{-x} + y^2$$

$$f_y = -2xe^{-2y} + e^{-x} + 2xy$$

$$f_x(0, 0) = 1 - 0 + 0 = 1$$

$$f_y(0, 0) = 0 + 1 + 0 = 1$$

21.

$$f(x, y) = 5x^4y^3 + 2xy$$

$$f_x = 5(4x^3)y^3 + 2y = 20x^3y^3 + 2y$$

$$f_y = 5x^4(3y^2) + 2x = 15x^4y^2 + 2x$$

$$f_{xx} = \frac{\partial}{\partial x}(f_x) = 20(3x^2)y^3 + 0 = 60x^2y^3$$

$$f_{yy} = \frac{\partial}{\partial y}(f_y) = 15x^4(2y) + 0 = 30x^4y$$

$$f_{xy} = \frac{\partial}{\partial y}(f_x) = 20x^3(3y^2) + 2(1) = 60x^3y^2 + 2$$

$$f_{yx} = \frac{\partial}{\partial x}(f_y) = 15(4x^3)y^2 + 2(1) = 60x^3y^2 + 2 = f_{xy}$$

23.

$$f(x, y) = e^{x^2y}$$

$$f_x = 2xye^{x^2y} \text{ and } f_y = x^2e^{x^2y}$$

$$f_{xx} = \frac{\partial}{\partial x}(f_x) = 2xy(e^{x^2y})(2xy) + e^{x^2y}(2y) = 2y(2x^2y + 1)e^{x^2y}$$

$$f_{yy} = \frac{\partial}{\partial y}(f_y)$$
$$= x^2(e^{x^2y})(x^2) = x^4e^{x^2y}$$
$$f_{xy} = \frac{\partial}{\partial y}(f_x)$$
$$= 2xy(e^{x^2y})(x^2) + e^{x^2y}(2x)$$
$$= 2x(x^2y+1)e^{x^2y}$$
$$f_{yx} = \frac{\partial}{\partial x}(f_y)$$
$$= x^2(e^{x^2y})(2xy) + e^{x^2y}(2x)$$
$$= 2x(x^2y+1)e^{x^2y} = f_{xy}$$

25. $f(s,t) = \sqrt{s^2+t^2} = (s^2+t^2)^{1/2}$

$$f_s = \frac{1}{2}(s^2+t^2)^{-1/2}(2s) = s(s^2+t^2)^{-1/2}$$
$$f_t = \frac{1}{2}(s^2+t^2)^{-1/2}(2t) = t(s^2+t^2)^{-1/2}$$
$$f_{ss} = s\left[-\frac{1}{2}(s^2+t^2)^{-3/2}(2s)\right] + (s^2+t^2)^{-1/2}(1)$$
$$= \frac{-s^2}{(s^2+t^2)^{3/2}} + \frac{1}{(s^2+t^2)^{1/2}}\frac{(s^2+t^2)}{(s^2+t^2)}$$
$$= \frac{t^2}{(s^2+t^2)^{3/2}}$$
$$f_{tt} = t\left[-\frac{1}{2}(s^2+t^2)^{-3/2}(2t)\right] + (s^2+t^2)^{-1/2}(1)$$
$$= \frac{s^2}{(s^2+t^2)^{3/2}}$$
$$f_{st} = \frac{\partial}{\partial t}(f_s) = s\left[-\frac{1}{2}(s^2+t^2)^{-3/2}(2t)\right]$$
$$= \frac{-st}{(s^2+t^2)^{3/2}}$$
$$f_{ts} = \frac{\partial}{\partial s}(f_t) = t\left[-\frac{1}{2}(s^2+t^2)^{-3/2}(2s)\right]$$
$$= \frac{-st}{(s^2+t^2)^{3/2}} = f_{st}$$

27. $D_1(p_1, p_2) = 500 - 6p_1 + 5p_2$
$D_2(p_1, p_2) = 200 + 2p_1 - 5p_2$

$$\frac{\partial D_1}{\partial p_2} = 5 \text{ and } \frac{\partial D}{\partial p_1} = 2$$

Since both partial derivatives are positive for all p_1 and p_2, the commodities are substitute commodities.

29.
$$D_1(p_1, p_2) = 3{,}000 + \frac{400}{p_1+3} + 50p_2$$
$$D_1(p_1, p_2) = 2{,}000 - 100p_1 + \frac{500}{p_2+4}$$

$$\frac{\partial D_1}{\partial p_2} = 50 \text{ and } \frac{\partial D_2}{\partial p_1} = -100$$

Since the partial derivaties are opposite in sign for all p_1 and p_2, the commodities are neither substitute nor complementary.

31.
$$D_1(p_1, p_2) = \frac{7p_2}{1+p_1^2}$$
$$D_2(p_1, p_2) = \frac{p_1}{1+p_2^2}$$

$$\frac{\partial D_1}{\partial p_2} = \frac{7}{1+p_1^2} > 0 \text{ and } \frac{\partial D_2}{\partial p_1} = \frac{1}{1+p_2^2}$$

Since both partial derivatives are positive for all p_1 and p_2, the commodities are substitute commdities.

33. $z = x^2 - y^2$

$$\frac{\partial z}{\partial x} = 2x \text{ and } \frac{\partial^2 z}{\partial x^2} = 2$$
$$\frac{\partial z}{\partial y} = -2y \text{ and } \frac{\partial^2 z}{\partial y^2} = -2$$

Since $\frac{\partial^2 z}{\partial x^2} + \frac{\partial^2 z}{\partial y^2} = 0$ the function satisfies Laplace's equation.

35.
$$z = xe^y - ye^x$$
$$\frac{\partial z}{\partial x} = e^y - ye^x \text{ and } \frac{\partial^2 z}{\partial x^2} = -ye^x$$
$$\frac{\partial z}{\partial y} = xe^y - e^x \text{ and } \frac{\partial^2 z}{\partial y^2} = xe^y$$

Since $\frac{\partial^2 z}{\partial x^2} + \frac{\partial^2 z}{\partial y^2} = -ye^x + xe^y \neq 0$ the function does not satisfy Laplace's equation.

37. The partial derivative

$$Q_K = \frac{\partial Q}{\partial K} = 30K^{-1}L^{1/3}$$
$$= \frac{30L^{1/3}}{K^{1/2}}$$

is the rate of change of the output with respect to the capital investment. This is an approximation to the additional number of units that will be produced each week if the capital investment is increased from K to $K+1$ while the size of the labor force is not changed. In particular, if the capital investment K is increased from 900 (thousand) to 901 (thousand) and the size of the labor force is $L = 1{,}000$, the resulting change in output is

$$\Delta Q = Q_K(900, 1000)$$
$$= \frac{30(1{,}000)^{1/3}}{(900)^{1/2}}$$
$$= \frac{30(10)}{30} = 10, \text{ or}$$

daily output will increase by 10 units.

39. **(a)**
$$Q_K(K, L) = 30K^{-2/3}L^{2/3}$$
$$Q_L(K, L) = 60K^{1/3}L^{-1/3}$$

So when $K = 5{,}495$ and $L = 4{,}587$,

$$Q_K(5495, 4587) = 30(5{,}495^{-2/3})(4{,}587^{2/3})$$
$$\approx 26.60$$
$$Q_L(5495, 4587) = 60(5{,}495^{1/3})(4{,}587^{-1/3})$$
$$\approx 63.72$$

(b) An increase of one unit capital ($1,000,000) results in an increase in output of 26.60 units, which is less than the 63.72 unit increase in output that results from a unit increase in the labor level. The government should encourage labor employment.

41. $F(L, r) = \frac{kL}{r^4}$

(a) $F(3.17, 0.085) = 60{,}727.24\ k$

$$\frac{\partial F}{\partial r} = \frac{k}{r^4} = 19{,}156.86\ k$$
$$\frac{\partial F}{\partial r} = -\frac{4kL}{r^5} = -2{,}857{,}752.58\ k$$

(b) $F(1.2L, 0.8r) = \frac{k(1.2L)}{(0.8r)^4} = 2.93F(L, r)$

$$\frac{\partial F}{\partial L}(1.2L, 0.8r) = 2.44\frac{\partial F}{\partial L}(L, r)$$
$$\frac{\partial F}{\partial r}(1.2L, 0.8r) = 3.66\frac{\partial F}{\partial r}(L, r)$$

43. $F(x, y) = 200 - 24\sqrt{x} + 4(0.1y + 3)^{3/2}$
$\frac{\partial F}{\partial y} = 6(0.1y + 3)^{1/2}(0.1) = 0.6(0.1y + 3)^{1/2}$
is the rate of change of demand with respect to the price of gasoline. When the selling price is kept constant,

$$\Delta F \approx \frac{\partial F}{\partial y}\Delta y$$

Since $y = 220$ cent and $\Delta = -1$,
$\Delta F \approx 0.6[0.1(220) + 3]^{1/3}(-1) = -3$
or demand will decrease by 3 bicycles.

45. $\frac{\partial V}{\partial R} = 2\pi RH$ is the rate of change of the volume with respect to the radius. When the height is kept constant,

$$\Delta V \approx \frac{\partial V}{\partial R}\Delta R$$

Since $R = 3$, $H = 12$, and $\Delta R = 1$,
$\Delta V \approx [2\pi(3)(12)](1) = 72\pi$, or an increase in volume of approximately 226 cubic cm.

47. **(a)** If the price x of the first lawnmower increases, the demand for that same lawnmower should fall. If the price y of the second (competing) lawnmower increases, the demand for the first lawnmower should increase.

(b) $D_x < 0$, $D_y > 0$

(c) With $D = a + bx + cy$, $D_x = b < 0$ and $D_y = c > 0$.

49. $P(x, y, u, v) = \dfrac{100xy}{xy + uv}$

$$P_x = \frac{(xy+uv)100y - 100xy^2}{(xy+uv)^2} = \frac{100uvy}{(xy+uv)^2}$$

$$P_y = \frac{(xy+uv)100x - 100x^2y}{(xy+uv)^2} = \frac{100uvx}{(xy+uv)^2}$$

$$P_u = -\frac{100xyv}{(xy+uv)^2},\ P_v = -\frac{100xyu}{(xy+uv)^2}$$

All of these partials measure the rate of change of percentage of total blood flow with respect to the quantities x, y, u, v respectively.

51.

$$\frac{\partial F}{\partial z} = \frac{c\pi x^2}{4}\left[\frac{1}{2}(y-z)^{-1/2}(-1)\right] = \frac{-c\pi x^2}{8\sqrt{y-z}}$$

is the rate of change of blood flow with respect to the pressure in the capillary. Since this rate is negative, the blood flow is decreasing.

53. $Q(K, L) = 120K^{1/2}L^{1/3}$

(a) $$Q_L = 120K^{1/2}\left(\frac{1}{3}L^{-2/3}\right) = 40K^{1/2}L^{-2/3}$$

$$Q_{LL} = -\frac{80}{3}K^{1/2}L^{-5/3}$$

$Q_{LL} < 0$; for a fixed level of capital investment, the effect on output of the addition of one worker hour is greater when the work force is small, than when it is large.

(b) $$Q_K = 60K^{-1/2}L^{1/3}$$

$$Q_{KK} = -30K^{-3/2}L^{1/3}$$

$Q_{KK} < 0$; for a fixed work force, the effect on output of the addition of $1,000 in capital investment is greater when the capital investment is small, than when it is large.

55. (a) To store the output function,
Press [y=] and
input $1{,}175x + 483L_1 + 3.1(x \wedge 2) * L_1 - 1.2(x \wedge 3) - 2.7(L_1 \wedge 2)$ for [y_1=].
From the home screen, input {71} [sto→] [2nd] L_1.
Use window dimensions
[0, 400]25 by [0, 250]25.
Use the value function under the calc menu and enter $x = 37$ to find $Q(37, 71) \approx 304{,}691$ units.
Use the value function again and enter $x = 38$ to find $Q(38, 71) \approx 317{,}310$ units.
From the home screen, input {72} [sto→] [2nd] L_1.
Use the value function under the calc menu and enter $x = 37$ to find
$Q(37, 72) \approx 309{,}031$ units.

(b) $Q_x(x, y) = 1{,}175 + 6.2xy - 3.6x^2$
To estimate the change in output when x is increased from 37 to 38 while y remains at 71, we find $Q_x(37, 71)$.
Press [y=].
Input $1{,}175 + 6.2xL_1 - 3.6x^2$ in [y_2=].
Deactivate $y_1 =$ so that only y_2 is activated.
From the home screen, input {71} [sto→] [2nd] L_1.
Use the value function under the calc menu and enter $x = 37$ to find $Q_x(37, 71) = 12{,}534$ units.
Thus, if the skilled workforce is increased from 37 to 38 and the unskilled remains constant at 71, the output is approximately increased by 12,534 units.
The actual change is $Q(38, 71) - Q(37, 71) = 317{,}310 - 304{,}691 = 12{,}619$ units.

(c) $Q_y(x, y) = 483 + 3.1x^2 - 5.4y$
To estimate the change in output when y is increased from 71 to 72 while x remains at 37, we find $Q_y(37, 71)$.
Press [y=].
Input $483 + 3.1x^2 - 5.4L_1$ for $y_3 =$.
Deactivate $y_1 =$ and $y_2 =$ so only $y_3 =$ is activated.
From the home screen, input {71} [sto→] [2nd] L_1.
Use the value function under the calc menu and enter $x = 37$ to find $Q_y(37, 71) \approx 4{,}344$ units.
Thus, if the unskilled workforce is increased from 71 to 72 and the skilled remains at 37, the output is approximately increased by 4,344 units. The actual change is $Q(37, 72) - Q(37, 71) = 309{,}031 - 304{,}691 = 4{,}340$ units.

57. $z = 2x + 3y;\ x = t^2;\ y = 5t$

$$\frac{dz}{dt} = \frac{\partial z}{\partial x} \cdot \frac{dx}{dt} + \frac{\partial z}{\partial y} \cdot \frac{dy}{dt}$$
$$= (2)(2t) + (3)(5)$$
$$= 4t + 15$$

59. $z = \dfrac{3x}{y}$; $x = t$; $y = t^2$

$$\frac{dz}{dt} = \frac{\partial z}{\partial x} \cdot \frac{dx}{dt} + \frac{\partial z}{\partial y} \cdot \frac{dy}{dt}$$
$$= \left(\frac{3}{y}\right)(1) + \left(-\frac{3x}{y^2}\right)(2t)$$
$$= \frac{3}{y} - \frac{6xt}{y^2}$$

61. $z = xy$; $x = e^{2t}$; $y = e^{-3t}$

$$\frac{dz}{dt} = \frac{\partial z}{\partial x} \cdot \frac{dx}{dt} + \frac{\partial z}{\partial y} \cdot \frac{dy}{dt}$$
$$= (y)(2e^{2t}) + (x)(-3e^{-3t})$$
$$= 2ye^{2t} - 3xe^{-3t}$$

63. $Q(x, y) = 10xy^{1/2}$

$$\Delta Q \approx \frac{\partial Q}{\partial x}\Delta x + \frac{\partial Q}{\partial y}\Delta y$$
$$= (10y^{1/2})\Delta x + \left(\frac{5x}{y^{1/2}}\right)\Delta y$$
$$= \left(10\sqrt{36}\right)(-3) + \left(\frac{5 \cdot 30}{\sqrt{36}}\right)(5)$$
$$= -55$$

or the number of units produced will decrease by 55.

65.
$$Q(x, y) = 200 - 10x^2 + 20xy$$
$$x(t) = 10 + 0.5t$$
$$y(t) = 12.8 + 0.2t^2$$

(a)
$$\frac{dQ}{dt} = \frac{\partial Q}{\partial x} \cdot \frac{dx}{dt} + \frac{\partial Q}{\partial y} \cdot \frac{dy}{dt}$$
$$= (-20x + 20y)(0.5) + (20x)(0.4t)$$

When $t = 4$, $x(4) = 12$ and $y(4) + 16$. So,

$$\frac{dQ}{dt} = [-20(12) + 20(16)](0.5) + [20(12)][0.4(4)]$$
$$= 424 \text{ units per month/month}$$

(b) When $t = 4$, $Q(12, 16) = 2{,}600$ so

$$100\frac{Q'(t)}{Q(t)} = 100\frac{424}{2{,}600} \approx 16.31\%$$

67. $Q(x, y) = 0.08x^2 + 0.12xy + 0.03y^2$

$$\Delta Q \approx \frac{\partial Q}{\partial x}\Delta x + \frac{\partial Q}{\partial y}\Delta y$$
$$= (0.16x + 0.12y)\Delta x + (0.12x + 0.06y)\Delta y$$
$$= [0.16(80) + 0.12(200)](0.5) + [0.12(80) + 0.06(200$$
$$= 61.6$$

or an increase of 61.6 units produced per day.

69. $Q(x, y) = 20x^{3/2}y$

$$\Delta Q \approx \frac{\partial Q}{\partial x}\Delta x + \frac{\partial Q}{\partial y}\Delta y$$
$$= (30x^{1/2}y)\Delta x + (20x^{3/2})\Delta y$$

Using $x = 36$ thousand, $y = 25$ thousand, $\Delta x = 0.5$ thousand and $\Delta y = -1$ thousand,

$$= \left[30\sqrt{36}(25)\right](0.5) + \left[20(36)^{3/2}\right](-1)$$
$$= -2{,}070$$

or a decrease in sales of 2,070 books.

71. (a) Cost = (area bottom)(cost per unit area)
+ (area top)(cost per unit area)
+(area sides)(cost per unit area)
+ (volume)(cost per unit volume)

$$C(R, H) = 0.0005\left[\pi R^2 + \pi R^2 + 2\pi RH\right]$$
$$+ 0.01(\pi R^2 H)$$
$$= 0.0005(2\pi)[R^2 + RH] + 0.01\pi R^2 H$$
$$= 0.001\pi[R^2 + RH + R^2 H]$$

(b)
$$\Delta C \approx \frac{\partial C}{\partial R}\Delta R + \frac{\partial C}{\partial H}\Delta H$$
$$= [0.001\pi(2R + H + 2RH)]\,\Delta R + \left[0.001\pi(R + R^2)\right]\Delta H$$

When $R = 3$, $H = 12$, $\Delta R = 0.3$ and $\Delta H = -0.2$,

$$\Delta C =\approx [0.001\pi(2\cdot 3 + 12 + 2\cdot 3\cdot 12)]\,(0.3) + [0.001\pi(3 + 3^2)](-0.2)$$
$$\approx 0.0773$$

or a decrease of 0.08 cents per can.

73. $x^2 + xy + y^3 = 1$

$$\frac{dy}{dx} = -\frac{f_x}{f_y} = -\frac{2x + y}{x + 3y^2}$$

When $x = -1$ and $y = 1$, the slope is $= -\dfrac{2(-1)+1}{-1+3(1)^2} = \dfrac{1}{2}$

The equation of the tangent line is

$$y - 1 = \frac{1}{2}(x + 1)$$
$$y = \frac{1}{2}x + \frac{3}{2}, \text{ or } x - 2y = -3$$

75. $U(x, y) = (x + 1)(y + 2) = xy + 2x + y + 2$

Need to find Δx so that $\Delta Q = 0$ when $\Delta y = -1$.

$$\Delta Q \approx \frac{\partial Q}{\partial x}\Delta x + \frac{\partial Q}{\partial y}\Delta y$$
$$0 = \frac{\partial Q}{\partial x}\Delta x + \frac{\partial Q}{\partial y}\Delta y$$
$$\Delta x = \left(-\frac{\partial Q}{\partial y}\Delta y\right) \Big/ \frac{\partial Q}{\partial x} = \frac{-(x+1)\Delta y}{y+2} = \frac{-(25+1)(-1)}{8+2} = 2.6$$

Increasing the first commodity by 2.6 units keeps utility from changing.

7.3 Optimizing Functions of Two Variables

1. $f(x, y) = 5 - x^2 - y^2$

$f_x = -2x$ so $f_x = 0$ when $x = 0$

$f_y = -2y$ so $f_y = 0$ when $y = 0$

and only critical point is (0, 0).

$$f_{xx} = -2,\ f_{yy} = -2,\ f_{xy} = 0$$
$$D = f_{xx}f_{yy} - (f_{xy})^2$$

For the point (0, 0),

$$D = (-2)(-2) - 0^2 > 0$$
$$f_{xx} < 0$$

So, (0, 0) is a relative maximum.

3. $f(x, y) = xy$

$f_x = y$ and $f_x = 0$ when $y = 0$

$f_y = x$ and $f_y = 0$ when $x = 0$

and only critical point is (0, 0).

$$f_{xx} = 0,\ f_{yy} = 0,\ f_{xy} = 1$$

For the point (0, 0),

$$D = (0)(0) - (1)^2 < 0$$

So, (0, 0) is a saddle point.

5. $f(x, y) = \dfrac{16}{x} + \dfrac{6}{y} + x^2 - 3y^2$

$$f_x = -\frac{16}{x^2} + 2x$$

So, $f_x = 0$ when $0 = -\dfrac{16}{x^2} + 2x$, or $x = 2$.

$$f_y = -\frac{6}{y^2} - 6y$$

So, $f_y = 0$ when $0 = -\dfrac{6}{y^2} - 6y$, or $y = -1$

and only critical point is $(2, -1)$.

$$f_{xx} = \frac{32}{x^3} + 2,\ f_{yy} = \frac{12}{y^3} - 6,\ f_{xy} = 0$$

For the point $(2, -1)$,

$$D = \left[\frac{32}{(2)^3} + 2\right]\left[\frac{12}{(-1)^3} - 6\right] - 0 < 0$$

So, $(2, -1)$ is a saddle point.

7. $f(x, y) = 2x^3 + y^3 + 3x^2 - 3y - 12x - 4$

$$f_x = 6x^2 + 6x - 12$$
$$= 6(x + 2)(x - 1)$$

So, $f_x = 0$ when $x = -2, 1$.

$$f_y = 3y^2 - 3$$
$$= 3(y + 1)(y - 1)$$

So, $f_y = 0$ when $y = -1, 1$ and the critical points are $(-2, -1)$, $(-2, 1)$, $(1, -1)$, and $(1, 1)$.

$$f_{xx} = 12x + 6,\ f_{yy} = 6y,\ f_{xy} = 0$$

For the point $(-2, -1)$,

$$D = [12(-2) + 6][6(-1)] - 0 > 0$$

and $f_{xx} < 0$, so $(-2, -1)$ is a relative maximum.
For the point $(-2, 1)$,

$$D = [12(-2) + 6][6(1)] - 0 < 0$$

So, $(-2, 1)$ is a saddle point.
For the point $(1, -1)$,

$$D = [12(1) + 6][6(-1)] - 0 < 0$$

So, $(1, -1)$ is a saddle point.
For the point$(1, 1)$,

$$D = [12(1) + 6][6(1)] > 0$$

and $f_{xx} > 0$, so $(1, 1)$ is a relative minimum.

9. $f(x, y) = x^3 + y^2 - 6xy + 9x + 5y + 2$

$$f_x = 3x^2 - 6y + 9$$

So, $f_x = 0$ when $0 = 3(x^2 - 2y + 3)$, or $0 = x^2 - 2y + 3$.

$$f_y = 2y - 6x + 5$$

So, $f_y = 0$ when $0 = 2y - 6x + 5$. Solving this system of equations by adding,

$$0 = x^2 - 6x + 8$$
$$= (x - 2)(x - 4)$$
$$\text{So, } x = 2, 4.$$

When $x = 2$, $0 = (2)^2 - 2y + 3$, or $y = \frac{7}{2}$.

When $x = 4$, $0 = (4)^2 - 2y + 3$, or $y = \frac{19}{2}$.

So, the critical points are $\left(2, \frac{7}{2}\right)$ and $\left(4, \frac{19}{2}\right)$

$$f_{xx} = 6x,\ f_{yy} = 2,\ f_{xy} = -6$$

For the point $\left(2, \frac{7}{2}\right)$,

$$D = 6(2)(2) - (-6)^2 < 0$$

So, $\left(2, \frac{7}{2}\right)$ is a saddle point.

For the point $\left(4, \frac{19}{2}\right)$,

$$D = 6(4)(2) - (-6)^2 > 0$$

and $f_{xx} > 0$, so $\left(4, \frac{19}{2}\right)$ is a relative minimum.

11. $f(x, y) = (x^2 + 2y^2)e^{1-x^2-y^2}$

$$f_x = (x^2 + 2y^2)(-2xe^{1-x^2-y^2}) + (e^{1-x^2-y^2})(2x)$$
$$= -2xe^{1-x^2-y^2}(x^2 + 2y^2 - 1)$$

So, $f_x = 0$ when $x = 0$ or $x^2 + 2y^2 - 1 = 0$

$$f_y = (x^2 + 2y^2)(-2ye^{1-x^2-y^2}) + (e^{1-x^2-y^2})(4y)$$
$$= -2ye^{1-x^2-y^2}(x^2 + 2y^2 - 2)$$

So, $f_y = 0$ when $y = 0$, or $x^2 + 2y^2 - 2 = 0$.
There are no solutions to the system of equations $x^2 + 2y^2 - 1 = 0$ and $x^2 + 2y^2 - 2 = 0$. Further, when $x = 0$, $f_y = 0$ when $0 = -2ye^{1-y^2}(2y^2 - 2)$, or $y = 0, -1, 1$. When $y = 0$, $f_x = 0$ when $0 = -2xe^{1-x^2}(x^2 - 1)$ or, $x = 0, -1, 1$. So, the critical points are $(-1, 0)$, $(0, 0)$, $(1, 0)$, $(0, -1)$ and $(0, 1)$.
Rewriting f_x as

$$f_x = -2e^{1-x^2-y^2}(x^3 + 2xy^2 - x)$$

$$f_{xx} = -2\left[e^{1-x^2-y^2}(3x^2 + 2y^2 - 1) + (x^3 + 2xy^2 - x)(-2xe^{1-x^2-y^2})\right]$$

$$f_{yy} = -2\left[e^{1-x^2-y^2}(x^2 + 6y^2 - 2) + (x^2y + 2y^3 - 2y)(-2ye^{1-x^2-y^2})\right]$$

$$f_{xy} = -2\left[e^{1-x^2-y^2}(4xy) + (x^3 + 2xy^2 - x)(-2ye^{1-x^2-y^2})\right]$$

For the point $(-1, 0)$,

$$D = (-4)(2) - 0 < 0$$

So, $(-1, 0)$ is a saddle point.
For the point $(0, 0)$,

$$D = (2e)(4e) - 0 > 0$$

and $f_{xx} > 0$, so $(0, 0)$ is a relative minimum.
For the point $(1, 0)$,

$$D - (-4)(2) - 0 < 0$$

So, $(1, 0)$ is a saddle point.
For the point $(0, -1)$,

$$D = (-2)(-8) - 0 > 0$$

and $f_{xx} < 0$, so $(0, -1)$ is a relative maximum.
For the point $(0, 1)$,

$$D = (-2)(-8) - 0 > 0$$

and $f_{xx} < 0$, so $(0, 1)$ is a relative maximum.

13. $f(x, y) = x^3 - 4xy + y^3$

$$f_x = 3x^2 - 4y$$

So, $f_x = 0$ when $0 = 3x^2 - 4y$, or $y = \dfrac{3x^2}{4}$.

$$f_y = -4x + 3y^2$$

So, $f_y = 0$ when $0 = -4x + 3y^2$

$$= -4x + 3\left(\frac{3x^2}{4}\right)^2$$

$$= \frac{27}{16}x^4 - 4x$$

$$= 4x\left(\frac{27}{64}x^3 - 1\right) = 0,$$

or $x = 0, \dfrac{4}{3}$.

When $x = 0$, $f_x = 0$ when $y = 0$.

When $x = \dfrac{4}{3}$, $f_x = 0$ when $0 = 3\left(\dfrac{4}{3}\right)^2 - 4y$,

or $y = \dfrac{4}{3}$.

So the critical points are $(0, 0)$ and $\left(\dfrac{4}{3}, \dfrac{4}{3}\right)$.

$$f_{xx} = 6x,\ f_{yy} = 6y,\ f_{xy} = -4$$

For the point $(0, 0)$,

$$D = 6(0)6(0) - (-4)^2 < 0$$

So, $(0, 0)$ is a saddle point.

For the point $\left(\dfrac{4}{3}, \dfrac{4}{3}\right)$,

$$D = 6\left(\frac{4}{3}\right)6\left(\frac{4}{3}\right) - (-4)^2 > 0$$

and $f_{xx} > 0$, so $\left(\dfrac{4}{3}, \dfrac{4}{3}\right)$ is a relative minimum.

15. $f(x, y) = e^{-(x^2+y^2-6y)} = e^{-x^2-y^2+6y}$

$$f_x = -2xe^{-x^2-y^2+6y}$$

So, $f_x = 0$ when $x = 0$.

$$f_y = (-2y + 6)e^{-x^2-y^2+6y} = -2(y - 3)e^{-x^2-y^2+6y}$$

So, $f_y = 0$ when $y = 3$. The only critical point is $(0, 3)$.

$$f_{xx} = -2\left[x\left(-2xe^{-x^2-y^2+6y}\right) + e^{-x^2-y^2+6y}(1)\right]$$

$$f_{yy} = -2\left[(y-3)(-2y+6)e^{-x^2-y^2+6y} + e^{-x^2-y^2+6y}(1)\right]$$

$$f_{xy} = -2\left[x(-2y+6)e^{-x^2-y^2+6y} + 0\right]$$

For the point (0, 3),

$$D = (-2e^9)(-2e^9) > 0$$

and $f_{xx} < 0$, so (0, 3) is a relative maximum.

17.

$$f(x, y) = \frac{1}{x^2 + y^2 + 3x - 2y + 1}$$

$$f_x = \frac{-(2x+3)}{(x^2 + y^2 + 3x - 2y + 1)^2}$$

So, $f_x = 0$ when $x = -\frac{3}{2}$.

$$f_y = \frac{-(2y-2)}{(x^2 + y^2 + 3x - 2y + 1)^2}$$

So, $f_y = 0$ when $y = 1$. The only critical point is $\left(-\frac{3}{2}, 1\right)$.

$$f_{xx} = \frac{1}{(x^2 + y^2 + 3x - 2y + 1)^4}\Big((x^2 + y^2 + 3x - 2y + 1)^2(-2) + (2x+3)\,[2(x^2 + y^2 + 3x - 2y + 1)(2x+3)]\Big)$$

$$f_{yy} = \frac{1}{(x^2 + y^2 + 3x - 2y + 1)^4}\Big((x^2 + y^2 + 3x - 2y + 1)^2(-2) + (2y-2) + [2(x^2 + y^2 + 3x - 2y + 1)(2y-2)]\Big)$$

$$f_{xy} = \frac{1}{(x^2 + y^2 + 3x - 2y + 1)^4}\Big(0 + (2x+3)\,[2(x^2 + y^2 + 3x - 2y + 1)(2y-2)]\Big)$$

For the point $\left(-\frac{3}{2}, 1\right)$,

$$D = (-4)(-4) - 0 > 0$$

and $f_{xx} < 0$, so $\left(-\frac{3}{2}, 1\right)$ is a relative maximum.

19.

$$f(x, y) = x \ln\left(\frac{y^2}{x}\right) + 3x - xy^2$$
$$= x(\ln y^2 - \ln x) + 3x - xy^2$$
$$= x \ln y^2 - x \ln x + 3x - xy^2$$

$$f_x = \ln y^2 - \left[x\left(\frac{1}{x}\right) + \ln x(1)\right] + 3 - y^2$$
$$= \ln y^2 - \ln x + 2 - y^2$$

So, $f_x = 0$ when $0 = 2\ln y - \ln x + 2 - y^3$

$$f_y = 2x\left(\frac{1}{y}\right) - 2xy = \frac{2x(1-y^2)}{y}$$

So, $f_y = 0$ when $x = 0$, $y = -1, 1$. We must reject $x = 0$, since f is undefined when $x = 0$.
When $y = -1$, $f_x = 0$ when

$$0 = \ln 1 - \ln x + 2 - 1$$
$$0 = 1 - \ln x$$
$$\ln x = 1, \text{ or } x = e.$$

When $y = 1$, $f_x = 0$ when
$0 = \ln 1 - \ln x + 2 - 1$, or $x = e$.
So, the critical points are $(e, -1)$ and $(e, 1)$.

$$f_{xx} = -\frac{1}{x},\ f_{yy} = -\frac{2x}{y^2} - 2x,\ f_{xy} = \frac{2}{y} - 2y$$

For the point $(e, -1)$,

$$D = \left(\frac{-1}{e}\right)(-4e) - 0 > 0$$

and $f_{xx} < 0$, so $(e, -1)$ is a relative maximum.
For the point $(e, 1)$,

$$D = \left(-\frac{1}{e}\right)(-4e) - 0 > 0$$

and $f_{xx} < 0$, so $(e, 1)$ is a relative maximum.

21. Profit = (profit from sales Duncan shirts)
+ (profit from sales O'Neal shirts)

$$P(x, y) = (x - 2)(40 - 50x + 40y) + (y - 2)(20 + 60x - 70y)$$

$$P_x = (x - 2)(-50) + (40 - 50x + 40y)(1) + (y - 2)(60) + 0 = 20(-5x + 5y + 1)$$

So, $P_x = 0$ when $0 = 20(-5x + 5y + 1)$, or $-5x + 5y + 1 = 0$.

$$P_y = (x - 2)(40) + 0 + (y - 2)(-70) + (20 + 60x - 70y)(1) = 20(5x - 7y + 4)$$

So, $P_y = 0$ when $0 = 20(5x - 7y + 4)$, or $0 = 5x - 7y + 4$.
Solving this system of equations by adding,

$$0 = -2y + 5, \text{ or } y = \frac{5}{2} = 2.5$$

When $y = 2.5$, $P_x = 0$ when $0 = -5x + 5(2.5) + 1$, or $x = 2.7$
So the critical point is (2.7, 2.5)

$$P_{xx} = -100,\ P_{yy} = -140,\ P_{xy} = 100$$

$$D = (-100)(-140) - (100)^2 > 0$$

and $P_{xx} < 0$
So, profit is maximized when Duncan shirts sell for \$2.70 and O'Neal shirts sell for \$2.50.

23. Let l, w, h be the dimensions of the box
Cost = (area) (cost per area)
Cost bottom = $(lw)(3)$
Cost top = $(lw)(5)$
Cost 4 sides = $2(lh)(1) + 2(wh)(1)$
$C = 8lw + 2lh + 2wh$
Since volume must be 32,
$32 = lwh$, or $h = \dfrac{32}{lw}$

$$C(l, w) = 8lw + 2l\left(\frac{32}{lw}\right) + 2w\left(\frac{32}{lw}\right) = 8lw + \frac{64}{w} + \frac{64}{l}$$

$$C_l = 8w - \frac{64}{l^2}$$

So, $C_l = 0$ when $0 = 8w - \dfrac{64}{l^2}$.

$$C_w = 8l - \frac{64}{w^2}$$

So, $C_w = 0$ when $0 = 8l - \dfrac{64}{w^2}$.
Solving each equation for w^2,

$$8w = \frac{64}{l^2}$$

$$w = \frac{8}{l^2},\ w^2 = \frac{64}{l^4}$$

$$8l = \frac{64}{w^2}$$

$$w^2 = \frac{8}{l}$$

So,

$$\frac{64}{l^4} = \frac{8}{l},\ 64l = 8l^4$$

$$8l(l^3 - 8) = 0, \quad \text{or } l = 2.$$

When $l = 2$, $w = \dfrac{8}{(2)^2} = 2$.
So, (2, 2) is the critical point.

$$C_{ll} = \frac{128}{l^3},\ C_{ww} = \frac{128}{w^3},\ C_{lw} = 8$$

$$D = (32)(32) - (8)^2 > 0 \text{ and } C_{ll} > 0$$

When $l = 2$ and $w = 2$, $h = \dfrac{32}{(2)(2)}$.
So, cost is minimized when the dimensions of the box are 2 ft × 2 ft × 8 ft.

25. Profit = revenue − cost

$$P(x, y) = [x(100 - x) + y(100 - y)] - [x^2 + xy + y^2] = -2x^2 - 2y^2 + 100x + 100y - xy$$

$$P_x = 4x + 100 - y$$

So, $P_x = 0$ when $0 = -4x + 100 - y$.

$$P_y = -4y + 100 - x$$

So, $P_y = 0$ when $0 = -4y + 100 - x$.
Solving this system of equations by multiplying the first equation by −4 and adding to second,

$$0 = 15x - 300, \text{ or } x = 20.$$

When $x = 20$, $P_x = 0$ when $0 = -4(20) + 100 - y$, or $y = 20$. So, the critical point is (20, 20).

$$P_{xx} = -4;\ P_{yy} = -4;\ P_{xy} = -1$$

$$D = (-4)(-4) - (-1)^2 > 0 \text{ and } P_{xx} < 0$$

So, profit is maximized when 20 gallons of each are produced.

27. $f(x, y) = C + xye^{1-x^2-y^2}$

$$f_x = y\left[x(-2xe^{1-x^2-y^2}) + e^{1-x^2-y^2}(1)\right]$$
$$= ye^{1-x^2-y^2}(-2x^2 + 1)$$

So, $f_x = 0$ when $y = 0$, or $x = \frac{\sqrt{2}}{2}$ (rejecting the negative solution).

$$f_y = x\left[y(-2ye^{1-x^2-y^2}) + e^{1-x^2-y^2}(1)\right]$$
$$= xe^{1-x^2-y^2}(-2y^2 + 1)$$

When $y = 0$, $f_y = 0$ when $x = 0$.

When $x = -\frac{\sqrt{2}}{2}$, $f_y = 0$ when $y = \frac{\sqrt{2}}{2}$.

When $x = \frac{\sqrt{2}}{2}$, $f_y = 0$ when $y = \frac{\sqrt{2}}{2}$.

Again rejecting the negative solutions, the critical points are (0, 0) and $\left(\frac{\sqrt{2}}{2}, \frac{\sqrt{2}}{2}\right)$. Rewriting f_x as

$$f_x = e^{1-x^2-y^2}(-2x^2y + y)$$
$$f_{xx} = (e^{1-x^2-y^2})(-4xy) + (-2x^2y + y)(-2xe^{1-x^2-y^2})$$

Similarly,

$$f_{yy} = (e^{1-x^2-y^2})(-4xy) + (-2xy^2 + x)(-2ye^{1-x^2-y^2})$$
$$f_{xy} = (e^{1-x^2-y^2})(-2x^2 + 1) + (-2x^2y + y)(-2ye^{1-x^2-y^2})$$

For the point (0, 0),

$$D = (0)(0) - (e)^2 < 0$$

So, the point (0, 0) does not correspond to the maximum.

For the point $\left(\frac{\sqrt{2}}{2}, \frac{\sqrt{2}}{2}\right)$,

$$D = (-2)(-2) - 0 > 0 \text{ and } f_{xx} < 0$$

So, $\frac{\sqrt{2}}{2}$ units of each stimuli maximizes performance.

29. $V_0 = xyz$, so $z = \frac{V_0}{xy}$ and

$$E(x, y) = \frac{k}{8m}\left(\frac{1}{x^2} + \frac{1}{y^2} + \frac{x^2y^2}{V_0}\right)$$
$$E_x = \frac{k}{8m}\left(-\frac{2}{x^3} + \frac{2xy^2}{V_0^2}\right)$$

So, $E_x = 0$ when $0 = -\frac{2}{x^3} + \frac{2xy^2}{V_0^2}$,

$$\text{or } x^2 = \frac{V_0}{\cdot y}.$$

$$E_y = \frac{k}{8m}\left(-\frac{2}{y^3} + \frac{2x^2y}{V_0^2}\right)$$

So, $E_y = 0$ when $0 = -\frac{2}{y^3} + \frac{2x^2y}{V_0^2}$,

$$\text{or } x^2 = \frac{V_0^2}{y^4}$$

$$\text{and } \frac{V_0}{y} = \frac{V_0^2}{y^4}, \text{ or } y = V_0^{1/3}.$$

When $y = V_0^{1/3}$, $x = \sqrt{\frac{V_0}{V_0^{1/3}}} = V_0^{1/3}$

and $z = \frac{V_0}{V_0^{1/3}V_0^{1/3}} = V_0^{1/3}$.

$$E_{xx} = \frac{k}{8m}\left(\frac{6}{x^4} + \frac{2y^2}{V_0^2}\right)$$

$$E_{yy} = \frac{k}{8m}\left(\frac{6}{y^4} + \frac{2x^2}{V_0^2}\right)$$

$$E_{xy} = \frac{k}{8m}\left(\frac{4xy}{V_0^2}\right)$$

$$D = \left(\frac{k}{mV_0^{4/3}}\right)\left(\frac{k}{mV_0^{4/3}}\right) - \left(\frac{k}{2mV_0^{4/3}}\right)^2 > 0$$

and $E_{xx} > 0$, so the ground state energy is maximized when $x = y = z = V_0^{1/3}$.

31. profit $=$ (profit from domestic market) $+$ profit from foreign market)

$$P(x, y) = x\left(60 - \frac{x}{5} + \frac{y}{20}\right) + y\left(50 - \frac{y}{10} + \frac{x}{20}\right)$$

$$= 50x - \frac{x^2}{5} + \frac{xy}{10} + 40y - \frac{y^2}{10}$$

$$P_x = 50 - \frac{2}{5}x + \frac{y}{10}$$

So, $P_x = 0$ when $0 = 50 - \frac{2}{5}x + \frac{y}{10} = 500 - 4x + y$.

$$P_y = \frac{x}{10} + 40 - \frac{y}{5}$$

So $P_y = 0$ when $0 = \frac{x}{10} + 40 - \frac{y}{5} = x + 400 - 2y$

Solving this system by multiplying the first equation by two and adding to the second,

$$0 = 1400 - 7x, \text{ or } x = 200.$$

When $x = 200$, $P_y = 0$ when $0 = 200 + 400 - 2y$, or $y = 300$.

$$P_{xx} = -\frac{2}{5},\ P_{yy} = -\frac{1}{5},\ P_{xy} = \frac{1}{10}$$

$$D = \left(-\frac{2}{5}\right)\left(-\frac{1}{5}\right) - \left(\frac{1}{10}\right)^2 > 0$$

and $P_{xx} < 0$

So, profit is maximized when 200 machines are supplied to the domestic market and 300 are supplied to the foreign market.

33. The square of the distance from $S(a, b)$ to each point is:

$$(a+5)^2 + (b-0)^2 = a^2 + 10a + 25 + b^2$$

$$(a-1)^2 + (b-7)^2 = a^2 - 2a + b^2 - 14b + 50$$

$$(a-9)^2 + (b-0)^2 = a^2 - 18a + 81 + b^2$$

$$(a-0)^2 + (b+8)^2 = a^2 + b^2 + 16b + 64$$

So, the sum of the distances is

$$f(a, b) = 4a^2 - 10a + 4b^2 + 2b + 220$$

$$f_a = 8a - 10, \text{ so } f_a = 0 \text{ when } a = \frac{5}{4}$$

$$f_b = 8b + 2, \text{ so } f_b = 0 \text{ when } b = -\frac{1}{4}$$

$$f_{aa} = 8,\ f_{bb} = 8,\ f_{ab} = 0 \text{ so,}$$

$$D = (8)(8) - 0 > 0 \text{ and } f_{aa} > 0$$

The sum is minimized at $\left(\frac{5}{4}, -\frac{1}{4}\right)$.

35. Since $p + q + r = 1$, $r = 1 - p - q$ and

$$P(p, q) = 2pq + 2p(1 - p - q) + 2(1 - p - q)q$$

$$= 2p - 2p^2 - 2pq + 2q - 2q^2$$

$$P_p = 2 - 4p - 2q$$

So, $p_p = 0$ when $0 = 2 - 4p - 2q$, or $0 = 1 - 2p - q$.

$$P_q = -2p + 2 - 4q$$

So, $P_q = 0$ when $0 = -2p + 2 - 4q$, or $0 = -p + 1 - 2q$.

Solving this system of equations by multiplying the first equation by negative two and adding to the second,

$$0 = -1 + 3p, \text{ or } p = \frac{1}{3}.$$

When $p = \frac{1}{3}$, $P_q = 0$ when $0 = -\frac{1}{3} + 1 - 2q$, or $q = \frac{1}{3}$.

$$P_{pp} = -4,\ P_{qq} = -4,\ P_{pq} = -2$$

$$D = (-4)(-4) - (-2)^2 > 0 \text{ and } P_{pp} < 0$$

So, so P is maximized when $p = \frac{1}{3}$, $q = \frac{1}{3}$, and $r = \frac{1}{3}$. The maximum is

$$P = 2\left(\frac{1}{3}\right)\left(\frac{1}{3}\right) + 2\left(\frac{1}{3}\right)\left(\frac{1}{3}\right) + 2\left(\frac{1}{3}\right)\left(\frac{1}{3}\right) = \frac{2}{3}$$

37. **(a)** The problem is to minimize the total time $T(x, y)$, where

$$T = \frac{\sqrt{(1.2)^2 + x^2}}{2} + \frac{\sqrt{(2.5)^2 + y^2}}{4} + \frac{4.3 - (x + y)}{6}$$

$$\frac{\partial T}{\partial x} = \frac{1}{2}\left[\frac{1}{2}\frac{2x}{\sqrt{(1.2)^2 + x^2}}\right] - \frac{1}{6}$$

$$\frac{\partial T}{\partial y} = \frac{1}{4}\left[\frac{1}{2}\frac{2y}{\sqrt{(2.5)^2 + y^2}}\right] - \frac{1}{6}$$

$$\frac{\partial T}{\partial x} = \frac{\partial T}{\partial y} = 0 \text{ when}$$

$$\frac{1}{2}\frac{x}{\sqrt{(1.2)^2 + x^2}} = \frac{1}{6} \text{ and } \frac{1}{4}\frac{y}{\sqrt{(2.5)^2 + y^2}} = \frac{1}{6}$$

which leads to $x = 0.424$ and $y = 2.236$.
In addition to his path, the "boundary" cases must also be considered. That is, a path where Tom moves directly to the river (perpendicular to the river), then Tom swims directly across the river (perpendicular to the river), and Mary runs to the finish. The second boundary path is along the diagonal connection S and F.

Case 1

$$x = 0,\ y = 0$$

$$\text{Time} = \frac{1.2}{2} + \frac{2.5}{4} + \frac{4.3}{6} \approx 1.942$$

Case 2

$$x = 0.424,\ y = 2.236$$

$$\text{Time} = \frac{1.273}{2} + \frac{3.354}{4} + \frac{1.64}{6} = 1.748$$

Case 3

$$x = 1.395,\ y = 2.905$$

$$\text{Time} = \frac{1.84}{2} + \frac{3.833}{4} + \frac{0}{6} = 1.878$$

The minimum time is when $x = 0.424$ miles and $y = 2.236$ miles.

(b) For the second team, the time is

$$T = \frac{\sqrt{(1.2)^2 + x^2}}{1.7} + \frac{\sqrt{(2.5)^2 + y^2}}{3.5} + \frac{4.3 - (x + y)}{6.3}$$

$$\frac{\partial T}{\partial x} = \frac{1}{1.7}\left[\frac{x}{\sqrt{(1.2)^2 + x^2}}\right] - \frac{1}{6.3}$$

$$\frac{\partial T}{\partial y} = \frac{1}{3.5}\left[\frac{y}{\sqrt{(2.5)^2 + y^2}}\right] - \frac{1}{6.3}$$

We must find when $\frac{\partial T}{\partial x} = \frac{\partial T}{\partial y} = 0$

Press [y =].

Input $\frac{\partial T}{\partial x}$ for $y_1 =$.

Use window dimensions [0, 2]0.5 by [−1, 2]0.5.

Press [graph].

Use the zero function under the calc menu to find $x \approx 0.3363$.

Repeat process for $\frac{\partial T}{\partial y}$ to find $y \approx 1.6704$.

Repeating the case scenarios as in part (a)

Case	x	y	Time
1	0	0	2.103
2	0.3363	1.6704	1.9562
3	1.395	2.905	2.177

Tom, Dick, and Mary will win by 0.208 hours (12.5 minutes)

(c) Writing Exercise—Answers will vary.

39. The goal is to maximize the livable space subject to a constraint on the surface area. Let s be the length along the floor, at each end, where a 6 foot tall person cannot stand. Then, the livable space is

$$L = 6(x - 2s)y$$

From similar triangles,

$$\frac{s}{6} = \frac{\frac{x}{2}}{\frac{\sqrt{3}}{2}x}$$

or, $s = \dfrac{6}{\sqrt{3}}$ and

$$L = 6\left(x - \frac{12}{\sqrt{3}}\right)y = 6xy - \frac{72}{\sqrt{3}}y$$

Since the surface area must be 500, the constraint is

$$500 = 2xy + 2\left(\frac{\sqrt{3}}{4}x^2\right)$$

and $\quad g(x) = 2xy + \dfrac{\sqrt{3}}{2}x^2$

$$L_x = 6y;\ L_y = 6x - \frac{72}{\sqrt{3}}$$

$$g_x = 2y + \sqrt{3}x;\ g_y = 2x$$

So, the three Lagrange equations are

$$6y = (2y + \sqrt{3}x)\lambda$$

$$6x - \frac{72}{\sqrt{3}} = 2x\lambda$$

$$2xy + \frac{\sqrt{3}}{2}x^2 = 500$$

Solving the second equation for λ and substituting into the first equation gives

$$6y = (2y + \sqrt{3}x)\left(3 - \frac{36}{\sqrt{3}x}\right)$$

$$6y = 6y + 3\sqrt{3}x - \frac{72y}{\sqrt{3}x} - 36$$

$$\frac{72}{\sqrt{3}x}y = 3\sqrt{3}x - 36$$

$$y = \frac{1}{8}x^2 - \frac{\sqrt{3}}{2}x$$

Substituting into the third equation gives

$$2x\left(\frac{1}{8}x^2 - \frac{\sqrt{3}}{2}x\right) + \frac{\sqrt{3}}{2}x^2 = 500$$

$$\frac{1}{4}x^3 - \sqrt{3}x^2 + \frac{\sqrt{3}}{2}x^2 = 500$$

$$x^3 - 4\sqrt{3}x^2 + 2\sqrt{3}x^2 = 2{,}000$$

$$x^3 - 2\sqrt{3}x^2 - 2{,}000 = 0$$

To use the calculator to solve

$$x^3 - 2\sqrt{3}x^2 - 2{,}000 = 0$$

Press [y=].
Input $x \wedge 3 - 2 * \sqrt{}(3) * x^2 - 2{,}000$ for $y_1 =$.
Use window dimensions [0, 20]5 by [−50, 500]150.
Press [graph].
Use the zero function under the calc menu to find $x = 13.866$.
When $x = 13.866$ feet,

$$y = \frac{1}{8}(13.866)^2 - \frac{\sqrt{3}}{2}(13.866)$$

$$\approx 12.025 \text{ feet.}$$

41. $f(x, y) = x^2 + y^2 - 4xy$,
$f_x = 2x - 4y = 0$ when $y = \dfrac{x}{2}$.
$f_y = 2y - 4x = 0$ when $y = 2x$.
So, (0, 0) is a critical point.
$f_{xx} = 2$, $f_{xy} = -4$, and $f_{yy} = 2$, so

$$D = 4 - (-4)^2 < 0$$

and (0, 0) is a saddle point.
The above is true but not asked for. If $x = 0$, $f(0, y) = y^2$ which is a parabola with a minimum at (0, 0) (in the vertical yz-plane). If $y = 0$, $f(x, 0) = x^2$ which is a parabola with a minimum at (0, 0) (in the vertical xz-plane).
If $y = x$, $f(x, x) = -2x^2$ which is a parabola with a maximum at (0, 0) (in the vertical plane passing through the z-axis and the line $y = x$ in the xy plane).

43. $f(x, y) = \dfrac{x^2 + xy + 7y^2}{x \ln y}$

To use the graphing utility to determine critical points of the function,

$$f_x = \frac{(x \ln y)(2x + y) - (x^2 + xy + 7y^2)(\ln y)}{(x \ln y)^2}$$

$$= \frac{\ln y[2x^2 + xy - x^2 - xy - 7y^2]}{x^2 \ln^2 y}$$

$$= \frac{x^2 - 7y^2}{x^2 \ln y}$$

$$f_y = \frac{(x \ln y)(x + 14y) - (x^2 + xy + 7y^2)\left(\frac{x}{y}\right)}{(x \ln y)^2}$$

$$= \frac{\underline{(xy \ln y)(x + 14y) - x(x^2 + xy + 7y^2)}}{\frac{y}{x^2(\ln y)^2}}$$

$$= \frac{x\left[(y \ln y)(x + 14y) - x^2 - xy - 7y^2\right]}{x^2 y(\ln y)^2}$$

$$= \frac{(y \ln y)(x + 14y) - x^2 - xy - 7y^2}{xy(\ln y)^2}$$

Next, $f_x = 0$ when $\dfrac{x^2 - 7y^2}{x^2 \ln y} = 0$, or $x^2 - 7y^2 = 0$.
$f_y = 0$ when
$(y \ln y)(x + 14y) - x^2 - xy - 7y^2 = 0$.
The critical points are found by solving the system

$$x^2 - 7y^2 = 0$$
$$(y \ln y)(x + 14y) - x^2 - xy - 7y^2 = 0$$

From the first equation, $x = \pm\sqrt{7}y$. Substitute $x = \sqrt{7}y$ into the second equation to obtain
$(y \ln y)(\sqrt{7}y + 14y) - 7y^2 - \sqrt{7}y^2 - 7y^2 = 0$
$y^2\left[(\ln y)(\sqrt{7} + 14) - 14 - \sqrt{7}\right] = 0$
Press [y =].
Input $x^2(\ln(x) * (14 + \sqrt{7}) - 14 - \sqrt{7})$ for $y_1 =$.
(Remember that we are actually solving for y.)
Use window dimensions $[-5, 10]1$ by $[-10, 10]1$.
Press [graph].
Using trace and zoom or the zero function under the calc menu to find the zeros are $y \approx 2.7182818 (y = e)$ and $y = 0$.

If $x = -\sqrt{7}y$, we also find the zeros to be $y = 0$ and $y = e$.
So, the critical points are $(\pm\sqrt{7}e, e)$.
The point $(0, 0)$ cannot be a critical point since $\ln 0$ is not defined.

45.

$$f(x, y) = 2x^4 + y^4 - 11x^2y + 18x^2$$
$$f_x = 8x^3 - 22xy + 36x$$
$$f_y = 4y^3 - 11x^2$$

The critical points are found by solving the system

$$2x(4x^2 - 11y + 18) = 0$$
$$4y^3 - 11x^2 = 0$$

Solving the first equation gives $2x = 0$, or $4x^2 - 11y + 18 = 0$.
If $2x = 0$, $x = 0$ and substituting this into the second equation gives $4y^3 = 0$, or $y = 0$. One critical point is $(0, 0)$.
To solve $4x^2 - 11y + 18 = 0$, solve the second equation to get $x^2 = \dfrac{4}{11}y^2$. and substitute. Then,

$$4\left(\frac{4}{11}y^3\right) - 11y + 18 = 0$$
$$\frac{16}{11}y^3 - 11y + 18 = 0$$

Press [y =].
Input $y = (16/11)x \wedge 3 - 11x + 18$ for $y_1 =$.
(Remember, we are actually solving for y.)
Use the window dimensions $[-10, 5]1$ by $[-10, 10]1$.
Press [graph].
Use trace and zoom or the zero function under the calc menu to find the zero is $y \approx -3.354$. We find we cannot use this value, however, since $x^2 = \dfrac{4}{11}(-3.354)^3$, $x^2 \approx -13.72$, which has no solution.
The only critical point is $(0, 0)$.

7.4 The Method of Least Squares

1. The sum $S(m, b)$ of the squares of the vertical distances from the three given points is
$S(m, b) = d_1^2 + d_2^2 + d_3^2 = (b-1)^2$
$+ (2m+b-3)^2 + (4m+b-2)^2$.
To minimize $S(m, b)$, set the partial derivatives
$\frac{\partial S}{\partial m} = 0$ and $\frac{\partial S}{\partial b} = 0$.

$$\frac{\partial S}{\partial m} = 2(2m+b-3)(2) + 2(4m+b-2)(4)$$
$$= 40m + 12b - 28 = 0$$
$$\frac{\partial S}{\partial b} = 2(b-1) + 2(2m+b-3) + 2(4m+b-2)$$
$$= 12m + 6b - 12 = 0.$$

Solve the resulting simplified equations $10m + 3b = 7$ and $6m + 3b = 6$ to get $m = \frac{1}{4}$ and $b = \frac{3}{2}$. So, the equation of the least-squares line is $y = \frac{1}{4}x + \frac{3}{2}$.

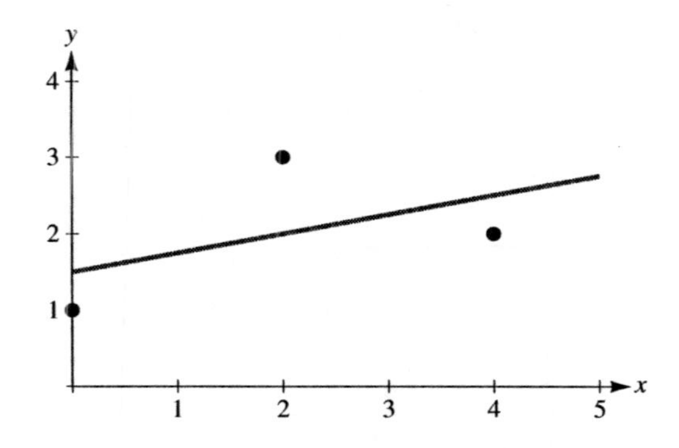

3. The sum $S(m, b)$ of the squares of the vertical distances from the four given points is
$S(m, b) = (m+b-2)^2 + (2m+b-4)^2$
$+(4m+b-4)^2 + (5m+b-2)^2$.
To minimize $S(m, b)$, set the partial derivatives
$\frac{\partial S}{\partial m} = 0$ and $\frac{\partial S}{\partial b} = 0$.
$$\frac{\partial S}{\partial m} = 2(m+b-2) + 2(2m+b-4)(2)$$
$$+ 2(4m+b-4)(4) + 2(5m+b-2)(5)$$
$$= 92m + 24b - 72 = 0$$
$$\frac{\partial S}{\partial b} = 2(m+b-2) + 2(2m+b-4)$$
$$+ 2(4m+b-4) + 2(5m+b-2)$$
$$= 24m + 8b - 24 = 0.$$

Solve the resulting simplified equations $23m + 6b = 18$ and $3m + b = 3$ to get $m = 0$ and $b = 3$. So, the equation of the least-squares line is $y = 3$.

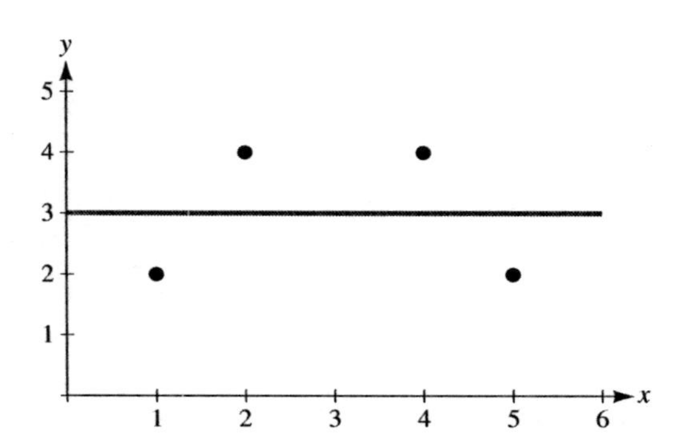

5.

x	y	xy	x^2
1	2	2	1
2	2	4	4
2	3	6	4
5	5	25	25
$\sum x = 10$	$\sum y = 12$	$\sum xy = 37$	$\sum x^2 = 34$

Using the formulas with $n = 4$,
$m = \frac{4(37) - 10(12)}{4(34) - (10)^2} = \frac{7}{9}$ and
$b = \frac{34(12) - 10(37)}{4(34) - (10)^2} = \frac{19}{18}$
So, the equation of the least-squares line is $y = \frac{7}{9}x + \frac{19}{18}$.

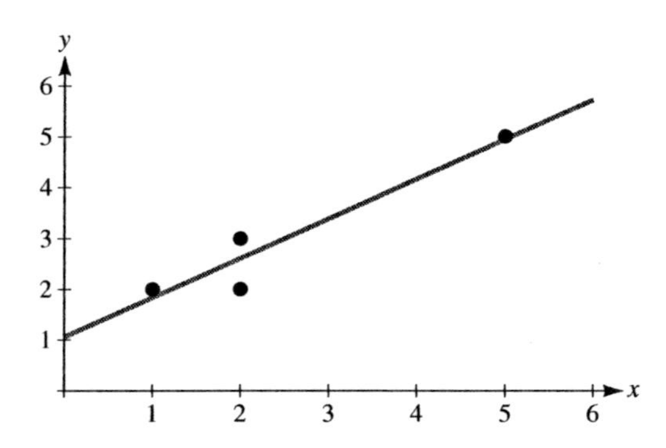

7.

x	y	xy	x^2
−2	5	−10	4
0	4	0	0
2	3	6	4
4	2	8	16
6	1	6	36
$\sum x = 10$	$\sum y = 15$	$\sum xy = 10$	$\sum x^2 = 60$

Using the formulas with $n = 5$,

$$m = \frac{5(10) - 10(15)}{5(60) - (10)^2} = -\frac{100}{200} = -\frac{1}{2} \text{ and}$$

$$b = \frac{60(15) - 10(10)}{5(60) - (10)^2} = \frac{800}{200} = 4$$

So, the equation of the least-squares line is $y = -\frac{1}{2}x + 4$.

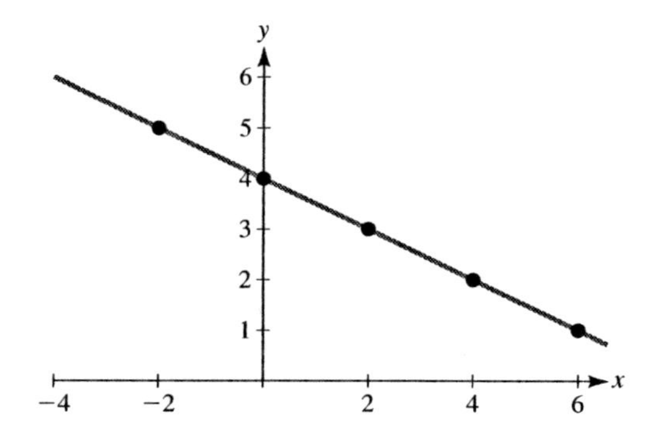

9.

x	y	xy	x^2
0	1	0	0
1	1.6	1.6	1
2.2	3	6.6	4.84
3.1	3.9	12.09	9.61
4	5	20	16
$\sum x = 10.3$	$\sum y = 14.5$	$\sum xy = 40.29$	$\sum x^2 = 31.45$

Using the formulas with $n = 5$,

$$m = \frac{5(40.29) - 14.5(10.3)}{5(31.45) - (10.3)^2} = \frac{52.10}{51.16} \approx 1.0184$$

$$b = \frac{31.45(14.5) - 10.3(40.29)}{5(31.45) - (10.3)^2} = \frac{41.038}{51.16}$$

≈ 0.8022. So, the equation of the least-squares line is
$y = 1.0184x + 0.8022$.

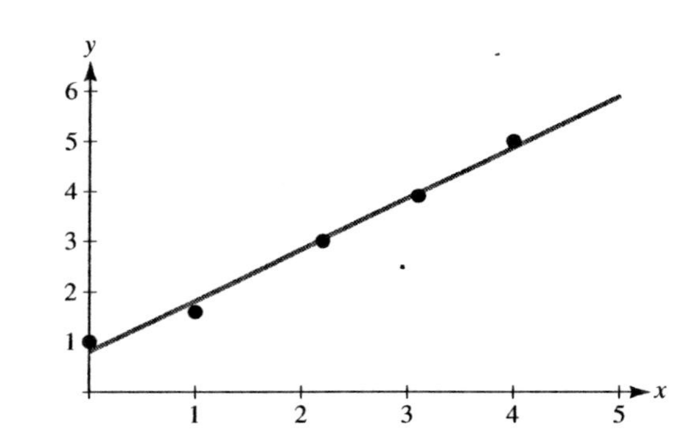

11.

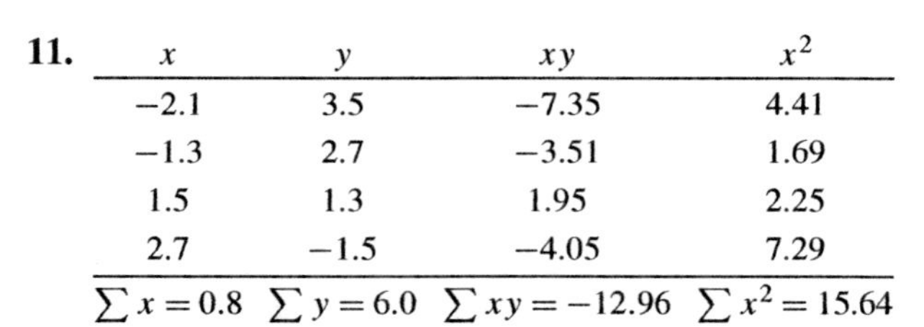

x	y	xy	x^2
−2.1	3.5	−7.35	4.41
−1.3	2.7	−3.51	1.69
1.5	1.3	1.95	2.25
2.7	−1.5	−4.05	7.29
$\sum x = 0.8$	$\sum y = 6.0$	$\sum xy = -12.96$	$\sum x^2 = 15.64$

Using the formulas with $n = 4$,

$$m = \frac{4(-12.96) - (0.8)(6.0)}{4(15.64) - (0.8)^2} = \frac{-56.64}{61.92} \approx -0.915$$

$$b = \frac{(15.64)(6.0) - (0.8)(-12.96)}{4(15.64) - (0.8)^2} = \frac{104.208}{61.92}$$

≈ 1.683.

So, the equation of the least-squares line is $y = -0.915x + 1.683$

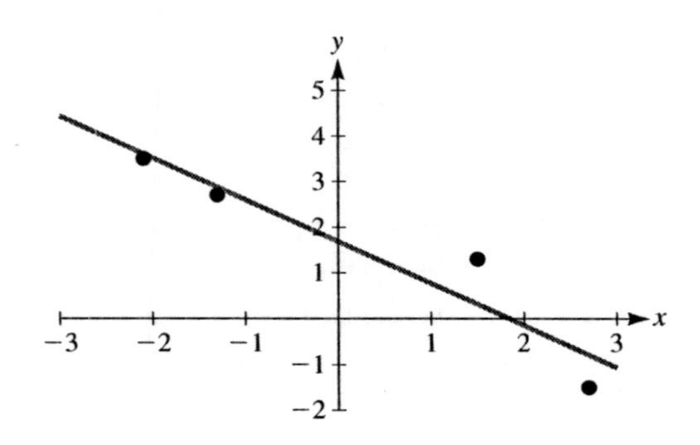

13. Since $y = Ae^{mx}$

$$\ln y = \ln A + \ln e^{mx} = \ln A + mx$$
$$= mx + \ln A$$

We can find the least-squares line, $Y = Mx + b$, using $Y = \ln y$. Then, use $M = m$ and $b = \ln A$.

x	$y = \ln y$	xy	x^2
1	2.75	2.75	1
3	2.83	8.49	9
5	2.91	14.55	25
7	3.00	21	49
10	3.11	31.1	100
$\sum x = 26$	$\sum y = 14.6$	$\sum xy = 77.89$	$\sum x^2 = 184$

Using the formulas with $n = 5$,
$$m = \frac{5(77.89) - (26)(14.6)}{5(184) - (26)^2} = \frac{9.85}{244} \approx 0.04 \text{ and}$$
$$b = \frac{(184)(14.6) - (26)(77.89)}{5(184) - (26)^2} = \frac{661.26}{244} \approx 2.710$$
For our exponential model, $y = Ae^{mx}$. Since $\ln A = b$,

$$A = e^b = e^{2.71} \approx 15.029$$

So, the exponential function that best fits the data is $y = 15.029e^{0.04x}$.

15. Since $y = Ae^{mx}$,

$$\ln y = \ln A + \ln e^{mx}$$
$$= mx + \ln A$$

We can find the least-squares line, $Y = Mx + b$, using $Y = \ln y$. Then, use $M = m$ and $b = \ln A$.

x	$y = \ln y$	xy	x^2
2	2.60	5.20	4
4	2.20	8.80	16
6	1.79	10.74	36
8	1.39	11.12	64
10	0.99	9.9	100
$\sum x = 30$	$\sum y = 8.97$	$\sum xy = 45.76$	$\sum x^2 = 220$

Using the formulas with $n = 5$,
$$m = \frac{5(45.76) - (30)(8.97)}{5(220) - (30)^2} = \frac{-40.3}{200} \approx -0.202$$
and
$$b = \frac{(220)(8.97) - (30)(45.76)}{5(220) - (30)^2} = \frac{600.6}{200} \approx 3.003$$
For our exponential model, $y = Ae^{mx}$. Since $\ln A = b$,

$$A = e^b = e^{3.003} \approx 20.15$$

So, the exponential function that best fits the data is $y = 20.15e^{-0.202x}$.

17. **(a)** Let x be the number of catalogs requested and y the number of applications received (both in units of 1,000). The given points (x, y) are plotted on the accompanying graph.

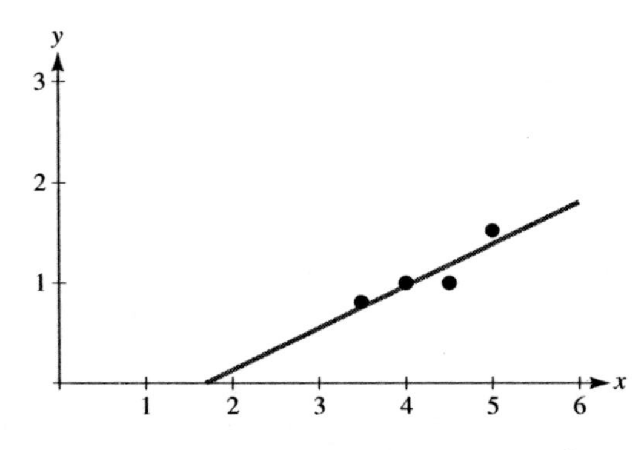

(b)

x	y	xy	x^2
4.5	1.0	4.5	20.25
3.5	0.8	2.8	12.25
4.0	1.0	4.0	16.00
5.0	1.5	7.5	25.00

$\sum x = 17.0$ $\sum y = 4.3$ $\sum xy = 18.8$ $\sum x^2 = 73.50$

Using the formulas with $n = 4$,
$m = \dfrac{4(18.8) - 17(4.3)}{4(73.5) - (17)^2} \approx 0.42$ and
$b = \dfrac{73.5(4.3) - 17(18.8)}{4(73.5) - (17)^2} \approx -0.71$
So, the equation of the least-squares line is $y = 0.42x - 0.71$.

(c) If 4,800 catalogs are requested by December 1, $x = 4.8$ and $y = 0.42(4.8) - 0.71 = 1.306$, which means that approximately 1,306 completed applications will be received by March 1.

19. (a)

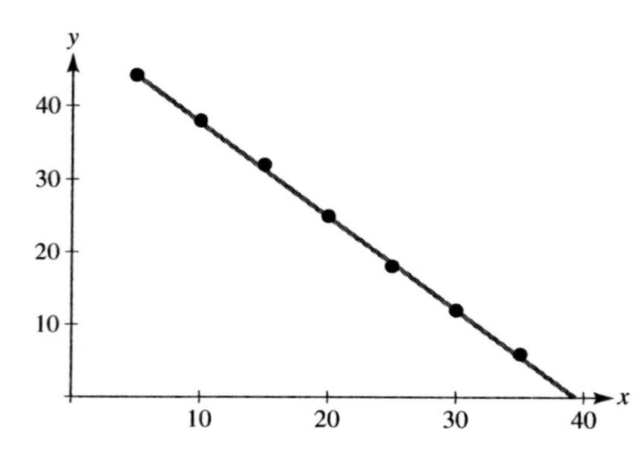

(b)

x	y	xy	x^2
5	44	220	25
10	38	380	100
15	32	480	225
20	25	500	400
25	18	450	625
30	12	360	900
35	6	210	1,225
$\sum x = 140$	$\sum y = 175$	$\sum xy = 2{,}600$	$\sum x^2 = 3{,}500$

Using the formulas with $n = 7$,
$m = \dfrac{7(2{,}600) - (140)(175)}{7(3{,}500) - (140)^2} = \dfrac{-6{,}300}{4{,}900}$
≈ -1.29 and

$b = \dfrac{(3{,}500)(175) - (140)(2{,}600)}{7(3{,}500) - (140)^2}$
$= \dfrac{248{,}500}{4{,}900} \approx 50.71$. So, the equation of the least-squares line is $y = -1.29x + 50.71$

(c) If 4,000 units are produced, $x = 40$ and

$$y = -1.29(40) + 50.71 = -0.89$$

Since this predicted price is negative, all 4,000 units cannot be sold at any price.

21. (a) Let x denote the number of hours after the polls open and y the corresponding percentage of registered voters that have already cast their ballots. Then

x	2	4	6	8	10
y	12	19	24	30	37

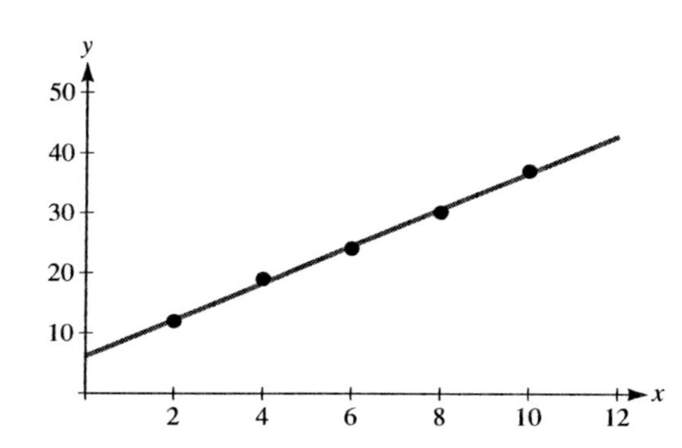

(b)

x	y	xy	x^2
2	12	24	4
4	19	76	16
6	24	144	36
8	30	240	64
10	37	370	100
$\sum x = 30$	$\sum y = 122$	$\sum xy = 854$	$\sum x^2 = 220$

Using the formulas with $n = 5$,
$m = \dfrac{5(854) - (30)(122)}{5(220) - (30)^2} = \dfrac{610}{200} = 3.05$ and
$b = \dfrac{(220)(122) - (30)(854)}{5(220) - (30)^2} = \dfrac{1{,}220}{200} = 6.10$
So, the equation of the least-squares line is $y = 3.05x + 6.10$

(c) When the polls close at 8:00 p.m., $x = 12$ and so $y = 3.05(12) + 6.1 = 42.7$, which means that approximately 42.7% of the registered voters can be expected to vote.

23. **(a)** Let x denote the number of decades after 1950 and y the corresponding population (in millions). Then,

x	0	1	2	3	4	5
y	150.7	179.3	203.2	226.5	248.7	291.4

Since $y = Ae^{mx}$,

$$\ln y = \ln A + \ln e^{mx}$$
$$\ln y = mx + \ln A$$

We can find the least-squares line, $Y = Mx + b$, using $Y = \ln y$. Then, use $M = m$ and $b = \ln A$.

x	$y = \ln y$	xy	x^2
0	5.02	0	0
1	5.19	5.19	1
2	5.31	10.62	4
3	5.42	16.26	9
4	5.52	22.08	16
5	5.67	28.35	25
$\sum x = 15$	$\sum y = 32.13$	$\sum xy = 82.5$	$\sum x^2 = 55$

Using the formulas with $n = 6$,
$m = \dfrac{6(82.5) - (15)(32.13)}{6(55) - (15)^2} = \dfrac{13.05}{105} \approx 0.124$
and
$b = \dfrac{(55)(32.13) - (15)(82.5)}{6(55) - (15)^2} = \dfrac{529.65}{105}$
≈ 5.044. For our exponential model, $P = Ae^{mx}$. Since $\ln A = b$,

$$A = e^b = e^{5.044} \approx 155.089$$

So, the exponential function that best fits the data is $P = 155.089e^{0.124x}$. So, the population is growing approximately 12.4% per decade.

(b) In the year 2005, $x = 5.5$ and $P = 155.089e^{0.124(5.5)} \approx 306.74$ million.

25. **(a)** Since $V(t) = Ae^{rt}$,

$$\ln V = \ln A + \ln e^{rt}$$
$$\ln V = rt + \ln A$$

We can find the least-squares line using $y = \ln V$. Then use $m = r$, $x = t$, and $b = \ln A$.

x	$y = \ln V$	xy	x^2
1	4.04	4.04	1
2	4.09	8.18	4
3	4.13	12.39	9
4	4.17	16.68	16
5	4.13	20.65	25
6	4.17	25.02	36
7	4.25	29.75	49
8	4.32	34.56	64
9	4.37	39.33	81
10	4.44	44.40	100
$\sum x = 55$	$\sum y = 42.11$	$\sum xy = 235.0$	$\sum x^2 = 385$

Using the formulas with $n = 10$,
$m = \dfrac{10(235.0) - (55)(42.11)}{10(385) - (55)^2} = \dfrac{33.95}{825}$
≈ 0.041 and

$b = \dfrac{(385)(42.11) - (55)(235.0)}{10(385) - (55)^2} = \dfrac{3{,}287.35}{825}$
≈ 3.985. For our exponential model, $V(t) = Ae^{rt}$. Since $\ln A = b$,

$$A = e^b = e^{3.985} \approx 53.785$$

So, the exponential function that best fits the data is $V(t) = 53.785e^{0.041t}$. Her account is growing at a rate of approximately 4.1% per year.

(b) When $t = 20$, $V(20) \approx 53.785e^{0.041(20)} \approx$ 122.1 thousand, or $122,100.

(c) To find t when $V(t) \approx 300$ thousand,

$$300 = 53.785e^{0.041t}$$
$$5.5778 = e^{0.041t}$$
$$\ln 5.5778 = 0.041t, \text{ or}$$
$$t \approx \frac{\ln 5.5778}{0.041} \approx 42 \text{ years}$$

(d) Using the two points named by Frank,

$$57 = Ae^{r(1)}$$
$$68 = Ae^{r(10)}$$

Solving the first for A and substituting in the second gives

$$68 = (57e^{-r})e^{10r}$$
$$1.19298 = e^{9r}$$
$$\ln 1.19298 = 9r$$

or $r \approx \frac{\ln 1.19298}{9} \approx 0.0196$ and $A = 57e^{-0.0196} \approx 55.89$. Frank's function fits the first and last data point, but may not be a good fit with the other data points. Frank's function would be less usable to predict other values.

27. (a) Let x denote the number of years after 1990 and y the corresponding DJIA. Then,

x	0	2	6	8	11	12
y	2,810	3,172	5,177	7,965	10,646	10,073

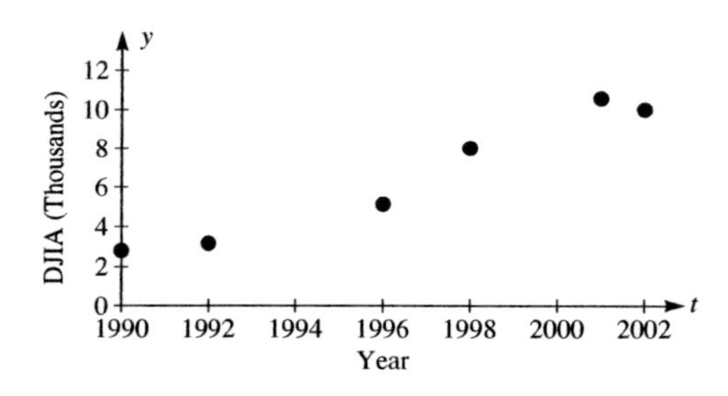

(b)

x	y	xy	x^2
0	2,810	0	0
2	3,172	6,344	4
6	5,177	31,062	36
8	7,965	63,720	64
11	10,646	117,106	121
12	10,073	120,876	144
$\sum x = 39$	$\sum y = 39{,}843$	$\sum xy = 339{,}108$	$\sum x^2 = 369$

Using the formulas with $n = 6$,

$$m = \frac{6(339{,}108) - (39)(39{,}843)}{6(369) - (39)^2} = \frac{480{,}771}{693} \approx 693.75$$

and

$$b = \frac{(369)(39{,}843) - (39)(339{,}108)}{6(369) - (39)^2} = \frac{1{,}476{,}855}{693} \approx 2{,}131.1.$$

So, the equation of the least-squares line is $y = 693.75x + 2{,}131.1$

(c) In 2003, when $x = 13$, $y = 693.75(13) + 2{,}131.1 \approx 11{,}150$; actual $= 8{,}608$.

(d) Writing exercise—Answers will vary.

29. (a) Let x denote the number of years after 1996 and y the corresponding GDP in billions of yuan. Then,

x	0	1	2	3	4	5
y	6,788	7,446	7,835	8,191	8,940	9,593

x	y	xy	x^2
0	6,788	0	0
1	7,446	7,446	1
2	7,835	15,670	4
3	8,191	24,573	9
4	8,940	35,760	16
5	9,593	47,965	25
$\sum x = 15$	$\sum y = 48{,}793$	$\sum xy = 131{,}414$	$\sum x^2 = 55$

Using the formulas with $n = 6$,

$$m = \frac{6(131{,}414) - (15)(48{,}793)}{6(55) - (15)^2} = \frac{56{,}589}{105} \approx 538.9$$

and

$$b = \frac{(55)(48{,}793) - (15)(131{,}414)}{6(55) - (15)^2} = \frac{712{,}405}{105} \approx 6{,}784.8.$$

So, the equation of the least-squares line is $y = 538.9x + 6{,}784.8$

(b) In the year 2005, when $x = 9$, $y = 538.9(9) + 6{,}784.8 \approx 11{,}635$ billion yuan is the predicted GDP.

31. (a) Let t denote the number of years after 1980 and N the corresponding number of reported AIDS cases. Then,

t	0	4	8	12	16	20
N	99	6,360	36,064	79,477	61,109	42,156

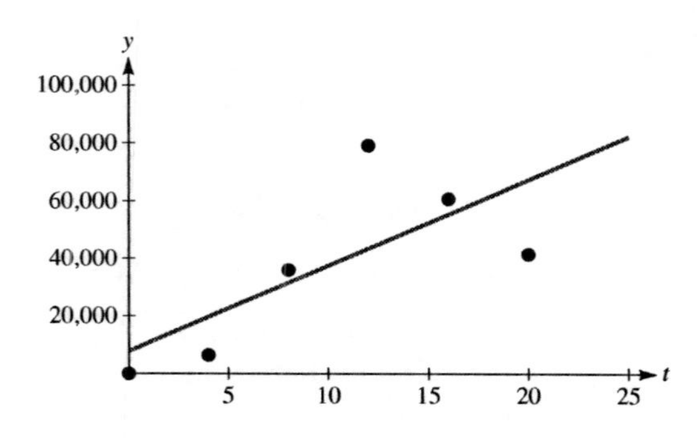

(b)

x	y	xy	x^2
0	99	0	0
4	6,360	25,440	16
8	36,064	288,512	64
12	79,477	953,724	144
16	61,109	977,744	256
20	42,156	843,120	400
$\sum x$ $= 60$	$\sum y$ $= 225{,}265$	$\sum xy$ $= 3{,}088{,}540$	$\sum x^2$ $= 880$

Using the formulas with $n = 6$,

$$m = \frac{6(3{,}088{,}540) - (60)(225{,}265)}{6(880) - (60)^2} = \frac{5{,}015{,}340}{1{,}680} \approx 2{,}985 \text{ and}$$

$$b = \frac{(880)(225{,}265) - (60)(3{,}088{,}540)}{6(880) - (60)^2} = \frac{12{,}920{,}800}{1{,}680} \approx 7{,}691.$$ So, the equation of the least-squares line is $N(t) = 2{,}985t + 7{,}691$.

(c) In 2005, when $t = 25$, $N(25) = 2{,}985(25) + 7{,}691 \approx 82{,}316$ reported cases is predicted.

(d) Writing exercise—Answers will vary.

33. (a)

$\ln W$	4.054	4.693	5.297	5.704	5.873	6.040	6.284	6.611
$\ln C$	1.668	2.617	3.645	4.358	4.649	4.905	5.276	5.766

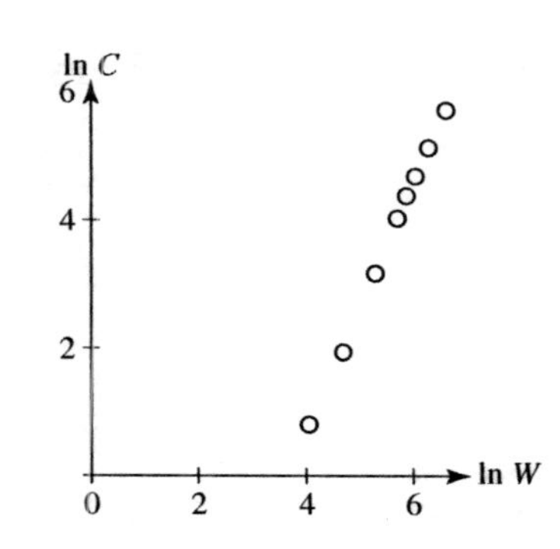

(b)

x	y	xy	x^2
4.054	1.668	6.762	16.435
4.693	2.617	12.282	22.024
5.297	3.645	19.308	28.058
5.704	4.358	24.858	32.536
5.873	4.649	27.304	34.492
6.040	4.905	29.626	36.482
6.284	5.276	33.154	39.489
6.611	5.766	38.119	43.705
$\sum x$ $= 44.556$	$\sum y$ $= 32.884$	$\sum xy$ $= 191.413$	$\sum x^2$ $= 253.221$

Using the formulas with $n = 8$,

$$m = \frac{8(191.413) - (44.556)(32.884)}{8(253.221) - (44.556)^2} \approx \frac{66.124}{40.531} \approx 1.631 \text{ and}$$

$$b = \frac{(253.221)(32.884) - (44.556)(191.413)}{8(253.221) - (44.556)^2} = \frac{-201.68}{40.531} \approx -4.976.$$ So, the equation of the least-squares line is $y = 1.631x - 4.976$

(c)

$$\ln C = 1.631 \ln W - 4.976$$

$$e^{\ln C} = e^{1.631 \ln W - 4.976}$$

$$C = e^{\ln W^{1.631}} e^{-4.976}$$

$$C = e^{-4.976} W^{1.631}$$

$$C(W) = 0.00690W^{1.631}$$

7.5 Constrained Optimization: The Method of Lagrange Multipliers

1.

$$f(x, y) = xy$$
$$g(x, y) = x + y$$
$$f_x = y;\ f_y = x;\ g_x = 1;\ g_y = 1$$

The three Lagrange equations are:

$$y = \lambda;\ x = \lambda;\ x + y = 1$$

From the first two equations, $x = y$ which, when substituted into the third equation gives
$2x = 1$, or $x = \frac{1}{2}$.
Since $x = y$, the corresponding value for y is
$y = \frac{1}{2}$. So, the constrained maximum is
$f\left(\frac{1}{2}, \frac{1}{2}\right) = \frac{1}{4}$.

3.

$$f(x, y) = x^2 + y^2$$
$$g(x, y) = xy$$
$$f_x = 2x;\ f_y = 2y;\ g_x = y;\ g_y = x$$

The three Lagrange equations are:

$$2x = \lambda y;\ 2y = \lambda x;\ xy = 1$$

Multiply the first equation by y and the second by x to get $2xy = \lambda y^2$ and $2xy = \lambda x^2$. Set the two expressions for $2xy$ equal to each other to get $\lambda y^2 = \lambda x^2$, $y^2 = x^2$, or $x = \pm y$. (Note that another solution of the equation $\lambda y^2 = \lambda x^2$ is $\lambda = 0$, which implies that $x = 0$ and $y = 0$, which is not consistent with the third equation.)
If $y = x$, the third equation becomes $x^2 = 1$, which implies that $x = \pm 1$ and $y = \pm 1$.
If $y = -x$, the third equation becomes $-x^2 = 1$, which has no solutions. So, the two points at which the constrained extrema can occur are $(1, 1)$ and $(-1, -1)$.
Since $f(1, 1) = 2$ and $f(-1, -1) = 2$, the constrained minimum is 2.

5.

$$f(x, y) = x^2 - y^2$$
$$g(x, y) = x^2 + y^2$$
$$f_x = 2x;\ f_y = -2y;\ g_x = 2x;\ g_y = 2y$$

The three Lagrange equations are:

$$2x = 2\lambda x;\ -2y = 2\lambda y;\ x^2 + y^2 = 4$$

From the first equation, either $\lambda = 1$ or $x = 0$. If $x = 0$, the third equation becomes $y^2 = 4$ or $y = \pm 2$.
From the second equation, either $\lambda = -1$ or $y = 0$. If $y = 0$, the third equation becomes $x^2 = 4$ or $x = \pm 2$.
If neither $x = 0$ nor $y = 0$, the first equation implies $\lambda = 1$ while the second equation implies $\lambda = -1$, which is impossible.
So, the only points at which the constrained extrema can occur are $(0, -2)$, $(0, 2)$, $(-2, 0)$, and $(2, 0)$.
Now, $f(0, -2) = -4$, $f(0, 2) = -4$, $f(-2, 0) = 4$, and $f(2, 0) = 4$. So, the constrained minimum is -4.

7.

$$f(x, y) = x^2 - y^2 - 2y$$
$$g(x, y) = x^2 + y^2$$
$$f_x = 2x;\ f_y = -2y - 2;\ g_x = 2x;\ g_y = 2y$$

The three Lagrange equations are:

$$2x = 2\lambda x;\ -2y - 2 = 2\lambda y;\ x^2 + y^2 = 1$$

From the first equation, either $\lambda = 1$ or $x = 0$. If $\lambda = 1$, the second equation becomes $2y - 2 = 2y$, $4y = -2$, or $y = -\frac{1}{2}$. From the third equation, $x^2 + \left(-\frac{1}{2}\right)^2 = 1$, or $x = \pm\frac{\sqrt{3}}{2}$.
If $x = 0$, the third equation becomes $0^2 + y^2 = 1$ or $y = \pm 1$. So, the only points at which the constrained extrema can occur are $\left(-\frac{\sqrt{3}}{2}, -\frac{1}{2}\right)$, $\left(\frac{\sqrt{3}}{2}, -\frac{1}{2}\right)$, $(0, -1)$, and $(0, 1)$. Now, $f\left(\frac{\sqrt{3}}{2}, -\frac{1}{2}\right) = f\left(-\frac{\sqrt{3}}{2}, -\frac{1}{2}\right) = \frac{3}{2}$, $f(0, -1) = 1$, and $f(0, 1) = -3$. So, the constrained maximum is $\frac{3}{2}$ and the constrained minimum is -3.

9. $f(x, y) = 2x^2 + 4y^2 - 3xy - 2x - 23y + 3$

$g(x, y) = x + y - 15 = 0$

$f_x = 4x - 3y - 2$

$f_y = 8y - 3x - 23$

$g_x = g_y = 1$

The three Lagrange equations are:

$$4x - 3y - 2 = \lambda$$
$$-3x + 8y - 23 = \lambda$$
$$x + y = 15$$

The first two lead to $7x - 11y = -21$.
Substitute $y = 15 - x$ to obtain $18x = 144$ or $x = 8$ and $y = 7$.
The constrained minimum is $f(8, 7) = -18$.

11. $f(x, y) = e^{xy}$

$g(x, y) = x^2 + y^2 - 4 = 0$

$f_x = ye^{xy}$, and $f_y = xe^{xy}$

$g_x = 2x$ and $g_y = 2y$

The three Lagrange equations are:

$$ye^{xy} = 2\lambda x$$
$$xe^{xy} = 2\lambda y$$
$$x^2 + y^2 - 4 = 0$$

Dividing the first two leads to $\frac{y}{x} = \frac{x}{y}$, or $x^2 = y^2$.
Substitute in $x^2 + y^2 = 4$ to obtain $x = \pm\sqrt{2}$ and $y = \pm\sqrt{2}$.
Now, $f(\sqrt{2}, -\sqrt{2}) = f(-\sqrt{2}, \sqrt{2}) = e^{-2}$ and $f(\sqrt{2}, \sqrt{2}) = f(-\sqrt{2}, -\sqrt{2}) = e^2$. So, the constrained maximum is e^2 and the constrained minimum is e^{-2}.

13. $f(x, y, z) = xyz$

$g(x, y, z) = x + 2y + 3z - 24 = 0$

$f_x = yz,\ f_y = xz$, and $f_z = xy$

$g_x = 1,\ g_y = 2$, and $g_z = 3$

The three Lagrange equations are:

$$yz = \lambda;\ xz = 2\lambda;\ xy = 3\lambda$$

Dividing the first two leads to $y = \frac{x}{2}$, dividing the first by the third leads to $z = \frac{x}{3}$.
Substitute in $x + 2y + 3z = 24$ to obtain $x = 8$, $y = 4$, and $z = \frac{8}{3}$.

The maximum is $f(8, 4, 8/3) = \frac{256}{3}$.

15. $f(x, y, z) = x + 2y + 3z$

$g(x, y.z) = x^2 + y^2 + z^2 - 16 = 0$

$f_x = 1,\ f_y = 2$, and $f_z = 3$

$g_x = 2x,\ g_y = 2y$, and $g_z = 2z$

The three Lagrange equations are:

$$1 = 2\lambda x;\ 2 = 2\lambda y;\ 3 = 2\lambda z$$

Dividing the first two leads to $y = 2x$, dividing the first by the third leads to $z = 3x$. Substitute in $x^2 + y^2 + z^2 = 16$ to obtain $x = \pm\frac{4}{\sqrt{14}}$, $y = \pm\frac{8}{\sqrt{14}}$, and $z = \pm\frac{12}{\sqrt{14}}$.
Now, $f\left(\frac{4}{\sqrt{14}}, \frac{8}{\sqrt{14}}, \frac{12}{\sqrt{14}}\right) = \frac{56}{\sqrt{14}} = 4\sqrt{14}$ and $f\left(\frac{-4}{\sqrt{14}}, \frac{-8}{\sqrt{14}}, \frac{-12}{\sqrt{14}}\right) = -\frac{56}{\sqrt{14}} = -4\sqrt{14}$.
So, the constrained maximum is $4\sqrt{14}$ and the constrained minimum is $-4\sqrt{14}$.

17. Let f denote the amount of fencing needed to enclose the pasture, x the side parallel to the river and y the sides perpendicular to the river. Then,

$$f(x, y) = x + 2y$$

The goal is to minimize this function subject to the constraint that the area $xy = 3{,}200$, so $g(x, y) = xy$. The partial derivatives are $f_x = 1$, $f_y = 2$, $g_x = y$, and $g_y = x$.
The three Lagrange equations are

$$1 = \lambda y;\ 2 = \lambda x;\ xy = 3{,}200$$

From the first equation, $\lambda = \frac{1}{y}$. From the second equation $\lambda = \frac{2}{x}$. Setting the two expressions for

λ equal to each other gives $\frac{1}{y} = \frac{2}{x}$ or $x = 2y$, and substituting this into the third equation yields $2y^2 = 3{,}200$, $y^2 = 1{,}600$, or $y = \pm 40$.
Only the positive value is meaningful in the context of this problem. So, $y = 40$, and (since $x = 2y$), $x = 80$. That is, to minimize the amount of fencing, the dimensions of the field should be 40 meters by 80 meters.

19. Let f denote the volume of the parcel. Then,

$$f(x, y) = x^2 y$$

The girth $4x$ plus the length y can be at most 108 inches. The goal is to maximize this function $f(x, y)$ subject to the constraint $4x + y = 108$, so $g(x, y) = 4x + y$.
The partial derivatives are $f_x = 2xy$, $f_y = x^2$, $g_x = 4$, and $g_y = 1$.
The three Lagrange equations are

$$2xy = 4\lambda; \; x^2 = \lambda; \; 4x + y = 108$$

From the first equation, $\lambda = \frac{xy}{2}$, which, combined with the second equation, gives $\frac{xy}{2} = x^2$ or $y = 2x$. (Another solution is $x = 0$, which is impossible in the context of this problem.)
Substituting $y = 2x$ into the third equation gives $6x = 108$ or $x = 18$, and since $y = 2x$, the corresponding value of y is $y = 36$.
So, the largest volume is $f(18, 36) = (18)^2(36) = 11{,}664$ cubic inches.

21. Let M denote the amount of metal used to construct the can. Then,

$$M(R, H) = 2\pi R^2 + 2\pi RH$$

The goal is to maximize this function $M(R, H)$ subject to the constraint that (volume) $\pi R^2 H = 6.89\pi$, so $g(R, H) = \pi R^2 H$. The partial derivatives are

$$M_R = 4\pi R + 2\pi H; \; M_H = 2\pi R,$$
$$g_R = 2\pi RH; \; g_H = \pi R^2$$

The three Lagrange equations are:

$$4\pi R + 2\pi H = 2\pi \lambda RH$$
$$2\pi R = \pi \lambda R^2$$
$$\pi R^2 H = 6.89\pi$$

The second equation leads to $\lambda = \frac{2}{R}$, which leads to $2R = H$, using the first equation. Using the third equation yields $H = \sqrt[3]{27.56} \approx 3.02$, and $R = \frac{H}{2} \approx 1.51$.
So, the amount of metal is minimized when the can's radius is 1.51 inches and its height is 3.02 inches.

23.
$$f(x, y) = 50x^{1/2}y^{3/2}$$

Since the constraint is $x + y = 8{,}000$, $g(x, y) = x + y$.

$$f_x = 25x^{-1/2}y^{3/2}; \; f_y = 75x^{1/2}y^{1/2};$$
$$g_x = 1; \; g_y = 1$$

The three Lagrange equations are:

$$25x^{-1/2}y^{3/2} = \lambda; \; 75x^{1/2}y^{1/2} = \lambda;$$
$$x + y = 8{,}000$$

Equating λ leads to $25x^{-1/2}y^{3/2} = 75x^{1/2}y^{1/2}$, or $y = 3x$.
Substituting in the constraint equation leads to $x = \$2{,}000$ for development and $y = \$6{,}000$ for promotion.

25. From problem 24, the three Lagrange equations were

$$20x^{-2/3}y^{2/3} = \lambda; \; 40x^{1/3}y^{-1/3} = \lambda;$$
$$x + y = 120$$

from which it was determined that the maximal output occurs when $x = 40$ and $y = 80$.
Substituting these values in the first equation gives

$$\lambda = 20(40)^{-2/3}(80)^{2/3} \approx 31.75$$

which implies that the maximal output will increase by approximately 31.75 units if the available money is increased by one thousand dollars and allocated optimally.

27. Let S denote the surface area of the bacterium. Then,

$$S(R, H) = 2\pi R^2 + 2\pi RH$$

The goal is to maximize this function subject to the constraint $\pi R^2 H = C$ (volume is fixed, C is a constant), so $g(R, H) = \pi R^2 H$.

$$S_R = 4\pi R + 2\pi H;\ S_H = 2\pi R;$$

$$g_R = 2\pi RH;\ g_H = \pi R^2$$

The three Lagrange equations are:

$$4\pi R + 2\pi H = 2\pi \lambda RH$$

$$2\pi R = \pi \lambda R^2$$

$$\pi R^2 H = C$$

The second equation leads to $\lambda = \frac{2}{R}$, which leads to $2R = H$, using the first equation.

29. The goal is to maximize $S(d_o, d_i) = d_o + d_i$ subject to the constraint $\frac{1}{d_o} + \frac{1}{d_i} = \frac{1}{L}$, so $g(d_o, d_i) = \frac{1}{d_o} + \frac{1}{d_i}$.

$$S_{d_o} = 1;\ S_{d_i} = 1;\ g_{d_o} = -\frac{1}{d_o^2};\ g_{d_i} = -\frac{1}{d_i^2}$$

The three Lagrange equations are:

$$1 = \lambda \cdot \frac{-1}{d_o^2};\ 1 = \lambda \cdot \frac{-1}{d_i^2};\ \frac{1}{d_o} + \frac{1}{d_i} = L$$

This leads to

$$\lambda = -(d_o^2);\ \lambda = -(d_i)^2, \text{ or } d_o = d_i$$

Substituting into the third equation,

$$d_o = d_i = 2L$$

and the maximum value of S is $4L$.

31. Let k be the cost per square cm of the bottom and sides. Then the cost of the top is $2k$ per square cm and the cost of the interior partitions is $\frac{2k}{3}$ per square cm. The goal is to minimize the cost of the box,

$$C(x, y) = k(x^2 + 4xy) + 2kx^2 + \frac{2k}{3}(2xy)$$

subject to the constraint $x^2 y = 800$, so $g(x, y) = x^2 y$.

$$C_x = 6kx + \frac{16}{3}ky$$

$$C_y = \frac{16}{3}kx,\ g_x = 2xy,\ g_y = x^2$$

The three Lagrange equations are:

$$6kx + \frac{16}{3}ky = 2\lambda xy;$$

$$\frac{16}{3}kx = \lambda x^2;\ x^2 y = 800$$

Solving the first two equations for λ and equating yields $x = \frac{8}{9}y$. Substituting into the third equation yields $\frac{9}{8}x^3 = 800$, or

$$x = 8.93 \text{ and } y = 10.04$$

33.

$$E(x, y, z) = \frac{k^2}{8m}\left(\frac{1}{x^2} + \frac{1}{y^2} + \frac{1}{z^2}\right)$$

$$g(x, y) = xyz$$

$$E_x = \frac{k^2}{8m}\left(-\frac{2}{x^3}\right)$$

$$E_y = \frac{k^2}{8m}\left(-\frac{2}{y^3}\right)$$

$$E_z = \frac{k^2}{8m}\left(-\frac{2}{z^3}\right)$$

$$g_x = yz,\ g_y = xz, \text{ and } g_z = xy$$

The three Lagrange equations are:

$$\frac{k^2}{8m}\left(-\frac{2}{x^3}\right) = \lambda yz$$

$$\frac{k^2}{8m}\left(-\frac{2}{y^3}\right) = \lambda xz$$

$$\frac{k^2}{8m}\left(-\frac{2}{z^3}\right) = \lambda xy$$

Dividing the first two leads to $y^2 = x^2$, or $y = x$; dividing the first by the third leads to

$z^2 = x^2$, or $z = x$. Substitute in $xyz = V_0$ to obtain $x = y = z = V_0^{1/3}$. The minimum E is $\frac{3k^2}{8m}V_0^{-2/3}$.

35. Let x denote the length of the shed, y the width, and z the height. The goal is to maximize the volume,

$$V = xyz$$

subject to the constraint

$$15xy + 12(2yz + xz) + 20xz = 8{,}000$$

so $g(x, y, z) = 15xy + 12(2yz + xz) + 20xz$.

$$f_x = yz,\ f_y = xz,\ \text{and}\ f_z = xy$$
$$g_x = 15y + 32z$$
$$g_y = 15x + 24z$$
$$g_z = 24y + 32x$$

The three Lagrange equations are:

$$yz = \lambda(15y + 32z)$$
$$xz = \lambda(15x + 24z)$$
$$xy = \lambda(24y + 32x)$$

Dividing the first two leads to $y = \frac{4x}{3}$ and dividing the first by the third leads to $z = \frac{5x}{8}$. Substitute in $15xy + 24yz + 32xz = 8{,}000$ to obtain

$$x = \frac{20\sqrt{3}}{3} \approx 11.55 \text{ ft}$$

$$y = \frac{80\sqrt{3}}{9} \approx 15.40 \text{ ft}$$

$$z = \frac{25\sqrt{3}}{6} \approx 7.22 \text{ ft}$$

37. From problem #36, $P_x = P_y = \lambda$. Using $P_y = \frac{6{,}400}{(y+2)^2} - 1{,}000$ and $y = 5$,

$$\lambda = P_y = \frac{64{,}000}{49} - 1{,}000 = 306.122$$

(for each $1,000)

Since the change in this promotion/development is $100, the corresponding change in profit is $30.61 (Remember that the Lagrange multiplier is the change in maximum profit for a 1 (thousand) dollar change in the constraint.)

39. **(a)** The goal is to maximize utility,

$$U(x, y) = 100x^{0.25}y^{0.75}$$

subject to the constraint $2x + 5y = 280$, so $g(x, y) = 2x + 5y$.

$$U_x = 25x^{-0.75}y^{0.75}$$
$$U_y = 75x^{0.25}y^{-0.25}$$
$$g_x = 2;\ g_y = 5$$

The three Lagrange equations are

$$25x^{-0.75}y^{0.75} = 2\lambda$$
$$75x^{0.25}y^{-0.25} = 5\lambda$$
$$2x + 5y = 280$$

Solving the first two equations for λ and equating yields

$$15x^{0.25}y^{-0.25} = 12.5x^{-0.75}y^{0.75},$$
$$\text{or } y = 1.2x.$$

Substituting in the third equation yields $x = 35$ and $y = 42$.

(b) $\lambda = (15)(35^{0.25})(42^{-0.25}) \approx 14.33$

which approximates the change in maximum utility due to an additional $1.00 in available funds.

41. $\lambda \approx \Delta u$ if $\Delta k = \$1$. Since

$$U(x, y) = x^{\alpha}y^{\beta},\ \alpha x^{\alpha-1}y^{\beta} = \lambda a$$

and $k = ax + by$, it follows that

$$\lambda = \frac{\alpha x^{\alpha-1} y^{\beta}}{a} = \frac{\alpha y^{\beta}}{a x^{1-\alpha}}$$

$$= \left(\frac{\alpha}{a}\right)\left(\frac{k\beta}{b}\right)^{\beta}\left(\frac{\alpha}{k\alpha}\right)^{1-\alpha}$$

$$= \left(\frac{\alpha}{a}\right)\left(\frac{k\beta}{b}\right)^{\beta}\left(\frac{\alpha}{k\alpha}\right)^{\beta}$$

$$= \left(\frac{\alpha}{a}\right)\left(\frac{k\beta a}{bk\alpha}\right)^{\beta} = \frac{\alpha\beta^{\beta}a^{\beta-1}}{\alpha^{\beta}b^{\beta}}$$

$$= \frac{\alpha^{\beta-1}\beta^{\beta}}{\alpha^{\beta-1}b^{\beta}} = \left(\frac{\alpha}{a}\right)^{\alpha}\left(\frac{\beta}{b}\right)^{\beta}$$

43. Let $Q(x, y)$ be the production level curve subject to $px + qy = k$. The three Lagrange equations then are $Q_x = \lambda p$, $Q_y = \lambda q$, and $px + qy = k$. From the first two equations $\dfrac{Q_x}{p} = \dfrac{Q_y}{q}$.

45. Need to find extrema of $Q(K, L) = 55[0.6K^{-1/4} + 0.4L^{-1/4}]^{-4}$ subject to $g(K, L) = 2K + 5L - 150 = 0$.

$$Q_K = -220\left[0.6K^{-1/4} + 0.4L^{-1/4}\right]^{-5}\left(-0.15K^{-5/4}\right)$$

$$Q_L = -220\left[0.6K^{-1/4} + 0.4L^{-1/4}\right]^{-5}\left(-0.1L^{-5/4}\right)$$

$$g_K = 2 \quad g_L = 5$$

The three Lagrange equations are

$$-220\left(0.6K^{-1/4} + 0.4L^{-1/4}\right)^{-5}\left(-0.15K^{-5/4}\right) = 2\lambda$$

$$-220\left(0.6K^{-1/4} + 0.4L^{-1/4}\right)^{-5}\left(-0.1L^{-5/4}\right) = 5\lambda$$

$$2K + 5L - 150 = 0$$

Solving the first two equations for λ gives

$$\frac{33K^{-5/4}\left(0.6K^{-1/4} + 0.4L^{-1/4}\right)^{-5}}{2} = \lambda$$

$$\frac{22L^{-5/4}\left(0.6K^{-1/4} + 0.4L^{-1/4}\right)^{-5}}{5} = \lambda$$

Setting these equal,

$$\frac{33K^{-5/4}\left(0.6K^{-1/4} + 0.4L^{-1/4}\right)^{-5}}{2} = \frac{22L^{-5/4}\left(0.6K^{-1/4} + 0.4L^{-1/4}\right)^{-5}}{5}$$

$$\frac{33K^{-5/4}}{2} = \frac{22L^{-5/4}}{5}$$

$$165L^{5/4} = 44K^{5/4}$$

$$L = \left(\frac{44K^{4/5}}{165}\right)^{4/5} = \left(\frac{44}{165}\right)^{4/5} K$$

Using the third equation,

$$2K + 5\left(\frac{44}{165}\right)^{4/5} K - 150 = 0$$

$$K \approx 40.14$$

$$L \approx \left(\frac{44}{165}\right)^{4/5} K \approx 13.89$$

$$Q(40.14, 13.89) = 55\left[0.6(40.14)^{-1/4} + 0.4(13.89)^{-1/4}\right]^{-4}$$

$$\approx 1395.4$$

47. $Q(K, L) = A\left[\alpha K^{-\beta} + (1-\alpha)L^{-\beta}\right]^{-1/\beta}$

$Q(sK, sL) = A\left[\alpha (sK)^{-\beta} + (1-\alpha)(sL)^{-\beta}\right]^{-1/\beta}$

$= A\left[\alpha s^{-\beta}K^{-\beta} + (1-\alpha)s^{-\beta}L^{-\beta}\right]^{-1/\beta}$

$= A(s^{-\beta})^{-1/\beta}\left[\alpha K^{-\beta} + (1-\alpha)L^{-\beta}\right]^{-1/;\beta}$

$= sA\left[\alpha K^{-\beta} + (1-\alpha)L^{-\beta}\right]^{-1/\beta}$

$= sQ(K, L)$

49. Need to find extrema of $f(x, y) = x - y$ subject to $g(x, y) = x^5 + x - z - y = 0$.

$$f_x = 1 \quad f_y = -1 \quad g_x = 5x^4 + 1 \quad g_y = -1$$

The three Lagrange equations are:

$$1 = \lambda(5x^4 + 1)$$

$$-1 = \lambda(-1)$$

$$x^5 + x - z - y = 0$$

From the second equation, $\lambda = 1$.
Then, from the first equation,

$$1 = 5x^4 + 1$$
$$5x^4 = 0$$
$$x = 0$$

Finally, from the third equation,

$$-2 - y = 0$$
$$y = -2$$

Therefore, a possible extremum occurs at the point $(0, -2)$. However, $f(1, 0) = 1$ and $f(-1, -4) = 3$, which shows $f(0, -2) = 2$ is not a local maximum or minimum point.
Press [y=].
Input $x \wedge 5 + x - 2$ for $y_1 =$ and input $x - L_1$ for $y_2 =$.
From the home screen, input {2, 1, 0, −1} [sto→] [2nd] L_1.
Use window dimensions [−4, 4]1 by [−4, 4]1.
Press [Graph].
From the graphs that the point $(0, -2)$ is an inflection point.

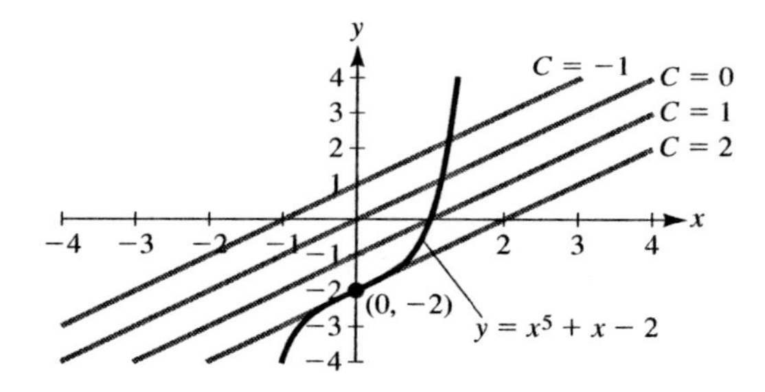

51. The goal is to maximize $P(K, L)$ subject to the constraint $C(K, L) = A$, so $g(K, L) = C(K, L)$. The three Lagrange equations are

$$\frac{\partial P}{\partial K} = \lambda\frac{\partial C}{\partial K}, \quad \frac{\partial P}{\partial L} = \lambda\frac{\partial C}{\partial L}, \quad C(K, L) = A$$

Divide the first two equations to eliminate λ and

$$\frac{\frac{\partial P}{\partial K}}{\frac{\partial P}{\partial L}} = \frac{\frac{\partial C}{\partial K}}{\frac{\partial C}{\partial L}} \text{ or } \frac{\frac{\partial P}{\partial K}}{\frac{\partial C}{\partial K}} = \frac{\frac{\partial P}{\partial L}}{\frac{\partial C}{\partial L}}$$

53. $F(x, y) = xe^{xy^2} + \frac{y}{x} + x\ln(x + y)$

(a)
$$0 = (x)\left[e^{xy^2}\left(x \cdot 2y\frac{dy}{dx} + y^2 \cdot 1\right)\right] + (e^{xy^2})(1) + \frac{x\frac{dy}{dx} - y \cdot 1}{x^2} + (x)\left[\frac{1}{x+y}\left(1 + \frac{dy}{dx}\right)\right] + \ln(x + y) \cdot 1$$

$$0 = 2x^2ye^{xy^2}\frac{dy}{dx} + xy^2e^{xy^2} + e^{xy^2} + \frac{1}{x}\frac{dy}{dx} - \frac{y}{x^2} + \frac{x}{x+y} + \frac{x}{x+y}\frac{dy}{dx} + \ln(x + y)$$

$$\frac{y}{x^2} - \frac{x}{x+y} - xy^2e^{xy^2} - e^{xy^2} + \ln(x + y) = \left(2x^2ye^{xy^2} + \frac{1}{x} + \frac{x}{x+y}\right)\frac{dy}{dx}$$

$$\frac{dy}{dx} = \frac{-xy^2e^{xy^2} - e^{xy^2} + \frac{y}{x^2} - \frac{x}{x+y} - \ln(x + y)}{2x^2ye^{xy^2} + \frac{1}{x} + \frac{x}{x+y}}$$

$$= -\frac{xy^2e^{xy^2} + e^{xy^2} - \frac{y}{x^2} + \frac{x}{x+y} + \ln(x + y)}{2x^2ye^{xy^2} + \frac{1}{x} + \frac{x}{x+y}}$$

(b)
$$F_x = xy^2e^{xy^2} + e^{xy^2} - \frac{y}{x^2} + \frac{x}{x+y} + \ln(x + y)$$

$$F_y = 2x^2ye^{xy^2} + \frac{1}{x} + \frac{x}{x+y}$$

$$\frac{dy}{dx} = -\frac{F_x}{F_y}$$

$$= -\frac{xy^2e^{xy^2} + e^{xy^2} - \frac{y}{x^2} + \frac{x}{x+y} + \ln(x + y)}{2x^2ye^{xy^2} + \frac{1}{x} + \frac{x}{x+y}}$$

55. Minimize $f(x, y) = \ln(x + 2y)$ subject to $xy + y = 5$.

$f(x, y) = \ln(x + 2y) \qquad g(x, y) = xy + y - 5 = 0$

$$f_x = \frac{1}{x + 2y} \qquad fy = \frac{2}{x + 2y}$$

$$g_x = y \qquad g_y = x + 1$$

The three Lagrange equations are:

$$\frac{1}{x + 2y} = \lambda y$$
$$\frac{2}{x + 2y} = \lambda(x + 1)$$
$$xy + y = 5$$

From the first equation, $\lambda = \frac{1}{y(x + 2y)}$.

From the second equation, $\lambda = \frac{2}{(x + 1)(x + 2y)}$.

Equating these two gives $(x + 1)(x + 2y) = 2y(x + 2y)$ or $y = \frac{(x + 1)}{2}$.

Substituting $y = \frac{1}{2}(x + 1)$ into the third equation,

$$\frac{1}{2}x(x + 1) + \frac{1}{2}(x + 1) = 5$$
$$x(x + 1) + (x + 1) = 10$$
$$(x + 1)[x + 1] = 10$$
$$(x + 1)^2 = 10$$

This gives $x \approx -4.1623$, $x \approx 2.1623$. This leads to the points $(-4.1623, -1.5811)$ and $(2.1623, 1.5811)$. We cannot use $(-4.1623, -1.5811)$ since this point leads $f(x, y)$ to be undefined.
Find $f(2.1623, 1.5811) = \ln[2.1623 + 2(1.5811)] \approx 1.6724$.

57. $f(x, y) = xe^{x^2-y}$ and

$g(x, y) = x^2 + 2y^2 - 1 = 0$

$$f_x = e^{x^2-y} + (x)(e^{x^2-y})(2x)$$
$$f_y = xe^{x^2-y}(-1) = -xe^{x^2-y}$$
$$f_x = (2x^2 + 1)(e^{x^2-y})$$
$$g_x = 2x \qquad g_y = 4y$$

The three Lagrange equations are:

$$(2x^2 + 1)(e^{x^2-y}) = \lambda(2x)$$
$$-xe^{x^2-y} = \lambda(4y)$$
$$x^2 + 2y^2 = 1$$

From the first equation, $\lambda = \frac{(2x^2 + 1)(e^{x^2-y})}{2x}$.

From the second equation, $\lambda = \frac{-xe^{x^2-y}}{4y}$.

Equating these and simplifying,

$$8x^2y + 4y = -2x^2$$
$$2x^2(1 + 4y) = -4y$$
$$x^2 = -\frac{2y}{4y + 1}$$

Substituting this into the third equation,

$$-\frac{2y}{4y + 1} + 2y^2 = 1$$
$$-2y + 2y^2(4y + 1) = 4y + 1$$
$$8y^3 + 2y^2 - 6y - 1 = 0$$

To solve, press [y =] and
input $8x \wedge 3 + 2x^2 - 6x - 1$ for $y_1 =$ (remember we are actually solving for y).
Use window dimensions [−4, 4]1 by [−4, 4]1.
Press [Graph].
Use the zero function under the calc menu to find that $y \approx -0.9184$, $y \approx -0.1636$, and $y \approx 0.832$.
We reject $y \approx -0.9184$ and $y \approx 0.832$ since these would result in x being undefined. If $y = -0.1636$, then $x = \pm 0.9729$.

The two points for consideration are (0.9729, −0.1636) and (−0.9729, −0.1636).
Press [y=].
Input $xe \wedge (x^2 - L_1)$ for $y_1 =$.
From the home screen, input $\{-0.1636\}$ [sto→] [2nd] L_1.
Press [Graph].
Use the value function under the calc menu to find $f(0.9729, -0.1636) \approx 2.952$ and $f(-0.9729, -0.1636) \approx -2.952$
The maximum point is (0.9729, −0.1636).

7.6 Double Integrals

1.

$$\int_0^1 \int_1^2 x^2 y \, dxdy$$
$$= \int_0^1 \left[\int_1^2 x^2 y \, dx \right] dy$$
$$= \int_0^1 \left[\frac{x^3}{3} y \Big|_1^2 \right] dy$$
$$= \int_0^1 \left[\frac{8}{3} y - \frac{1}{3} y \right] dy$$
$$= \frac{7}{6} y^2 \Big|_0^1 = \frac{7}{6}$$

3.

$$\int_0^{\ln 2} \int_{-1}^0 2xe^y \, dxdy$$
$$= \int_0^{\ln 2} \left[\int_{-1}^0 2xe^y \, dx \right] dy$$
$$= \int_0^{\ln 2} \left[x^2 e^y \Big|_{-1}^0 \right] dy$$
$$= \int_0^{\ln 2} \left[-e^y \right] \, dy = -e^y \Big|_0^{\ln 2} = -1.$$

5.

$$\int_1^3 \int_0^1 \frac{2xy}{x^2+1} \, dxdy$$
$$= \int_1^3 \left[\int_0^1 \frac{2xy}{x^2+1} \, dx \right] dy$$
$$= \int_1^3 \left[y \ln(x^2+1) \Big|_0^1 \right] dy$$
$$= \int_1^3 y \ln 2 dy = \ln 2 \left(\frac{1}{2} \right) y^2 \Big|_1^3 = 4 \ln 2$$

7.

$$\int_0^4 \int_{-1}^1 x^2 y \, dxdy$$
$$= \int_0^4 \left[\int_{-1}^1 x^2 y \, dx \right] dx$$
$$= \int_0^4 \left[\frac{y^2}{2} x^2 \Big|_{-1}^1 \right] dx = 0$$

9.

$$\int_2^3 \int_1^2 \frac{x+y}{xy} \, dy \, dx$$
$$= \int_2^3 \int_1^2 \left[\frac{1}{y} + \frac{1}{x} \right] dy \, dx$$
$$= \int_2^3 \left[\ln(y) + \frac{y}{x} \right] \Big|_1^2 dx$$
$$= (x \ln 2 + \ln x) \Big|_2^3 = \ln 2 + \ln \frac{3}{2} = \ln 3$$

11.

$$\int_0^4 \int_0^{\sqrt{x}} x^2 y \, dy \, dx = \int_0^4 \left[\int_0^{\sqrt{x}} x^2 y \, dy \right] dx$$
$$= \int_0^4 \left[\frac{x^2 y^2}{z} \Big|_0^{\sqrt{x}} \right] dx = \int_0^4 \frac{x^3}{2} \, dx$$
$$= \frac{x^4}{8} \Big|_0^4 = 32$$

13. $$\int_0^1 \int_{y-1}^{1-y} (2x+y)\,dx\,dy$$
$$= \int_0^1 \left[\int_{y-1}^{1-y} (2x+y)\,dx \right] dy$$
$$= \int_0^1 \left[(x^2+xy)\Big|_{y-1}^{1-y} \right] dy$$
$$= \int_0^1 \Big[[(1-y)^2 + (1-y)y]$$
$$-[(y-1)^2 + (y-1)y] \Big]\,dy$$
$$= \int_0^1 2y - 2y^2\,dy = y^2 - \frac{2y^3}{3}\Big|_0^1 = \frac{1}{3}$$

15. $$\int_0^1 \int_0^4 \sqrt{xy}\,dy\,dx = \int_0^1 \left[\int_0^4 x^{\frac{1}{2}} y^{\frac{1}{2}}\,dy \right] dx$$
$$= \int_0^1 \left[\frac{2x^{\frac{1}{2}} y^{\frac{3}{2}}}{3} \Big|_0^4 \right] dx = \int_0^1 \frac{16x^{\frac{1}{2}}}{3}\,dx$$
$$= \frac{32x^{\frac{3}{2}}}{9}\Big|_0^1 = \frac{32}{9}$$

17. $$\int_1^e \int_0^{\ln x} xy\,dydx = \int_1^e \left[\int_0^{\ln x} xy\,dy \right] dx$$
$$= \int_1^e \left[\frac{xy^2}{2}\Big|_0^{\ln x} \right] dx = \int_1^e \frac{x(\ln x)^2}{2}\,dx$$

Using integration by parts with

$$u = (\ln x)^2 \quad \text{and} \quad dV = \frac{x}{2}\,dx$$
$$= \frac{x^2}{4}(\ln x)^2\Big|_1^e - \int_1^e \frac{x}{2} \ln x\,dx$$
$$= \frac{e^2}{4} - \int_1^e \frac{x}{2} \ln x\,dx$$

Using integration by parts again, with

$$u = \ln x \quad \text{and} \quad dV = \frac{x}{2}\,dx$$

$$= \frac{e^2}{4} - \left[\frac{x^2}{4} \ln x\Big|_1^e - \int_1^e \frac{x}{4}\,dx \right]$$
$$= \frac{e^2}{4} - \left[\left(\frac{x^2}{4} \ln x - \frac{x^2}{8} \right)\Big|_1^e \right]$$
$$= \frac{e^2}{4} - \left[\left(\frac{e^2}{4} - \frac{e^2}{8} \right) - \left(0 - \frac{1}{8} \right) \right] = \frac{e^2-1}{8}$$

19. Solving $x^2 = 3x$ yields $x = 0$ and $x = 3$. Similarly, after solving each equation for x, $\sqrt{y} = \frac{y}{3}$ when $y = 0$ and $y = 9$. So, R can be described in terms of vertical cross sections by $0 \le x \le 3$ and $x^2 \le y \le 3x$ and in terms of horizontal cross sections by $0 \le y \le 9$ and $\frac{y}{3} \le x \le \sqrt{y}$.

21. The given points form a rectangle. So, R can be described in terms of vertical cross sections by $-1 \le x \le 2$ and $1 \le y \le 2$ and in terms of horizontal cross sections by $1 \le y \le 2$ and $-1 \le x \le 2$.

23. Solving $\ln x = 0$ yields $x = 1$, with the second boundary given as $x = e$. Similarly, solving $y = \ln x$ for x yields $x = e^y$, with the second boundary given as $y = 0$. So, R can be described in terms of vertical cross sections by $1 \le x \le e$ and $0 \le y \le \ln x$ and in terms of horizontal cross sections by $0 \le y \le 1$ and $e^y \le x \le e$.

25. $$\iint_R 3xy^2\,dA = \int_{-1}^0 \int_{-1}^2 3xy^2\,dxdy$$
$$= \int_{-1}^0 \left[\int_{-1}^2 3xy^2\,dx \right] dy = \int_{-1}^0 \left[\frac{3x^2y^2}{2}\Big|_{-1}^2 \right] dy$$
$$= \int_{-1}^0 \frac{9y^2}{2}\,dy = \frac{3y^3}{2}\Big|_{-1}^0 = \frac{3}{2}$$

Note: problem can be equivalently worked as

$$\int_{-1}^2 \int_{-1}^0 3xy^2\,dydx.$$

27. Since the line joining the points (0, 0) and (1, 1) is $y = x$,

$$\iint_R xe^y\,dA = \int_0^1 \int_0^x xe^y\,dydx$$

$$= \int_0^1 \left[\int_0^x xe^y\,dy\right] dx = \int_0^1 \left[xe^y\big|_0^x\right] dx$$

$$= \int_0^1 (xe^x - x)\,dx = \int_0^1 xe^x\,dx - \int_0^1 x\,dx$$

$$= \int_0^1 xe^x\,dx - \frac{x^2}{2}\bigg|_0^1 = \int_0^1 xe^x\,dx - \frac{1}{2}$$

Using integration by parts with

$$u = x \quad \text{and} \quad dV = e^x\,dx$$

$$= xe^x\big|_0^1 - \int_0^1 e^x\,dx - \frac{1}{2}$$

$$= (xe^x - e^x)\big|_0^1 - \frac{1}{2} = \frac{1}{2}$$

Note: problem can be equivalently worked as $\int_0^1 \int_0^y xe^y\,dxdy$.

29. Solving $x^2 = 2x$ yields $x = 0$ and $x = 2$, so

$$\iint_R (2y - x)\,dA = \int_0^2 \int_{x^2}^{2x} (2y - x)\,dydx$$

$$= \int_0^2 \left[\int_{x^2}^{2x} (2y - x)\,dy\right] dx$$

$$= \int_0^2 \left[(y^2 - xy)\big|_{x^2}^{2x}\right] dx$$

$$= \int_0^2 \left[[(2x)^2 - x(2x)] - [(x^2)^2 - x(x^2)]\right] dx$$

$$= \int_0^2 (2x^2 - x^4 + x^3)\,dx$$

$$= \left(\frac{2x^3}{3} - \frac{x^5}{5} + \frac{x^4}{4}\right)\bigg|_0^2 = \frac{44}{15}$$

Note: problem can be equivalently worked as

$$\int_0^4 \int_{\sqrt{y}}^{\frac{y}{2}} (2y - x)\,dxdy.$$

31. The line joining the points $(-1, 0)$ and $(0, 1)$ is $y = x + 1$, or $x = y - 1$. Similarly, the line joining the points $(0, 1)$ and $(1, 0)$ is $y = 1 - x$, or $x = 1 - y$. So,

$$\iint_R (2x + 1)\,dA = \int_0^1 \int_{y-1}^{1-y} (2x + 1)\,dxdy$$

$$= \int_0^1 \left[\int_{y-1}^{1-y} (2x + 1)\,dx\right] dy$$

$$= \int_0^1 \left[(x^2 + x)\big|_{y-1}^{1-y}\right] dy$$

$$= \int_0^1 \Big[[(1 - y)^2 + (1 - y)]$$

$$-[(y - 1)^2 + (y - 1)]\Big]\,dy$$

$$= \int_0^1 (2 - 2y)\,dy = (2y - y^2)\big|_0^1 = 1$$

33. After solving each equation for x, $2y = -y$ when $y = 0$, with the other boundary given as $y = 2$. So,

$$\iint_R \frac{1}{y^2 + 1}\,dA = \int_0^2 \int_{-y}^{2y} \frac{1}{y^2 + 1}\,dxdy$$

$$= \int_0^2 \left[\int_{-y}^{2y} \frac{1}{y^2 + 1}\,dx\right] dy = \int_0^2 \left[\frac{x}{y^2 + 1}\bigg|_{-y}^{2y}\right] dy$$

$$= \int_0^2 \left[\left[\frac{2y}{y^2 + 1} - \frac{-y}{y^2 + 1}\right]\right] dy = \int_0^2 \frac{3y}{y^2 + 1}\,dy$$

$$= 3\int_0^2 \frac{y}{y^2 + 1}\,dy$$

Using substitution with $u = y^2 + 1$,

$$= 3\int_1^5 \frac{1}{u} \cdot \frac{1}{2}\,du = \frac{3}{2}\int_1^5 \frac{1}{u}\,du$$

$$= \frac{3}{2}\left(\ln|u|\big|_1^5\right) = \frac{3}{2}(\ln 5 - \ln 1) = \frac{3\ln 5}{2}$$

35. After solving each equation for x, $y^{\frac{1}{3}} = y$ when $y = 0$ and $y = 1$. So,

$$\iint_R 12x^2e^{y^2}\,dA = \int_0^1\int_y^{y^{\frac{1}{3}}} 12x^2e^{y^2}\,dxdy$$

$$= \int_0^1\left[\int_y^{y^{\frac{1}{3}}} 12x^2e^{y^2}\,dx\right]dy$$

$$= \int_0^1\left[4x^3e^{y^2}\Big|_y^{y^{\frac{1}{3}}}\right]dy$$

$$= \int_0^1 (4ye^{y^2} - 4y^3e^{y^2})\,dy$$

$$= 4\int_0^1 ye^{y^2}dy - 4\int_0^1 y^2(ye^{y^2})\,dy$$

Using substitution for the first integral with $u = y^2$, and using integration by parts for the second integral with

$$u = y^2 \quad\text{and}\quad dV = ye^{y^2}\,dy$$

(where solving for V requires substitution as well)

$$= 4\int_0^1 \frac{e^4}{2}\,du - 4\left[\frac{y^2}{2}e^{y^2}\Big|_0^1 - \int_0^1 ye^{y^2}\,dy\right]$$

$$= 4\left(\frac{1}{2}e^{y^2}\Big|_0^1\right) - 4\left[\left(\frac{y^2}{2}e^{y^2}\Big|_0^1\right) - \frac{1}{2}e^{y^2}\Big|_0^1\right]$$

$$= 2\left(e^{y^2} - y^2e^{y^2} + e^{y^2}\right)\Big|_0^1$$

$$= 2\left(2e^{y^2} - y^2e^{y^2}\right)\Big|_0^1 = 2(e-2)$$

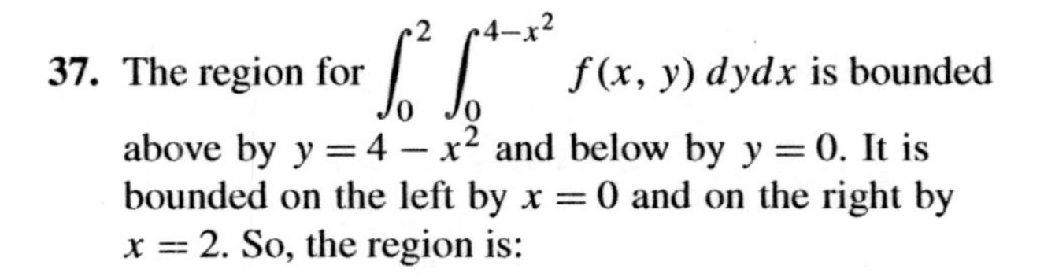

37. The region for $\int_0^2\int_0^{4-x^2} f(x, y)\,dydx$ is bounded above by $y = 4 - x^2$ and below by $y = 0$. It is bounded on the left by $x = 0$ and on the right by $x = 2$. So, the region is:

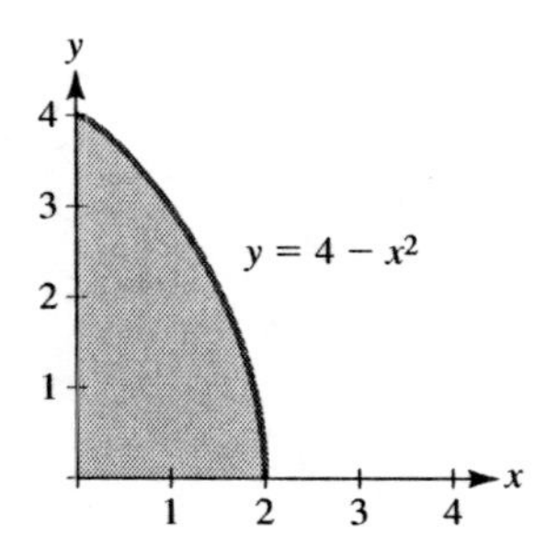

Reversing the integration yields

$$\int_0^4\int_0^{\sqrt{4-y}} f(x, y)\,dxdy.$$

39. The region for $\int_0^1\int_{x^3}^{\sqrt{x}} f(x, y)\,dydx$ is bounded above by $y = \sqrt{x}$ and below by $y = x^3$. It is bounded on the left by $x = 0$ and on the right by $x = 1$. So, the region is:

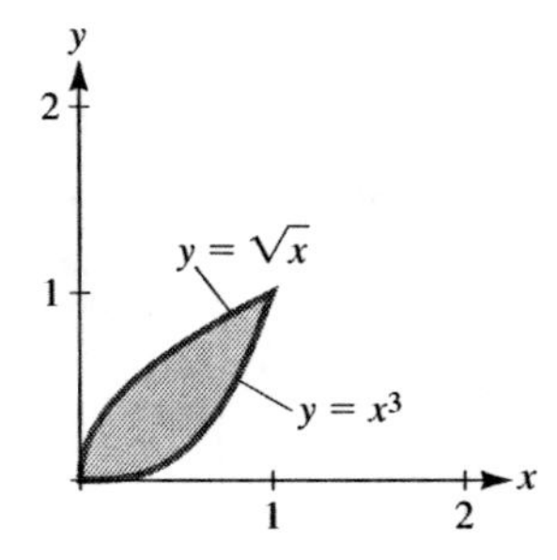

Reversing the integration yields

$$\int_0^1\int_{y^2}^{y^{\frac{1}{3}}} f(x, y)\,dxdy.$$

41. The region for $\int_1^{e^2}\int_{\ln x}^{2} f(x, y)\,dydx$ is bounded above by $y = 2$ and below by $y = \ln x$. It is bounded on the left by $x = 1$ and on the right by $x = e^2$. So, the region is:

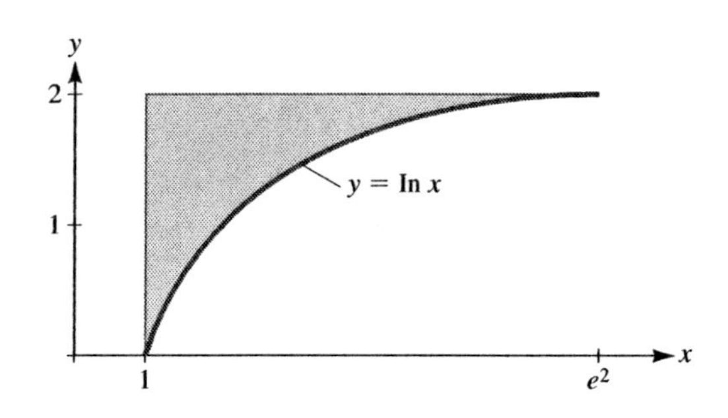

Reversing the integration yields

$$\int_0^2 \int_1^{e^y} f(x, y)\, dxdy.$$

43. The region for $\int_{-1}^{1} \int_{x^2+1}^{2} f(x, y)\, dydx$ is bounded above by $y = 2$ and below by $y = x^2 + 1$. It is bounded on the left by $x = -1$ and on the right by $x = 1$. So, the region is:

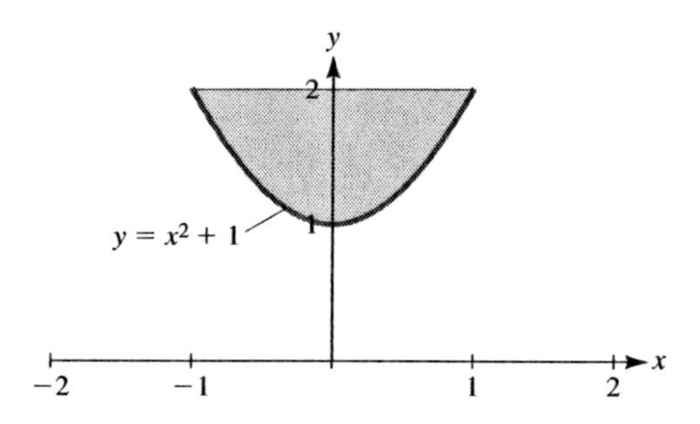

Reversing the integration yields

$$\int_1^2 \int_{-\sqrt{y-1}}^{\sqrt{y-1}} f(x, y)\, dxdy.$$

45. The line joining the points $(-4, 0)$ and $(2, 6)$ is $y = x + 4$, which the bottom boundary being $y = 0$. So, the area of R is

$$\int_{-4}^{2} \int_0^{x+4} (1)\, dydx = \int_{-4}^{2} \left[\int_0^{x+4} 1\, dy\right] dx$$

$$= \int_{-4}^{2} \left[y\Big|_0^{x+4}\right] dx = \int_{-4}^{2} (x + 4)\, dx$$

$$= \left(\frac{x^2}{2} + 4x\right)\Bigg|_{-4}^{2} = 18$$

47. Solving $\frac{1}{2}x^2 = 2x$ yields $x = 0$ and $x = 4$. So, the area of R is

$$\int_0^4 \int_{\frac{x^2}{2}}^{2x} (1)\, dydx = \int_0^4 \left[\int_{\frac{x^2}{2}}^{2x} 1\, dy\right] dx$$

$$= \int_0^4 \left[y\Big|_{\frac{x^2}{2}}^{2x}\right] dx = \int_0^4 \left(2x - \frac{x^2}{2}\right) dx$$

$$= \left(x^2 - \frac{x^3}{6}\right)\Bigg|_0^4 = \frac{16}{3}$$

49. Solving $x^2 - 4x + 3 = 0$ yields $x = 1$ and $x = 3$. So, the area of R is

$$\int_1^3 \int_0^{x^2-4x+3} (1)\, dydx$$

$$= \int_1^3 \left[\int_{x^2-4x+3}^{0} 1\, dy\right] dx$$

$$= \int_1^3 \left[y\Big|_{x^2-4x+3}^{0}\right] dx$$

$$= \int_1^3 (-x^2 + 4x - 3)\, dx$$

$$= \left(-\frac{x^3}{3} + 2x^2 - 3x\right)\Bigg|_1^3 = \frac{4}{3}$$

51. Solving $\ln x = 0$ yields $x = 1$, with the other boundary given as $x = e$. So, the area of R is

$$\int_1^e \int_0^{\ln x} (1)\, dydx = \int_1^e \left[\int_0^{\ln x} 1\, dy\right] dx$$

$$= \int_1^e \left[y\Big|_0^{\ln x}\right] dx = \int_1^e \ln x\, dx$$

Using integration by parts with

$$u = \ln x \quad \text{and} \quad dV = dx$$
$$= x \ln x\Big|_1^e - \int_1^e 1\,dx$$
$$= (x \ln x - x)\Big|_1^e = 1$$

53. After solving each equation for x, $\sqrt{4-y} = \dfrac{y}{3}$ when $y = -12$ and $y = 3$. However, the region is also bounded by $y = 0$, making the limits $y = 0$ and $y = 3$. So, the area of R is

$$\int_0^3 \int_{\frac{y}{3}}^{\sqrt{4-y}} (1)\,dxdy = \int_0^3 \left[\int_{\frac{y}{3}}^{\sqrt{4-y}} 1\,dy\right] dx$$
$$= \int_0^3 \left[x\Big|_{\frac{y}{3}}^{\sqrt{4-y}}\right] dy = \int_0^3 \left(\sqrt{4-y} - \frac{4}{3}\right) dy$$
$$= \int_0^3 \sqrt{4-y}\,dy - \int_0^3 \frac{y}{3}\,dy$$
$$= \int_0^3 \sqrt{4-y}\,dy - \frac{y^2}{6}\Big|_0^3 = \int_0^3 \sqrt{4-y}\,dy - \frac{3}{2}$$

Using substitution with $u = 4 - y$,

$$= \int_4^1 u^{\frac{1}{2}} - du - \frac{3}{2} = -\int_4^1 u^{\frac{1}{2}}\,du - \frac{3}{2}$$
$$= \int_1^4 u^{\frac{1}{2}}du - \frac{3}{2} = \frac{2}{3}u^{\frac{3}{2}}\Big|_1^4 - \frac{3}{2} = \frac{19}{6}$$

55.
$$V = \int_0^1 \int_0^2 (6 - 2x - 2y)\,dydx$$
$$= \int_0^1 \left[\int_0^2 (6 - 2x - 2y)\,dy\right] dx$$
$$= \int_0^1 \left[(6y - 2xy - y^2)\Big|_0^2\right] dx$$
$$= \int_0^1 (8 - 4x)\,dx$$
$$= (8x - 2x^2)\Big|_0^1 = 6$$

Note: problem can be equivalently worked as $\int_0^2 \int_0^1 (6 - 2x - 2y)\,dxdy$.

57.
$$V = \int_1^2 \int_1^3 \frac{1}{xy}\,dydx$$
$$= \int_1^2 \left[\int_1^3 \frac{1}{x} \cdot \frac{1}{y}\,dy\right] dx$$
$$= \int_1^2 \left[\frac{1}{x} \ln|y|\Big|_1^3\right] dx = \int_1^2 \frac{1}{x} \ln 3\,dx$$
$$= \left[\ln 3 \ln|x|\right]_1^2 = (\ln 3)(\ln 2)$$

Note: problem can be equivalently worked as $\int_1^3 \int_1^2 \dfrac{1}{xy}\,dxdy$.

59.
$$V = \int_0^1 \int_0^2 xe^{-y}\,dydx$$
$$= \int_0^1 \left[\int_0^2 xe^{-y}\,dy\right] dx$$
$$= \int_0^1 \left[-xe^{-y}\Big|_0^2\right] dx = \int_0^1 (-xe^{-2} + x)\,dx$$
$$= \int_0^1 (1 - e^{-2})x\,dx = (1 - e^{-2})\frac{x^2}{2}\Big|_0^1$$
$$= \frac{e^2 - 1}{2e^2} = \frac{1}{2}\left(1 - \frac{1}{e^2}\right)$$

61. After solving both equations for x, $y = 2 - y$ when $y = 1$, with the other boundary given as $y = 0$. So,

$$V = \int_0^1 \int_y^{2-y} (2x + y)\,dxdy$$
$$= \int_0^1 \left[\int_y^{2-y} (2x + y)\,dx\right] dy$$
$$= \int_0^1 \left[(x^2 + xy)\Big|_y^{2-y}\right] dy$$
$$= \int_0^1 \Big[\left[(2-y)^2 + (2-y)y\right] - \left[(y)^2 + (y)y\right]\Big] dy$$
$$= \int_0^1 (4 - 2y - 2y^2)\,dy$$
$$= \left(4y - y^2 - \frac{2y^3}{3}\right)\Big|_0^1 = \frac{7}{3}$$

63. Solving $8 - x^2 = x^2$ yields $x = -2$ and $x = 2$. So,

$$V = \int_{-2}^{2}\int_{x^2}^{8-x^2} (x+1)\,dydx$$
$$= \int_{-2}^{2}\left[\int_{x^2}^{8-x^2} (x+1)\,dy\right] dx$$
$$= \int_{-2}^{2}\left[(x+1)y\Big|_{x^2}^{8-x^2}\right] dx$$
$$= \int_{-2}^{2}\left[(x+1)(8-x^2) - (x+1)(x^2)\right] dx$$
$$= \int_{-2}^{2}(8 + 8x - 2x^2 - 2x^3)\,dx$$
$$= \left(8x + 4x^2 - \frac{2x^3}{3} - \frac{x^4}{2}\right)\Big|_{-2}^{2} = \frac{64}{3}$$

65. The area of the rectangular region is 15.

$$f_{av} = \frac{1}{15}\int_{-2}^{3}\int_{-1}^{2} xy(x-2y)\,dydx$$
$$= \frac{1}{15}\int_{-2}^{3}\left(\frac{x^2y^2}{2} - \frac{2xy^3}{3}\right)\Big|_{-1}^{2} dx$$
$$= \frac{1}{15}\int_{-2}^{3}\left(2x^2 - \frac{16x}{3} - \frac{x^2}{2} - \frac{2x}{3}\right) dx$$
$$= \frac{1}{15}\int_{-2}^{3}(1.5x^2 - 6x)dx$$
$$= \frac{1}{15}\left(\frac{x^3}{2} - 3x^2\right)\Big|_{-2}^{3} = \frac{1}{6}$$
$$\approx 0.1667$$

67. The area of the rectangular region is 2.

$$f_{av} = \frac{1}{2}\int_{0}^{2}\int_{0}^{1} xye^{x^2y}\,dxdy$$

Using substitution with $u = x^2y$,

$$= \frac{1}{2}\int_{0}^{2}\left[\int_{0}^{y} e^u \cdot \frac{1}{2}\,du\right] dy$$
$$= \frac{1}{4}\int_{0}^{2}\left(e^u\Big|_{0}^{y}\right) dy$$
$$= \frac{1}{4}\int_{0}^{2}(e^y - 1)\,dy$$
$$= \frac{1}{4}(e^y - y)\Big|_{0}^{2} = \frac{1}{4}(e^2 - 3) \approx 1.0973$$

69. The area of the rectangular region is $\frac{3}{2}$. The line joining the points (0, 0) and (3, 1) is $y = \frac{x}{3}$.

$$f_{av} = \frac{1}{\frac{3}{2}}\int_{0}^{3}\int_{\frac{x}{3}}^{1} 6xy\,dydx$$
$$= \frac{2}{3}\int_{0}^{3}\left[3xy^2\Big|_{\frac{x}{3}}^{1}\right] dx$$
$$= \frac{2}{3}\int_{0}^{3}\left(3x - \frac{x^3}{3}\right) dx$$
$$= \frac{2}{3}\left(\frac{3x^2}{2} - \frac{x^4}{12}\right)\Big|_{0}^{3} = \frac{9}{2}$$

Note: problem can be equivalently worked as $\int_0^1\int_0^{3y} 6xy\,dxdy$.

71. The area of the given region is

$$\int_{-2}^{2} 4 - x^2\,dx = \left(4x - \frac{x^3}{3}\right)\Big|_{-2}^{2} = \frac{32}{3}$$
$$f_{av} = \frac{32}{3}\int_{-2}^{2}\int_{0}^{4-x^2} x\,dydx$$
$$= \frac{32}{3}\int_{-2}^{2}(xy)\Big|_{0}^{4-x^2} dx = \frac{1}{16}\int_{-2}^{2}(4x - x^3)\,dx$$
$$= \frac{32}{3}\left(2x^2 - \frac{x^4}{4}\right)\Big|_{-2}^{2} = 0$$

73. $$\int_{1}^{3}\int_{2}^{5} \frac{\ln xy}{y}\,dydx = \int_{1}^{3}\int_{2}^{5} \ln xy \cdot \frac{1}{y}\,dydx$$

Using substitution with $u = \ln xy$,

$$= \int_1^3 \left[\int_{\ln 2x}^{\ln 5x} u\, du \right] dx = \int_1^3 \left(\frac{u^2}{2} \right) \Big|_{\ln 2x}^{\ln 5x} dx$$

$$= \frac{1}{2} \int_1^3 (\ln^2 5x - \ln^2 2x)\, dx$$

$$= \frac{1}{2} \int_1^3 (\ln 5x + \ln 2x)(\ln 5x - \ln 2x)\, dx$$

$$= \frac{1}{2} \int_1^3 (\ln 10x^2) \left(\ln \frac{5}{2} \right) dx$$

$$= \frac{\ln 2.5}{2} \int_1^3 \ln 10x^2\, dx$$

Using integration by parts with

$$u = \ln 10x^2 \quad \text{and} \quad dV = dx$$

$$du = \frac{2}{x}\, dx \qquad V = x$$

$$= \frac{\ln 2.5}{2} \left[x \ln 10x^2 \Big|_1^3 - \int_1^3 x \cdot \frac{2}{x} dx \right]$$

$$= \frac{\ln 2.5}{2} \left[x \ln 10x^2 - 2x \right]_1^3$$

$$= \frac{\ln 2.5}{2} [(3 \ln 90 - 6) - (\ln 10 - 2)] \approx 3.297$$

75. $$\int_0^1 \int_0^1 x^3 e^{x^2 y}\, dy\, dx = \int_0^1 \int_0^1 x e^{x^2 y} x^2\, dy\, dx$$

Using substitution with $u = x^2 y$,

$$= \int_0^1 \int_0^{x^2} x e^u\, du\, dx = \int_0^1 \left[x \left(e^u \Big|_0^{x^2} \right) \right] dx$$

$$= \int_0^1 (x e^{x^2} - x e^0)\, dx = \int_0^1 x e^{x^2}\, dx - \int_0^1 x\, dx$$

Using substitution with $u = x^2$,

$$= \frac{1}{2} \int_0^1 e^u\, du - \int_0^1 x\, dx$$

$$= \frac{1}{2} \left(e^u \Big|_0^1 \right) - \frac{x^2}{2} \Big|_0^1$$

$$= \frac{1}{2}(e^1 - e^0) - \left(\frac{1}{2} - 0 \right) \approx 0.859$$

77. $$Q_{av} = \frac{1}{35} \int_0^7 \int_0^5 (2x^3 + 3x^2 y + y^3)\, dx dy$$

$$= \frac{1}{35} \int_0^7 (0.5x^4 + x^3 y + x y^3) \Big|_0^5 dy$$

$$= \frac{1}{7} ((0.5)(125y) + (0.5)(25y^2) + 0.25y^4) \Big|_0^7$$

$$= \frac{943}{4} = 235.75$$

79. $$P(xy) = \int_{70}^{89} \int_{100}^{125} [(x - 30)(70 + 5x - 4y) + (y - 40)(80 - 6x + 7y)]\, dx\, dy$$

$$= \int_{70}^{89} \int_{100}^{125} [5x^2 + 7y^2 + 160x - 10xy - 80y - 5{,}300]\, dx\, dy$$

$$= \int_{70}^{89} [1.6667x^3 + 7xy^2 + 80x^2 - 5x^2 y - 80xy - 5{,}300x] \Big|_{100}^{125} dy$$

$$= \int_{70}^{89} [1{,}909{,}218.75 + 175y^2 - 30{,}125y]\, dy$$

$$= [1{,}906{,}041.67y + 58.33y^3 - 15{,}062.5y^2] \Big|_{70}^{89}$$

$$= 1.1826(10^7)$$

The area is $(125 - 100)(89 - 70) = 475$.

The average profit is $\frac{1.1826(10^7)}{475} = 24{,}896.8$ or \$2,489,800.

81. $$E(x, y) = \frac{90}{5{,}280}(2x + y^2) \text{ miles}$$

$$E_{av} = \frac{0.01705}{12} \int_0^3 \int_0^4 (2x + y^2)\, dx dy$$

$$= 0.00142 \int_0^3 (16 + 4y^2)\, dy$$

$$= 0.00142(16y + 1.333y^3) \Big|_0^3 = 630 \text{ ft.}$$

83. Value $= \int_{-1}^{1}\int_{-1}^{1}(300+x+y)e^{-0.01x}\,dxdy$

$$= \int_{-1}^{1}\int_{-1}^{1}[(300+y)e^{-0.01x}+xe^{--0.01x}]\,dxdy$$

$$= \int_{-1}^{1}\left[\begin{array}{l}\frac{(300+y)}{-0.01}e^{-0.01x}\\ -100xe^{-0.01x}-10{,}000e^{-0.01x}\end{array}\right]\Bigg|_{x=-1}^{x=1} dy$$

$$= \int_{-1}^{1}\left[\begin{array}{l}39{,}900e^{0.01}-40{,}100e^{-0.01}\\ +(100e^{0.01}-100e^{-0.01})y\end{array}\right] dy$$

$$= 79{,}800e^{0.01}-80{,}200e^{-0.01}=1{,}200.007$$

or roughly 1.2 million dollars.

85. (a)

$$S_{av} = \frac{0.0072}{(142)(76.8)}\int_{3.2}^{80}\int_{38}^{180}W^{0.425}H^{0.725}\,dHdW$$

$$= \frac{0.0072}{(142)(76.8)}\int_{3.2}^{80}W^{0.425}\left(\int_{38}^{180}H^{0.725}\,dH\right)dW$$

$$= \frac{0.0072}{(142)(76.8)}\int_{3.2}^{80}W^{0.425}\left(\frac{H^{1.725}}{1.725}\Bigg|_{38}^{180}\right)dW$$

$$\approx \frac{0.0072}{(142)(76.8)}\int_{3.2}^{80}W^{0.425}(4195.71)\,dW$$

$$\approx 0.00277\int_{3.2}^{80}W^{0.425}\,dW$$

$$0.00277\left(\frac{W^{1.425}}{1.425}\Bigg|_{3.2}^{80}\right)$$

$$\approx 0.00277(357.802)\approx 0.991 \text{ sq meters}$$

(b) No. It could only be interpreted as the person's average surface area from birth until his/her adult weight and height was first reached.

87. Solving $4-x^2=0$ yields $x=-2$ and $x=2$. So,

$$V = \int_{-2}^{2}\int_{0}^{4-x^2}(20-x^2-y^2)dy\,dx$$

$$= \int_{-2}^{2}\left[\left(20y-x^2y-\frac{y^3}{3}\right)_0^{4-x^2}\right]dx$$

$$= \int_{-2}^{2}\left[20(4-x^2)-x^2(4-x^2)-\frac{(4-x^2)^3}{3}\right]dx$$

$$= \int_{-2}^{2}\left[80-20x^2-4x^2+x^4-\frac{64-48x^2+12x^4-x^6}{3}\right]$$

$$= \int_{-2}^{2}\left[\frac{176}{3}-8x^2-3x^4+\frac{x^6}{3}\right]dx$$

$$= \left[\frac{176}{3}x-\frac{8}{3}x^3-\frac{3}{5}x^5+\frac{1}{21}x^7\right]_{-2}^{2}$$

$$= \left[\left(\frac{352}{3}-\frac{64}{3}-\frac{96}{5}+\frac{128}{21}\right)-\left(-\frac{352}{3}+\frac{64}{3}+\frac{96}{5}-\frac{128}{21}\right)\right]$$

$$= \frac{17408}{105}\approx 165.79\,m^3$$

89.

$$E = \int_{-2}^{2}\int_{-2}^{2}\left(1-\frac{1}{9}\left(x^2+y^2\right)\right)dydx$$

$$= \int_{-2}^{2}\left[y-\frac{1}{9}\left(x^2y+\frac{y^3}{3}\right)\right]\Bigg|_{y=-2}^{y=2}dx$$

$$= \left(4x-\frac{4x^3}{27}-\frac{16x}{27}\right)\Bigg|_{x=-2}^{x=2}$$

$$= 2\left(8-\frac{64}{27}\right)=\frac{304}{27}$$

91.

$$f_{av} = \frac{1}{2}\int_{1}^{2}\int_{1}^{3}xy\ln\left(\frac{y}{x}\right)dydx$$

$$= \frac{1}{2}\int_{1}^{2}\int_{1}^{3}(xy\ln y-xy\ln x)\,dydx$$

$$= \frac{1}{2}\int_{1}^{2}\left(\frac{xy^2}{2}\ln y-\frac{xy^2}{4}-\frac{xy^2}{2}\ln x\right)\Bigg|_{1}^{3}dx$$

$$= \frac{1}{2}\int_{1}^{2}\left[x\left(\frac{9}{2}\ln 3-2\right)-4x\ln x\right]dx$$

$$= \frac{1}{2}\left[\frac{x^2}{2}\left(\frac{9}{2}\ln 3-2\right)-2x^2\ln x+x^2\right]\Bigg|_{1}^{2}$$

$$= \frac{27}{8}\ln 3-4\ln 2$$

Checkup for Chapter 7

1. (a) $f(x, y) = x^3 + 2xy^2 - 3y^4$
The domain is the set of all real pairs (x, y).

$$f_x = 3x^2 + 2y^2;\ f_y = 4xy - 12y^3;$$
$$f_{xx} = 6x;\ f_{yx} = 4y$$

(b) $f(x, y) = \dfrac{2x + y}{x - y}$
The domain is the set of all real pairs (x, y) such that $x - y \neq 0$, or $y \neq x$.

$$f_x = \frac{(x - y)(2) - (2x + y)(1)}{(x - y)^2} = -\frac{3y}{(x - y)^2}$$
$$f_y = \frac{(x - y)(1) - (2x + y)(-1)}{(x - y)^2} = \frac{3x}{(x - y)^2}$$
$$f_{xx} = (-3y) - 2(x - y)^{-3}(1) = \frac{6y}{(x - y)^3}$$
$$f_{yx} = \frac{(x - y)^2(3) - (3x)2(x - y)(1)}{(x - y)^4}$$
$$= -\frac{3(x + y)}{(x - y)^3}$$

(c) $f(x, y) = e^{2x-y} + \ln(y^2 - 2x)$
The domain of e^{2x-y} is the set of all real pairs (x, y), but the domain of $\ln(y^2 - 2x)$ is the set of all real pairs such that $y^2 - 2x > 0$, or $y^2 > 2x$.

$$f_x = 2e^{2x-y} - \frac{2}{y^2 - 2x}$$
$$f_y = -e^{2x-y} + \frac{2y}{y^2 - 2x}$$
$$f_{xx} = 4e^{2x-y} + 2(y^2 - 2x)^{-2}(-2)$$
$$= 4e^{2x-y} - \frac{4}{(y^2 - 2x)^2}$$
$$f_{yx} = -2e^{2x-y} - 2y(y^2 - 2x)^{-2}(-2)$$
$$= -2e^{2x-y} + \frac{4y}{(y^2 - 2x)^2}$$

2. (a) $f(x, y) = x^2 + y^2$
Level curves are of the form $x^2 + y^2 = C$, which are circles having the origin as their center and radius $\sqrt{C}$, and also the single point $(0, 0)$, when $C = 0$.

(b) $f(x, y) = x + y^2$
Level curves are of the form $x + y^2 = C$, which are parabolas having a horizontal axis, opening to the left, and a vertex on the x-axis.

3. (a)
$$f(x, y) = 4x^3 + y^3 - 6x^2 - 6y^2 + 5$$
$$f_x = 12x^2 - 12x = 12x(x - 1)$$
$$f_x = 0 \text{ when } x = 0, 1$$
$$f_y = 3y^2 - 12y = 3y(y - 4)$$
$$f_y = 0 \text{ when } y = 0, 4$$

So, the critical points are $(0, 0)$, $(0, 4)$, $(1, 0)$, and $(1, 4)$.

$$f_{xx} = 24x - 12;\ f_{yy} = 6y - 12;\ f_{xy} = 0$$

For the point $(0, 0)$,

$$D = [24(0) - 12][6(0) - 12] - 0 > 0$$
$$\text{and } f_{xx} < 0$$

So, $(0, 0)$ is a relative maximum.
For the point $(0, 4)$,

$$D = [24(0) - 12][6(4) - 12] - 0 < 0$$

So, $(0, 4)$ is a saddle point.
For the point $(1, 0)$,

$$D = [24(1) - 12][6(0) - 12] - 0 < 0$$

So, $(1, 0)$ is a saddle point.
For the point $(1, 4)$,

$$D = [24(1) - 12][6(4) - 12] - 0 > 0$$
$$\text{and } f_{xx} > 0$$

So, $(1, 4)$ is a relative minimum.

(b)
$$f(x, y) = x^2 - 4xy + 3y^2 + 2x - 4y$$
$$f_x = 2x - 4y + 2$$
$$f_x = 0 \text{ when } 2x - 4y = -2$$

$$f_y = -4x + 6y - 4$$

$$f_y = 0 \text{ when } -4x + 6y = 4$$

Solving this system of equations, by multiplying the first by two and adding to the second, gives $y = 0$, and $x = -1$. So, the only critical point is $(-1, 0)$.

$$f_{xx} = 2;\ f_{yy} = 6;\ f_{xy} = -4$$

$$D = (2)(6) - (-4)^2 < 0$$

So, $(-1, 0)$ is a saddle point.

(c) $f(x, y) = xy - \dfrac{1}{y} - \dfrac{1}{x}$

$$f_x = y + \frac{1}{x^2}$$

$$f_x = 0 \text{ when } y = -\frac{1}{x^2},\ \text{or } y^2 = \frac{1}{x^4}$$

$$f_y = y + \frac{1}{x^2}$$

$$f_y = 0 \text{ when } 0 = x + \frac{1}{y^2}$$

$$0 = x + x^4$$
$$0 = x(x^3 + 1)$$

or, $x = -1$ (rejecting $x = 0$ since f undefined for $x = 0$) and $y = -1$. So, the only critical point is $(-1, -1)$.

$$f_{xx} = -\frac{2}{x^3};\ f_{yy} = -\frac{2}{y^3};\ f_{xy} = 1$$

$$D = \left[-\frac{2}{(-1)^3}\right]\left[-\frac{2}{(-1)^3}\right] - (1)^2 > 0$$

$$\text{and } f_{xx} > 0$$

So, $(-1, -1)$ is a relative minimum.

4. (a) $f(x, y) = x^2 + y^2$
$g(x, y) = x + 2y$

$$f_x = 2x;\ f_y = 2y;\ g_x = 1;\ g_y = 2$$

The three Lagrange equations are

$$2x = \lambda;\ 2y = 2\lambda;\ x + 2y = 4$$

Equating λ from the first two equations gives

$$2x = y$$

Substituting in the third equation gives $x = \dfrac{4}{5}$. Then, $y = \dfrac{8}{5}$ and the minimum value of the function is $f\left(\dfrac{4}{5}, \dfrac{8}{5}\right) = \dfrac{16}{5}$.

(b) $f(x, y) = xy^2$
$g(x, y) = 2x^2 + y^2$

$$f_x = y^2;\ f_y = 2xy;\ g_x = 4x;\ g_y = 2y$$

The three Lagrange equations are

$$y^2 = 4\lambda x;\ 2xy = 2\lambda y$$
$$2x^2 + y^2 = 6$$

Solving the first two equations for λ and equating gives $y^2 = 4x^2$.
Substituting into the third equation gives $x = -1, 1$. When $x = -1$, $y = -2$ or 2. When $x = 1$, $y = -2$ or 2. So, the critical points are $(-1, -2)$, $(-1, 2)$, $(1, -2)$, and $(1, 2)$.

$$f(-1, -2) = f(-1, 2) = -4 \text{ and}$$
$$f(1, -2) = f(1, 2) = 4$$

So, the maximum value of f is 4, and the minimum value of f is -4.

5. (a)

$$\int_{-1}^{3}\int_{0}^{2} x^3 y\, dx\, dy$$
$$= \int_{-1}^{3}\left(\frac{x^4 y}{4}\Big|_0^2\right) dy = \int_{-1}^{3} 4y\, dy = (2y^2)\Big|_{-1}^{3}$$
$$= 16$$

(b)

$$\int_{0}^{2}\int_{-1}^{1} x^2 e^{xy}\, dx\, dy$$
$$= \int_{-1}^{1}\int_{0}^{2} x^2 e^{xy}\, dy\, dx$$
$$= \int_{-1}^{1}\int_{0}^{2} x e^{xy} x\, dy\, dx$$

Using substitution with $u = xy$,

$$= \int_{-1}^{1} x \int_{0}^{2x} e^u \, du \, dx$$

$$= \int_{-1}^{1} x \left(e^u \Big|_0^{2x} \right) dx$$

$$= \int_{-1}^{1} x(e^{2x} - e^0) \, dx$$

$$= \int_{-1}^{1} (xe^{2x} - x) \, dx$$

$$= \int_{-1}^{1} xe^{2x} \, dx - \int_{-1}^{1} x \, dx$$

Using integration by parts with

$$u = x \quad \text{and} \quad dV = e^{2x} \, dx$$

$$du = dx \qquad V = \frac{1}{2} e^{2x}$$

$$= \frac{x}{2} e^{2x} \Big|_{-1}^{1} - \int_{-1}^{1} \frac{1}{2} e^{2x} \, dx - \int_{-1}^{1} x \, dx$$

$$= \left(\frac{x}{2} e^{2x} - \frac{1}{4} e^{2x} - \frac{x^2}{2} \right) \Big|_{-1}^{1}$$

$$= \left(\frac{1}{2} e^2 - \frac{1}{4} e^2 - \frac{1}{2} \right) - \left(-\frac{1}{2} e^{-2} - \frac{1}{4} e^{-2} - \frac{1}{2} \right)$$

$$= \frac{1}{4} e^2 + \frac{3}{4} e^{-2} = \frac{1}{4}(e^2 + 3e^{-2})$$

$$= \frac{1}{4} \left(e^2 + \frac{3}{e^2} \right) = \frac{1}{4} \left(\frac{e^4 + 3}{e^2} \right)$$

$$= \frac{e^4 + 3}{4e^2}$$

(c) $$\int_{1}^{2} \int_{1}^{y} \frac{y}{x} \, dx \, dy = \int_{1}^{2} y \left(\int_{1}^{y} \frac{1}{x} \, dx \right) dy$$

$$= \int_{1}^{2} y \left(\ln |x| \, \Big|_1^y \right) dy = \int_{1}^{2} y \ln y \, dy$$

Using integration by parts with

$$u = \ln y \text{ and } dV = y \; dy$$

$$= \frac{y^2}{2} \ln y \Big|_1^2 - \int_{1}^{2} \frac{y}{2} \, dy$$

$$= \left(\frac{y^2}{2} \ln y - \frac{y^2}{4} \right) \Big|_1^2 = (2 \ln 2 - 1) - \left(0 - \frac{1}{4} \right)$$

$$= 2 \ln 2 - \frac{3}{4}$$

(d) $$\int_{0}^{2} \int_{0}^{2-x} xe^{-y} \, dy \, dx = \int_{0}^{2} (x - e^{-y}) \Big|_0^{2-x} dx$$

$$= \int_{0}^{2} -xe^{x-2} + x \, dx$$

Using integration by parts with

$$u = -x \text{ and } dV = e^{x-2} \, dx$$

$$= -xe^{x-2} \Big|_0^2 - \int_{0}^{2} -e^{x-2} \, dx + \int_{0}^{2} x \, dx$$

$$= -xe^{x-2} \Big|_0^2 + \int_{0}^{2} e^{x-2} \, dx + \int_{0}^{2} x \, dx$$

$$= \left(-xe^{x-2} + e^{x-2} + \frac{x^2}{2} \right) \Big|_0^2$$

$$= (-2e^0 + e^0 + 2) - (0 + e^{-2} + 0)$$

$$= 1 - \frac{1}{e^2} = \frac{e^2 - 1}{e^2}$$

6. $Q(K, L) = 120K^{3/4}L^{1/4}$

$Q_K = \dfrac{90L^{1/4}}{K^{1/4}}$; $Q_L = \dfrac{30K^{3/4}}{L^{3/4}}$

When $K = 1{,}296$ thousand dollars and $L = 20{,}736$ worker-hours,

$$Q_K = 180 \quad \text{and} \quad Q_L = 3.75$$

7. $U(x, y) = \ln(x^2\sqrt{y})$; $g(x, y) = 20x + 30y$

$$U_x = \frac{1}{x^2\sqrt{y}} \cdot 2x\sqrt{y} = \frac{2}{x}$$

$$U_y = \frac{1}{x^2\sqrt{y}} \cdot \frac{1}{2} x^2 y^{-1/2} = \frac{1}{2y}$$

$$g_x = 20; \; g_y = 30$$

The three Lagrange equations are

$$\frac{2}{x} = 20\lambda; \quad \frac{1}{2y} = 30\lambda;$$

$$20x + 30y = 300$$

Solving the first two equations for λ and equating gives $x = 6y$. Substituting in the third equation gives $y = 2$, so $x = 12$. Everett should buy 12 DVDs and 2 video games.

8. $E = 0.05(xy - 2x^2 - y^2 + 95x + 20y)$

$$E_x = 0.05(y - 4x + 95)$$

$$E_x = 0 \text{ when } 4x - y = 95$$

$$E_y = 0.05(x - 2y + 20)$$

$$E_y = 0 \text{ when } -x + 2y = 20$$

Solving the system of equations by multiplying the first by two and adding to the second gives $x = 30$ units of A, so $y = 25$ units of B.
Since the combined dosage is less than 60 units, there will not be a risk of side effects. Further, this is an equivalent dosage of $E(30, 25) = 83.75$ units, it will be effective.

9. The area of the given region is

$$\int_1^2 \frac{1}{y}\,dy = \ln|y|\Big|_1^2 = \ln 2$$

$$T_{AV} = \frac{1}{\ln 2}\int_1^2 \int_0^{\frac{1}{y}} 10ye^{-xy}\,dx\,dy$$

Using substitution with $u = -xy$ and $du = -y\,dx$,

$$= \frac{1}{\ln 2}\int_1^2 \left[-10e^{-xy}\Big|_0^{\frac{1}{y}}\right] dy$$

$$= \frac{1}{\ln 2}\int_1^2 10(1 - e^{-1})\,dy$$

$$= \frac{1}{\ln 2}\left[10(1 - e^{-1})y\Big|_1^2\right]$$

$$= \frac{10(1 - e^{-1})}{\ln 2}\left(y\Big|_1^2\right) = \frac{10(1 - e^{-1})}{\ln 2}(2 - 1)$$

$$= \frac{10(1 - \frac{1}{e})}{\ln 2}\ ^\circ\text{C}$$

10. Let x denote the year of operation and y the corresponding profit, in millions of dollars.

(a)

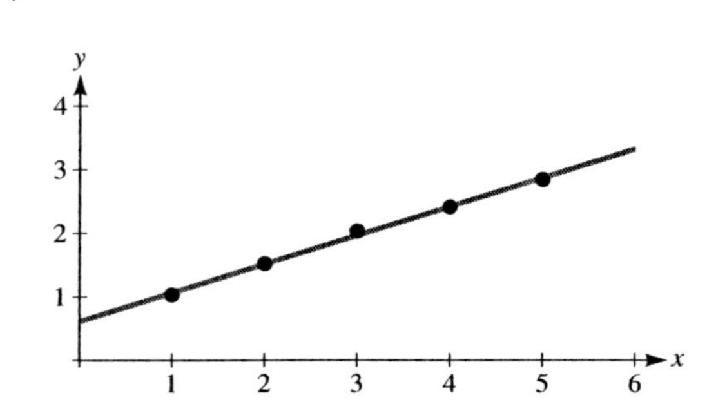

(b)

x	y	xy	x^2
1	1.03	1.03	1
2	1.52	3.04	4
3	2.03	6.09	9
4	2.41	9.64	16
5	2.84	14.20	25
$\sum x = 15$	$\sum y = 9.83$	$\sum xy = 34.00$	$\sum x^2 = 55$

Using the formulas with $n = 5$,

$$m = \frac{5(34) - (15)(9.83)}{5(55) - (15)^2} = \frac{22.55}{50} \approx 0.451$$

$$b = \frac{(55)(9.83) - (15)(34)}{5(55) - (15)^2} = \frac{30.65}{50} \approx 0.613$$

So, the equation of the least squares line is $y = 0.451x + 0.613$.

(c) When $x = 6$, $y = 0.451(6) + 0.613 = 3.319$ so the prediction is \$3,319,000.

Review Problems

1. $f(x, y) = 2x^3y + 3xy^2 + \frac{y}{x}$

$$f_x = 6x^2y + 3y^2 - \frac{y}{x^2}$$

$$f_y = 2x^3 + 6xy + \frac{1}{x}$$

3. $f(x, y) = xye^{xy}$

$$f_x = (xy)(e^{xy} \cdot y) + (e^{xy})(y)$$
$$= ye^{xy}(xy + 1)$$
$$f_y = (xy)(e^{xy} \cdot x) + (e^{xy})(x)$$
$$= xe^{xy}(xy + 1)$$

5. $$f(x, y) = \ln \frac{xy}{x + 3y} = \ln x + \ln y - \ln(x + 3y)$$

$$f_x = \frac{1}{x} - \frac{1}{x + 3y} = \frac{3y}{x(x + 3y)}$$

$$f_y = \frac{1}{y} - \frac{3}{x + 3y} = \frac{x}{y(x + 3y)}$$

7. $Q = 40K^{1/3}L^{1/2}$

The marginal product of capital is

$$\frac{\partial Q}{\partial K} = \frac{40}{3}K^{-2/3}L^{1/2} = \frac{40L^{1/2}}{3K^{2/3}}$$

which is approximately the change ΔQ in output due to one (thousand dollar) unit increase in capital. When $K = 125$ (thousand) and $L = 900$,

$$\Delta Q \approx \frac{\partial Q}{\partial K} = \frac{40(900)^{1/2}}{3(125)^{2/3}} = 16 \text{ units}$$

9. **(a)** When $f = 2$, the level curve $x^2 - y = 2$ is a parabola, with vertical axis, opening up, and having the vertex $(0, -2)$.
When $f = -2$, the level curve $x^2 - y = -2$ is a parabola, with vertical axis, opening up, and having the vertex $(0, 2)$.

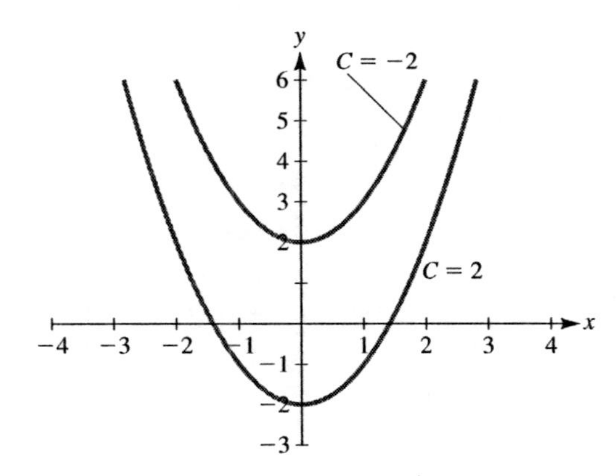

(b) When $f = 0$, the level curve is $6x + 2y = 0$, or $y = -3x$, which is a line through the origin with slope -3. When $f = 1$, the level curve is $6x + 2y = 1$, or $y = -3x + \frac{1}{2}$, which is the same line translated up $\frac{1}{2}$ a unit. When $f = 2$, the level curve is $6x + 2y = 2$, or $y = -3x + 1$, which is the same line translated up one unit.

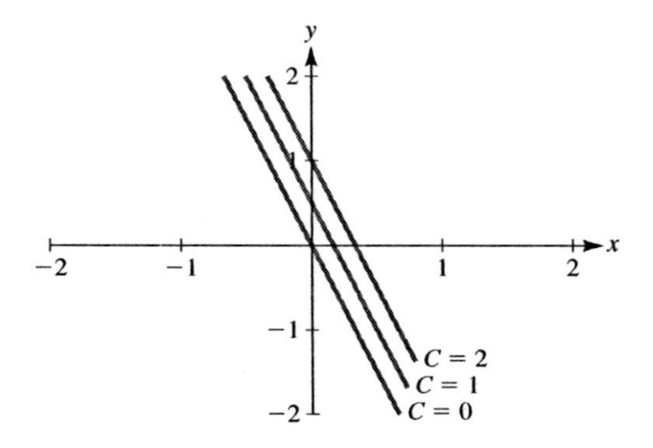

11. $$Q(x, y) = 60x^{1/3}y^{2/3}$$

For any value of x, the slope of the level curve $Q = k$ is an approximation of the change in unskilled labor y that should be made to offset a one-unit increase in skilled labor x so that the level of output will remain constant. So,

$$\Delta Q = \text{change in unskilled labor}$$
$$\approx \frac{dQ}{dx} = -\frac{Q_x}{Q_y}$$
$$= -\frac{20x^{-2/3}y^{2/3}}{40x^{1/3}y^{-1/3}} = -\frac{y}{2x}$$

When $x = 10$ and $y = 40$,

$$\Delta Q \approx \frac{dQ}{dx} = -\frac{40}{2(10)} = -2$$

That is, the level of unskilled labor should be decreased by approximately 2 workers.

13. $$f(x, y) = x^2 + y^3 + 6xy - 7x - 6y$$
$$f_x = 2x + 6y - 7$$
$$f_y = 3y^2 + 6x - 6$$

To find the critical points, set $f_x = 0$ and $f_y = 0$.

So, $2x+6y-7=0$ and $3y^2+6x-6=0$,
or $2x+6y-7=0$ and $2x+y^2-2=0$.
Subtracting the two equations gives $y^2-6y+5=0$,
$(y-1)(y-5)=0$, or $y=1$ and $y=5$.
When $y=1$, the first equation gives $x=\frac{1}{2}$ and when $y=5$, the first equation gives $x=-\frac{23}{2}$.

So, the critical points of f are $\left(\frac{1}{2}, 1\right), \left(-\frac{23}{2}, 5\right)$.

Since $f_{xx}=2$, $f_{yy}=6y$, and $f_{xy}=6$,

$$D=f_{xx}f_{yy}-(f_{xy})^2=(2)(6y)-36=12(y-3)$$

For the point $\left(\frac{1}{2}, 1\right)$,

$$D=12(-2)=-24<0$$

and f has a saddle point at $\left(\frac{1}{2}, 1\right)$.

For the point $\left(-\frac{23}{2}, 5\right)$,

$$D=12(2)=24>0$$

$$\text{and } f_{xx}>0$$

So, f has a relative minimum at $\left(-\frac{23}{2}, 5\right)$.

15. $f(x, y)=xe^{2x^2+5xy+2y^2}$

$$\begin{aligned} f_x &= (x)\left[e^{2x^2+5xy+2y^2}(4x+5y)\right] \\ &\quad + (e^{2x^2+5xy+2y^2})(1) \\ &= e^{2x^2+5xy+2y^2}\left[x(4x+5y)+1\right] \\ &= e^{2x^2+5xy+2y^2}(4x^2+5xy+1) \end{aligned}$$

$$f_x=0 \text{ when } 4x^2+5xy+1=0$$

$$f_y=x\left[e^{2x^2+5xy+2y^2}(5x+4y)\right]$$

$$f_y=0 \text{ when } x(5x+4y)e^{2x^2+5xy+2y^2}=0$$

So, $f_y=0$ when $x=0$ and when $5x+4y=0$, or

$$x=-\frac{4}{5}y$$

When $x=0$, substituting into $f_x=0$ yields no solution.
When $x=-\frac{4}{5}y$, $f_x=0$ when

$$0=4\left(-\frac{4}{5}y\right)^2+5\left(-\frac{4}{5}y\right)(y)+1$$

$$\text{or, } y=\pm\frac{5}{6}$$

$$\text{When } y=\frac{5}{6},\ x=-\frac{4}{5}\left(\frac{5}{6}\right)=-\frac{2}{3}$$

$$\text{When } y=-\frac{5}{6},\ x=-\frac{4}{5}\left(-\frac{5}{6}\right)=\frac{2}{3}$$

So, the critical points are $\left(-\frac{2}{3}, \frac{5}{6}\right)$ and $\left(\frac{2}{3}, -\frac{5}{6}\right)$.

$$\begin{aligned} f_{xx} &= (e^{2x^2+5xy+2y^2})(8x+5y) \\ &\quad + (4x^2+5xy+1)[e^{2x^2+5xy+2y^2}(4x+5y)] \\ f_{yy} &= x[(e^{2x^2+5xy+2y^2})(4) \\ &\quad + (5x+4y)e^{2x^2+5xy+2y^2}(5x+4y)] \\ f_{xy} &= (e^{2x^2+5xy+2y^2})(5x) \\ &\quad + (4x^2+5xy+1)e^{2x^2+5xy+2y^2}(5x+4y) \end{aligned}$$

For the point $\left(-\frac{2}{3}, \frac{5}{6}\right)$,

$$D\approx(-0.7076)(-2.0218)-(-1.6174)^2<0$$

So, $\left(-\frac{2}{3}, \frac{5}{6}\right)$ is a saddle point.

For the point $\left(\frac{2}{3}, -\frac{5}{6}\right)$,

$$D\approx(0.7076)(2.0218)-(1.6174)^2<0$$

So, $\left(\frac{2}{3}, -\frac{5}{6}\right)$ is also a saddle point.

17. The goal is to maximize the area of a rectangle

$$A(l, w)=lw$$

subject to the comstraint $2l+2w=k$, where k is some positive constant. So, $g(l, w)=2l+2w$.

$$A_l = w;\ A_w = l;\ g_l = 2;\ g_w = 2$$

The three Lagrange equations are

$$w = 2\lambda;\ l = 2\lambda;\ 2l + 2w = k$$

Solving the first two for λ and equating gives $\frac{w}{2} = \frac{l}{2}$, or $w = l$. So, the rectangle having the greatest area is a square.

19. From problem 18, the profit function is

$$P(x, y) = \frac{50y}{y+2} + \frac{20x}{x+5} - x - y$$

The constraint is $x + y = 11$ thousand dollars, so $g(x, y) = x + y$.

$$P_x = \frac{100}{(x+5)^2} - 1;\ P_y = \frac{100}{(y+2)^2} - 1;$$

$g_x = 1;\ g_y = 1$
The three Lagrange equations are

$$\frac{100}{(x+5)^2} - 1 = \lambda$$
$$\frac{100}{(y+2)^2} - 1 = \lambda$$
$$x + y = 11$$

From the first two equations,

$$(x+5)^2 = (y+2)^2$$

or $y = x + 3$ (rejecting the negative solution). Substituting into the third equation gives $x = 4$, and the corresponding value of y is 7.
So, to maximize profit, \$4,000 should be spent on development and \$7,000 should be spent on promotion.

21.
$$f(x, y) = \frac{12}{x} + \frac{18}{y} + xy$$

Suppose y is fixed (say at $y = 1$), then f is very large when x is quite small.
f is also large when x is large, with smaller values of f occurring between these extremes. The same reasoning applies to y when x is fixed.

$$f_x = -\frac{12}{x^2} + y;\ f_y = -\frac{18}{y^2} + x$$

To find the critical points, set $f_x = 0$ and $f_y = 0$. Then $y = \frac{12}{x^2}$ and $x = \frac{18}{y^2}$. Substituting leads to

$$y = \frac{12}{x^2} = \frac{12}{\left(\frac{18}{y^2}\right)^2} = \frac{12y^4}{18^2}$$

or $y = 0$ (which is not in the domain of the function) and $12y^3 = 18^2$, $y^3 = 27$, $y = 3$. The corresponding value for $x = \frac{18}{3^2} = 2$. So, the critical point of f is $(2, 3)$.

$$f_{xx} = \frac{24}{x^3};\ f_{yy} = \frac{36}{y^3};\ f_{xy} = 1$$

For the point $(2, 3)$,
$D = \frac{(24)(36)}{(2^3)(3^3)} - 1 > 0$ and $f_{xx}(2, 3) > 0$, so the minimum is $f(2, 3) = 18$.

23.
$$\int_0^1 \int_0^2 e^{-x-y}\, dydx$$
$$= \int_0^1 \int_0^2 e^{-x}e^{-y}\, dydx$$
$$= \int_0^1 (-e^{-x}e^{-y})\Big|_0^2\, dx$$
$$= \int_0^1 (-e^{-x}e^{-2} + e^{-x})\, dx$$
$$= (1 - e^{-2}) \int_0^1 e^{-x}\, dx$$
$$= (1 - e^{-2})(-e^{-x})\Big|_0^1$$
$$= (1 - e^{-2})(-e^{-1} + 1) = 0.5466$$

25.

$$\int_0^1 \int_{-1}^1 xe^{2y}\,dydx$$
$$= \int_0^1 \left(\frac{1}{2}\right) xe^{2y}\Big|_{-1}^1 dx$$
$$= \int_0^1 \left(\frac{xe^2}{2} - \frac{xe^{-2}}{2}\right) dx$$
$$= \frac{e^2 - e^{-2}}{2} \int_0^1 x\,dx$$
$$= \frac{e^2 - e^{-2}}{4} \approx 1.8134$$

27.

$$I = \int_1^e \int_1^e (\ln x + \ln y)\,dydx$$
$$= \int_1^e \left[y(\ln x) + (y \ln y - y)\right]\Big|_1^e dx$$
$$= \int_1^e [(e-1)\ln x + 1]\,dx$$
$$= [(e-1)(x \ln x - x) + x]\,|_1^e = 3.4366$$

29.

$$\int_1^2 \int_0^x e^{\frac{y}{x}}\,dy\,dx = \int_1^2 \int_0^x e^{\frac{1}{x}y}\,dy\,dx$$
$$= \int_1^2 xe^{\frac{y}{x}}\Big|_0^x dx = \int_1^2 xe - x\,dx$$
$$= \int_1^2 (e-1)x\,dx = (e-1)\frac{x^2}{2}\Big|_1^2$$
$$= (e-1)\left(2 - \frac{1}{2}\right) = \frac{3}{2}(e-1)$$

31.

$$\int\int_R (x+2y)dA$$
$$= \int_0^1 \int_{-2}^2 (x+2y)\,dydx$$
$$= \int_0^1 (xy + y^2)\Big|_{-2}^2 dx$$
$$= \int_0^1 4x\,dx = 2x^2\Big|_0^1 = 2$$

33.

$$V = \int_1^2 \int_2^3 xe^{-y}\,dydx$$
$$= \int_1^2 (-xe^{-y})\Big|_2^3 dx$$
$$= \int_1^2 (x)(e^{-2} - e^{-3})\,dx$$
$$= (e^{-2} - e^{-3})\frac{3}{.2} = 0.1283$$

35. The sum of the three numbers is $x + y + z = 20$, so $z = 20 - x - y$. Their product is

$$P = xyz = xy(20 - x - y) = 20xy - x^2y - xy^2$$
$$P_x = 20y - 2xy - y^2$$
$$P_x = 0 \text{ when } y(20 - 2x - y) = 0$$
$$P_y = 20x - x^2 - 2xy$$
$$P_y = 0 \text{ when } x(20 - x - 2y) = 0$$

Since the numbers must be positive, reject the solution $x = 0$ or $y = 0$. Solving the system of equations by multiplying the first by -2 and adding to the second gives $x = \frac{20}{3}$. When $x = \frac{20}{3}$, $20 - \frac{20}{3} - 2y = 0$, or $y = \frac{20}{3}$. Then, $z = 20 - \frac{20}{3} - \frac{20}{3} = \frac{20}{3}$. So, the product is maximized when $x = y = z = \frac{20}{3}$.

37. Using the hint in the problem, let D denote the square of the distance from the origin to the surface. Then,

$$D = x^2 + y^2 + z^2$$

Since $y^2 - z^2 = 10$, $y^2 = 10 + z^2$ and

$$D = x^2 + 10 + 2z^2$$
$$D_x = 2x, \text{ so } D_x = 0 \text{ when } x = 0$$
$$D_z = 4z, \text{ so } D_z = 0 \text{ when } z = 0$$

When $z = 0$, $y^2 = 10$ or $y = \pm\sqrt{10}$.
So, the critical points are $(0, -\sqrt{10}, 0)$ and $(0, \sqrt{10}, 0)$.

$$D_{xx} = 2, \; D_{zz} = 4, \; D_{xz} = 0$$

For the point $(0, -\sqrt{10}, 0)$,

$$D = (2)(4) - 0 > 0$$
$$\text{and } D_{xx} > 0$$

So, $(0, -\sqrt{10}, 0)$ is a relative minimum.
For the point $(0, \sqrt{10}, 0)$,

$$D > 0 \text{ and } D_{xx} > 0$$

So, it is also a relative minimum. The square of the distance, using either point, is

$$D = 0 + 10 + 0 = 10$$

So, the minimum distance $= \sqrt{10}$.

39. (a) Let x denote the monthly advertising expenditure and y the corresponding sales (both measured in units of \$1,000). Then

x	3	4	7	9	10
y	78	86	138	145	156

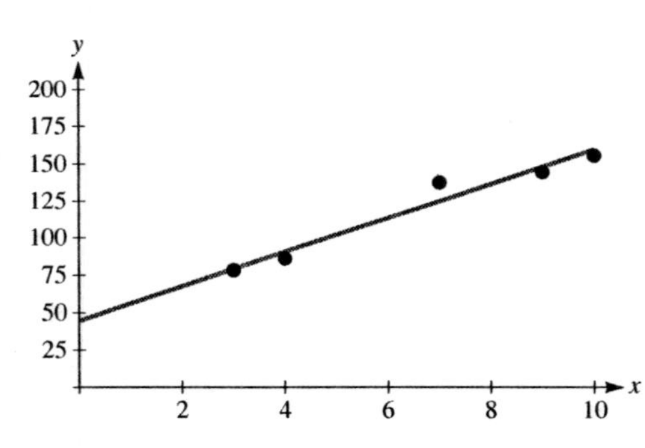

(b)

x	y	xy	x^2
3	78	234	9
4	86	344	16
7	138	966	49
9	145	1,305	81
10	156	1,560	100
$\sum x = 33$	$\sum y = 603$	$\sum xy = 4{,}409$	$\sum x^2 = 255$

Using the formulas with $n = 5$,

$$m = \frac{5(4{,}409) - 33(603)}{5(255) - (33)^2} = 11.54$$
$$b = \frac{255(603) - 33(4{,}409)}{5(255) - (33)^2} = 44.45$$

So, the equation of the least-squares line is

$$y = 11.54x + 44.45$$

(c) $y = 11.54(5) + 44.45 = 102.15$ thousand, or \$102,150.

41.
$$Q(x, y) = 200 - 10x^2 + 20y$$
$$x(t) = 5 + 0.02t$$
$$y(t) = 6 + 0.4\sqrt{t}$$
$$\frac{dQ}{dt} = \frac{\partial Q}{\partial x} \cdot \frac{dx}{dt} + \frac{\partial Q}{\partial y} \cdot \frac{dy}{dt}$$
$$= (-20x)(0.02) + (20)\left(\frac{0.2}{\sqrt{t}}\right)$$

When $t = 9$, $x(9) = 5.18$ and

$$\frac{dQ}{dt} = -20(5.18)(0.02) + 20\left(\frac{0.2}{\sqrt{9}}\right)$$
$$\approx -0.739$$

or demand is decreasing at a rate of 3/4 quart per month.

43.
$$p(x, y) = \frac{1}{4}x^{1/3}y^{1/2}$$
$$x = 129 - \sqrt{8t}$$
$$y = 15.60 + 0.2t$$
$$Q = \frac{4184}{p}$$

$$\frac{dQ}{dt} = \frac{dQ}{dp} \cdot \frac{dp}{dt}$$
where $\frac{dp}{dt} = \frac{\partial p}{\partial x} \cdot \frac{dx}{dt} + \frac{\partial p}{\partial y} \cdot \frac{dy}{dt}$

$$\frac{dQ}{dt} = -\frac{4184}{p^2}\left[\left(\frac{1}{12}x^{-2/3}y^{1/2}\right)\left(-\frac{\sqrt{8}}{2\sqrt{t}}\right) + \left(\frac{1}{8}x^{1/3}y^{-1/2}\right)(0.2)\right]$$

When $t = 2$, $x = 125$, $y = 16$ and $p(125, 16) = 5$ so

$$\frac{dQ}{dt} = -\frac{4184}{(5)^2}\left[\left(\frac{1}{12}\cdot\frac{1}{25}\cdot 4\right)\left(-\frac{\sqrt{8}}{2\sqrt{2}}\right) + \left(\frac{1}{8}\cdot 5\cdot\frac{1}{4}\right)(0.2)\right]$$

≈ -3.00

or demand is decreasing at a rate of 3 pies per week.

45. $Q(E, T) = 125E^{2/3}T^{1/2}$

$$\frac{dQ}{dt} = \frac{\partial Q}{\partial E}\cdot\frac{dE}{dt} + \frac{\partial Q}{\partial T}\cdot\frac{dT}{dt}$$

$$= \left(\frac{250}{3}E^{-1/3}T^{1/2}\right)\left(\frac{1}{11}\right) + \left(\frac{125}{2}E^{2/3}T^{-1/2}\right)(-0.21)$$

$$= \left[\frac{250}{3}(151)^{-1/3}(10)^{1/2}\right]\left(\frac{1}{11}\right) + \left[\frac{125}{2}(151)^{2/3}(10)^{-1/2}\right](-0.21)$$

≈ -113.19

or decreasing at a rate of 113 units per day.

47. $N(r, s) = 40e^{-r/2}e^{-s/3}$

$$\text{Pollution} = \int_2^3\int_1^2 40e^{-r/2}e^{-s/3}ds\, dr$$

$$= \int_2^3 \left(40e^{-r/2}\cdot -3e^{s/3}\Big|_1^2\right) dr$$

$$= -120\int_2^3 \left[e^{-r/2}\left(e^{-2/3} - e^{-1/3}\right)\right] dr$$

$$= -120\left(e^{-2/3} - e^{-1/3}\right)\int_2^3 e^{-r/2}dr$$

$$= -120\left(e^{-2/3} - e^{1/3}\right)\left[-2e^{-r/2}\Big|_2^3\right]$$

$$= 240\left(e^{2/3} - e^{-1/3}\right)\left(e^{-3/2} - e^{-1}\right)$$

≈ 7.056 units

49. With $Q = x^a y^b$, $Q_x = ax^{a-1}y^b$ and $Q_y = bx^a y^{b-1}$.

$$xQ_x + yQ_y = x(ax^{a-1}y^b) + y(bx^a y^{b-1})$$
$$= (a+b)x^a y^b = (a+b)Q$$

If $b = 1 - a$, then $xQ_x + yQ_y = (a+b)Q = Q$.

Chapter 8

Differential Equations

8.1 Introduction to Differential Equations

1.
$$\frac{dy}{dx} = 3x^2 + 5x - 6$$
$$y = \int \frac{dy}{dx}\,dx$$
$$y = \int (3x^2 + 5x - 6)\,dx$$
$$= x^3 + \frac{5}{2}x^2 - 6x + C.$$

3. Separate the variables
$$\frac{dy}{dx} = 3y$$
$$\frac{1}{y}\,dy = 3\,dx$$

and integrate
$$\int \frac{1}{y}\,dy = \int 3\,dx,$$
$$\ln|y| = 3x + C_1,$$
$$|y| = e^{3x+C_1} = e^{C_1}e^{3x}, \text{ or } y = Ce^{3x}$$

where C is the constant $\pm e^{C_1}$.

5. Separate the variables of
$$\frac{dy}{dx} = e^y$$
$$\frac{1}{e^y}\,dy = dx$$

and integrate
$$\int e^{-y}\,dy = \int dx,$$
$$-e^{-y} = x + C_1 \text{ or } e^{-y} = C - x$$

where C is the constant $-C_1$. So,
$$\ln e^{-y} = \ln(C - x),$$
$$-y = \ln(C - x), \text{ or } y = -\ln(C - x)$$

7. Separate the variables of
$$\frac{dy}{dx} = \frac{x}{y}$$
$$y\,dy = x\,dx$$

and integrate
$$\int y\,dy = \int x\,dx,$$
$$\frac{y^2}{2} = \frac{x^2}{2} + C_1 \text{ or } y^2 = x^2 + C$$

$y = \pm\sqrt{x^2 + C}$, where C is the constant $2C_1$.

9. Separate the variables of
$$\frac{dy}{dx} = \sqrt{xy} = \sqrt{x}\sqrt{y}$$
$$\frac{1}{\sqrt{y}}\,dy = \sqrt{x}\,dx$$

and integrate
$$\int y^{-1/2}\,dy = \int x^{1/2}\,dx$$
$$2y^{1/2} = \frac{2}{3}x^{3/2} + C_1$$
$$y = \left(\frac{1}{3}x^{3/2} + C\right)^2$$

where C is the constant $2C_1$.

11. Separate the variables of

$$\begin{aligned}\frac{dy}{dx} &= \frac{y}{x-1}\\ \frac{1}{y}\,dy &= \frac{1}{x-1}\,dx\end{aligned}$$

and integrate

$$\begin{aligned}\int \frac{1}{y}\,dy &= \int \frac{1}{x-1}\,dx\\ \ln|y| &= \ln|x-1| + C_1\\ \ln|y| - \ln|x-1| &= C_1\\ \ln\frac{|y|}{|x-1|} &= C_1\\ \frac{|y|}{|x-1|} &= e^{C_1}\\ |y| &= e^{C_1}|x-1|\\ y &= \pm e^{C_1}|x-1|\\ y &= C|x-1|\end{aligned}$$

where C is the constant $\pm e^{C_1}$.

13. Separate the variables of

$$\begin{aligned}\frac{dy}{dx} &= \frac{y+3}{(2x-5)^6}\\ \frac{1}{y+3}\,dy &= \frac{1}{(2x-5)^6}\,dx\end{aligned}$$

and integrate

$$\begin{aligned}\int \frac{1}{y+3}\,dy &= \int (2x-5)^{-6}\,dx\\ \ln|y+3| &= -\frac{1}{10}(2x-5)^{-5} + C_1\\ |y+3| &= e^{-1/10(2x-5)^{-5}+C_1}\\ |y+3| &= e^{C_1}e^{-1/10(2x-5)^{-5}}\\ y+3 &= \pm e^{C_1}e^{-1/10(2x-5)^{-5}}\\ y &= -3 + Ce^{-1/10(2x-5)^{-5}}\end{aligned}$$

where C is the constant $\pm e^{C_1}$.
Note: $\ln|y+3| = -\frac{1}{10}(2x-5)^{-5} + C_1$
$\ln|y+3|^{10} = -(2x-5)^{-5} + C_1$.

15. Separate the variables of

$$\begin{aligned}\frac{dx}{dt} &= \frac{xt}{2t+1}\\ \frac{1}{x}\,dx &= \frac{t}{2t+1}\,dt\end{aligned}$$

and integrate

$$\int \frac{1}{x}\,dx = \int \frac{t}{2t+1}\,dt$$

using substitution with $u = 2t+1$,

$$\begin{aligned}\int \frac{1}{x}\,dx &= \frac{1}{2}\int \frac{\frac{u-1}{2}}{u}\,du\\ &= \frac{1}{4}\int \frac{u-1}{u}\,du\\ &= \frac{1}{4}\int 1 - \frac{1}{u}\,du\\ \ln|x| &= \frac{1}{4}u - \frac{1}{4}\ln|u| + C_1\\ &= \frac{1}{4}(2t+1) - \frac{1}{4}\ln|2t+1| + C_1\\ &= \frac{t}{2} - \frac{1}{4}\ln|2t+1| + C_2\\ &= \frac{t}{2} + \ln(2t+1)^{-1/4} + C_2\end{aligned}$$

where C_2 is the constant $\frac{1}{4} + C_1$.

$$\begin{aligned}|x| &= e^{t/2+\ln(2t+1)^{-1/4}+C_2}\\ |x| &= e^{t/2}(2t+1)^{-1/4}\cdot e^{C_2}\\ x &= \frac{\pm e^{C_2}e^{t/2}}{(2t+1)^{1/4}}\\ x &= \frac{Ce^{t/2}}{(2t+1)^{1/4}}\end{aligned}$$

where C is the constant $\pm e^{C_2}$.

17. Separate the variables of

$$\begin{aligned}\frac{dy}{dx} &= xe^{x-y} = x\cdot\frac{e^x}{e^y}\\ e^y\,dy &= xe^x\,dx\end{aligned}$$

and integrate

$$\int e^y\,dy = \int xe^x\,dx$$

Let

$$u = x \quad \text{and} \quad dV = e^x\,dx$$
$$du = dx \qquad V = e^x$$
$$\int e^y\,dy = xe^x - \int e^x\,dx$$
$$e^y = xe^x - e^x + C_1$$
$$y = \ln\left(xe^x - e^x + C_1\right)$$

19. Separate the variables of

$$\frac{dy}{dt} = y\ln\sqrt{t} = y\ln t^{1/2} = y\frac{1}{2}\ln t$$
$$\frac{1}{y}\,dy = \frac{1}{2}\ln t\,dt$$

and integrate

$$\int \frac{1}{y}\,dy = \frac{1}{2}\int \ln t\,dt$$

Let

$$u = \ln t \quad \text{and} \quad dV = dt$$
$$du = \frac{1}{2}\,dt \qquad V = t$$
$$\int \frac{1}{y}\,dy = \frac{1}{2}\left[t\ln t - \int t\cdot\frac{1}{t}\,dt\right]$$
$$\ln|y| = \frac{1}{2}\left[t\ln t - t\right] + C_1$$
$$|y| = e^{t/2(\ln t - 1) + C_1}$$
$$|y| = e^{C_1}\cdot e^{t/2(\ln t - 1)}$$
$$y = \pm e^{C_1}\cdot e^{t/2(\ln t - 1)}$$
$$y = Ce^{t(\ln t - 1)/2}$$

where C is the constant $\pm e^{C_1}$.

21.

$$\frac{dy}{dx} = e^{5x}$$
$$\int \frac{dy}{dx}\,dx = \int e^{5x}\,dx$$
$$y = \frac{1}{5}e^{5x} + C$$

Since $y = 1$ when $x = 0$,

$$1 = \frac{1}{5}e^0 + C, \text{ or } C = \frac{4}{5}$$

So,

$$y = \frac{1}{5}e^{5x} + \frac{4}{5}$$

23.

$$\frac{dy}{dx} = \frac{x}{y^2}$$
$$y^2\,dy = x\,dx$$
$$\int y^2\,dy = \int x\,dx$$
$$\frac{y^3}{3} = \frac{x^2}{2} + C_1$$
$$y^3 = \frac{3}{2}x^2 + C_2$$

where C_2 is the constant $3C_1$,

$$y = \left(\frac{3}{2}x^2 + C_2\right)^{1/3}$$

since $y = 3$ when $x = 2$,

$$3 = \left[\frac{3}{2}(2)^2 + C_2\right]^{1/3}$$
$$3 = \left(6 + C_2\right)^{1/3}$$
$$27 = 6 + C_2, \text{ or } C_2 = 21$$

So,

$$y = \left(\frac{3}{2}x^2 + 21\right)^{1/3}$$
$$= \left(\frac{3x^2 + 42}{2}\right)^{1/3}$$

25.

$$\frac{dy}{dx} = y^2(4-x)^{1/2}$$

$$\frac{1}{y^2}\,dy = (4-x)^{1/2}\,dx$$

$$\int y^{-2}\,dy = \int (4-x)^{1/2}\,dx$$

$$\frac{y^{-1}}{-1} = \frac{-2}{3}(4-x)^{3/2} + C_1$$

$$\frac{1}{y} = \frac{2}{3}(4-x)^{3/2} - C_1$$

Since $y = 2$ when $x = 4$,

$$\frac{1}{2} = \frac{2}{3}(0) - C_1, \text{ or } C_1 = -\frac{1}{2}$$

$$\frac{1}{y} = \frac{2}{3}(4-x)^{3/2} + \frac{1}{2} = \frac{4(4-x)^{3/2}+3}{6}$$

$$y = \frac{6}{4(4-x)^{3/2}+3}$$

27.

$$\frac{dy}{dt} = \frac{y+1}{t(y-1)}$$

$$\frac{y-1}{y+1}\,dy = \frac{1}{t}\,dt$$

$$\left(1 - \frac{2}{y+1}\right)\,dy = \frac{1}{t}\,dt$$

$$y - 2\ln|y+1| = \ln|t| + C_1$$

Since $y = 2$ when $t = 1$,

$$2 - 2\ln 3 = 0 + C_1,$$

$$\text{or } C_1 = 2(1 - \ln 3)$$

$$y - 2\ln|y+1| = \ln|t| + 2(1 - \ln 3)$$

29. Let Q denote the number of bacteria. Then, $\frac{dQ}{dt}$ is the rate of change of Q, and since this rate of change is proportional to Q,

$$\frac{dQ}{dt} = kQ$$

where k is a positive constant of proportionality.

31. Let Q denote the investment. Then $\frac{dQ}{dt}$ is the rate of change of Q, and since this rate of change is equal to 7% of the size of Q,

$$\frac{dQ}{dt} = 0.07Q$$

33. Let P denote the population, Then $\frac{dP}{dt}$ is the rate of change of P, and since this rate of change is the constant 500,

$$\frac{dP}{dt} = 500$$

35. Let T_m = temperature of the surrounding medium

$T(t)$ = object's temperature at time t

Then, $\frac{dT}{dt}$ is the rate of change of T and since this rate is proportional to $T_m - T$,

$$\frac{dT}{dt} = k\left(T_m - T\right)$$

37. Let F = total number of facts and

$R(t)$ = number of facts recalled at time t.

Then, $\frac{dR}{dt}$ is the rate of change of R and since this rate is proportional to $F - R$,

$$\frac{dR}{dt} = k(F - R)$$

39. Let C = number of people involved and

$P(t)$ = number of people implicated at time t.

Then, $\frac{dP}{dt}$ is the rate of change of P and since this rate is proportional to $(P)(C - P)$,

$$\frac{dP}{dt} = kP(C - P)$$

41. If $y = Ce^{kx}$, the derivative of y is

$$\frac{dy}{dx} = Ce^{kx}\cdot k = kCe^{kx} = ky,$$

the given differential equation.

43.

$$y = C_1e^x + C_2xe^x$$
$$\frac{dy}{dx} = C_1e^x + C_2(xe^x + e^x)$$
$$= (C_1 + C_2)e^x + C_2xe^x$$
$$\frac{d^2y}{dx^2} = (C_1 + C_2)e^x + C_2(xe^x + e^x)$$
$$= (C_1 + 2C_2)e^x + C_2xe^x$$
$$\frac{d^2y}{dx^2} - 2\frac{dy}{dx} + y = (C_1 + 2C_2)e^x + C_2xe^x$$
$$- 2C_1e^x - 2C_2xe^x - 2C_2e^x$$
$$+ C_1e^x + C_2xe^x$$
$$= (C_1 + 2C_2 - 2C_1 - 2C_2 + C_1)e^x$$
$$+ (C_2 - 2C_2 + C_2)xe^x$$
$$= 0 \cdot e^x + 0 \cdot xe^x = 0$$

45. Rate revenue changes = (# barrels)(rate selling price changes).

$$\frac{dR}{dt} = 400(56 + 0.04t), \text{ where } t \text{ is in months.}$$
$$\text{Revenue} = \int_0^{24} 400(56 + 0.04t)\, dt$$
$$= 400(56t + 0.02t^2)\Big|_0^{24}$$
$$= 400\left[\left(56(24) + 0.02(24)^2\right) - 0\right]$$
$$\approx \$542{,}208$$

47. (a) rate salt flows out

$$= (\text{salt/gal) flowing out})(\text{gal/min flowing out})$$
$$= \left(\frac{S(t)}{200}\right)(5) = \frac{S(t)}{40} \text{ gal/min}$$

(b) $\dfrac{dS}{dt}$ = (rate salt enters) − (rate salt leaves)

$$= (\text{salt/gal flowing in})(\text{gal/min flowing in}) - \frac{S}{40}$$
$$= (0)(5) - \frac{S}{40} = -\frac{S}{40}$$

(c)

$$\frac{dS}{dt} = -\frac{S}{40}$$
$$\int \frac{1}{S} dS = -\frac{1}{40}\int dt$$
$$\ln|S| = -\frac{1}{40}t + C_1$$
$$|S| = e^{-1/40t + C_1}$$
$$|S| = e^{C_1} \cdot e^{-1/40t}$$
$$S = \pm e^{C_1} \cdot e^{-1/40t}$$
$$S = Ce^{-t/40}$$

When $t = 0$, $S(0) = (2\text{lbs/gal})(200\text{gal}) = 400\text{lbs}$, so $400 = Ce^0$, or $C = 400$. So, $S(t) = 400e^{-t/40}$.

49. Let $P(t)$ = number of infected residents and C = total number of susceptible residents. We need to maximize the rate at which residents become infected, or

$$\frac{dP}{dt} = kP(C - P)$$

So, $$\frac{d^2P}{dt^2} = k\left[P\left(-\frac{dP}{dt}\right) + (C - P)\left(\frac{dP}{dt}\right)\right]$$
$$= k(C - 2P)\frac{dP}{dt}$$
$$\frac{d^2P}{dt^2} = 0 \quad \text{when} \quad 0 = k(C - 2P),$$
$$\left(\text{eliminating when } \frac{dP}{dt} = 0\right)$$

or, $P = \dfrac{C}{2}$.

When $0 < P < \dfrac{C}{2}$, $\dfrac{d^2P}{dt^2} > 0$, so $\dfrac{dP}{dt}$ is increasing

$P > \dfrac{C}{2}$, $\dfrac{d^2P}{dt^2} < 0$, so $\dfrac{dP}{dt}$ is decreasing

Therefore, $\dfrac{dP}{dt^2}$ is a maximum when $P = \dfrac{C}{2}$.

51.

$$\frac{dp}{dt} = k(1 - p)$$

where k is a constant of proportionality

$$\int \frac{dp}{1-p} = \int k\,dt$$
$$-\ln|1-p| = kt + C_1$$
$$\ln|1-p| = -kt - C_1$$
$$|1-p| = e^{-kt-C_1}$$
$$|1-p| = e^{-kt} \cdot e^{-C_1}$$
$$1-p = \pm e^{-C_1} e^{-kt}$$
$$1-p = Ce^{-kt}$$
$$p(t) = 1 - Ce^{-kt}$$

When $t = 0$, $p(0) = 0$, so

$$0 = 1 - Ce^0, \quad \text{or} \quad C = 1$$
$$\text{and} \quad p(t) = 1 - e^{-kt}.$$

Further, when $t = 8$, $p(8) = 0.05$, so

$$0.05 = 1 - e^{-8k}$$
$$e^{-8k} = 0.95$$
$$-8k = \ln 0.95, \quad \text{or} \quad k = -\frac{\ln 0.95}{8}$$

$$\text{and} \quad p(t) = 1 - e^{-\left(-\frac{\ln 0.95}{8}\right)t}$$
$$= 1 - e^{\left(\frac{\ln 0.95}{8}\right)t} = 1 - e^{\ln(0.95)^{\frac{1}{8}t}} = 1 - (0.95)^{\frac{t}{8}}$$

53. Let P be the number of people involved and Q the number of people implicated.

$$\frac{dQ}{dt} = KQ(P-Q)$$
$$\int \frac{dQ}{Q(P-Q)} = \int k\,dt$$

Using the method of partial fractions,

$$\frac{1}{Q(P-Q)} = \frac{A}{Q} + \frac{B}{P-Q}$$
$$1 = A(P-Q) + BQ$$

Since this must be true for all values of P and Q, choose $Q = 0$. Then,

$$1 = AP, \quad \text{or} \quad A = \frac{1}{P}$$

Choosing $Q = P$,

$$1 = BP, \quad \text{or} \quad B = \frac{1}{P}$$

So,

$$\frac{1}{Q(P-Q)} = \frac{1}{PQ} + \frac{1}{P(P-Q)}$$

and

$$\int \frac{dQ}{Q(P-Q)} = \int \frac{dQ}{PQ} + \int \frac{dQ}{P(P-Q)}$$

Since P is a constant,

$$= \frac{1}{P}\int \frac{dQ}{Q} + \frac{1}{P}\int \frac{dQ}{P-Q}$$

Now,

$$\frac{1}{P}\int \frac{dQ}{Q} + \frac{1}{P}\int \frac{dQ}{P-Q} = \int k\,dt$$
$$\frac{1}{P}\ln|Q| - \frac{1}{P}\ln|P-Q| = kt + C_1$$
$$\ln|Q| - \ln|P-Q| = Pkt + PC_1$$
$$\ln\left|\frac{Q}{P-Q}\right| = Pkt + PC_1$$
$$\left|\frac{Q}{P-Q}\right| = e^{Pkt+PC_1}$$
$$\left|\frac{Q}{P-Q}\right| = e^{Pkt} \cdot e^{PC_1}$$
$$\frac{Q}{P-Q} = \pm e^{PC_1} \cdot e^{Pkt}$$
$$\frac{Q}{P-Q} = Ce^{Pkt}$$

Rather than solving for Q at this point, is easier to substitute known values first. When $t = 0$, $Q = 7$ so

$$\frac{7}{P-7} = Ce^0, \quad \text{or} \quad C = \frac{7}{P-7}$$

and

$$\frac{Q}{P-Q} = \left(\frac{7}{P-7}\right)e^{Pkt}$$

Further, when $t = 3$, $Q = 16$ so

$$\frac{16}{P-16} = \left(\frac{7}{P-7}\right) e^{3Pk}$$

and, when $t = 6$, $Q = 28$ so

$$\frac{28}{P-28} = \left(\frac{7}{P-7}\right) e^{6PK}$$
$$= \left(\frac{7}{P-7}\right) \left(e^{3Pk}\right)^2$$

From above,

$$\frac{16}{P-16} = \left(\frac{7}{P-7}\right) e^{3Pk}$$
$$\frac{16(P-7)}{7(P-16)} = e^{3Pk}$$

Substituting,

$$\frac{28}{P-28} = \left(\frac{7}{P-7}\right) \left[\frac{16(P-7)}{7(P-16)}\right]^2$$
$$\frac{28}{P-28} = \frac{256(P-7)}{7(P-16)^2}$$
$$196(P-16)^2 = 256(P-7)(P-28)$$
$$196(P^2 - 32P + 256) = 256(P^2 - 35P + 196)$$
$$0 = 60P^2 - 2688P$$
$$0 = P(60P - 2688)$$

Rejecting the solution $P = 0$, $P \approx 45$ people.

55. (a)

$$\frac{dR}{dS} = \frac{k}{s}$$
$$\int dR = \int \frac{k}{s}\, ds$$
$$R = k \ln|S| + C$$

When $S = S_0$, $R = 0$ so

$$0 = k \ln|S_0| + C, \text{ or}$$
$$C = -k \ln|S_0|$$

and

$$R = k \ln|S| - k \ln|S_0|$$
$$R = k(\ln|S| - \ln|S_0|)$$
$$R = k \ln\left|\frac{S}{S_0}\right|$$

(b)

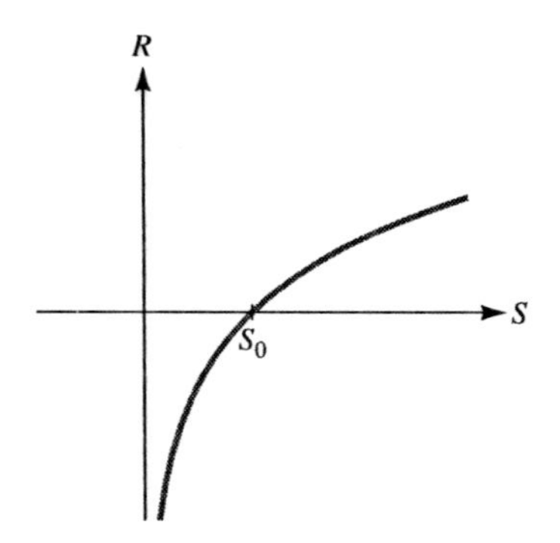

57. (a)

$$\frac{dp}{dt} = -\frac{1}{2}(S - D)$$
$$= -\frac{1}{2}\left[(80 + 3p) - (120 - 2p)\right]$$
$$= -\frac{1}{2}(-40 + 5p)$$
$$\int \frac{dp}{5p - 40} = \int -\frac{1}{2} dt$$

Using substitution with $u = 5p - 40$ and $\frac{1}{5} du = dp$,

$$\frac{1}{5} \ln|5p - 40| = -\frac{1}{2} t + C_1$$
$$\ln|5p - 40| = -\frac{5}{2} t + C_2$$
$$|5p - 40| = e^{-\frac{5}{2}t} + e^{C_2}$$
$$|5p - 40| = e^{-\frac{5}{2}t} \cdot e^{C_2}$$
$$5p - 40 = \pm e^{C_2} \cdot e^{-\frac{5}{2}t}$$
$$5p - 40 = C_3 e^{-\frac{5}{2}t}$$
$$p = \frac{C_3 e^{-\frac{5}{2}t} + 40}{5} = Ce^{-\frac{5}{2}t} + 8$$

When $t = 0$, $p = 5$, so $5 = Ce^0 + 8$, or $C = -3$ and $p(t) = 8 - 3e^{-\frac{5}{2}t}$

(b) For $S = D$,
$80 + 3p = 120 - 2p$ or $pe = \$8$

(c) $\lim_{t\to\infty}\left(8 - 3e^{-\frac{5}{2}t}\right) = 8 - 0 = 8$

59. $\dfrac{dS}{dt} = \dfrac{aS}{b + cS + S^2}$

$$\int \frac{b + cS + S^2}{aS}\,dS = \int dt$$

$$\int \left(\frac{b}{aS} + \frac{c}{a} + \frac{S}{a}\right) dS = \int dt$$

$$\frac{b}{a}\ln S + \frac{c}{a}S + \frac{S^2}{2a} = t + k$$

61.

$$\frac{dP}{dx} = k\frac{P}{x}$$

$$\int \frac{dP}{P} = \int \frac{k}{x}\,dx$$

$$\ln P = k\ln x + C_1$$

$$P = e^{k\ln x + C_1}$$

$$P = e^{C_1}e^{\ln x^k}$$

$$P(x) = Cx^k$$

8.2 First-Order Linear Differential Equations

1.

$$\frac{dy}{dx} + \left(\frac{3}{x}\right)y = x$$

The integrating factor is

$$I(x) = e^{\int \frac{3}{x}dx} = e^{3\ln x} = e^{\ln x^3} = x^3$$

$$y = \frac{1}{x^3}\left[\int x^3 \cdot x\,dx + C\right]$$

$$y = \frac{1}{x^3}\left[\frac{x^5}{5} + C\right] = \frac{x^2}{5} + \frac{C}{x^3}$$

3.

$$\frac{dy}{dx} + \left(\frac{1}{2x}\right)y = \sqrt{x}e^x$$

The integrating factor is

$$I(x) = e^{\int \frac{1}{2x}dx} = e^{\frac{1}{2}\ln x} = e^{\ln x^{\frac{1}{2}}} = x^{\frac{1}{2}}$$

$$y = \frac{1}{x^{\frac{1}{2}}}\left[\int x^{\frac{1}{2}} \cdot x^{\frac{1}{2}}e^x\,dx + C\right]$$

Using integration by parts with

$$u = x \text{ and } dV = e^x\,dx$$

$$y = \frac{1}{x^{\frac{1}{2}}}\left[xe^x - \int e^x\,dx + C\right]$$

$$y = \frac{1}{x^{\frac{1}{2}}}[xe^x - e^x + C]$$

$$y = x^{\frac{1}{2}}e^x - \frac{e^x}{x^{\frac{1}{2}}} + \frac{C}{x^{\frac{1}{2}}}, \text{ or}$$

$$y = \sqrt{x}e^x - \frac{e^x}{\sqrt{x}} + \frac{C}{\sqrt{x}} = \frac{1}{\sqrt{x}}(xe^x - e^x + C)$$

5. Dividing every term by x^2,

$$\frac{dy}{dx} + \left(\frac{1}{x}\right)y = \frac{2}{x^2}$$

The integrating factor is

$$I(x) = e^{\int \frac{1}{x}dx} = e^{\ln x} = x$$

$$y = \frac{1}{x}\left[\int x \cdot \frac{2}{x^2}\,dx + C\right]$$

$$y = \frac{1}{x}\,[2\ln|x| + C]$$

$$y = \frac{2}{x}\ln|x| + \frac{C}{x}$$

7.

$$\frac{dy}{dx} + \left(\frac{2x+1}{x}\right)y = e^{-2x}$$

The integrating factor is

$$I(x) = e^{\int \frac{2x+1}{x}\,dx} = e^{\int 2+\frac{1}{x}\,dx}$$
$$= e^{2x+\ln x} = e^{2x}\cdot e^{\ln x} = xe^{2x}$$
$$y = \frac{1}{xe^{2x}}\left[\int xe^{2x}\cdot e^{-2x}\,dx + C\right]$$
$$y = \frac{1}{xe^{2x}}\left[\int x\,dx + C\right]$$
$$y = \frac{1}{xe^{2x}}\left[\frac{x^2}{2} + C\right]$$
$$y = \frac{xe^{-2x}}{2} + \frac{Ce^{-2x}}{x} = \frac{x}{2e^{2x}} + \frac{C}{xe^{2x}}$$

9.
$$\frac{dy}{dx} = \frac{1+xy}{1+x}$$
$$(1+x)\frac{dy}{dx} = 1+xy$$
$$(1+x)\frac{dy}{dx} - xy = 1$$
$$\frac{dy}{dx} + \left(-\frac{x}{1+x}\right)y = \frac{1}{1+x}$$

The integrating factor is

$$I(x) = e^{\int -\frac{x}{1+x}\,dx}$$

Using substitution with $u = 1+x$ and $-x = 1-u$,

$$e^{\int -\frac{x}{1+x}\,dx} = e^{\int \frac{1-u}{u}\,du}$$
$$= e^{\int \frac{1}{u}-1\,du}$$
$$= e^{\ln u - u} = e^{\ln(1+x)}\cdot e^{-(1+x)}$$
$$= (1+x)e^{-(1+x)}$$

$$y = \frac{1}{(1+x)e^{-(1+x)}}\left[\int (1+x)e^{-(1+x)}\cdot\frac{1}{1+x}\,dx + C_1\right]$$
$$y = \frac{1}{(1+x)e^{-(1+x)}}\left[\int e^{-(1+x)}dx + C_1\right]$$
$$y = \frac{1}{(1+x)e^{-(1+x)}}\left[-e^{-(1+x)} + C_1\right]$$
$$y = -\frac{1}{1+x} + \frac{C_1}{(1+x)e^{-(1+x)}}$$
$$y = -\frac{1}{1+x} + \frac{C_1e^{1+x}}{1+x}$$
$$y = \frac{C_1e^{1+x}-1}{1+x} = \frac{1}{1+x}(C_1e\cdot e^x - 1)$$
$$= \frac{1}{1+x}(Ce^x - 1)$$

11.
$$\frac{dx}{dt} + \left(\frac{1}{1+t}\right)x = t$$

The integrating factor is

$$I(t) = e^{\int \frac{1}{1+t}\,dt}$$
$$= e^{\ln(1+t)} = 1+t$$
$$x = \frac{1}{1+t}\left[\int (1+t)t\,dt + C\right]$$
$$x = \frac{1}{1+t}\left[\int t+t^2\,dt + C\right]$$
$$x = \frac{1}{1+t}\left[\frac{t^2}{2} + \frac{t^3}{3} + C\right]$$

13.
$$\frac{dy}{dx} + \left(\frac{t}{t+1}\right)y = t+1$$

The integrating factor is

$$I(t) = e^{\int \frac{t}{t+1}\,dt}$$

Using substitution with $u = t+1$ and $t = u-1$,

$$e^{\int \frac{t}{t+1}\,dt} = e^{\int \frac{u-1}{u}\,du} = e^{\int 1-\frac{1}{u}\,du}$$
$$= e^{(1+t)-\ln(1+t)}$$
$$= e^{(1+t)}\cdot e^{-\ln(1+t)} = e^{(1+t)}\cdot e^{\ln\left(\frac{1}{1+t}\right)} = \frac{e^{1+t}}{1+t}$$

$$y = \frac{1+t}{e^{1+t}}\left[\int \frac{e^{1+t}}{1+t} \cdot t + 1\, dt + C_1\right]$$
$$y = \frac{1+t}{e^{1+t}}\left[\int e^{1+t}\, dt + C_1\right]$$
$$y = \frac{1+t}{e^{1+t}}\left[e^{1+t} + C_1\right]$$
$$y = 1 + t + \frac{C_1(1+t)}{e^{1+t}}$$
$$y = 1 + t + C_1(1+t)e^{-(1+t)}$$
$$y = 1 + t + C_1e^{-1}(1+t)e^{-t} = 1 + t + C(1+t)e^{-t}$$
$$= (1+t)(1+Ce^{-t})$$

15. Dividing every term by x,

$$\frac{dy}{dx} + \left(-\frac{2}{x}\right)y = 2x^2$$

The integrating factor is

$$I(x) = e^{\int -\frac{2}{x}\, dx} = e^{-2\ln x}$$
$$= e^{\ln x^{-2}} = \frac{1}{x^2}$$
$$y = x^2\left[\int \frac{1}{x^2} \cdot 2x^2\, dx + C\right]$$
$$y = x^2\left[\int 2\, dx + C\right]$$
$$y = x^2[2x + C] = 2x^3 + Cx^2$$

Since $y = 2$ when $x = 1$,

$$2 = 2 + C, \text{ or } C = 0$$

So,
$y = 2x^3$

17. $\dfrac{dy}{dx} + xy = x + e^{-\frac{x^2}{2}}$
The integrating factor is

$$I(x) = e^{\int x\, dx} = e^{\frac{x^2}{2}}$$
$$y = \frac{1}{e^{\frac{x^2}{2}}}\left[\int e^{\frac{x^2}{2}}\left(x + e^{-\frac{x^2}{2}}\right) dx + C\right]$$
$$y = e^{-\frac{x^2}{2}}\left[\int \left(xe^{\frac{x^2}{2}} + 1\right) dx + C\right]$$

Using substitution with $u = \frac{x^2}{2}$ and $du = x\, dx$,

$$y = e^{-\frac{x^2}{2}}\left[\int e^u\, du + \int 1\, dx + C\right]$$
$$y = e^{-\frac{x^2}{2}}\left[e^{\frac{x^2}{2}} + x + C\right]$$
$$y = 1 + xe^{-\frac{x^2}{2}} + Ce^{-\frac{x^2}{2}}$$

Since $y = -1$ when $x = 0$,

$$-1 = 1 + 0 + Ce^0, \text{ or } C = -2$$

So,

$$y = 1 + xe^{-\frac{x^2}{2}} - 2e^{-\frac{x^2}{2}} = 1 + (x-2)e^{-\frac{x^2}{2}}$$

19. $$\frac{dy}{dx} + \left(\frac{1}{x}\right)y = \frac{1}{x^2}$$

The integrating factor is

$$I(x) = e^{\int \frac{1}{x}\, dx} = e^{\ln x} = x$$
$$y = \frac{1}{x}\left[\int x \cdot \frac{1}{x^2}\, dx + C\right]$$
$$y = \frac{1}{x}\left[\int \frac{1}{x}\, dx + C\right]$$
$$y = \frac{1}{x}[\ln x + C] = \frac{\ln x + C}{x}$$

Since $y = -2$ when $x = 1$,

$$-2 = \ln 1 + C, \text{ or } C = -2$$

So,

$$y = \frac{\ln x - 2}{x}$$

21. (a) Separating the variables,

$$\frac{dy}{dx} + 3y = 5$$
$$\frac{dy}{dx} = 5 - 3y$$
$$\frac{dy}{5-3y} = dx$$
$$\int \frac{dy}{5-3y} = \int dx$$

Using substitution with $u = 5 - 3y$,

$$-\frac{1}{3}\ln|5-3y| = x + C_1$$
$$\ln|5-3y| = -3x - 3C_1$$
$$|5-3y| = e^{-3x-3C_1}$$
$$5 - 3y = \pm e^{-3C_1}\cdot e^{-3x}$$
$$5 - 3y = Ce^{-3x}$$
$$y = \frac{5 - Ce^{-3x}}{3}$$

(b)

$$\frac{dy}{dx} + (3)y = 5$$

The integrating factor is

$$I(x) = e^{\int 3\,dx} = e^{3x}$$
$$y = e^{-3x}\left[\int e^{3x}\cdot 5\,dx + C\right]$$
$$y = e^{-3x}\left[\frac{5}{3}e^{3x} + C\right] = \frac{5}{3} + Ce^{-3x}$$
$$y = \frac{5 + Ce^{-3x}}{3}$$

Note: the constants in (a) and (b) will have opposite signs.

23. (a) Separating the variables,

$$\frac{dy}{dx} + \frac{y}{x+1} = \frac{2}{x+1}$$
$$(x+1)\frac{dy}{dx} + y = 2$$
$$(x+1)\frac{dy}{dx} = 2 - y$$
$$\frac{dy}{2-y} = \frac{dx}{x+1}$$
$$\int\frac{dy}{2-y} = \int\frac{dx}{x+1}$$

$$-\ln|2-y| = \ln|x+1| + C_1$$
$$-C_1 = \ln|x+1| + \ln|2-y|$$
$$-C_1 = \ln|(x+1)(2-y)|$$
$$e^{-C_1} = |(x+1)(2-y)|$$
$$\pm e^{-C_1} = (x+1)(2-y)$$
$$C = (x+1)(2-y)$$
$$\frac{C}{x+1} = 2 - y$$
$$y = 2 - \frac{C}{x+1}$$

(b)

$$\frac{dy}{dx} + \left(\frac{1}{x+1}\right)y = \frac{2}{x+1}$$

The integrating factor is

$$I(x) = e^{\int\frac{1}{x+1}dx} = e^{\ln(x+1)} = x + 1$$
$$y = \frac{1}{x+1}\left[\int(x+1)\cdot\frac{2}{x+1}\,dx + C_1\right]$$
$$y = \frac{1}{x+1}\left[\int 2\,dx + C_1\right]$$
$$y = \frac{1}{x+1}[2x + C_1] = \frac{2x + C_1}{x+1}$$

Letting $C_1 = 2 - C$,

$$y = \frac{2x + 2 - C}{x+1} = \frac{2(x+1) - C}{x+1} = 2 - \frac{C}{x+1}$$

25. Since $\frac{dy}{dx}$ = slope of lines tangent to the graph of $y = f(x)$,

$$\frac{dy}{dx} = x + y$$
$$\frac{dy}{dx} + (-1)y = x$$

The integrating factor is

$$I(x) = e^{\int -1\,dx} = e^{-x}$$
$$y = e^{x}\left[\int e^{-x}\cdot x\,dx + C\right]$$

Using integration by parts with

$$u = x \text{ and } dV = e^{-x}\,dx$$

$$y = e^x\left[-xe^{-x} - \int -e^{-x}\,dx + C\right]$$
$$y = e^x\left[-xe^{-x} - e^{-x} + C\right]$$
$$y = -x - 1 + Ce^x$$

Since $y = 2$ when $x = -1$,

$$2 = 1 - 1 + Ce^{-1}, \text{ or } C = 2e$$

So,

$$y = -x - 1 + 2e^{1+x}$$

27. Since $\frac{dp}{dt}$ = rate of change,

$$\frac{dp}{dt} = 0.01p + 1000t$$
$$\frac{dp}{dt} + (-0.01)p = 1000t$$

The integrating factor is

$$I(t) = e^{\int -0.01\,dt} = e^{-0.01t}$$
$$p = e^{0.01t}\left[\int e^{-0.01t}\cdot 1000t\,dt + C\right]$$

Using integration by parts with

$$u = t \text{ and } dV = e^{-0.01t}\,dt$$
$$p = e^{0.01t}\left[1000\left(-100te^{-0.01t} - \int -100e^{-0.01t}\,dt\right) + C\right]$$
$$p = e^{0.01t}\left[1000\left(-100te^{-0.01t} - 10{,}000e^{-0.01t}\right) + C\right]$$
$$p(t) = -100{,}000t - 10{,}000{,}000 + Ce^{0.01t}$$

Since $p = 200{,}000$ when $t = 0$,

$$200{,}000 = 0 - 10{,}000{,}000 + C, \text{ or } C = 10{,}200{,}000$$

So,

$$p(t) = -100{,}000t - 10{,}000{,}000 + 10{,}200{,}000e^{0.01t}$$

Nine months from now,

$$p(9) = -100{,}000(9) - 10{,}000{,}000 + 10{,}200{,}000e^{0.01(9)}$$

or approximately $260,578.

29. Annually, the interest earned by the balance V in the account is rV. The other source of change is the annual deposit D. So,

$$\frac{dV}{dt} = rV + D$$

(a)
$$\frac{dV}{dt} = rV + D$$
$$\frac{dV}{dt} + (-r)V = D$$

The integrating factor is

$$I(t) = e^{\int -r\,dt} = e^{-rt}$$
$$V = e^{rt}\left[\int e^{-rt}\cdot D\,dt + C\right]$$
$$V = e^{rt}\left[-\frac{D}{r}e^{-rt} + C\right]$$
$$V(t) = -\frac{D}{r} + Ce^{rt}$$

When $t = 0$, $V(0) = 0$ and

$$0 = -\frac{D}{r} + C, \text{ or } C = \frac{D}{r}$$

So,

$$V(t) = -\frac{D}{r} + \frac{D}{r}e^{rt} = \frac{D}{r}(e^{rt} - 1)$$

(b) With $D = 8{,}000$, $r = 0.04$, and $t = 20$

$$V(20) = \frac{8{,}000}{0.04}\left(e^{0.04(20)} - 1\right)$$

or approximately $245,108.

(c) If $V = 800{,}000$, $r = 0.05$, and $t = 30$

$$800{,}000 = \frac{D}{0.05}\left(e^{0.05(30)} - 1\right)$$
$$800{,}000 = 20D(e^{1.5} - 1), \text{ or}$$
$$D \approx 11{,}489$$

So, he should deposit approximately $11,489 per year.

31. Using the result of part (a) in problem 29:

$$V(t) = -\frac{D}{r} + Ce^{rt}$$

When $t = 0$, $V(0) = A$ the amount of his parent's deposit, so

$$A = -\frac{800}{0.04} + Ce^{0}$$
$$C = A + 20{,}000$$

So,

$$V(t) = -20{,}000 + (A + 20{,}000)e^{0.04t}$$
$$V(4) = -20{,}000 + Ae^{0.16} + 20{,}000e^{0.16}$$

The balance after 4 years must be \$8,000 so

$$8{,}000 = -20{,}000 + Ae^{0.16} + 20{,}000e^{0.16}$$

or, $A \approx \$3{,}860.03$.

33. (a)

$$\frac{dS}{dt} = \begin{bmatrix}\text{rate salt}\\ \text{enters}\end{bmatrix} - \begin{bmatrix}\text{rate salt}\\ \text{leaves}\end{bmatrix}$$
$$\frac{dS}{dt} = \begin{pmatrix}\text{amt salt}\\ \text{per gallon}\end{pmatrix}\begin{pmatrix}\text{rate gallons}\\ \text{enter}\end{pmatrix} - \begin{pmatrix}\text{amt salt}\\ \text{per gallon}\end{pmatrix}\begin{pmatrix}\text{rate gallons}\\ \text{leave}\end{pmatrix}$$
$$\frac{dS}{dt} = 0 - 3\begin{pmatrix}\text{amt salt}\\ \text{per gallon}\end{pmatrix}$$

Since $S(t)$ is the total amount of salt in the tank, the tank originally holds 40 gallons, and the tank loses a net 2 gal/min,

$$\begin{matrix}\text{amt salt}\\ \text{per gallon}\end{matrix} = \frac{S(t)}{40 - 2t}$$

So,
$$\frac{dS}{dt} = -\frac{3S}{40 - 2t}$$
$$\frac{dS}{dt} + \left(\frac{3}{40 - 2t}\right)S = 0$$

The integrating factor is

$$I(t) = e^{\int \frac{3}{40-2t}\,dt}$$
$$= e^{-\frac{3}{2}\ln(40-2t)} = (40 - 2t)^{-\frac{3}{2}}$$
$$S(t) = (40 - 2t)^{\frac{3}{2}}[0 + C] = (40 - 2t)^{\frac{3}{2}}C$$

Initially, there are 5 pounds of salt in the tank, so

$$5 = (40)^{\frac{3}{2}}C, \quad \text{or } C \approx 0.0197642$$

and

$$S(t) = 0.0197642(40 - 2t)^{\frac{3}{2}}$$
$$= 0.05590(20 - t)^{\frac{3}{2}}$$

(b) For $S = 4$,

$$4 = 0.05590(20 - t)^{\frac{3}{2}}$$
$$(71.5542)^{\frac{2}{3}} = 20 - t, \quad \text{or}$$
$$t \approx 2.76 \text{ minutes}$$

(c) Since the tank originally holds 40 gallons and is losing a net 2 gal/min, it takes 20 minutes to drain the tank. Choosing $t = 19.99$ minutes,

$$S = 0.05590(20 - 19.99)^{\frac{3}{2}}$$
$$S \approx 0.000056 \text{ lbs}$$

Or, essentially no salt remains.

35. (a) Let $g(t)$ be the amount of poisonous gas in the closet after t minutes.

$$\frac{dg}{dt} = \begin{pmatrix}\text{rate gas}\\ \text{enters}\end{pmatrix} - \begin{pmatrix}\text{rate gas}\\ \text{leaves}\end{pmatrix}$$
$$\frac{dg}{dt} = \begin{pmatrix}\text{amt gas}\\ \text{per ft}^3\end{pmatrix}\begin{pmatrix}\text{rate gas}\\ \text{enters}\end{pmatrix} - \begin{pmatrix}\text{amt gas}\\ \text{per ft}^3\end{pmatrix}\begin{pmatrix}\text{rate gas}\\ \text{leaves}\end{pmatrix}$$
$$\frac{dg}{dt} = (1.00)(0.2) - \left(\frac{g(t)}{2{,}000}\right)(0.2)$$
$$\frac{dg}{dt} + \left(\frac{1}{10{,}000}\right)g = 0.2$$

The integrating factor is

$$I(t) = e^{\int \frac{1}{10{,}000}\,dt} = e^{\frac{t}{10{,}000}}$$
$$g = e^{-\frac{t}{10{,}000}}\left[\int e^{\frac{t}{10{,}000}} \cdot 0.2dt + C\right]$$
$$g = e^{-\frac{t}{10{,}000}}\left[2{,}000e^{\frac{t}{10{,}000}} + C\right]$$
$$g(t) = 2{,}000 + Ce^{-\frac{t}{10{,}000}}$$

When $t = 0$, there is no poisonous gas in the room, so
$0 = 2{,}000 + Ce^0$, or $C = -2{,}000$ and
$g(t) = 2{,}000 - 2{,}000e^{-\frac{t}{10{,}000}}$

(b) The air becomes hazardous when

$$g(t) = (0.01)(2{,}000) = 20$$
$$20 = 2{,}000 - 2{,}000e^{-\frac{t}{10{,}000}}$$
$$e^{-\frac{t}{10{,}000}} = 0.99$$
$$-\frac{t}{10{,}000} = \ln 0.99$$
$$t \approx 100.5 \text{ minutes}$$

37. Let $A(t)$ be the amount of money in the account at time t.

$$\begin{matrix}\text{net rate of} \\ \text{change of } A\end{matrix} = \begin{pmatrix}\text{rate interest} \\ \text{is added}\end{pmatrix} - \begin{pmatrix}\text{rate money} \\ \text{is withdrawn}\end{pmatrix}$$
$$\frac{dA}{dt} = 0.05A - 5{,}000$$
$$\frac{dA}{dt} + (-0.05)A = -5{,}000$$

The integrating factor is

$$I(t) = e^{\int -0.05\,dt} = e^{-0.05t}$$
$$A = e^{0.05t}\left[\int e^{-0.05t} \cdot -5{,}000dt + C\right]$$
$$A = e^{0.05t}[100{,}000e^{-0.05t} + C]$$
$$A(t) = 100{,}000 + Ce^{0.05t}$$

When $t = 0$, the amount is his initial investment, A_I.

$$A_I = 100{,}000 + C, \text{ or } C = A_I - 100{,}000$$
$$A(t) = 100{,}000 + (A_I - 100{,}000)e^{0.05t}$$

After ten years, the account will be depleted, so

$$0 = 100{,}000 + (A_I - 100{,}000)e^{0.05(10)}$$
$$A_I e^{0.5} = 100{,}000e^{0.5} - 100{,}000$$
$$A_I = 100{,}000 - 100{,}000e^{-0.5}$$
$$A_I \approx 39{,}347$$

So, he should initially deposit approximately \$39,347.

39. **(a)**
$$\frac{dA}{dt} = \begin{pmatrix}\text{rate drug} \\ \text{enters}\end{pmatrix} - \begin{pmatrix}\text{rate drug} \\ \text{leaves}\end{pmatrix}$$
$$= \begin{pmatrix}\text{amt drug} \\ \text{per mg}\end{pmatrix}\begin{pmatrix}\text{rate mg} \\ \text{enter}\end{pmatrix} - \begin{pmatrix}\text{rate drug} \\ \text{leaves}\end{pmatrix}$$
$$= (1.00)(3) - kA$$
$$\frac{dA}{dt} = 3 - kA$$

where k is a constant of proportionality.

$$\frac{dA}{dt} + kA = 3$$

The integrating factor is

$$I(t) = e^{\int k\,dt} = e^{kt}$$
$$A = e^{-kt}\left[\int e^{kt} \cdot 3\,dt + C\right]$$
$$A = e^{-kt}\left[\frac{3}{k}e^{kt} + C\right]$$
$$A(t) = \frac{3}{k} + Ce^{-kt}$$

When $t = 0$, $A(0) = 0$ so

$$0 = \frac{3}{k} + C, \text{ or } C = -\frac{3}{k}$$
$$A(t) = \frac{3}{k} - \frac{3}{k}e^{-kt}$$

(b) When $t = 1$, $A(1) = 2.3$ so

$$2.3 = \frac{3}{k} - \frac{3}{k}e^{-k}$$
$$\frac{3}{k} - \frac{3}{k}e^{-k} - 2.3 = 0$$

Press [y=]. We will use the variable x for k on the calculator.
Enter $(3/x) - (3 * (e \wedge -x)/x) - 2.3$ for $y_1 =$.
Use window dimensions $[-5, 5]1$ by $[-5, 5]1$ and press [Graph].
Use the zero function under the calc menu.
Enter $x = 0$ for the left bound and $x = 1$ for the right bound and $x = 0.5$ for the guess. We find that $k \approx 0.5572$.
When $t = 8$

$$A(8) = \frac{3}{0.5572} - \frac{3}{0.5572}e^{-8(0.5572)}$$
$$A(8) \approx 5.32 \text{ mg}$$

So, after eight hours, there will be approximately 5.32 mg of the drug in the patient's bloodstream.

$$\text{As } t \to \infty,\ A(t) \to \frac{3}{k}, \text{ or } 5.384 \text{ mg.}$$

(c) For a steady-state level of 9 mg, need

$$\frac{\text{rate infused}}{k} = 9$$

or an infusion rate of approximately 5.015 mg/hr.

41. (a) Since $S = D$

$$34 + 3p + 2\frac{dp}{dt} = 25 - 2p + 5\frac{dp}{dt}$$
$$\frac{dp}{dt} = \frac{5p+9}{3}$$
$$\int \frac{dp}{5p+9} = \int \frac{1}{3}dt$$

Using substitution with $u = 5p + 9$ and $\frac{1}{5}du = dp$

$$\frac{1}{5}\ln|5p+9| = \frac{1}{3}t + C_1$$
$$\ln|5p+9| = \frac{5}{3}t + C_2$$
$$|5p+9| = e^{\frac{5}{3}t + C_2}$$
$$|5p+9| = e^{\frac{5}{3}t} \cdot e^{C_2}$$
$$5p + 9 = \pm e^{C_2} \cdot e^{\frac{5}{3}t}$$
$$5p + 9 = C_3 e^{\frac{5}{3}t}$$
$$p = \frac{C_3 e^{\frac{5}{3}t} - 9}{5} = Ce^{\frac{5}{3}t} - \frac{9}{5}$$

When $t = 0$, $p = 8$, so $8 = Ce^0 - \frac{9}{5}$, or $C = \frac{49}{5}$

and
$$p(t) = \frac{49}{5}e^{\frac{5}{3}t} - \frac{9}{5}$$

(b) $\lim_{t\to\infty} \frac{49}{5}e^{\frac{5}{3}t} - \frac{9}{5} = \infty$, so the pricing of this commodity is unstable.

8.3 Additional Applications of Differential Equations

1. $xy^2 = C$
Differentiating implicitly,

$$x \cdot 2y\frac{dy}{dx} + y^2 \cdot 1 = 0$$
$$\frac{dy}{dx} = -\frac{y^2}{2xy} = -\frac{y}{2x}$$

where $\frac{dy}{dx}$ represents the slope at any given point on any curve of the form $xy^2 = C$. The slope of any orthogonal trajectory curve is the negative reciprocal, or

$$\frac{dY}{dX} = \frac{2x}{y} = \frac{2X}{Y}$$

Since $x = X$ and $y = Y$ where the orthogonal trajectory curve intersects the original curve.

$$Y\,dY = 2X\,dx$$
$$\int Y\,dY = \int 2X\,dx$$
$$\frac{1}{2}Y^2 = X^2 + C_1, \text{ or}$$
$$Y^2 - 2X^2 = C$$

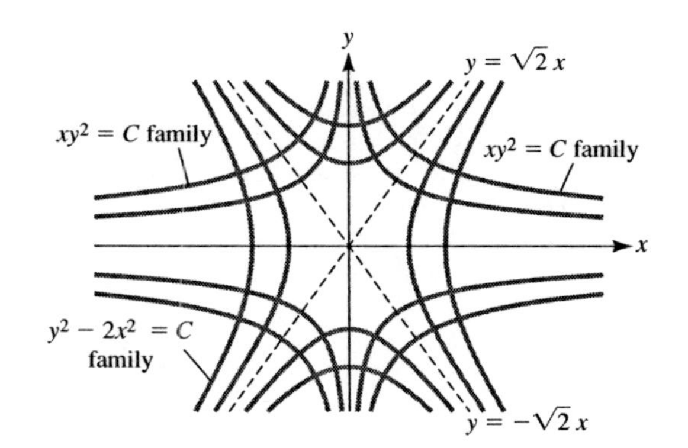

3. $y = Cx^2$
Differentiating,

$$\frac{dy}{dx} = 2Cx$$

where $\frac{dy}{dx}$ represents the slope at any given point on any curve of the form $y = Cx^2$.
Since $y = Cx^2$, $C = \frac{y}{x^2}$ and

$$\frac{dy}{dx} = \frac{2y}{x}$$

The slope of any orthogonal trajectory curve is the negative reciprocal, or

$$\frac{dY}{dX} = -\frac{x}{2y} = -\frac{X}{2Y}$$

since $x = X$ and $y = Y$ where the orthogonal trajectory curve intersects the original curve.

$$\int 2Y\,dY = \int -X\,dX$$

$$Y^2 = -\frac{1}{2}X^2 + C_1$$

$$X^2 + 2Y^2 = K$$

where $K = 2C_1$

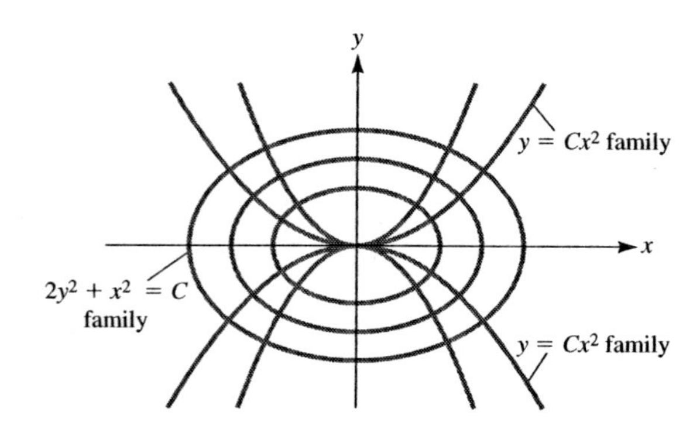

5. $S(t) = 2 + 3p(t);\; D(t) = 10 - p(t)$

(a)

$$\frac{dp}{dt} = k[D(t) - S(t)]$$

$$\frac{dp}{dt} = 0.02[8 - 4p]$$

$$\frac{dp}{dt} + (0.08)p = 0.16$$

The integrating factor is

$$I(t) = e^{\int 0.08\,dt} = e^{0.08t}$$

$$p = e^{-0.08t}\left[\int e^{0.08t} \cdot 0.16\,dt + C\right]$$

$$p = e^{-0.08t}[2e^{0.08t} + C]$$

$$p(t) = 2 + Ce^{-0.08t}$$

When $t = 0$, $p_0 = 1$ and

$$1 = 2 + Ce^0, \text{ or } C = -1$$

So,

$$p(t) = 2 - e^{-0.08t}$$

Note: The integration may also be done using separation of variables.

(b) $p(4) = 2 - e^{-0.08(4)} \approx \$1.27$ per unit

(c) $\lim_{t\to\infty} p(t) = \lim_{t\to\infty} (2 - e^{-0.08t}) = 2$

The price levels to approximately \$2 per unit.

7. $S(t) = 2 + p(t),\; D(p) = 3 + 7e^{-t}$

(a)

$$\frac{dp}{dt} = k[D(t) - S(t)]$$

$$\frac{dp}{dt} = 0.02[1 + 7e^{-t} - p]$$

$$\frac{dp}{dt} + (0.02)p = 0.02 + 0.14e^{-t}$$

The integrating factor is

$$I(t) = e^{\int 0.02\,dt} = e^{0.02t}$$

$$p = e^{-0.02t}\left[\int e^{0.02t}(0.02 + 0.14e^{-t})\,dt + C\right]$$

$$p = e^{-0.02t}\left[\int (0.02e^{0.02t} + 0.14e^{-0.98t})\,dt + C\right]$$

$$p = e^{-0.02t}\left[e^{0.02t} - \frac{1}{7}e^{-0.98t} + C\right]$$

$$p(t) = 1 - \frac{1}{7}e^{-t} + Ce^{-0.02t}$$

When $t = 0$, $p_0 = 4$ and

$$4 = 1 - \frac{1}{7}e^{0} + Ce^{0}, \text{ or } C = \frac{22}{7}$$

So,

$$p(t) = 1 - \frac{1}{7}e^{-t} + \frac{22}{7}e^{-0.02t}$$

(b)
$$p(4) = 1 - \frac{1}{7}e^{-4} + \frac{22}{7}e^{-0.02(4)}$$
$$\approx \$3.90 \text{ per unit}$$

(c)
$$\lim_{t\to\infty} p(t) = \lim_{t\to\infty}\left(1 - \frac{1}{7}e^{-t} + \frac{22}{7}e^{-0.02t}\right) = 1$$

The price levels to approximately $1.00 per unit.

9. $\frac{dQ}{dt} = kQ(t)$, where k is a constant of proportionality.

Separating the variables,

$$\frac{dQ}{Q} = k\,dt$$

$$\int \frac{dQ}{Q} = \int k\,dt$$

$$\ln Q = kt + C_1$$

$$Q = e^{kt+C_1}$$

$$Q = e^{C_1}\cdot e^{kt}$$

$$Q(t) = Ce^{kt}$$

When $t = 0$, $Q(0) = Q_0$ so

$$Q(t) = Q_0e^{kt}$$

When $t = 3$, $Q(3) = \frac{1}{2}Q_0$ and

$$\frac{1}{2}Q_0 = Q_0e^{3k}$$

$$\frac{1}{2} = e^{3k}$$

$$\ln 0.5 = 3k$$

$$\frac{\ln 0.5}{3} = k$$

So,

$$Q(t) = Q_0e^{\frac{\ln 0.5}{3}t}$$

When $Q(t) = \frac{1}{4}Q_0$,

$$\frac{1}{4}Q_0 = Q_0e^{\frac{\ln 0.5}{3}t}$$

$$\frac{1}{4} = e^{\frac{\ln 0.5}{3}t}$$

$$\ln 0.25 = \frac{\ln 0.5}{3}t$$

$$t = \frac{3\ln 0.25}{\ln 0.5} = 6$$

It takes six minutes for $\frac{3}{4}$ of the sugar to dissolve. Note: Since it takes 3 minutes for half of the original amount to dissolve, it takes three more minutes for half of the remaining half to dissolve. Thus, it takes six total minutes for $\frac{3}{4}$ of the sugar to dissolve.

11. Using Newton's law of cooling,

$$\frac{dT}{dt} = k(T - T_m)$$

$$\frac{dT}{dt} = k(T + S)$$

Separating the variables,

$$\frac{dT}{T+5} = k\,dt$$

$$\int \frac{dT}{T+5} = \int k\,dt$$

$$\ln(T+5) = kt + C_1$$

$$T + 5 = e^{kt+C_1}$$

$$T + 5 = e^{C_1}\cdot e^{kt}$$

$$T(t) = Ce^{kt} - 5$$

When $t = 0$, $T(0) = T_0$ and

$$T_0 = Ce^0 - 5, \text{ or } C = T_0 + 5$$

and

$$T(t) = (T_0 + 5)e^{kt} - 5$$

When $t = 10$, $T(10) = 70$ and

$$70 = (T_0 + 5)e^{10k} - 5$$

When $t = 20$, $T(20) = 50$ and

$$50 = (T_0 + 5)e^{20k} - 5$$

Noting that $e^{20k} = (e^{10k})^2$, solve the first equation for e^{10k} and substitute into the second equation.

$$e^{10k} = \frac{75}{(T_0 + 5)}$$

$$50 = (T_0 + 5)\left(\frac{75}{T_0 + 5}\right)^2 - 5$$

$$55 = \frac{5{,}625}{T_0 + 5}, \text{ or } T_0 \approx 97.27°\text{C}$$

13. Using Newton's law of cooling,

$$\frac{dT}{dt} = k(T - T_m)$$

$$\frac{dT}{dt} = k(T - 50)$$

Separating the variables,

$$\frac{dT}{T - 50} = k\,dt$$

$$\int \frac{dT}{T - 50} = \int k\,dt$$

$$\ln|T - 50| = kt + C_1$$

$$|T - 50| = e^{kt + C_1}$$

$$T - 50 = \pm e^{C_1} \cdot e^{kt}$$

$$T(t) = Ce^{kt} + 50$$

At the time of death, $t = 0$, the body's temperature was 98.6°F and

$$98.6 = Ce^0 + 50, \text{ or } C = 48.6$$

and $T(t) = 48.6e^{kt} + 50$.

Let t_0 be the time when the body is found. Then,

$$80 = 48.6e^{kt_0} + 50$$

Also, when $t = t_0 + \frac{1}{3}$,

$$78 = 48.6e^{k(t_0 + \frac{1}{3})} + 50$$

$$78 = 48.6e^{\frac{k}{3}}e^{kt_0} + 50$$

Solving the first equation for $48.6e^{kt_0}$ and substituting into the second equation,

$$78 = e^{\frac{k}{3}}(30) + 50$$

$$\frac{28}{30} = e^{\frac{k}{3}}$$

$$\ln\frac{14}{15} = \frac{k}{3}, \text{ or } k = 3\ln\frac{14}{15}$$

Now,

$$30 = 48.6e^{kt_0}$$

$$\frac{30}{48.6} = e^{(3\ln\frac{14}{15})t_0}$$

$$\ln\frac{30}{48.6} = \left(3\ln\frac{14}{15}\right)t_0$$

$$t_0 = \frac{\ln\frac{30}{48.6}}{3\ln\frac{14}{15}} \approx 2.33$$

When the body was found at 3:00 p.m., approximately 2 hours and 20 minutes had passed since death. So, the time of death was 12:40 p.m.

15.

$$\frac{dA}{dt} = \begin{bmatrix}\text{rate drug}\\\text{enters}\end{bmatrix} - \begin{bmatrix}\text{rate drug}\\\text{leaves}\end{bmatrix}$$

$$\frac{dA}{dt} = \begin{pmatrix}\text{amt drug}\\\text{per cc}\end{pmatrix}\begin{pmatrix}\text{rate cc's}\\\text{enter}\end{pmatrix} - \begin{pmatrix}\text{amt drug}\\\text{per cc}\end{pmatrix}\begin{pmatrix}\text{rate cc's}\\\text{leave}\end{pmatrix}$$

$$\frac{dA}{dt} = (.16)(6) - \frac{A(t)}{200}(6)$$

$$\frac{dA}{dt} + \frac{3}{100}A = 0.96$$

The integrating factor is

$$I(t) = e^{\int \frac{3}{100}\,dt} = e^{0.03t}$$

$$A = e^{-0.03t}\left[\int e^{0.03t}(0.96)\,dt + C\right]$$

$$A = e^{-0.03t}\left[0.96\left(\frac{1}{0.03}e^{0.03t}\right) + C\right]$$

$$A(t) = 32 + Ce^{-0.03t}$$

When $t = 0$, $A(0) = 0$ and

$$0 = 32 + Ce^{0}, \text{ or } C = -32$$

and

$$A(t) = 32 - 32e^{-0.03t} = 32(1 - e^{-0.03t})$$

17.

$$\frac{da}{dt} = \begin{bmatrix}\text{rate drug}\\\text{enters}\end{bmatrix} - \begin{bmatrix}\text{rate drug}\\\text{disbursed}\end{bmatrix}$$

$$= \begin{pmatrix}\text{amt drug}\\\text{per cc}\end{pmatrix}\begin{pmatrix}\text{rate cc's}\\\text{enter}\end{pmatrix} - \begin{pmatrix}\text{amt drug}\\\text{per cc}\end{pmatrix}\begin{pmatrix}\text{rate cc's}\\\text{disbursed}\end{pmatrix}$$

$$\frac{da}{dt} = (c_0)(A) - \left(\frac{a(t)}{V}\right)(A)$$

$$\frac{da}{dt} + \left(\frac{A}{V}\right)a = Ac_0$$

The integrating factor is

$$I(t) = e^{\int \frac{A}{V}\,dt} = e^{\frac{A}{V}t}$$

$$a = e^{-\frac{A}{V}t}\left[\int e^{\frac{A}{V}t}\cdot Ac_0 + C\right]$$

$$a = e^{-\frac{A}{V}t}\left[Vc_0e^{\frac{A}{V}t} + C\right]$$

$$a(t) = Vc_0 + Ce^{-\frac{A}{V}t}$$

When $t = 0$, there is no drug in the organ, so $0 = Vc_0 + Ce^{0}$, or $C = -Vc_0$
and
$a(t) = Vc_0 - Vc_0e^{-\frac{A}{V}t}$
Since a is the amount of drug in the organ, $\frac{a}{V}$ is the concentration of the drug. To find t when the concentration is L,

$$\frac{a}{V} = c_0 - c_0e^{-\frac{A}{V}t}$$

$$L = c_0 - c_0e^{-\frac{A}{V}t}$$

$$\frac{L}{c_0} = 1 - e^{-\frac{A}{V}t}$$

$$e^{-\frac{A}{V}t} = 1 - \frac{L}{c_0}$$

$$-\frac{A}{V}t = \ln\left(1 - \frac{L}{c_0}\right)$$

$$t = -\frac{V}{A}\ln\left(1 - \frac{L}{c_0}\right) \text{ seconds}$$

19. (a)

$$\frac{dN}{dt} = \begin{bmatrix}\text{rate nutrient}\\\text{enters}\end{bmatrix} - \begin{bmatrix}\text{rate nutrient}\\\text{leaves}\end{bmatrix}$$

$$\frac{dN}{dt} = \begin{pmatrix}\text{amt nutrient}\\\text{per gallon}\end{pmatrix}\begin{pmatrix}\text{rate gallons}\\\text{enter}\end{pmatrix} - \begin{pmatrix}\text{amt nutrient}\\\text{per gallon}\end{pmatrix}\begin{pmatrix}\text{rate gallons}\\\text{leave}\end{pmatrix}$$

$$\frac{dN}{dt} = \left(\frac{0.2te^{-\frac{t}{40}}}{3{,}000}\right)(75) - \left(\frac{N}{3{,}000}\right)(75)$$

$$\frac{dN}{dt} + \left(\frac{1}{40}\right)N = 0.005te^{-\frac{t}{40}}$$

The integrating factor is

$$I(t) = e^{\int \frac{1}{40}\,dt} = e^{\frac{t}{40}}$$

$$N = e^{-\frac{t}{40}}\left[\int e^{\frac{t}{40}}\cdot 0.005te^{-\frac{t}{40}}\,dt + C\right]$$

$$N = e^{-\frac{t}{40}}\left[\int 0.005t\,dt + C\right]$$

$$N(t) = e^{-\frac{t}{40}}\left[\frac{0.005}{2}t^2 + C\right]$$

When $t = 0$, $N(0) = 0$ and

$$0 = e^{0}[0 + C], \text{ or } C = 0$$

So, $\quad N(t) = 0.0025t^2e^{-\frac{t}{40}}$.

(b) The maximum concentration occurs when $N'(t) = 0$, or

$$N'(t) = \left(\frac{15}{2}t^2\right)\left(-\frac{1}{40}e^{-\frac{t}{40}}\right) + (e^{-\frac{t}{40}})(15t)$$

$$0 = -\frac{3}{16}t^2e^{-\frac{t}{40}} + 15te^{-\frac{t}{40}}$$

$$0 = te^{-\frac{t}{40}}\left(-\frac{3}{16}t + 15\right)$$

$$t = 80 \text{ (rejecting } t = 0, \text{ when } N(0) = 0)$$

When $0 \le t < 80$, $N'(t) > 0$. When $t > 80$, $N'(t) < 0$. Therefore $\frac{N(80)}{3{,}000}$ is the maximum concentration.

$$N(80) = 0.0025(80)^2e^{-\frac{80}{40}} = \frac{16}{e^2}$$

The maximum concentration $= \dfrac{16}{3{,}000e^2}$

≈ 0.000722 ounces per gallon

21.

$$\frac{dP}{dt} = rP - E(t)$$

$$\frac{dP}{dt} = 0.03P - 10t$$

$$\frac{dP}{dt} + (-0.03)P = -10t$$

The integrating factor is

$$I(t) = e^{\int -0.03\,dt} = e^{-0.03t}$$

$$P = e^{0.03t}\left[\int e^{-0.03t}(-10t)\,dt + C\right]$$

Using integration by parts with

$$u = -10t \text{ and } dV = e^{-0.03t}\,dt$$

$$P = e^{0.03t}\left[\frac{1{,}000}{3}te^{-0.03t} + \frac{100{,}000}{9}e^{-0.03t} + C\right]$$

$$P(t) = \frac{1{,}000}{3}t + \frac{100{,}000}{9} + Ce^{0.03t}$$

When $t = 0$, $P(0) = 100{,}000$ and

$$100{,}000 = 0 + \frac{100{,}000}{9} + Ce^0,$$

or $C = \dfrac{800{,}000}{9}$ and

$$P(t) = \frac{1{,}000}{3}t + \frac{100{,}000}{9} + \frac{800{,}000}{9}e^{0.03t}$$

23.

$$\frac{dP}{dt} = rP + I(t)$$

$$\frac{dP}{dt} = 0.02P + 100e^{-t}$$

$$\frac{dP}{dt} + (-0.02)P = 100e^{-t}$$

The integrating factor is

$$I(t) = e^{\int -0.02\,dt} = e^{-0.02t}$$

$$P = e^{0.02t}\left[\int e^{-0.02t}(100e^{-t})\,dt + C\right]$$

$$P = e^{0.02t}\left[\int 100e^{-1.02t}\,dt + C\right]$$

$$P = e^{0.02t}\left[-\frac{100}{1.02}e^{-1.02t} + C\right]$$

$$P(t) = -\frac{100}{1.02}e^{-t} + Ce^{0.02t}$$

When $t = 0$, $P(0) = 300{,}000$ and

$$300{,}000 = -\frac{100}{1.02}e^0 + Ce^0,$$

or $C \approx 300{,}098$ and

$$P(t) \approx -98.04e^{-t} + 300{,}098e^{0.02t}$$

25. $\dfrac{dN}{dt} = kN(60 - N)$

$$\int \frac{dN}{N(60-N)} = \int k\,dt$$

Using formula 6 in the integration table,

$$\frac{1}{60}\ln\left|\frac{N}{60-N}\right| = kt + C_1$$
$$\ln\left|\frac{N}{60-N}\right| = 60kt + C_2$$
$$\left|\frac{N}{60-N}\right| = e^{60kt+C_2}$$
$$\left|\frac{N}{60-N}\right| = e^{60kt}\cdot e^{c_2}$$
$$\frac{N}{60-N} = \pm e^{c_2}\cdot e^{60kt}$$
$$\frac{N}{60-N} = C_3e^{60kt}$$
$$N = (60C_3 - C_3N)e^{60kt}$$
$$Ne^{-60kt} = 60C_3 - C_3N$$
$$N\left(e^{-60kt} + C_3\right) = 60C_3$$
$$N = \frac{60C_3}{C_3 + e^{-60kt}} = \frac{60}{1+\frac{1}{C_3}e^{-60kt}}$$
$$N(t) = \frac{60}{1+Ce^{-60kt}}$$

When $t = 0$, $N(0) = 2$, so
$2 = \frac{60}{1+Ce^0}$, or $C = 29$
and
$N(t) = \frac{60}{1+29e^{-60kt}}$
When $t = 1$, $N(1) = 3$, so

$$3 = \frac{60}{1+29e^{-60k}}$$
$$1 + 29e^{-60k} = 20$$
$$e^{-60k} = \frac{19}{29}$$
$$-60k = \ln\frac{19}{29}$$
$$k = \frac{\ln\frac{19}{29}}{-60} \approx 0.00705$$

and

$$N(t) = \frac{60}{1+29e^{-0.423t}}$$

Twenty people will know his identity when

$$20 = \frac{60}{1+29e^{-0.423t}}$$
$$1 + 29e^{0.423t} = 3$$
$$e^{0.423t} = \frac{2}{29}$$
$$-0.423t = \ln\frac{2}{29}$$
$$t = \frac{\ln\frac{2}{29}}{-0.423} \approx 6.3 \text{ days}$$

The spy is a plucked duck.

27. (a)
$$\frac{da}{dt} = \left[\begin{pmatrix}\text{rate drug}\\ \text{enters}\end{pmatrix} - \begin{pmatrix}\text{rate drug}\\ \text{disbursed}\end{pmatrix}\right]$$
$$= \begin{pmatrix}\text{amt drug}\\ \text{per unit}\end{pmatrix}\begin{pmatrix}\text{rate units}\\ \text{enter}\end{pmatrix} - \begin{pmatrix}\text{amt drug}\\ \text{per unit}\end{pmatrix}\begin{pmatrix}\text{rate units}\\ \text{disbursed}\end{pmatrix}$$
$$\frac{da}{dt} = (1.00)(5) - \left(\frac{a(t)}{800+2t}\right)(4)$$
$$\frac{da}{dt} + \left(\frac{4}{800+2t}\right)a = 5$$

The integrating factor is
$I(t) = e^{\int \frac{4}{800+2t}dt}$
Using substitution with $u = 800 + 2t$ and $\frac{1}{2}du = dt$,

$$I(t) = e^{2\ln(800+2t)} = (800+2t)^2$$
$$a = (800+2t)^{-2}\left[\int (800+2t)^2\cdot 5dt + C\right]$$

Using substitution with $u = 800 + 2t$ and $\frac{1}{2}du = dt$,

$$a = (800+2t)^{-2}\left[\frac{5}{2}\frac{(800+2t)^3}{3} + C\right]$$

$$a(t) = \frac{5}{6}(800+2t) + C(800+2t)^{-2}$$

When $t = 0$, $a(0) = 50$, so

$$50 = \frac{5}{6}(800) + \frac{C}{(800)^2}$$

or, $C = -\frac{1{,}184{,}000{,}000}{3}$

and

$$a(t) = \frac{5}{6}(800+2t) - \frac{1{,}184{,}000{,}000}{3(800+2t)^2}$$

Since a is the amount of drug in the organ, $\frac{a}{800+2t}$ is the concentration of the drug. So,

$$c(t) = \frac{5}{6} - \frac{1{,}184{,}000{,}000}{3(800+2t)^3}$$

$$c(t) = \frac{5}{6} - \frac{148{,}000{,}000}{3(400+t)^3}$$

(b) $0.13 = \frac{5}{6} - \frac{148{,}000{,}000}{3(400+t)^3}$

$$\frac{148{,}000{,}000}{3(400+t)^3} = 0.70\bar{3}$$

$$70{,}142{,}183 \approx (400+t)^3$$

$$400 + t \approx 412.407$$

$$t \approx 12.4 \text{ hours}$$

29. A chemotactic organism follows a path that is orthogonal to the curves of constant nutrient concentration, $3x^2 + y^2 = C$. Differentiating implicitly,

$$6x + 2y\frac{dy}{dx} = 0$$

$$\frac{dy}{dx} = -\frac{3x}{y}$$

where $\frac{dy}{dx}$ represents the slope of the nutrient curves. The slope of the orthogonal trajectory curves is the negative reciprocal, or

$$\frac{dY}{dX} = \frac{y}{3x} = \frac{Y}{3X}$$

Since $x = X$ and $y = Y$ where the orthogonal trajectory curve intersects the nutrient curve.

$$\frac{dY}{Y} = \frac{1}{3X}\,dX$$

$$\int \frac{dY}{Y} = \int \frac{1}{3X}\,dX$$

$$\ln|Y| = \frac{1}{3}\ln|X| + C_1$$

$$|Y| = e^{\frac{1}{3}\ln|X| + C_1}$$

$$Y = \pm e^{C_1}\cdot e^{\frac{1}{3}\ln|X|}$$

$$Y = C|X|^{\frac{1}{3}}$$

If the organism is introduced at the point (1, 1)

$$1 = C|1|^{\frac{1}{3}}, \text{ or } C = 1$$

So, the organism travels along the orthogonal trajectory curve $Y = |X|^{\frac{1}{3}}$.

31. Let $M(t)$ be the value of the money market account after t years and $S(t)$ the value of the stock fund.

$$\frac{dM}{dt} = 0.02M - 0.25M = -0.23M$$

$$\frac{dM}{M} = -0.23\,dt$$

$$\ln M = -0.23t + C_1$$

$$M = e^{-0.23t + C_1}$$

$$M(t) = Ce^{-0.23t}$$

When $t = 0$, $M(0) = 30{,}000$ and

$$30{,}000 = Ce^0, \text{ or } C = 30{,}000$$

So, $M(t) = 30{,}000e^{-0.23t}$

Now, $\frac{dS}{dt} = 0.1S + 0.25M$

$$\frac{dS}{dt} = 0.1S + 0.25(30{,}000e^{-0.23t})$$

$$\frac{dS}{dt} + (-0.1)S = 7{,}500e^{-0.23t}$$

The integrating factor is

$$I(t) = e^{\int -0.1\,dt} = e^{-0.1t}$$

$$S = e^{0.1t}\left[\int e^{-0.1t}\cdot 7{,}500e^{-0.23t}\,dt + C\right]$$
$$S = e^{0.1t}\left[\int 7{,}500e^{-0.33t}\,dt + C\right]$$
$$S = e^{0.1t}\left[\frac{7{,}500}{-0.33}e^{-0.33t} + C\right]$$
$$S(t) = \frac{7{,}500}{-0.33}e^{-0.23t} + Ce^{0.1t}$$

When $t = 0$, $S(0) = 0$ and

$$0 = \frac{7{,}500}{-0.33}e^{0} + Ce^{0}, \text{ or } C = \frac{7{,}500}{0.33}$$

So,

$$S(t) = \frac{7{,}500}{-0.33}e^{-0.23t} + \frac{7{,}500}{0.33}e^{0.1t}$$

After five years,

$$M(5) = 30{,}000e^{-0.23(5)} \approx \$9{,}499$$
$$S(5) = \frac{7{,}500}{-0.33}e^{-0.23(5)} + \frac{7{,}500}{0.33}e^{0.1(5)} \approx \$30{,}275$$

33. Working in units of thousand people,

$$\frac{dN}{dt} = 0.04(250 - N)(N - 2.5)$$

(a) Separating the variables,

$$\int \frac{dN}{(250 - N)(N - 2.5)} = \int 0.04\,dt$$

Using formula 6 on page 458, with $u = 250 - N$ and $-du = dN$,

$$-\int \frac{du}{u(247.5 - u)} = \int 0.04\,dt$$
$$-\frac{1}{247.5}\ln\left|\frac{250 - N}{N - 2.5}\right| = 0.04t + C_1$$
$$-\ln\left|\frac{250 - N}{N - 2.5}\right| = 9.9t + 247.5C_1$$
$$\ln\left|\frac{N - 2.5}{250 - N}\right| = 9.9t + C_2$$
$$\left|\frac{N - 2.5}{250 - N}\right| = e^{9.9t + C_2}$$
$$\frac{N - 2.5}{250 - N} = \pm e^{C_2}\cdot e^{9.9t}$$
$$\frac{N - 2.5}{250 - N} = C_3e^{9.9t}$$

When $t = 0$, $N(0) = N_0$ and

$$\frac{N_0 - 2.5}{250 - N_0} = C_3e^{0}$$
$$\text{or} \quad C_3 = \frac{N_0 - 2.5}{250 - N_0}$$

So,

$$\frac{N - 2.5}{250 - N} = \left(\frac{N_0 - 2.5}{250 - N_0}\right)e^{9.9t}$$

To simplify the algebra, rename the right-hand side K. Then,

$$\frac{N - 2.5}{250 - N} = K$$
$$N - 2.5 = 250K - KN$$
$$N = \frac{250K + 2.5}{1 + K}$$

Substituting back for K,

$$N = \frac{250\left(\frac{N_0 - 2.5}{250 - N_0}\right)e^{9.9t} + 2.5}{1 + \frac{N_0 - 2.5}{250 - N_0}e^{9.9t}}$$
$$= \frac{\frac{250e^{9.9t}(N_0 - 2.5) + 2.5(250 - N_0)}{250 - N_0}}{\frac{250 - N_0 + (N_0 - 2.5)e^{9.9t}}{250 - N_0}}$$
$$N(t) = \frac{250e^{9.9t}(N_0 - 2.5) + 2.5(250 - N_0)}{250 - N_0 + (N_0 - 2.5)e^{9.9t}}$$

(b) If $N_0 = 5$ thousand people,

$$N(t) = \frac{250e^{9.9t}(2.5) + 2.5(245)}{245 + (2.5)e^{9.9t}}$$
$$= \frac{625e^{9.9} + 612.5}{245 + 2.5e^{9.9t}}$$

and

$$\lim_{t\to\infty} N(t) = \lim_{t\to\infty} \frac{625e^{9.9} + 612.5}{245 + 2.5e^{9.9}}$$
$$= \lim_{t\to\infty} \frac{e^{9.9}(625 + 612.5e^{-9.9})}{e^{9.9}(245e^{-9.9} + 2.5)}$$
$$= \frac{625}{2.5} = 250 \text{ thousand people,}$$

or the entire susceptible population.

(c) If $N_0 = 1$ thousand people,

$$N(t) = \frac{250e^{9.9t}(-1.5) + 2.5(249)}{249 + (-1.5)e^{9.9t}}$$

So, $N(t) = 0$ when

$$0 = -375e^{9.9t} + 622.5$$
$$e^{9.9t} = 1.66$$
$$t = \frac{\ln 1.66}{9.9} \approx 0.051$$

So, the epidemic dies out after 0.051 days.

(d) In general,

$$N(t) = \frac{N_s e^{k(N_s - m)t}(N_0 - m) + m(N_s - N_0)}{N_s - N_0 + (N_0 - m)e^{k(N_s - m)t}}$$

So, $N(t) = 0$ when

$$0 = N_s e^{k(N_s - m)t}(N_0 - m) + m(N_s - N_0)$$
$$N_s e^{k(N_s - m)t}(N_0 - m) = m(N_0 - N_s)$$
$$e^{k(N_s - m)t} = \frac{m(N_0 - N_s)}{N_s(N_0 - m)}$$
$$k(N_s - m)t = \ln\left[\frac{m(N_0 - N_s)}{N_s(N_0 - m)}\right]$$
$$t = \frac{\ln\left[\frac{m(N_0 - N_s)}{N_s(N_0 - m)}\right]}{k(N_s - m)} = \frac{\ln\left[\frac{m(N_s - N_0)}{N_s(m - N_0)}\right]}{k(N_s - m)}$$

Since $N_s > m$ and since k is a positive constant, the denominator is always positive.
Since $N_s > N_0$ and m is a positive constant, the numerator of the logarithm is positive.
If $N_0 > m$, the denominator of the logarithm would be negative and the logarithm would be undefined (giving no answer for t).
If $N_0 < m$, the logarithm is defined and such a time t exists.

35. **(a)** Setting $\frac{dR}{dt} = 0$,
$63R - 3RS = 0$
or, $S_e = 21$ thousand squirrels
Setting $\frac{dS}{dt} = 0$,
$265 - RS = 0$
$546 - 21R = 0$
or, $R_e = 26$ thousand rabbits

(b) $\frac{dR}{dS} = \frac{63R - 3RS}{26S - RS} = \frac{R(63 - 3S)}{S(26 - R)}$

$$\int \frac{26 - R}{R} dR = \int \frac{63 - 3S}{S} dS$$
$$\int \left(\frac{26}{R} - 1\right) dR = \int \left(\frac{63}{S} - 3\right) dS$$

$26 \ln R - R = 63 \ln S - 3S + C$
When $t = 0$, $R = S = 4$ thousand, so
$26 \ln 4 - 4 = 63 \ln 4 - 12 + C$
or, $C \approx -43.3$
and
$26 \ln R - R = 63 \ln S - 3S - 43.3$

37. $$\frac{dD}{dt} = -kD; \quad \frac{dS}{dt} = 2kS$$

(a) Separating the variables,

$$\frac{dD}{D} = -k\, dt$$
$$\int \frac{dD}{D} = \int -k\, dt$$
$$\ln D = -kt + C_1$$
$$D = e^{-kt + C_1}$$
$$D(t) = Ce^{-kt}$$

When $t = 0$, $D(0) = 50$ and

$$50 = Ce^0, \text{ or } C = 50$$

So, $D(t) = 50e^{-kt}$

Similarly,

$$\frac{dS}{S} = 2k\,dt$$
$$\int \frac{dS}{S} = \int 2k\,dt$$
$$\ln S = 2kt + C_1$$
$$S = e^{2kt + C_1}$$
$$S(t) = Ce^{2kt}$$
$$5 = Ce^0, \text{ or } C = 5$$

So, $S(t) = 5e^{2kt}$

Now,

$$D(10) = 50e^{-10k}$$
$$S(10) = 5e^{20k}$$
$$50e^{-10k} = 5e^{20k}$$
$$50 = 5e^{30k}$$
$$10 = e^{30k}$$
$$\ln 10 = 30k$$
$$k = \frac{\ln 10}{30}$$

(b) $D(t) = 50e^{-\frac{\ln 10}{30}t} \approx 50e^{-0.07675t}$

$$S(t) = 5e^{2\left(\frac{\ln 10}{30}\right)t} = 5e^{\frac{\ln 10}{15}t} \approx 5e^{0.15351t}$$

(c) $D(10) = S(10) = 50e^{-0.0768(10)}$

$$D(10) \approx 23.197$$

At equilibrium, approximately 23 units are supplied and demanded.

39. (a) Since the pollutant enters the first lake all at once and no more enters,

$$\frac{dP_1}{dt} = 0 - \begin{bmatrix}\text{rate pollutant}\\ \text{leaves}\end{bmatrix}$$
$$= 0 - \begin{pmatrix}\text{amt pollutant}\\ \text{per gallon}\end{pmatrix}\begin{pmatrix}\text{rate gallons}\\ \text{leave}\end{pmatrix}$$
$$= -\left(\frac{P_1}{700{,}000}\right)(1{,}500)$$
$$= -\frac{3}{1{,}400}P_1$$

Separating the variables,

$$\int \frac{dP_1}{P_1} = \int -\frac{3}{1{,}400}\,dt$$
$$\ln P_1 = -\frac{3}{1{,}400}t + C_1$$
$$P_1 = e^{-\frac{3}{1{,}400}t + C_1}$$
$$P_1(t) = Ce^{-\frac{3}{1{,}400}t}$$

When $t = 0$, all of the pollutant is dumped into the first lake, so $P_1(0) = 2{,}000$ and

$$2{,}000 = Ce^0, \text{ or } C = 2{,}000$$

So, $P_1(t) = 2{,}000e^{-\frac{3}{1{,}400}t}$.

(b) $$\frac{dP_2}{dt} = aP_1 - bP_2$$

Since $\dfrac{dP_2}{dt} = \begin{bmatrix}\text{rate pollutant}\\ \text{enters}\end{bmatrix} - \begin{bmatrix}\text{rate pollutant}\\ \text{leaves}\end{bmatrix}$

$$= \begin{pmatrix}\text{amt pollutant}\\ \text{per gallon}\end{pmatrix}\begin{pmatrix}\text{rate gallons}\\ \text{enter}\end{pmatrix} - \begin{pmatrix}\text{amt pollutant}\\ \text{per gallon}\end{pmatrix}\begin{pmatrix}\text{rate gallons}\\ \text{leave}\end{pmatrix}$$
$$= \left(\frac{P_1}{700{,}000}\right)(1{,}500) - \left(\frac{P_2}{400{,}000}\right)(1{,}500)$$
$$\frac{dP_2}{dt} = \frac{3}{1{,}400}P_1 - \frac{3}{800}P_2$$

So, $a = \dfrac{3}{1{,}400}$ and $b = \dfrac{3}{800}$.

$$\frac{dP_2}{dt} + \left(\frac{3}{800}\right)P_2 = \frac{3}{1{,}400}P_1$$
$$\frac{dP_2}{dt} + \left(\frac{3}{800}\right)P_2 = \frac{3}{1{,}400}\left(2{,}000e^{-\frac{3}{1{,}400}t}\right)$$
$$\frac{dP_2}{dt} + \left(\frac{3}{800}\right)P_2 = \frac{30}{7}e^{-\frac{3}{1{,}400}t}$$

The integrating factor is

$$I(t) = e^{\int \frac{3}{800}\,dt} = e^{\frac{3}{800}t}$$

$$P_2 = e^{-\frac{3}{800}t}\left[\int e^{\frac{3}{800}t}\cdot\frac{30}{7}e^{-\frac{3}{1,400}t}\,dt + C\right]$$

$$P_2 = e^{-\frac{3}{800}t}\left[\int \frac{30}{7}e^{\frac{9}{5,600}t}\,dt + C\right]$$

$$P_2 = e^{-\frac{3}{800}t}\left[\frac{30}{7}\cdot\frac{5,600}{9}e^{\frac{9}{5,600}t} + C\right]$$

$$P_2(t) = \frac{8,000}{3}e^{-\frac{3}{1,400}t} + Ce^{-\frac{3}{800}t}$$

When $t = 0$, $P_2(0) = 0$ and

$$0 = \frac{8,000}{3} + C,\ \text{or } C = -\frac{8,000}{3}$$

So, $P_2(t) = \dfrac{8,000}{3}\left(e^{-\frac{3}{1,400}t} - e^{-\frac{3}{800}t}\right).$

(c) To find the maximum pollutant in the second lake, find the critical point(s) from the derivative. Since

$$\begin{aligned}\frac{dP_2}{dt} &= \frac{3}{1,400}P_1 - \frac{3}{800}P_2\\ &= \frac{3}{1,400}\left(2,000e^{-\frac{3}{1,400}t}\right)\\ &\quad - \frac{3}{800}\left(\frac{8,000}{3}\left[e^{-\frac{3}{1,400}t} - e^{-\frac{3}{800}t}\right]\right)\\ &= \frac{30}{7}e^{-\frac{3}{1,400}t} - 10e^{-\frac{3}{1,400}t} + 10e^{-\frac{3}{800}t}\\ &= -\frac{40}{7}e^{-\frac{3}{1,400}t} + 10e^{-\frac{3}{800}t}\end{aligned}$$

$\dfrac{dP_2}{dt} = 0$ when

$$\frac{40}{7}e^{-\frac{3}{1,400}t} = 10e^{-\frac{3}{800}t}$$

$$\frac{4}{7}e^{-\frac{3}{1,400}t} = e^{-\frac{3}{800}t}$$

$$\ln\left(\frac{4}{7}e^{-\frac{3}{1,400}t}\right) = \ln e^{-\frac{3}{800}t}$$

$$\ln\frac{4}{7} - \frac{3}{1,400}t = -\frac{3}{800}t$$

$$\frac{9}{5,600}t = -\ln\frac{4}{7}$$

$$\frac{9}{5,600}t = \ln\frac{7}{4}$$

$$t = \frac{5,600}{9}\ln\frac{7}{4}$$

The maximum in the second lake occurs approximately 348.2 hours after the pollutant is dumped. The maximum amount in the second lake is approximately

$$P_2(348.2) = \frac{8,000}{3}\left(e^{-\frac{3}{1,400}(348.2)} - e^{-\frac{3}{800}(348.2)}\right)$$

$$\approx 542 \text{ pounds}$$

41. $$\frac{dP}{dt} = kP^{1+c}$$

(a) Separating the variables,

$$P^{-(1+c)}\,dP = k\,dt$$

$$\int P^{-(1+c)}\,dP = \int k\,dt$$

$$\frac{P^{-(1+c)+1}}{-(1+c)+1} = kt + C_1$$

$$\frac{P^{-c}}{-c} = kt + C_1$$

$$P^{-c} = -ckt - cC_1$$

$$P^{-c} = -ckt - C_2$$

When $t = 0$, $P(0) = P_0$ and

$$P_0^{-c} = 0 - C_2,\ \text{or } C_2 = -P_0^{-c}$$

So,

$$P^{-c} = -ckt + P_0^{-c}$$

$$P(t) = (-ckt + P_0^{-c})^{-\frac{1}{c}} = \frac{1}{\left(\frac{1}{P_0^c} - ckt\right)^{\frac{1}{c}}}$$

(b) In order for

$$\lim_{t\to t_a}\left(\frac{1}{P_0^c} - ckt\right)^{-\frac{1}{c}} = \infty$$

need to have

$$\lim_{t\to t_a} \frac{1}{P_0^c} - ckt = 0$$

$$\frac{1}{P_0^c} - ckt_a = 0$$

$$\frac{1}{P_0^c} = ckt_a$$

$$t_a = \frac{1}{P_0^c ck}$$

(c)

$$P(t) = \frac{P_0}{(1 - ckt\ P_0^c)^{\frac{1}{c}}}$$

Since $c = 0.02$ and $P_0 = P(0) = 2$,

$$P(t) = \frac{2}{[1 - 0.02kt(2)^{0.02}]^{\frac{1}{0.02}}}$$

$$\approx \frac{2}{(1 - 0.020279kt)^{50}}$$

Now, $P(2) = 4$, so

$$4 \approx \frac{2}{\left(1 - 0.020279k(2)\right)^{50}}$$

$$4^{0.02} \approx \frac{2^{0.02}}{1 - 0.040558k}$$

$$1 - 0.040558k \approx \frac{2^{0.02}}{4^{0.02}}$$

$$1 - 0.040558k \approx \frac{2^{0.02}}{2^{0.04}}$$

$$1 - 0.040558k \approx 2^{-0.02}$$

$$k \approx 0.3394$$

So,

$$P(t) \approx \frac{2}{[1 - 0.020279(0.3394)t]^{50}}$$

$$P(t) \approx \frac{2}{[1 - 0.00688t]^{50}}$$

(d) Using the result of part (b),

$$t_a = \frac{1}{P_0^c\ ck}$$

$$= \frac{1}{2^{0.02}(0.02)(0.3394)}$$

$$\approx 145.3 \text{ months}$$

8.4 Approximate Solutions of Differential Equations

1. To sketch solutions of the differential equation $y' = x + y$, start by drawing the family of parallel lines $x + y = m$, or $y = -x + m$, for several values of m. Along each such line, place small segments of slope m. This slope field approximates the solution curves, with the solution through the point (1, 1) drawn in.

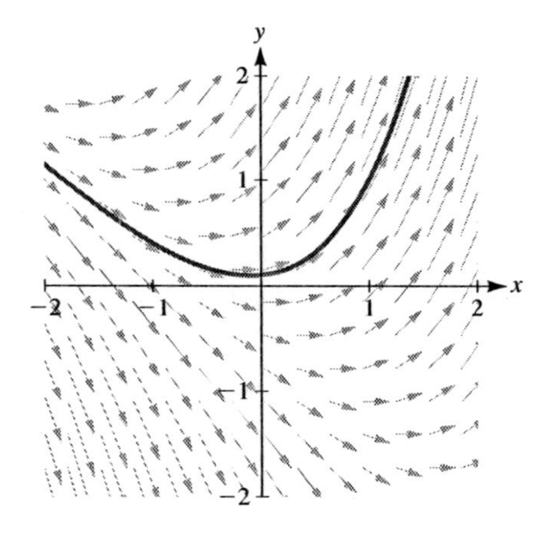

3. To sketch solutions of the differential equation $y' = y^2$, start by drawing the family of lines $y^2 = m$, for several values of $m \geq 0$. Note that for each value of $m \neq 0$, the lines are the pairs $y = \sqrt{m}$ and $y = -\sqrt{m}$. Along each such line, place small segments of slope m. This slope field approximates the solution curves, with the solution through the point (0, 1) drawn in.

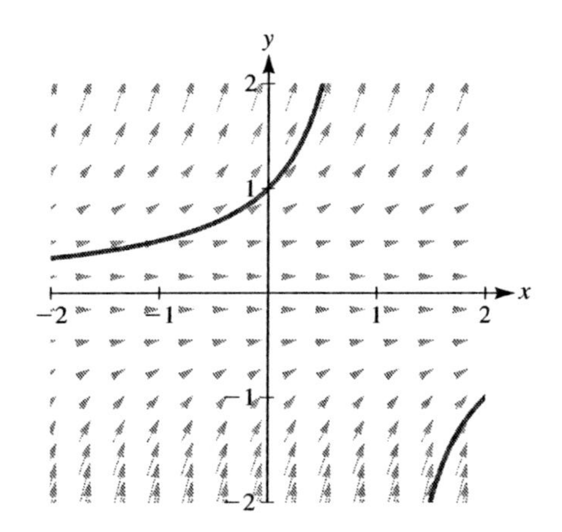

5. To sketch solutions of the differential equation $y' = y(2 - y)$, start by drawing the family of lines $y(2 - y) = m$, or $y^2 - 2y + m = 0$, for several values of $m \leq 1$ (so there are real solutions to the quadratic). Again, for $m \neq 1$, note that the lines are pairs of the form $y = \dfrac{-2 \pm \sqrt{4 - 4m}}{2}$. Along each such line, place small segments of slope m. This slope field approximates the solution curves, with the solution through the point $(0, -1)$ drawn in.

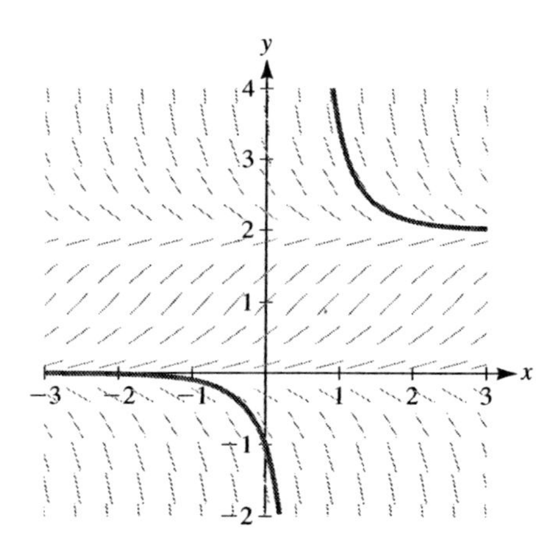

7. To sketch solutions of the differential equation $y' = x^2 + y^2$, start by drawing the family of circles $x^2 + y^2 = m$, for several values of $m > 0$. Along each such circle, place small segments of slope m. Also sketch the graph of $x^2 + y^2 = 0$, which is the origin, and draw a horizontal segment for a slope of zero. This slope field approximates the solution curves, with the solution through the point $(1, -1)$ drawn in.

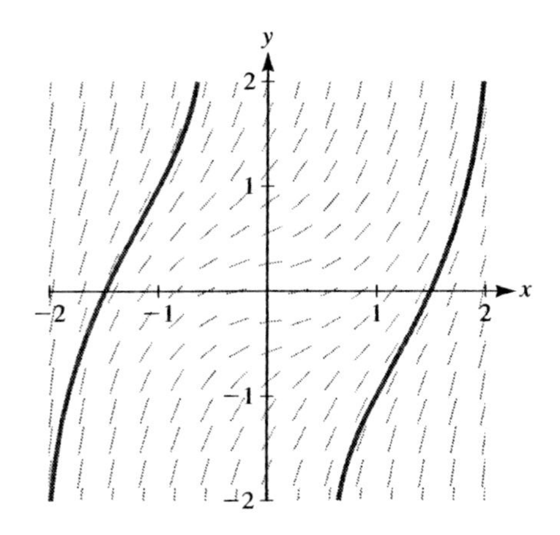

9. To sketch solutions of the differential equation $y' = \dfrac{2y - 3x}{x + y}$, start by drawing the family of lines $\dfrac{2y - 3x}{x + y} = m$, or $y = \dfrac{3 + m}{2 - m}x$, for several values of $m \neq 2$. Along each such line, place small segments of slope m. This slope field approximates the solution curves, with the solution through the point $(0, 1)$ drawn in.

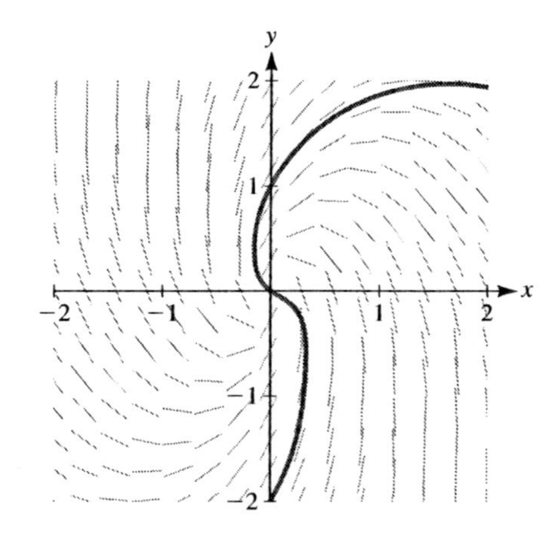

11. To sketch solutions of the differential equation $y' = e^{xy}$, start by drawing the family of curves $e^{xy} = m$, or $y = \dfrac{\ln m}{x}$, for several values of $m > 0$.

Along each such pair of curves, place small segments of slope m. This slope field approximates the solution curves, with the solution through the point (0, 0) drawn in.

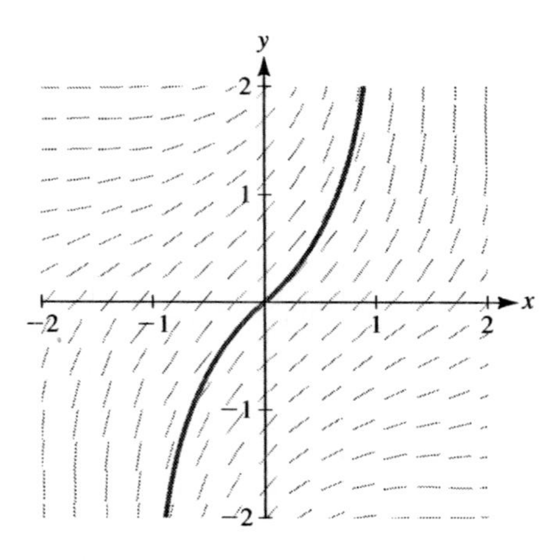

13. Using Euler's method with $f(x, y) = 2x + y$,

k	x_k	y_k	$f(x_k, y_k)$
0	0	1	1
1	0.2	1.2	1.6
2	0.4	1.52	2.32
3	0.6	1.984	3.184
4	0.8	2.6208	4.2208
5	1.0	3.46496	

Plotting the points (x_k, y_k) and connecting successive points with line segments yields the Euler solution to the differential equation for the given initial value.

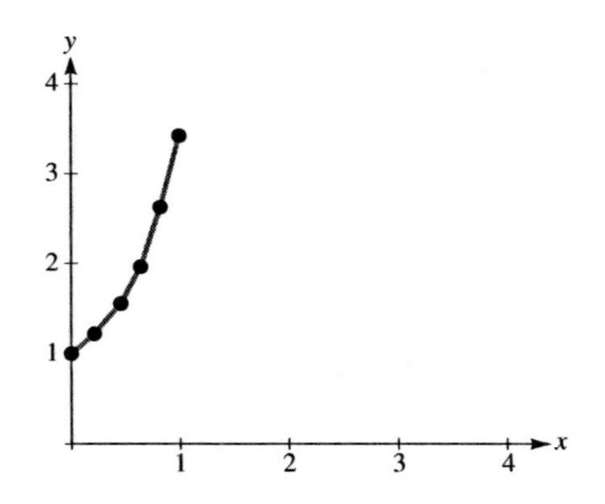

15. Using Euler's method with $f(x, y) = \dfrac{y}{x} + x$,

k	x_k	y_k	$f(x_k, y_k)$
0	1	0	1
1	1.2	0.2	1.36667
2	1.4	0.47333	1.73810
3	1.6	0.82095	2.11309
4	1.8	1.24357	2.49087
5	2.0	1.74175	

Plotting the points (x_k, y_k) and connecting successive points with line segments yields the Euler solution to the differential equation for the given initial value.

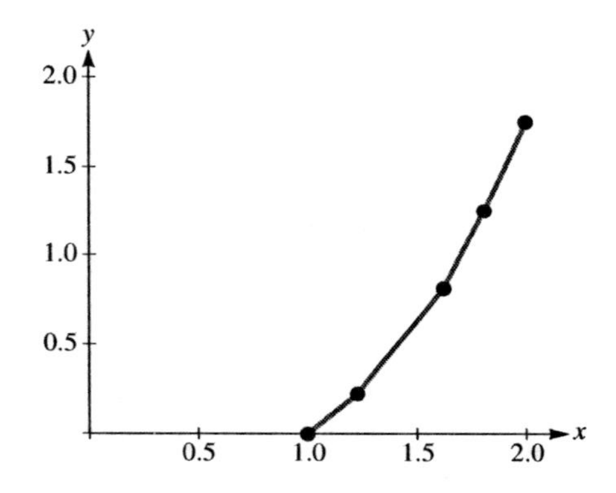

17. Using Euler's method with $f(x, y) = \sqrt{x} - y$,

k	x_k	y_k	$f(x_k, y_k)$
0	1	1	0
1	1.5	1	0.22474
2	2.0	1.11237	0.30184
3	2.5	1.26329	0.31785
4	3.0	1.42222	0.30984
5	3.5	1.57713	0.29370
6	4.0	1.72398	0.27602
7	4.5	1.86199	0.25933
8	5.0	1.99166	

Plotting the points (x_k, y_k) and connecting successive points with line segments yields the Euler solution to the differential equation for the given initial value.

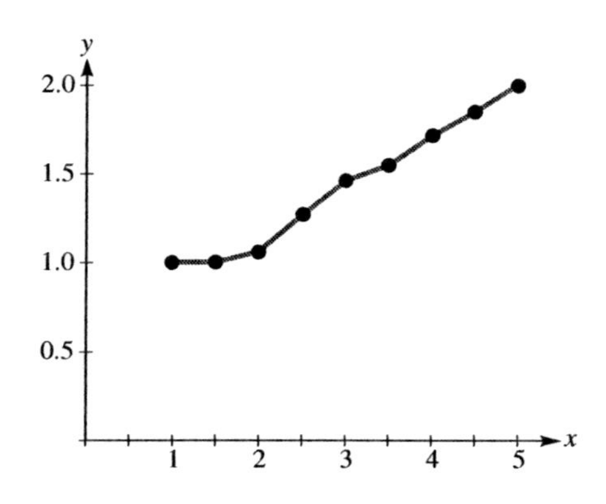

19. Using Euler's method with $f(x, y) = \dfrac{x^2 - y^2}{xy}$,

k	x_k	y_k	$f(x_k, y_k)$
0	1	1	0
1	1.4	1	0.68571
2	1.8	1.27429	0.70462
3	2.2	1.55613	0.70642
4	2.6	1.83870	0.70684
5	3.0	2.12144	

Plotting the points (x_k, y_k) and connecting successive points with line segments yields the Euler solution to the differential equation for the given initial value.

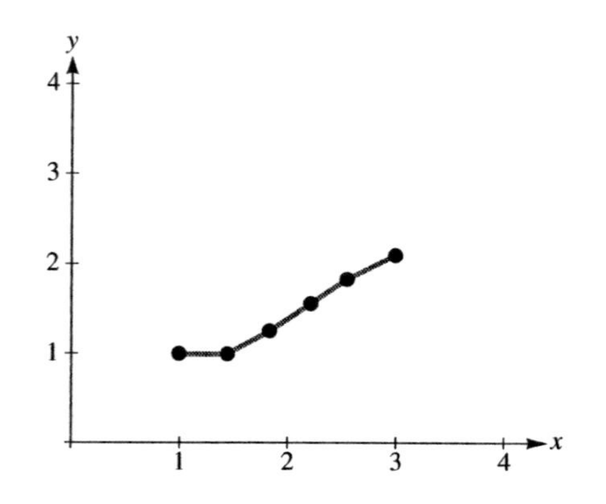

21. Using Euler's method with $f(x, y) = x + 2y$,

k	x_k	y_k	$f(x_k, y_k)$
0	0	1	2
1	0.2	1.4	3
2	0.4	2.0	4.4
3	0.6	2.88	6.36
4	0.8	4.152	9.104
5	1.0	5.9728	

The estimate for $y(1)$ is 5.9728.

23. Using Euler's method with $f(x, y) = \dfrac{x - y}{2x + y}$,

k	x_k	y_k	$f(x_k, y_k)$
0	0	1	−1
1	0.2	0.8	−0.5
2	0.4	0.7	−0.2
3	0.6	0.66	−0.03226
4	0.8	0.6535	0.64987
5	1.0	0.6665	

The estimate for $y(1)$ is 0.6665.

25. Using Euler's method with $f(x, y) = x^2 + 2y^2$,

k	x_k	y_k	$f(x_k, y_k)$
0	0	0	0
1	0.4	0	0.16
2	0.8	0.064	0.64819
3	1.2	0.32328	1.64902
4	1.6	0.98289	4.49213
5	2.0	2.77974	

The estimate for $y(2)$ is 2.77974.

27. Using Euler's method with $f(t, L) = \dfrac{\sqrt{1 + t^2}}{t}$,

k	t_k	L_k	$f(t_k, L_k)$
0	1	0	1.41421
1	1.5	0.70711	1.20185
2	2.0	1.30804	1.11803
3	2.5	1.86706	1.07703
4	3.0	2.40558	1.05409
5	3.5	2.93263	1.04002
6	4.0	3.45264	

The estimate for $L(4)$ is 3.45264.

29. **(a)** Using Euler's method with $f(t, L) = 1.03(5.3 - L)$,

k	t_k	L_k	$f(t_k, L_k)$
0	0	0.2	5.253
1	0.4	2.3012	3.08876
2	0.8	3.53671	1.81619
3	1.2	4.26318	1.06792
4	1.6	4.69035	0.62793
5	2	4.94153	

In two months, the length of the fish will be approximately 4.94 inches.

(b) Separating the variables,

$$\int \frac{dL}{5.3 - L} = \int 1.03\, dt$$
$$-\ln|5.3 - L| = 1.03t + C_1$$
$$\ln|5.3 - L| = -1.03t - C_1$$
$$|5.3 - L| = e^{-1.03t - C_1}$$
$$|5.3 - L| = C_2 e^{-1.03t}$$
$$5.3 - L = \pm C_2 e^{-1.03t}$$
$$5.3 - L = C_3 e^{-1.03t}$$
$$L(t) = 5.3 - C_3 e^{-1.03t}$$

Using the condition that $L(0) = 0.2$,

$$0.2 = 5.3 - C_3, \text{ or } C_3 = 5.1$$
$$L(t) = 5.3 - 5.1e^{-1.03t}$$
$$L(2) = 5.3 - 5.1e^{-1.03(2)}$$
$$\approx 4.65 \text{ inches}$$

The approximation using Euler's method is off by about 6%.

31. Using Euler's method with $f(K, Q) = 0.2Q + K^{\frac{1}{2}}$,

k	K_k	Q_k	$f(K_k, Q_k)$
0	2	500	101.41421
1	3	601.4142	122.01489
2	4	723.4291	146.68582
3	5	870.1149	

When the capital expenditure is $5,000 approximately 870 units will be produced.

8.5 Difference Equations

1.
$$y_0 = 1$$
$$y_1 = 3y_0 = 3(1) = 3$$
$$y_2 = 3y_1 = 3(3) = 9$$
$$y_3 = 3y_2 = 3(9) = 27$$
$$y_4 = 3y_3 = 3(27) = 81$$

3.
$$y_0 = 1$$
$$y_1 = (y_0)^2 = (1)^2 = 1$$
$$y_2 = (y_1)^2 = (1)^2 = 1$$
$$y_3 = (y_2)^2 = (1)^2 = 1$$
$$y_4 = (y_3)^2 = (1)^2 = 1$$

5.
$$y_0 = 1$$
$$y_1 = 1$$
$$y_2 = y_1 + y_0 = 1 + 1 = 2$$
$$y_3 = y_2 + y_1 = 2 + 1 = 3$$
$$y_4 = y_3 + y_2 = 3 + 2 = 5$$

7.
$$y_0 = 1$$
$$y_1 = \frac{y_0}{y_0 + 2} = \frac{1}{1 + 2} = \frac{1}{3}$$
$$y_2 = \frac{y_1}{y_1 + 2} = \frac{\frac{1}{3}}{\frac{1}{3} + 2} = \frac{1}{7}$$
$$y_3 = \frac{y_2}{y_2 + 2} = \frac{\frac{1}{7}}{\frac{1}{7} + 2} = \frac{1}{15}$$
$$y_4 = \frac{y_3}{y_3 + 2} = \frac{\frac{1}{15}}{\frac{1}{15} + 2} = \frac{1}{31}$$

9.
$$ny_n + (n - 1)y_{n-1} = 2n - 3; \quad y_n = A + \frac{B}{n}$$

$$n\left(A+\frac{B}{n}\right)+(n-1)\left(A+\frac{B}{n-1}\right)=2n-3$$
$$nA+B+An-A+\frac{nB}{n-1}-\frac{B}{n-1}=2n-3$$
$$2An+\frac{nB-B}{n-1}+B-A=2n-3$$
$$2An+\frac{B(n-1)}{n-1}+B-A=2n-3$$
$$2An+2B-A=2n-3$$

So,

$$2An=2n, \quad \text{or} \quad A=1$$
$$2B-A=-3$$
$$2B-1=-3$$
$$B=-1$$

11.

$$y_n=-y_{n-1}$$
$$y_n=(-1)y_{n-1}+0$$

Since $a=-1$, $b=0$, and $y_0=1$,

$$y_n=(-1)^n(1)+\left(\frac{1-(-1)^n}{1-(-1)}\right)(0)$$
$$y_n=(-1)^n$$

13.

$$y_n=\frac{1}{2}y_{n-1}+1$$

Since $a=\frac{1}{2}$, $b=1$, and $y_0=0$,

$$y_n=\left(\frac{1}{2}\right)^n(0)+\left(\frac{1-\left(\frac{1}{2}\right)^n}{1-\frac{1}{2}}\right)(1)$$
$$y_n=\frac{1-\left(\frac{1}{2}\right)^n}{\frac{1}{2}}=2\left[1-\left(\frac{1}{2}\right)^n\right]$$
$$y_n=2-2\left(\frac{1}{2}\right)^n=2-\left(\frac{1}{2}\right)^{n-1}=2-\frac{1}{2^{n-1}}$$

15.

$$y_n-y_{n-1}=\frac{1}{8}y_{n-1}+\frac{1}{4}$$
$$y_n=\frac{9}{8}y_{n-1}+\frac{1}{4}$$

Since $a=\frac{9}{8}$, $b=\frac{1}{4}$, and $y_0=0$,

$$y_n=\left(\frac{9}{8}\right)^n(0)+\left(\frac{1-\left(\frac{9}{8}\right)^n}{1-\frac{9}{8}}\right)\frac{1}{4}$$
$$y_n=\left(\frac{1-\left(\frac{9}{8}\right)^n}{-\frac{1}{8}}\right)\frac{1}{4}=-2\left[1-\left(\frac{9}{8}\right)^n\right]$$
$$y_n\doteq 2\left(\frac{9}{8}\right)^n-2$$

17.

$$y_n-y_{n-1}=\frac{1}{10}y_{n-1}$$
$$y_n=\frac{11}{10}y_{n-1}+0$$

Since $a=\frac{11}{10}$, $b=0$, and $y_0=10$,

$$y_n=\left(\frac{11}{10}\right)^n(10)+\left(\frac{1-\left(\frac{11}{10}\right)^n}{1-\frac{11}{10}}\right)(0)$$
$$y_n=\frac{11^n}{10^{n-1}}$$

19. Using the interest equation $A=P(1+rt)$ with $r=0.05$ and $t=1$,

$$y_n=1.05y_{n-1}+50; \quad y_0=1{,}000$$

Now, $a=1.05$ and $b=50$, so

$$y_n=(1.05)^n(1{,}000)+\left(\frac{1-(1.05)^n}{1-1.05}\right)(50)$$
$$y_n=1{,}000(1.05)^n-1{,}000[1-(1.05)^n]$$
$$y_n=1{,}000\left[(1.05)^n-(1-(1.05)^n)\right]$$
$$y_n=1{,}000[2(1.05)^n-1]$$
$$y_n=2{,}000(1.05)^n-1{,}000$$

21. (a)

$$y_n=1.08y_{n-1}+-2{,}000$$

Since $a=1.08$, $b=-2{,}000$, and $y_0=50{,}000$

$$y_n = (1.08)^n(50{,}000) + \left(\frac{1-(1.08)^n}{1-1.08}\right)(-2{,}000)$$
$$y_n = 50{,}000(1.08)^n + 25{,}000[1-(1.08)^n]$$
$$y_n = 25{,}000[2(1.08)^n + (1-(1.08)^n)]$$
$$y_n = 25{,}000[(1.08)^n + 1]$$
$$y_n = 25{,}000(1.08)^n + 25{,}000$$

(b) Need

$$50{,}000 = (1.08)^n(50{,}000) + \left(\frac{1-(1.08)^n}{1-1.08}\right)(-h)$$

for all values of n.

$$50{,}000 = (1.08)^n(50{,}000) + 12.5h[1-(1.08)^n]$$
$$50{,}000 - 50{,}000(1.08)^n = 12.5h[1-(1.08)^n]$$
$$50{,}000[1-(1.08)^n] = 12.5h[1-(1.08)^n]$$
$$\frac{50{,}000[1-(1.08)^n]}{12.5[1-(1.08)^n]} = h$$

or, $h = 4{,}000$

23. **(a)** $S_{n-1} + S_n = k(P - S_n) - 0.25S_n$, or $S_n - S_{n-1} = k(P - S_{n-1}) - 0.2S_{n-1}$.

(b) $$S_n = (0.8 - k)S_{n-1} + kP$$

Since $S_0 = 500$, $S_1 = 500 - (0.2)(500) + 700 = 1{,}100$ and $P = 10{,}000$

$$S_1 = (0.8-k)S_0 + kP$$
$$1{,}100 = (0.8-k)500 + 10{,}000k,$$

or $k = \frac{7}{95}$

So,

$$S_n = \left(0.8 - \frac{7}{95}\right)S_{n-1} + \frac{14{,}000}{19}$$

Using $a = 0.8 - \frac{7}{95}$, $b = \frac{14{,}000}{19}$ and $S_0 = 500$,

$$S_n = \left(0.8 - \frac{7}{95}\right)^n 500 + \left[\frac{1-\left(0.8-\frac{7}{95}\right)^n}{1-\left(0.8-\frac{7}{95}\right)}\right]\left(\frac{7}{95}\right)(10{,}000)$$

$$S_n \approx (0.72632)^n 500 + 2{,}692.3491\left[1-\left(0.8-\frac{7}{95}\right)^n\right]$$

In the long run,

$$\lim_{n\to\infty} S_n = 0 + 2{,}692.3491(1-0)$$

approximately 2,692 people will be sick.

25. **(a)**
$$C_n - C_{n-1} = -0.25C_{n-1} + 50$$
$$C_n = 0.75C_{n-1} + 50$$

Using $a = 0.75$, $b = 50$ and $C_0 = 300$,

$$C_n = (0.75)^n 300 + \left[\frac{1-(0.75)^n}{1-0.75}\right]50$$
$$C_n = (0.75)^n 300 + 200[1-(0.75)^n]$$
$$C_n = (0.75)^n 100 + 200$$

(b) When $n = 10$,

$$C_{10} = (0.75)^{10}100 + 200 \approx 205.631$$

approximately 206 cars will be in the fleet. In the long run,

$$\lim_{n\to\infty} C_n = 0 + 200$$

the fleet will consist of 200 cars.

27. Using

$$A = \frac{rD}{1-(1+r)^{-n}}$$

with a monthly interest rate of $\frac{0.08}{12}$ for $n = 60$ months,

$$A = \frac{\frac{0.08}{12}(22{,}000)}{1-\left(1+\frac{0.08}{12}\right)^{-60}} \approx 446.08$$

So, his monthly payment must be approximately \$446.08.

29. Let S_n be the total amount in Evelyn's savings account after n months. Then,

$$S_n - S_{n-1} = \frac{0.05}{12} S_{n-1} + A$$
$$S_n = \left(1 + \frac{0.05}{12}\right) S_{n-1} + A$$

Using $a = 1 + \frac{0.05}{12}$, $b = A$ and $S_0 = 0$,

$$S_n = \left(1 + \frac{0.05}{12}\right)^n (0) + \frac{1 - \left(1 + \frac{0.05}{12}\right)^n}{1 - \left(1 + \frac{0.05}{12}\right)} A$$
$$S_n = -240A\left[1 - \left(1 + \frac{0.05}{12}\right)^n\right]$$

Need $S_n = 7{,}000$ when $t = 36$ months, so

$$7{,}000 = -240A\left[1 - \left(1 + \frac{0.05}{12}\right)^{36}\right]$$
$$A \approx 180.63$$

So, Evelyn's monthly deposit should be approximately \$180.63.

31.

$$I_n - I_{n-1} = cI_{n-1} - h$$
$$I_n = (1 + c)I_{n-1} - h$$

Using $a = 1 + c$, $b = -h$

$$I_n = (1 + c)^n I_0 + \frac{1 - (1 + c)^n}{1 - (1 + c)}(-h)$$
$$I_n = (1 + c)^n I_0 + \frac{h}{c}\left[1 - (1 + c)^n\right]$$
$$I_n = \left(I_0 - \frac{h}{c}\right)(1 + c)^n + \frac{h}{c}$$

33. (a) If $P_m = P_{m+1} = L$, then

$$P_{m+1} - P_m = L - L = 0$$
$$kP_m\left(1 - \frac{1}{K}P_m\right) = 0$$

Since $k \neq 0$ and $P_m \neq 0$,

$$1 - \frac{1}{K}P_m = 0$$
$$P_m = K$$

So, $K = L$. Now,

$$P_{m+2} - P_{m+1} = kP_{m+1}\left(1 - \frac{1}{L}P_{m+1}\right)$$
$$P_{m+2} = P_{m+1} + kP_{m+1}\left(1 - \frac{1}{L}P_{m+1}\right)$$
$$= L + Lk\left(1 - \frac{1}{L} \cdot L\right)$$
$$= L$$
$$P_{m+3} = P_{m+2} + kP_{m+2}\left(1 - \frac{1}{L}P_{m+2}\right)$$
$$= L + Lk\left(1 - \frac{1}{L} \cdot L\right)$$
$$= L$$

Since this is the pattern for all $P_n \geq P_m$,

$$P_n = L \text{ for all } n \geq m.$$

(b)

$$P_n - P_{n-1} = kP_{n-1}\left(1 - \frac{1}{K}P_{n-1}\right)$$
$$P_1 - P_0 = kP_0\left(1 - \frac{1}{K}P_0\right)$$
$$5 = 50k\left(1 - \frac{50}{K}\right)$$
$$1 = 10k\left(1 - \frac{50}{K}\right)$$
$$k = \frac{1}{10\left(1 - \frac{50}{K}\right)}$$
$$k = \frac{K}{10(K - 50)}$$

Now,

$$P_2 - P_1 = kP_1\left(1 - \frac{1}{K}P_1\right)$$
$$5 = 55k\left(1 - \frac{55}{K}\right)$$
$$1 = 11k\left(1 - \frac{55}{K}\right)$$
$$1 = 11k\left(\frac{K-55}{K}\right)$$

Substituting,

$$1 = 11\left(\frac{K}{10(K-50)}\right)\left(\frac{K-55}{K}\right)$$
$$1 = \frac{11(K-55)}{10(K-50)}$$
$$10K - 500 = 11K - 605$$
$$K = L = 105$$
$$k = \frac{105}{10(105-50)} = \frac{21}{110}$$

(c) $$\frac{dP}{dt} = kP\left(1 - \frac{P}{K}\right)$$

Separating the variables,

$$\int \frac{dP}{P\left(1-\frac{P}{K}\right)} = \int k\,dt$$
$$\int \frac{dP}{P\left(1-\frac{1}{K}P\right)} = \int k\,dt$$

Using formula 6 on page 458, with $u = P$ and $du = dP$,

$$\frac{1}{1}\ln\left|\frac{P}{1-\frac{1}{K}P}\right| = kt + C_1$$
$$\left|\frac{KP}{K-P}\right| = e^{kt+C_1}$$
$$\frac{KP}{K-P} = \pm e^{C_1}e^{kt}$$
$$\frac{KP}{K-P} = Ce^{kt}$$

Since $P(0) = 50$,

$$\frac{50K}{K-50} = Ce^0$$
$$\frac{50K}{K-50} = C$$

So,

$$\frac{KP}{K-P} = \frac{50K}{K-50}e^{kt}$$

Also, $P(1) = 55$ so

$$\frac{55K}{K-55} = \frac{50K}{K-50}e^{k\cdot 1}$$
$$\frac{55K(K-50)}{50K(K-55)} = e^k$$
$$\frac{11(K-50)}{10(K-55)} = e^k$$
$$k = \ln\left[\frac{11(K-50)}{10(K-55)}\right]$$

So,

$$\frac{KP}{K-P} = \frac{50K}{K-50}e^{\ln\left[\frac{11(K-50)}{10(K-55)}\right]t}$$
$$\frac{KP}{K-P} = \frac{50K}{K-50}\cdot\left(e^{\ln\left[\frac{11(K-50)}{10(K-55)}\right]}\right)^t$$
$$\frac{KP}{K-P} = \frac{50K}{K-50}\left[\frac{11(K-50)}{10(K-55)}\right]^t$$
$$\frac{P}{K-P} = \frac{50}{K-50}\left[\frac{11(K-50)}{10(K-55)}\right]^t$$

Now, $P(2) = 60$ so

$$\frac{60}{K-60} = \frac{50}{K-50}\left[\frac{11(K-50)}{10(K-55)}\right]^2$$
$$\frac{60}{K-60} = \frac{121(K-50)}{2(K-55)^2}$$
$$120(K-55)^2 = 121(K-50)(K-60)$$
$$120K^2 - 13{,}200K + 363{,}000$$
$$= 121K^2 - 13{,}310K + 363{,}000$$
$$0 = K^2 - 110K$$
$$0 = K(K-110)$$
$$K = 110 \quad (\text{rejecting } K = 0)$$

So,

$$K = L = 110$$
$$k = \ln\left[\frac{11(110-50)}{10(110-55)}\right] = \ln\left(\frac{6}{5}\right)$$

35. (a) The amount of money coming into the bank at the beginning of the nth day which is already new currency is, working in million dollars,

$$\left(\frac{C_{n-1}}{5{,}000}\right) 18$$

So,

$$C_n = C_{n-1} + \left[18 - \left(\frac{C_{n-1}}{5{,}000}\right) 18\right]$$
$$C_n = \frac{4{,}982}{5{,}000} C_{n-1} + 18$$

Using $a = \dfrac{4{,}982}{5{,}000}$, $b = 18$, and $C_0 = 0$,

$$C_n = \left(\frac{4{,}982}{5{,}000}\right)^n (0) + \frac{1 - \left(\frac{4{,}982}{5{,}000}\right)^n}{1 - \frac{4{,}982}{5{,}000}}(18)$$
$$C_n = 5{,}000\left[1 - \left(\frac{4{,}982}{5{,}000}\right)^n\right]$$

(b) Need 0.9 (5,000) million to be new currency, so

$$4500 = 5{,}000\left[1 - \left(\frac{4{,}982}{5{,}000}\right)^n\right]$$
$$0.9 = 1 - \left(\frac{4{,}982}{5{,}000}\right)^n$$
$$\left(\frac{4{,}982}{5{,}000}\right)^n = 0.1$$
$$n \ln\left(\frac{4{,}982}{5{,}000}\right) = \ln 0.1$$
$$n = \frac{\ln 0.1}{\ln\left(\frac{4{,}982}{5{,}000}\right)}$$
$$n \approx 638.45 \text{ years}$$

(c) Solving the same problem using a first-order linear differential equation,

$$\frac{dC}{dt} = 18 - \left(\frac{C}{5{,}000}\right) 18$$
$$\frac{dC}{dt} = -\frac{18}{5{,}000}C + 18$$
$$\frac{dC}{dt} + \left(\frac{18}{5{,}000}\right) C = 18$$

The integrating factor is

$$I(t) = e^{\int \frac{18}{5{,}000}\, dt} = e^{\frac{18t}{5{,}000}}$$
$$C(t) = e^{-\frac{18t}{5{,}000}}\left[\int e^{\frac{18t}{5{,}000}} \cdot 18\, dt + c\right]$$
$$C(t) = e^{-\frac{18t}{5{,}000}}\left[5{,}000 e^{\frac{18t}{5{,}000}} + c\right]$$
$$C(t) = 5{,}000 + ce^{-\frac{18t}{5{,}000}}$$

When $t = 0$, $c(0) = 0$ so

$$0 = 5{,}000 + ce^0, \quad \text{or } c = -5{,}000$$

and

$$C(t) = 5{,}000 - 5{,}000e^{-\frac{18t}{5{,}000}}$$
$$4{,}500 = 5{,}000\left(1 - e^{-\frac{18t}{5{,}000}}\right)$$
$$0.9 = 1 - e^{-\frac{18t}{5{,}000}}$$
$$e^{-\frac{18t}{5{,}000}} = 0.1$$
$$-\frac{18t}{5{,}000} = \ln 0.1$$
$$t = \frac{-5{,}000 \ln 0.1}{18}$$
$$t \approx 639.61 \text{ years}$$

The resultant times are very close.

(d) Writing Exercise—Answers will vary.

37. $S_n = 3p_{n-1} - 2;\ D_n = -p_n + 1.3;\ p_0 = 1$

(a) Using $a = 1.3$, $b = 1$, $c = -2$ and $d = 3$, the equilibrium price is

$$p_e = \frac{1.3 + 2}{1 + 3} = 0.825$$

(b) Using $p_n = \left(-\dfrac{d}{b}\right) p_{n-1} + \left(\dfrac{a-c}{b}\right)$,

$$p_n = \left(-\frac{3}{1}\right) p_{n-1} + \left(\frac{1.3+2}{1}\right)$$
$$p_n = -3p_{n-1} + 3.3$$

Using $a = -3$, $b = 3.3$, and $p_0 = 1$,

$$p_n = (-3)^n(1) + \frac{1-(-3)^n}{1-(-3)}(3.3)$$
$$p_n = (-3)^n + 0.825[1-(-3)^n]$$
$$p_n = 0.175(-3)^n + 0.825$$

(c)

n	p_n	S_n	D_n
1	0.3	−1.1	1
2	2.4	5.2	−1.1
3	−3.9	−13.7	5.2

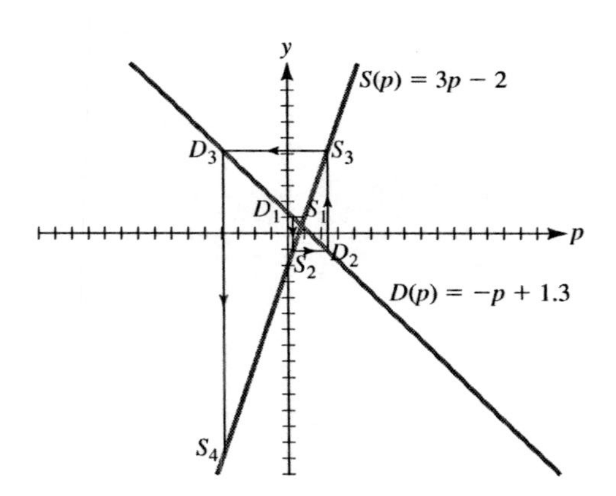

39. $S_n = 2p_{n-1} - 1$; $D_n = -2p_n + 7$; $p_0 = -3$

(a) Using $a = 7$, $b = 2$, $c = -1$, and $d = 2$, the equilibrium price is

$$p_e = \frac{7+1}{2+2} = 2$$

(b) Using $p_n = \left(-\frac{d}{b}\right) p_{n-1} + \left(\frac{a-c}{b}\right)$

$$p_n = \left(-\frac{2}{2}\right) p_{n-1} + \left(\frac{7-(-1)}{2}\right)$$
$$p_n = -p_{n-1} + 4$$

Using $a = -1$, $b = 4$, and $p_0 = 3$,

$$p_n = (-1)^n(3) + \frac{1-(-1)^n}{1-(-1)}(4)$$
$$p_n = 3(-1)^n + 2[1-(-1)^n]$$
$$p_n = (-1)^n + 2$$

(c)

n	p_n	S_n	D_n
1	1	1	5
2	3	5	1
3	1	1	5

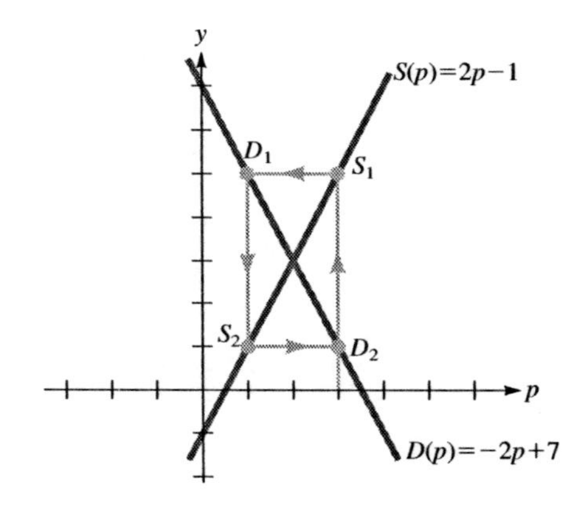

Checkup for Chapter 8

1. **(a)**

$$\frac{dy}{dx} = \frac{xy}{x^2+1}$$

Separating the variables,

$$\int \frac{dy}{y} = \int \frac{x}{x^2+1}\,dx$$

Using substitution for the right side integral with $u = x^2 + 1$ and $\frac{1}{2}du = x\,dx$,

$$\ln|y| = \frac{1}{2}\ln(x^2+1) + C_1$$
$$\ln|y| - \ln(x^2+1)^{\frac{1}{2}} = C_1$$

$$\ln\left|\frac{y}{\sqrt{x^2+1}}\right| = C_1$$

$$\frac{y}{\sqrt{x^2+1}} = e^{C_1}$$

$$y = C\sqrt{x^2+1}$$

(b)

$$\frac{dy}{dx} = \frac{2y}{x} + x^3$$

$$\frac{dy}{dx} + \left(-\frac{2}{x}\right)y = x^3$$

The integrating factor is

$$I(x) = e^{\int -\frac{2}{x}\,dx} = e^{-2\ln x} = e^{\ln x^{-2}} = \frac{1}{x^2}$$

$$y = x^2\left[\int \frac{1}{x^2}\cdot x^3\,dx + C\right]$$

$$y = x^2\left[\frac{x^2}{2} + C\right]$$

$$y = \frac{x^4}{2} + Cx^2$$

(c)

$$\frac{dy}{dx} = xe^{y-x}$$

$$\frac{dy}{dx} = xe^{-x}\cdot e^{y}$$

Separating the variables,

$$\int e^{-y}\,dy = \int xe^{-x}\,dx$$

Using integration by parts for the right side integral,

$$-e^{-y} = -xe^{-x} - \int -e^{-x}\,dx$$

$$-e^{-y} = -xe^{-x} - e^{-x} + C$$

$$e^{-y} = xe^{-x} + e^{-x} - e^{-x}e^{x}C$$

$$e^{-y} = e^{-x}(x + 1 - Ce^{x})$$

$$-y = \ln[e^{-x}(x + 1 - Ce^{x})]$$

$$-y = -x + \ln\left|x + 1 - Ce^{x}\right|$$

$$y = x - \ln\left|x + 1 - Ce^{x}\right|$$

2. (a)

$$\frac{dy}{dx} = -\frac{2}{x^2y}$$

Separating the variables,

$$\int y\,dy = \int -\frac{2}{x^2}\,dx$$

$$\frac{y^2}{2} = \frac{2}{x} + C_1$$

$$y^2 = \frac{4}{x} + C$$

When $x = -1$, $y = 1$ so

$$1 = -4 + C, \text{ or } C = 5$$

$$\text{and}\quad y^2 = \frac{4}{x} + 5$$

$$y = \sqrt{\frac{4}{x} + 5} = \sqrt{\frac{4 + 5x}{x}}$$

(b)

$$\frac{dy}{dx} - \left(\frac{1}{x}\right)y = 5$$

The integrating factor is

$$I(x) = e^{\int \frac{1}{x}\,dx} = e^{\ln x} = x$$

$$y = \frac{1}{x}\left[\int x\cdot 5\,dx + C\right]$$

$$y = \frac{1}{x}\left[\frac{5x^2}{2} + C\right]$$

$$y = \frac{5x}{2} + \frac{C}{x}$$

When $x = 1$, $y = 0$ so

$$0 = \frac{5}{2} + C, \text{ or } C = -\frac{5}{2}$$

$$y = \frac{5}{2}\left(x - \frac{1}{x}\right)$$

(c)

$$\frac{dy}{dx} + \left(\frac{x}{x^2+1}\right)y = x$$

The integrating factor, using substitution with $u = x^2 + 1$ and $\frac{1}{2}du = x\,dx$, is

$$I(x) = e^{\int \frac{x}{x^2+1}\,dx} = (x^2+1)^{\frac{1}{2}}$$

$$y = \frac{1}{(x^2+1)^{\frac{1}{2}}}\left[\int (x^2+1)^{\frac{1}{2}} \cdot x\,dx + C\right]$$

Using substitution again for the integration,

$$y = \frac{1}{(x^2+1)^{\frac{1}{2}}}\left[\frac{1}{3}(x^2+1)^{\frac{3}{2}} + C\right]$$

$$y = \frac{1}{3}(x^2+1) + \frac{C}{(x^2+1)^{\frac{1}{2}}}$$

When $x = 0$, $y = 0$ so

$$0 = \frac{1}{3} + C, \text{ or } C = -\frac{1}{3}$$

$$y = \frac{1}{3}\left[(x^2+1) - \frac{1}{(x+1)^{\frac{1}{2}}}\right]$$

3. **(a)**

$$y_n = -\frac{1}{2}y_{n-1}$$

Using $a = -\frac{1}{2}$, $b = 0$ and $y_0 = -1$,

$$y_n = \left(-\frac{1}{2}\right)^n(-1) + \frac{1-\left(-\frac{1}{2}\right)}{1-\left(-\frac{1}{2}\right)}(0)$$

$$y_n = -1\left(-\frac{1}{2}\right)^n = -1(-2)^{-n}$$

$$= -\left(-\frac{1}{2}\right)^n$$

(b)

$$y_n - y_{n-1} = \frac{1}{3}y_{n-1}$$

$$y_n = \frac{4}{3}y_{n-1}$$

Using $a = \frac{4}{3}$, $b = 0$ and $y_0 = 2$,

$$y_n = \left(\frac{4}{3}\right)^n(2) + \frac{1-\left(\frac{4}{3}\right)^n}{1-\left(\frac{4}{3}\right)}(0)$$

$$y_n = 2\left(\frac{4}{3}\right)^n$$

(c)

$$y_n - y_{n-1} = \frac{1}{4}y_{n-1} - \frac{1}{2}$$

$$y_n = \frac{5}{4}y_{n-1} - \frac{1}{2}$$

Using $a = \frac{5}{4}$, $b = -\frac{1}{2}$ and $y_0 = 0$,

$$y_n = \left(\frac{5}{4}\right)^n(0) + \frac{1-\left(\frac{5}{4}\right)^n}{1-\left(\frac{5}{4}\right)}\left(-\frac{1}{2}\right)$$

$$y_n = 2\left[1-\left(\frac{5}{4}\right)^n\right]$$

4. **(a)** To sketch solutions of the differential equation $y' = 2x - y$, start by drawing the family of parallel lines $2x - y = m$, or $y = 2x - m$, for several values of m. Along each such line, place small segments of slope m. This slope field approximates the solution curves, with the solution through the point (0, 0) drawn in.

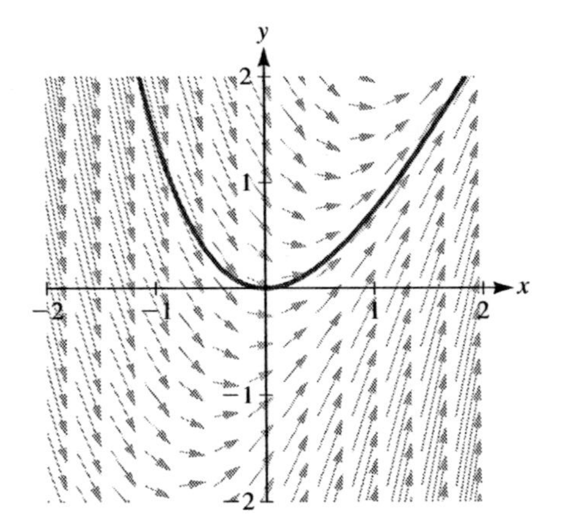

(b) To sketch solutions of the differential equation $y' = x^2 - y^2$, start by drawing the family of hyperboles $x^2 - y^2 = m$ for several values of m. Along each such hyperbola, place small segments of slope m. This slope field approximates the solution curves, with the solution through the point (1, 1) drawn in.

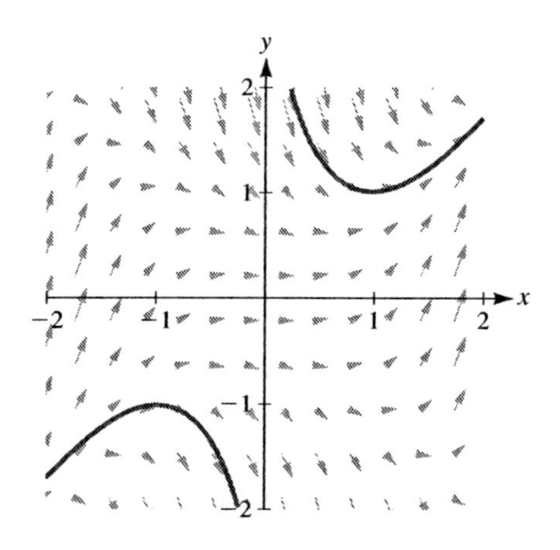

(c) To sketch solutions of the differential equation $y' = \frac{y}{x+1}$, start by drawing the family of lines $\frac{y}{x+1} = m$, or $y = m(x+1)$, for several values of m. Along each such line, place small segments of slope m. This slope field approximates the solution curves, with the solution through the point (0, 0) drawn in.

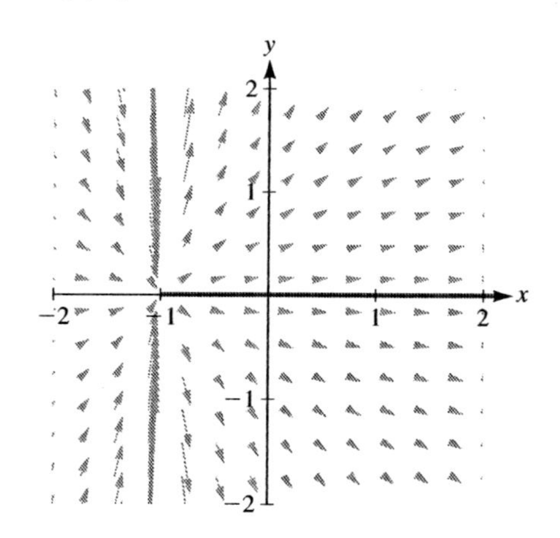

5.
$$\frac{dP}{dt} = -0.11P$$

6.
$$\frac{dP}{dt} = k\left[S(p) - D(p)\right]$$

where k is a constant of proportionality.

7. $\frac{dA}{dt} = kA$, where k is a constant of proportionality.

Separating the variables,

$$\int \frac{dA}{A} = \int k\,dt$$
$$\ln|A| = kt + C_1$$
$$|A| = e^{kt+C_1}$$
$$A = \pm e^{C_1} \cdot e^{kt}$$
$$A(t) = Ce^{kt}$$

When $t = 0$, $A(0) = A_0$ so

$$A_0 = Ce^0, \quad \text{or } C = A_0$$

and

$$A(t) = A_0 e^{kt}$$

When $t = 4$, $A(4) = \frac{1}{2}A_0$ so

$$\frac{1}{2}A_0 = A_0 e^{4k}$$
$$\ln\frac{1}{2} = 4k, \quad \text{or } k = \frac{\ln\frac{1}{2}}{4}$$

and

$$A(t) = A_0 e^{\left(\frac{\ln\frac{1}{2}}{4}\right)t}$$

For the drug to remain effective, need $A(2) =$ 1,800 mg, so

$$1{,}800 = A_0 e^{\left(\frac{\ln\frac{1}{2}}{4}\right)(2)}$$
$$1{,}800 = A_0 e^{\ln\left(\frac{1}{2}\right)^{\frac{1}{2}}}$$
$$1{,}800 = A_0\left(\frac{1}{2}\right)^{\frac{1}{2}}$$
$$A_0 = 1{,}800\left(\frac{1}{2}\right)^{-\frac{1}{2}} = 1{,}800\sqrt{2} \approx 2{,}546 \text{ mg}$$

8. $\frac{dV}{dt} = kV^2$, where k is a constant of proportionality.
Separating the variables,

$$\int \frac{dV}{V^2} = \int k\, dt$$
$$-\frac{1}{V} = kt + C$$
$$V(t) = -\frac{1}{kt + C}$$

When $t = 0$, $V(0) = 10{,}000$ so

$$10{,}000 = -\frac{1}{C}, \text{ or } C = -0.0001$$

and

$$V(t) = -\frac{1}{kt - 0.0001}$$

When $t = 10$, $V(10) = 20{,}000$ so

$$20{,}000 = -\frac{1}{10k - 0.0001}, \text{ or } k = 0.000005$$

and

$$V(t) = -\frac{1}{0.000005t - 0.0001}$$
$$V(t) = -\frac{200{,}000}{t - 20}$$

Need t when $V(t) = 50{,}000$ so

$$50{,}000 = -\frac{200{,}000}{t - 20}$$
$$t - 20 = -4$$
$$t = 16 \text{ years}$$

9.

$$\frac{dA}{dt} = \begin{pmatrix}\text{rate salt}\\ \text{enters}\end{pmatrix} - \begin{pmatrix}\text{rate salt}\\ \text{leaves}\end{pmatrix}$$
$$\frac{dA}{dt} = \begin{pmatrix}\text{salt per}\\ \text{gallon}\end{pmatrix}\begin{pmatrix}\text{rate gallons}\\ \text{enter}\end{pmatrix} - \begin{pmatrix}\text{salt per}\\ \text{gallon}\end{pmatrix}\begin{pmatrix}\text{rate gallons}\\ \text{leave}\end{pmatrix}$$
$$\frac{dA}{dt} = (2)(4) = \left(\frac{A(t)}{200 + 3t}\right)(1)$$
$$\frac{dA}{dt} = 8 - \frac{A}{200 + 3t}$$
$$\frac{dA}{dt} + \left(\frac{1}{200 + 3t}\right)A = 8$$

The integrating factor is

$$I(t) = e^{\int \frac{1}{200+3t}\, dt} = e^{\frac{1}{3}\ln(200+3t)}$$
$$= e^{\ln(200+3t)^{\frac{1}{3}}} = (200 + 3t)^{\frac{1}{3}}$$
$$A = (200 + 3t)^{-\frac{1}{3}}\left[\int (200 + 3t)^{\frac{1}{3}} \cdot 8\, dt + C\right]$$

Using substitution with $u = 200 + 3t$ and $\frac{1}{3}\, du = dt$,

$$A = (200 + 3t)^{-\frac{1}{3}}\left[2(200 + 3t)^{\frac{4}{3}} + C\right]$$
$$A(t) = 2(200 + 3t) + C(200 + 3t)^{-\frac{1}{3}}$$

When $t = 0$, $A(0) = 75$ so

$$75 = 2(200) + C(200)^{-\frac{1}{3}}, \text{ or}$$
$$C = -325(200)^{\frac{1}{3}}$$

and

$$A(t) = 400 + 6t - 325\left(\frac{200}{200 + 3t}\right)^{\frac{1}{3}}$$

When the tank is full,

$$200 + 3t = 500$$
$$t = 100 \text{ minutes}$$

$$A(100) = 400 + 6(100) - 325\left(\frac{200}{200 + 3(100)}\right)^{\frac{1}{3}}$$
$$A(100) \approx 760.54 \text{ lbs.}$$

10. $S_n = 20 + p_{n-1}$; $D_n = 50 - 2p_n$; $p_0 = 20$

(a) Using $a = 50$, $b = 2$, $c = 20$ and $d = 1$, the equilibrium price is

$$p_e = \frac{50 - 20}{2 + 1} = 10$$

(b) Using $p_n = \left(-\frac{d}{b}\right)p_{n-1} + \left(\frac{a - c}{b}\right)$

$$p_n = \left(-\frac{1}{2}\right)p_{n-1} + \left(\frac{50 - 20}{2}\right)$$
$$p_n = -\frac{1}{2}p_{n-1} + 15$$

Using $a = -\frac{1}{2}$, $b = 15$ and $p_0 = 20$,

$$p_n = \left(-\frac{1}{2}\right)^n (20) + \frac{1-\left(-\frac{1}{2}\right)^n}{1-\left(-\frac{1}{2}\right)}(15)$$

$$p_n = 20\left(-\frac{1}{2}\right)^n + 10\left[1-\left(-\frac{1}{2}\right)^n\right]$$

$$p_n = 10\left(-\frac{1}{2}\right)^n + 10 = 10\left[1+\left(-\frac{1}{2}\right)^n\right]$$

(c)

n	p_n	S_n	D_n
1	5	25	40
2	12.5	32.5	25
3	8.75	28.75	32.5

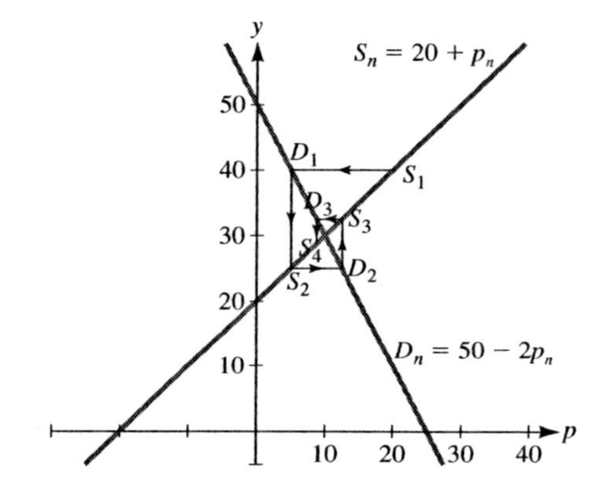

Review Problems

1.

$$\frac{dy}{dx} = x^3 - 3x^2 + 5,$$

$$y = \int (x^3 - 3x^2 + 5)\,dx$$

$$= \frac{x^4}{4} - x^3 + 5x + C.$$

3. Separate the variables of

$$\frac{dy}{dx} = k(80 - y)$$

and integrate to get

$$\int \frac{1}{80-y}\,dy = \int k\,dx,$$

$$-\ln|80 - y| = kx + C_1,$$

$$|80 - y| = e^{-kx-C_1} = e^{-C_1}e^{-kx},$$

$$80 - y = Ce^{-kx}, \text{ or } y = 80 - Ce^{-kx}$$

where $C = \pm e^{-C_1}$

5.

$$\frac{dy}{dx} = e^{x+y}$$

Using the hint and separating the variables,

$$\frac{dy}{dx} = e^x e^y$$

$$\int e^{-y}\,dy = \int e^x\,dx$$

$$-e^{-y} = e^x + C$$

$$e^{-y} = -e^x - C$$

$$-y = \ln(-e^x - C)$$

$$y = -\ln(-e^x - C)$$

7.

$$\frac{dy}{dx} + \left(\frac{4}{x}\right)y = e^{-x}$$

The integrating factor is

$$I(x) = e^{\int \frac{4}{x}\,dx} = e^{4\ln x} = e^{\ln x^4} = x^4$$

$$y = \frac{1}{x^4}\left[\int x^4 \cdot e^{-x}\,dx + C\right]$$

Using an integral table reduction formula (or integration by parts four times)

$$y = \frac{1}{x^4}\left[\frac{1}{-1}x^4e^{-x} - \frac{4}{-1}\int x^3e^{-x}\,dx + C\right]$$

$$y = \frac{1}{x^4}\left[-x^4e^{-x} + 4\left(\frac{1}{-1}x^3e^{-x} - \frac{3}{-1}\int x^2e^{-x}\,dx\right) + C\right]$$

$$y = \frac{1}{x^4}\left[-x^4e^{-x} - 4x^3e^{-x} + 12\left(\frac{1}{-1}x^2e^{-x} - \frac{2}{-1}\int xe^{-x}\,dx\right) + C\right]$$

$$y = \frac{1}{x^4}\left[-x^4e^{-x} - 4x^3e^{-x} - 12x^2e^{-x} + 24\left(\frac{1}{-1}xe^{-x} - \frac{1}{-1}\int e^{-x}\,dx\right) + C\right]$$

$$y = \frac{1}{x^4}\left[-x^4e^{-x} - 4x^3e^{-x} - 12x^2e^{-x} - 24xe^{-x} - 24e^{-x} + C\right]$$

$$y = -e^{-x} - \frac{4}{x}e^{-x} - \frac{12}{x^2}e^{-x} - \frac{24}{x^3}e^{-x} - \frac{24}{x^4}e^{-x} + \frac{C}{x^4}$$

$$y = \frac{C}{x^4} - e^{-x}\left(1 + \frac{4}{x} + \frac{12}{x^2} + \frac{24}{x^3} + \frac{24}{x^4}\right)$$

9.
$$\frac{dy}{dx} + \left(\frac{2}{x+1}\right)y = \frac{4}{x+2}$$

The integrating factor is

$$I(x) = e^{\int \frac{2}{x+1}\,dx} = e^{2\ln(x+1)} = e^{\ln(x+1)^2} = (x+1)^2$$

$$y = \frac{1}{(x+1)^2}\left[\int (x+1)^2 \cdot \frac{4}{x+2}\,dx + C\right]$$

$$y = \frac{1}{(x+1)^2}\left[\int \frac{4(x^2+2x+1)}{x+2}\,dx + C\right]$$

$$y = \frac{1}{(x+1)^2}\left[4\int \frac{x(x+2)}{x+2} + \frac{1}{x+2}\,dx + C\right]$$

$$y = \frac{1}{(x+1)^2}\left[4\left(\frac{x^2}{2} + \ln|x+2|\right) + C\right]$$

$$y = \frac{1}{(x+1)^2}\left[2x^2 + 4\ln|x+2| + C\right]$$

11.
$$\frac{dy}{dx} = \frac{\ln x}{xy}$$

Separating the variables,

$$\int y\,dy = \int \frac{\ln x}{x}\,dx$$

Using substitution with $u = \ln x$ and $du = \frac{1}{x}\,dx$,

$$\frac{y^2}{2} = \frac{(\ln x)^2}{2} + C_1$$

$$y^2 = (\ln x)^2 + 2C_1$$

$$y^2 = (\ln x)^2 + C$$

$$y = \pm\sqrt{(\ln x)^2 + C}$$

13.
$$\frac{dy}{dx} = 5x^4 - 3x^2 - 2,$$

$$y = \int (5x^4 - 3x^2 - 2)\,dx = x^5 - x^3 - 2x + C.$$

Since $y = 4$ when $x = 1$, $4 = 1 - 1 - 2 + C$, or $C = 6$.

So,

$$y = x^5 - x^3 - 2x + 6$$

15.
$$\frac{dy}{dx} + (-0.06)y = 0$$

The integrating factor is

$$I(x) = e^{\int -0.06\,dx} = e^{-0.06x}$$

$$y = e^{0.06x}\left[\int e^{-0.06x} \cdot 0\,dx + C\right]$$

$$y = Ce^{0.06x}$$

Since $y = 100$ when $x = 0$,

$$100 = Ce^0, \text{ or } C = 100$$

So,

$$y = 100e^{0.06x}$$

17.
$$\frac{dy}{dx} + \left(-\frac{5}{x}\right)y = x^2$$

The integrating factor is

$$I(x) = e^{\int -\frac{5}{x}} = e^{-5\ln x} = e^{\ln x^{-5}} = \frac{1}{x^5}$$

$$y = x^5\left[\int \frac{1}{x^5}\cdot x^2\,dx + C\right]$$
$$y = x^5\left[-\frac{1}{2x^2} + C\right]$$
$$y = -\frac{1}{2}x^3 + Cx^5$$

Since $y = 4$ when $x = -1$,

$$4 = \frac{1}{2} - C, \text{ or } C = -\frac{7}{2}$$
$$y = -\frac{1}{2}x^3 - \frac{7}{2}x^5$$

19. $$\frac{dy}{dx} + (-x)y = e^{\frac{x^2}{2}}$$

The integrating factor is

$$I(x) = e^{\int -x\,dx} = e^{-\frac{x^2}{2}}$$
$$y = e^{\frac{x^2}{2}}\left[\int e^{-\frac{x^2}{2}}\cdot e^{\frac{x^2}{2}}\,dx + C\right]$$
$$y = e^{\frac{x^2}{2}}[x + C] = xe^{\frac{x^2}{2}} + Ce^{\frac{x^2}{2}}$$

Since $y = 4$ when $x = 0$,

$$4 = 0 + Ce^0, \text{ or } C = 4$$
$$y = (x+4)e^{\frac{x^2}{2}}$$

21. $$\frac{dy}{dx} = y \ln\sqrt{x}$$

Separating the variables,

$$\int \frac{dy}{y} = \int \ln\sqrt{x}\,dx$$
$$\int \frac{dy}{y} = \frac{1}{2}\int \ln x\,dx$$

Using integration by parts with $u = \ln x$ and $dV = dx$,

$$\ln|y| = \frac{1}{2}\left[x\ln x - \int \frac{1}{x}\cdot x\,dx\right] + C_1$$
$$\ln|y| = \frac{1}{2}[x\ln x - x] + C_1$$
$$\ln|y| = \ln x^{\frac{x}{2}} - \frac{x}{2} + C_1$$
$$|y| = e^{\ln x^{\frac{x}{2}} - \frac{x}{2} + C_1}$$
$$|y| = e^{\ln x^{\frac{x}{2}}}\cdot e^{-\frac{x}{2}}\cdot e^{C_1}$$
$$|y| = C_2x^{\frac{x}{2}}\cdot e^{-\frac{x}{2}}$$
$$y = \pm C_2x^{\frac{x}{2}}\cdot e^{-\frac{x}{2}}$$
$$y = C_3x^{\frac{x}{2}}e^{-\frac{x}{2}}$$

Since $y = -4$ when $x = 1$,

$$-4 = C_3(1)^{\frac{1}{2}}e^{\left(-\frac{1}{2}\right)}$$
$$C_3 = -4e^{\frac{1}{2}}$$
$$y = -4e^{\frac{1}{2}}x^{\frac{x}{2}}e^{-\frac{x}{2}}$$
$$y = -4x^{\frac{x}{2}}e^{\frac{(1-x)}{2}}$$

23. $$\frac{dy}{dx} = \frac{xy}{\sqrt{1-x^2}}$$
$$\frac{dy}{y} = \frac{x\,dx}{\sqrt{1-x^2}}$$

Using substitution with $u = 1 - x^2$,

$$\ln|\,y\,| = -\sqrt{1-x^2} + C$$

Since $y = 2$ when $x = 0$, $C = 1 + \ln 2$ and

$$\ln|\,y\,| = -\sqrt{1-x^2} + 1 + \ln 2$$
$$\ln\left|\frac{y}{2}\right| = 1 - \sqrt{1-x^2}$$
$$y = 2e^{1-\sqrt{1-x^2}}$$

25. $$y_n = y_{n-1} + 2$$

Using $a = 1$, $b = 2$ and $y_0 = 1$,

$$y_n = 1 + 2n$$

27. $$y_n - y_{n-1} = 3$$
$$y_n = y_{n-1} + 3$$

Using $a = 1$, $b = 3$ and $y_0 = 1$,

$$y_n = 1 + 3n$$

29.

$$y_n - 5y_{n-1} = 2$$
$$y_n = 5y_{n-1} + 2$$

Using $a = 5$, $b = 2$ and $y_0 = 2$,

$$y_n = (5)^n(2) + \frac{1-(5)^n}{1-5}(2)$$
$$y_n = 2(5)^n - \frac{1}{2}[1-(5)^n]$$
$$y_n = \frac{5}{2}(5)^n - \frac{1}{2}$$
$$y_n = \frac{1}{2}(5)^{n+1} - \frac{1}{2} = \frac{1}{2}\left[(5)^{n+1} - 1\right]$$

31. To sketch solutions of the differential equation $y' = 3x - y$, start by drawing the family of parallel lines $3x - y = m$, or $y = 3x - m$, for several values of m. Along each such line, place small segments of slope m. This slope field approximates the solution curves, with the solution through the point (1, 0) drawn in.

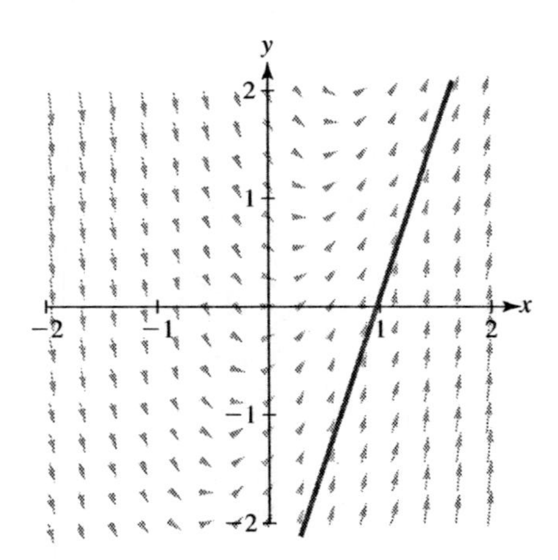

33. To sketch solutions of the differential equation $y' = x + y^2$, start by drawing the family of parabolas $x + y^2 = m$, or $x = -y^2 + m$, for several values of m. Along each such line, place small segments of slope m. This slope field approximates the solution curves, with the solution through the point (1, 1) drawn in.

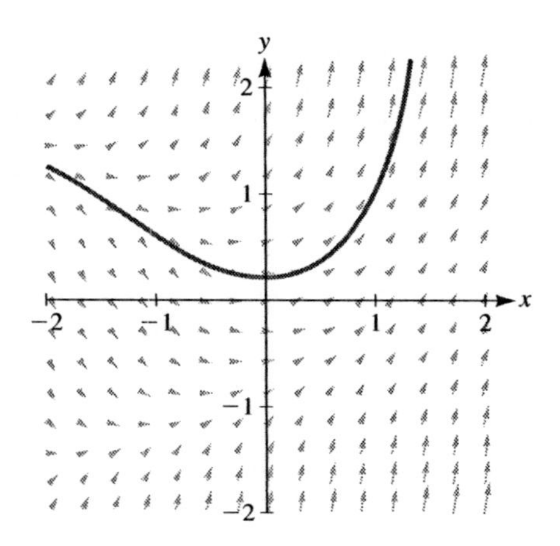

35. To sketch solutions of the differential equation $y' = \frac{y}{x} + \frac{x}{y}$, start by drawing the family of curves $\frac{y}{x} + \frac{x}{y} = m$, for several values of m. Along each such line, place small segments of slope m. This slope field approximates the solution curves, with the solution through the point (1, 0) drawn in.

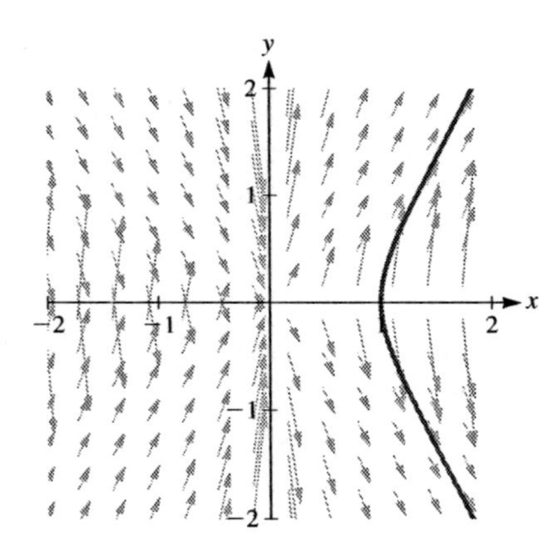

37. Using Euler's method with $f(x, y) = 3x - y$,

k	x_k	y_k	$f(x_k, y_k)$
0	0	0	0
1	0.1	0	0.3
2	0.2	0.03	0.57
3	0.3	0.087	0.813
4	0.4	0.1683	1.0317
5	0.5	0.27147	

Plotting the points (x_k, y_k) and connecting successive points with line segments yields the Euler solution to the differential equation for the given initial value.

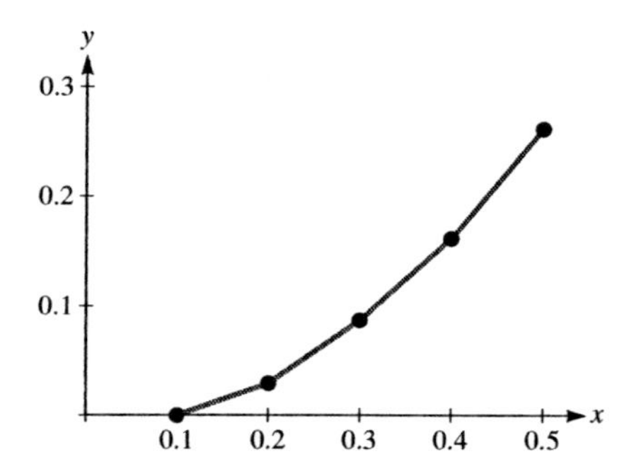

39. Using Euler's method with $f(x, y) = \frac{y}{x} + y$,

k	x_k	y_k	$f(x_k, y_k)$
0	1	1	2
1	1.2	1.4	2.56667
2	1.4	1.91333	3.28000
3	1.6	2.56933	4.17516
4	1.8	3.40437	

Plotting the points (x_k, y_k) and connecting successive points with line segments yields the Euler solution to the differential equation for the given initial value.

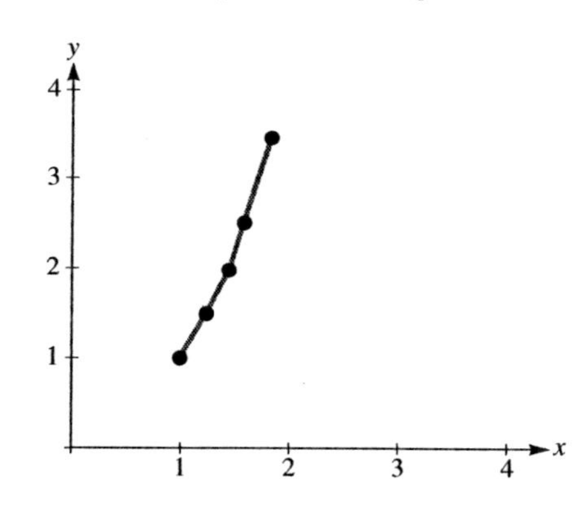

41. Using Euler's method with $f(x, y) = x + 4y$,

k	x_k	y_k	$f(x_k, y_k)$
0	0	1	4
1	0.2	1.8	7.4
2	0.4	3.28	13.52
3	0.6	5.984	24.536
4	0.8	10.8912	44.3648
5	1.0	19.76416	

The estimate for $y(1)$ is 19.76416.

43. Using Euler's method with $f(x, y) = e^{xy}$,

k	x_k	y_k	$f(x_k, y_k)$
0	0	0	1
1	0.4	0.4	1.17351
2	0.8	0.86940	2.00476
3	1.2	1.67130	7.43028
4	1.6	4.64341	1684.9000
5	2.0	678.59971	

The estimate for $y(2)$ is 678.5997.

45.

$$\frac{dN}{dt} = k(A - N)$$

where k is a constant of proportionality. Separating the variables,

$$\int \frac{dN}{A - N} = \int k\,dt$$

$$-\ln(A - N) = kt + C_1$$

$$\ln(A - N) = -kt - C_1$$

$$A - N = e^{-kt - C_1}$$

$$A - N = Ce^{-kt}$$

$$N(t) = A - Ce^{-kt}$$

When $t = 0$, $N(0) = 0$ and

$$0 = A - Ce^0, \text{ or } C = A$$

and

$$N(t) = A - Ae^{-kt} = A(1 - e^{-kt})$$

47. Let $Q(t)$ denote the number of pounds of salt in the tank after t minutes.
Then $\frac{dQ}{dt}$ is the rate of change of salt with respect to time (measured in pounds per minute).

$$\frac{dQ}{dt} = \text{(rate at which salt enters)}$$
$$- \text{(rate at which salt leaves)}$$
$$= \frac{\text{pounds entering}}{\text{gallon}} \frac{\text{gallons entering}}{\text{minute}}$$
$$- \frac{\text{pounds leaving}}{\text{gallon}} \frac{\text{gallons leaving}}{\text{minute}}.$$

Now, $\frac{\text{pounds leaving}}{\text{gallon}}$

$$= \frac{\text{pounds of salt in the tank}}{\text{gallons of brine in the tank}}$$
$$= \frac{Q}{200}.$$

So, $\frac{dQ}{dt} = -\frac{Q}{200}(4) = -\frac{Q}{50}$.
Separate the variables and integrate to get

$$\int \frac{1}{Q} dQ = -\int \frac{1}{50}\, dt,$$
$$\ln |\, Q \,| = -\frac{t}{50} + C_1,$$
$$Q = e^{C_1} e^{-t/50} = Ce^{-t/50},$$

where $C = e^{C_1}$. Since there are initially 600 pounds of salt in the tank (3 pounds of salt per gallon times 200 gallons), $600 = Q(0) = C$. So,

$$Q(t) = 600e^{-t/50}$$

The amount of salt in the tank after 100 minutes is $Q(100) = 600e^{-2} = 81.2012$ pounds.

49.
$$\frac{dP}{dt} = kP(10 - P)$$

where k is a constant of proportionality.
Separating the variables,

$$\int \frac{dP}{P(10-P)} = \int k\, dt$$

Using integral tables,

$$\frac{1}{10} \ln \left| \frac{P}{10-P} \right| = kt + C_1$$
$$\ln \left(\frac{P}{10-P} \right) = 10kt + 10C_1$$
$$\frac{P}{10-P} = e^{10kt + 10C_1}$$
$$\frac{P}{10-P} = Ce^{10kt}$$

When $t = 0$, $P = 4$ and

$$\frac{2}{3} = Ce^0, \text{ or } C = \frac{2}{3}$$
$$\frac{P}{10-P} = \frac{2}{3}e^{10kt}$$

Further, when $t = 4$, $P = 7.4$ so

$$\frac{7.4}{10-7.4} = \frac{2}{3}e^{10k(4)}$$
$$\frac{111}{26} = e^{40k}$$
$$\ln \frac{111}{26} = 40k$$
$$k = \frac{\ln \frac{111}{26}}{40} \approx 0.036286$$
$$\frac{P}{10-P} = \frac{2}{3}e^{0.36286t}$$
$$\frac{3}{2}P = 10e^{0.36286t} - Pe^{0.36286t}$$
$$\frac{3}{2}P + Pe^{0.36286t} = 10e^{0.36286t}$$
$$\left(\frac{3}{2} + e^{0.36286t}\right) P = 10e^{0.36286t}$$
$$P(t) = \frac{10e^{0.36286t}}{\frac{3}{2} + e^{0.36286t}} = \frac{20e^{0.36286t}}{3 + 2e^{0.36286t}}$$

When $t = 10$,

$$P(10) = \frac{20e^{0.36286(10)}}{3 + 2e^{0.36286(10)}} \approx 9.617 \text{ million}$$

51. (a)
$$\frac{dC}{dt} = R - kC$$

Separating the variables,

$$\int \frac{dC}{R - kC} = \int dt$$

$$-\frac{1}{k} \ln|R - kC| = t + C_1$$

$$\ln|R - kC| = -kt - kC_1$$

$$|R - kC| = e^{-kt - kC_1}$$

$$|R - kC| = e^{-kC_1} \cdot e^{-kt}$$

$$R - kC = \pm e^{-kC_1} e^{-kt}$$

$$R - kC = C_2 e^{-kt}$$

$$C(t) = \frac{R - C_2 e^{-kt}}{k}$$

When $t = 0$, $C(0) = C_0$ and

$$C_0 = \frac{R - C_2 e^0}{k}, \text{ or } C_2 = R - kC_0$$

and

$$C(t) = \frac{R - (R - kC_0)e^{-kt}}{k}$$

$$C(t) = \frac{R}{k} + \left(C_0 - \frac{R}{k}\right) e^{-kt}$$

(b)

$$\lim_{t \to \infty} \frac{R}{k} + \left(C_0 - \frac{R}{k}\right) e^{-kt} = \frac{R}{k}$$

(c)

$$\frac{1}{2} C_0 = \frac{R}{k} + \left(C_0 - \frac{R}{k}\right) e^{-kt}$$

$$\frac{\frac{1}{2}C_0 - \frac{R}{k}}{C_0 - \frac{R}{k}} = e^{-kt}$$

$$\frac{\frac{C_k - 2R}{2k}}{\frac{C_0 k - R}{k}} = e^{-kt}$$

$$\frac{C_0 k - 2R}{2(C_0 k - R)} = e^{-kt}$$

$$\ln\left(\frac{C_0 k - 2R}{2(C_0 k - R)}\right) = -kt$$

$$t = -\frac{1}{k} \ln\left(\frac{C_0 k - 2R}{2(C_0 k - R)}\right)$$

$$t = -\frac{1}{k} \ln\left[\frac{1}{2}\left(\frac{C_0 k - R - R}{C_0 k - R}\right)\right]$$

$$t = -\frac{1}{k} \ln\left[\frac{1}{2}\left(1 - \frac{R}{C_0 k - R}\right)\right]$$

$$t = -\frac{1}{k} \ln\left[\frac{1}{2}\left(1 + \frac{R}{R - C_0 k}\right)\right]$$

53.

$$E_n = 1.05 E_{n-1}$$

Using $a = 1.05$, $b = 0$ and $E_0 = 100$,

$$E_n = (1.05)^n (100) + \frac{1 - (1.05)^n}{1 - (1.05)}(0)$$

$$E_n = 100(1.05)^n$$

55. (a)

$$p_n = 0.3 p_{n-1}$$

Using $a = 0.3$ and $b = 0$,

$$p_n = (0.3)^n p_0 + \frac{1 - (0.3)^n}{1 - (0.3)}(0)$$

$$p_n = (0.3)^n p_0$$

(b) The probability that a defective unit receives its imperfection during one of the stages of production is 100%, or 1.

$$1 = \sum_{i=0}^{\infty} (0.3)^i p_0$$

$$1 = \frac{1}{1 - 0.3} p_0, \text{ or } p_0 = 0.7$$

So, $p_n = 0.7(0.3)^n$

and $p_3 = 0.7(0.3)^3 = 0.0189$, or 1.89%

57.

$$P_n = \frac{bP_{n-1}}{a + P_{n-1}}$$

(a)

$$P_{n-1} = \frac{bP_{n-2}}{a + P_{n-2}}$$

So, $$\frac{b}{a} P_{n-1} = \frac{b^2 P_{n-2}}{a(a + P_{n-2})}$$

Now,

$$P_n = \frac{bP_{n-1}}{a + P_{n-1}} = \frac{b\left(\frac{bP_{n-2}}{a+P_{n-2}}\right)}{a + \left(\frac{bP_{n-2}}{a+P_{n-2}}\right)}$$

$$= \frac{\frac{b^2 P_{n-2}}{a+P_{n-2}}}{\frac{a(a+P_{n-2})+bP_{n-2}}{a+P_{n-2}}} = \frac{b^2 P_{n-2}}{a(a + P_{n-2}) + bP_{n-2}}$$

Comparing P_n and $\frac{b}{a}P_{n-1}$, note that the numerators are the same, but the denominator of P_n is larger (since $b > 0$ and $P_{n-2} > 0$).

$$\text{So,} \quad P_n < \frac{b}{a}P_{n-1}$$

$$P_n < \frac{b}{a}\left(\frac{b}{a}P_{n-2}\right)$$

$$P_n < \frac{b}{a}\left[\frac{b}{a}\left(\frac{b}{a}P_{n-3}\right)\right]$$

$$\vdots$$

$$P_n < \left(\frac{b}{a}\right)^n P_0$$

(b)

$$\lim_{n\to\infty} P_n < \lim_{n\to\infty}\left[\left(\frac{b}{a}\right)^n P_0\right]$$

Since $a > b$, $\lim_{n\to\infty}\left(\frac{b}{a}\right)^n = 0$ and $\lim_{n\to\infty} P_n = 0$.

(c) Let $L = \lim_{n\to\infty} P_n$. Then

$$\lim_{n\to\infty} \frac{bP_{n-1}}{a + P_{n-1}} = L$$

But, $\lim_{n\to\infty} P_n = \lim_{n\to\infty} P_{n-1} = L$. So,

$$\lim_{n\to\infty} \frac{bP_{n-1}}{a + P_{n-1}} = \frac{bL}{a + L} = L$$

$$bL = aL + L^2$$

$$0 = L^2 + (a - b)L$$

$$0 = L[L + (a - b)]$$

So, $L = 0$ or $L = b - a$. Since $P_0 > 0$ and the colony grows, $L = 0$ is rejected and

$$\lim_{n\to\infty} P_n = b - a$$

59. (a) $R_1 = 1$ is the initial pair of rabbits, which is now one month old.
$R_2 = 1$ since the initial pair is just now old enough to reproduce.
$R_3 = 2$ since the first pair remains and have reproduced a second pair, but the second pair isn't old enough to reproduce.
$R_4 = 3$ since the initial pair remains, they have reproduced again, and the second pair is just now old enough to reproduce.
$R_n = R_{n-1} + R_{n-2}$ since the number of rabbit pairs after n months = number pairs last month + number new rabbits = R_{n-1} + number of newly reproductive pairs = $R_{n-1} + R_{n-2}$.

(b)

$$R_n = R_{n-1} + R_{n-2}$$

Since $R_n = \lambda^n$

$$\lambda^n = \lambda^{n-1} + \lambda^{n-2}$$

Dividing both sides by λ^{n-2},

$$\lambda^2 = \lambda + 1$$

$$\lambda^2 - \lambda - 1 = 0$$

$$\lambda = \frac{1 \pm \sqrt{1+4}}{2} = \frac{1 \pm \sqrt{5}}{2}$$

(c)

$$R_n = A\left(\frac{1+\sqrt{5}}{2}\right)^n + B\left(\frac{1-\sqrt{5}}{2}\right)^n$$

Since $R_1 = 1$,

$$1 = A\left(\frac{1+\sqrt{5}}{2}\right) + B\left(\frac{1-\sqrt{5}}{2}\right)$$

$$2 = A\left(1+\sqrt{5}\right) + B\left(1-\sqrt{5}\right)$$

$$B = \frac{2 - A(1+\sqrt{5})}{1-\sqrt{5}}$$

Since $R_2 = 1$,

$$1 = A\left(\frac{1+\sqrt{5}}{2}\right)^2 + B\left(\frac{1-\sqrt{5}}{2}\right)^2$$

$$1 = A\left(\frac{1+\sqrt{5}}{2}\right)^2 + \left[\frac{2 - A(1+\sqrt{5})}{1-\sqrt{5}}\right]\left(\frac{1-\sqrt{5}}{2}\right)^2$$

$$4 = A(1+\sqrt{5})^2 + [2 - A(1+\sqrt{5})](1-\sqrt{5})$$

$$4 = 6A + 2A\sqrt{5} + 2 - A(1+\sqrt{5}) - 2\sqrt{5} + \sqrt{5}A(1+\sqrt{5})$$

$$4 = 6A + 2\sqrt{5}A + 2 - A - \sqrt{5}A - 2\sqrt{5} + \sqrt{5}A + 5A$$

$$4 = 10A + 2\sqrt{5}A - 2\sqrt{5} + 2$$

$$2 + 2\sqrt{5} = 10A + 2\sqrt{5}A$$

$$1 + \sqrt{5} = 5A + \sqrt{5}A$$

$$A = \frac{1+\sqrt{5}}{5+\sqrt{5}}$$

$$= \frac{1+\sqrt{5}}{5+\sqrt{5}} \cdot \frac{5-\sqrt{5}}{5-\sqrt{5}}$$

$$= \frac{4\sqrt{5}}{20} = \frac{\sqrt{5}}{5} = \frac{1}{\sqrt{5}}$$

$$B = \frac{2 - \frac{1}{\sqrt{5}}(1+\sqrt{5})}{1-\sqrt{5}}$$

$$= \frac{1 - \frac{1}{\sqrt{5}}}{1-\sqrt{5}} \cdot \frac{1+\sqrt{5}}{1+\sqrt{5}}$$

$$= \frac{\sqrt{5} - \frac{1}{\sqrt{5}}}{-4} = \frac{\frac{4}{\sqrt{5}}}{-4} = -\frac{1}{\sqrt{5}}$$

(d) Writing Exercise—Answers will vary.

Chapter 9

Infinite Series and Taylor Series Approximations

9.1 Infinite Series

1.

$$\frac{1}{3}+\frac{1}{9}+\frac{1}{27}+\frac{1}{81}+\cdots$$
$$=\frac{1}{3}+\frac{1}{3^2}+\frac{1}{3^3}+\frac{1}{3^4}+\cdots$$
$$=\sum_{n=1}^{\infty}\frac{1}{3^n}$$

Note: there are additional ways to write this summation.

3.

$$\frac{1}{2}+\frac{2}{3}+\frac{3}{4}+\frac{4}{5}+\cdots$$
$$=\frac{1}{1+1}+\frac{2}{2+1}+\frac{3}{3+1}+\frac{4}{4+1}+\cdots$$
$$=\sum_{n=1}^{\infty}\frac{n}{n+1}$$

Again, there are additional ways to write this summation.

5.

$$\frac{1}{2}-\frac{4}{3}+\frac{9}{4}-\frac{16}{5}+\cdots$$
$$=\frac{1}{2}+\frac{-4}{3}+\frac{9}{4}+\frac{-16}{5}+\cdots$$
$$=\frac{(-1)^2(1)^2}{1+1}+\frac{(-1)^3(2)^2}{2+1}+\frac{(-1)^4(3)^2}{3+1}$$
$$+\frac{(-1)^5(4)^2}{4+1}+\cdots$$
$$=\sum_{n=1}^{\infty}\frac{(-1)^{n+1}(n)^2}{n+1}=\sum_{n=1}^{\infty}(-1)^{n+1}\frac{n^2}{n+1}$$

Again, there are additional ways to write this summation.

7. $\sum_{n=1}^{\infty}\frac{1}{2^n}$

$$S_4=\frac{1}{2^1}+\frac{1}{2^2}+\frac{1}{2^3}+\frac{1}{2^4}$$
$$=\frac{1}{2}+\frac{1}{4}+\frac{1}{8}+\frac{1}{16}=\frac{15}{16}$$

9. $\sum_{n=1}^{\infty}\frac{(-1)^n}{n}$

$$S_4=\frac{(-1)^1}{1}+\frac{(-1)^2}{2}+\frac{(-1)^3}{3}+\frac{(-1)^4}{4}$$
$$=-1+\frac{1}{2}-\frac{1}{3}+\frac{1}{4}=-\frac{7}{12}$$

11. $\sum_{n=0}^{\infty}\left(\frac{4}{5}\right)^n$

Since $\left|\frac{4}{5}\right| < 1$ and the series starts with $n = 0$,

$$= \frac{1}{1-\frac{4}{5}} = \frac{1}{\frac{1}{5}} = 5$$

13. $$\sum_{n=}^{\infty}\frac{2}{3^n} = \sum_{n=0}^{\infty} 2\left(\frac{1}{3}\right)^n$$

Since $\left|\frac{1}{3}\right| < 1$ and the series starts with $n = 0$,

$$= \frac{2}{1-\frac{1}{3}} = \frac{2}{\frac{2}{3}} = 2\cdot\frac{3}{2} = 3$$

15. $$\sum_{n=1}^{\infty}\left(\frac{3}{2}\right)^n = \frac{3}{2} + \left(\frac{3}{2}\right)^2 + \left(\frac{3}{2}\right)^3 + \left(\frac{3}{2}\right)^4 + \cdots$$

$$= \frac{3}{2}\left[1 + \left(\frac{3}{2}\right) + \left(\frac{3}{2}\right)^2 + \left(\frac{3}{2}\right)^3 + \cdots\right]$$

$$= \sum_{n=0}^{\infty}\frac{3}{2}\left(\frac{3}{2}\right)^n$$

Since $\left|\frac{3}{2}\right| > 1$ and the series starts with $n = 0$, the series diverges.

17.

$$\sum_{n=2}^{\infty}\frac{3}{(-4)^n} = 3\sum_{n=2}^{\infty}\left(-\frac{1}{4}\right)^n$$

$$= 3\left[\left(-\frac{1}{4}\right)^2 + \left(-\frac{1}{4}\right)^3 + \left(-\frac{1}{4}\right)^4 + \left(-\frac{1}{4}\right)^5 + \cdots\right]$$

$$= 3\left(-\frac{1}{4}\right)^2\left[1 + \left(-\frac{1}{4}\right) + \left(-\frac{1}{4}\right)^2 + \left(-\frac{1}{4}\right)^3 + \cdots\right]$$

$$= \frac{3}{16}\sum_{n=0}^{\infty}\left(-\frac{1}{4}\right)^n$$

Since $\left|-\frac{1}{4}\right| < 1$ and the series starts with $n = 0$,

$$= \frac{3}{16}\left[\frac{1}{1-\left(-\frac{1}{4}\right)}\right] = \frac{3}{16}\left[\frac{1}{\frac{5}{4}}\right] = \frac{3}{16}\cdot\frac{4}{5} = \frac{3}{20}$$

19. $$\sum_{n=1}^{\infty} 5(0.9)^n$$

$$= 5\left[0.9 + (0.9)^2 + (0.9)^3 + (0.9)^4 + \cdots\right]$$

$$= 5(0.9)\left[1 + (0.9) + (0.9)^2 + (0.9)^3 + \cdots\right]$$

$$= 4.5\sum_{n=0}^{\infty}(0.9)^n$$

Since $|0.9| < 1$ and the series starts with $n = 0$,

$$= 4.5\left[\frac{1}{1-0.9}\right] = 4.5\left[\frac{1}{0.1}\right] = 45$$

21. $$\sum_{n=1}^{\infty}\frac{3n}{4^{n+2}} = \sum_{n=1}^{\infty}\frac{3^n}{4^2\cdot 4^n} = \frac{1}{16}\sum_{n=1}^{\infty}\left(\frac{3}{4}\right)^n$$

$$= \frac{1}{16}\left[\frac{3}{4} + \left(\frac{3}{4}\right)^2 + \left(\frac{3}{4}\right)^3 + \left(\frac{3}{4}\right)^4 + \cdots\right]$$

$$= \frac{1}{16}\left(\frac{3}{4}\right)\left[1 + \left(\frac{3}{4}\right) + \left(\frac{3}{4}\right)^2 + \left(\frac{3}{4}\right)^3 + \cdots\right]$$

$$= \frac{3}{64}\sum_{n=0}^{\infty}\left(\frac{3}{4}\right)^n$$

Since $\left|\frac{3}{4}\right| < 1$ and the series starts with $n = 0$,

$$= \frac{3}{64}\left[\frac{1}{1-\frac{3}{4}}\right] = \frac{3}{64}\left[\frac{1}{\frac{1}{4}}\right] = \frac{3}{64}\cdot 4 = \frac{3}{16}$$

23. $$\sum_{n=0}^{\infty}\frac{4^{n+1}}{5^{n-1}} = \sum_{n=0}^{\infty}\frac{4^1\cdot 4^n}{5^{-1}\cdot 5^n} = 20\sum_{n=0}^{\infty}\left(\frac{4}{5}\right)^n$$

Since $\left|\frac{4}{5}\right| < 1$ and the series starts with $n = 0$,

$$= 20\left[\frac{1}{1-\frac{4}{5}}\right] = 20\left[\frac{1}{\frac{1}{5}}\right] = 20\cdot 5 = 100$$

25. $\sum_{k=1}^{\infty} 2a_k = 2\sum_{k=1}^{\infty} a_k = 2.5 = 10$

27. $$= \sum_{k=1}^{\infty}\left[\frac{1}{2^k} - b_k\right] = \sum_{k=1}^{\infty}\frac{1}{2^k} - \sum_{k=1}^{\infty} b_k$$
$$= \sum_{k=1}^{\infty}\left(\frac{1}{2}\right)^k - \sum_{k=1}^{\infty} b_k$$

In the first term, $\left|\frac{1}{2}\right| < 1$ but series starts with $k = 1$. Rewriting,

$$= \sum_{k=0}^{\infty}\left(\frac{1}{2}\right)^k - 1 - \sum_{k=1}^{\infty} b_k$$
$$= \frac{1}{1-\frac{1}{2}} - 1 - (-3) = 2 - 1 + 3 = 4$$

29. $$\sum_{k=0}^{\infty}\frac{1}{(k+1)(k+2)} = \sum_{k=0}^{\infty}\left[\frac{1}{k+1} - \frac{1}{k+2}\right]$$
$$S_n = \left(\frac{1}{1}-\frac{1}{2}\right) + \left(\frac{1}{2}-\frac{1}{3}\right) + \left(\frac{1}{3}-\frac{1}{4}\right) + \cdots + \left(\frac{1}{n} - \frac{1}{n+1}\right) + \left(\frac{1}{n+1} - \frac{1}{n+2}\right)$$
$$S_n = 1 - \frac{1}{n+2}$$
$$\lim_{n\to\infty} S_n = \lim_{n\to\infty}\left(1 - \frac{1}{n+2}\right) = 1$$

31. $$\sum_{k=1}^{\infty}\frac{1}{(2k-1)(2k+1)} = \sum_{k=1}^{\infty}\frac{1}{2}\left[\frac{1}{2k-1} - \frac{1}{2k+1}\right]$$
$$= \frac{1}{2}\sum_{k=1}^{\infty}\left[\frac{1}{2k-1} - \frac{1}{2k+1}\right]$$
$$S_n = \frac{1}{2}\left[\left(\frac{1}{1}-\frac{1}{3}\right) + \left(\frac{1}{3}-\frac{1}{5}\right) + \left(\frac{1}{5}-\frac{1}{7}\right) + \cdots + \left(\frac{1}{2(n-1)-1} - \frac{1}{2(n-1)+1}\right) + \left(\frac{1}{2n-1} - \frac{1}{2n+1}\right)\right]$$
$$S_n = \frac{1}{2}\left[\left(1-\frac{1}{3}\right) + \left(\frac{1}{3}-\frac{1}{5}\right) + \left(\frac{1}{5}-\frac{1}{7}\right) + \cdots + \left(\frac{1}{2n-3} - \frac{1}{2n-1}\right) + \left(\frac{1}{2n-1} - \frac{1}{2n+1}\right)\right]$$
$$S_n = \frac{1}{2}\left(1 - \frac{1}{2n+1}\right)$$
$$\lim_{n\to\infty} S_n = \frac{1}{2}\lim_{n\to\infty}\left(1 - \frac{1}{2n+1}\right)$$
$$= \frac{1}{2}(1-0) = \frac{1}{2}$$

33. $$0.3333\cdots = \frac{3}{10} + \frac{3}{100} + \frac{3}{1000} + \frac{3}{10{,}000} + \cdots$$
$$= \frac{3}{10} + \frac{3}{10^2} + \frac{3}{10^3} + \frac{3}{10^4} + \cdots$$
$$= \frac{3}{10}\left[1 + \frac{1}{10} + \frac{1}{10^2} + \frac{1}{10^3} + \cdots\right]$$
$$= \frac{3}{10}\sum_{n=0}^{\infty}\left(\frac{1}{10}\right)^n$$

Since $\left|\frac{1}{10}\right| < 1$ and the series starts with $n = 0$,

$$= \frac{3}{10}\left[\frac{1}{1-\frac{1}{10}}\right] = \frac{3}{10}\left[\frac{1}{\frac{9}{10}}\right] = \frac{3}{10}\cdot\frac{10}{9} = \frac{1}{3}$$

35. $$0.252525\cdots = \frac{25}{100} + \frac{25}{10{,}000} + \frac{25}{1{,}000{,}000} + \cdots$$
$$= \frac{25}{100} + \frac{25}{(100)^2} + \frac{25}{(100)^3} + \cdots$$
$$= \frac{25}{100}\left[1 + \frac{1}{100} + \frac{1}{100^2} + \cdots\right]$$
$$= \frac{25}{100}\sum_{n=0}^{\infty}\left(\frac{1}{100}\right)^n$$

Since $\left|\frac{1}{100}\right| < 1$ and the series starts with $n = 0$,

$$= \frac{25}{100}\left[\frac{1}{1-\frac{1}{100}}\right] = \frac{25}{100}\left[\frac{1}{\frac{99}{100}}\right] = \frac{25}{100}\cdot\frac{100}{99} = \frac{25}{99}$$

37. Of the 50 billion dollar tax rebate, 0.92 (50) billion will remain after one year. After the second year, 92% of this amount, or 0.92 $[0.92(50)] = (0.92)^2(50)$ will remain. So, the amount spent is

$$(50)(0.92) + (50)(0.92)^2 + 50(0.92)^3 + \cdots$$
$$= 50(0.92)\left[1 + (0.92) + (0.92)^2 + \cdots\right]$$
$$= 46\sum_{n=0}^{\infty}(0.92)^n$$

Since $|0.92| < 1$ and the series starts with $n = 0$,

$$= 46\left[\frac{1}{1-0.92}\right] = 46\left[\frac{1}{0.08}\right] = 575$$

The total amount of spending generated will be 575 billion dollars. Including the government's 50 billion, the total is 625 billion dollars.

39. When $n = 0$, a payment of \$1,000 is made and the present value of that payment is \$1,000. When $n = 1$, another payment of \$1,000 is made but the present value of that money is $1000e^{-0.04}$ (using $P = Be^{-rt}$). When $n = 2$, the present value of the payment is $1000e^{-0.04(2)}$. The present value of the entire investment will be

$$1000 + 1000e^{-0.04(1)} + 1000e^{-0.04(2)} + 1000e^{-0.04(3)} + \cdots$$

Since $e^{-0.04t} = (e^{-0.04})^t$,

$$= 1000 + 1000(e^{-0.04})^1 + 1000(e^{-0.04})^2 + 1000(e^{-0.04})^3 + \cdots$$
$$= 1000\left[1 + (e^{-0.04})^1 + (e^{-0.04})^2 + (e^{-0.04})^3 + \cdots\right]$$
$$= 1000\sum_{n=0}^{\infty}(e^{-0.04})^n$$

Since $\left|e^{-0.04}\right| < 1$ and the series starts with $n = 0$,

$$= 1000\left[\frac{1}{1-e^{-0.04}}\right] \approx 25{,}503$$

The present value of the investment is approximately \$25,503.

41. The present value of the first payment is $A(1+r)^{-1}$. The present value of the second payment is $A(1+r)^{-2}$. In general, the present value of the nth payment is $A(1+r)^{-n}$. The present value of the investment is

$$\frac{A}{1+r} + \frac{A}{(1+r)^2} + \frac{A}{(1+r)^3} + \cdots$$
$$= \frac{A}{1+r}\left[1 + \frac{1}{1+r} + \frac{1}{(1+r)^2} + \cdots\right]$$
$$= \frac{A}{1+r}\sum_{n=0}^{\infty}\frac{1}{(1+r)^n}$$
$$= \frac{A}{1+r}\sum_{n=0}^{\infty}\left(\frac{1}{1+r}\right)^n$$

Since $1 + r > 1$, $\left|\frac{1}{1+r}\right| < 1$. Further, the series starts with $n = 0$, so

$$= \frac{A}{1+r}\left[\frac{1}{1-\frac{1}{1+r}}\right]$$
$$= \frac{A}{1+r}\left[\frac{1}{\frac{r}{1+r}}\right]$$
$$= \frac{A}{1+r}\left[\frac{1+r}{r}\right] = \frac{A}{r}$$

43. After 1 day, the fraction of the first injection remaining is

$$20e^{-1/2}$$

After 2 days, the fraction of the first injection remaining is $20e^{-2/2}$ and the fraction of the second injection is $20e^{-1/2}$, or

$$20e^{-1/2} + 20e^{-2/2}$$

After 3 days, the fraction of the first injection is $20e^{-3/2}$, the fraction of the second injection is $20e^{-2/2}$, and the fraction of the third injection is $20e^{-1/2}$, or

$$20e^{-1/2} + 20e^{-2/2} + 20e^{-3/2}$$

So, the total amount of drug remaining as the injections continue is

$$20e^{-1/2} + 20e^{-2/2} + 20e^{-3/2} + \cdots$$
$$= 20e^{-1/2(1)} + 20e^{-1/2(2)} + 20e^{-1/2(3)} + \cdots$$

Since $e^{-1/2t} = \left(e^{-1/2}\right)^t$,

$$= 20\left(e^{-1/2}\right)^1 + 20\left(e^{-1/2}\right)^2 + 20\left(e^{-1/2}\right)^3 + \cdots$$
$$= 20e^{-1/2}\left[1 + \left(e^{-1/2}\right)^1 + \left(e^{-1/2}\right)^2 + \cdots\right]$$
$$= 20e^{-1/2}\sum_{n=0}^{\infty}\left(e^{-1/2}\right)^n$$

Since $\left|e^{-1/2}\right| < 1$ and the series starts with $n = 0$,

$$= 20e^{-1/2}\left[\frac{1}{1 - e^{-1/2}}\right] = \frac{20}{e^{1/2} - 1}$$

Approximately 30.83 units of the drug will remain in the patient's system.

45. After 1 day, the amount of toxin in a person's body is

$$(1 - q)d$$

After 2 days, the amount of toxin remaining from the first day is $(1 - q)\left[(1 - q)d\right]$ and the amount from the second day is $(1 - q)d$, for a total of

$$(1 - q)d + (1 - q)^2 d$$

After 3 days, the amount from the first day is $(1 - q)\left[(1 - q)^2 d\right]$, the amount from the second day is $(1 - q)\left[(1 - q)d\right]$, and the amount from the third day is $(1 - q)d$, for a total of

$$(1 - q)d + (1 - q)^2 d + (1 - q)^3 d$$

So, the total amount of toxin in a person's body is

$$(1 - q)d + (1 - q)^2 d + (1 - q)^3 d + \cdots$$
$$= (1 - q)d\left[1 + (1 - q) + (1 - q)^2 + \cdots\right]$$
$$= (1 - q)d\sum_{n=0}^{\infty}(1 - q)^n$$

Since $|1 - q| < 1$ and the series starts with $n = 0$,

$$= (1 - q)d\left[\frac{1}{1 - (1 - q)}\right]$$
$$= (1 - q)d\left[\frac{1}{q}\right] = \frac{1 - q}{q}d$$

47. **(a)** Since $s = 16t^2$, $t = \sqrt{\frac{s}{16}}$. The ball first falls a distance of 16 feet, so $t_1 = \sqrt{\frac{16}{16}} = 1$. The height of the ball after rebounding is (0.36)(16). The time to fall this distance is $t_2 = \sqrt{\frac{(0.36)(16)}{16}} = \sqrt{0.36}$. Assume it takes the same amount of time to rebound to this height as it does to fall from this height. So far, the total time is $1 + 2\sqrt{0.36}$. The height os the ball when it rebounds again is $(0.36)\left[(0.36)(16)\right] = (0.36)^2(16)$. The time to fall from this position is $t_3 = \sqrt{\frac{(0.36)^2(16)}{16}}$, which is again doubled to account for the time to rebound to this position. The total time the ball travels is

$$1 + 2(0.36)^{1/2} + 2(0.36)^{2/2} + 2(0.36)^{3/2} + \cdots$$
$$= 1 + 2(0.36)^{1/2}\left[1 + (0.36)^{1/2} + (0.36)^{2/2} + \cdots\right]$$

Since $(0.36)^{n/2} = \left[(0.36)^{1/2}\right]^n$

$$= 1 + 2(0.36)^{1/2}\sum_{n=0}^{\infty}\left[(0.36)^{1/2}\right]^n$$

Since $|(0.36)^{1/2}| < 1$ and the series starts with $n = 0$,

$$= 1 + 2(0.36)^{1/2}\left[\frac{1}{1 - (0.36)^{1/2}}\right]$$
$$= 1 + (1.2)(2.5) = 4$$

So, the total time the ball will travel is 4 seconds.

(b) From above, $t_1 = \sqrt{\frac{H}{16}} = \frac{H^{1/2}}{4}$, $t_2 = \sqrt{\frac{rH}{16}} = \frac{r^{1/2}H^{1/2}}{4}$, $t_3 = \sqrt{\frac{r^2 H}{16}} = \frac{rH^{1/2}}{4}$ and the total time is

$$\frac{H^{1/2}}{4}+2\left(\frac{r^{1/2}H^{1/2}}{4}\right)+2\left(\frac{r^{2/2}H^{1/2}}{4}\right)$$
$$+2\left(\frac{r^{3/2}H^{1/2}}{4}\right)+\cdots$$
$$=\frac{H^{1/2}}{4}+\frac{r^{1/2}H^{1/2}}{2}\left[1+r^{1/2}+r^{3/2}+\cdots\right]$$

Since $r^{n/2}=\left(r^{1/2}\right)^n$,

$$=\frac{H^{1/2}}{4}+\frac{r^{1/2}H^{1/2}}{2}\sum_{n=0}^{\infty}\left(r^{1/2}\right)^n$$

Now, $|r^{1/2}|<1$ and the series starts with $n=0$, so

$$=\frac{H^{1/2}}{4}+\frac{r^{1/2}H^{1/2}}{2}\left(\frac{1}{1-r^{1/2}}\right)$$
$$=\frac{H^{1/2}}{4}\left[1+\frac{2r^{1/2}}{1-r^{1/2}}\right]$$
$$=\frac{H^{1/2}}{4}\left[\frac{1-r^{1/2}}{1-r^{1/2}}+\frac{2r^{1/2}}{1-r^{1/2}}\right]$$
$$=\frac{H^{1/2}}{4}\left[\frac{1+r^{1/2}}{1-r^{1/2}}\right]$$

(c) Since the total time is 2.25 seconds,

$$2.25=\frac{\sqrt{9}}{4}\left[\frac{1+r^{1/2}}{1-r^{1/2}}\right]$$
$$3=\frac{1+r^{1/2}}{1-r^{1/2}}$$
$$3-3r^{1/2}=1+r^{1/2},\quad\text{or } r=0.25.$$

When first dropped, the ball travels 9 feet. It rebolunds to a height of (0.25)(9) feet, which is traveled twice (rising and falling). It next rebounds to a height of $(0.25)[(0.25)(9)]=(0.25)^2(9)$ feet, which is also traveled twice. The total distance traveled is

$$=9+2(0.25)(9)+2(0.25)^2(9)+2(0.25)^3(9)+\cdots$$
$$=9+18(0.25)\left[1+(0.25)+(0.25)^2+\cdots\right]$$
$$=9+4.5\sum_{n=0}^{\infty}(0.25)^n$$

Since $|0.25|<1$ and the series starts with $n=0$,

$$=9+4.5\left(\frac{1}{1-0.25}\right)=15\text{ feet}$$

49. After 1 year, $1.7\times10^9m^3$ of the gas are consumed. During the second year, $1.074(1.7\times10^9)m^3$ are consumed, for a total of

$$1.7\times10^9+1.074(1.7\times10^9)$$

During the third year, another $1.074\left[1.074(1.7\times10^9)\right]m^3$ are consumed, for a total of

$$1.7\times10^9+1.074(1.7\times10^9)$$
$$+1.074\left[1.074(1.7\times10^9)\right]$$

The total consumption for n years is

$$S_n=1.7\times10^9+1.074(1.7\times10^9)+(1.074)^2(1.7\times10^9)$$
$$+\cdots+(1.074)^{n-1}(1.7\times10^9)$$

Now,

$$1.074S_n=1.074(1.7\times10^9)+(1.074)^2(1.7\times10^9)$$
$$+(1.074)^3(1.7\times10^9)+\cdots+(1.074)^n(1.7\times10^9)$$

So,

$$1.074S_n-S_n=-1.7\times10^9+(1.074)^n(1.7\times10^9)$$
$$S_n\frac{1.7\times10^9\left[(1.074)^n-1\right]}{0.074}$$

Need to fine n when

$$\frac{1.7 \times 10^9 \left[(1.074)^n - 1\right]}{0.074} = 3 \times 10^{11}$$

$$(1.074)^n - 1 \approx 0.13059 \times 10^2$$

$$(1.074)^n \approx 14.059$$

$$n \ln 1.074 \approx \ln 14.059$$

$$n \approx \frac{\ln 14.059}{\ln 1.074} \approx 37.03$$

The gas reserves will be consumed approximately 37 years later, in 2037.

51. Since the initial dosage differs from subsequent injections, consider them separately. For the 800 mg injection, after 1 four-hour time period, 0.7(800) mg remain. After 2 four-hour time periods, 0.7[0.7(800)] or $(0.7)^2(800)$ mg remain. After n four-hour time periods, $(0.7)^n(800)$ mg remain.
Considering the 400 mg injections, after 1 four-hour time period, the first such injection has just been given. After 2 four-hour time periods, 0.7(400) mg remain from the first injection and the second injection has just been given. After 3 four-hour time periods, the amount remaining is 0.7[0.7(400)], or $(0.7)^2(400)$ from the first injection, (0.7)(400) from the second injection, and the next injection has just been given. Combining all injections, the total remaining after n four-hour time periods is

$$S_n = 400 + 0.7(400) + (0.7)^2(400) + (0.7)^3(400) + \cdots + (0.7)^{n-1}(400) + (0.7)^n(800)$$

Now,

$$0.7S_n = (0.7)(400) + (0.7)^2(400) + (0.7)^3(400) + \cdots + (0.7)^n(400) + (0.7)^{n+1}(800)$$

So,

$$S_n - 0.7S_n = 400 + (0.7)^n(800) - (0.7)^n(400) - (0.7)^{n+1}(800)$$

$$0.3S_n = 400 + 400(0.7)^n - 800(0.7)(0.7)^n$$

$$0.3S_n = 400 - 160(0.7)^n$$

$$Sn = \frac{400 - 160(0.7)^n}{0.3}$$

Need to find first value of n when $S_n \geq 1000$ units.

$$S_1 = 960$$

$$S_2 = 1072$$

Scelerat has until the second injection is given, or 8 hours.

53. **(a)** Press [y =]. Enter $300 * e^{(-0.2*x)}$ for $y_1 =$. Press [2nd] [quit]. Press [vars] and select 1:Function under the y-vars menu. Select 1: y_1 and press [enter]. There should be y_1 displayed on the home screen. At the curser, insert (1) and press [enter]. This evaluates the function at $x = 1$ and the display should read 245.619 . . . To store this value as A, press [sto→] [alpha] A and enter.

(b) Press [vars] and select function under the Y-vars menu. Enter y_1. Insert (2) and press enter. Press [+] [alpha] A and enter. The display should read 446.715 . . . Press [sto→] [alpha] B to store this result as B.

(c) Repeat this process for $x = 3$, 4, and 5. The final display should be 856.521 . . .

55. $S_n = a + ar + ar^2 + ar^3 + \cdots + ar^{n-1}$

(a) When $r = 1$, $S_n = na$

$$\lim_{n\to\infty} S_n = \lim_{n\to\infty} na,$$

which does not exist when $a \neq 0$. So, the series diverges.

(b) When $r = -1$,
$S_n = a$ when n is odd
$S_n = 0$ when n is even
$\lim_{n\to\infty} S_n$ does not exist, as no single value is approached (bounces back and forth between a and zero), so the series diverges.

9.2 Tests for Convergence

1.

$$\sum_{k=1}^{\infty} \frac{2k}{k+5}$$

$$\lim_{k\to\infty} a_k \lim_{k\to\infty} \frac{2k}{k+5} = \lim_{k\to\infty} \frac{2}{1+\frac{5}{k}} = \frac{2}{1+0} = 2$$

Since $\lim_{k\to\infty} a_k \neq 0$, the series diverges.

3. $$\sum_{k=1}^{\infty} 1+(-1)^k$$

$$\lim_{k\to\infty} a_k = \lim_{k\to\infty} 1+(-1)^k$$

When k is even, $a_k = 2$. When k is odd, $a_k = 0$. Since no single value is approached as $k \to \infty$ (bounces back and forth between 2 and 0), $\lim_{k\to\infty} a_k$ does not exist and the series diverges.

5. $$\sum_{k=1}^{\infty} \frac{2}{3k+1}$$

For $n \geq 1$, $f(n) = \dfrac{2}{3n+1}$ is continuous, positive and decreasing, so the integral test applies.

$$\int_1^{\infty} \frac{2}{3n+1} dn = 2 \lim_{b\to\infty} \int_1^b \frac{1}{3n+1} dn$$

$$= \frac{2}{3} \lim_{b\to\infty} \int_4^{3b+1} \frac{1}{u} du$$

$$= \frac{2}{3} \lim_{b\to\infty} \left[\ln u \Big|_4^{3b+1} \right]$$

$$= \frac{2}{3} \lim_{b\to\infty} [\ln(3b+1) - \ln 4]$$

Since $\lim\limits_{b\to\infty} \ln(3b+1)$ does not exist, the improper integral diverges and the series must diverge.

7. $$\sum_{k=1}^{\infty} \frac{k}{3+k^2}$$

For $n \geq 1$, $f(n) = \dfrac{n}{3+n^2}$ is continuous, positive and decreasing, so the integral test applies.

$$\int_1^{\infty} \frac{n}{3+n^2} dn = \lim_{b\to\infty} \int_1^b \frac{n}{3+n^2} dn$$

Using substitution with $u = 3+n^2$ and $\frac{1}{2}\, du = n\, dn$,

$$\frac{1}{2} du = n\, dn,$$

$$= \frac{1}{2} \lim_{b\to\infty} \int_1^{3+b^2} \frac{1}{u} du$$

$$= \frac{1}{2} \lim_{b\to\infty} \left[\ln u \Big|_4^{3+b^2} \right]$$

$$= \frac{1}{2} \lim_{b\to\infty} \left[\ln(3+b^2) - \ln 4 \right]$$

Since $\lim\limits_{b\to\infty} \ln(3+b^2)$ does not exist, the improper integral diverges and the series must diverge.

9. $\displaystyle\sum_{k=1}^{\infty} \frac{1}{k^{5/6}}$ is a p-series with $p = \frac{5}{6}$. Since $\frac{5}{6} \leq 1$, the series diverges.

11. $\displaystyle\sum_{k=1}^{\infty} \frac{1}{k^2\sqrt{k}} = \sum_{k=1}^{\infty} \frac{1}{k^{5/2}}$ is a p-series with $p = \frac{5}{2}$. Since $\frac{5}{2} > 1$, the series converges.

13. $$\sum_{k=1}^{\infty} \frac{3}{2^k+1}$$

$$\frac{3}{2^k+1} < \frac{3}{2^k} \text{ for all } k \geq 1$$

$$\sum_{k=1}^{\infty} \frac{3}{2^k} = \frac{3}{2} + \frac{3}{2^2} + \frac{3}{2^3} +$$

$$= \frac{3}{2}\left[1 + \frac{1}{2} + \frac{1}{2^2} + \cdots\right]$$

$$= \frac{3}{2} \sum_{k=0}^{\infty} \left(\frac{1}{2}\right)^n = \frac{3}{2}\left[\frac{1}{1-\frac{1}{2}}\right] = 3$$

Since $\displaystyle\sum_{k=1}^{\infty} \frac{3}{2k}$ converges, $\displaystyle\sum_{k=1}^{\infty} \frac{3}{2k+1}$ must also converge.

15. $$\sum_{k=2}^{\infty} \frac{1}{k^{1/2}-1}$$

$$\frac{1}{k^{1/2}-1} < \frac{1}{k^{1/2}} \text{ for all } k \geq 2$$

$$\sum_{k=2}^{\infty} \frac{1}{k^{1/2}} = \sum_{k=1}^{\infty} \frac{1}{k^{1/2}} - 1$$

$\sum_{k=1}^{\infty} \frac{1}{k^{1/2}}$ is a p-series with $p = \frac{1}{2}$. Since $\frac{1}{2} \leq 1$, this series diverges. In turn, $\sum_{k=2}^{\infty} \frac{1}{k^{1/2}}$ also diverges. By the comparison test, $\sum_{k=2}^{\infty} \frac{1}{\sqrt{k}-1}$ also diverges.

17. $$\sum_{k=1}^{\infty} \frac{2^k}{k^3}$$

Using the ratio test,

$$\begin{aligned} L &= \lim_{k\to\infty} \left| \frac{\frac{2^{k+1}}{(k+1)^3}}{\frac{2^k}{k^3}} \right| \\ &= \lim_{k\to\infty} \frac{k^3 2^{k+1}}{(k+1)^3 2^k} = \lim_{k\to\infty} \frac{2k^3}{(k+1)^3} \\ &= 2 \lim_{k\to\infty} \left(\frac{k}{k+1} \right)^3 \\ &= 2 \lim_{k\to\infty} \left(\frac{1}{1+\frac{1}{k}} \right)^3 = 2 \end{aligned}$$

Since $L > 1$, the series diverges.

19. $$\sum_{k=1}^{\infty} \frac{(-3)^k}{k \cdot 2^{2k}}$$

Using the ratio test,

$$\begin{aligned} L &= \lim_{k\to\infty} \left| \frac{\frac{(-3)^{k+1}}{(k+1)2^{2(k+1)}}}{\frac{(-3)^k}{k \cdot 2^{2k}}} \right| \\ &= \lim_{k\to\infty} \left| \frac{(-3)^{k+1} \cdot k \cdot 2^{2k}}{(-3)^k (k+1) 2^{2k+2}} \right| \\ &= \lim_{k\to\infty} \left| \frac{-3 \cdot k}{2^2(k+1)} \right| = \frac{3}{4} \lim_{k\to\infty} \left(\frac{k}{k+1} \right) = \frac{3}{4} \end{aligned}$$

Since $L < 1$, the series converges.

21. $\sum_{k=3}^{\infty} \frac{k}{\ln k}$

Since k increases more rapidly than $\ln k$,

$$\lim_{k\to\infty} a_n = \lim_{k\to\infty} \frac{k}{\ln k} \neq 0$$

and the series diverges.

23. $\sum_{k=1}^{\infty} \left(\frac{\pi}{2} \right)^k$

$$\lim_{k\to\infty} a_n = \lim_{k\to\infty} \left(\frac{\pi}{2} \right)^k \neq 0$$

So, the series diverges.

25. $\sum_{k=1}^{\infty} \frac{1}{(3k)^{3/2}}$

Using the comparison test,

$$\frac{1}{(3k)^{3/2}} < \frac{1}{k^{3/2}} \text{ for all } k \geq 1$$

$\sum_{k=1}^{\infty} \frac{1}{k^{3/2}}$ is a p-series with $p = \frac{3}{2}$. Since $\frac{3}{2} > 1$, this series converges and the original series must converge as well.

27. $\sum_{k=3}^{\infty} \frac{1}{k\sqrt{\ln k}}$

Using the integral test,

$$\int_3^{\infty} \frac{1}{n\sqrt{\ln n}}\, dn = \lim_{b\to\infty} \int_3^b \frac{1}{n\sqrt{\ln n}}\, dn$$

Using substitution with $u = \ln n$ and $du = \frac{1}{n}\, dn$,

$$\begin{aligned} &= \lim_{b\to\infty} \int_{\ln 3}^{\ln b} u^{-1/2}\, du = \lim_{b\to\infty} \left(2u^{1/2} \Big|_{\ln 3}^{\ln b} \right) \\ &= 2 \lim_{b\to\infty} \left(\sqrt{\ln b} - \sqrt{\ln 3} \right) \end{aligned}$$

This limit does not exist. Since the improper integral diverges, the series diverges.

29. $\sum_{k=1}^{\infty} \frac{e^{1/k}}{k^2}$ Using the integral test,

$$\int_1^{\infty} \frac{e^{1/n}}{n^2}\, dn = \lim_{b\to\infty} \int_1^b \frac{e^{1/n}}{n^2}\, dn$$

Using substitution with $u = \frac{1}{n}$ and $-du = \frac{1}{n^2}\,dn$,

$$= \lim_{b\to\infty} \int_1^{1/b} -e^u\,du = \lim_{b\to\infty} \int_{1/b}^1 e^u\,du$$

$$= \lim_{b\to\infty} \left(e^u\Big|_{1/b}^1\right) = \lim_{b\to\infty} \left(e^1 - e^{1/b}\right)$$

$$= e - 1$$

Since the improper integral converges, the series converges.

31. $\displaystyle\sum_{k=2}^{\infty} \frac{\ln k}{e^k}$

Using the integral test,

$$\int_2^{\infty} \frac{\ln n}{e^n}\,dn = \lim_{b\to\infty} \int_2^b \frac{\ln n}{e^n}\,dn$$

Using integration by parts with $u = \ln n$ and $du = \frac{1}{n}\,dn$,

$$= \lim_{b\to\infty} \left[-e^{-n} \ln n\Big|_2^b - \int_2^b -\frac{1}{n} e^{-n}\,dn \right]$$

$$= \lim_{b\to\infty} \left[-e^{-n} \ln n\Big|_2^b + \int_2^b \frac{1}{ne^n}\,dn \right]$$

Now, using the comparison test since

$$\frac{1}{ne^n} < \frac{1}{e^n} \text{ for } n \geq 2$$

$$\lim_{b\to\infty} \int_2^b e^{-n}\,dn = \lim_{b\to\infty} \left[-e^{-n}\Big|_2^b \right]$$

$$= \lim_{b\to\infty} \left[-e^{-b} + e^{-2} \right] = 0 + e^{-2} = \frac{1}{e^2}$$

Since $\int_2^{\infty} e^{-n}\,dn$ converges, $\int_2^{\infty} \frac{1}{ne^n}\,dn$ also converges. Further,

$$\lim_{b\to\infty} \left[-e^{-n} \ln n\Big|_2^b \right]$$

$$= \lim_{b\to\infty} \left[-e^{-b} \ln b + e^{-2} \ln 2 \right] = \frac{\ln 2}{e^2}$$

So the series converges.

33. $\displaystyle\sum_{k=1}^{\infty} \frac{1}{k^{\sqrt{2}}}$

This is a p-series with $p = \sqrt{2}$. Since $\sqrt{2} > 1$, the series converges.

35. $\displaystyle\sum_{k=1}^{\infty} e^{1/k}$

$$\lim_{k\to\infty} a_k = \lim_{k\to\infty} e^{1/k} = 1$$

Since $\lim_{k\to\infty} a_k \neq 0$, the series diverges.

37. $\displaystyle\sum_{k=1}^{\infty} \frac{1}{k^2 2^k}$

(1) $\displaystyle\frac{1}{k^2 2^k} < \frac{1}{1^k}$ for all ≥ 1

$$\sum_{k=1}^{\infty} \frac{1}{2^k} = \sum_{k=1}^{\infty} \left(\frac{1}{2}\right)^k = \frac{1}{1-\frac{1}{2}} = 2$$

So, this series converges.

(2) $\displaystyle\frac{1}{k^2 2^k} < \frac{1}{k^2}$ for all $k \geq 1$

$\displaystyle\sum_{k=1}^{\infty} \frac{1}{k^2}$ is a p-series with $p = 2$. Since $2 > 1$, this series converges.

39. $\displaystyle\sum_{k=1}^{\infty} \frac{1}{(1+2k)^p}$

When $p = 1$, using the integral test,

$$\int_1^{\infty} \frac{1}{1+2n}\,dn = \lim_{b\to\infty} \int_1^b \frac{1}{1+2n}\,dn$$

Using substitution with $u = 1 + 2n$ and $\frac{1}{2}\,du = dn$,

$$= \frac{1}{2} \lim_{b\to\infty} \int_3^{1+2b} \frac{1}{u}\,du = \frac{1}{2} \lim_{b\to\infty} \left(\ln u\Big|_3^{1+2b} \right)$$

$$= \frac{1}{2} \lim_{b\to\infty} [\ln(1+2b) - \ln 3]$$

Since this limit does not exist, the series diverges.
When $p \neq 1$, using the integral test,

$$\int_1^{\infty} \frac{1}{(1+2n)^p}\,dn = \lim_{b\to\infty}\int_1^b \frac{1}{(1+2n)^p}\,dn$$
$$= \frac{1}{2}\lim_{b\to\infty}\int_3^{1+2b} u^{-p}\,du = \frac{1}{2}\lim_{b\to\infty}\left(\frac{u^{-p+1}}{-p+1}\Big|_3^{1+2b}\right)$$
$$= \frac{1}{2}\lim_{b\to\infty}\left[\frac{(1+2b)^{-p+1}}{-p+1} - \frac{3^{-p+1}}{-p+1}\right]$$

When $p > 1$, $-p+1 < 0$ and

$$= \frac{1}{2}\lim_{b\to\infty}\left[\frac{1}{(-p+1)(1+2b)^{p-1}} - \frac{3^{-p+1}}{-p+1}\right]$$
$$= \frac{1}{2}\left[0 - \frac{3^{-p+1}}{-p+1}\right] = \frac{-3^{-p+1}}{2(-p+1)} = \frac{3^{-p+1}}{2(p-1)}$$

This integral converges, so the series converges. When $p < 1$, $-p+1 > 0$. Since

$$\lim_{b\to\infty}\frac{(1+2b)^{-p+1}}{-p+1} \text{ does not exist,}$$

this integral diverges, as does the series.

41. $\displaystyle\sum_{k=1}^{\infty}\frac{c^k}{k^c}$

When $c = 1$,

$$= \sum_{k=1}^{\infty}\frac{1}{k}$$

This is a p-series with $p = 1$. So, the series diverges. Using the ratio test,

$$L = \lim_{k\to\infty}\left|\frac{\frac{c^{k+1}}{(k+1)^c}}{\frac{c^k}{k^c}}\right|$$
$$= \lim_{k\to\infty}\frac{k^c c^{k+1}}{c^k(k+1)^c} = \lim_{k\to\infty}\frac{ck^c}{(k+1)^c}$$
$$= c\lim_{k\to\infty}\left(\frac{k}{k+1}\right)^c = c\lim_{k\to\infty}\left(\frac{1}{1+\frac{1}{k}}\right)^c$$
$$= c\cdot 1 = c$$

When $c > 1$, the series diverges. When $c < 1$, the series converges.

43. $\displaystyle\sum_{n=0}^{\infty}\frac{1}{3n+1}$

Using the integral test,

$$\int_0^{\infty}\frac{1}{3n+1}\,dn = \lim_{b\to\infty}\int_0^b \frac{1}{3n+1}\,dn$$

Using substitution with $u = 3n+1$ and $\frac{1}{3}\,du = dn$,

$$= \frac{1}{3}\lim_{b\to\infty}\int_1^{3b+1}\frac{1}{u}\,du = \frac{1}{3}\lim_{b\to\infty}\left[\ln u\Big|_1^{3b+1}\right]$$
$$= \frac{1}{3}\lim_{b\to\infty}\left[\ln(3b+1) - \ln 1\right]$$

Since $\lim_{b\to\infty}\ln(3b+1)$ does not exist, the integral diverges, as does the series.

45. The probability that the first board will break because it is weaker than all of the preceeding boards is 1. The probability that the second board will break because it is weaker than all of the preceeding boards is $\frac{1}{2}$. The probability that the third board will break because it is weaker than all of the preceeding boards is $\frac{1}{3}$. This pattern continues, making the sum from 1 to 100 boards

$$= 1 + \frac{1}{2} + \frac{1}{3} + \frac{1}{4} + \cdots + \frac{1}{99} + \frac{1}{100}$$
$$\approx 5.19$$

So, approximately 5 boards out of the 100 will break.

47. Need to find if series converges or diverges, for the series

$$\sum_{k=1}^{\infty}\frac{T}{k} = T\sum_{k=1}^{\infty}\frac{1}{k}$$

This is a p-series with $p = 1$, so the series diverges. The racer will not finish the race in a finite time.

49. $\displaystyle\sum_{k=1}^{\infty}\frac{k+2}{\sqrt[3]{k^5+k^2}}$

Let $a_k = \dfrac{k+2}{\sqrt[3]{k^5+k^2}}$ and $b_k = \dfrac{k}{k^{5/3}} = \dfrac{1}{k^{2/3}}$.

$$\lim_{k\to\infty}\frac{a_k}{b_k}=\lim_{k\to\infty}\frac{k+2}{\sqrt[3]{k^5+k^2}}\cdot\frac{k^{2/3}}{1}$$

$$=\lim_{k\to\infty}\frac{k^{5/3}+2k^{2/3}}{\sqrt[3]{k^5+k^2}}=\lim_{k\to\infty}\frac{1+\frac{2}{k}}{\sqrt[3]{1+\frac{1}{k^3}}}$$

$$=\frac{1+0}{\sqrt[3]{1+0}}=1$$

Since the limit is a positive finite number, both series converge or both diverge.

$$\sum_{k=1}^{\infty}\frac{1}{k^{2/3}} \text{ is a } p\text{-series with } p=2/3.$$

Since $\frac{2}{3}<1$, this series diverges and both must diverge.

51. $\displaystyle\sum_{k=1}^{\infty}\frac{1+2^k}{3+5^k}$

Let $a_k=\dfrac{1+2^k}{3+5^k}$ and $b_k=\dfrac{2^k}{5^k}$.

$$\lim_{k\to\infty}\frac{a_k}{b_k}=\lim_{k\to\infty}\frac{(1+2^k)}{(3+5^k)}\cdot\frac{5^k}{2^k}$$

$$=\lim_{k\to\infty}\frac{5^k+10^k}{3(2^k)+10^k}$$

$$=\lim_{k\to\infty}\frac{\left(\frac{1}{2}\right)^k+1}{3\left(\frac{1}{5}\right)^k+1}=\frac{0+1}{3(0)+1}=1$$

Since the limit is a positive finite number, both series converge or both diverge.

$$\sum_{k=1}^{\infty}\frac{2^k}{5^k}=\sum_{k=1}^{\infty}\left(\frac{2}{5}\right)^k$$

This geometric series converges, so both series converge.

53. $\displaystyle\sum_{n=3}^{\infty}\frac{1}{(\ln n)^2}$

Since $\sqrt{n}>\ln n$,

$$\sum_{n=3}^{\infty}\frac{1}{(\sqrt{n})^2}<\sum_{n=3}^{\infty}\frac{1}{(\ln n)^2}$$

Now,

$$\sum_{n=3}^{\infty}\frac{1}{(\sqrt{n})^2}=\sum_{n=3}^{\infty}\frac{1}{n}=\sum_{n=3}^{\infty}\frac{1}{n}-\left(1+\frac{1}{2}\right)$$

Since $\displaystyle\sum_{n=1}^{\infty}\frac{1}{n}$ is a p-series with $p=1$, the series diverges, as does $\displaystyle\sum_{n=3}^{\infty}\frac{1}{n}=\sum_{n=1}^{\infty}\frac{1}{n}-\frac{3}{2}$. So, the original series must also diverge.

9.3 Functions as Power Series; Taylor Series

1. $\displaystyle\sum_{k=0}^{\infty}5^kx^k$

Using the ratio test,

$$L=\lim_{k\to\infty}\left|\frac{5^{k+1}x^{k+1}}{5^kx^k}\right|=\lim_{k\to\infty}|5x|=|5x|$$

The series converges if $|5x|<1$, or $|x|<\frac{1}{5}$. The radius of convergence is $R=\frac{1}{5}$ and the interval of abhsolute convergence is $-\frac{1}{5}<x<\frac{1}{5}$.

3. $\displaystyle\sum_{k=1}^{\infty}\frac{x^k}{k^{1/2}}$

Using the ratio test,

$$L=\lim_{k\to\infty}\left|\frac{\frac{x^{k+1}}{(k+1)^{1/2}}}{\frac{x^k}{k^{1/2}}}\right|=\lim_{k\to\infty}\left|\frac{k^{1/2}x^{k+1}}{(k+1)^{1/2}x^k}\right|$$

$$=\lim_{k\to\infty}\left|\frac{k^{1/2}x}{(k+1)^{1/2}}\right|=|x|$$

The series converges if $|x|<1$. The radius of convergence is $R=1$ and the interval of absolute convergence is $-1<x<1$.

5. $\displaystyle\sum_{k=1}^{\infty}\frac{2^{2k}x^k}{k^2}$

Using the ratio test,

$$L = \lim_{k\to\infty}\left|\frac{\frac{2^{2(k+1)}x^{k+1}}{(k+1)^2}}{\frac{2^{2k}x^k}{k^2}}\right| = \lim_{k\to\infty}\left|\frac{2^{2k+2}x^{k+1}k^2}{2^{2k}x^k(k+1)^2}\right|$$

$$= \lim_{k\to\infty}\left|\frac{4xk^2}{(k+1)^2}\right| = |4x|$$

The series converges when $|4x| < 1$, or $|x| < \frac{1}{4}$. The radius of convergence is$R = \frac{1}{4}$ and the interval of absolute convergence is $-\frac{1}{4} < x < \frac{1}{4}$.

7. $\displaystyle\sum_{k=1}^{\infty} \frac{s^k x^k}{k!}$

Using the ratio test,

$$L = \lim_{k\to\infty}\left|\frac{\frac{5^{k+1}x^{k+1}}{(k+1)!}}{\frac{5^k x^k}{k!}}\right| = \lim_{k\to\infty}\left|\frac{5^{k+1}x^{k+1}k!}{5^k x^k(k+1)!}\right|$$

$$= \lim_{k\to\infty}\left|\frac{5x}{k+1}\right| = 0 \text{ for all } x$$

This series converges for all x, so the radius of convergence is $R = \infty$ and the interval of absolute convergence is the entire x-axis.

9. $f(x) = \dfrac{x}{1+x}$

Using $\dfrac{1}{1-x} = \displaystyle\sum_{n=0}^{\infty} x^n$ for $|x| < 1$, replace x by $-x$ to get

$$\frac{1}{1+x} = \sum_{n=0}^{\infty}(-x)^n = \sum_{n=0}^{\infty}(-1)^n x^n$$

Now, the interval of convergence is $|-x| < 1$, or $|x| < 1$.
Next, multiply by x to get

$$\frac{x}{1+x} = \sum_{n=0}^{\infty}(-1)^n x^{n+1} = \sum_{n=1}^{\infty}(-1)^{n+1}x^n$$

The interval of convergence remains $|x| < 1$, or $-1 < x < 1$.

11. $f(x) = \dfrac{x^2}{1-x^2}$

Using $\dfrac{1}{1-x} = \displaystyle\sum_{n=0}^{\infty} x^n$ for $|x| < 1$, replace x by x^2 to get

$$\frac{1}{1-x^2} = \sum_{n=0}^{\infty}(x^2)^n = \sum_{n=0}^{\infty} x^{2n}$$

Now, the interval of convergence is $|x^2| < 1$, or $-1 < x < 1$.
Next, multiply by x^2 to get

$$\frac{x^2}{1-x^2} = \sum_{n=0}^{\infty} x^2 \cdot x^{2n} = \sum_{n=0}^{\infty} x^{2(n+1)}$$

The interval of convergence remains $-1 < x < 1$.

13. $f(x) = \ln(2+x)$ Since

$$\ln(2+x) = \int \frac{1}{2+x}\,dx$$

first find a power series for $\dfrac{1}{2+x}$. To use

$$\frac{1}{1-x} = \sum_{n=0}^{\infty} x^n$$

rewrite

$$\frac{1}{2+x} = \frac{1}{2\left[1-\left(-\frac{1}{2}x\right)\right]}$$

Replacing x by $-\frac{1}{2}x$,

$$\frac{1}{2+x} = \frac{1}{2}\sum_{n=0}^{\infty}\left(-\frac{1}{2}x\right)^n$$

$$= \frac{1}{2}\sum_{n=0}^{\infty}\frac{(-1)^n}{2^n}x^n$$

$$= \sum_{n=0}^{\infty}\frac{(-1)^n}{2^{n+1}}x^n$$

So,

$$\ln(2+x) = \int_0^{\infty} \frac{1}{2} - \frac{1}{4}x + \frac{1}{8}x^2 - \frac{1}{16}x^3 + \cdots\, dx$$

$$= C + \frac{1}{2}x - \frac{1}{8}x^2 + \frac{1}{24}x^3 - \frac{1}{64}x^4 + \cdots$$

$$= C + \sum_{n=0}^{\infty} \frac{(-1)^n}{(n+1)2^{n+1}} x^{n+1}$$

$$= C + \sum_{n=1}^{\infty} \frac{(-1)^{n+1}}{n2^n} x^n$$

Since $\ln(2+x) = \ln 2$ when $x = 0$,

$$= \ln 2 + \sum_{n=1}^{\infty} \frac{(-1)^{n+1}}{n2^n} x^n$$

Since the modified geometric series converges for $\left|-\frac{1}{2}x\right| < 1$, the interval of convergence for the power series is $\left|\frac{1}{2}x\right| < 1$, or $-2 < x < 2$.

15. For $f(x) = e^{3x}$ about $a = 0$, the Taylor coeficients are

$$f(x) = e^{3x} \qquad f(0) = 1 \qquad a_0 = \frac{1}{0!}$$

$$f'(x) = 3e^{3x} \qquad f'(0) = 3 \qquad a_1 = \frac{3}{1!}$$

$$f''(x) = 9e^{3x} \qquad f''(0) = 9 \qquad a_2 = \frac{9}{2!}$$

$$f'''(x) = 27e^{3x} \qquad f'''(0) = 27 \qquad a_3 = \frac{27}{3!}$$

$$a_n = \frac{3^n}{n!}$$

The corresponding Taylor series is

$$f(x) = \sum_{n=0}^{\infty} \frac{3^n}{n!} x^n$$

17. For $f(x) = \frac{1}{2}(e^x + e^{-x})$ about $a = 0$, the Taylor coefficients are

$$f(x) = \frac{1}{2}(e^x + e^{-x}) \qquad f(0) = 1 \quad a_0 = \frac{1}{0!}$$

$$f'(x) = \frac{1}{2}(e^x - e^{-x}) \qquad f'(0) = 0 \quad a_1 = \frac{0}{1!}$$

$$f''(x) = \frac{1}{2}(e^x + e^{-x}) \qquad f''(0) = 1 \quad a_2 = \frac{1}{2!}$$

$$f'''(x) = \frac{1}{2}(e^x - e^{-x}) \qquad f'''(0) = 0 \quad a_3 = \frac{0}{3!}$$

$$a_n = \frac{1}{n!} \text{ when } n \text{ is even}$$

$$a_n = 0 \text{ when } n \text{ is odd}$$

So, $a_{2n} = \dfrac{1}{(2n)!}$

The corresponding Taylor series is

$$f(x) = \sum_{n=0}^{\infty} \frac{1}{(2n)!} x^{2n}$$

19. For $f(x) = e^{-3x}$ about $a = -1$, the Taylor coefficients are

$$f(x) = e^{-3x} \qquad f(-1) = e^3 \quad a_0 = \frac{e^3}{0!}$$

$$f'(x) = -3e^{-3x} \qquad f'(-1) = -3e^3 \quad a_1 = \frac{-3e^3}{1!}$$

$$f''(x) = 9e^{-3x} \qquad f''(-1) = 9e^3 \quad a_2 = \frac{9e^3}{2!}$$

$$f'''(x) = -27e^{-3x} \qquad f'''(-1) = -27e^3 \quad a_3 = \frac{-27e^3}{3!}$$

$$a_n = \frac{(-1)^n(3)^n e^3}{n!}$$

The corresponding Taylor series is

$$\sum_{n=0}^{\infty} \frac{(-1)^n(3)^n e^3}{n!} [x - (-1)]^n$$

$$= \sum_{n=0}^{\infty} \frac{(-3)^n e^3}{n!} (x+1)^n$$

21. For $f(x) = \ln(2x)$ about $a = \frac{1}{2}$, the Taylor coefficients are

$$f(x) = \ln(2x) \qquad f\left(\frac{1}{2}\right) = 0 \qquad a_0 = \frac{0}{0!} = \frac{0(2^0)}{0!}$$

$$f'(x) = \frac{1}{x} \qquad f'\left(\frac{1}{2}\right) = 2 \qquad a_1 = \frac{2}{1!} = \frac{1(2^1)}{1!}$$

$$f''(x) = \frac{-1}{x^2} \qquad f''\left(\frac{1}{2}\right) = -4 \qquad a_2 = \frac{-4}{2!} = \frac{-1(2^2)}{2!}$$

$$f'''(x) = \frac{2}{x^3} \qquad f'''\left(\frac{1}{2}\right) = 16 \qquad a_3 = \frac{16}{3!} = \frac{2(2^3)}{3!}$$

$$f^{iv}(x) = \frac{-6}{x^4} \qquad f^{iv}\left(\frac{1}{2}\right) = -96 \qquad a_4 = \frac{-96}{4!} = \frac{-3 \cdot 2(2^4)}{41!}$$

Ignoring the first coefficient, which is zero,

$$a_n = \frac{(-1)^{n+1}(n-1)!(2^n)}{n!}$$

$$a_n = \frac{(-1)^{n+1}(2^n)}{n}$$

The corresponding Taylor series is

$$\sum_{n=1}^{\infty} \frac{(-1)^{n+1}(2^n)}{n} \left(x - \frac{1}{2}\right)^n$$

23. For $f(x) = \dfrac{x}{1+x}$ about $a = 2$, the Taylor coefficients are

$$f(x) = \frac{x}{1+x} \qquad f(2) = \frac{2}{3} \qquad a_0 = \frac{\frac{2}{3}}{0!} = \frac{2}{3}$$

$$f'(x) = \frac{1}{(1+x)^2} \qquad f1(2) = \frac{1}{9} \qquad a_1 = \frac{\frac{1}{9}}{1!} = \frac{1!}{(3^{1+1})1!}$$

$$f''(x) = \frac{-2}{(1+x)^3} \qquad f''(2) = \frac{-2}{27} \qquad a_2 = \frac{\frac{-2}{27}}{2!} = \frac{-(2!)}{(3^{2+1})2!}$$

$$f'''(x) = \frac{6}{(1+x)^4} \qquad f''' = \frac{6}{81} \qquad a_3 = \frac{\frac{6}{81}}{3!} = \frac{3!}{(3^{3+1})3!}$$

Excluding a_0,

$$a_n = \frac{(-1)^{n+1}}{3^{n+1}} = \left(-\frac{1}{3}\right)^{n+1}$$

The corresponding Taylor series is

$$f(x) = \frac{2}{3} + \sum_{n=1}^{\infty} \left(-\frac{1}{3}\right)^{n+1} (x-2)^n$$

25. $f(x) = 3x^2 e^{x^3}$ about $x = 0$

Using

$$e^x = \sum_{n=0}^{\infty} \frac{x^n}{n!}$$

$$e^{x^3} = \sum_{n=0}^{\infty} \frac{(x^3)^n}{n!} = \sum_{n=0}^{\infty} \frac{x^{3n}}{n!}$$

(a) Multiplying by $3x^2$ gives

$$3x^2 e^{x^3} = 3x^2 \sum_{n=0}^{\infty} \frac{x^{3n}}{n!} = \sum_{n=0}^{\infty} \frac{3x^{3n+2}}{n!}$$

(b) Differentiating term-by-term,

$$e^{x^3} = 1 + x^3 + \frac{x^6}{2} + \frac{x^9}{6} + \frac{x^{12}}{24} + \frac{x^{15}}{120} + \cdots$$

$$\frac{d}{dx}(e^{x^3}) = 3x^2 + 3x^5 + \frac{3}{2}x^8 + \frac{3}{6}x^{11} + \frac{3}{24}x^{14} + \cdots$$

$$\frac{d}{dx}(e^{x^3}) = \sum_{n=0}^{\infty} \frac{3x^{3n+2}}{n!}$$

(Note that $\dfrac{d}{dx}(e^{x^3}) = 3x^2 e^{x^3}$)

27. $\int x^2 e^{-x^2}\, dx$ about $x = 0$

Using

$$e^x = \sum_{n=0}^{\infty} \frac{x^n}{n!}$$

$$e^{(-x^2)} = \sum_{n=0}^{\infty} \frac{(-x^2)^n}{n!} = \sum_{n=0}^{\infty} \frac{(-1)^n x^{2n}}{n!}$$

Multiplying by x^2 gives

$$x^2 e^{-x^2} = \sum_{n=0}^{\infty} \frac{(-1)^n x^{2(n+1)}}{n!}$$

Integrating term-by-term,

$$\int x^2 e^{-x}\,dx = \int x^2 - x^4 + \frac{x^6}{2} - \frac{x^8}{6} + \frac{x^{10}}{24} - \cdots\,dx$$
$$= \frac{x^3}{3} - \frac{x^5}{5} + \frac{x^7}{7(2)} - \frac{x^9}{9(6)} + \frac{x^{11}}{11(24)} - \cdots$$
$$= \frac{x^3}{3(0!)} - \frac{x^5}{5(1!)} + \frac{x^7}{7(2!)} - \frac{x^9}{9(3!)} + \frac{x^{11}}{11(4!)} - \cdots$$
$$= \sum_{n=0}^{\infty} \frac{(-1)^n x^{2n+3}}{(2n+3)n!}$$

29. $\sqrt{1.2}; n = 3$
Using $f(x) = \sqrt{x} = x^{1/2}$ about $x = 1$, the Taylor coefficients are

$$f(x) = x^{1/2} \qquad f(1) = 1 \quad a_0 = \frac{1}{0!} = 1$$
$$f'(x) = \frac{1}{2x^{1/2}} \qquad f'(1) = \frac{1}{2} \quad a_1 = \frac{\frac{1}{2}}{1!} = \frac{1}{2}$$
$$f''(x) = \frac{-1}{4x^{3/2}} \qquad f''(1) = -\frac{1}{4} \quad a_2 = \frac{\frac{-1}{4}}{2!} = -\frac{1}{8}$$
$$f'''(x) = \frac{3}{8x^{5/2}} \qquad f'''(1) = \frac{3}{8} \quad a_3 = \frac{\frac{3}{8}}{3!} = \frac{1}{16}$$

Using degree 3 with $x - 1 = 1.2 - 1 = 0.2$,

$$P_3(x) = 1(0.2)^0 + \frac{1}{2}(0.2) - \frac{1}{8}(0.2)^2 + \frac{1}{16}(0.2)^3$$
$$= 1.0955$$

31. $\ln(0.7), n = 5$
Using $f(x) = \ln x$ about $x = 1$ (see Example 9.3.7),

$$\ln x = \sum_{n=1}^{\infty} \frac{(-1)^{n+1}}{n}(x-1)^n$$
$$= (x-1) - \frac{1}{2}(x-1)^2 + \frac{1}{3}(x-1)^3 - \cdots$$

For degree 5 with $x - 1 = 0.7 - 1 = -0.3$,

$$P_s(x) = (-0.3) - \frac{1}{2}(-0.3)^2 + \frac{1}{3}(-0.3)^3$$
$$- \frac{1}{4}(-0.3)^4 + \frac{1}{5}(-0.3)^5$$
$$\approx -0.3565$$

33. $\frac{1}{\sqrt{e}} = e^{-1/2}; n = 4$
Using $f(x) = e^x$ about $x = 0$ (see Example 9.3.6),

$$e^x = \sum_{n=0}^{\infty} \frac{x^n}{n!}$$
$$= 1 + x + \frac{x^2}{2} + \frac{x^3}{6} + \frac{x^4}{24} + \cdots$$

For degree 4 with $x = -\frac{1}{2}$,

$$P_4(x) = 1 - \frac{1}{2} + \frac{1}{8} - \frac{1}{48} + \frac{1}{384}$$
$$\approx 0.6068$$

35. $\ln \sqrt[3]{4} = \ln 4^{1/3} = \ln\left(\frac{1}{2}\right)^{-2/3} = -\frac{2}{3}\ln 0.5; n = 5$
Using $f(x) = \ln x$ about $x = 1$,

$$\ln x = \sum_{n=1}^{\infty} \frac{(-1)^{n+1}}{n}(x-1)^n$$
$$= (x-1) - \frac{1}{2}(x-1)^2 + \frac{1}{3}(x-1)^3 - \cdots$$

For degree 5 with $x - 1 = 0.5 = -0.5$,

$$P_5(x) = -\frac{2}{3}\left[(-0.5) - \frac{1}{2}(-0.5)^2 + \frac{1}{3}(-0.5)^3 - \frac{1}{4}(-0.5)^4 + \frac{1}{5}(-0.5)^5\right]$$
$$\approx 0.4590$$

37. $\int_{-0.2}^{0.1} e^{-x^2}\,dx; n = 4$
Use

$$e^x = \sum_{n=0}^{\infty} \frac{x^n}{n!}$$
$$= 1 + x + \frac{x^2}{2} + \frac{x^3}{6} + \frac{x^4}{24} + \cdots$$

Substituting $-x^2$ for x gives

$$e^{-x^2} = 1 + (-x^2) + \frac{(-x^2)^2}{2} + \frac{(-x^2)^3}{6} + \frac{(-x^2)^4}{24} + \cdots$$
$$= 1 - x^2 + \frac{x^4}{2} - \frac{x^6}{6} + \frac{x^8}{24} - \cdots$$

$$P_4(x) = 1 - x^2 + \frac{x^4}{2}$$

$$\int_{-0.2}^{0.1} e^{-x^2}\,dx \approx \int_{-0.2}^{0.1} 1 - x^2 + \frac{x^4}{2}\,dx$$

$$= \left(x - \frac{x^3}{3} + \frac{x^5}{10}\right)\Big|_{-0.2}^{0.1}$$

$$= \left(0.1 - \frac{(0.1)^3}{3} + \frac{(0.1)^5}{10}\right) - \left(-0.2 - \frac{(-0.2)^3}{3} + \frac{(-0.2)^5}{10}\right)$$

$$\approx 0.2970$$

39. $\displaystyle\int_{-1/2}^{0} \frac{1}{1-x^3}dx;\ n = 9$

Use $\displaystyle\frac{1}{1-x} = \sum_{n=0}^{\infty} x^n$

$$= 1 + x + x^2 + x^3 + x^4 + \cdots$$

Substituting x^3 for x gives

$$= 1 + x^3 + (x^3)^2 + (x^3)^3 + (x^3)^4 + \cdots$$
$$= 1 + x^3 + x^6 + x^9 + x^{12} + \cdots$$

$$P_9(x) = 1 + x^3 + x^6 + x^9$$

$$\int_{-1/2}^{0} \frac{1}{1-x^3}\,dx \approx \int_{1/2}^{0} 1 + x^3 + x^6 + x^9\,dx$$

$$= \left(x + \frac{x^4}{4} + \frac{x^7}{7} + \frac{x^{10}}{10}\right)\Big|_{1/2}^{0}$$

$$= 0 - \left(-\frac{1}{2} + \frac{\left(-\frac{1}{2}\right)^4}{4} + \frac{\left(-\frac{1}{2}\right)^7}{7} + \frac{\left(\frac{1}{2}\right)^{10}}{10}\right)$$

$$\approx 0.4854$$

41. $f(x) = 1 - e^{-x};\ c = 0$

Use

$$e^x = \sum_{n=0}^{\infty} \frac{x^n}{n!} = 1 + x + \frac{x^2}{2} + \frac{x^3}{6} + \frac{x^4}{24} + \cdots$$

Substituting $-x$ for x gives

$$e^{-x} = 1 - x + \frac{(-x)^2}{2} + \frac{(-x)^3}{6} + \frac{(-x)^4}{24} + \cdots$$
$$= 1 - x + \frac{x^2}{2} - \frac{x^3}{6} + \frac{x^4}{24} - \cdots$$

$$P_0(x) = 1 - 1 = 0$$

$$P_1(x) = 1 - (1 - x) = x$$

$$P_2(x) = 1 - \left(1 - x + \frac{x^2}{2}\right) = x - \frac{x^2}{2}$$

$$P_3(x) = 1 - \left(1 - x + \frac{x^2}{2} - \frac{x^3}{6}\right) = x - \frac{x^2}{2} + \frac{x^3}{6}$$

Press [y=]. Enter $y = 1 - e \wedge (-x)$ for y_1, $y = 0$ for $y_2 =$, $y = x$ for $y_3 =$, $y = x - (x^2/2)$ for $y_4 =$, and $y = x - (x^2/2) + (x \wedge (3)/6)$ for $y_5 =$. Use window dimensions $[-3, 3]1$ by $[-4, 2]1$. Press [graph].

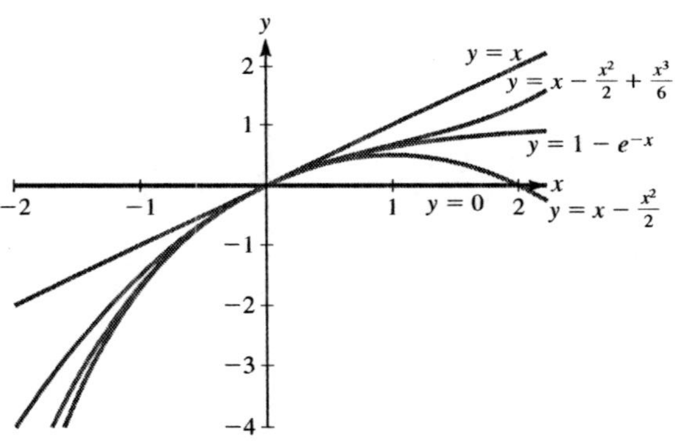

43. $f(x) = \dfrac{\ln x}{x};\ c = 1$

$$P_0(x) = f(1) = 0$$

$$P_1(x) = f(1) + f'(1)(x-1) = x - 1$$

$$P_1(x) = f(1) + f'(1)(x-1) + \frac{f''(1)}{2!}(x-1)^2$$
$$= (x-1) - \frac{3}{2}(x-1)^2$$

$$P_3(x) = f(1) + f'(1)(x-1) + \frac{f''(1)}{2!}(x-1)^2 + \frac{f'''(1)}{3!}(x-1)^3$$
$$= (x-1) - \frac{3}{2}(x-1)^2 + \frac{11}{6}(x-1)^3$$

Press [y=]. Enter $y = \ln(x)/x$ for $y_1 =$, $y = 0$ for $y_2 =$, $y = x - 1$ for $y_3 =$, $y = x - 1 - \frac{3}{2} * (x-1)^2$

for $y_4 =$, and $y = x - 1 - \frac{3}{2} * (x-1)^2 + \frac{11}{6} * (x-1)^3$ for $y_5 =$
Use window dimensions [0, 6]1 by [−8, 4]1. Press graph

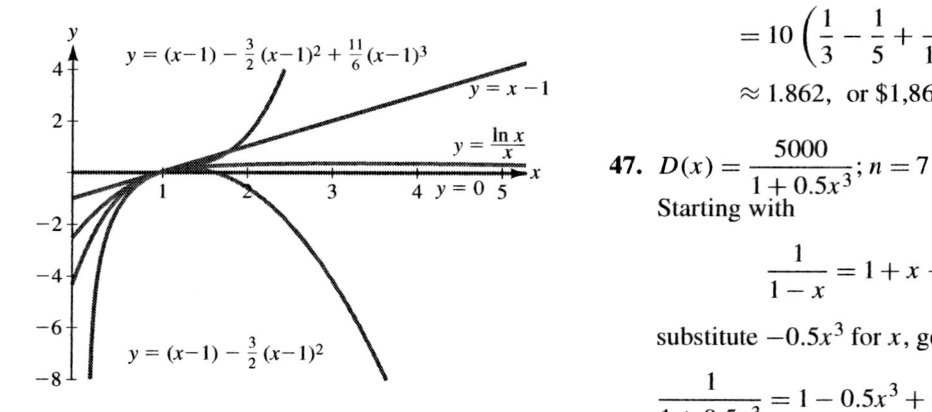

45. $P'(x) = 10x^2e^{-x^2}; n = 8$
Use

$$e^x = \sum_{n=0}^{\infty} \frac{x^n}{n!}$$
$$= 1 + x + \frac{x^2}{2} + \frac{x^3}{6} + \frac{x^4}{24} + \cdots$$

Substituting $-x^2$ for x gives

$$e^{-x^2} = 1 - x^2 + \frac{x^4}{2} - \frac{x^6}{6} + \frac{x^8}{24} - \cdots$$

Multiplying by $10x^2$ gives

$$10x^2e^{-x^2} = 10\left(x^2 - x^4 + \frac{x^6}{2} - \frac{x^8}{6} + \frac{x^{10}}{24} - \cdots\right)$$
$$P_8(x) = 10\left(x^2 - x^4 + \frac{x^6}{2} - \frac{x^8}{6}\right)$$

So, the integral needed is

$$\int_0^1 10\left(x^2 - x^4 + \frac{x^6}{2} - \frac{x^8}{6}\right) dx$$
$$= 10\left(\frac{x^3}{3} - \frac{x^5}{5} + \frac{x^7}{14} - \frac{x^9}{54}\right)\Big|_0^1$$
$$= 10\left(\frac{1}{3} - \frac{1}{5} + \frac{1}{14} - \frac{1}{54}\right) = \frac{352}{189}$$
$$\approx 1.862, \text{ or } \$1{,}862$$

47. $D(x) = \dfrac{5000}{1 + 0.5x^3}; n = 7$
Starting with

$$\frac{1}{1-x} = 1 + x + x^2 + x^3 + \cdots$$

substitute $-0.5x^3$ for x, getting

$$\frac{1}{1 + 0.5x^3} = 1 - 0.5x^3 + (-0.5x^3)^2 + (-0.5x^3)^3 + \cdots$$
$$= 1 - 0.5x^3 + 0.25x^6 - 0.125x^9 + \cdots$$

Multiplying by $2\pi x \cdot 5000$ gives

$$2\pi x \cdot \frac{5000}{1 + 0.5x^3}$$
$$= 10{,}000\pi\left(x - 0.5x^4 + 0.25x^7 + \cdots\right)$$

So, the integral needed is

$$10{,}000\pi \int_0^1 x - 0.5x^4 + 0.25x^7\, dx$$
$$= 10{,}000\pi\left(\frac{x^2}{2} - 0.1x^5 + \frac{0.25}{8}x^8\right)\Big|_0^1$$
$$= 10{,}000\pi\left(\frac{1}{2} - 0.1 + \frac{0.25}{8}\right) \approx 13{,}548 \text{ people}$$

49. $P_{ave} = \dfrac{1}{10}\displaystyle\int_0^{10} \frac{1}{1 + e^{-1.2t}} dt$.

(a)
$$P(t) = \frac{1}{1 + e^{-1.2t}}$$
$$P'(t) = \frac{1.2e^{-1.2t}}{(1 + e^{-1.2t})^2}$$
$$P''(t) = \frac{-1.44e^{-1.2t}(1 - e^{-1.2t})}{(1 + e^{-1.2t})^3}$$

For $c = 0$,

$$P(0) = \frac{1}{2} \qquad a_0 = \frac{\frac{1}{2}}{0!} = \frac{1}{2}$$

$$P'(0) = \frac{1.2}{4} \qquad a_1 = \frac{\frac{3}{10}}{1!} = \frac{3}{10}$$

$$P''(0) = 0 \qquad a_2 = \frac{0}{2!} = 0$$

$$P_2 = \frac{1}{2} + \frac{3}{10}t$$

$$\begin{aligned} P_{ave} &\approx \frac{1}{10}\int_0^{10}\left(\frac{1}{2} + \frac{3}{10}t\right) dt \\ &= \frac{1}{10}\left[\frac{1}{2}t + \frac{3}{20}t^2\Big|_0^{10}\right] \\ &= \frac{1}{10}\,[(5 + 15) - 0] = 2 \end{aligned}$$

For $c = 5$,

$$P(5) \approx 0.00753 \qquad a_0 \approx \frac{0.99753}{0!} = 0.99753$$

$$P'(5) \approx 0.0029598 \qquad a_1 \approx \frac{0.0029598}{1!} = 0.0029598$$

$$P''(5) \approx -0.0035342 \quad a_2 \approx \frac{-0.0035342}{2!} = -0.0017671$$

$$P_2 \approx 0.99753 + 0.0029598(t - 5) - 0.0017671(t - 5)^2$$

$$\begin{aligned} P_{ave} &\approx \frac{1}{10}\int_0^{10} [0.99753 + 0.0029598(t - 5) \\ &\quad - 0.0017671(t - 5)^2]dt \\ &= \frac{1}{10}\left[0.99753t + 0.0014799(t - 5)^2 - 0.000589(t - 5)^3\Big|_0^{10}\right] \\ &\approx \frac{1}{10}\,[(9.9753 + 0.036998 - 0.073625) \\ &\quad -(0 + 0.036998 + 0.073625)] \\ &\approx 0.9828 \end{aligned}$$

For $c = 10$,

$$P(10) \approx 0.0999994 \qquad a_0 \approx \frac{0.999994}{0!} = 0.999994$$

$$P'(10) \approx 0.0000074 \qquad a_1 \approx \frac{0.0000074}{1!} = 0.0000074$$

$$P''(10) \approx -0.0000088 \quad a_2 \approx \frac{-0.0000088}{2!} = -0.0000044$$

$$P_2 \approx 0.999994 + 0.0000074(t - 10) - 0.0000044(t - 10)^2$$

$$\begin{aligned} P_{ave} &\approx \frac{1}{10}\int_0^{10} [0.999994 + 0.0000074(t - 10) \\ &\quad - 0.0000044(t - 10)^2]dt \\ &\approx \frac{1}{10}\left[0.999994t + 0.0000037(t - 10)^2 - 0.00000147(t - 10)^3\Big|_0^{10}\right] \\ &\approx \frac{1}{10}\,[(9.99994 + 0 - 0) - (0 + 0.00037 + 0.00147)] \\ &\approx 0.99981 \end{aligned}$$

(b) The best estimate is when $c = 5$.

Checkup for Chapter 9

1. **(a)**

$$\begin{aligned} \sum_{n=0}^{\infty}\frac{(-3)^n}{5^{n+1}} &= \sum_{n=0}^{\infty}\frac{(-3)^n}{5 \cdot 5^n} \\ &= \frac{1}{5}\sum_{n=0}^{\infty}\left(-\frac{3}{5}\right)^n \end{aligned}$$

Since $\left|-\frac{3}{5}\right| < 1$, this is a geometric series with sum

$$= \frac{1}{5} \cdot \frac{1}{1 + \frac{3}{5}} = \frac{1}{8}$$

(b) $\displaystyle\sum_{n=1}^{\infty}\frac{2^{2n}}{3^{n-1}}$

Since $\displaystyle\lim_{n\to\infty} a_n = \lim_{n\to\infty}\frac{2^{2n}}{3^{n-1}} \neq 0$, this series diverges (The ratio test also works here.)

2. **(a)** $\displaystyle\sum_{n=1}^{\infty}\frac{1}{n^{3/2}}$

Since this is a p-series with $p = \frac{3}{2}$, the series converges.

(b) $\sum_{n=1}^{\infty} \frac{3^n}{n!}$

Using the ratio test,

$$L = \lim_{n\to\infty} \left| \frac{\frac{3^{n+1}}{(n+1)!}}{\frac{3^n}{n!}} \right|$$

$$= \lim_{n\to\infty} \left(\frac{3^{n+1}}{(n+1)!} \cdot \frac{n!}{3^n} \right)$$

$$= \lim_{n\to\infty} \left(\frac{3}{n+1} \right) = 0$$

Since $L < 1$, the series converges.

3. **(a)** $\sum_{n=1}^{\infty} \frac{n(n^2+1)}{100n^3+9}$

Since $\lim_{n\to\infty} a_n$

$$= \lim_{n\to\infty} \frac{n(n^2+1)}{100n^3+9}$$

$$= \lim_{n\to\infty} \frac{n^3+n}{100n^3+9}$$

$$= \lim_{n\to\infty} \frac{1+\frac{1}{n^2}}{100+\frac{9}{n^3}} = \frac{1}{100} \neq 0$$

the series diverges.

(b) $\sum_{n=2}^{\infty} \frac{\ln n}{\sqrt{n}}$

Using the comparison test,

$$\sum_{n=2}^{\infty} \frac{1}{\sqrt{n}} < \sum_{n=2}^{\infty} \frac{\ln n}{\sqrt{n}}$$

Now,

$$\sum_{n=2}^{\infty} \frac{1}{\sqrt{n}} = \sum_{n=1}^{\infty} \frac{1}{\sqrt{n}} - 1$$

Since $\sum_{n=1}^{\infty} \frac{1}{n^{1/2}}$ is a p-series with $p \leq 1$, the series diverges, as does

$$\sum_{n=2}^{\infty} \frac{1}{n^{1/2}} = \sum_{n=1}^{\infty} \frac{1}{n^{1/2}} - 1$$

So, the original series must also diverge.

4. **(a)** $\sum_{n=1}^{\infty} \frac{1}{\sqrt{n}} = \sum_{n=1}^{\infty} \frac{1}{n^{1/2}}$

This is a p-series with $p = \frac{1}{2}$, so it diverges.

(b) $\sum_{n=1}^{\infty} \sqrt{n}e^{-n} = \sum_{n=1}^{\infty} \frac{\sqrt{n}}{e^n}$

Using the ratio test,

$$L = \left| \frac{\frac{\sqrt{n+1}}{e^{n+1}}}{\frac{\sqrt{n}}{e^n}} \right| = \lim_{n\to\infty} \left| \frac{\sqrt{n+1}}{e^{n+1}} \cdot \frac{e^n}{\sqrt{n}} \right|$$

$$= \lim_{n\to\infty} \left| \frac{\sqrt{n+1}}{e\sqrt{n}} \right| = \lim_{n\to\infty} \left| \frac{1}{e} \left(\frac{n+1}{n} \right)^{1/2} \right|$$

$$= \lim_{n\to\infty} \left| \frac{1}{e} \left(\frac{1+\frac{1}{n}}{1} \right)^{1/2} \right|$$

$$= \frac{1}{e} < 1$$

So, this series converges.

(c) $\sum_{n=1}^{\infty} \frac{n^2}{2^n}$

Using the ratio test,

$$L = \lim_{n\to\infty} \left| \frac{\frac{(n+1)^2}{2^{n+1}}}{\frac{n^2}{2^n}} \right| = \lim_{n\to\infty} \left| \frac{(n+1)^2}{2^{n+1}} \cdot \frac{2^n}{n^2} \right|$$

$$= \lim_{n\to\infty} \left| \frac{(n+1)^2}{2n^2} \right| = \lim_{n\to\infty} \left| \frac{1}{2} \left(\frac{n+1}{n} \right)^2 \right|$$

$$= \lim_{n\to\infty} \left| \frac{1}{2} \left(\frac{1+\frac{1}{n}}{1} \right)^2 \right| = \frac{1}{2} < 1$$

So, this series converges.

(d) $\sum_{n=0}^{\infty} \frac{1}{n^2+1} = 1 + \sum_{n=1}^{\infty} \frac{1}{n^2+1}$

Using the comparison test,

$$\sum_{n=1}^{\infty}\frac{1}{n^2+1}<\sum_{n=1}^{\infty}\frac{1}{n^2}$$

Now, $\sum_{n=1}^{\infty}\frac{1}{n^2}$ is a p-series with $p=2$, so it converges. So, $\sum_{n=1}^{\infty}\frac{1}{n^2+1}$ also converges and

$$1+\sum_{n=1}^{\infty}\frac{1}{n^2+1}=\sum_{n=0}^{\infty}\frac{1}{n^2+1}$$

converges.

5. **(a)** $\sum_{n=1}^{\infty}\frac{x^n}{n+1}$

Using the ratio test,

$$L=\lim_{n\to\infty}\left|\frac{\frac{x^{n+1}}{n+2}}{\frac{x^n}{n+1}}\right|=\lim_{n\to\infty}\left|\frac{x^{n+1}}{n+2}\cdot\frac{n+1}{x^n}\right|$$

$$=\lim_{n\to\infty}\left|x\left(\frac{n+1^2}{n+2}\right)\right|=\lim_{n\to\infty}\left|x\left(\frac{1+\frac{1}{n}}{1+\frac{2}{n}}\right)\right|$$

$$=|x|$$

Since the series converges when $L<1$, need $|x|<1$. So, the interval of absolute convergence is $-1<x<1$.

(b) $\sum_{n=1}^{\infty}\frac{(2x)^n}{n!}$

Using the ratio test,

$$L=\lim_{n\to\infty}\left|\frac{\frac{(2x)^{n+1}}{(n+1)!}}{\frac{(2x)^n}{n!}}\right|=\lim_{n\to\infty}\left|\frac{(2x)^{n+1}}{(n+1)!}\cdot\frac{n!}{(2x)^n}\right|$$

$$=\lim_{n\to\infty}\left|\frac{2x}{n+1}\right|=0$$

Since $L<0$ for all x, the interval of absolute convergence is all real x.

6. **(a)** $f(x)=\frac{x}{1+x^2}$

Using $\frac{1}{1-x}=\sum_{n=0}^{\infty}x^n$,

Substitute $-x^2$ for x to get

$$\frac{1}{1+x^2}=\sum_{n=0}^{\infty}(-x^2)^n=\sum_{n=0}^{\infty}(-1)^n x^{2n}$$

Multiply by x gives

$$\frac{x}{1+x^2}=\sum_{n=0}^{\infty}(-1)^n x^{2n+1}$$

(b) $f(x)=\frac{5}{2+3x}$

Rewrite the function as

$$=\frac{5}{2}\left(\frac{1}{1+\frac{3}{2}x}\right)$$

Using $\frac{1}{1-x}=\sum_{n=0}^{\infty}x^n$, substitute $-\frac{3}{2}x$ for x

$$\frac{1}{1+\frac{3}{2}x}=\sum_{n=0}^{\infty}\left(-\frac{3}{2}x\right)^n=\sum_{n=0}^{\infty}\left(-\frac{3}{2}\right)^n x^n$$

Multiplying by $\frac{5}{2}$ gives

$$\frac{5}{2}\left(\frac{1}{1+\frac{3}{2}x}\right)=\frac{5}{2+3x}=\sum_{n=0}^{\infty}\left(\frac{5}{2}\right)\left(-\frac{3}{2}\right)^n x^n$$

7. **(a)** $f(x)=e^x+e^{-x}$; about $x=0$

$$f(x)=e^x+e^{-x}\qquad f(0)=2\quad a_0=\frac{2}{0!}$$

$$f'(x)=e^x-e^{-x}\qquad f'(0)=0\quad a_1=\frac{0}{1!}$$

$$f''(x)=e^x+e^{-x}\qquad f''(0)=2\quad a_2=\frac{2}{2!}$$

$$f'''(x)=e^x-e^{-x}\qquad f'''(0)=0\quad a_3=\frac{0}{3!}$$

$$a_n=\frac{2}{n!}\text{ when } n \text{ is even}$$

$$a_n=0\text{ when } n \text{ is odd}$$

So, $a_{2n}=\frac{2}{(2n)!}$ The corresponding Taylor series is

$$\sum_{n=0}^{\infty}\frac{2}{(2n)!}x^{2n}$$

(b) $g(x) = \ln(x^2 + 1)$; about $x = 0$
Using the fact that

$$\ln(x^2 + 1) = \int \frac{2x}{1 + x^2}\, dx$$

first find a power series for $\dfrac{2x}{1+x^2}$ and then integrate term-by-term.
Since

$$\frac{1}{1-x} = 1 + x + x^2 + x^3 + \cdots$$

substituting $-x^2$ for x gives

$$\frac{1}{1+x^2} = 1 + (-x^2) + (-x^2)^2 + (-x^2)^3 + \cdots$$
$$= 1 - x^2 + x^4 - x^6 + \cdots$$

Multiplying by $2x$ gives

$$\frac{2x}{1+x^2} = 2(x - x^3 + x^5 - x^7 + \cdots)$$

So,

$$\ln(x^2 + 1) = 2\int x - x^3 + x^5 - x^7 + \cdots\, dx$$

$$\ln(x^2 + 1) = 2\left[\frac{x^2}{2} - \frac{x^4}{4} + \frac{x^6}{6} - \frac{x^8}{8} + \cdots\right]$$
$$= x^2 - \frac{x^4}{2} + \frac{x^6}{3} - \frac{x^8}{4} + \cdots$$
$$= \sum_{n=1}^{\infty} \frac{(-1)^{n+1}}{n} x^{2n}$$

8. After 1 day, the fraction of the first injection remaining is $25e^{-1/3}$.
After 2 days, the fraction of the first injection remaining is $25e^{-2/3}$ and the fraction of the second injection is $25e^{-1/3}$, or

$$25e^{-1/3} + 25e^{-2/3}$$

After 3 days, the fraction of the first injection is $25e^{-3/3}$, the fraction of the second injection is $25e^{-2/3}$, and the fraction of the third injection is $25e^{-1/3}$, or

$$25e^{-1/3} + 25e^{-2/3} + 25e^{-3/3}$$

So, the total amount remaining as the injections continue is

$$25e^{-1/3} + 25e^{-2/3} + 25e^{-3/3} + \cdots$$
$$= 25e^{-1/3(1)} + 25e^{-1/3(2)} + 25e^{-1/3(3)} + \cdots$$

Since $e^{-1/3(t)} = \left(e^{-1/3}\right)^t$,

$$= 25\left(e^{-1/3}\right)^1 + 25\left(e^{-1/3}\right)^2 + 25\left(e^{-1/3}\right)^3 + \cdots$$
$$= 25e^{-1/3}\left[1 + \left(e^{-1/3}\right) + \left(e^{-1/3}\right)^2 + \cdots\right]$$
$$= 25^{-1/3} \sum_{n=0}^{\infty} \left(e^{-1/3}\right)^n$$

Since $|e^{-1/3}| < 1$ and the series starts with $n = 0$,

$$= 25e^{-1/3}\left[\frac{1}{1 - e^{-1/3}}\right] = \frac{25}{e^{1/3} - 1}$$

approximately 63.2 units of the drug will remain in the patient's bloodstream.

9. Using the present value formula $P = Be^{-rt}$, the present value of the first withdrawel is $5{,}000e^{-0.4(1)}$. The present value of the second withdrawel is $5{,}000e^{-0.04(2)}$. In general, the present value of the nth withdrawal is $5{,}000e^{-0.04(n)}$. Adding the values together, the present amount of money needed is

$$P = \sum_{n=1}^{\infty} 5{,}000e^{-0.04n}$$

Since $e^{-0.04n} = (e^{-0.04})^n$,

$$= 5{,}000(e^{-0.04})^1 + 5{,}000(e^{-0.04})^2 + 5{,}000(e^{-0.04})^3 + \cdots$$
$$= 5{,}000e^{-0.04}\left[1 + (e^{-0.04}) + (e^{-0.04})^2 + \cdots\right]$$
$$= 5{,}000(e^{-0.04}) \sum_{n=0}^{\infty} (e^{-0.04})^n$$

Since $|e^{-0.04}| < 1$ and the series starts with $n = 0$,

$$= 5{,}000(e^{-0.04})\left[\frac{1}{1-e^{-0.04}}\right]$$
$$= \frac{5{,}000}{e^{0.04}-1} \approx \$122{,}516.67$$

10. To find the power series for $f(x) = \frac{1}{1+x^2}$, start with

$$\frac{1}{1-x} = \sum_{n=0}^{\infty} x^n$$

Substituting $-x^2$ for x gives

$$\frac{1}{1+x^2} = \sum_{n=0}^{\infty}(-x^2)^n = \sum_{n=0}^{\infty}(-1)^n x^{2n}$$
$$= 1 - x^2 + x^4 - x^6 + \cdots$$

Integrating term-by-term,

$$\int_0^1 (1 - x^2 + x^4 - x^6 + \cdots)\,dx$$
$$= \left(x - \frac{x^3}{3} + \frac{x^5}{5} - \frac{x^7}{7} + \cdots\right)\Big|_0^1$$
$$= \left(1 - \frac{1}{3} + \frac{1}{5} - \frac{1}{7} + \frac{1}{9} - \cdots\right) - 0$$
$$= \sum_{n=0}^{\infty} \frac{(-1)^n}{2n+1}$$

A partial sum of this series, based on the desired accuracy, will give an estimate of the integral.

Review Problems

1. $\displaystyle\sum_{n=0}^{\infty} \frac{(-2)^n}{5^{n+1}}$

Writing $(-2)^n$ as $\frac{(-2)^{n+1}}{-2}$,

$$= \sum_{n=0}^{\infty} -\frac{1}{2}\left(-\frac{2}{5}\right)^{n+1} = -\frac{1}{2}\sum_{n=0}^{\infty}\left(-\frac{2}{5}\right)^{n+1}$$
$$= -\frac{1}{2}\left[\sum_{n=0}^{\infty}\left(-\frac{2}{5}\right)^n - 1\right] = \frac{1}{2} - \frac{1}{2}\sum_{n=0}^{\infty}\left(-\frac{2}{5}\right)^n$$

This series is geometric with $|r| = \frac{2}{5}$, so the sum is

$$\frac{1}{2} - \frac{1}{2}\left[\frac{1}{1+\frac{2}{5}}\right] = \frac{1}{7}$$

3. $\displaystyle\sum_{n=0}^{\infty} \frac{13}{(-5)^n} = 13\sum_{n=0}^{\infty} \frac{1}{(-1)^n(5)^n} = 13\sum_{n=0}^{\infty} \frac{(-1)^n}{(5)^n}$

$$= 13\sum_{n=0}^{\infty}\left(-\frac{1}{5}\right)^n$$

This series is geometric with $|r| = \frac{1}{5}$, so the sum is

$$= 13\left[\frac{1}{1+\frac{1}{5}}\right] = \frac{65}{6}$$

5. $\displaystyle\sum_{n=0}^{\infty} e^{-0.5n} = \sum_{n=0}^{\infty}(e^{-0.5})^n$

This series is geometric with $|r| = e^{-0.5}$, so the sum is

$$= \frac{1}{1-e^{-0.5}} = \frac{e^{0.5}}{e^{0.5-1}} = \frac{\sqrt{e}}{\sqrt{e}-1}$$

7. $\displaystyle\sum_{n=0}^{\infty}\left(\frac{2}{3}\right)^n + \left(\frac{3}{2}\right)^n$

Using the comparison test,

$$\sum_{n=0}^{\infty}\left(\frac{3}{2}\right)^n < \sum_{n=0}^{\infty}\left(\frac{2}{3}\right)^n + \left(\frac{3}{2}\right)^n$$

Since $\displaystyle\sum_{n=0}^{\infty}\left(\frac{3}{2}\right)^n$ is a geometric series with $|r| = \frac{3}{2}$, it diverges. So the original series must also diverge.

9. $\displaystyle\sum_{n=0}^{\infty} \frac{1}{2n+1}$

Using the integral test,

$$\int_0^{\infty} \frac{1}{2x+1}\,dx = \lim_{b\to\infty}\int_0^b \frac{1}{2x+1}\,dx$$

Using substitution with $u = 2x+1$ and $\frac{1}{2}\,du = dx$

$$= \lim_{b\to\infty} \frac{1}{2}\int_1^{2b+1} \frac{1}{u}\,du = \frac{1}{2}\lim_{b\to\infty}\left(\ln u\Big|_1^{2b+1}\right)$$

$$= \frac{1}{2}\lim_{b\to\infty} [\ln(2b+1) - 0] = \infty$$

Since the improper integral diverges, the series diverges.

11. $\sum_{n=1}^{\infty} \ln\left(2 + \frac{1}{n}\right)$

Since $\lim_{n\to\infty} a_n = \lim_{n\to\infty} \ln\left(2+\frac{1}{n}\right) = \ln 2 \neq 0$, this series diverges.

13. $\sum_{n=2}^{\infty} \frac{\ln\sqrt{n}}{\sqrt{n}}$

Using the integral test,

$$\int_2^{\infty} \frac{\ln\sqrt{x}}{\sqrt{x}}\,dx = \lim_{b\to\infty}\int_2^b \frac{\ln\sqrt{x}}{\sqrt{x}}\,dx$$

$$= \lim_{b\to\infty}\int_2^b \frac{\frac{1}{2}\ln x}{x^{1/2}}\,dx$$

$$= \frac{1}{2}\lim_{b\to\infty}\int_2^b \frac{\ln x}{x^{1/2}}\,dx$$

Using integration by parts, with $u = \ln x$ and $dV = x^{-1/2}\,dx$,

$$= \frac{1}{2}\lim_{b\to\infty}\left[2x^{1/2}\ln x\Big|_2^b - \int_2^b \frac{1}{x}\cdot 2x^{1/2}\,dx\right]$$

$$= \frac{1}{2}\lim_{b\to\infty}\left[2x^{1/2}\ln x\Big|_2^b - 2\int_2^b x^{-1/2}\,dx\right]$$

$$= \frac{1}{2}\lim_{b\to\infty}\left(2x^{1/2}\ln x - 4x^{1/2}\right)\Big|_2^b$$

$$= \frac{1}{2}\lim_{b\to\infty} 2\left(x^{1/2}\ln x - 2x^{1/2}\right)\Big|_2^b$$

$$= \lim_{b\to\infty}\left(x^{1/2}\ln x - 2x^{1/2}\right)\Big|_2^b$$

$$= \lim_{b\to\infty}\left[\left(b^{1/2}\ln b - 2b^{1/2}\right) - \left(\sqrt{2}\ln 2 - 2\sqrt{2}\right)\right]$$

Now,

$\lim_{b\to\infty}\left(\sqrt{2}\ln 2 - 2\sqrt{2}\right)$ is finite and

$$\lim_{b\to\infty}\left(b^{1/2}\ln b - 2b^{1/2}\right)$$

$$= \lim_{b\to\infty}\left[b^{1/2}(\ln b - 2)\right] = \infty,$$

Since the improper integral diverges, the series diverges.

15. $\sum_{n=1}^{\infty} \frac{n^3}{3^n}$

Using the ratio test,

$$L = \lim_{n\to\infty}\left|\frac{\frac{(n+1)^3}{3^{n+1}}}{\frac{n^3}{3^n}}\right| = \lim_{n\to\infty}\left|\frac{(n+1)^3}{3^{n+1}}\cdot\frac{3^n}{n^3}\right|$$

$$= \lim_{n\to\infty}\left|\frac{1}{3}\left(\frac{n+1}{n}\right)^3\right| = \lim_{n\to\infty}\left|\frac{1}{3}\left(\frac{1+\frac{1}{n}}{1}\right)^3\right| = \frac{1}{3}$$

Since $L < 1$, the series converges.

17. $\sum_{n=1}^{\infty} \frac{(-3)^{2n}}{n!}$

Using the ratio test,

$$L = \lim_{n\to\infty}\left|\frac{\frac{(-3)^{2(n+1)}}{(n+1)!}}{\frac{(-3)^{2n}}{n!}}\right| = \lim_{n\to\infty}\left|\frac{(-3)^{2n+2}}{(n+1)!}\cdot\frac{n!}{(-3)^{2n}}\right|$$

$$= \lim_{n\to\infty}\left|\frac{(-3)^2}{n+1}\right| = 0$$

Since $L < 1$, the series converges.

19. $\sum_{n=1}^{\infty} \frac{(2x)^{2n}}{3^{n-1}}$

Using the ratio test,

$$L = \lim_{n\to\infty}\left|\frac{\frac{(2x)^{2(n+1)}}{3^n}}{\frac{(2x)^{2n}}{3^{n-1}}}\right| = \lim_{n\to\infty}\left|\frac{(2x)^{2n+2}}{3^n}\cdot\frac{3^{n-1}}{(2x)^{2n}}\right|$$

$$= \lim_{n\to\infty}\left|\frac{(2x)^2}{3}\right| = \left|\frac{4}{3}x^2\right|$$

The series converges when $L < 1$, or

$$\left|\frac{4}{3}x^2\right| < 1$$
$$\frac{4}{3}\left|x^2\right| < 1$$
$$\left|x^2\right| < \frac{3}{4}$$
$$|x| < \frac{\sqrt{3}}{2}$$

So, the interval of absolute convergence is $-\frac{\sqrt{3}}{2} < x < \frac{\sqrt{3}}{2}$.

21. $\sum_{n=1}^{\infty} e^n x^{n-1}$

Using the ratio test,

$$L = \lim_{n\to\infty}\left|\frac{e^{n+1}x^n}{e^n x^{n-1}}\right| = \lim_{n\to\infty}|ex| = e|x|$$

The series converges when $L < 1$, or

$$e|x| < 1$$
$$|x| < \frac{1}{e}$$

So, the interval of absolute convergence is $-\frac{1}{e} < x < \frac{1}{e}$.

23. $f(x) = e^x - e^{-x}$; about $x = 0$

Use

$$e^x = \sum_{n=0}^{\infty}\frac{x^n}{n!}$$

Substituting $-x$ for x gives

$$e^{-x} = \sum_{n=0}^{\infty}\frac{(-x)^n}{n!} = \sum_{n=0}^{\infty}\frac{(-1)^n x^n}{n!}$$

So, $e^x - e^{-x}$

$$= \sum_{n=0}^{\infty}\frac{x^n}{n!} - \sum_{n=0}^{\infty}\frac{(-1)^n x^n}{n!}$$
$$= \left(1 + x + \frac{x^2}{2!} + \frac{x^3}{3!} + \cdots\right) - \left(1 - x + \frac{x^2}{2!} - \frac{x^3}{3!} + \cdots\right)$$
$$= 2x + 2\left(\frac{x^3}{3!}\right) + 2\left(\frac{x^5}{5!}\right) + \cdots$$
$$= 2\left[x + \left(\frac{x^3}{3!}\right) + \left(\frac{x^5}{5!}\right) + \cdots\right]$$
$$= 2\sum_{n=0}^{\infty}\frac{x^{2n+1}}{(2n+1)!}$$

25. $f(x) = \ln\left(\frac{x+1}{2x+1}\right) = \ln(x+1) - \ln(2x+1)$

Using the facts that

$$\ln(x+1) = \int \frac{1}{1+x}\,dx$$
$$\ln(2x+1) = \int \frac{2}{1+2x}\,dx$$

find a power series for $\frac{1}{1+x}$ and $\frac{2}{1+2x}$, then integrate term-by-term. Since

$$\frac{1}{1-x} = 1 + x + x^2 + x^3 + \cdots$$

substituting $-x$ for x gives

$$\frac{1}{1+x} = 1 + (-x) + (-x)^2 + (-x)^3 + \cdots$$
$$= 1 - x + x^2 - x^3 + \cdots$$

and substituting $-2x$ for x gives

$$\frac{1}{1+2x} = 1 + (-2x) + (-2x)^2 + (-2x)^3 + \cdots$$
$$= 1 - 2x + 4x^2 - 8x^3 + \cdots$$

Multiplying by 2 gives

$$\frac{2}{1+2x} = 2 - 4x + 8x^2 - 16x^3 + \cdots$$

So, $\ln(x+1) - \ln(2x+1)$

$$= \int \left(1 - x + x^2 - x^3 + \cdots\right) - \left(2 - 4x + 8x^2 - 16x^3 + \cdots\right) dx$$

$$= \int -1 + 3x - 7x^2 + 15x^3 + \cdots \, dx$$

$$= -x + \frac{3}{2}x^2 - \frac{7}{3}x^3 + \frac{15}{4}x^4 + \cdots$$

$$= \sum_{n=1}^{\infty} \frac{(-1)^n(2^n - 1)}{n} x^n$$

27. $51.34747\cdots = 51.3 + 0.04747\cdots$
Now,

$$0.04747\cdots = \frac{47}{1000} + \frac{47}{100{,}000} + \frac{47}{10{,}000{,}000} + \cdots$$

$$= \frac{47}{1000}\left[1 + \frac{1}{100} + \frac{1}{(100)^2} + \frac{1}{(100)^3} + \cdots\right]$$

$$= \frac{47}{1000}\sum_{n=0}^{\infty}\left(\frac{1}{100}\right)^n$$

This is a geometric series with $|r| = \frac{1}{100}$, so the sum is

$$\frac{47}{1000}\left[\frac{1}{1 - \frac{1}{100}}\right] = \frac{47}{990}$$

adding back the 51.3 gives

$$\frac{513}{10} + \frac{47}{990} = \frac{50{,}834}{990} = \frac{25{,}417}{495}$$

29. When first dropped, the ball travels a distance H. It rebounds to a height of $0.75H$ and then drops that same distance. Next, it rebounds to a height of $0.75(0.75H)$, or $(0.75)^2H$, and then drops that same distance. In general, the ball rebounds to a height of $(0.75)^nH$ and then drops that same distance. The total distance traveled is

$$H + 2(0.75)H + 2(0.75)^2H + 2(0.75)^3H + \cdots$$

$$= H + 2(0.75)H[1 + (0.75) + (0.75)^2 + \cdots]$$

$$= H + 1.5H\sum_{n=0}^{\infty}(0.75)^n$$

This is a geometric series with $|r| = 0.75$, so the sum is

$$= H + 1.5H\left[\frac{1}{1 - 0.75}\right] = H + 6H = 7H$$

Since the total distance traveled is given as 70 feet, $H = 10$ feet.

31. Using the present value formula $P = Be^{-rt}$, the present value of the second payment is $5{,}000e^{-0.05(1)}$. The present value of the third payment is $5{,}000e^{-0.05(2)}$. In general, the present value of the $(n+1)^{\text{th}}$ payment is $5{,}000e^{-0.05(n)}$. Adding the values together, the present value of the investment is

$$P = \sum_{n=1}^{\infty} 5{,}000e^{-0.05n}$$

Since $e^{-0.05n} = (e^{-0.05})^n$,

$$= 5{,}000(e^{-0.05})^1 + 5{,}000(e^{-0.05})^2 + 5{,}000(e^{-0.05})^3 + \cdots$$

$$= 5{,}000e^{-0.05}\left[1 + (e^{-0.05}) + (e^{-0.05})^2 + \cdots\right]$$

$$= 5{,}000e^{-0.05}\sum_{n=0}^{\infty}(e^{-0.05})^n$$

Since $|e^{-0.05}| < 1$ and the series starts with $n = 0$,

$$= 5{,}000e^{-0.05}\left[\frac{1}{1 - e^{-0.05}}\right]$$

$$= \frac{5{,}000}{e^{0.05} - 1} \approx \$97{,}520.84$$

Adding the \$5,000 payment which is made immediately, the present value of the investment is approximately \$102,521.

33. After 1 day, the fraction of the first injection remaining is $25e^{-k}$. After 2 days, the fraction of the first injection remaining is $25e^{-2k}$ and the fraction of the second injection is $25e^{-k}$. After 3 days, the fraction of the first injection is $25e^{-3k}$, the fraction of the second injection is $25e^{-2k}$, and the fraction of the third injection is $25e^{-k}$. So, the total amount remaining as the injections continue is

$$25e^{-k} + 25e^{-2k} + 25e^{-3k} + \cdots$$
$$= 25e^{-k\cdot 1} + 25e^{-k\cdot 2} + 25e^{-k\cdot 3} + \cdots$$

Since $e^{-k(n)} = (e^{-k})^n$,

$$= 25(e^{-k})^1 + 25(e^{-k})^2 + 25(e^{-k})^3 + \cdots$$
$$= 25e^{-k}\left[1 + (e^{-k} + (e^{-k})^2 + \cdots\right]$$
$$= 25e^{-k}\sum_{n=0}^{\infty}(e^{-k})^n$$

Since $|e^{-k}| < 1$ and the series starts with $n = 0$,

$$= 25e^{-k}\left[\frac{1}{1-e^{-k}}\right]$$
$$= \frac{25}{e^k - 1}$$

Since 70 units eventually accumulate in the patient's bloodstream,

$$70 = \frac{25}{e^k - 1}$$
$$e^k - 1 = \frac{25}{70} = \frac{5}{14}$$
$$e^k = \frac{19}{14}$$
$$k = \ln\left(\frac{19}{14}\right) \approx 0.3054$$

35. **(a)** Since the book contains a total of T words, the sum of the word counts must be T.

(b) $$\sum_{k=1}^{\infty}\frac{T}{k(k+1)} = T\sum_{k=1}^{\infty}\frac{1}{k(k+1)}$$
$$= T(1) = T$$

(c) Writing Exercise—Answers will vary.

37. $f(x) = \dfrac{1}{(1-x)^2}$; about $a = 2$

$$f(x) = \frac{1}{(1-x)^2} \quad f(2) = 1 \quad a_0 = \frac{1}{0!} = \frac{(-1)^0(1!)}{0!}$$
$$f'(x) = \frac{2}{(1-x)^3} \quad f'(2) = -2 \quad a_1 = \frac{-2}{1!} = \frac{(-1)(2!)}{1!}$$
$$f''(x) = \frac{6}{(1-x)^4} \quad f''(2) = 6 \quad a_2 = \frac{6}{2!} = \frac{(-1)^2(3!)}{2!}$$
$$a_n = \frac{(-1)^n(n+1)!}{n!} = (-1)^n(n+1)$$

The corresponding Taylor series is

$$\sum_{n=0}^{\infty}(-1)^n(n+1)(x-2)^n$$

39. $f(x) = x\ln x$, about $a = 1$

$$f(x) = x\ln x \quad f(1) = 0 \quad a_0 = \frac{0}{0!} = 0$$
$$f'(x) = 1 + \ln x \quad f'(1) = 1 \quad a_1 = \frac{1}{1!} = 1$$
$$f''(x) = \frac{1}{x} \quad f''(1) = 1 \quad a_2 = \frac{1}{2!} = \frac{1}{2}$$
$$f'''(x) = -\frac{1}{x^2} \quad f'''(1) = -1 \quad a_3 = \frac{-1}{3!} = -\frac{1}{6}$$
$$f'v(x) = \frac{2}{x^3} \quad f'^v(1) = 2 \quad a_4 = \frac{2}{4!} = \frac{1}{12}$$
$$fv(x) = \frac{6}{x^4} \quad f^v(1) = -6 \quad a_5 = \frac{-6}{5!} = -\frac{1}{20}$$

Starting with $n = 2$,

$$a_n = \frac{(-1)^n}{n(n-1)}$$

The corresponding Taylor series is

$$0 + (x-1) + \sum_{n=2}^{\infty}\frac{(-1)^n}{n(n+1)}(x-1)^n$$

41. $\sqrt{0.9}$, $n = 3$
Using $f(x) = \sqrt{x} = x^{1/2}$ about $x = 1$, the Taylor coefficients are

$f(x) = x^{1/2} \qquad f(1) = 1 \quad a_0 = \frac{1}{0!} = 1$

$f'(x) = \frac{1}{2x^{1/2}} \qquad f'(1) = \frac{1}{2} \quad a_1 = \frac{1/2}{1!} = \frac{1}{2}$

$f''(x) = \frac{-1}{4x^{3/2}} \qquad f''(1) = -\frac{1}{4} \quad a_2 = \frac{-1/4}{2!} = -\frac{1}{8}$

$f'''(x) = \frac{3}{8x^{5/2}} \qquad f'''(1) = \frac{3}{8} \quad a_3 = \frac{3/8}{3!} = \frac{1}{16}$

Using degree 3 with $x - 1 = 0.9 - 1 = -0.1$,

$$P_3 = (-0.1)^0 + \frac{1}{2}(-0.1) - \frac{1}{8}(-0.1)^2 + \frac{1}{16}(-0.1)^3$$
$$\approx 0.9487$$

43. **(a)** The time until the bee meets the first train is $\frac{1000}{60+30} = \frac{1000}{90}$ seconds. The new distance the bee travels to again meet a train is

$$1000 - \left(\frac{1000}{90}\right)(60) = 1000\left(1 - \frac{2}{3}\right)$$

The time to meet the train is

$$\frac{1000\left(1 - \frac{2}{3}\right)}{90}$$

The new distance the bee travels to again meet a train is

$$1000\left(1 - \frac{2}{3}\right) - \frac{1000\left(1 - \frac{2}{3}\right)}{90}(60)$$
$$= 1000\left(1 - \frac{2}{3}\right)\left[1 - \frac{2}{3}\right] = 1000\left(1 - \frac{2}{3}\right)^2$$

The time to meet the train is

$$\frac{1000\left(1 - \frac{2}{3}\right)^2}{90}$$

The total time the bee travels will be

$$\frac{1000}{90} + \frac{1000}{90}\left(1 - \frac{2}{3}\right) + \frac{1000}{90}\left(1 - \frac{2}{3}\right)^2 + \cdots$$
$$= \frac{1000}{90}\left[1 + \frac{1}{3} + \left(\frac{1}{3}\right)^2 + \left(\frac{1}{3}\right)^3 + \cdots\right]$$
$$= \frac{1000}{90}\sum_{n=0}^{\infty}\left(\frac{1}{3}\right)^n$$

Since $\left|\frac{1}{3}\right| < 1$ and the series starts with $n = 0$, the sum is

$$= \frac{1000}{90}\left[\frac{1}{1 - \frac{1}{3}}\right] = \frac{1000}{90}\left(\frac{3}{2}\right) = \frac{50}{3} \text{ seconds}$$

Since the bee flies at 60 ft/sec, it will travel a total distance of

$$(60)\left(\frac{50}{3}\right) = 1{,}000 \text{ ft}$$

(b) The easiest way to work the problem is to realize that each train will travel 500 feet when they collide. Traveling at a rate of 30 ft/sec, this will take $\frac{50}{3}$ seconds.

(c) Consider each train separately since they will travel the same distance, 90 feet.
Since $a(t) = \frac{dV}{dt} = -5$,

$$v(t) = -5t + V_0$$

where $V_0 = 30$ ft/sec.
Since $v(t) = \frac{ds}{dt}$,

$$S(t) = -\frac{5}{2}t^2 + 30t + S_0$$

where $S_0 = 0$. Want to know the time it takes the train to travel 90 feet, or

$$90 = -\frac{5}{2}t^2 + 30t$$

$$5t^2 - 60t + 180 = 0$$

$$t = \frac{60 \pm \sqrt{3600 - 4(5)(180)}}{2(5)} = 6 \text{ seconds}$$

Before the deceleration begins, each train has traveled $500 - 90 = 410$ feet at a rate of 30 ft/sec. The time for this to occur is $\frac{41}{3}$ seconds, so the total time the bee flies is $\frac{59}{3}$ seconds. So the bee travels

$$(60)\left(\frac{59}{3}\right) = 1,\ 180 \text{ feet}$$

45. $f(x) = \dfrac{1}{\sqrt{1-x^2}}$; about $x = 0$

(a)

$$f(x) = \frac{1}{\sqrt{1-x^2}} \qquad f(0) = 1 \quad a_0 = \frac{1}{0!} = 1$$

$$f'(x) = \frac{x}{(1-x^2)^{3/2}} \qquad f'(0) = 0 \quad a_1 = \frac{0}{1!} = 0$$

$$f''(x) = \frac{1}{(1-x^2)^{5/2}} \qquad f''(0) = 1 \quad a_2 = \frac{1}{2!} = \frac{1}{2}$$

$$f'''(x) = \frac{5x}{(1-x^2)^{7/2}} \qquad f'''(0) = 0 \quad a_3 = \frac{0}{3!} = 0$$

$$P_0(x) = a_0x^0 = 1$$

$$P_1(x) = a_0x^0 + a_1x^1 = 1$$

$$P_2(x) = a_0x^0 + a_1x^1 + a_2x^2 = 1 + \frac{1}{2}x^2$$

$$P_3(x) = a_0x^0 + a_1x^1 + a_2x^2 + a_3x^3 = 1 + \frac{1}{2}x^2$$

Press [y=]. Since $P_0 = P_1$, enter $y = 1$ for $y_1 =$. Since $P_2 = P_3$, enter $y = 1 + \frac{1}{2} * x^2$ for $y_2 =$. Use window dimensions [0, 1]0.2 by [0.5, 2.2]0.2. Press [graph]

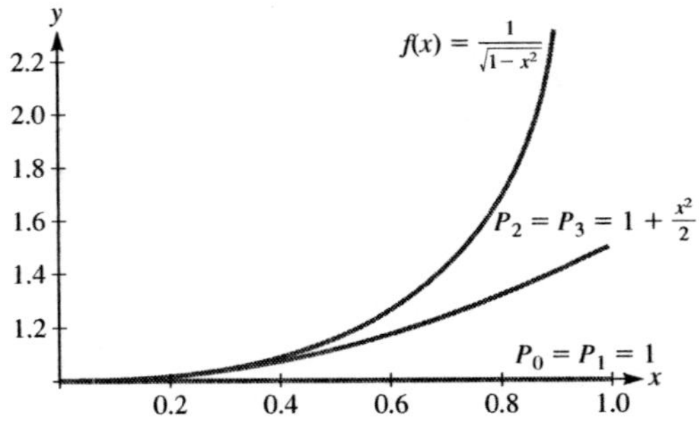

(b)

$$\int_0^{1/2} 1 + \frac{1}{2}x^2\,dx = \left(x + \frac{1}{6}x^3\right)\Big|_0^{1/2}$$

$$= \frac{1}{2} + \frac{1}{48} = \frac{25}{48}$$

47. **(a)** $e^x = \displaystyle\sum_{n=0}^{\infty} \frac{x^n}{n!}$

$$P_2 = 1 + x + \frac{1}{2}x^2$$

$$\frac{S(t) - S(0)}{R(t)} + 1 = 1 + x + \frac{1}{2}x^2$$

$$0 = \frac{1}{2}x^2 + x - \frac{S(t) - S(0)}{R(t)}$$

$$x = \frac{-1 \pm \sqrt{1 + (4)\left(\frac{1}{2}\right)\left(\frac{S(t)-S(0)}{R(t)}\right)}}{2\left(\frac{1}{2}\right)}$$

$$x = -1 + \sqrt{1 + 2\frac{S(t) - S(0)}{R(t)}}$$

Need to find t where $x = \frac{\ln 2}{\lambda}t$, so

$$t = \frac{\lambda}{\ln 2}\left[-1 + \sqrt{1 + 2\frac{S(t) - S(0)}{R(t)}}\right]$$

(b)

$$t = \frac{48.6 \times 10^9}{\ln 2}\left[-1 + \sqrt{1 + 2\frac{0.0004S_0 - 0}{0.05S_0}}\right]$$

$$= \frac{48.6 \times 10^9}{\ln 2}\left[-1 + \sqrt{1 + 2\frac{0.0004}{0.05}}\right]$$

$$\approx 5.587 \times 10^8 \text{ years}$$

(c) Writing Exercise—Answers will vary.

Chapter 10

Probability and Calculus

10.1 Discrete Random Variables

1. Discrete, since only possible values are heads and tails.

3. Continuous, since weight is a continuum of values.

5. Continuous, since living area is a continuum of values.

7. Discrete, since there are a specific, countable number of players from which values can come.

9. Discrete, since there are a specific, countable number of passengers from which values can come.

11. Since counting the number of 6's thrown in the rolls, two sample space is {0, 1, 2}.

13. Since counting the number of even numbers thrown in two rolls, the sample space is {0, 1, 2}.

15.
$$E(X) = 0\left(\frac{1}{8}\right) + 1\left(\frac{1}{8}\right) + 2\left(\frac{1}{4}\right) + 4\left(\frac{1}{2}\right) = \frac{21}{8} = 2.625$$

$$\begin{aligned}\text{Var}(X) &= \left(0 - \frac{21}{8}\right)^2\left(\frac{1}{8}\right) + \left(1 - \frac{21}{8}\right)^2\left(\frac{1}{8}\right) + \left(2 - \frac{21}{8}\right)^2\left(\frac{1}{4}\right) + \left(4 - \frac{21}{8}\right)^2\left(\frac{1}{2}\right)\\ &= \frac{441}{512} + \frac{169}{512} + \frac{25}{256} + \frac{121}{128} = \frac{1144}{512}\\ &= \frac{143}{64} \approx 2.234\end{aligned}$$

$$\sigma(X) = \sqrt{\text{Var}(X)} = \sqrt{\frac{143}{64}} = \frac{\sqrt{143}}{8} \approx 1.495$$

17.
$$E(X) = 0\left(\frac{1}{7}\right) + 2\left(\frac{3}{7}\right) + 4\left(\frac{2}{7}\right) + 6\left(\frac{1}{7}\right) = \frac{20}{7} \approx 2.857$$

$$\begin{aligned}\text{Var}(X) &= \left(0 - \frac{20}{7}\right)^2\left(\frac{1}{7}\right) + \left(2 - \frac{20}{7}\right)^2\left(\frac{3}{7}\right) + \left(4 - \frac{20}{7}\right)^2\left(\frac{2}{7}\right) + \left(6 - \frac{20}{7}\right)^2\left(\frac{1}{7}\right)\\ &= \frac{400}{343} + \frac{108}{343} + \frac{128}{343} + \frac{484}{343} = \frac{1120}{343}\\ &= \frac{160}{49} \approx 3.265\end{aligned}$$

$$\sigma(X) = \sqrt{\text{Var}(X)} = \sqrt{\frac{160}{49}} = \frac{4}{7}\sqrt{10} \approx 1.807$$

19. (a)
$X(TTTT) = 4$; $X(HTTT) = 3$; $X(THTT) = 3$; $X(TTHT) = 3$; $X(TTTH) = 3$; $X(HHTT) = 2$; $X(HTHT) = 2$; $X(HTTH) = 2$; $X(THHT) = 2$; $X(THTH) = 2$; $X(TTHH) = 2$; $X(HHHT) = 1$; $X(HHTH) = 1$; $X(HTHH) = 1$; $X(THHH) = 1$; $X(HHHH) = 0$

(b) $HTTT, THTT, TTHT, TTTH$

(c) All events

(d)
$$P(X = 0) = \frac{1}{16}$$
$$P(X \le 3) = \frac{15}{16}$$
$$P(X > 3) = \frac{1}{16}$$

21.

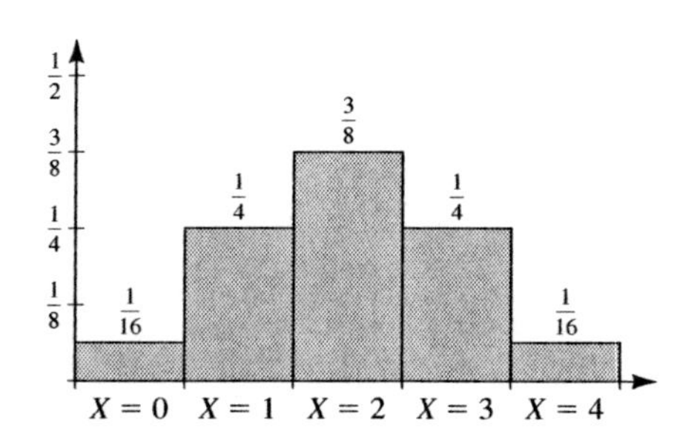

$$E(X) = 0\left(\frac{1}{16}\right) + 1\left(\frac{1}{4}\right) + 2\left(\frac{3}{8}\right) + 3\left(\frac{1}{4}\right) + 4\left(\frac{1}{16}\right) = 2$$

$$\begin{aligned}\text{Var}(X) &= (0-2)^2\left(\frac{1}{16}\right) + (1-2)^2\left(\frac{1}{4}\right) + (2-2)^2\left(\frac{3}{8}\right) \\ &\quad + (3-2)^2\left(\frac{1}{4}\right) + (4-2)^2\left(\frac{1}{16}\right) \\ &= \frac{1}{4} + \frac{1}{4} + 0 + \frac{1}{4} + \frac{1}{4} = 1\end{aligned}$$

$$\sigma(X) = \sqrt{\text{Var}(X)} = 1$$

23.

$$P(X = 1{,}000{,}000) = \frac{1}{100{,}000} = 0.00001$$

$$P(X = 100{,}000) = \frac{5}{100{,}000} = 0.00005$$

$$P(X = 10{,}000) = \frac{50}{100{,}000} = 0.00005$$

$$P(X = 0) = \frac{99{,}944}{100{,}000} = 0.99944$$

A fair price for a ticket would be its expected value, or

$$\begin{aligned}E(X) &= 1{,}000{,}000(0.00001) + 100{,}000(0.00005) \\ &\quad + 10{,}000(0.0005) + 0(0.99944) \\ &= \$20.00\end{aligned}$$

25. Since there are three possible outcomes, namely a green, black or red number, and since only red wins, $X(\text{red}) = 1$, $X(\text{black}) = -1$ and $X(\text{green}) = -1$.

Further $P(X = 1) = \frac{18}{37}$, $P(X = -1) = \frac{1}{37} + \frac{18}{37}$. So, the expected value is

$$\begin{aligned}E(X) &= 1\left(\frac{18}{37}\right) - 1\left(\frac{1}{37}\right) - 1\left(\frac{18}{37}\right) \\ &= -\frac{1}{37}, \text{ or an approximate loss of \$0.03}\end{aligned}$$

27. (a)

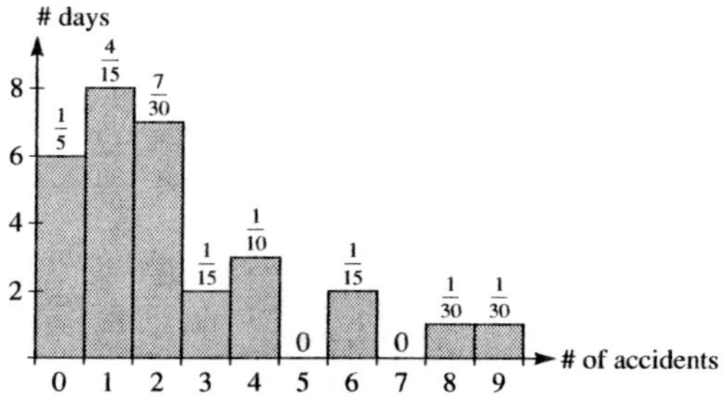

(b)

$$\begin{aligned}E(X) &= 0\left(\frac{1}{5}\right) + 1\left(\frac{4}{15}\right) + 2\left(\frac{7}{30}\right) \\ &\quad + 3\left(\frac{1}{15}\right) + 4\left(\frac{1}{10}\right) \\ &\quad + 5(0) + 6\left(\frac{1}{15}\right) + 7(0) + 8\left(\frac{1}{30}\right) + 9\left(\frac{1}{30}\right) \\ &= 0 + \frac{4}{15} + \frac{7}{15} + \frac{3}{15} + \frac{4}{10} + 0 + \frac{6}{15} + 0 + \frac{4}{15} \\ &\quad + \frac{9}{30} = \frac{69}{30} = \frac{23}{10} = 2.3\end{aligned}$$

On any given day, can expect approximately 2 accidents.

(c)

$$\begin{aligned}\mathrm{Var}(X) &= \left(0-\tfrac{23}{10}\right)^2\left(\tfrac{1}{5}\right)+\left(1-\tfrac{23}{10}\right)^2\left(\tfrac{4}{15}\right)\\ &+\left(2-\tfrac{23}{10}\right)^2\left(\tfrac{7}{30}\right)\\ &+\left(3-\tfrac{23}{10}\right)^2\left(\tfrac{1}{15}\right)+\left(4-\tfrac{23}{10}\right)^2\left(\tfrac{1}{10}\right)\\ &+\left(5-\tfrac{23}{10}\right)^2(0)\\ &+\left(6-\tfrac{23}{10}\right)^2\left(\tfrac{1}{15}\right)+\left(7-\tfrac{23}{10}\right)^2(0)\\ &+\left(8-\tfrac{23}{10}\right)^2\left(\tfrac{1}{30}\right)+\left(9-\tfrac{23}{10}\right)^2\left(\tfrac{1}{30}\right)\\ &=\frac{529}{500}+\frac{676}{1{,}500}+\frac{63}{3{,}000}+\frac{49}{1{,}500}+\frac{289}{1{,}000}+0\\ &+\frac{1{,}369}{1{,}500}+0+\frac{3{,}249}{3{,}000}+\frac{4{,}489}{3{,}000}=\frac{16{,}030}{3{,}000}=\frac{1{,}603}{300}\\ &\approx 5.3433\end{aligned}$$

$$\sigma(X)=\sqrt{\mathrm{Var}(X)}=\sqrt{\frac{1603}{300}}=\frac{\sqrt{4809}}{30}\approx 2.3116$$

29. $E(X)=0(0.961)+1(0.015)+2(0.009)+3(0.005)$
$\quad +4(0.004)+5(0.004)+6(0.002)=0.096$

On average, would not expect any people to be bumped.

31. (a) $X(\text{win})=w$; $X(\text{lose})=-1$

$P(X=w)=p$

$P(X=-1)=1-p$

The expected value is

$$E(X)=wp-1(1-p)=wp-1+p$$

(b) Want $E(X)=0$, or

$$0=wp-1+p$$

$$\frac{1-p}{p}=w$$

Fair odds are $\dfrac{1-p}{p}$ to 1.

33. $w=\dfrac{1-p}{p}$

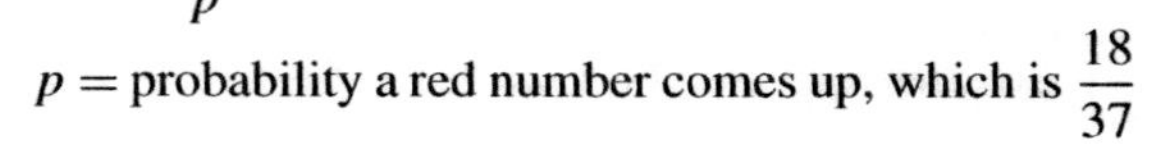

p = probability a red number comes up, which is $\dfrac{18}{37}$

$$w=\frac{1-\frac{18}{37}}{\frac{18}{37}}=\frac{19}{18}$$

Fair odds are $\dfrac{19}{18}$ to 1.

35. $P(1)=(1-0.8)^0(0.8)=0.8$
$P(2)=(1-0.8)^1(0.8)=0.16$
$P(3)=(1-0.8)^2(0.8)=0.032$
$P(4)=(1-0.8)^3(0.8)=0.0064$
$P(5)=(1-0.8)^4(0.8)=0.00128$
$P(6)=(1-0.8)^5(0.8)=0.000256$
$P(7)=(1-0.8)^6(0.8)=0.0000512$
$P(8)=(1-0.8)^7(0.8)=0.0000102$

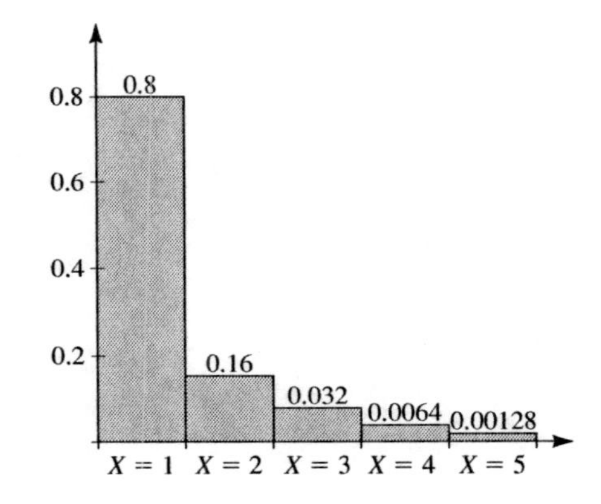

37. (a) $P(X=5)=(1-0.07)^4(0.07)$
$\quad\approx 0.05236$

(b) $P(X\le 5)=P(X=1)+P(X=2)$
$\quad +P(X=3)+P(X=4)+P(X=5)$
$\quad =0.07+(1-0.07)(0.07)$
$\quad +(1-0.07)^2(0.07)+(1-0.07)^3(0.07)$
$\quad +(1-0.07)^4(0.07)$
$\quad \approx 0.3043$

(c) $P(X\ge 5)=1-P(X<5)$
$\quad =1-0.3043+P(X=5)$
$\quad \approx 0.7481$

(d) $E(X) = \frac{1}{p} = \frac{1}{0.07} = \frac{100}{7} \approx 14$ chips

39. (a) The probability that the tenth page will jam is

$$P(X = 10) = (0.9995)^9(0.0005) \approx 0.000498$$

(b) $E(X) = \frac{1}{0.0005} = 2{,}000$
On the average, would expect the first paper jam on page 2,000 of a large document.

(c) $P(X > 25) = 1 - P(X \le 25)$

$$= 1 - \sum_{n=1}^{25} (0.9995)^{n-1}(0.0005)$$

$$= 1 - 0.0005 \sum_{n=1}^{25} (0.9995)^{n-1}$$

$\sum_{n=1}^{25} (0.9995)^{n-1} = S_{25}$, where

$$S_{25} = 1 + 0.9995 + (0.9995)^2 + \cdots + (0.9995)^{24}$$

$$0.9995S_{25} = 0.9995 + (0.9995)^2 + \cdots + (0.9995)^{24} + (0.9995)^{25}$$

Subtracting,

$$0.0005S_{25} = 1 - (0.9995)^{25}$$

$$S_{25} = \frac{1 - (0.9995)^{25}}{0.0005}$$

So,

$$1 - 0.0005 \sum_{n=1}^{25} (0.9995)^{n-1}$$

$$= 1 - 0.0005 \left[\frac{1 - (0.9995)^{25}}{0.0005}\right]$$

$$= (0.9995)^{25} \approx 0.9876$$

41. (a)

$$P(X = 5) = (1 - 0.04)^4(0.04) \approx 0.0340$$

(b)

$$P(X \le 5) = P(X = 1) + P(X = 2) + P(X = 3) + P(X = 4) + P(X = 5)$$

$$= 0.04 + (1 - 0.04)(0.04) + (1 - 0.04)^2(0.04) + (1 - 0.04)^3(0.04) + (1 - 0.04)^4(0.04)$$

$$\approx 0.1846$$

(c) $P(X > 12) = \sum_{n=13}^{\infty} (1 - 0.04)^{n-1}(0.04)$

$$= (0.96)^{12}(0.04) + (0.96)^{13}(0.04) + (0.96)^{14}(0.04) + \cdots$$

$$= (0.96)^{12}(0.04)\left[1 + 0.96 + (0.96)^2 + (0.96)^3 + \cdots\right]$$

$$= (0.96)^{12}(0.04) \sum_{n=0}^{\infty} (0.96)^n$$

$$= (0.96)^{12}(0.04)\left(\frac{1}{1 - 0.96}\right)$$

$$\approx 0.6127$$

(d) $E(X) = \frac{1}{P} = \frac{1}{0.04} = 25$ students

43. $P(X \ge 100) = \sum_{n=100}^{\infty} (0.98)^{n-1}(0.02)$

$$= (0.98)^{99}(0.02) + (0.98)^{100}(0.02) + (0.98)^{101}(0.02) + \cdots$$

$$= (0.98)^{99}(0.02)\left[1 + 0.98 + (0.98)^2 + (0.98)^3 + \cdots\right]$$

$$= (0.98)^{99}(0.02) \sum_{n=0}^{\infty} (0.98)^n$$

$$= (0.98)^{99}(0.02)\left(\frac{1}{1 - 0.98}\right) \approx 0.1353$$

45. $E(X) = \sum_{k=1}^{\infty} X_k p(X_k)$

$$\sum_{k=1}^{\infty} k(1 - p)^{k-1}p$$

$$= p \sum_{k=1}^{\infty} k(1 - p)^{k-1}$$

$$= p\left[1 + 2(1 - p)^2 + 3(1 - p)^3 + 4(1 - p)^4 + \cdots\right]$$

$$= p\left[\frac{1}{(1 - [1 - p])^2}\right] = p\left(\frac{1}{p^2}\right) = \frac{1}{p}$$

47. $\sum_{n=1}^{\infty} P(X=n) = \sum_{n=1}^{\infty} \frac{a}{n^b} = a \sum_{n=1}^{\infty} \frac{1}{n^b} = 1$

So, $a = \frac{1}{\zeta(b)}$ where $\zeta(b) = \sum_{n=1}^{\infty} \frac{1}{n^b}$

10.2 Continuous Random Variables

1. Since $f(x) \geq 0$ for all x, the first condition is met. Checking the second condition,

$$\int_{-\infty}^{\infty} f(x)\,dx = \int_0^{\infty} \frac{10}{(x+10)^2}\,dx$$

$$= \lim_{N\to\infty} \int_0^N \frac{10}{(x+10)^2}\,dx = 10 \lim_{N\to\infty} \int_0^N \frac{1}{(x+10)^2}\,dx$$

Using substitution with $u = x + 10$ and $du = dx$,

$$= 10 \lim_{N\to\infty} \int_{10}^{N+10} \frac{1}{u^2}\,du$$

$$= 10 \lim_{N\to\infty} \left(-\frac{1}{u}\Big|_{10}^{N+10}\right)$$

$$= 10 \lim_{N\to\infty} \left(-\frac{1}{N+10} + \frac{1}{10}\right) = 10\left(0 + \frac{1}{10}\right) = 1$$

The third condition is also met, so f *is* a probability density function.

3. Since $f(x) \geq 0$ for all x, the first condition is met. Checking the second condition,

$$\int_{-\infty}^{\infty} f(x)\,dx = \int_0^{\infty} xe^{-x}\,dx$$

$$= \lim_{N\to\infty} \int_0^N xe^{-x}\,dx$$

Using integration by parts with $u = x$ and $dV = e^{-x}\,dx$,

$$= \lim_{N\to\infty} \left[-xe^{-x}\Big|_0^N - \int_0^N 1 \cdot -e^{-x}\,dx\right]$$

$$= \lim_{N\to\infty} \left[-xe^{-x}\Big|_0^N + \int_0^N e^{-x}\,dx\right]$$

$$= \lim_{N\to\infty} \left[\left(-xe^{-x} - e^{-x}\right)\Big|_0^N\right]$$

$$= \lim_{N\to\infty} \left[\left(-Ne^{-N} - e^{-N}\right) - (0-1)\right]$$

$$= (0-0) - (0-1) = 1$$

The third condition is also met, so f *is* a probability density function.

5. The first condition is not met. For example $f(-1) = \frac{3}{2}(-1)^2 + 2(-1) = -\frac{1}{2}$. Since it is not the case that $f(x) \geq 0$ for all x, f is *not* a probability density function.

7. **(a)** $P(2 \leq x \leq 5) = \int_2^5 \frac{1}{3}\,dx$

$$= \frac{x}{3}\Big|_2^5 = 1$$

Note: $\int_{-\infty}^{\infty} f(x)\,dx = \int_2^5 f(x)\,dx$ in this problem, so needn't even integrate to conclude that the probability is 1.

(b) $P(3 \leq x \leq 4) = \int_3^4 \frac{1}{3}\,dx$

$$= \frac{x}{3}\Big|_3^4 = \frac{1}{3}$$

(c) $P(X \geq 4) = \int_4^5 \frac{1}{3}\,dx$

$$= \frac{x}{3}\Big|_4^5 = \frac{1}{3}$$

Note: can also calculate as $1 - \int_0^4 \frac{1}{3}\,dx$.

9. **(a)** $P(0 \leq x \leq 4) = \int_0^4 \frac{1}{8}(4-x)\,dx$

$$= \int_0^4 \frac{1}{2} - \frac{1}{8}x\,dx = \left(\frac{x}{2} - \frac{x^2}{16}\right)\Big|_0^4$$

$$= (2-1) - 0 = 1$$

Note: $\int_{-\infty}^{\infty} f(x)\,dx = \int_0^4 f(x)\,dx$ in this problem, so needn't even integrate to conclude that the probability is 1.

(b) $P(2 \le x \le 3) = \int_2^3 \frac{1}{8}(4-x)\,dx$

$$= \left(\frac{x}{2} - \frac{x^2}{16}\right)\Big|_2^3 = \left(\frac{3}{2} - \frac{9}{16}\right) - \left(1 - \frac{1}{4}\right)$$

$$= \frac{3}{16}$$

(c) $P(X \ge 1) = \int_1^4 \frac{1}{8}(4-x)\,dx$

$$= \left(\frac{x}{2} - \frac{x^2}{16}\right)\Big|_1^4 = (2-1) - \left(\frac{1}{2} - \frac{1}{16}\right) = \frac{9}{16}$$

Note: can also calculate as $1 - \int_0^1 \frac{1}{8}(4-x)\,dx$.

11. **(a)** $P(1 \le x < \infty) = \int_1^{\infty} \frac{3}{x^4}\,dx$

$$= 3 \lim_{N\to\infty} \int_1^N \frac{1}{x^4}\,dx = 3 \lim_{N\to\infty} \left(-\frac{1}{3x^3}\Big|_1^N\right)$$

$$= 3 \lim_{N\to\infty} \left(-\frac{1}{3N^3} + \frac{1}{3}\right) = 3\left(0 + \frac{1}{3}\right) = 1$$

Note: $\int_{-\infty}^{\infty} f(x)\,dx = \int_1^{\infty} \frac{3}{x^4}\,dx$ in this problem, So needn't even integrate to conclude that the probability is 1.

(b) $P(1 \le x \le 2) = \int_1^2 \frac{3}{x^4}\,dx$

$$= 3\left(-\frac{1}{3x^3}\Big|_1^2\right) = 3\left(-\frac{1}{24} + \frac{1}{3}\right) = \frac{7}{8}$$

(c) $P(X \ge 2) = 1 - \int_1^2 \frac{3}{x^4}\,dx$

$$= 1 - \frac{7}{8} = \frac{1}{8}$$

13. **(a)** $P(X \ge 0) = \int_0^{\infty} 2xe^{-x^2}\,dx$

$$= \lim_{N\to\infty} \int_0^N 2xe^{-x^2}\,dx$$

Using substitution with $u = -x^2$ and $-du = 2x\,dx$,

$$= \lim_{N\to\infty} -\int_0^{-N^2} e^u\,du = \lim_{N\to\infty} \left(-e^u\Big|_0^{-N^2}\right)$$

$$= \lim_{N\to\infty} \left(-e^{-N^2} + e^0\right) = 0 + 1 = 1$$

Note: $\int_{-\infty}^{\infty} f(x)\,dx = \int_0^{\infty} f(x)\,dx$ in this problem, so needn't even integrate to conclude that the probability is 1.

(b) $P(1 \le x \le 2) = \int_1^2 2xe^{-x^2}\,dx$

$$= -e^u\Big|_{-1}^{-4} = -e^{-4} + e^{-1} = \frac{1}{e} - \frac{1}{e^4}$$

(c) $P(X \le 2) = \int_0^2 2xe^{-x^2}\,dx$

$$= -e^u\Big|_0^{-4} = -e^{-4} + 1 = 1 - \frac{1}{e^4}$$

15. **(a)** $P(X \ge 4) = \int_4^7 \frac{3}{28} + \frac{3}{x^2}\,dx$

$$= \left(\frac{3}{28}X - \frac{3}{x}\right)\Big|_4^7 = \left(\frac{3}{4} - \frac{3}{7}\right) - \left(\frac{3}{7} - \frac{3}{4}\right) = \frac{9}{14}$$

(b) $P(X < 5) = \int_3^5 \frac{3}{28} + \frac{3}{x^2}\,dx$

$$= \left(\frac{3}{28}x - \frac{3}{x}\right)\Big|_3^5 = \left(\frac{15}{28} - \frac{3}{5}\right) - \left(\frac{9}{28} - 1\right) = \frac{43}{70}$$

(c) $P(4 \le X \le 6) = \int_4^6 \frac{3}{28} + \frac{3}{x^2}\,dx$

$$= \left(\frac{3}{28}x - \frac{3}{x}\right)\Big|_4^6 = \left(\frac{9}{14} - \frac{1}{2}\right) - \left(\frac{3}{7} - \frac{3}{4}\right) = \frac{13}{28}$$

17. The probability density function for this situation is

$$f(x) = \begin{cases} \frac{1}{20} & 0 \le x \le 20 \\ 0 & \text{otherwise} \end{cases}$$

(a) $P(X \ge 8) = \int_8^{20} \frac{1}{20}\,dx$

$$= \frac{x}{20}\Big|_8^{20} = \frac{3}{5}$$

(b) $P(2 \le x \le 5) = \int_2^5 \frac{1}{20}\,dx$

$$= \frac{x}{20}\Big|_2^5 = \frac{3}{20}$$

19. **(a)** $P(50 \le x \le 60) = \int_{50}^{60} 0.01e^{-0.01x}\,dx$

$$= -e^{-0.01x}\Big|_{50}^{60} = -e^{-0.6} + e^{-0.5} \approx 0.0577$$

(b) $P(0 \le x \le 60) = \int_0^{60} 0.01e^{-0.01x}\,dx$

$$= -e^{-0.01x}\Big|_0^{60} = -e^{-0.6} + e^0 \approx 0.4512$$

(c) $P(X > 60) = 1 - P(0 \le X \le 60)$
$\approx 1 - 0.4512 = 0.5488$

21. **(a)** $P(X \ge 8) = 1 - \int_0^8 \frac{1}{4}e^{-x/4}\,dx$

$$1 - \left(-e^{-x/4}\Big|_0^8\right) = 1 - \left(-e^{-2} + e^0\right) = e^{-2} = \frac{1}{e^2}$$

(b) $P(1 \le X \le 5) = \int_1^5 \frac{1}{4}e^{-x/4}\,dx$

$$= -e^{-x/4}\int_1^5 = -e^{-5/4} + e^{-1/4}$$

$$= \frac{1}{e^{1/4}} - \frac{1}{e^{5/4}} \approx 0.4923$$

23. **(a)** $P(X \le 5) = \int_0^5 \frac{1}{16}xe^{-x/4}\,dx$
Using integration by parts with $u = x$ and $dV = e^{-x/4}\,dx$,

$$= \frac{1}{16}\left[-4xe^{-x/4}\Big|_0^5 - \int_0^5 1 \cdot -4e^{-x/4}\,dx\right]$$

$$= \frac{1}{16}\left[-4xe^{-x/4}\Big|_0^5 + 4\int_0^5 e^{-x/4}\,dx\right]$$

$$= \frac{1}{16}\left[\left(-4xe^{-x/4} - 16e^{-x/4}\right)\Big|_0^5\right]$$

$$= -\frac{1}{4}\left[\left(xe^{-x/4} + 4e^{-x/4}\right)\Big|_0^5\right]$$

$$-\frac{1}{4}\left[(x+4)e^{-x/4}\Big|_0^5\right]$$

$$= -\frac{1}{4}\left[9e^{-5/4} - 4e^0\right] = 1 - \frac{9}{4e^{5/4}} \approx 0.3554$$

(b) $P(X \ge 10) = 1 - \int_0^{10} \frac{1}{16}xe^{-x/4}\,dx$

$$= 1 - \left(-\frac{1}{4}\left[(x+4)e^{-x/4}\Big|_0^{10}\right]\right)$$

$$= 1 - \left(-\frac{1}{4}\left[14e^{-5/2} - 4e^0\right]\right) = 1 + \frac{7}{2}e^{-5/2} - 1$$

$$= \frac{7}{2e^{5/2}} \approx 0.2873$$

25. $P(X \le 12) = \int_0^{12} 0.08e^{-0.08t}\,dt$

$$= -e^{-0.08t}\Big|_0^{12} = -e^{-0.96} + e^0$$

≈ 0.6171

27. **(a)** The probability density function for this situation is

$$f(t) = \begin{cases} \frac{1}{60} & 0 \le t \le 60 \\ 0 & \text{otherwise} \end{cases}$$

(b) $P(X \ge 45) = \int_{45}^{60} \frac{1}{60}\,dt$

$$= \frac{t}{60}\Big|_{45}^{60} = 1 - \frac{3}{4} = \frac{1}{4}$$

(c) $P(5 \le x \le 15) = \int_5^{15} \frac{1}{60}\,dt$

$$= \frac{t}{60}\Big|_5^{15} = \frac{1}{4} - \frac{1}{12} = \frac{1}{6}$$

29. The probability density function for this situation is

$$f(t) = \begin{cases} 0.5e^{-0.5} & x \geq 0 \\ 0 & x<0 \end{cases}$$

$$P(X > 2) = 1 - \int_0^2 0.5e^{-0.5}\,dx$$

$$= 1 - \left(-e^{-0.5}\Big|_0^2\right)$$

$$= 1 + \left(e^{-0.5}\Big|_0^2\right) = 1 + \left(e^{-1} - e^0\right)$$

$$= e^{-1} = \frac{1}{e} \approx 0.3679$$

31. (a)

$$\int_1^2 \int_0^2 \frac{1}{6} e^{-\frac{x}{2}} e^{-\frac{y}{3}}\,dy\,dx$$

$$= \int_1^2 \frac{1}{6} e^{-x/2} \left(\int_0^2 e^{-y/3}\,dy \right) dx$$

$$= \int_1^2 \frac{1}{6} e^{-x/2} \left(-3e^{-y/3}\Big|_0^2 \right) dx$$

$$= -\frac{1}{2} \int_1^2 e^{-x/3} \left(e^{-y/3}\Big|_0^2 \right) dx$$

$$= -\frac{1}{2} \int_1^2 e^{-x/2} \left(e^{-2/3} - e^0 \right) dx$$

$$= -\frac{1}{2} \left(e^{-2/3} - 1 \right) \int_1^2 e^{-x/2}\,dx$$

$$= -\frac{1}{2} \left(e^{-2/3} - 1 \right) \left(-2e^{-x/2}\Big|_1^2 \right)$$

$$= \left(e^{-2/3} - 1 \right) \left(e^{-x/2}\Big|_1^2 \right)$$

$$= \left(e^{-2/3} - 1 \right) \left(e^{-1} - e^{1/2} \right) \approx 0.1161$$

(b) Since $x + y \leq 1$, we want

$$\int_0^1 \int_0^{1-x} \frac{1}{6} e^{-x/2} e^{-y/3}\,dy\,dx$$

$$= -\frac{1}{2} \int_0^1 e^{-x/2} \left(e^{-y/3}\Big|_0^{1-x} \right) dx$$

$$= -\frac{1}{2} \int_0^1 e^{-x/2} \left(e^{\frac{x-1}{3}} - e^0 \right) dx$$

$$= -\frac{1}{2} \int_0^1 e^{\frac{-x-2}{6}} - e^{-x/2}\,dx$$

$$= -\frac{1}{2} \left[\left(-6e^{\frac{-x-2}{6}} + 2e^{-x/2} \right) \Big|_0^1 \right]$$

$$= \left(3e^{\frac{-x-2}{6}} - e^{-x/2} \right) \Big|_0^1 = \left(3e^{-1/2} - e^{-1/2} \right)$$

$$- \left(3e^{-1/3} - e^0 \right)$$

$$\approx 0.06347$$

33. (a)

$$\int_0^3 \int_0^3 \frac{1}{12} e^{-x/4} e^{-y/3}\,dy\,dx$$

$$= \frac{1}{12} \int_0^3 e^{-x/4} \left(-3e^{-y/3} \right) \Big|_0^3 dx$$

$$= -\frac{1}{4} \int_0^3 e^{-x/4} \left(e^{-1} - e^0 \right) dx$$

$$= -\frac{1}{4} \left(\frac{1}{e} - 1 \right) \int_0^3 e^{-x/4}\,dx$$

$$= -\frac{1}{4} \left(\frac{1}{e} - 1 \right) \left(-4e^{-x/4}\Big|_0^3 \right)$$

$$= \left(\frac{1}{e} - 1 \right) \left(e^{-3/4} - e^0 \right) \approx 0.3335$$

(b) This is the complement of the value found in part (a), or

$$1 - 0.3335 = 0.6665$$

35. Using the constraint $x + y < 50$, need to find

$$1 - \int_0^{50} \int_0^{50-x} \frac{1}{500} e^{-x/10} e^{-y/50}\,dy\,dx$$

$$= 1 - \frac{1}{500}\int_0^{50} e^{-x/10}\left(-50e^{-y/50}\right)\Big|_0^{50-x} dx$$
$$= 1 + \frac{1}{10}\int_0^{50} e^{-x/10}\left(e^{-y/50}\Big|_0^{50-x}\right) dx$$
$$= 1 + \frac{1}{10}\int_0^{50} e^{-x/10}\left(e^{-y/50}\Big|_0^{50-x}\right) dx$$
$$= 1 + \frac{1}{10}\int_0^{50} e^{-x/10}\left(e^{\frac{x-50}{50}} - e^0\right) dx$$
$$= 1 + \frac{1}{10}\int_0^{50} e^{\frac{-4x-50}{50}} - e^{-x/10}\, dx$$
$$= 1 + \frac{1}{10}\left(-\frac{25}{2}e^{\frac{-4x-50}{50}} + 10e^{-x/10}\right)\Big|_0^{50}$$
$$= 1 + \frac{1}{10}\left[\left(-\frac{25}{2}e^{-5} + 10e^{-5}\right) - \left(-\frac{25}{2}e^{-1} + 10e^0\right)\right]$$
$$\approx 0.458$$

37. Need to find m such that

$$\int_0^m \frac{1}{40}\,dt = \frac{1}{2}$$
$$\frac{t}{40}\Big|_0^m = \frac{1}{2}$$
$$\frac{m}{40} - 0 = \frac{1}{2}$$
$$m = 20 \text{ seconds}$$

39. Need to find m such that

$$\int_a^m \frac{1}{b-a}\,dt = \frac{1}{2}$$
$$\frac{t}{b-a}\Big|_a^m = \frac{1}{2}$$
$$\frac{m}{b-a} - \frac{a}{b-a} = \frac{1}{2}$$
$$m - a = \frac{1}{2}(b-a)$$
$$m = \frac{1}{2}b - \frac{1}{2}a + a$$
$$m = \frac{1}{2}b + \frac{1}{2}a$$
$$m = \frac{b+a}{2}$$

41. **(a)** $$r(x) = \int_x^\infty f(x)\,dx = 1 - \int_0^x f(x)\,dx$$

(b) $$r(10) = 1 - \int_0^{10} 20e^{-20x}\,dx$$
$$= 1 - 20\left(-\frac{1}{20}e^{-20x}\Big|_0^{10}\right)$$
$$= 1 + \left(e^{-20x}\Big|_0^{10}\right)$$
$$= 1 + \left(e^{-200} - e^0\right) = e^{-200}$$

43. **(a)**

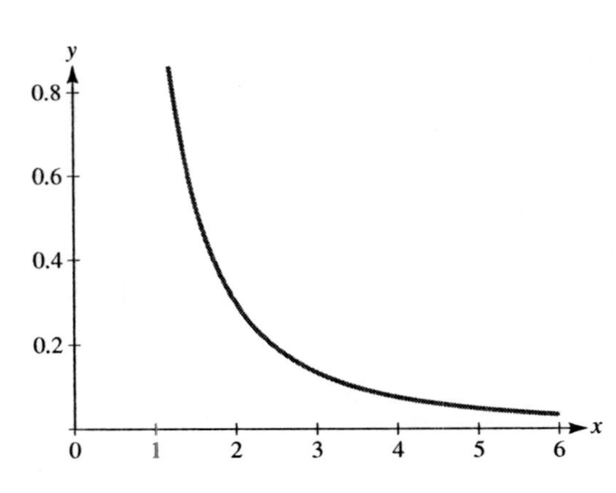

(b) $$\int_{10}^{20} \frac{1.2}{x^2}\,dx = 1.2\int_{10}^{20} \frac{1}{x^2}\,dx$$
$$= 1.2\left(-\frac{1}{x}\Big|_{10}^{20}\right) = 1.2\left(-\frac{1}{20} + \frac{1}{10}\right)$$
$$= 0.06$$

(c) Writing Exercise—Answers will vary.

10.3 Expected Value and Variance of Continuous Random Variables

1.

$$E(X) = \int_{-\infty}^{\infty} x f(x)\,dx$$
$$= \int_2^5 \frac{1}{3}x\,dx = \frac{1}{6}x^2\Big|_2^5$$
$$= \frac{1}{6}(25-4) = \frac{7}{2}$$
$$\text{Var}(X) = \int_{-\infty}^{\infty} x^2 f(x)\,dx - [E(X)]^2$$
$$= \int_2^5 \frac{1}{3}x^2\,dx - \left(\frac{7}{2}\right)^2$$
$$= \frac{1}{9}x^3\Big|_2^5 - \frac{49}{4} = \frac{1}{9}(125-8) = \frac{3}{4}$$

3.

$$E(X) = \int_{-\infty}^{\infty} x f(x)\,dx$$
$$= \int_0^4 \frac{1}{8}x(4-x)\,dx = \frac{1}{8}\int_0^4 4x - x^2\,dx$$
$$= \frac{1}{8}\left(2x^2 - \frac{x^3}{3}\right)\Big|_0^4 = \frac{1}{8}\left[\left(32 - \frac{64}{3}\right) - 0\right] = \frac{4}{3}$$
$$\text{Var}(X) = \int_{-\infty}^{\infty} x^2 f(x)\,dx - [E(x)]^2$$
$$= \int_0^4 \frac{1}{8}x^2(4-x)\,dx - \left(\frac{4}{3}\right)^2$$
$$= \frac{1}{8}\int_0^4 4x^2 - x^3\,dx - \frac{16}{9}$$
$$= \frac{1}{8}\left(\frac{4}{3}x^3 - \frac{x^4}{4}\right)\Big|_0^4 - \frac{16}{9} = \frac{8}{9}$$

5.

$$E(X) = \int_{-\infty}^{\infty} x f(x)\,dx$$
$$= \lim_{N\to\infty}\int_1^N \frac{3}{x^3}\,dx = 3\lim_{N\to\infty}\left(-\frac{1}{2x^2}\Big|_1^N\right)$$
$$= 3\lim_{N\to\infty}\left(-\frac{1}{2N^2} + \frac{1}{2}\right) = 3\left(0 + \frac{1}{2}\right) = \frac{3}{2}$$
$$\text{Var}(X) = \int_{-\infty}^{\infty} x^2 f(x)\,dx - [E(X)]^2$$
$$= \lim_{N\to\infty}\int_1^N \frac{3}{x^2}\,dx - \left(\frac{3}{2}\right)^2$$
$$= 3\lim_{N\to\infty}\left(-\frac{1}{x}\Big|_1^N\right) - \frac{9}{4}$$
$$= 3\lim_{N\to\infty}\left(-\frac{1}{N} + 1\right) - \frac{9}{4}$$
$$= 3(0+1) - \frac{9}{4} = \frac{3}{4}$$

7.

$$E(X) = \int_{-\infty}^{\infty} x f(x)\,dx$$
$$= \lim_{N\to\infty}\int_0^N 4xe^{-4x}\,dx$$
$$= 4\lim_{N\to\infty}\int_0^N xe^{-4x}\,dx$$

Using integration by parts with $u = x$ and $dV = e^{-4x}\,dx$

$$= 4\lim_{N\to\infty}\left[-\frac{x}{4}e^{-4x}\Big|_0^N - \int_0^N 1\cdot -\frac{1}{4}e^{-4x}\,dx\right]$$
$$= 4\lim_{N\to\infty}\left[-\frac{x}{4}e^{-4x}\Big|_0^N + \frac{1}{4}\int_0^N e^{-4x}\,dx\right]$$
$$= 4\lim_{N\to\infty}\left[-\frac{x}{4}e^{-4x} - \frac{1}{16}e^{-4x}\right]\Big|_0^N$$
$$= 4\lim_{N\to\infty}\left[\left(-\frac{N}{4}e^{-4N} - \frac{1}{16}e^{-4N}\right) - \left(0 - \frac{1}{16}e^0\right)\right]$$
$$= 4\left[(0-0) - \left(0 - \frac{1}{16}\right)\right] = \frac{1}{4}$$

$$\text{Var}(X) = \int_{-\infty}^{\infty} x^2 f(x)\,dx - [E(X)]^2$$
$$= \lim_{N\to\infty} \int_0^N 4x^2e^{-4x}\,dx - \left(\frac{1}{4}\right)^2$$
$$= 4 \lim_{N\to\infty} \int_0^N x^2e^{-4x}\,dx - \frac{1}{16}$$

Using integration by parts twice,

$$= 4 \lim_{N\to\infty} \left[-\frac{x^2}{4}e^{-4x}\Big|_0^N - \int_0^N 2x \cdot -\frac{1}{4}e^{-4x}\,dx \right] - \frac{1}{16}$$
$$= 4 \lim_{N\to\infty} \left[-\frac{x^2}{4}e^{-4x}\Big|_0^N + \frac{1}{2}\int_0^N xe^{-4x}\,dx \right] - \frac{1}{16}$$
$$= 4 \lim_{N\to\infty} \left[-\frac{x^2}{4}e^{-4x}\Big|_0^N \right.$$
$$\left. + \frac{1}{2}\left(-\frac{x}{4}e^{-4x}\Big|_0^N - \int_0^N 1 \cdot -\frac{1}{4}e^{-4x}\,dx \right) \right] - \frac{1}{16}$$
$$= 4 \lim_{N\to\infty} \left[-\frac{x^2}{4}e^{-4x}\Big|_0^N - \frac{x}{8}e^{-4x}\Big|_0^N + \frac{1}{8}\int_0^N e^{-4x}\,dx \right] - \frac{1}{16}$$
$$= 4 \lim_{N\to\infty} \left[-\frac{x^2}{4}e^{-4x} - \frac{x}{8}e^{-4x} - \frac{1}{32}e^{-4x} \right]\Big|_0^N - \frac{1}{16}$$
$$= 4 \lim_{N\to\infty} \left[\left(-\frac{N^2}{4}e^{-4x} - \frac{N}{8}e^{-4N} - \frac{1}{32}e^{-4N} \right) \right.$$
$$\left. - \left(0 - 0 - \frac{1}{32}e^0 \right) \right] - \frac{1}{16}$$
$$= 4\left[(0-0-0) - \left(0 - 0 - \frac{1}{32} \right) \right] - \frac{1}{16} = \frac{1}{16}$$

9. $$E(X) = \int_{-\infty}^{\infty} xf(x)\,dx$$
$$= \int_0^1 20x(x^3 - x^4)\,dx = 20\int_0^1 x^4 - x^5\,dx$$
$$= 20\left(\frac{x^5}{5} - \frac{x^6}{6} \right)\Big|_0^1 = 20\left[\left(\frac{1}{5} - \frac{1}{6} \right) - 0 \right] = \frac{2}{3}$$

$$\text{Var}(X) = \int_{-\infty}^{\infty} x^2 f(x)\,dx - [E(X)]^2$$
$$= \int_0^1 20x^2(x^3 - x^4)\,dx - \left(\frac{2}{3}\right)^2$$
$$= 20\int_0^1 x^5 - x^6\,dx - \frac{4}{9}$$
$$= 20\left(\frac{x^6}{6} - \frac{x^7}{7} \right)\Big|_0^1 - \frac{4}{9}$$
$$= 20\left[\left(\frac{1}{6} - \frac{1}{7} \right) - 0 \right] - \frac{4}{9} = \frac{2}{63}$$

11. $$E(X) = \int_{-\infty}^{\infty} xf(x, y)\,dy\,dx$$
$$= \int_0^{\infty}\int_0^{\infty} \frac{1}{2}xe^{-x/2}e^{-y}\,dy\,dx$$
$$= \int_0^{\infty} \frac{1}{2}xe^{-x/2}\left(\int_0^{\infty} e^{-y}\,dy \right) dx$$
$$= \int_0^{\infty} \frac{1}{2}xe^{-x/2}\left[\lim_{N\to\infty} \left(-e^{-y}\Big|_0^N \right) \right] dx$$
$$= \int_0^{\infty} \frac{1}{2}xe^{-x/2}\left[\lim_{N\to\infty} \left(-e^{-N} + e^0 \right) \right] dx$$
$$= \int_0^{\infty} \frac{1}{2}xe^{-x/2}(1)\,dx = \frac{1}{2}\lim_{N\to\infty}\int_0^N xe^{-x/2}\,dx$$

Using integration by parts with $u = x$ and $dV = e^{-x/2}\,dx$

$$= \frac{1}{2}\lim_{N\to\infty}\left[-2xe^{-x/2}\Big|_0^N - \int_0^N 1 \cdot -2e^{-x/2}\,dx \right]$$
$$= \frac{1}{2}\lim_{N\to\infty}\left[-2xe^{-x/2}\Big|_0^N + 2\int_0^N e^{-x/2}\,dx \right]$$
$$= \frac{1}{2}\lim_{N\to\infty}\left[-2xe^{-x/2} - 4e^{-x/2} \right]\Big|_0^N$$
$$= \frac{1}{2}\lim_{N\to\infty}\left[\left(-2Ne^{-N/2} - 4e^{-N/2} \right) - \left(0 - 4e^0 \right) \right]$$
$$= \frac{1}{2}[(0-0) - (0-4)] = 2$$

$$E(Y) = \int_0^\infty \int_0^\infty \frac{1}{2} y e^{-y} e^{-x/2}\, dx\, dy$$
$$= \int_0^\infty \frac{1}{2} y e^{-y} \left(\int_0^\infty e^{-x/2}\, dx \right) dy$$
$$= \int_0^\infty \frac{1}{2} y e^{-y} \left[\lim_{N\to\infty} \left(-2e^{-x/2} \Big|_0^N \right) \right] dy$$
$$= \int_0^\infty \frac{1}{2} y e^{-y} \left[\lim_{N\to\infty} \left(-2e^{-N/2} + 2e^0 \right) \right] dy$$
$$= \int_0^\infty \frac{1}{2} y e^{-y}(2)\, dy = \lim_{N\to\infty} \int_0^N y e^{-y}\, dy$$

Using integration by parts with $u = y$ and $dV = e^{-y}\, dy$

$$= \lim_{N\to\infty} \left[-ye^{-y} \Big|_0^N - \int_0^N 1 \cdot -e^{-y}\, dy \right]$$
$$= \lim_{N\to\infty} \left[-ye^{-y} \Big|_0^N + \int_0^N e^{-y}\, dy \right]$$
$$= \lim_{N\to\infty} \left[-ye^{-y} - e^{-y} \right] \Big|_0^N$$
$$= \lim_{N\to\infty} \left[\left(-Ne^{-N} - e^{-N} \right) - \left(0 - e^0 \right) \right]$$
$$= (0 - 0) - (0 - 1) = 1$$

13. $$E(X) = \int_{-\infty}^\infty x f(x, y)\, dy\, dx$$
$$= \int_0^\infty \int_0^\infty xye^{-x-y}\, dy\, dx$$
$$= \int_0^\infty \int_0^\infty xye^{-x}e^{-y}\, dy\, dx$$
$$= \int_0^\infty xe^{-x} \left[\lim_{N\to\infty} \int_0^N ye^{-y}\, dy \right] dx$$

Using integration by parts with $u = y$ and $dV = e^{-y}\, dy$,

$$= \int_0^\infty xe^{-x} \left[\lim_{N\to\infty} \left(-ye^{-y} \Big|_0^N - \int_0^\infty 1 \cdot -e^{-y}\, dy \right) \right] dx$$
$$= \int_0^\infty xe^{-x} \left[\lim_{N\to\infty} \left(-ye^{-y} \Big|_0^N + \int_0^N e^{-y}\, dy \right) \right] dx$$
$$= \int_0^\infty xe^{x} \left[\lim_{N\to\infty} \left(-ye^{-y} - e^{-y} \right) \Big|_0^N \right] dx$$
$$= \int_0^\infty xe^{x} \left[\lim_{N\to\infty} \left((-Ne^{-N} - e^{-N}) - (0 - e^0) \right) \right] dx$$
$$= \int_0^\infty xe^{x}(1)\, dx = \lim_{N\to\infty} \int_0^N xe^{x}\, dx$$

Using integration by parts with $u = x$ and $dV = e^x\, dx$,

$$= \lim_{N\to\infty} \left[xe^x \Big|_0^N - \int_0^N 1e^x\, dx \right]$$
$$= \lim_{N\to\infty} \left[\left(xe^x - e^x \right) \Big|_0^N \right]$$
$$= \lim_{N\to\infty} \left[\left(Ne^N - e^N \right) - \left(0 - e^0 \right) \right] = 1$$

$$E(Y) = \int_{-\infty}^\infty y f(x, y)\, dx\, dy$$
$$= \int_0^\infty \int_0^\infty y^2 e^{-x-y}\, dx\, dy$$
$$= \int_0^\infty \int_0^\infty y^2 e^{-x} e^{-y}\, dx\, dy$$
$$= \int_0^\infty y^2 e^{-y} \left[\lim_{N\to\infty} \int_0^N e^{-x}\, dx \right] dy$$
$$= \int_0^\infty y^2 e^{-y} \left[\lim_{N\to\infty} \left(-e^{-x} \Big|_0^N \right) \right] dy$$
$$= \int_0^\infty y^2 e^{-y} \left[\lim_{N\to\infty} \left(-e^{-N} + e^0 \right) \right] dy$$
$$= \int_0^\infty y^2 e^{-y}(1)\, dy = \lim_{N\to\infty} \int_0^N y^2 e^{-y}\, dy$$

Using integration by parts twice,

$$= \lim_{N\to\infty}\left[-y^2e^{-y}\Big|_0^N - \int_0^N 2y\cdot -e^{-y}\,dy\right]$$

$$= \lim_{N\to\infty}\left[-y^2e^{-y}\Big|_0^N + 2\int_0^N ye^{-y}\,dy\right]$$

$$= \lim_{N\to\infty}\left[-y^2e^{-y}\Big|_0^N + 2\left(-ye^{-y}\Big|_0^N - \int_0^N 1\cdot -e^{-y}\,dy\right)\right]$$

$$= \lim_{N\to\infty}\left[-y^2e^{-y}\Big|_0^N - 2ye^{-y}\Big|_0^N + 2\int_0^N e^{-y}\,dy\right]$$

$$= \lim_{N\to\infty}\left[-y^2e^{-y} - 2ye^{-y} - 2e^{-y}\right]\Big|_0^N$$

$$= \lim_{N\to\infty}\left[\left(-N^2e^{-N} - 2Ne^{-N} - 2e^{-N}\right) - (0 - 0 - 2e^0)\right]$$

$$= (0-0-0) - (0-0-2) = 2$$

15. $$E(X) = \int_{-\infty}^{\infty} xf(x)\,dx$$

$$= \int_3^7 x\left(\frac{3}{28} + \frac{3}{x^2}\right)dx$$

$$= \int_3^7 \frac{3}{28}x + \frac{3}{x}\,dx$$

$$= \left(\frac{3}{56}x^2 + 3\ln x\right)\Big|_3^7$$

$$= \left(\frac{147}{56} + 3\ln 7\right) - \left(\frac{27}{56} + 3\ln 3\right) \approx 4.6848 \text{ years}$$

17. $$E(X) = \int_{-\infty}^{\infty} xf(x)\,dx$$

$$= \int_0^{20} \frac{1}{20}x\,dx = \frac{x^2}{40}\Big|_0^{20} = 10 \text{ minutes}$$

19. $$E(X) = \int_{-\infty}^{\infty} xf(x)\,dx$$

$$= \lim_{N\to\infty}\int_0^N 0.05xe^{-0.05x}\,dx$$

$$= 0.05\lim_{N\to\infty}\int_0^N xe^{-0.05x}\,dx$$

Using integration by parts with $u = x$ and $dV = e^{-0.05x}\,dx$,

$$= 0.05\lim_{N\to\infty}\left[-20xe^{-0.05x}\Big|_0^N - \int_0^N 1\cdot -20e^{-0.05x}\,dx\right]$$

$$= 0.05\lim_{N\to\infty}\left[-20xe^{-0.05x}\Big|_0^N + 20\int_0^N e^{-0.05x}\,dx\right]$$

$$= 0.05\lim_{N\to\infty}\left[-20xe^{-0.05x} - 400e^{-0.05x}\right]\Big|_0^N$$

$$= 0.05\lim_{N\to\infty}\left[\left(-20Ne^{-0.05N} - 400e^{-0.05N}\right) - (0 - 400e^0)\right]$$

$$= 0.05\,[(0-0) - (0-400)] = 20 \text{ months}$$

21. (a) Need the expected value to be 5, so

$$E(X) = \int_{-\infty}^{\infty} xf(x)\,dx$$

$$5 = \lim_{N\to\infty}\int_0^N \lambda xe^{-\lambda x}\,dx$$

$$5 = \lambda\lim_{N\to\infty}\int_0^N xe^{-\lambda x}\,dx$$

Using integration by parts with $u = x$ and $dV = e^{-\lambda x}\,dx$,

$$5 = \lambda\lim_{N\to\infty}\left[-\frac{x}{\lambda}e^{-\lambda x}\Big|_0^N - \int_0^N 1\cdot -\frac{1}{\lambda}e^{-\lambda x}\,dx\right]$$

$$5 = \lambda\lim_{N\to\infty}\left[-\frac{x}{\lambda}e^{-\lambda x}\Big|_0^N + \frac{1}{\lambda}\int_0^N e^{-\lambda x}\,dx\right]$$

$$5 = \lambda\lim_{N\to\infty}\left[-\frac{x}{\lambda}e^{-\lambda x} - \frac{1}{\lambda^2}e^{-\lambda x}\right]\Big|_0^N$$

$$5 = \lambda\lim_{N\to\infty}\left[\left(-\frac{N}{\lambda}e^{-\lambda N} - \frac{1}{\lambda^2}e^{-\lambda N}\right) - \left(0 - \frac{1}{\lambda^2}e^0\right)\right]$$

$$5 = \lambda\left[(0-0) - \left(0 - \frac{1}{\lambda^2}\right)\right]$$

$$5 = \frac{1}{\lambda}, \text{ or } \lambda = \frac{1}{5}$$

(b) $$\int_0^2 \frac{1}{5}e^{-x/5}\,dx = \frac{1}{5}\int_0^2 e^{-x/5}\,dx$$

$$= \frac{1}{5}\left(-5e^{-x/5}\Big|_0^2\right) = -\left(e^{-x/5}\Big|_0^2\right)$$

$$= -\left(e^{-2/5} - e^0\right) = 1 - e^{-2/5} \approx 0.3297$$

(c)

$$1-\int_0^7 \frac{1}{5}e^{-x/5}\,dx = 1+\left(e^{-x/5}\Big|_0^7\right)$$
$$= 1+\left(e^{-7/5}-e^0\right) = e^{-7/5} \approx 0.2466$$

23. (a) To be a probability density function, need

$$\lim_{N\to\infty}\int_0^N Ae^{-bx}\,dx = 1$$
$$A\lim_{N\to\infty}\int_0^N e^{-bx}\,dx = 1$$
$$A\lim_{N\to\infty}\left(-\frac{1}{b}e^{-bx}\Big|_0^N\right) = 1$$
$$-\frac{A}{b}\lim_{N\to\infty}\left(e^{-bN}-e^0\right) = 1$$
$$-\frac{A}{b}(0-1) = 1$$
$$A = b$$

Also, know that $P(X \geq 1) = 0.9$, so

$$1-\int_0^1 Ae^{-bx}\,dx = 0.9$$
$$\int_0^1 Ae^{-Ax}\,dx = 0.1$$
$$A\left(-\frac{1}{A}e^{-Ax}\Big|_0^1\right) = 0.1$$
$$-e^{-A}+e^0 = 0.1$$
$$e^{-A} = 0.9$$
$$A = -\ln 0.9$$
$$A = b \approx 0.1054$$

(b) To find $P(X \geq 10)$, use

$$1-\int_0^{10} 0.1054e^{-0.1054x}\,dx$$
$$= 1-0.1054\left(\frac{1}{-0.1054}e^{-0.1054x}\Big|_0^{10}\right)$$
$$= 1+\left(e^{-1.054}-e^0\right) \approx 0.3485$$

or approximately 35% of the sample population.

(c) To find when $P(X \leq t) = 0.9$, use

$$\int_0^t 0.1054e^{-0.1054x}\,dx = 0.9$$
$$0.1054\left(\frac{1}{-0.1054}e^{-0.1054x}\Big|_0^t\right) = 0.9$$
$$-e^{-0.1054t}+e^0 = 0.9$$
$$e^{-0.1054t} = 0.1$$
$$-0.1054t = \ln 0.1$$
$$t = \frac{\ln 0.1}{-0.1054} \approx 21.846$$

So, it takes approximately 21.8 weeks.

(d)

$$E(X) = \lim_{N\to\infty}\int_0^N 0.1054xe^{-0.1054x}\,dx$$
$$= 0.1054\lim_{N\to\infty}\int_0^N xe^{-0.1054x}\,dx$$

Using integration by parts with $u = x$ and $dV = e^{-0.1054x}\,dx$,

$$= 0.1054\lim_{N\to\infty}\left[-\frac{x}{0.1054}e^{-0.1054x}\Big|_0^N - \int_0^N 1\cdot -\frac{1}{0.1054}e^{-0.1054x}\,dx\right]$$
$$= -\lim_{N\to\infty}\left[xe^{-0.1054x}\Big|_0^N - \int_0^N e^{-0.1054x}\,dx\right]$$
$$= -\lim_{N\to\infty}\left[xe^{-0.1054x} + \frac{1}{0.1054}\left(e^{-0.1054x}\Big|_0^N\right)\right]$$
$$= -\lim_{N\to\infty}\left[xe^{-0.1054x} + \frac{1}{0.1054}e^{-0.1054x}\right]\Big|_0^N$$
$$= -\lim_{N\to\infty}\left[\left(Ne^{-0.1054N} + \frac{1}{0.1054}e^{-0.1054N}\right) - \left(0+\frac{1}{0.1054}e^0\right).\right]$$
$$= -\left[(0+0)-\left(0+\frac{1}{0.1054}\right)\right]$$
$$= \frac{1}{0.1054} \approx 9.49 \text{ weeks}$$

(e)
$$\text{Var}(X) = \lim_{N\to\infty} \int_0^N 0.1054x^2 e^{-0.1054x}\,dx - (E(X))^2$$
$$= 0.1054 \lim_{N\to\infty} \int_0^N x^2 e^{-0.1054x}\,dx - (9.49)^2$$

Using integration by parts twice,

$$0.1054 \lim_{N\to\infty} \int_0^N x^2 e^{-0.1054x}\,dx$$
$$= 0.1054 \lim_{N\to\infty} \left[-\frac{x^2}{0.1054} e^{-0.1054x}\Big|_0^N - \int_0^N 2x\left(-\frac{1}{0.1054} e^{-0.1054x}\right) dx \right]$$
$$= -\lim_{N\to\infty} \left[x^2 e^{-0.1054x}\Big|_0^N - 2\int_0^N x e^{-0.1054x}\,dx \right]$$
$$= -\lim_{N\to\infty} \left[x^2 e^{-0.1054x}\Big|_0^N - 2\left(-\frac{x}{0.1054} e^{-0.1054x}\Big|_0^N - \int_0^N 1\cdot -\frac{1}{0.1054} e^{-0.1054x}\,dx \right)\right]$$
$$= -\lim_{N\to\infty} \left[x^2 e^{-0.1054x} + \frac{2x}{0.1054} e^{-0.1054x}\Big|_0^N - \frac{2}{0.1054}\int_0^N e^{-0.1054x}\,dx \right] - (9.49)^2$$
$$= -\lim_{N\to\infty} \left[x^2 e^{-0.1054x} + \frac{2x}{0.1054} e^{-0.1054x} + \frac{2}{(0.1054)^2} e^{-0.1054x} \right]\Big|_0^N$$
$$= -\lim_{N\to\infty} \left[\left(N^2 e^{-0.1054N} + \frac{2N}{0.1054} e^{-0.1054N} + \frac{2}{(0.1054)^2} e^{0.1054N} \right) - \left(0 + 0 + \frac{2}{(0.1054)^2} e^0 \right) \right]$$
$$= -\left[(0+0+0) - \left(0 + 0 + \frac{2}{(0.1054)^2} \right) \right]$$
$$= \frac{2}{(0.1054)^2} \approx 180.03$$

Remembering to subtract $[E(X)]^2 = (9.49)^2$,

$$\text{Var}(X) \approx 89.97$$

25.
$$E(X) = \int_{-\infty}^{\infty} x f(x)\,dx$$
$$= \lim_{N\to\infty} \int_0^N 0.5x e^{0.5x}\,dx$$
$$= \lim_{N\to\infty} 0.5 \int_0^N x e^{-0.5x}\,dx$$

Using integration by parts with $u = x$ and $dV = e^{0.5x}\,dx$,

$$= \lim_{N\to\infty} 0.5 \left[-2xe^{-0.5x}\Big|_0^N 1\cdot -2e^{-0.5x}\,dx \right]$$
$$= \lim_{N\to\infty} 0.5 \left[-2xe^{-0.5x}\Big|_0^N + 2\int_0^N e^{-0.5x}\,dx \right]$$
$$= \lim_{N\to\infty} 0.5 \left[-2x^{0.5x} - 4e^{-0.5x} \right]\Big|_0^N$$
$$= \lim_{N\to\infty} \left[-xe^{-0.5x} - 2e^{-0.5x} \right]\Big|_0^N$$
$$= \lim_{N\to\infty} \left[\left(-Ne^{-0.5N} - 2e^{-0.5N} \right) - (0 - 2e^0) \right]$$
$$= (0-0) - (0-2) = 2 \text{ hours}$$

27.
$$E(X) = \int_{-\infty}^{\infty} x f(x)\,dx$$
$$= \int_a^b x\left(\frac{1}{b-a}\right) dx$$
$$= \frac{1}{b-a}\int_a^b x\,dx$$
$$= \frac{1}{b-a}\left(\frac{x^2}{2}\Big|_a^b\right)$$
$$= \frac{1}{b-a}\left(\frac{b^2}{2} - \frac{a^2}{2}\right) = \frac{1}{b-a}\left(\frac{b^2-a^2}{2}\right)$$
$$= \frac{1}{b-a}\cdot\frac{(b-a)(b+a)}{2} = \frac{b+a}{2}$$

29.

$$E(X)=\int_{-\infty}^{\infty} xf(x)\,dx$$
$$=\lim_{N\to\infty}\int_0^N \lambda x e^{-\lambda x}\,dx$$
$$=\lim_{N\to\infty}\lambda\int_0^N xe^{-\lambda x}\,dx$$

Using integration by parts with $u=x$ and $dV=e^{-\lambda x}\,dx$,

$$=\lim_{N\to\infty}\lambda\left[-\frac{x}{\lambda}e^{-\lambda x}\Big|_0^N-\int_0^N 1\cdot -\frac{1}{\lambda}e^{-\lambda x}\,dx\right]$$
$$=\lim_{N\to\infty}\lambda\left[-\frac{x}{\lambda}e^{-\lambda x}\Big|_0^N+\frac{1}{\lambda}\int_0^N e^{-\lambda x}\,dx\right]$$
$$=\lim_{N\to\infty}\lambda\left[-\frac{x}{\lambda}e^{-\lambda x}-\frac{1}{\lambda^2}e^{-\lambda x}\right]\Big|_0^N$$
$$=\lim_{N\to\infty}\left[-xe^{-\lambda x}-\frac{1}{\lambda}e^{-\lambda N}\right]\Big|_0^N$$
$$=\lim_{N\to\infty}\left[\left(-Ne^{-\lambda N}-\frac{1}{\lambda}e^{-\lambda N}\right)-\left(0-\frac{1}{\lambda}e^{0}\right)\right]$$
$$=(0-0)-\left(0-\frac{1}{\lambda}\right)=\frac{1}{\lambda}$$

31. **(a)** $E(X)=\frac{1}{\lambda}=\frac{1}{2}$

So,

$$f(x)=\begin{cases}2e^{-2x} & x\ge 0\\ 0 & x<0\end{cases}$$

(b)

$$1-\int_0^1 2e^{-2x}\,dx$$
$$=1-2\int_0^1 e^{-2x}\,dx$$
$$=1-2\left(-\frac{1}{2}e^{-2x}\Big|_0^1\right)$$
$$=1+\left(e^{-2x}\Big|_0^1\right)=1+e^{-2}-e^{0}=e^{-2}\approx 0.1353$$

(c)

$$\int_{1/4}^{1/2} 2e^{-2x}\,dx=2\int_{1/4}^{1/2} e^{-2x}\,dx$$
$$=2\left(-\frac{1}{2}e^{-2x}\Big|_{1/4}^{1/2}\right)=-e^{-2x}\Big|_{1/4}^{1/2}$$
$$=-e^{-1}+e^{-1/2}\approx 0.2387$$

33.

$$E(X)=\int_{-\infty}^{\infty}\int_{-\infty}^{\infty} xf(x,y)\,dy\,dx$$
$$=\int_0^{\infty}\int_0^{\infty}\frac{x}{12}e^{-x/4}e^{-y/3}\,dy\,dx$$
$$=\int_0^{\infty}\frac{x}{12}e^{-x/4}\left(\int_0^{\infty}e^{-y/3}\,dy\right)dx$$
$$=\int_0^{\infty}\frac{x}{12}e^{-x/4}\left(\lim_{N\to\infty}\int_0^N e^{-y/3}\,dy\right)dx$$
$$=\int_0^{\infty}\frac{x}{12}e^{-x/4}\left(\lim_{N\to\infty}-3e^{-y/3}\Big|_0^N\right)dx$$
$$=\int_0^{\infty}\frac{x}{12}e^{-x/4}\left(\lim_{N\to\infty}\left[-3e^{-N/3}+3e^{0}\right]\right)dx$$
$$=\int_0^{\infty}\frac{x}{12}e^{-x/4}(3)\,dx$$
$$=\lim_{N\to\infty}\frac{1}{4}\int_0^N xe^{-x/4}\,dx$$

Using integration by parts with $u=x$ and $dV=e^{-x/4}\,dx$,

$$=\lim_{N\to\infty}\frac{1}{4}\left[-4xe^{-x/4}\Big|_0^N-\int_0^N 1\cdot -4e^{-x/4}\,dx\right]$$
$$=\lim_{N\to\infty}\frac{1}{4}\left[-4xe^{-x/4}\Big|_0^N+4\int_0^N e^{-x/4}\,dx\right]$$
$$=\lim_{N\to\infty}\frac{1}{4}\left[-4xe^{-x/4}-16e^{-x/4}\right]\Big|_0^N$$
$$=\lim_{N\to\infty}\left[-xe^{-x/4}-4e^{-x/4}\right]\Big|_0^N$$
$$=\lim_{N\to\infty}\left[\left(-Ne^{-N/4}-4e^{-N/4}\right)-(0-4e^{0})\right]=4$$

The patient in bed 107A can expect to stay 4 days.

$$E(Y) = \int_{-\infty}^{\infty}\int_{-\infty}^{\infty} yf(x, y)\,dx\,dy$$

$$= \int_0^{\infty}\int_0^{\infty} \frac{y}{12}e^{-x/4}e^{-y/3}\,dx\,dy$$

$$= \int_0^{\infty} \frac{y}{12}e^{-y/3}\left(\int_0^{\infty} e^{-x/4}\,dx\right)dy$$

$$= \int_0^{\infty} \frac{y}{12}e^{-y/3}\left(\lim_{N\to\infty}\int_0^{N} e^{-x/4}\,dx\right)dy$$

$$= \int_0^{\infty} \frac{y}{12}e^{-y/3}\left(\lim_{N\to\infty} -4e^{-x/4}\Big|_0^N\right)dy$$

$$= \int_0^{\infty} \frac{y}{12}e^{-y/3}\left(\lim_{N\to\infty}\left[-4e^{-N/4}+4e^{0}\right]\right)dy$$

$$= \int_0^{\infty} \frac{y}{12}e^{-y/3}(4)\,dy = \lim_{N\to\infty}\frac{1}{3}\int_0^{N} ye^{-y/3}\,dy$$

Using integration by parts with $u = y$ and $dV = e^{-y/3}\,dy$,

$$= \lim_{N\to\infty}\frac{1}{3}\left[-3ye^{-y/3}\Big|_0^N - \int_0^N 1\cdot -3e^{-y/3}\,dy\right]$$

$$= \lim_{N\to\infty}\frac{1}{3}\left[-3ye^{-y/3}\Big|_0^N + 3\int_0^N e^{-y/3}\,dy\right]$$

$$= \lim_{N\to\infty}\frac{1}{3}\left[-3ye^{-y/3} - 9e^{-y/3}\right]\Big|_0^N$$

$$= \lim_{N\to\infty}\left[-ye^{-y/3} - 3e^{-y/3}\right]\Big|_0^N$$

$$= \lim_{N\to\infty}\left[-Ne^{-N/3} - 3e^{-N/3} - (0 - 3e^{0})\right] = 3$$

The patient in bed 107b can expect to stay 3 days.

35. (a) Need $x + y < 8$, so

$$\int_0^8\int_0^{8-x} \frac{1}{5120}(x^2 + xy + 2y^2)\,dy\,dx$$

$$= \frac{1}{5,120}\int_0^8 \left(x^2y + \frac{x}{2}y^2 + \frac{2}{3}y^3\right)\Big|_0^{8-x}\,dx$$

$$= \frac{1}{5,120}\int_0^8 \left[x^2(8-x) + \frac{x}{2}(8-x)^2 + \frac{2}{3}(8-x)^3\right] - 0\,dx$$

$$= \frac{1}{5,120}\int_0^8 -\frac{7}{6}x^3 + 16x^2 - 96x + \frac{1,024}{3}\,dx$$

$$= \frac{1}{5,120}\left(-\frac{7}{24}x^4 + \frac{16}{3}x^3 - 48x^2 + \frac{1024}{3}x\right)\Big|_0^8$$

$$= \frac{1}{5,120}\left[\left(-\frac{3,584}{3} + \frac{8,192}{3} - 3,072 + \frac{8,192}{3}\right) - 0\right] = \frac{7}{30}$$

(b) $$E(X) = \int_{-\infty}^{\infty}\int_{-\infty}^{\infty} xf(x, y)\,dy\,dx$$

$$= \int_0^8\int_0^8 \frac{x}{5,120}\left(x^2 + xy + 2y^2\right)\,dy\,dx$$

$$= \int_0^8 \frac{x}{5,120}\left(\int_0^8 x^2 + xy = 2y^2\,dy\right)dx$$

$$= \int_0^8 \frac{x}{5,120}\left(x^2y + \frac{x}{2}y^2 + \frac{2}{3}y^3\right)\Big|_0^8\,dx$$

$$= \int_0^8 \frac{x}{5,120}\left[\left(8x^2 + 32x + \frac{1,024}{3}\right) - 0\right]dx$$

$$= \frac{1}{5,120}\int_0^8 8x^3 + 32x^2 + \frac{1,024}{3}x\,dx$$

$$= \frac{1}{5,120}\left(2x^4 + \frac{32}{3}x^3 + \frac{512}{3}x^2\right)\Big|_0^8$$

$$= \frac{1}{5,120}\left[\left(8,192 + \frac{16,384}{3} + \frac{32,768}{3}\right) - 0\right] = 4.8$$

To produce a unit of A, 4.8 is the expected number of worker-hours.

$$E(Y) = \int_0^8\int_0^8 \frac{y}{5,120}\left(x^2 + xy + 2y^2\right)\,dx\,dy$$

$$= \int_0^8 \frac{y}{5,120}\left(\int_0^8 x^2 + xy + 2y^2\,dx\right)dy$$

$$= \int_0^8 \frac{y}{5,120}\left(\frac{1}{3}x^3 + \frac{y}{2}x^2 + 2y^2x\right)\Big|_0^8\,dy$$

$$= \int_0^8 \frac{y}{5,120}\left[\left(\frac{512}{3}y + 32y + 16y^2\right) - 0\right]dy$$

$$= \frac{1}{5,120} \int_0^8 \frac{512}{3}y + 32y^2 + 16y^3\,dy$$
$$= \frac{1}{5,120} \left(\frac{25b}{3}y^2 + \frac{32}{3}y^3 + 4y^4 \right) \Big|_0^8$$
$$= \frac{1}{5,120} \left[\left(\frac{16,384}{3} + \frac{16,384}{3} + 16,384 \right) - 0 \right]$$
$$= \frac{16}{3}$$

(c)
$$\text{cost} = \left(\frac{24}{5}\right)(18) + \left(\frac{16}{3}\right)(20)$$
$$\approx \$193.07$$

10.4 Normal and Poisson Probability Distributions

1.
$$f(x) = \frac{1}{2\sqrt{2\pi}} e^{-\frac{x^2}{8}}$$
$$= \frac{1}{2\sqrt{2\pi}} e^{-\frac{(x-0)^2}{2(2)^2}}$$

So, the expected value is 0, the standard deviation is 2, and the variance is 4.

3.
$$f(x) = \frac{1}{\sqrt{6\pi}} e^{-\frac{(x+1)^2}{6}}$$
$$= \frac{1}{\sqrt{3}\sqrt{2\pi}} e^{\frac{-(x+1)^2}{2(\sqrt{3})^2}}$$

So, the expected value is -1, the standard deviation is $\sqrt{3}$, and the variance is 3.

5. Using the table,

$$P(Z \le 1.24) = 0.8925$$

7. Since $P(Z > 3.49) \approx 0$

$$(P(Z \le 4.26) \approx P(Z \le 3.49) \approx 1$$

9. $P(-1 \le Z \le 1) = P(Z \le 1) - P(Z \le -1)$

Now,

$$P(Z \le -1) = P(Z \ge 1)$$
$$= 1 - P(Z \le 1)$$

So,

$$P(-1 \le Z \le 1) = P(Z \le 1) - [1 - P(Z \le 1)]$$
$$= 2P(Z \le 1) - 1$$

Since $P(Z \le 1) = 0.8413$,
$= 2(0.8413) - 1 = 0.6826$

11. Using the table,

$$P(Z \le b) = 0.8413$$

when $b = 1.00$

13. Using the table,

$$P(Z < b) = 0.8643$$

when $b = 1.10$

15. $\mu = 60$, $\sigma = 4$

$$Z = \frac{x - 60}{4}$$

(a) $P(X \ge 68) = P(Z \ge 2)$
$$= 1 - P(Z \le 2)$$
$$= 1 - 0.9772 = 0.0228$$

(b) $P(X \ge 60) = P(Z \ge 0) = 0.5$

(c) $P(56 \le X \le 64) = P(-1 \le Z \le 1)$
$$= P(Z \le 1) - P(Z \le -1)$$

where $P(Z \le -1) = 1 - P(Z \le 1)$
$$= P(Z \le 1) - [1 - P(Z \le 1)]$$
$$= 2P(Z \le 1) - 1$$
$$= 2(0.8413) - 1 = 0.6826$$

(d) $P(X \le 40) = P(Z \le -5)$
Since $P(X \le -3.49) = P(X \ge 3.49) \approx 0$,
$$P(Z \le -5) \approx 0$$

17. $\mu = 4$, $\sigma = 0.5$

$$Z = \frac{x-4}{0.5}$$

(a) $P(X \geq x) = 0.0228$ is the same as

$$P\left(Z \geq \frac{x-4}{0.5}\right) = 0.0228$$

To use the table,

$$P\left(Z \geq \frac{x-4}{0.5}\right) = 1 - P\left(Z \leq \frac{x-4}{0.5}\right)$$
$$= 1 - 0.0228 = 0.9772$$

From the table,

$$\frac{x-4}{0.5} = 2.00$$
$$x = 5$$

(b) $P(X \leq x) = 0.1587$ is the same as

$$P\left(Z \leq \frac{x-4}{0.5}\right) = 0.1587$$

Since the smallest value in the table is 0.5, find

$$P\left(Z \geq \frac{x-4}{0.5}\right) = 1 - 0.1587 = 0.8413$$

From the table,

$$\frac{x-4}{0.5} = 1$$
$$x = 3.5$$

By symmetry,

$$P\left(Z \leq \frac{x-4}{0.5}\right) = 0.1587$$

occurs for $x = -3.5$

19. $\mu = 15$, $\sigma = 4$

$$Z = \frac{x-15}{4}$$

(a)
$$\begin{aligned} P(11 \leq X \leq 19) &= P(-1 \leq Z \leq 1) \\ &= P(Z \leq 1) - P(Z \leq -1) \\ &= P(Z \leq 1) - [1 - P(Z \leq 1)] \\ &= 2P(Z \leq 1) - 1 \end{aligned}$$

Using the table,

$$\approx 2(0.8413) - 1 = 0.6826$$

(b)
$$\begin{aligned} P(7 \leq X \leq 23) &= P(-2 \leq Z \leq 2) \\ &= P(Z \leq 2) - P(Z \leq -2) \\ &= P(Z \leq 2) - [1 - P(Z \leq 2)] \\ &= 2P(Z \leq 2) - 1 \end{aligned}$$

Using the table,

$$2(0.9772) - 1 = 0.9544$$

21. $\lambda = 3$

(a)
$$\begin{aligned} P(X = 1) &= \frac{3^1}{1!}e^{-3} \\ &= 3e^{-3} \approx 0.149 \end{aligned}$$

(b)
$$\begin{aligned} P(X = 3) &= \frac{3^3}{3!}e^{-3} \\ &= \frac{9}{2}e^{-3} \approx 0.224 \end{aligned}$$

(c)
$$\begin{aligned} P(X = 4) &= \frac{3^4}{4!}e^{-3} \\ &= \frac{27}{8}e^{-3} \approx 0.168 \end{aligned}$$

(d)
$$\begin{aligned} P(X = 8) &= \frac{3^8}{8!}e^{-3} \\ &= \frac{729}{4480}e^{-3} \approx 0.008 \end{aligned}$$

23.
$$\begin{aligned} P(X = 0) &= \frac{2^0}{0!}e^{-2} \\ &= e^{-2} \approx 0.135 \\ P(X = 1) &= \frac{2^1}{1!}e^{-2} \\ &= 2e^{-2} \approx 0.271 \\ P(X = 2) &= \frac{2^2}{2!}e^{-2} \\ &= 2e^{-2} \approx 0.271 \\ P(X = 3) &= \frac{2^3}{3!}e^{-2} \\ &= \frac{4}{3}e^{-2} \approx 0.180 \end{aligned}$$

$$P(X=4) = \frac{2^4}{4!}e^{-2}$$
$$= \frac{2}{3}e^{-2} \approx 0.090$$
$$P(X=5) = \frac{2^5}{5!}e^{-2}$$
$$= \frac{4}{15}e^{-2} \approx 0.036$$
$$P(X=6) = \frac{2^6}{6!}e^{-2}$$
$$= \frac{8}{90}e^{-2} \approx 0.012$$

Since approximately 99.7% of the probability lies within three standard deviations ($3\sqrt{2} \approx 4.2$) of the mean, will assume $P(X = x) \approx 0$ for $x > 6$.

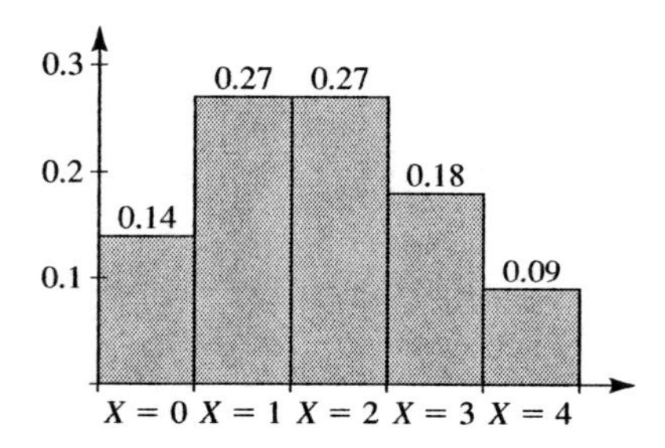

25. **(a)** False; this is only true for a standard normal distribution, where $\mu = 0$.

(b) True; when $\mu = 0$, probability is distributed symmetrically about $z = 0$.

27. $\mu = 50, \sigma = 5$

Approximately 68.3% of adult basset hounds have weight within one standard deviation of the mean, or

$$50 - 5 \le w \le 50 + 5$$
$$45 \le w \le 55$$

Approximately 95.4% of adult basset hounds have weight within two standard deviations of the mean, or

$$50 - 2(5) \le w \le 50 + 2(5)$$
$$40 \le w \le 60$$

Approximately 99.7% of adult basset hounds have weight within three standard deviations of the mean, or

$$50 - 3(5) \le w \le 50 + 3(5)$$
$$35 \le w \le 65$$

29. $\mu = 42{,}500, \sigma = 3{,}000$

$$Z = \frac{x - 42{,}500}{3{,}000}$$

$P(41{,}000 \le X \le 44{,}000)$

$$= P(-0.5 \le Z \le 0.5)$$
$$= P(Z \le 0.5) - P(Z \le -0.5)$$
$$= P(Z \le 0.5) - P(Z \ge 0.5)$$
$$= P(Z \le 0.5) - [1 - P(Z \le 0.5)]$$
$$= 2P(Z \le 0.5) - 1$$

Using the table,

$$= 2(0.6915) - 1 = 0.3830$$

31. $\mu = 1, \sigma = 0.01$

$$Z = \frac{x - 1}{0.01}$$

(a) $P(0.98 \le X \le 1.02) = P(-2 \le Z \le 2)$

$$= P(Z \le 2) - P(Z \le -2)$$
$$= P(Z \le 2) - P(Z \ge 2)$$
$$= P(Z \le 2) - [1 - P(Z \le 2)]$$
$$= 2P(Z \le 2) - 1$$

Using the table,

$$= 2(0.9772) - 1 = 0.9544$$

(b) $P(X < 0.97) = P(Z < -3)$

$$= P(Z > 3) = 1 - P(Z < 3)$$

Using the table,

$$= 1 - 0.9987 = 0.0013$$

33. $\mu = 5,\ Z = \dfrac{x - \mu}{\sigma}$

$$P(4.9 \le X \le 5.1) = 0.99$$

Now,

$$P(4.9 \le X \le 5.1) = P\left(\frac{-0.1}{\sigma} \le Z \le \frac{0.1}{\sigma}\right)$$
$$= P\left(Z \le \frac{0.1}{\sigma}\right) - P\left(Z \le -\frac{0.1}{\sigma}\right)$$
$$= P\left(Z \le \frac{0.1}{\sigma}\right) - P\left(Z \ge -\frac{0.1}{\sigma}\right)$$
$$= P\left(Z \le \frac{0.1}{\sigma}\right) - \left[1 - P\left(Z \le -\frac{0.1}{\sigma}\right)\right]$$
$$= 2P\left(Z \le \frac{0.1}{\sigma}\right) - 1$$

So,

$$2P\left(Z \le \frac{0.1}{\sigma}\right) - 1 = 0.99$$
$$= P\left(Z \le \frac{0.1}{\sigma}\right) = 0.995$$

Using the table,

$$\frac{0.1}{\sigma} = 2.575$$
$$\sigma \approx 0.0388$$

35. $\mu = 150, \sigma = 25$

$$Z = \frac{x - 150}{25}$$
$$P(X > 180) = P(Z > 1.2)$$
$$= 1 - P(Z < 1.2)$$

Using the table,

$$= 1 - 0.8849 = 0.1151$$

37. There are seats available for $\frac{200}{1000}$ or 0.2 of the members. Since $\mu = 40$ and $\sigma = 5$,

$$Z = \frac{x - 40}{5}$$

Need

$$P\left(Z > \frac{x - 40}{5}\right) = 0.2$$

Now,

$$P\left(Z > \frac{x - 40}{5}\right) = 1 - P\left(Z < \frac{x - 40}{5}\right)$$

So,

$$1 - P\left(Z < \frac{x - 40}{5}\right) = 0.2$$
$$P\left(Z < \frac{x - 40}{5}\right) = 0.8$$

Using the table,

$$\frac{x - 40}{5} = 0.84$$
$$x = 44.2 \text{ years old}$$

39. $\mu = 24.5, \sigma = 5.25$

$$Z = \frac{x - 24.5}{5.25}$$

(a) $P(X > 32) = P\left(Z > \frac{10}{7}\right)$

$$= 1 - P\left(Z < \frac{10}{7}\right)$$

Using the table with $z \approx 1.43$,

$$= 1 - 0.9236 = 0.0764, \text{ or } 7.64\%$$

(b) To find x such that $P(X > x) = 0.9$, find $P(X < x) = 0.1$. Now,

$$P(X < x) = P\left(Z < \frac{x - 24.5}{5.25}\right)$$

Since the smallest value in the table is 0.5, find

$$P\left(Z > \frac{x - 24.5}{5.25}\right) = 1 - 0.1 = 0.9$$

From the table,

$$\frac{x - 24.5}{5.25} \approx 1.28$$
$$x = 31.22$$

By symmetry, the desired value of x is the same distance from the mean on the opposite side. So,

$$x = 24.5 - (31.22 - 24.5)$$
$$= 17.78 \text{ ounces}$$

41. $\mu = 175, \sigma = 35$

$$Z = \frac{x - 175}{35}$$

(a) $P(X > 225) = P\left(Z > \frac{10}{7}\right)$

$$= 1 - P\left(Z < \frac{10}{7}\right)$$

Using the table with $z \approx 1.43$,

$$= 1 - 0.9236 = 0.0764, \text{ or } 7.64\%$$

(b) To find x such that $P(X > x) = 0.98$, find $P(X < x) = 0.02$. Now,

$$P(X < x) = P\left(Z < \frac{x - 175}{35}\right)$$

Since the smallest value in the table is 0.5, find

$$P\left(Z > \frac{x - 175}{35}\right) = 1 - 0.02 = 0.98$$

From the table,

$$\frac{x - 175}{35} \approx 2.05$$

$$x = 246.75$$

By symmetry, the desired value of x is the same distance from the mean on the opposite side. So,

$$x = 175 - (246.75 - 175)$$
$$= 103.25 \text{ mg/dL}$$

43. $\mu = 266, \sigma = 16$

$$Z = \frac{x - 266}{16}$$

(a) $P(X > 290) = P(Z > 1.5)$

$$= 1 - P(Z < 1.5)$$

Using the table,

$$= 1 - 0.9332 = 0.0668$$

(b) $P(X < 260) = P\left(Z < -\frac{3}{8}\right)$ Since the table starts with $z = 0$, find

$$P\left(Z > \frac{3}{8}\right) = 1 - P\left(Z < \frac{3}{8}\right)$$

Using the table, averaging the values for $z = 0.37$ and $z = 0.38$,

$$= 1 - 0.6462 = 0.3538$$

(c) $P(260 \le X \le 280) = P\left(-\frac{3}{8} \le Z \le \frac{7}{8}\right)$

$$= P\left(Z \le \frac{7}{8}\right) - P\left(Z \le -\frac{3}{8}\right)$$

Using the table, averaging the values for $z = 0.87$ and 0.88, and using the value from part (b),

$$= 0.8092 - 0.3538 = 0.4554$$

45. $\mu = 10, \sigma = 0.05$

$$Z = \frac{x - 10}{0.05}$$

$$P(9.9 \le X \le 10.15) = P(-2 \le Z \le 3)$$
$$= P(Z \le 3) - P(Z \le -2)$$

Since the table starts with $z = 0$, find

$$P(Z \ge 2) = 1 - P(Z \le 2)$$

Using the table,

$$P(Z \le 3) - [1 - P(Z \le 2)]$$
$$= 0.9987 - [1 - 0.9772] = 0.9759$$

47. $\mu = -1.6, \sigma = 8$

$$Z = \frac{x + 1.6}{8}$$

(a) $P(0 \le X \le 10) = P(0.2 \le Z \le 1.45)$

$$= P(Z \le 1.45) - P(Z \le 0.2)$$

Using the table,

$$= 0.9265 - 0.5793 = 0.3472$$
$$P(-5 \le X \le 5) = P(-0.425 \le Z \le 0.825)$$
$$= P(Z \le 0.825) - P(Z \le -0.425)$$

Since the table starts with $z = 0$, find

$$P(Z \ge 0.425) = 1 - P(Z \le 0.425)$$

Using the table, averaging values,

$$P(Z \le 0.825) - [1 - P(Z \le 0.425)]$$
$$= 0.7953 - [1 - 0.6646] = 0.4599$$

(b) $P(X > 5) = P(Z > 0.825)$

$$= 1 - P(Z \le 0.825)$$

Using a result from part (a),

$$= 1 - 0.7953 = 0.2047$$

(c) $P(X < 5) = 1 - P(X > 5)$
Using the result of part (a),

$$= 1 - 0.2047 = 0.7953$$

49. Machine shop A: $\mu = 0.2$, $\sigma = 0.010$

$$Z = \frac{x - 0.2}{0.010}$$

$$\begin{aligned} P(0.18 \le X \le 0.22) &= P(-2 \le Z \le 2) \\ &= P(Z \le 2) - P(Z \le -2) \\ &= P(Z \le 2) - P(Z \ge 2) \\ &= P(Z \le 2) - [1 - P(Z \le 2)] \\ &= 2P(Z \le 2) - 1 \end{aligned}$$

Using the table,

$$= 2(0.9772) - 1 = 0.9544$$

For every 1000 washers, approximately 954 will be useable. The cost per useable washer is approximately 0.105 cents.
Machine shop B: $\mu = 0.2$, $\sigma = 0.011$

$$Z = \frac{x - 0.2}{0.011}$$

$$\begin{aligned} P(0.18 \le X \le 0.22) &\approx P(-1.818 \le Z \le 1.818) \\ &= P(Z \le 1.818) - P(Z \le -1.818) \\ &= P(Z \le 1.818) - P(Z \ge 1.818) \\ &= P(Z \le 1.818) - [1 - P(Z \le 1.818)] \\ &= 2P(Z \le 1.818) - 1 \end{aligned}$$

Using the table with $z = 1.82$,

$$= 2(0.9656) - 1 = 0.9312$$

For every 1,000 washers, approximately 931 will be useable. The cost per washer is approximately 0.097 cents.
So, shop B offers the better deal.

51. $\mu = 11$, $\sigma = 2$

$$Z = \frac{x - 11}{2}$$

(a) $P(10 \le X \le 18) = P\left(-\frac{1}{2} \le Z \le \frac{7}{2}\right)$

$$\begin{aligned} &P(Z \le 3.5) - P(Z \le -0.5) \\ &P(Z \le 3.5) - P(Z \ge 0.5) \\ &P(Z \le 3.5) - [1 - P(Z \le 0.5)] \\ &P(Z \le 3.5) + P(Z \le 0.5) - 1 \end{aligned}$$

Using the table,

$$= 1 + 0.6915 - 1 = 0.6915$$

(b) $P(X \le 9) = P(Z \le -1)$

$$\begin{aligned} &= P(Z \ge 1) \\ &= 1 - P(Z \le 1) \end{aligned}$$

Using the table,

$$= 1 - 0.8413 = 0.1587$$

Without the drug, approximately 16% of the population would recover in 9 days, so it's not overly impressive.

(c) $P(X \le 6) = P\left(Z \le -\frac{5}{2}\right)$

$$\begin{aligned} &= P(Z \ge 2.5) \\ &= 1 - P(Z \le 2.5) \end{aligned}$$

Using the table,

$$= 1 - 0.9938 = 0.0062$$

Without the drug, approximately 0.6% of the population would recover in 6 days, so now the results are impressive.

(d) Writing Exercise—Answers will vary.

53. (a) $E(X) = \lambda$, so $\lambda = 6.1$

(b)

$$\begin{aligned} P(X = 0) &= \frac{6.1^0}{0!} e^{-6.1} \\ &= e^{-6.1} \approx 0.00224 \end{aligned}$$

(c)

$$\begin{aligned} P(X = 3) &= \frac{6.1^3}{3!} e^{-6.1} \\ &= \frac{226.981}{6} e^{-6.1} \approx 0.0848 \end{aligned}$$

(d) $P(X < 6) = P(X = 0) + P(X = 1)$
$+P(X = 2) + P(X = 3) + P(X = 4)$
$+P(X = 5)$

$$\left[\frac{6.1^0}{0!}+\frac{6.1^1}{1!}+\frac{6.1^2}{2!}+\frac{6.1^3}{3!}+\frac{6.1^4}{!4}+\frac{6.1^5}{5!}\right]e^{-6.1}$$
$$\approx 0.4298$$

55. When $n = 100{,}000$ then $\lambda = 3$.

$$\begin{aligned} P(X \geq 1) &= 1 - P(X = 0) \\ &= 1 - \frac{3^0}{0!}e^{-3} \\ &= 1 - e^{-3} \approx 0.9502 \end{aligned}$$

When $n = 1{,}000{,}000$ then $\lambda = 30$

$$\begin{aligned} P(x \geq 1) &= 1 - P(x = 0) \\ &= 1 - \frac{30^0}{0!}e^{-30} \\ &= 1 - e^{-30} \approx 1 \end{aligned}$$

Approximately 5% more likely in the larger population.

57. (a) $P(X = 0) = 0.05$

$$\begin{aligned} \frac{\lambda^0}{0!}e^{-\lambda} &= 0.05 \\ e^{-\lambda} &= 0.05 \\ -\lambda &= \ln 0.05 \\ \lambda &= -\ln 0.05 \approx 3 \end{aligned}$$

(b) $P(X = 5) = \frac{3^5}{5!}e^{-3} \approx 0.10$

(c) $P(X \leq 5) = P(X = 0) + P(X = 1) + P(X = 2) + P(X = 3) + P(X = 4) + P(X = 5)$

$$= \left[\frac{3^0}{0!}+\frac{3^1}{1!}+\frac{3^2}{2!}+\frac{3^3}{3!}+\frac{3^4}{4!}+\frac{3^5}{5!}\right]e^{-3}$$
$$\approx 0.92$$

(d)

$$\begin{aligned} P(X \geq 1) &= 1 - P(X = 0) \\ &= 1 - \frac{3^0}{0!}e^{-3} \\ &= 1 - e^{-3} \approx 0.95 \end{aligned}$$

(e) $E(X) = \lambda = 3$ hits

59. $f(x) = \frac{1}{\sqrt{2\pi}}e^{-x^2/2}$

To find the maximum, get the critical point(s) from the derivative.

$$f'(x) = \frac{1}{\sqrt{2\pi}}e^{-x^2/2}(-x)$$

$$f'(x) = 0 \text{ when}$$

$$-\frac{x}{\sqrt{2\pi}}e^{-x^2/2} = 0$$

or, when $x = 0$. Noting that when

$x < 0,\ f'(x) > 0$ function increasing

$x > 0,\ f'(x) < 0$ function decreasing

So, the absolute maximum occurs when $x = 0$.
To find the inflection points, get the critical point(s) from the second derivative.

$$\begin{aligned} f''(x) &= \left(-\frac{x}{\sqrt{2\pi}}\right)\left(-xe^{-x^2/2}\right) + \left(e^{-x^2/2}\right)\left(-\frac{1}{\sqrt{2\pi}}\right) \\ &= \frac{1}{\sqrt{2\pi}}e^{-x^2/2}\left(x^2 - 1\right) \end{aligned}$$

$f''(x) = 0$ when $x = \pm 1$

Noting that when

$$\begin{aligned} x < -1,&\ f''(x) > 0 \\ -1 < x < 1,&\ f''(x) < 0 \\ x > 1,&\ f''(x) > 0 \end{aligned}$$

Inflection points exist for both $x = -1$ and $x = 1$.

61. Show $P(\mu - 2\sigma \leq X \leq \mu + 2\sigma) = 0.954$ Using the simplest case of a standard normal distribution (since all normal distributions can be translated to a standard normal distribution), show

$$P(-2 \leq Z \leq 2) = 0.95$$

Now,

$$\begin{aligned} P(-2 \leq Z \leq 2) &= P(Z \leq 2) - P(Z \leq -2) \\ &= P(Z \leq 2) - P(Z \geq 2) \\ &= P(Z \leq 2) - [1 - P(Z \leq 2)] \\ &= 2P(Z \leq 2) - 1 \end{aligned}$$

Using the table,

$$= 2(0.9772) - 1 = 0.9544$$

or approximately 95.4%

Similarly,

$$\begin{aligned} P(-3 \le Z \le 3) &= P(Z \le 3) - P(Z \le -3) \\ &= P(Z \le 3) - P(Z \ge 3) \\ &= P(Z \le 3) - [1 - P(Z \le 3)] \\ &= 2P(Z \le 3) - 1 \\ &= 2(0.9987) - 1 = 0.9974 \end{aligned}$$

or approximately 99.7%

Checkup for Chapter 10

1.

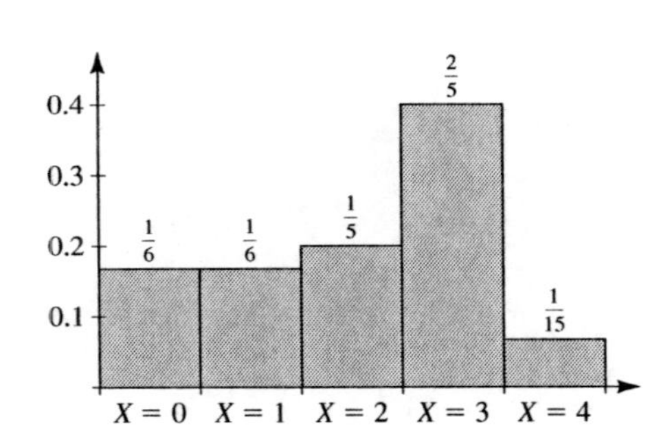

$$\begin{aligned} E(X) &= x_0p(x_0) + x_1p(x_1) + x_2p(x_2) \\ &\quad + x_3p(x_3) + x_4p(x_4) \\ &= 0\left(\frac{1}{6}\right) + 1\left(\frac{1}{6}\right) + 2\left(\frac{1}{5}\right) + 3\left(\frac{2}{5}\right) + 4\left(\frac{1}{15}\right) \\ &\approx 2.03 \end{aligned}$$

$$\begin{aligned} \text{Var}(x) &= (0 - 2.03)^2\left(\frac{1}{6}\right) + (1 - 2.03)^2\left(\frac{1}{6}\right) \\ &\quad + (2 - 2.03)^2\left(\frac{1}{5}\right) \\ &\quad + (3. - 2.03)^2\left(\frac{2}{5}\right) + (4 - 2.03)^2\left(\frac{1}{15}\right) \\ &\approx 1.50 \end{aligned}$$

$$\sigma(x) = \sqrt{1.50} \approx 1.22$$

2. **(a)**

$$\begin{aligned} \int_{-\infty}^{\infty} f(x)\,dx &= \int_0^1 \frac{4}{3}x^{1/3}\,dx \\ &= \frac{4}{3}\int_0^1 x^{1/3}\,dx = \frac{4}{3}\left(\frac{3}{4}x^{4/3}\Big|_0^1\right) \\ &= x^{4/3}\Big|_0^1 = 1 \end{aligned}$$

Yes, f *is* a probability density function.

(b)

$$\begin{aligned} \int_{-\infty}^{\infty} f(x)\,dx &= \int_0^{\infty} 2e^{-x/2}\,dx \\ &= \lim_{N\to\infty} 2\int_0^N e^{-x/2}\,dx \\ &= \lim_{N\to\infty} 2\left(-2e^{-x/2}\Big|_0^N\right) \\ &= -4\lim_{N\to\infty}\left(e^{-x/2}\Big|_0^N\right) \\ &= -4\lim_{N\to\infty}\left(e^{-N/2} - e^0\right) \\ &= -4(0 - 1) = 4 \end{aligned}$$

No, f is *not* a probability density function.

3. **(a)**

$$\begin{aligned} P(2 \le X \le 3) &= \int_2^3 \frac{1}{3}e^{-x/3}\,dx \\ &= \frac{1}{3}\left(-3e^{-x/3}\Big|_2^3\right) = e^{-x/3}\Big|_2^3 \\ &= -e^{-1} + e^{-2/3} \approx 0.1455 \end{aligned}$$

(b)

$$\begin{aligned} P(X \ge 3) &= 1 - P(X \le 3) \\ &= 1 - \int_0^3 \frac{1}{3}e^{-x/3}\,dx \\ &= 1 + \left(e^{-x/3}\Big|_0^3\right) = 1 + \left(e^{-1} - e^0\right) \\ &= \frac{1}{e} \approx 0.3679 \end{aligned}$$

4.

$$E(X) = \int_{-\infty}^{\infty} x f(x)\,dx$$
$$= \int_0^6 x \cdot \frac{1}{36}(6x - x^2)\,dx$$
$$= \frac{1}{36}\int_0^6 6x^2 - x^3\,dx$$
$$= \frac{1}{36}\left(2x^3 - \frac{x^4}{4}\right)\Big|_0^6$$
$$= \frac{1}{36}(432 - 324) = 3$$
$$\text{Var}(x) = \int_{-\infty}^{\infty} x^2 f(x)\,dx - [E(X)]^2$$
$$= \int_0^6 x^2 \cdot \frac{1}{36}(6x - x^2)\,dx - 9$$
$$= \frac{1}{36}\int_0^6 6x^3 - x^4\,dx - 9$$
$$= \frac{1}{36}\left(\frac{3}{2}x^4 - \frac{x^5}{5}\right)\Big|_0^6 - 9$$
$$= \frac{1}{36}(1944 - 1555.2) - 9 = 1.8$$

5. **(a)** $P(Z \geq 0.85) = 1 - P(Z \leq 0.85)$ Using the table,

$$= 1 - 0.8023 = 0.1977$$

(b) $P(-1.30 \leq Z \leq 3.25)$

$$= P(Z \leq 3.25) - P(Z \leq -1.30)$$

Since the table starts with $z = 0$, find

$$P(Z \geq 1.30) = 1 - P(Z \leq 1.30)$$

So,

$$P(Z \leq 3.25) - P(Z \leq -1.30)$$
$$= P(Z \leq 3.25) - [1 - P(Z \leq 1.30)]$$

Using the table,

$$= 0.9994 - [1 - 0.9032] = 0.9026$$

6. $\mu = 12, \sigma = 2$

$$Z = \frac{x - 12}{2}$$

(a) $P(10 \leq X \leq 14) = P(-1 \leq Z \leq 1)$
$= P(Z \leq 1) - P(Z \leq -1)$
Since the table starts with $z = 0$, find

$$P(Z \geq 1) = 1 - P(Z \leq 1)$$

So,

$$P(Z \leq 1) - P(Z \leq -1)$$
$$= P(Z \leq 1) - [1 - P(Z \leq 1)]$$
$$= 2P(Z \leq 1) - 1$$

Using the table,

$$= 2(0.8413) - 1 = 0.6826$$

(b) $P(8 \leq X \leq 16) = P(-2 \leq Z \leq 2)$

$$= P(Z \leq 2) - P(Z \leq -2)$$
$$= P(Z \leq 2) - [1 - P(Z \leq 2)]$$
$$= 2P(Z \leq 2) - 1$$
$$= 2(0.9772) - 1 = 0.9544$$

7.

$$f(x) = \begin{cases} \frac{1}{3} & 1 \leq X \leq 4 \\ 0 & \text{otherwise} \end{cases}$$
$$P(X \geq 2) = \int_2^4 \frac{1}{3}\,dx$$
$$= \frac{1}{3}x\Big|_2^4 = \frac{4}{3} - \frac{2}{3} = \frac{2}{3} \approx 0.6667$$

8. $P(X > 5) = 1 - P(X \leq 5)$

$$= 1 - P(X = 1) - P(X = 2) - P(X = 3) - P(X = 4) - P(X = 5)$$
$$= 1 - (1 - 0.05)^0(0.05) - (1 - 0.05)^1(0.05) - (1 - 0.05)^2(0.05)$$
$$- (1 - 0.05)^3(0.05) - (1 - 0.05)^4(0.05)$$
$$\approx 0.7738$$

9.

$$f(x) = \begin{cases} 0.5e^{-0.5x} & x \geq 0 \\ 0 & x < 0 \end{cases}$$

(a) $P(X \geq 3) = 1 - P(X \leq 3)$

$$= 1 - \int_0^3 0.5e^{-0.5x}\,dx$$
$$= 1 - 0.5\left(\frac{1}{-0.5}e^{-0.5x}\Big|_0^3\right)$$
$$= 1 + \left(e^{-0.5x}\Big|_0^3\right)$$
$$= 1 + \left(e^{-1.5} - e^0\right) = e^{-1.5} \approx 0.2231$$

(b)
$$E(X) = \int_{-\infty}^{\infty} xf(x)\,dx$$
$$= \int_0^{\infty} 0.5xe^{-0.5x}\,dx$$
$$= \lim_{N\to\infty} 0.5 \int_0^N xe^{-0.5x}\,dx$$

Using integration by parts with $u = x$ and $dV = e^{-0.5x}\,dx$,

$$= \lim_{N\to\infty} 0.5\left[-2xe^{-0.5x}\Big|_0^N - \int_0^N 1 \cdot -2e^{-0.5x}\,dx\right]$$
$$= \lim_{N\to\infty} 0.5\left[-2xe^{-0.5x}\Big|_0^N + 2\int_0^N e^{-0.5x}\,dx\right]$$
$$= \lim_{N\to\infty} 0.5\left[-2xe^{-0.5x} - 4e^{-0.5x}\right]\Big|_0^N$$
$$= \lim_{N\to\infty}\left[-xe^{-0.5x} - 2e^{-0.5x}\right]\Big|_0^N$$
$$= \lim_{N\to\infty}\left[\left(-Ne^{-0.5N} - 2e^{-0.5N}\right) - \left(0 - 2e^0\right)\right]$$
$$= (0-0) - (0-2) = 2 \text{ years}$$

This expected value represents the life expectancy of those who receive the drug.

10. $\mu = 150, \sigma = 24$

$$Z = \frac{x - 150}{24}$$
$$P(X \geq x) = 0.2$$

Now,

$$P(X \geq x) = P\left(Z \geq \frac{x-150}{24}\right)$$
$$= 1 - P\left(Z \leq \frac{x-150}{24}\right)$$

So,

$$0.2 = 1 - P\left(Z \leq \frac{x-150}{24}\right)$$
$$P\left(Z \leq \frac{x-150}{24}\right) = 0.8$$

Using the table,

$$\frac{x-150}{24} \approx 0.84$$
$$x \approx 170.16$$

So, the minimum qualifying score is approximately 170 points.

11. $P(X \geq 5) = 1 - P(X < 5)$
$= 1 - P(X = 0) - P(X = 1) - P(X = 2) - P(X = 3) - P(X = 4)$
Since $E(X) = \lambda = 5$,

$$= 1 - \frac{5^0}{0!}e^{-5} - \frac{5^1}{1!}e^{-5} - \frac{5^2}{2!}e^{-5} - \frac{5^3}{3!}e^{-5} - \frac{5^4}{4!}e^{-5}$$
$$= 1 - \left(1 + 5 + \frac{25}{2} + \frac{125}{6} + \frac{625}{24}\right)e^{-5}$$
$$\approx 0.5595$$

12. **(a)** This is the complement of the probability that both components last at least 5 years, or

$$1 - \int_5^{\infty}\int_5^{\infty} 0.1e^{-x/2}e^{-y/5}\,dy\,dx$$
$$= 1 - \int_5^{\infty} 0.1e^{-x/2}\left(\int_5^{\infty} e^{-y/5}\,dy\right)dx$$
$$= 1 - \int_5^{\infty} 0.1e^{-x/2}\left(\lim_{N\to\infty}\int_5^N e^{-y/5}\,dy\right)dx$$
$$= 1 - \int_5^{\infty} 0.1e^{-x/2}\left(\lim_{N\to\infty}\left[-5e^{-y/5}\right]\Big|_5^N\right)dx$$

$$= 1 - \int_5^{\infty} -0.5e^{-x/2}\left(\lim_{N\to\infty}\left[e^{-y/5}\right]\Big|_5^N\right)dx$$
$$= 1 - \int_5^{\infty} -0.5e^{-x/2}\left(\lim_{N\to\infty}\left[e^{-N/5} - e^{-1}\right]\right)dx$$
$$= 1 - \int_5^{\infty} -0.5e^{-x/2}\left(-e^{-1}\right)dx$$
$$= 1 - 0.5e^{-1}\int_5^{\infty} e^{-x/2}\,dx$$
$$= 1 - 0.5e^{-1}\lim_{N\to\infty}\int_5^{N} e^{-x/2}\,dx$$
$$= 1 - 0.5e^{-1}\lim_{N\to\infty}\left(-2e^{-x/2}\Big|_5^N\right)$$
$$= 1 + e^{-1}\lim_{N\to\infty}\left(e^{-x/2}\Big|_5^N\right)$$
$$= 1 + e^{-1}\lim_{N\to\infty}\left(^{-N/2} - e^{-5/2}\right)$$
$$= 1 + e^{-1}\left(-e^{-5/2}\right) = 1 - e^{-7/2} \approx 0.9698$$

(b) $E(X) = \int_{-\infty}^{\infty}\int_{-\infty}^{\infty} xf(x, y)\,dy\,dx$

$$= \int_0^{\infty}\int_0^{\infty} 0.1xe^{-x/2}e^{-y/5}\,dy\,dx$$
$$= \int_0^{\infty} 0.1xe^{-x/2}\left(\int_0^{\infty} e^{-y/5}\,dy\right)dx$$
$$= \int_0^{\infty} 0.1xe^{-x/2}\left(\lim_{N\to\infty}\int_0^{N} e^{-y/5}\,dy\right)dx$$
$$= \int_0^{\infty} 0.1xe^{-x/2}\left(\lim_{N\to\infty}\left[-5e^{-y/5}\right]\Big|_0^N\right)dx$$
$$= \int_0^{\infty} -0.5xe^{-x/2}\left(\lim_{N\to\infty}\left[e^{-y/5}\right]\Big|_0^N\right)dx$$
$$= \int_0^{\infty} -0.5xe^{-x/2}\left(\lim_{N\to\infty}\left[e^{-N/5} - e^{0}\right]\right)dx$$
$$= \int_0^{\infty} -0.5xe^{-x/2}(-1)\,dx$$
$$= \lim_{N\to\infty} 0.5\int_0^{N} xe^{-x/2}\,dx$$

Using integration by parts with $u = x$ and $dV = e^{-x/2}\,dx$,

$$= \lim_{N\to\infty} 0.5\left[-2xe^{-x/2}\Big|_0^N - \int_0^N 1.-2e^{-x/2}\,dx\right]$$
$$= \lim_{N\to\infty} 0.5\left[-2xe^{-x/2}\Big|_0^N + 2\int_0^N e^{-x/2}\,dx\right]$$
$$= \lim_{N\to\infty} 0.5\left[-2xe^{-x/2} - 4e^{-x/2}\right]\Big|_0^N$$
$$= \lim_{N\to\infty}\left[-xe^{-x/2} - 2e^{-x/2}\right]\Big|_0^N$$
$$= \lim_{N\to\infty}\left[\left(-N^{-N/2} - 2^{-N/2}\right) - \left(-0 - 2e^{0}\right)\right]$$
$$= (0 - 0) - (0 - 2) = 2 \text{ years}$$

$$E(Y) = \int_{-\infty}^{\infty}\int_{-\infty}^{\infty} yf(x, y)\,dx\,dy$$
$$= \int_0^{\infty}\int_0^{\infty} 0.1ye^{-x/2}e^{-y/5}\,dx\,dy$$
$$= \int_0^{\infty} 0.1ye^{-y/5}\left(\int_0^{\infty} e^{-x/2}\,dx\right)dy$$
$$= \int_0^{\infty} 0.1ye^{-y/5}\left(\lim_{N\to\infty}\int_0^{N} e^{-x/2}\,dx\right)dy$$
$$= \int_0^{\infty} 0.1ye^{-y/5}\left(\lim_{N\to\infty}\left[-2e^{-x/2}\right]\Big|_0^N\right)dy$$
$$= \int_0^{\infty} -0.2ye^{-y/5}\left(\lim_{N\to\infty}\left[e^{-x/2}\right]\Big|_0^N\right)dy$$
$$= \int_0^{\infty} -0.2ye^{-y/5}\left(\lim_{N\to\infty}\left[^{-N/2} - e^{0}\right]\right)dy$$
$$= \int_0^{\infty} -0.2ye^{-y/5}(-1)\,dy$$
$$= \lim_{N\to\infty} 0.2\int_0^{N} ye^{-y/5}\,dy$$

Using integration by parts with $u = y$ and $dV = e^{-y/5}\,dy$,

$$= \lim_{N\to\infty} 0.2\left[-5ye^{-y/5}\Big|_0^N - \int_0^N 1\cdot -5e^{-y/5}\,dy\right]$$

$$= \lim_{N\to\infty} 0.2\left[-5ye^{-y/5}\Big|_0^N + 5\int_0^N e^{-y/5}\,dy\right]$$

$$= \lim_{N\to\infty} 0.2\left[-5ye^{-y/5} - 25e^{-y/5}\right]\Big|_0^N$$

$$= \lim_{N\to\infty}\left[-ye^{-y/5} - 5e^{-y/5}\right]\Big|_0^N$$

$$= \lim_{N\to\infty}\left[\left(-Ne^{-N/5} - 5e^{-N/5}\right) - \left(0 - 5e^{0}\right)\right]$$

$$= (0-0) - (0-5) = 5 \text{ years}$$

The second component lasts $2\frac{1}{2}$ times longer than the first component.

Review Problems

1.

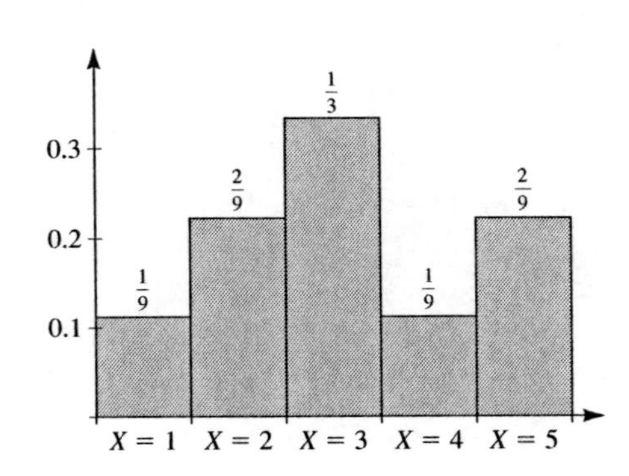

$$E(X) = x_1p(x_1) + x_2p(x_2) + x_3p(x_3)$$
$$x_4p(x_4) + x_5p(x_5)$$
$$= 1\left(\frac{1}{9}\right) + 2\left(\frac{2}{9}\right) + 3\left(\frac{1}{3}\right) + 4\left(\frac{1}{9}\right) + 5\left(\frac{2}{9}\right)$$
$$= \frac{28}{9} \approx 3.11$$

$$\text{Var}(x) = (x_1-\mu)^2p(x_1) + (x_2-\mu)^2p(x_2) + (x_3-\mu)^2p(x_3)$$
$$+ (x_4-\mu)^2p(x_4) + (x_5-\mu)p(x_5)$$
$$= \left(1-\frac{28}{9}\right)^2\left(\frac{1}{9}\right) + \left(2-\frac{28}{9}\right)^2\left(\frac{2}{9}\right) + \left(3-\frac{28}{9}\right)^2\left(\frac{1}{3}\right)$$
$$+ \left(4-\frac{28}{9}\right)^2\left(\frac{1}{9}\right) + \left(5-\frac{28}{9}\right)^2\left(\frac{2}{9}\right)$$
$$= \frac{361}{729} + \frac{200}{729} + \frac{1}{243} + \frac{64}{729} + \frac{578}{729}$$
$$= \frac{134}{81} \approx 1.65$$

$$\sigma(x) = \sqrt{1.65} \approx 1.29$$

3. Since the values of X are any non-negative integer, X *is* a discrete random variable. Since no fruit fly can lay eggs indefinitely, it is a finite discrete random variable.

5. Since the values of X are amounts of money, which at most is measured by dollars and cents, X *is* a discrete random variable. Since there are a limited number of players on a team, there are a limited number of outcomes so X *is* a finite discrete random variable.

7. $$P(2 \le X \le 7) = \int_2^7 f(x)\,dx$$
$$= \int_3^7 \frac{1}{5}\,dx = \frac{x}{5}\Big|_3^7$$
$$= \frac{7}{5} - \frac{3}{5} = \frac{4}{5} = 0.8$$

$$P(X \ge 5) = \int_5^8 \frac{1}{8}\,dx$$
$$= \frac{x}{5}\Big|_5^8 = \frac{8}{5} - \frac{5}{5} = \frac{3}{5} = 0.6$$

9.

$$P(X \le 1) = \int_0^1 0.75x(2x - x^2)\,dx$$
$$= 0.75 \int_0^1 2x^2 - x^3\,dx$$
$$= 0.75\left(\frac{2}{3}x^3 - \frac{x^4}{4}\right)\Big|_0^1$$
$$= 0.75\left[\left(\frac{2}{3} - \frac{1}{4}\right) - 0\right]$$
$$= \frac{5}{16} = 0.3125$$

$$P(1 \le X \le 3) = \int_1^3 f(x)\,dx$$
$$= \int_1^2 0.75x(2x - x^2)\,dx$$
$$= 0.75\left(\frac{2}{3}x^3 - \frac{x^4}{4}\right)\Big|_1^2$$
$$= 0.75\left[\left(\frac{16}{3} - 4\right) - \left(\frac{2}{3} - \frac{1}{4}\right)\right]$$
$$= \frac{11}{16} = 0.6875$$

11.

$$E(X) = \int_{-\infty}^{\infty} xf(x)\,dx$$
$$= \int_0^2 0.75x^2(2x - x^2)\,dx$$
$$= 0.75 \int_0^2 2x^3 - 4^4\,dx$$
$$= 0.75\left(\frac{x^4}{2} - \frac{x^5}{5}\right)\Big|_0^2$$
$$= 0.75\left[\left(8 - \frac{32}{5}\right) - 0\right]$$
$$= \frac{6}{5} = 1.2$$

$$\text{Var}(X) = \int_{-\infty}^{\infty} x^2 f(x)\,dx - [E(X)]^2$$
$$= \int_0^2 0.75x^3(2x - x^2)\,dx - \left(\frac{6}{5}\right)^2$$
$$= 0.75 \int_0^2 2x^4 - x^5\,dx - \frac{36}{25}$$
$$= 0.75\left(\frac{2}{5}x^5 - \frac{x^6}{6}\right)\Big|_0^2 - \frac{36}{25}$$
$$= 0.75\left[\left(\frac{64}{5} - \frac{64}{6}\right) - 0\right] - \frac{36}{25}$$
$$= \left(\frac{3}{4}\right)\left[\frac{64}{30}\right] - \frac{36}{25} = \frac{8}{5} - \frac{36}{25} = \frac{4}{25} = 0.16$$

13. Need $\int_{-\infty}^{\infty} f(x)\,dx = 1$. Now,

$$\int_0^{\infty} cxe^{-x/4}\,dx = \lim_{N\to\infty} c\int_0^N xe^{-x/4}\,dx$$

Using integration by parts with $u = x$ and $dV = e^{-x/4}\,dx$,

$$= c\lim_{N\to\infty}\left[-4xe^{-x/4}\Big|_0^N - \int_0^N 1\cdot -4e^{-x/4}\,dx\right]$$
$$= c\lim_{N\to\infty}\left[-4xe^{-x/4}\Big|_0^N + 4\int_0^N e^{-x/4}\,dx\right]$$
$$= c\lim_{N\to\infty}\left[-4xe^{-x/4} - 16e^{-x/4}\right]\Big|_0^N$$
$$= -4c\lim_{N\to\infty}\left[xe^{-x/4} + 4e^{-x/4}\right]\Big|_0^N$$
$$= -4c\lim_{N\to\infty}\left[\left(Ne^{-N/4} + 4e^{-N/4}\right) - (0 + 4e^0)\right]$$
$$= -4c\,[(0 + 0) - 4] = 16c$$

So, $16c = 1$ or $c = \frac{1}{16}$.

15. $\mu = 7, \sigma = 2$

$$Z = \frac{x - 7}{2}$$

$$P(X \ge 9) = P(Z \ge 1)$$
$$= 1 - P(Z \le 1)$$

Using the table,

$$= 1 - 0.8413 = 0.1587$$

17.
$$\begin{aligned} E(X) &= x_1p(x_1) + x_2p(x_2) + x_3p(x_3) \\ &= 0.25(0.6) + 0.1(0.25) - 0.1(0.15) \\ &= 0.16 \end{aligned}$$

She should expect a 16% increase from this stock.

19.
$$\begin{aligned} E(X) &= x_1p(x_1) + x_2p(x_2) + x_3p(x_3) + x_4p(x_4) \\ &= 100(0.08) + 20(0.12) + 5(0.20) + 0(0.6) \\ &= \$11.40 \end{aligned}$$

21. $\mu = 80, \sigma = 7$

$$Z = \frac{x - 80}{7}$$

(a) $P(X > 90) = P\left(Z > \frac{10}{7}\right)$

$$= 1 - P\left(Z < \frac{10}{7}\right)$$

Using the table with $z = 1.43$,

$$= 1 - 0.9236 = 0.0764$$

In a group of 200,

$$200(0.0764) \approx 15$$

Approximately 15 children weigh more than 90 lbs.

(b) $P(X < 70) = P(Z < -\frac{10}{7})$

$$= P\left(Z > \frac{10}{7}\right)$$

$$= 1 - P\left(Z < \frac{10}{7}\right)$$

Using the result of part (a), approximately 15 children weigh less than 70 lbs. Note: since 70 and 90 are the same distance from the mean, by symmetry, $P(X < 70) = P(X > 90)$

(c) Since

$$f(x) = \frac{1}{\sigma\sqrt{2\pi}}e^{(-x-\mu)^2/2\sigma^2}$$

Here,

$$f(x) = \frac{1}{7\sqrt{2\pi}}e^{-(x-80)^2/98}$$

$$f(80) = \frac{1}{7\sqrt{2\pi}}e^0$$

$$= \frac{1}{7\sqrt{2\pi}} \approx 0$$

The probability is zero that any particular child will weigh exactly 80 pounds, so no children weigh exactly 80 pounds.

23. $\mu = 0.03, \sigma = 0.0015$

$$Z = \frac{x - 0.03}{0.0015}$$

$$P(X < 0.025) = P\left(Z < -\frac{10}{3}\right)$$

$$P\left(Z > \frac{10}{3}\right)$$

$$1 - P\left(Z < \frac{10}{3}\right)$$

Using the table with $z = 3.33$,

$$= 1 - 0.9996 = 0.0004$$

25. **(a)**

$$\begin{aligned} P(X \le 10) &= \int_0^{10} 0.4e^{-0.4t}\,dt \\ &= 0.4\int_0^{10} e^{-0.4t}\,dt \\ &= 0.4\left(\frac{1}{-0.4}e^{-0.4t}\Big|_0^{10}\right) \\ &= -e^{-0.4t}\Big|_0^{10} = -e^{-4} + e^0 \\ &= 1 - \frac{1}{e^4} \approx 0.9817 \end{aligned}$$

(b) $P(X > 20) = 1 - P(X \le 20)$

$$\begin{aligned} &= 1 - \int_0^{20} 0.4e^{-0,4t}\,dt \\ &= 1 + \left(e^{-0.4t}\Big|_0^{20}\right) \\ &= 1 + \left(e^{-8} - e^0\right) = e^{-8} \approx 0.000333 \end{aligned}$$

(c)

$$E(X) = \int_{-\infty}^{\infty} xf(x)\,dx$$
$$= \int_{0}^{\infty} 0.4te^{-0.4t}\,dt$$
$$= \lim_{N\to\infty} 0.4 \int_{0}^{N} te^{-0.4t}\,dt$$

Using integration by parts with $u = t$ and $dV = e^{-0.4t}\,dt$,

$$= \lim_{N\to\infty} 0.4 \left[2.5te^{-0.4t}\Big|_0^N - \int_0^N 1\cdot -2.5e^{-0.4t}\,dt \right]$$
$$= \lim_{N\to\infty} 0.4 \left[-2.5te^{-0.4t}\Big|_0^N + 2.5 \int_0^N e^{-0.4t}\,dt \right]$$
$$= \lim_{N\to\infty} 0.4 \left[-2.5te^{-0.4t} - 6.25e^{-0.4t} \right]\Big|_0^N$$
$$= \lim_{N\to\infty} \left[-te^{-0.4t} - 2.5e^{-0.4t} \right]\Big|_0^N$$
$$= \lim_{N\to\infty} \left[\left(-Ne^{-0.4N} - 2.5e^{-0.4N}\right) - \left(0 - 2.5e^0\right) \right]$$
$$= (0 - 0) - (0 - 2.5) = 2.5 \text{ years}$$

27. (a)

$$P(X < 10) = \int_0^{10} 0.25xe^{-x/2}\,dx$$
$$= 0.25 \int_0^{10} xe^{-x/2}\,dx$$

Using integration, by parts with $u = x$ and $dV = e^{-x/2}\,dx$,

$$= 0.25 \left[-2xe^{-x/2}\Big|_0^{10} - \int_0^{10} 1\cdot -2e^{-x/2}\,dx \right]$$
$$= 0.25 \left[-2xe^{-x/2}\Big|_0^{10} + 2 \int_0^{10} e^{-x/2}\,dx \right]$$
$$= 0.25 \left[-2xe^{-x/2} - 4e^{-x/2} \right]\Big|_0^{10}$$
$$= -0.5 \left[xe^{-x/2} + 2e^{-x/2} \right]\Big|_0^{10}$$
$$= -0.5 \left[\left(10e^{-5} + 2e^{-5}\right) - (0 + 2e^0) \right]$$
$$= -0.5 \left[12e^{-5} - 2 \right] \approx 0.9596$$

(b)

$$E(X) = \int_{-\infty}^{\infty} xf(x)\,dx$$
$$= \int_0^{\infty} 0.25x^2e^{-x/2}\,dx$$
$$= \lim_{N\to\infty} 0.25 \int_0^N x^2e^{-x/2}\,dx$$

Using integration by parts twice,

$$= 0.25 \lim_{N\to\infty} \left[-2x^2e^{-x/2}\Big|_0^N - \int_0^N 2x\cdot -2e^{-x/2}\,dx \right]$$
$$= 0.25 \lim_{N\to\infty} \left[-2x^2e^{-x/2}\Big|_0^N + 4 \int_0^N xe^{-x/2}\,dx \right]$$
$$= 0.25 \lim_{N\to\infty} \left[-2x^2e^{-x/2}\Big|_0^N + 4\left(-2xe^{-x/2}\Big|_0^N - \int_0^N 1\cdot -2e^{-x/2}\,dx \right) \right]$$
$$= 0.25 \lim_{N\to\infty} \left[-2x^2e^{-x/2} - 8xe^{-x/2}\Big|_0^N + 8 \int_0^N e^{-x/2}\,dx \right]$$
$$= 0.25 \lim_{N\to\infty} \left[-2x^2e^{-x/2} - 8xe^{-x/2} - 16e^{-x/2} \right]\Big|_0^N$$
$$= 0.25 \lim_{N\to\infty} \left[\left(-2N^2e^{-N/2} - 8Ne^{-N/2} - 16e^{-N/2}\right) - \left(0 - 0 - 16e^0\right) \right]$$
$$= 0.25\,[(0 - 0 - 0) - (0 - 0 - 16)] = 4 \text{ feet}$$

29. (a) $P(X = 10{,}000) = 1 - 0.02 = 0.98$
$P(x = -90{,}000) = 0.02$
These are the only possibilities.

(b)
$$E(X) = x_1p(x_1) + x_2p(x_2)$$
$$= 10{,}000(0.98) - 90{,}000(0.02)$$
$$= 8{,}000$$

Expected value of each policy is a profit of \$8,000.

(c) $16{,}000 = c(0.98) + (c - 100{,}000)(0.02)$
$18{,}000 = c$
The company should charge \$18,000 per policy.

31. (a)

$$P(X \le 10) = 2P(10 \le X \le 20)$$

$$\int_0^{10} \lambda e^{-\lambda x}\,dx = 2\int_{10}^{20} \lambda e^{-\lambda x}\,dx$$

$$-e^{-\lambda x}\Big|_0^{10} = -2e^{-\lambda x}\Big|_{10}^{20}$$

$$-e^{10\lambda} + e^0 = -2\left(e^{-20\lambda} - e^{-10\lambda}\right)$$

$$-e^{10\lambda} + 1 = -2e^{-20\lambda} + 2e^{-10\lambda}$$

$$2e^{-20\lambda} - 3e^{-10\lambda} + 1 = 0$$

$$2(e^{-10\lambda})^2 - 3(e^{10\lambda}) + 1 = 0$$

$$e^{-10\lambda} = \frac{3 \pm \sqrt{9 - 4(2)(1)}}{2(2)}$$

$$e^{-10\lambda} = \frac{3 \pm 1}{4}$$

$e^{-10\lambda} = \frac{1}{2}$ (rejecting $e^{-10\lambda} = 1$ which leads to $\lambda = 0$).

$$-10\lambda = \ln\frac{1}{2}$$

$$\lambda \frac{\ln\frac{1}{2}}{-10} = \frac{-\ln\frac{1}{2}}{10} = \frac{\ln 2}{10}$$

(b)

$$P(X \le 5) = \int_0^5 \frac{\ln 2}{10} e^{-\frac{\ln 2}{10}x}\,dx$$

$$= -e^{\frac{-\ln 2}{10}x}\Big|_0^5 = -e^{\frac{\ln 2}{2}} + e^0$$

$$= -e^{\frac{\ln\frac{1}{2}}{2}} + 1$$

$$= -e^{\ln(\frac{1}{2})^{1/2}} + 1$$

$$= -\sqrt{\frac{1}{2}} + 1 = 1 - \frac{\sqrt{2}}{2}$$

(c)

$$E(X) = \int_{-\infty}^{\infty} xf(x)\,dx$$

$$= \int_0^{\infty} \frac{\ln 2}{10} x e^{-\frac{\ln 2}{10}x}\,dx$$

$$= \lim_{N\to\infty} \frac{\ln 2}{10} \int_0^N x e^{-\frac{\ln 2}{10}x}\,dx$$

Using integration by parts with $u = x$ and $dV = e^{-\frac{\ln 2}{10}x}\,dx$,

$$= \lim_{N\to\infty} \frac{\ln 2}{10}\left[-\frac{10x}{\ln 2}e^{-\frac{\ln 2}{10}x}\Big|_0^N - \int_0^N 1 \cdot -\frac{10}{\ln 2}e^{-\frac{\ln 2}{10}x}\,dx\right]$$

$$\lim_{N\to\infty} \frac{\ln 2}{10}\left[-\frac{10x}{\ln 2}e^{-\frac{\ln 2}{10}x}\Big|_0^N + \frac{10}{\ln 2}\int_0^N e^{-\frac{\ln 2}{10}x}\,dx\right]$$

$$\lim_{N\to\infty}\left[-xe^{-\frac{\ln 2}{10}x}\Big|_0^N + \int_0^N e^{-\frac{\ln 2}{10}x}\,dx\right]$$

$$= \lim_{N\to\infty}\left[-xe^{-\frac{\ln 2}{10}x} - \frac{10}{\ln 2}e^{-\frac{\ln 2}{10}x}\right]\Big|_0^N$$

$$= \lim_{N\to\infty}\left[\left(-Ne^{-\frac{\ln 2}{10}N} - \frac{10}{\ln 2}e^{-\frac{\ln 2}{10}N}\right) - \left(0 - \frac{10}{\ln 2}e^0\right)\right]$$

$$= (0 - 0) - \left(0 - \frac{10}{\ln 2}\right) = \frac{10}{\ln 2} \approx 14.2 \text{ weeks}$$

33. $\mu = 12, \sigma = 0.05$

$$Z = \frac{x - 12}{0.05}$$

(a) $P(X \ge 11.8) = P(Z \ge -4) = 1 - P(Z \le -4)$
Since the table starts with $z = 0$,

$$P(Z \le -4) = P(Z \ge 4) = 1 - P(Z \le 4)$$

So,

$$P(Z \ge -4) = 1 - [1 - P(Z \le 4)] = P(Z \le 4)$$

Since 99.7% of probability is between $\mu - 3\sigma$ and $\mu + 3\sigma$, or -3 and 3,

$$P(Z \le 4) \approx 1$$

(b) $P(X \le x) = P\left(Z \le \frac{x - 12}{0.05}\right)$
Want $P\left(Z \le \frac{x-12}{0.05}\right) = 0.95$
Using the table,

$$\frac{x - 12}{0.05} \approx 1.645$$

$$x \approx 12.08$$

35. $\mu = 72.3, \sigma = 16.4$

$$Z = \frac{x - 72.3}{16.4}$$

$P(50 \le X \le 75) \approx P(-1.36 \le Z \le 0.16)$
$= P(Z \le 0.16) - P(Z \le -1.36)$
Since the table starts with $z = 0$,

$$P(Z \le -1.36) = P(Z \ge 1.36) \\ = 1 - P(Z \le 1.36)$$

So,

$$P(-1.36 \le Z \le 0.16) \\ = P(Z \le 0.16) - [1 - P(Z \le 1.36)]$$

Using the table,

$$= 0.5636 - [1 - 0.9131] \\ = 0.4767$$

In a class of 82 students, approximately $82(0.4767) \approx 39$ will have scores between 50 and 75.

37. **(a)** $P(X \ge 5) = 1 - P(X < 5)$

$$= 1 - \int_4^5 0.75(x-4)(6-x)\,dx$$
$$= 1 - 0.75 \int_4^5 10x - 24 - x^2\,dx$$
$$= 1 - 0.75\left(5x^2 - 24x - \frac{x^3}{3}\right)\Big|_4^5$$
$$= 1 - 0.75\left[\left(125 - 120 - \frac{125}{3}\right) - \left(80 - 96 - \frac{64}{3}\right)\right]$$
$$= 0.5$$

(b) $E(X) = \int_{-\infty}^{\infty} xf(x)\,dx$

$$= \int_4^6 0.75x(x-4)(6-x)\,dx$$
$$= 0.75 \int_4^6 10x^2 - 24x - x^3\,dx$$
$$= 0.75\left(\frac{10}{3}x^3 - 12x^2 - \frac{x^4}{4}\right)\Big|_4^6$$
$$= 0.75\left[(720 - 432 - 324) - \left(\frac{640}{3} - 192 - 64\right)\right]$$
$$= 5$$

39. $E(X) = \lambda$, so $\lambda = 4$

(a)
$$P(X = 4) = \frac{4^4}{4!}e^{-4} \\ = \frac{32}{3}e^{-4} \approx 0.1954$$

(b) $P(X = 0) = \frac{4^0}{0!}e^{-4} = e^{-4} \approx 0.01832$

(c) $P(X < 4) = P(X = 0) + P(X = 1)$
$+ P(X = 2) + P(X = 3)$

$$= \left(\frac{4^0}{0!} + \frac{4^1}{1!} + \frac{4^2}{2!} + \frac{4^3}{3!}\right)e^{-4}$$
$$= \left(1 + 4 + 8 + \frac{32}{3}\right)e^{-4}$$
$$\approx 0.4335$$

41. **(a)** For a 24 hour period,

$$E(X) = \lambda = 2.4$$
$$P(X = 0) = \frac{2.4^0}{0!}e^{-2.4} = e^{-2.4} \approx 0.0907$$

(b) For a 12 hour period,

$$E(X) = \lambda = 1.2$$
$$P(X \ge 1) = 1 - P(X = 0) \\ = 1 - \frac{1.2^0}{0!}e^{-1.2} \\ = 1 - e^{-1.2} \approx 0.6988$$

(c) For a 1 hour period,

$$E(X) = \lambda = 0.1$$
$$P(X=0) = \frac{0.1^0}{0!}e^{-0.1} = e^{-0.1} \approx 0.9048$$

43. (a)

$$\int_{-\infty}^{\infty}\int_{-\infty}^{\infty} f(x, y)\, dy\, dx$$
$$= \int_0^1\int_0^1 0.4(2x+3y)\, dy\, dx$$
$$= 0.4\int_0^1\int_0^1 2x+3y\, dy\, dx$$
$$= 0.4\int_0^1 \left(2xy + \frac{3}{2}y^2\Big|_0^1\right) dx$$
$$= 0.4\int_0^1 \left[\left(2x+\frac{3}{2}\right) - 0\right] dx$$
$$= 0.4\left(x^2 + \frac{3}{2}x\Big|_0^1\right)$$
$$= 0.4\left[\left(1+\frac{3}{2}\right) - 0\right] = 1$$

Also, $f(x, y) \geq 0$ for $0 \leq x \leq 1$ and $0 \leq y \leq 1$. So, f is a probability density function.

(b)

$$\int_0^{0.5}\int_{0.5}^1 0.4(2x+3y)\, dy\, dx$$
$$= 0.4\int_0^{0.5}\left(2xy + \frac{3}{2}y^2\Big|_{0.5}^1\right) dx$$
$$= 0.4\int_0^{0.5}\left[\left(2x+\frac{3}{2}\right) - (x+0.375)\right] dx$$
$$= 0.4\int_0^{0.5} x + 1.125\, dx$$
$$= 0.4\left(\frac{x^2}{2} + 1.125x\right)\Big|_0^{0.5}$$
$$= 0.4\,[(0.125 + 0.5625) - 0]$$
$$= 0.275$$

(c)

$$\int_{0.8}^1\int_{0.8}^1 0.4(2x+3y)\, dy\, dx$$
$$= 0.4\int_{0.8}^1\left(2xy + \frac{3}{2}y^2\Big|_{0.8}^1\right) dx$$
$$= 0.4\int_{0.8}^1\left[\left(2x+\frac{3}{2}\right) - (1.6x + 0.96)\right] dx$$
$$= 0.4\int_{0.8}^1 0.4x + 0.54\, dx$$
$$= 0.4\left(0.2x^2 + 0.54x\Big|_{0.8}^1\right)$$
$$= 0.4\,[(0.2 + 0.54) - (0.128 + 0.432)]$$
$$= 0.072$$

(d)

$$\int_0^1\int_0^{1-x} 0.4(2x+3y)\, dy\, dx$$
$$= 0.4\int_0^1\left(2xy + \frac{3}{2}y^2\Big|_0^{1-x}\right) dx$$
$$= 0.4\int_0^1\left[2x(1-x) + \frac{3}{2}(1-x)^2\right] dx$$
$$= 0.4\int_0^1 \frac{3}{2} - x - \frac{1}{2}x^2\, dx$$
$$= 0.4\left(\frac{3}{2}x - \frac{x^2}{2} - \frac{x^3}{6}\right)\Big|_0^1$$
$$= 0.4\left[\left(\frac{3}{2} - \frac{1}{2} - \frac{1}{6}\right) - 0\right]$$
$$\approx 0.333$$

45. (a)

$$\int_0^5\int_0^5 0.125e^{-x/4}e^{-y/2}\, dy\, dx$$
$$= 0.125\int_0^5 e^{-x/4}\left(\int_0^5 e^{-y/2}\, dy\right) dx$$
$$= 0.125\int_0^5 e^{-x/4}\left(-2e^{-y/2}\Big|_0^5\right) dx$$

$$= -0.25 \int_0^5 e^{-x/4}\left(e^{-5/2} - e^0\right) dx$$

$$= -0.25\left(e^{-5/2} - 1\right)\int_0^5 e^{-x/4}dx$$

$$= -0.25\left(e^{-5/2} - 1\right)\left(-4e^{-x/4}\Big|_0^5\right)$$

$$= \left(e^{-5/2} - 1\right)\left(e^{-5/4} - 1\right) \approx 0.6549$$

(b)

$$\int_0^8 \int_0^{8-x} 0.125e^{-x/4}e^{-y/2}\, dy\, dx$$

$$= 0.125 \int_0^8 e^{-x/4}\left(\int_0^{8-x} e^{-y/2}\, dy\right) dx$$

$$= 0.125 \int_0^8 e^{-x/4}\left(-2e^{-y/2}\Big|_0^{8-x}\right) dx$$

$$= -0.25 \int_0^8 e^{-x/4}\left(e^{\frac{x-8}{2}} - e^0\right) dx$$

$$= -0.25 \int_0^8 e^{\frac{x-16}{4}} e^{-x/4}\, dx$$

$$= -0.25\left(4e^{\frac{x-16}{4}} + 4e^{-x/4}\right)\Big|_0^8$$

$$= -1\left(e^{\frac{x-16}{4}} + e^{-x/4}\right)\Big|_0^8$$

$$= -1\left[\left(e^{-2} + e^{-2}\right) - \left(e^{-4} + e^0\right)\right]$$

$$= -1\left[2e^{-2} - e^{-4} - 1\right] \approx 0.7476$$

47. (a)

$$\int_{1/3}^1 1\, dy = y\Big|_{1/3}^1 = 1 - \frac{1}{3} = \frac{2}{3}$$

(b) Since the chance of arriving after the tram is the same as the chance of arriving before the tram, the probability is $\frac{1}{2}$.

(c) Want the tram to arrive after the tourist, or $X \geq Y$. When the tourist arrives within the first $\frac{2}{3}$ of a given hour, need the tram to arrive within $\frac{1}{3}$ of an hour of the tourist, or $X \leq Y + \frac{1}{3}$. So, $Y \leq X \leq Y + \frac{1}{3}$. Since the combined times need to be within the one hour total, $0 \leq 4Y + \frac{1}{3} \leq 1$. This limits Y as ranging from 0 to $\frac{2}{3}$. So, when the tourist arrives within the first $\frac{2}{3}$ of an hour, the p;robability of connecting is

$$\int_0^{2/3} \int_y^{y+1/3} 1\, dx\, dy$$

$$= \int_0^{2/3}\left(\int_y^{y+1/3}\right) dy = \int_0^{2/3} \frac{1}{3}\, dy$$

$$= \frac{y}{3}\Big|_0^{2/3} = \frac{2}{9}$$

Similarly, when the tourist arrives in the last $\frac{1}{3}$ of a given hour, the probability of connecting is

$$\int_{2/3}^1 \int_y^1 1\, dx\, dy$$

$$= \int_{2/3}^1 \left(x\Big|_y^1\right) dy = \int_{2/3}^1 1 - y\, dy$$

$$= y - \frac{y^2}{2}\Big|_{2/3}^1 = \frac{1}{18}$$

So, the total probability of connecting is $\frac{2}{9} + \frac{1}{18} = \frac{5}{18}$.

49. (a)

$$P(X \leq 2) = \int_0^2 \frac{5.36}{(x + 5.36)^2}\, dx$$

Using substitution with $u = x + 5.36$ and $du = dx$,

$$= 5.36 \int_{5.36}^{7.36} \frac{1}{u^2}\, du$$

$$= 5.36\left(-\frac{1}{u}\Big|_{5.36}^{7.36}\right)$$

$$= 5.36\left(-\frac{1}{7.36} + \frac{1}{5.36}\right)$$

$$= 1 - \frac{5.36}{7.36} \approx 0.2717$$

(b) $P(X \geq 10) = 1 - P(X \leq 10)$

$$= 1 - \int_0^{10} \frac{5.36}{(x+5.36)^2}\,dx$$

$$= 1 - 5.36\left(-\frac{1}{u}\Big|_{5.36}^{15.36}\right)$$

$$= 1 - 5.36\left(-\frac{1}{15.36} + \frac{1}{5.36}\right)$$

$$= 1 + \frac{5.36}{15.36} - 1 = \frac{5.36}{15.36} \approx 0.3490$$

(c) $E(X) = \int_{-\infty}^{\infty} xf(x)\,dx$

$$= \int_0^{\infty} \frac{5.36x}{(x+5.36)^2}\,dx$$

$$= 5.36 \lim_{N\to\infty} \int_0^N \frac{x}{(x+5.36)^2}\,dx$$

Using formula #3 in the short table of integrals,

$$= 5.36 \lim_{N\to\infty}\left[\frac{5.36}{5.36+x} + \ln|5.36+x|\right]\Big|_0^N$$

$$= 5.36 \lim_{N\to\infty}\left[\left(\frac{5.36}{5.36+N} + \ln|5.36+N|\right) - (1+\ln 5.36)\right]$$

$$= \infty$$

This distribution does not have a finite expected value.

(d) Writing Exercise—Answers will vary.

Chapter 11

Trigonometric Functions

11.1 The Trigonometric Functions

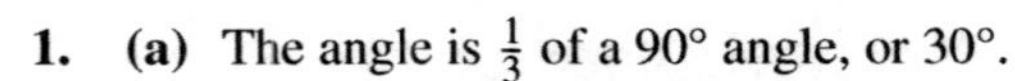

1. **(a)** The angle is $\frac{1}{3}$ of a 90° angle, or 30°.

(b) The angle is $\frac{2}{3}$ of a 90° angle, or 60°.

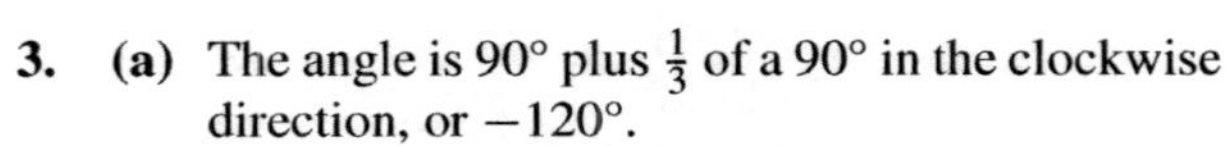

3. **(a)** The angle is 90° plus $\frac{1}{3}$ of a 90° in the clockwise direction, or $-120°$.

(b) The angle is 360° plus $\frac{1}{3}$ of a 90° angle, or 390°.

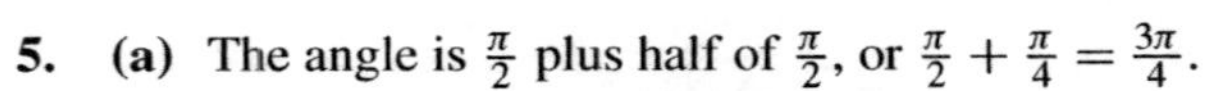

5. **(a)** The angle is $\frac{\pi}{2}$ plus half of $\frac{\pi}{2}$, or $\frac{\pi}{2} + \frac{\pi}{4} = \frac{3\pi}{4}$.

(b) The angle is two-thirds of $\frac{\pi}{2}$, or $\frac{\pi}{3}$.

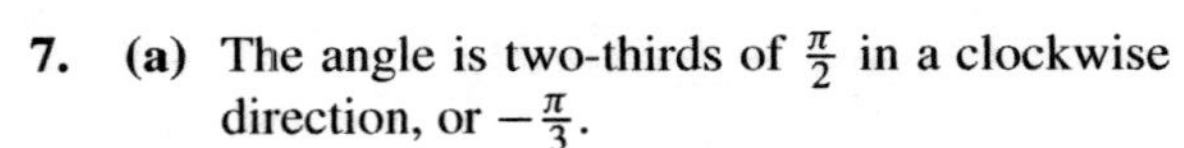

7. **(a)** The angle is two-thirds of $\frac{\pi}{2}$ in a clockwise direction, or $-\frac{\pi}{3}$.

(b) The angle is $\frac{\pi}{2}$ plus two-thirds of $\frac{\pi}{2}$, or $\frac{\pi}{2} + \frac{\pi}{3} = \frac{5\pi}{6}$.

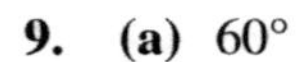

9. **(a)** 60°

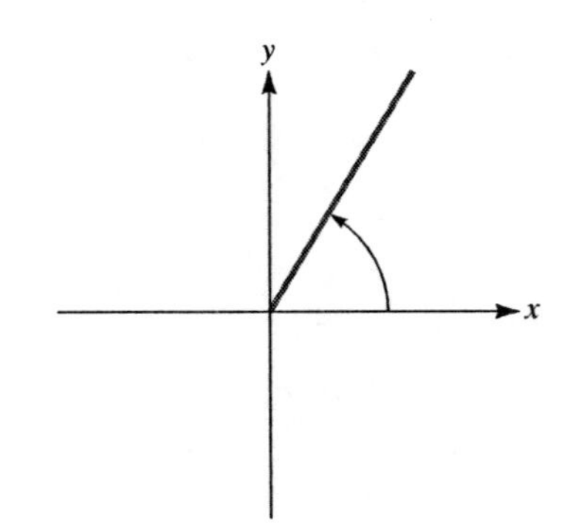

(b) 120°

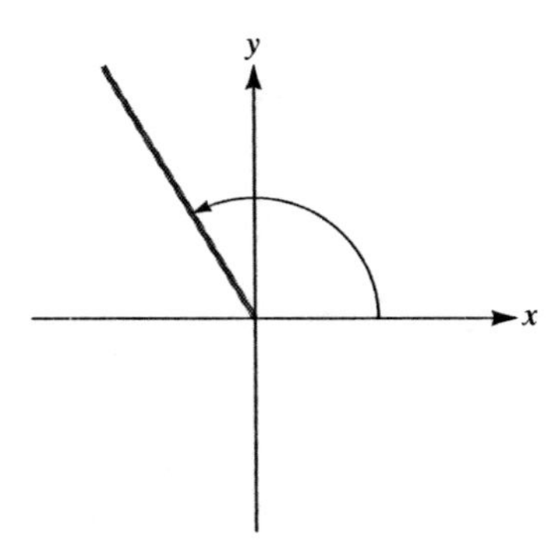

11. **(a)** 45°

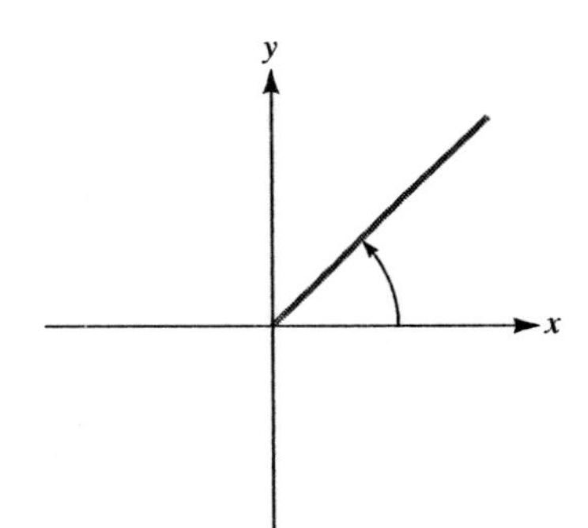

(b) $-150°$

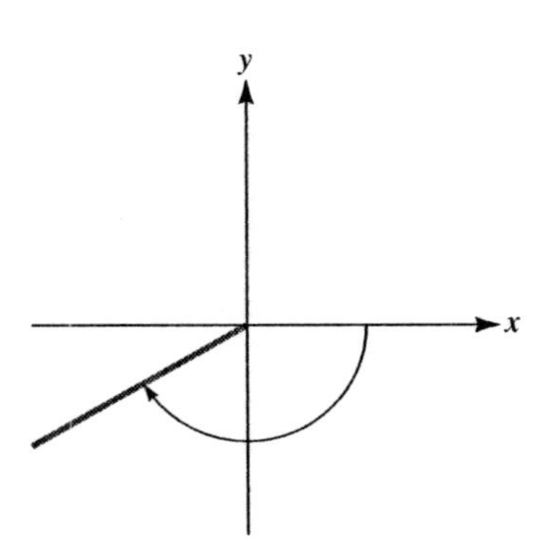

13. **(a)** $\dfrac{3\pi}{4}$

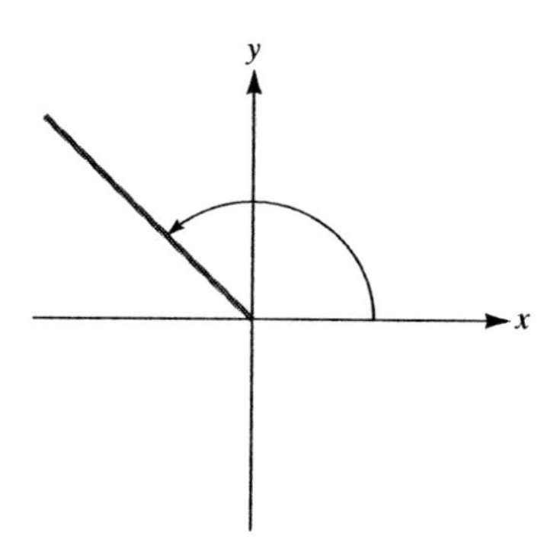

(b) $-\dfrac{2\pi}{3}$

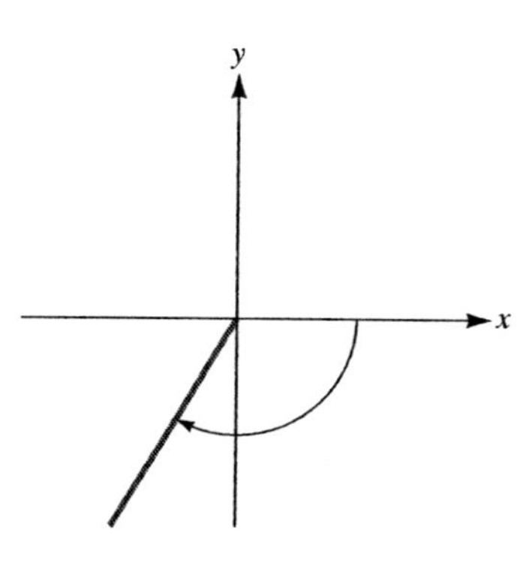

15. **(a)** $-\dfrac{5\pi}{3}$

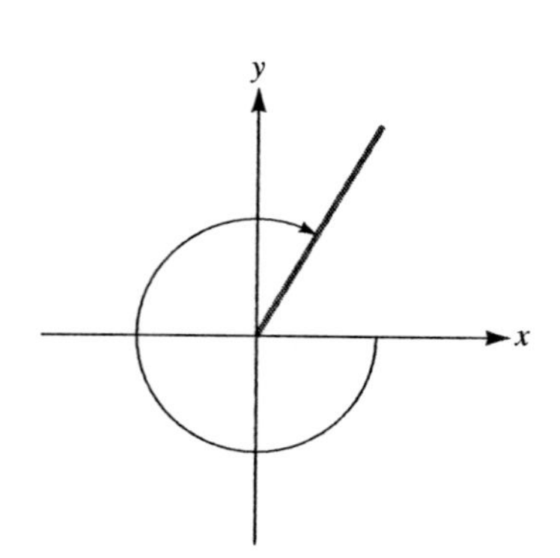

(b) $\dfrac{3\pi}{2}$

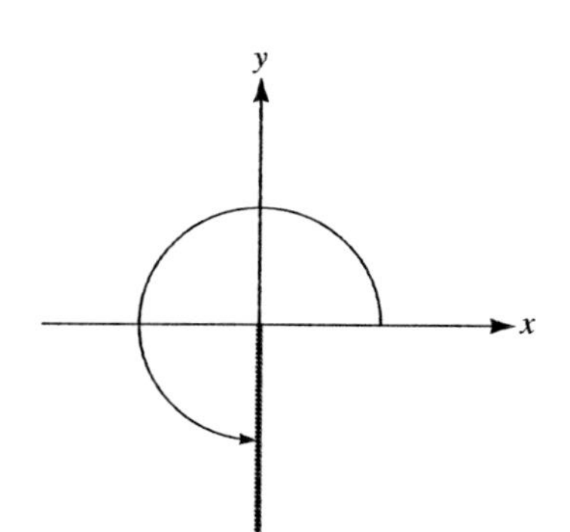

17. **(a)** $15° \cdot \dfrac{\pi}{180°} = \dfrac{\pi}{12}$

(b) $-240° \cdot \dfrac{\pi}{180°} = -\dfrac{4\pi}{3}$

19. **(a)** $135° \cdot \dfrac{\pi}{180°} = \dfrac{3\pi}{4}$

(b) $540° \cdot \dfrac{\pi}{180°} = 3\pi$

21. **(a)** $\dfrac{5\pi}{6} \cdot \dfrac{180°}{\pi} = 150°$

(b) $-\dfrac{\pi}{12} \cdot \dfrac{180°}{\pi} = -15°$

23. **(a)** $3\pi \cdot \dfrac{180°}{\pi} = 540°$

(b) $1 \cdot \dfrac{180°}{\pi} = \dfrac{180°}{\pi}$

25. $\cos \dfrac{7\pi}{2} = \cos \left(\dfrac{3\pi}{2} + 2\pi \right) = \cos \dfrac{3\pi}{2} = 0$

27.
$$\sin \left(-\frac{7\pi}{2} \right) = -\sin \left(\frac{7\pi}{2} \right) = -\sin \left(\frac{3\pi}{2} + 2\pi \right)$$
$$= -\sin \frac{3\pi}{2} = -(-1) = 1$$

29.
$$\cot \frac{5\pi}{2} = \frac{\cos \frac{5\pi}{2}}{\sin \frac{5\pi}{2}} = \frac{\cos \left(\frac{\pi}{2} + 2\pi \right)}{\sin \left(\frac{\pi}{2} + 2\pi \right)} = \frac{\cos \frac{\pi}{2}}{\sin \frac{\pi}{2}}$$
$$= \frac{0}{1} = 0$$

31.
$$\csc\left(-\frac{7\pi}{2}\right) = \frac{1}{\sin\left(-\frac{7\pi}{2}\right)} = \frac{1}{-\sin\frac{7\pi}{2}}$$
$$= \frac{1}{-\sin\left(\frac{3\pi}{2}+2\pi\right)} = \frac{1}{-\sin\frac{3\pi}{2}}$$
$$= \frac{1}{-(-1)} = 1$$

33. $\cos\left(-\frac{2\pi}{3}\right) = \cos\frac{2\pi}{3} = -\frac{1}{2}$

35.
$$\sin\left(-\frac{7\pi}{6}\right) = -\sin\frac{7\pi}{6} = -\sin\left(\pi+\frac{\pi}{6}\right)$$
$$= -\left(\sin\pi\cos\frac{\pi}{6} + \cos\pi\sin\frac{\pi}{6}\right)$$
$$= -\left(0 - 1\cdot\frac{1}{2}\right) = \frac{1}{2}$$

37.
$$\cot\frac{\pi}{3} = \frac{\cos\frac{\pi}{3}}{\sin\frac{\pi}{3}} = \frac{\frac{1}{2}}{\frac{\sqrt{3}}{2}} = \frac{1}{\sqrt{3}} = \frac{\sqrt{3}}{3}$$

39.
$$\tan\left(-\frac{\pi}{4}\right) = \frac{\sin\left(-\frac{\pi}{4}\right)}{\cos\left(-\frac{\pi}{4}\right)} = \frac{-\sin\frac{\pi}{4}}{\cos\frac{\pi}{4}} = \frac{-\frac{\sqrt{2}}{2}}{\frac{\sqrt{2}}{2}} = -1$$

41.
$$\csc\frac{2\pi}{3} = \frac{1}{\sin\frac{2\pi}{3}} = \frac{1}{\frac{\sqrt{3}}{2}} = \frac{2}{\sqrt{3}} = \frac{2\sqrt{3}}{3}$$

43.
$$\sec\frac{5\pi}{4} = \frac{1}{\cos\frac{5\pi}{4}} = \frac{1}{\cos\left(\pi+\frac{\pi}{4}\right)}$$
$$= \frac{1}{\cos\pi\cos\frac{\pi}{4} - \sin\pi\sin\frac{\pi}{4}}$$
$$= \frac{1}{(-1)\left(\frac{\sqrt{2}}{2}\right) - 0} = -\frac{2}{\sqrt{2}} = -\sqrt{2}$$

45. Using a right triangle having an adjacent side of length 1, an opposite side of length $\sqrt{3}$, and a hypotenuse of length 2,
$$\cos\theta = \frac{\text{adjacent}}{\text{hypotenuse}} \quad \text{or} \quad \cos\frac{\pi}{3} = \frac{1}{2}$$
$$\sin\theta = \frac{\text{opposite}}{\text{hypotenuse}} \quad \text{or} \quad \sin\frac{\pi}{3} = \frac{\sqrt{3}}{2}$$

47.
$$\sin\frac{7\pi}{6} = \sin\left(\pi+\frac{\pi}{6}\right)$$
$$\sin\pi\cos\frac{\pi}{6} + \cos\pi\sin\frac{\pi}{6}$$
$$= 0 + (-1)\left(\frac{1}{2}\right) = -\frac{1}{2}$$
$$\cos\frac{7\pi}{6} = \cos\left(\pi+\frac{\pi}{6}\right)$$
$$= \cos\pi\cos\frac{\pi}{6} - \sin\pi\sin\frac{\pi}{6}$$
$$= (-1)\left(\frac{\sqrt{3}}{2}\right) - 0 = -\frac{\sqrt{3}}{2}$$
$$\sin\frac{5\pi}{4} = \sin\left(\pi+\frac{\pi}{4}\right)$$
$$= \sin\pi\cos\frac{\pi}{4} + \cos\pi\sin\frac{\pi}{4}$$
$$= 0 + (-1)\left(\frac{\sqrt{2}}{2}\right) = -\frac{\sqrt{2}}{2}$$
$$\cos\frac{5\pi}{4} = \cos\left(\pi+\frac{\pi}{4}\right)$$
$$= \cos\pi\cos\frac{\pi}{4} - \sin\pi\sin\frac{\pi}{4}$$
$$= (-1)\left(\frac{\sqrt{2}}{2}\right) - 0 = -\frac{\sqrt{2}}{2}$$
$$\sin\frac{4\pi}{3} = \sin\left(\pi+\frac{\pi}{3}\right)$$
$$= \sin\pi\cos\frac{\pi}{3} + \cos\pi\sin\frac{\pi}{3}$$
$$= 0 + (-1)\left(\frac{\sqrt{3}}{2}\right) = -\frac{\sqrt{3}}{2}$$
$$\cos\frac{4\pi}{3} = \cos\left(\pi+\frac{\pi}{3}\right)$$
$$= \cos\pi\cos\frac{\pi}{3} - \sin\pi\sin\frac{\pi}{3}$$
$$= (-1)\left(\frac{1}{2}\right) - 0 = -\frac{1}{2}$$

$$\sin\frac{3\pi}{2} = -1$$

$$\cos\frac{3\pi}{2} = 0$$

$$\begin{aligned}\sin\frac{5\pi}{3} &= \sin\left(\pi + \frac{2\pi}{3}\right)\\ &= \sin\pi\cos\frac{2\pi}{3} + \cos\pi\sin\frac{2\pi}{3}\\ &= 0 + (-1)\left(\frac{\sqrt{3}}{2}\right) = -\frac{\sqrt{3}}{2}\end{aligned}$$

$$\begin{aligned}\cos\frac{5\pi}{3} &= \cos\left(\pi + \frac{2\pi}{3}\right)\\ &= \cos\pi\cos\frac{2\pi}{3} - \sin\pi\sin\frac{2\pi}{3}\\ &= (-1)\left(-\frac{1}{2}\right) - 0 = \frac{1}{2}\end{aligned}$$

$$\begin{aligned}\sin\frac{7\pi}{4} &= \sin\left(\pi + \frac{3\pi}{4}\right)\\ &= \sin\pi\cos\frac{3\pi}{4} + \cos\pi\sin\frac{3\pi}{4}\\ &= 0 + (-1)\left(\frac{\sqrt{2}}{2}\right) = -\frac{\sqrt{2}}{2}\end{aligned}$$

$$\begin{aligned}\cos\frac{7\pi}{4} &= \cos\left(\pi + \frac{3\pi}{4}\right)\\ &= \cos\pi\cos\frac{3\pi}{4} - \sin\pi\sin\frac{3\pi}{4}\\ &= (-1)\left(-\frac{\sqrt{2}}{2}\right) - 0 = \frac{\sqrt{2}}{2}\end{aligned}$$

$$\begin{aligned}\sin\frac{11\pi}{6} &= \sin\left(\pi + \frac{5\pi}{6}\right)\\ &= \sin\pi\cos\frac{5\pi}{6} + \cos\pi\sin\frac{5\pi}{6}\\ &= 0 + (-1)\left(\frac{1}{2}\right) = -\frac{1}{2}\end{aligned}$$

$$\begin{aligned}\cos\frac{11\pi}{6} &= \cos\left(\pi + \frac{5\pi}{6}\right)\\ &= \cos\pi\cos\frac{5\pi}{6} - \sin\pi\sin\frac{5\pi}{6}\\ &= (-1)\left(-\frac{\sqrt{3}}{2}\right) - 0 = \frac{\sqrt{3}}{2}\end{aligned}$$

$$\sin 2\pi = 0$$

$$\cos 2\pi = 1$$

49.

$$\csc\theta = \frac{5}{3}$$

$$\csc\theta = \frac{1}{\sin\theta} = \frac{1}{\frac{\text{opposite}}{\text{hypotenuse}}} = \frac{\text{hypotenuse}}{\text{opposite}}$$

Using a right triangle having a hypotenuse of length 5, an opposite side of length 3, and an adjacent side of length $\sqrt{25-9} = 4$,

$$\tan\theta = \frac{\text{opposite}}{\text{adjacent}} = \frac{3}{4}$$

51.

$$\cos\theta = \frac{3}{5}$$

$$\cos\theta = \frac{\text{adjacent}}{\text{hypotenuse}}$$

Using a right triangle having a hypotenuse of length 5, an adjacent side of length 3, and an opposite side of $\sqrt{25-9} = 4$,

$$\tan\theta = \frac{\text{opposite}}{\text{adjacent}} = \frac{4}{3}$$

53.

$$\cot\theta = \frac{4}{3}$$

$$\cot\theta = \frac{\text{adjacent}}{\text{opposite}}$$

Using a right triangle having an adjacent side of length 4, an opposite side of length 3, and a hypotenuse of length $\sqrt{16+9} = 5$,

$$\sec\theta = \frac{1}{\cos\theta} = \frac{1}{\frac{\text{adjacent}}{\text{hypotenuse}}} = \frac{\text{hypotenuse}}{\text{adjacent}} = \frac{5}{4}$$

55.

$$3\sin^2\theta\cos 2\theta = 2; 0 \le \theta \le 2\pi$$
$$3\sin^2\theta + (\cos^2\theta - \sin^2\theta) = 2$$
$$2\sin^2\theta + \cos^2\theta = 2$$
$$2\sin^2\theta + (1 - \sin^2\theta) = 2$$
$$\sin^2\theta + 1 = 2$$
$$\sin^2\theta = 1$$
$$\sin\theta = 1$$
$$\theta = \frac{\pi}{2}, \frac{3\pi}{2}$$

57.

$$\cos 2\theta = \cos\theta; 0 \le \theta \le \pi$$
$$\cos^2\theta - \sin^2\theta = \cos\theta$$
$$\cos^2\theta - (1 - \cos^2\theta) = \cos\theta$$
$$2\cos^2\theta - 1 = \cos\theta$$
$$2\cos^2\theta - \cos\theta - 1 = 0$$
$$(2\cos\theta + 1)(\cos\theta - 1) = 0$$
$$2\cos\theta + 1 = 0 \text{ or } \cos\theta - 1 = 0$$
$$\cos\theta = -\frac{1}{2} \text{ or } \cos\theta = 1$$
$$\theta = \frac{2\pi}{3}, 0$$

59.

$$3\cos^2\theta - \sin^2\theta = 2; 0 \le \theta \le \pi$$
$$3\cos^2\theta - (1 - \cos^2\theta) = 2$$
$$4\cos^2\theta - 1 = 2$$
$$\cos^2\theta = \frac{3}{4}$$
$$\cos\theta = \pm\frac{\sqrt{3}}{2}$$
$$\theta = \frac{\pi}{6}, \frac{5\pi}{6}$$

61.

$$\cos^2\theta - \sin^2\theta + \cos\theta = 0; 0 \le \theta \le \pi$$
$$\cos^2\theta - (1 - \cos^2\theta) + \cos\theta = 0$$
$$2\cos^2\theta + \cos\theta - 1 = 0$$
$$(2\cos\theta - 1)(\cos\theta + 1) = 0$$
$$2\cos\theta - 1 = 0 \text{ or } \cos\theta + 1 = 0$$
$$\cos\theta = \frac{1}{2} \text{ or } \cos\theta = -1$$
$$\theta = \frac{\pi}{3}, \pi$$

63. $E(t) = E_0 \sin\left(\frac{2\pi t}{28}\right)$

(a) $$E(21\cdot 365) = E_0\sin\left(\frac{2\pi}{28}\cdot 21\cdot 365\right) = -E_0$$

(be sure to calculate using radian mode)

(b) $P(t) = P_0\sin\left(\frac{2\pi t}{23}\right)$

$$P(21\cdot 365) = P_0\sin\left(\frac{2\pi}{23}\cdot 21\cdot 365\right) \approx 0.998P_0$$

(c) The maximum emotional level occurs when

$$\sin\left(\frac{2\pi t}{28}\right) = 1$$
$$\frac{2\pi t}{28} = \frac{\pi}{2} + 2\pi k,$$

where k is a non-negative integer

$$t = 7 + 28k$$

The maximum physical level occurs when

$$\sin\left(\frac{2\pi t}{23}\right) = 1$$
$$\frac{2\pi t}{23} = \frac{\pi}{2} + 2\pi k$$
$$t = \frac{23}{4} + 23k$$

For both levels to be maximum simultaneously, need $7 + 28k = \frac{23}{4} + 23k$

$$5k = -\frac{5}{4}, \text{ or } k = -\frac{1}{4}$$

Since k can only be a non-negative integer, both maximums cannot occur simultaneously.

65. $P(t) = A + B \cos kt$

(a) When $t = 0$, $P(0) = 130$ so

$$130 = A + B$$
$$B = 130 - A$$

and

$$P(t) = A + (130 - A)\cos kt$$

The function is minimized when $kt = \pi$, so $P\left(\frac{\pi}{k}\right) = 82$ and

$$82 = A + (130 - A)(-1)$$

or, $A = 106$, $B = 24$, and

$$P(t) = 106 + 24 \cos kt$$

Since the pulse rate is 79 heartbeats per minute and the maximum blood pressure occurs every 2π radians,

$$kt = 2\pi$$
$$t = \frac{2\pi}{k}$$
$$\frac{79}{60} = \frac{1}{\frac{2\pi}{k}}$$
$$k = \frac{79\pi}{30} \approx 8.2729$$

(b)

$$P(60) = 106 + 24\cos\left(\frac{79\pi}{30} \cdot 60\right)$$
$$= 130$$

67.

$$\sin\left(\frac{\pi}{2} - \theta\right) = \sin\left(\frac{\pi}{2} + (-\theta)\right)$$
$$= \sin\frac{\pi}{2}\cos(-\theta) + \cos\frac{\pi}{2}\sin(-\theta)$$
$$= \sin\frac{\pi}{2}\cos\theta + cos\frac{\pi}{2} - \sin\theta$$
$$= (1)\cos\theta - 0 = \cos\theta$$

69. On a circle of radius 1 centered about the origin, an angle θ corresponds to a point on the circle with coordinates $(x, y) = (\cos\theta, \sin\theta)$. For $0 \le \theta < 2\pi$, this correspondence is one-to-one. Need to show that for an angle θ with corresponding coordinates (x_0, y_0), the coordinates corresponding to the angle $\frac{\pi}{2} - \theta$ must be (y_0, x_0). Considering quadrant by quadrant,

Quadrant I

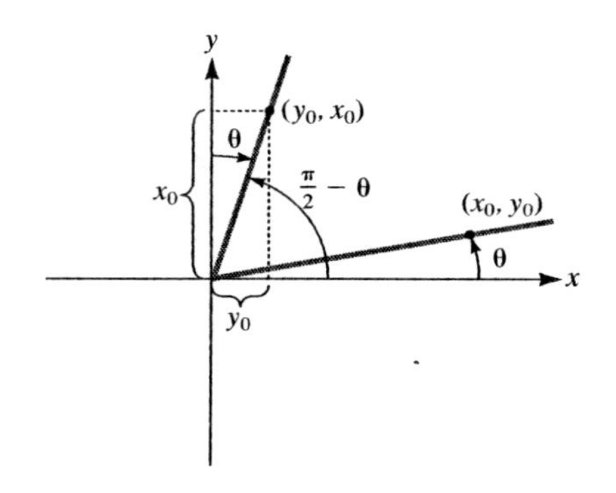

Quadrant II

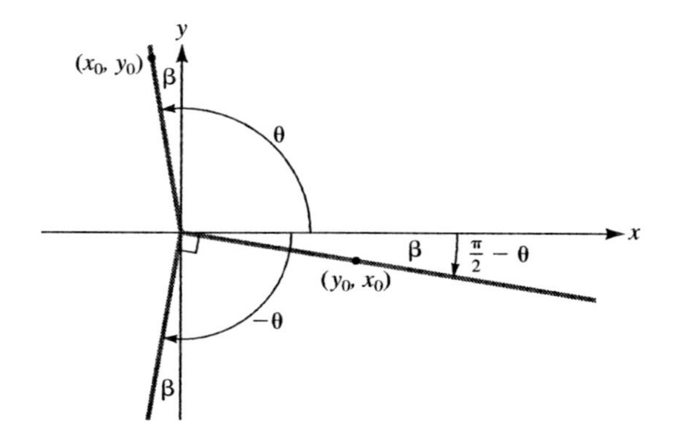

Quadrant III

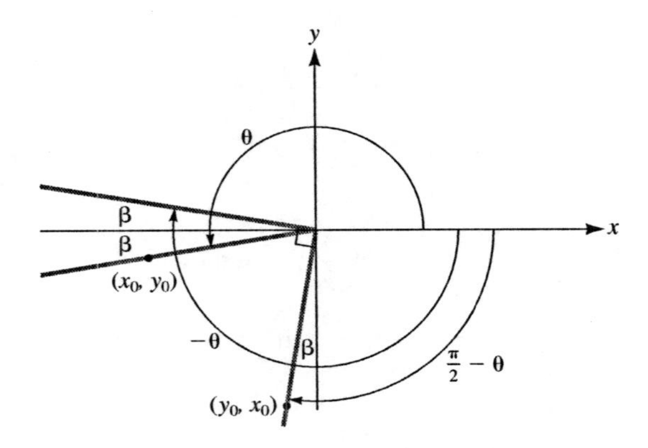

Quadrant IV

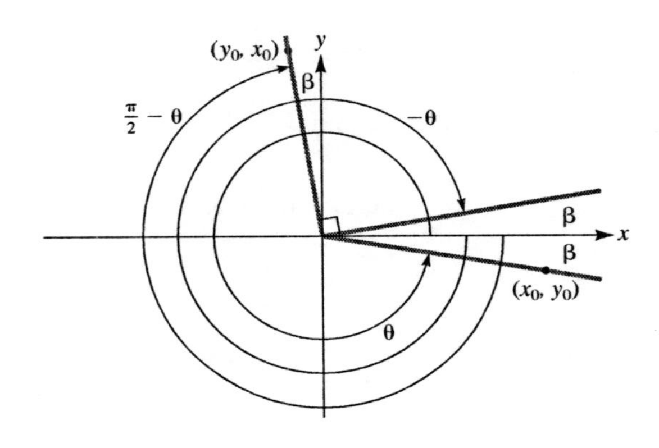

11.2 Differentiation and Integration of Trigonometric Functions

1.
$$f(t) = \sin 3t$$
$$f'(t) = \cos 3t \cdot \frac{d}{dt}(3t) = 3\cos 3t$$

3.
$$f(t) = \sin(1-2t)$$
$$f'(t) = \cos(1-2t) \cdot \frac{d}{dt}(1-2t) = -2\cos(1-2t)$$

5.
$$f(t) = \cos(t^3+1)$$
$$f'(t) = -\sin(t^3+1) \cdot \frac{d}{dt}(t^3+1) = -3t^2\sin(t^3+1)$$

7.
$$f(t) = \cos^2\left(\frac{\pi}{2}-t\right) = \left[\cos\left(\frac{\pi}{2}-t\right)\right]^2$$
$$f'(t) = 2\left[\cos\left(\frac{\pi}{2}-t\right)\right]\frac{d}{dt}\left[\cos\left(\frac{\pi}{2}-t\right)\right]$$
$$= 2\cos\left(\frac{\pi}{2}-t\right)\cdot -\sin\left(\frac{\pi}{2}-t\right)\cdot\frac{d}{dt}(-t)$$
$$= 2\cos\left(\frac{\pi}{2}-t\right)\sin\left(\frac{\pi}{2}-t\right)$$

9.
$$f(x) = \cos(1+3x)^2$$
$$f'(x) = -\sin(1+3x)^2 \cdot \frac{d}{dt}(1+3x)^2$$
$$= \left[-\sin(1+3x)^2\right][2(1+3x)(3)]$$
$$= -6(1+3x)\sin(1+3x)^2$$

11. $f(u) = e^{-u/2}\cos 2\pi u$
$$f'(u) = e^{-u/2}\cdot\frac{d}{du}(\cos 2\pi u) + \cos 2\pi u\cdot\frac{d}{du}\left(e^{-u/2}\right)$$
$$f'(u) = e^{-u/2}\cdot -\sin 2\pi u\frac{d}{du}(2\pi u) + \cos 2\pi u\cdot -\frac{1}{2}e^{-u/2}$$
$$= -2\pi e^{-u/2}\sin 2\pi u - \frac{1}{2}e^{-u/2}\cos 2\pi u.$$
$$= e^{-u/2}\left[-2\pi\sin 2\pi u - \frac{1}{2}\cos 2\pi u\right]$$

13.
$$f(t) = \frac{\sin t}{1+\sin t}$$
$$f'(t) = \frac{(1+\sin t)\frac{d}{dt}(\sin t) - (\sin t)\frac{d}{dt}(1+\sin t)}{(1+\sin t)^2}$$
$$f'(t) = \frac{(1+\sin t)(\cos t) - (\sin t)(\cos t)}{(1+\sin t)^2}$$
$$f'(t) = \frac{\cos t}{(1+\sin t)^2}$$

15.
$$f(t) = \tan(1 - t^3)$$
$$f'(t) = \sec^2(1 - t^3)\frac{d}{dt}(1 \cdot t^3)$$
$$= -3t^2 \sec^2(1 - t^3)$$

17. $f(t) = \sec\left(\frac{\pi}{2} - 2\pi t\right)$
$$f'(t) = \sec\left(\frac{\pi}{2} - 2\pi t\right)\tan\left(\frac{\pi}{2} - 2\pi t\right) \cdot \frac{d}{dt}\left(\frac{\pi}{2} - 2\pi t\right)$$
$$f'(t) = -2\pi \sec\left(\frac{\pi}{2} - 2\pi t\right)\tan\left(\frac{\pi}{2} - 2\pi t\right)$$

19.
$$f(t) = \ln \sin^2 t$$
$$f'(t) = \frac{1}{\sin^2 t} \cdot \frac{d}{dt}(\sin^2 t)$$
$$= \frac{1}{\sin^2 t} \cdot \frac{d}{dt}\left[(\sin t)^2\right]$$
$$= \frac{1}{\sin^2 t} \cdot 2(\sin t)\frac{d}{dt}(\sin t)$$
$$= \frac{1}{\sin^2 t} \cdot 2 \sin t \cos t = 2 \cot t$$

21. $\int \sin\left(\frac{t}{2}\right) dt$

Using substitution with $u = \frac{t}{2}$ and $du = \frac{1}{2}\, dt$, or $2\, du = dt$
$$= 2\int \sin u\, du = 2(-\cos u) + C$$
$$= -2\cos\left(\frac{t}{2}\right) + C$$

23. $\int x \cos(1 - 3x^2)\, dx$

Using substitution with $u = 1 - 3x^2$ and $du = -6x\, dx$, or $-\frac{1}{6}\, du = x\, dx$,
$$= -\frac{1}{6}\int \cos u\, du = -\frac{1}{6}\sin u + C$$
$$= -\frac{1}{6}\sin(1 - 3x^2) + C$$

25. $\int x \cos(2x)\, dx$

Need to use integration by parts using $u = x$ and $dV = \cos(2x)\, dx$. To integrate
$$\int dV = \int \cos(2x)\, dx$$
use substitution with $u = 2x$ and $du = 2\, dx$, or $\frac{1}{2}\, du = dx$
$$\int dV = \frac{1}{2}\int \cos u\, du$$
$$= \frac{1}{2}\sin u = \frac{1}{2}\sin(2x)$$
So,
$$\int x \cos(2x)\, dx = \frac{x}{2}\sin(2x) - \frac{1}{2}\int \sin(2x)\, dx$$
Using substitution again, with $u = 2x$ and $\frac{1}{2}\, du = dx$
$$= \frac{x}{2}\sin(2x) - \frac{1}{4}\int \sin u\, du$$
$$= \frac{x}{2}\sin(2x) + \frac{1}{4}\cos(2x) + C$$

27. $\int \sin x \cos x\, dx$

Using substitution with $u = \sin x$ and $du = cosx\, dx$,
$$= \int u\, du = \frac{u^2}{2} + C = \frac{1}{2}\sin^2 x + C$$

29. $\int x \sin x\, dx$

Using integrfation by parts with $u = x$ and $dV = \sin x\, dx$
$$= -x\cos x - \int -\cos x\, dx$$
$$= -x\cos x + \sin x + C$$

31. $f(t) = 2 + 1.5\sin\left[\frac{\pi}{2}(t - 1)\right]$

period: $\frac{2\pi}{p} = \frac{\pi}{2}$, or $p = 4$
amplitude: $b = 1.5$
phase shift: $d = 1$
Vertical shift: $a = 2$

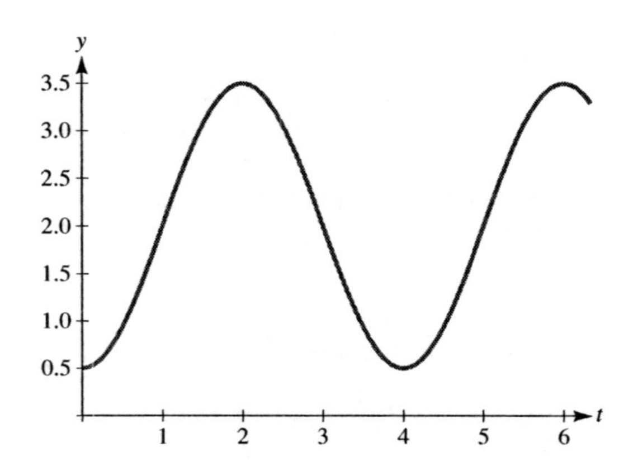

33. $f(t) = 3 - \cos\left[2\left(t - \frac{\pi}{2}\right)\right]$

period: $\dfrac{2\pi}{p} = 2$, or $p = \pi$
amplitude: $b = -1$
phase shift: $d = \frac{\pi}{2}$
Vertical shift: $a = 3$

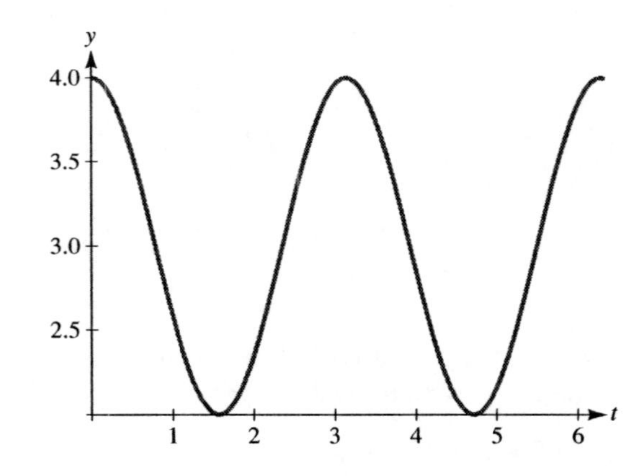

35. For $0 \le x \le \frac{1}{3}$, the graph of $y = \cos(\pi x)$ is above the graph of $y = \sin\left(\frac{\pi}{2}x\right)$. (Check the values of each function for x-values between 0 and $\frac{1}{3}$ to confirm.) So, the area is

$$\int_0^{1/3} \cos(\pi x) - \sin\left(\frac{\pi}{2}x\right) dx$$
$$= \int_0^{1/3} \cos(\pi x)\, dx - \int_0^{1/3} \sin\left(\frac{\pi}{2}x\right) dx$$

Using substitution for the first integral with $u = \pi x$ and $\frac{1}{\pi}\, du = dx$, and for the second integral with $u = \frac{\pi}{2}x$ and $\frac{2}{\pi}\, du = dx$,

$$= \frac{1}{\pi}\int_0^{\pi/3} \cos u\, du - \frac{2}{\pi}\int_0^{\pi/6} \sin u\, du$$
$$= \frac{1}{\pi}\left(\sin u\Big|_0^{\pi/3}\right) - \frac{2}{\pi}\left(-\cos u\Big|_0^{\pi/6}\right)$$
$$= \frac{1}{\pi}\left(\sin u\Big|_0^{\pi/3}\right) + \frac{2}{\pi}\left(\cos u\Big|_0^{\pi/6}\right)$$
$$= \frac{1}{\pi}\left(\frac{\sqrt{3}}{2} - 0\right) + \frac{2}{\pi}\left(\frac{\sqrt{3}}{2} - 1\right)$$
$$= \frac{3\sqrt{3} - 4}{2\pi}$$

37. For $0 \le x \le 1$, the graph of $y = \sin(\pi x)$ is above the graph of $y = x^2 - x$. (Check the values of each function for x-values between 0 and 1 to confirm.) So, the area is

$$\int_0^1 \left[\sin(\pi x) - (x^2 - x)\right] dx$$
$$= \int_0^1 \sin(\pi x) dx - \int_0^1 (x^2 - x) dx$$

Using substitution for the first integral with $u = \pi x$ and $\frac{1}{\pi} du = dx$,

$$= \frac{1}{\pi}\int_0^{\pi} \sin u\, du - \int_0^1 (x^2 - x) dx$$
$$= \frac{1}{\pi}\left(-\cos u\Big|_0^{\pi}\right) - \left(\frac{x^3}{3} - \frac{x^2}{2}\right)\Big|_0^1$$
$$= \frac{1}{\pi}(1 + 1) - \left[\left(\frac{1}{3} - \frac{1}{2}\right) - 0\right]$$
$$= \frac{2}{\pi} + \frac{1}{6}$$

39. Using $\int \pi r^2 dr$ with $r = \sqrt{\sin x}$,
Volume of solid

$$= \int_0^{\pi} \pi \left(\sqrt{\sin x}\right)^2 dx$$
$$= \pi \int_0^{\pi} \sin x \, dx$$
$$= \pi \left(-\cos x \Big|_0^{\pi}\right)$$
$$= \pi(1+1) = 2\pi$$

41. Using $\int \pi r^2 dr$ with $r = \sec x$,
Volume of solid

$$= \int_{-\pi/3}^{\pi/3} \pi(\sec x)^2 dx$$
$$= \int_{-\pi/3}^{\pi/3} \sec^2 x \, dx$$
$$= \pi \left(\tan x \Big|_{-\pi/3}^{\pi/3}\right)$$
$$= \pi\left(\sqrt{3}+\sqrt{3}\right) = 2\sqrt{3}\pi$$

43. $\dfrac{dy}{dt} = \dfrac{\sin(2t)}{y+1}$
Separating the variables,

$$\int y+1 \, dy = \int \sin(2t) \, dt$$

Using substitution for the first integral with $u = y+1$ and $du = dy$, and for the second integral with $w = 2t$ and $\frac{1}{2} dw = dt$,

$$\int u \, du = \frac{1}{2}\int \sin w \, dw$$
$$\frac{u^2}{2} = \frac{1}{2}(-\cos w) + C$$
$$\frac{(y+1)^2}{2} = -\frac{1}{2}\cos(2t) + C_1$$
$$(y+1)^2 = -\cos(2t) + 2C_1$$
$$(y+1)^2 = -\cos(2t) + C$$

When $t = 0$, $y = 1$ so

$$(1+1)^2 = -\cos 0 + C$$
$$4 = -1 + C, \text{ or } C = 5$$

So,

$$(y+1)^2 = -\cos(2t) + 5$$
$$y + 1 = \sqrt{5 - \cos(2t)}$$
$$y = -1 + \sqrt{5 - \cos(2t)}$$

(rejecting the negative solution since $y = 1$ when $t = 0$)

45. $\dfrac{dy}{dx} = \dfrac{\sin x}{\cos y}$
Separating the variables,

$$\int \cos y \, dy = \int \sin x \, dx$$
$$\sin y = -\cos x + C$$

When $x = \pi$, $y = 0$ so
$0 = 1 + C$, or $C = -1$
So,
$\sin y = -\cos x - 1$

47.

$$T(t) = 60 + 12\sin\left(\frac{\pi t}{10} - \frac{5}{6}\right)$$
$$\frac{dT}{dt} = 12\left[\cos\left(\frac{\pi t}{10} - \frac{5}{6}\right)\right]\left[\frac{\pi}{10}\right]$$
$$\frac{dT}{dt} = \frac{6\pi}{5}\cos\left(\frac{\pi t}{10} - \frac{5}{6}\right)$$

At noon,

$$\frac{dT}{dt} = \frac{6\pi}{5}\cos\left(\frac{\pi(12)}{10} - \frac{5}{6}\right)$$
$$\approx -3.69$$

So, the temperature is getting cooler, decreasing at a rate of approximately 3.69 degrees per hour.

49. **(a)** Press [y=]. Enter $50\sin(0.3x+1) + 10\cos(0.6x+2)$ for $y_1 =$.
Use window dimensions $[0, 3\pi]\frac{\pi}{2}$ by $[0, 45]10$. Press [graph]. To find the x-intercept, use the zero function under the calc menu to find the x-intercept at $x \approx 7.8$. To find the y-intercept, use the value function under the calc menu. Enter $x = 0$ to find the y-intercept at $y \approx 38$.

(b) To find the area of the garden, we will need $\int_0^{7.8} y_1$. Use the $\int f(x)\,dx$ function under the calc menu with $x = 0$ as the lower limit and

$x = 7.8$ as the upper limit. We see the area is about 244.73.

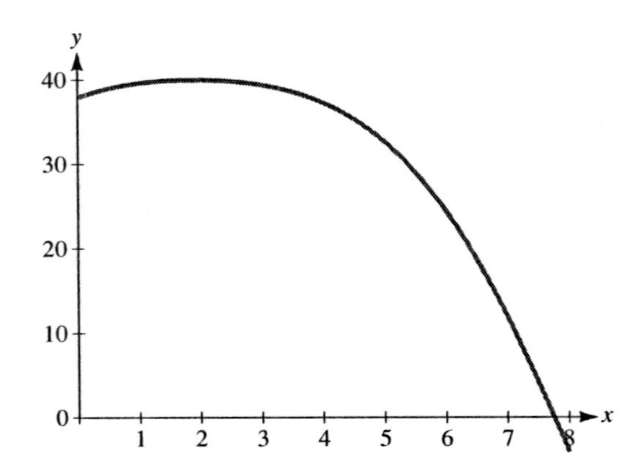

51. $\dfrac{dB}{dt} = \sqrt{t} + \sin\left(\dfrac{\pi t}{4}\right)$

Separating the variables

$$\int dB = \int t^{1/2} + \sin\left(\frac{\pi t}{4}\right) dt$$

Using substitution with $u = \dfrac{\pi t}{4}$ and $\dfrac{4}{\pi}\, du = dt$,

$$\int dB = \int t^{1/2}\, dt + \frac{4}{\pi}\int \sin u\, du$$

$$B(t) = \frac{2}{3}t^{3/2} - \frac{4}{\pi}\cos\left(\frac{\pi t}{4}\right) + C$$

When $t = 0$, $B(0) = B_o$ so

$$B_o = 0 - \frac{4}{\pi}(1) + C, \text{ or}$$

$$C = B_o + \frac{4}{\pi} \text{ and}$$

$$B(t) = \frac{2}{3}t^{3/2} - \frac{4}{\pi}\cos\left(\frac{\pi t}{4}\right) + B_o + \frac{4}{\pi}$$

When $t = 24$,

$$\begin{aligned} B(24) &= \frac{2}{3}(24)^{3/2} - \frac{4}{\pi}\cos\left(\frac{\pi(24)}{4}\right) + B_o + \frac{4}{\pi} \\ &= \frac{2}{3}\left(2\sqrt{6}\right)^3 - \frac{4}{\pi}(1) + B_o + \frac{4}{\pi} \\ &= 32\sqrt{6} + B_o \end{aligned}$$

53. $T(t) = 98.1 + 0.6\cos\left[\frac{\pi(t-15)}{24}\right]$

(a) To find the absolute maximum and absolute minimum on the interval $0 \le t \le 24$, first find the relative minimum(s) and maximum(s), if any.

$$\begin{aligned} T'(t) &= 0 + (0.6) - \sin\left[\frac{\pi(t-15)}{24}\right] \cdot \frac{\pi}{24} \\ &= -\frac{\pi}{40}\sin\left[\frac{\pi(t-15)}{24}\right] \end{aligned}$$

$T'(t) = 0$ when

$$\sin\left[\frac{\pi(t-15)}{24}\right] = 0$$

$$\frac{\pi(t-15)}{24} = 0 + \pi k$$

where k is any integer.
Since $0 \le t \le 24$, the only solution is $t = 15$. The absolute maximum and absolute minimum must occur among $T(0)$, $T(15)$ and $T(24)$.

$$T(0) \approx 97.87$$
$$T(15) = 98.7$$
$$T(24) \approx 98.33$$

So, the maximum temperature is 98.7°and occurs when $t = 15$, or 3 p.m. The minimum temperature is approximately 97.87°and occurs when $t = 0$, or at midnight.

(b) At noon, when $t = 12$,

$$\begin{aligned} T'(12) &= -\frac{\pi}{40}\sin\left[\frac{\pi(12-15)}{24}\right] \\ &= -\frac{\pi}{40}\sin\left(-\frac{\pi}{8}\right) \\ &\approx 0.0301 \end{aligned}$$

His body temperature is changing at the rate of approximately 0.03 degrees per hour.

(c) To find his maximum rate, find the critical point(s) of the second derivative.

$$T''(t) = -\frac{\pi}{40}\cos\left[\frac{\pi(t-15)}{24}\right]\left(\frac{\pi}{24}\right)$$

$T''(t) = 0$ when

$$\cos\left[\frac{\pi(t-15)}{24}\right] = 0$$

$$\frac{\pi(t-15)}{24} = \frac{\pi}{2} + k\pi$$

where k is any integer.
Since $0 \leq t \leq 24$, the only solution is $t = 3$. So, his temperature is increasing most rapidly at 3 a.m.

55. (a) The rate of change of sales is the derivative, or

$$S'(t) = 0 + 11.2\left[\cos\left(\frac{\pi t}{5}\right)\right]\left(\frac{\pi}{5}\right) + 2.3\left[-\sin\left(\frac{\pi t}{3}\right)\right]\left(\frac{\pi}{3}\right)$$

$$S'(t) = \frac{11.2\pi}{5}\cos\left(\frac{\pi t}{5}\right) - \frac{2.3\pi}{3}\sin\left(\frac{\pi t}{3}\right)$$

When $t = 6$,

$$S'(6) = \frac{11.2\pi}{5}\cos\left(\frac{\pi \cdot 6}{5}\right) - \frac{2.3\pi}{3}\sin\left(\frac{\pi \cdot 6}{3}\right)$$

$s'(6) \approx -5.69$

Sales will be decreasing at a rate of approximately 5,693 units per month.

(b)

$$S_{AV} = \frac{1}{12-0}\int_0^{12} 27.5 + 11.2\sin\left(\frac{\pi t}{5}\right) + 2.3\cos\left(\frac{\pi t}{3}\right)dt$$

$$= \frac{1}{12}\left[27.5t - \frac{(11.2)(5)}{\pi}\cos\left(\frac{\pi t}{5}\right) + \frac{(2.3)(3)}{\pi}\sin\left(\frac{\pi t}{3}\right)\right]\Big|_0^{12}$$

$$\approx \frac{1}{12}[(330 - 5.5083 + 0) - (0 - 17.8254 + 0)]$$

$$\approx 28.526$$

or approximately 28,526 units per month.

57. $\tan\theta = \dfrac{12}{x}$

$$(\sec^2\theta)\left(\frac{d\theta}{dt}\right) = -\frac{12}{x^2}\cdot\frac{dx}{dt}$$

When $x = 9$, the hypotenuse $= \sqrt{(12)^2 + (9)^2}$ and $\sec\theta = \dfrac{\text{hypotenuse}}{\text{adjacent}} = \dfrac{15}{9}$, so

$$\frac{d\theta}{dt} = \frac{-\frac{12}{(9)^2}\cdot -5}{\left(\frac{15}{9}\right)^2} = \frac{4}{15}\text{ radians/ sec}$$

59. Let x be the horizontal distance between the observer and the plane. Then,

$$\tan\theta = \frac{3}{x}$$

$$(\sec^2\theta)\left(\frac{d\theta}{dt}\right) = -\frac{3}{x^2}\cdot\frac{dx}{dt}$$

When $x = 4$, the hypotenuse $= \sqrt{(3)^2 + (4)^2}$ and

$$\sec\theta = \frac{\text{hypotenuse}}{\text{adjacent}} = \frac{5}{4},\text{ so}$$

$$\frac{d\theta}{dt} = \frac{-\frac{3}{(4)^2}\cdot -500}{\left(\frac{5}{4}\right)^2} = 60\text{ radians/hr}$$

61. By definition,

$$\frac{d}{dt}(\cos t) = \lim_{h\to 0}\frac{\cos(t+h) - \cos t}{h}$$

Using the addition formula for cosine,

$$= \lim_{h\to 0}\frac{\cos t\cos h - \sin t\sin h - \cos t}{h}$$

$$= \lim_{h\to 0}\frac{\cos t(\cos h - 1) - \sin t\sin h}{h}$$

$$= \lim_{h\to 0}\cos t\left(\frac{\cos h - 1}{h}\right) - \sin t\left(\frac{\sin h}{h}\right)$$

Since $\lim_{h\to 0}\dfrac{\cos h - 1}{h} = 0$ and $\lim_{h\to 0}\dfrac{\sin h}{h} = 1$,

$$\frac{d}{dt}(\cos t) = -\sin t$$

63. Press [y=]. Enter $(\cos(x) - 1)/x$ as $y_1 =$. Use window dimensions $[-\pi, \pi]\pi/4$ by $[-1, 1]0.5$

and press [graph]. To evaluate $\lim_{h\to 0} \frac{\cos x - 1}{x}$, we will approach $x = 0$ from both the left and the right. Use the [trace] and zoom-in feature to see $\lim_{h\to 0} \frac{\cos x - 1}{x} = 0.$

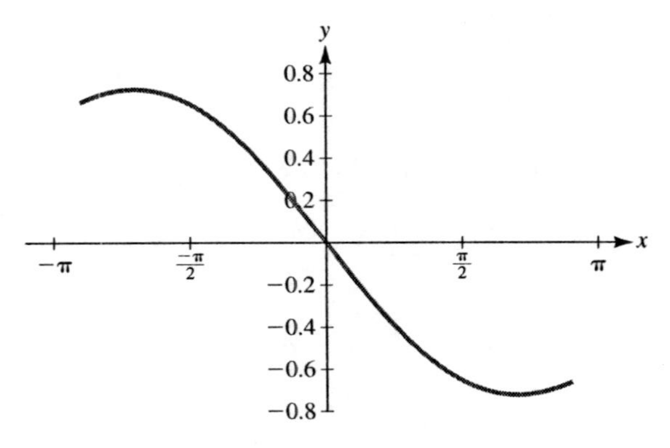

65. **(a)** When $f(t) = a + b\sin\left[\frac{2\pi}{p}(t-d)\right]$,

$$f(t+p) = a + b\sin\left[\frac{2\pi}{p}(t+p-d)\right]$$
$$= a + b\sin\left[\frac{2\pi}{p}(t-d) + 2\pi\right]$$

Using the addition formula for sine,

$$= a + b\left[\sin\frac{2\pi}{p}(t-d)\cos 2\pi + \sin 2\pi\cos\frac{2\pi}{p}(t-d)\right]$$

Since $\sin 2\pi = 0$ and $\cos 2\pi = 1$,

$$= a + b\sin\left[\frac{2\pi}{p}(t-d)\right] = f(t)$$

(b) When $f(t) = a + b\cos\left[\frac{2\pi}{p}(t-d)\right]$,

$$f(t+p) = a + b\cos\left[\frac{2\pi}{p}(t+p-d)\right]$$
$$= a + b\cos\left[\frac{2\pi}{p}(t-d) + 2\pi\right]$$

Using the addition formula for cosine,

$$= a + b\left[\cos\frac{2\pi}{p}(t-d)\cos 2\pi - \sin\frac{2\pi}{p}(t-d)\sin 2\pi\right]$$

Since $\cos 2\pi = 1$ and $\sin 2\pi = 0$,

$$= a + b\cos\left[\frac{2\pi}{p}(t-d)\right] = f(t)$$

11.3 Additional Applications Involving Trigonometric Functions

1. Using $CS = \int_0^{q_o} D(q)\,dq - p_o q_0$ and $p_o = D(q_o)$,

$$P_0 = D(3) = 10 - 2(3) + \sin\frac{\pi(3)}{5} = 4$$
$$CS \approx \int_0^3 \left[10 - 2q + \sin\left(\frac{\pi q}{3}\right)\right] dq - (4)(3)$$
$$\approx \int_0^3 (10 - 2q)\,dq + \int_0^3 \sin\left(\frac{\pi q}{3}\right) dq - 12$$

Using substitution for the second integral with $u = \pi q/3$ and $\frac{3}{\pi}\,du = dq$,

$$\approx \int_0^3 (10-2q)dq + \frac{3}{\pi}\int_0^\pi \sin u\,du - 12$$
$$\approx (10q - q^2)\Big|_0^3 + \frac{3}{\pi}\left(-\cos u\Big|_0^\pi\right) - 12$$
$$\approx (10q - q^2)\Big|_0^3 - \frac{3}{\pi}\left(\cos u\Big|_0^\pi\right) - 12$$
$$\approx [(30-9) - 0] - \frac{3}{\pi}(-1-1) - 12$$
$$\approx \$10.91$$

Note: Be sure your calculator is in radian mode.

3. Using $CS = \int_0^{q_o} D(q)\,dq - p_o q_o$ and $P_o = D(q_o)$,

$$p_0 = D(2) = 15 - 5(2) + \sin(2\pi)\cos(2\pi)$$
$$= 5$$

$$CS = \int_0^2 [15 - 5q + \sin(\pi q)\cos(\pi q)]\,dq - (2)(5)$$
$$= \int_0^2 (15 - 5q)dq + \int_0^2 \sin(\pi q)\cos(\pi q)dq - 10$$

Using substitution for the second integral with $u = \sin \pi q$ and $\frac{1}{\pi}du = \cos \pi q\, dq$,

$$= \int_0^2 (15 - 5q)dq + \int_0^0 \sin u\, du - 10$$
$$= \left(15q - \frac{5q^2}{2}\right)\Big|_0^2 + 0 - 10$$
$$= [(30 - 10) - 0] - 10 = \$10$$

5. Using $PS = p_0q_0 - \int_0^{q_0} S(q)dq$ and $p_0 = S(q_0)$,

$$p_0 = S(4) = 3 + 2(4) + 0.5\cos(4\pi)$$
$$= 11.50$$

$$PS = (11.50)(4) - \int_0^4 [3 + 2q + 0.5\cos(\pi q)]\,dq$$
$$= 46 - \left[\int_0^4 (3 + 2q)dq + 0.5\int_0^4 \cos(\pi q)dq\right]$$

Using substitution for the second integral with $u = \pi q$ and $\frac{1}{\pi}du = dq$,

$$= 46 - \left[\int_0^4 (3 + 2q)dq + \frac{1}{2\pi}\int_0^{4\pi} \cos u\, du\right]$$
$$= 46 - \left[(3q + q^2)\Big|_0^4 + \frac{1}{2\pi}\left(\sin u\Big|_0^{4\pi}\right)\right]$$
$$= 46 - \left[(12 + 16) - 0 + \frac{1}{2\pi}(\sin 4\pi - \sin 0)\right]$$
$$= \$18$$

7. Using $PS = p_0q_0 - \int_0^{q_0} S(q)dq$ and $p_0 = S(q_0)$,

$$p_0 = S(3) = 1 + 5(3) + 2\sin\left(\frac{3\pi}{2}\right)\cos\left(\frac{3\pi}{2}\right)$$
$$= 16$$

$$PS = (16)(3) - \int_0^3 \left[1 + 5q + 2\sin\left(\frac{\pi q}{2}\right)\cos\left(\frac{\pi q}{2}\right)\right] dq$$
$$= 48 - \left[\int_0^3 (1 + 5q)dq + 2\int_0^3 \sin\left(\frac{\pi q}{2}\right)\cos\left(\frac{\pi q}{2}\right) dq\right]$$

Using substitution for the second integral with $u = \sin\left(\frac{\pi q}{2}\right)$ and $\frac{2}{\pi}dq = \cos\left(\frac{\pi q}{2}\right) dq$,

$$= 48 - \left[\int_0^3 (1 + 5q)dq + \frac{4}{\pi}\int_0^{-1} u\, du\right]$$
$$= 48 - \left[\int_0^3 (1 + 5q)dq - \frac{4}{\pi}\int_{-1}^{0} u\, du\right]$$
$$= 48 - \left[\left(q + \frac{5q^2}{2}\right)\Big|_0^3 - \frac{4}{\pi}\left(\frac{u^2}{2}\Big|_{-1}^0\right)\right]$$
$$= 48 - \left[\left(q + \frac{5q^2}{2}\right)\Big|_0^3 - \frac{2}{\pi}\left(u^2\Big|_{-1}^0\right)\right]$$
$$= 48 - \left[\left(3 + \frac{45}{2}\right) - 0 - \frac{2}{\pi}(0 - 1)\right]$$
$$= 48 - \left[3 + \frac{45}{2} + \frac{2}{\pi}\right] \approx \$21.86$$

Note: Be sure your calculator is in radian mode.

9. Using $FV = e^{rT}\int_0^T f(t)e^{-rt}dt$,

$$FV = e^{(0.04)(5)}\int_0^5 2{,}000\cos\left(\frac{t}{5}\right) e^{-0.04t}dt$$
$$= 2{,}000e^{0.2}\int_0^5 \cos\left(\frac{t}{5}\right) e^{-0.04t}dt$$

Using the formula for $\int e^{au}\cos(bu)du$,

$$= 2000e^{0.2}\left(\frac{e^{-0.04t}}{0.0016+0.04}\left[-0.04\cos\left(\frac{t}{5}\right)+0.2\sin\left(\frac{t}{5}\right)\right]\Bigg|_0^5\right)$$

$$= \frac{2000}{0.0416}e^{0.2}\left(e^{-0.2}[-0.04\cos 1+0.2\sin 1]\right.$$
$$\left.-e^0[-0.04\cos 0+0.2\sin 0]\right)$$

$$\approx \frac{2000}{0.0416}e^{0.2}\left(e^{-0.2}[-0.02161+0.16829]\right.$$
$$-[-0.04+0])$$

$$\approx \frac{2000}{0.0416}\left(0.14668+0.04e^{0.2}\right)$$
$$\approx \$9401$$

Using $PV = \int_0^T f(t)e^{-rt}dt$,

$$PV = \int_0^5 2000\cos\left(\frac{t}{5}\right)e^{-0.04t}dt$$

$$= 2000\int_0^5 \cos\left(\frac{t}{5}\right)e^{-0.04t}dt$$

Using the formula for $\int e^{au}\cos(bu)du$,

$$= 2000\left(\frac{e^{-0.04t}}{0.0016+0.04}\left[-0.04\cos\left(\frac{t}{5}\right)+0.2\sin\left(\frac{t}{5}\right)\right]\Bigg|_0^5\right)$$

$$= \frac{2000}{0.0416}\left(e^{-0.2}[-0.04\cos 1+0.2\sin 1]\right.$$
$$\left.-e^0[-0.04\cos 0+0.2\sin 0]\right)$$

$$= \frac{2000}{0.0416}\left(0.14668e^{-0.2}+0.04\right)$$
$$\approx \$7,697$$

11. Using $FV = e^{rT}\int_0^T f(t)e^{-rt}dt$,

$$FV = e^{(0.03)(3)}\int_0^3 3000\sin(\pi t)e^{-0.03t}dt$$

$$= 3000e^{0.09}\int_0^3 \sin(\pi t)e^{-0.03t}dt$$

Using the formula for $\int e^{au}\sin(bu)du$,

$$= 3000e^{0.09}\left(\frac{e^{-0.03t}}{0.0009+\pi^2}[-0.03\sin(\pi t)-\pi\cos(\pi t)]\Big|_0^3\right)$$

$$= \frac{3000}{0.0009+\pi^2}e^{0.09}\left(e^{-0.09}[0+\pi]-e^0[0-\pi]\right)$$

$$= \frac{3000}{0.0009+\pi^2}\left[\pi+\pi e^{0.09}\right]$$
$$\approx \$2,000$$

Using $PV = \int_0^T f(t)e^{-rt}dt$,

$$PV = \int_0^3 3000\sin(\pi t)e^{-0.03t}dt$$

$$= 3000\int_0^3 \sin(\pi t)e^{-0.03t}dt$$

Using the formula for $\int e^{au}\sin(bu)du$,

$$= 3000\left(\frac{e^{-0.03t}}{0.0009+\pi 2}[-0.03\sin(\pi t)-\pi\cos(\pi t)]\Big|_0^3\right)$$

$$= \frac{3000}{0.0009+\pi^2}\left(e^{-0.09}[0+\pi)-e^0[0-\pi]\right)$$

$$= \frac{3000}{0.0009+\pi^2}\left(\pi e^{-0.09}+\pi\right)$$
$$\approx \$1,828$$

13. $x(t) = \sin 2t + \cos 2t$

(a) Press [y=]. Enter $\sin(2x)+\cos(2x)$ for $y_1 =$. Note we using the variable x in place of t. Use window dimensions $[0, \pi/2]\pi/4$ by $[-1, 1.6]0.2$ Press [graph].

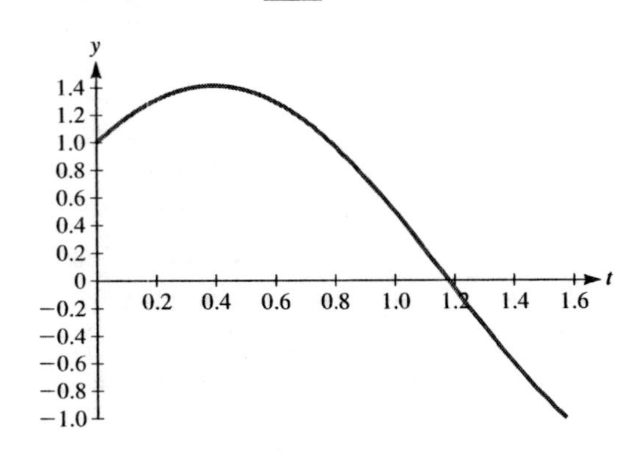

(b) To find the absolute max/min on $0 \le t \le \frac{\pi}{2}$, find the critical point(s) from the derivative.

$$x'(t) = 2\cos 2t - 2\sin 2t$$
$$0 = 2(\cos 2t - \sin 2t)$$
$$0 = \cos 2t - \sin 2t$$
$$\sin 2t = \cos 2t$$

For $0 \le t \le \frac{\pi}{2}$,
$2t = \frac{\pi}{4}$, or $t = \frac{\pi}{8}$
Comparing

$$x\left(\frac{\pi}{8}\right) = \sqrt{2}$$
$$x(0) = 1$$
$$x\left(\frac{\pi}{2}\right) = -1$$

So, the maximum occurs when $t = \frac{\pi}{8}$ and the minimum occurs when $t = \frac{\pi}{2}$.

15. $x(t) = 5\sin 3t + 2\cos t$

(a) Press [y=]. Enter $5\sin(3x) + 2\cos(x)$ for $y_1 =$. Note we are using the variable x in place of t. Use window dimensions $[0, \pi]\pi/6$ by $[-6, 8]2$. Press [graph].

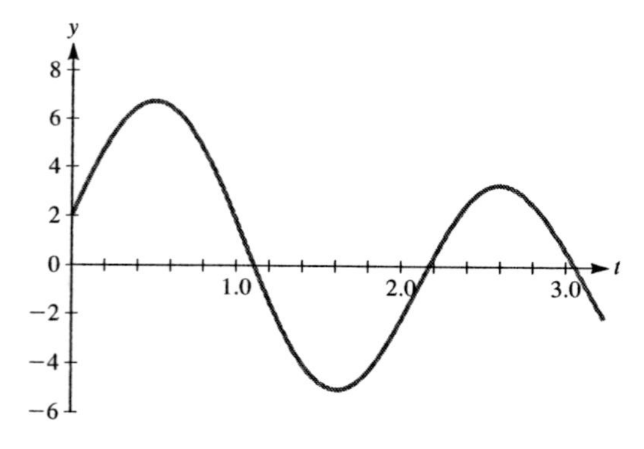

(b) To find the absolute max/min on $0 \le t \le \pi$, find the critical point(s) from the derivative.

$$x'(t) = 15\cos 3t - 2\sin t$$
$$0 = 15\cos 3t - 2\sin t$$

To find t for which x is maximized, use the maximum function under the calc menu with $x = 0.4$ as the left bound, $x = 0.6$ as the right bound, and $x = 0.5$ as the guess. We see that x is maximized for $t \approx 0.502$. Similarly, use the minimum function under the calc menu with $x = 1.5$ as the left bound, $x = 1.7$ as the right bound, and $x = 1.6$ as the guess. We see that x is minimized for $t \approx 1.615$.
Comparing

$$x(0.502) \approx 6.743$$
$$x(1.615) \approx -5.045$$
$$x(0) = 2$$
$$x(\pi) = -2$$

So, the maximum occurs when $t \approx 0.502$ and the minimum occurs when $t \approx 1.615$.

17. $x(t) = e^{-t}(\sin t + \cos t)$

(a) Press [y=]. Enter $e \wedge (-x) * (\sin(x) + \cos(x))$ for $y_1 =$. Note we are using the variable x in place of t. Use window dimensions $\left[\frac{\pi}{2}, \frac{3\pi}{2}\right]\frac{\pi}{12}$ by $[-0.04, 0.22]0.02$

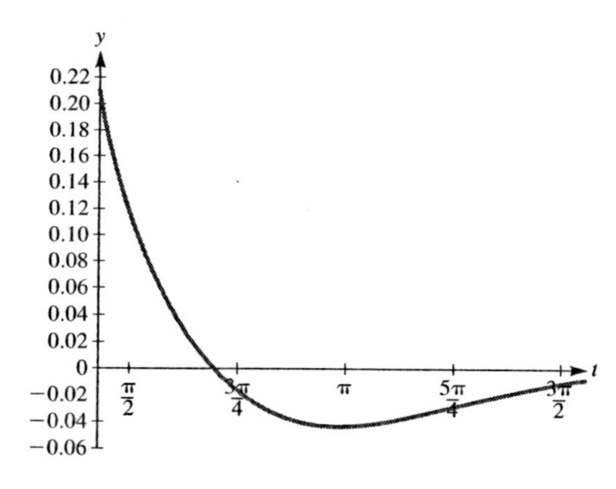

(b) To find the absolute max/min on $\frac{\pi}{2} \le t \le \frac{3\pi}{2}$, find the critical point(s) from the derivative.

$$x'(t) = e^{-t}(\cos t - \sin t) + (\sin t + \cos t)(-e^{-t})$$
$$0 = e^{-t}[\cos t - \sin t - \sin t - \cos t]$$
$$0 = e^{-t}(-2\sin t)$$
$$\sin t = 0, \text{ or } t = \pi$$

Comparing,

$$x(\pi) = -e^{-\pi} \approx -0.0432$$
$$x\left(\frac{\pi}{2}\right) = e^{-\pi/2} \approx 0.2079$$
$$x\left(\frac{3\pi}{2}\right) = -e^{-3\pi/2} \approx -0.0090$$

So, the maximum occurs when $t = \frac{\pi}{2}$ and the minimum occurs wshen $t = \pi$.

19. $f(t) = \cot t - \sqrt{2} \csc t$
To find the absolute max/min on $0 < t < \pi$, find the critical point(s) from the derivative.

$$f'(t) = -\csc^2 t - \sqrt{2}(-\cot t \csc t)$$
$$f'(t) = -\csc^2 t + \sqrt{2}(\cos t \csc^2 t)$$
$$0 = -\csc^2 t\left(1 - \sqrt{2}\cos t\right)$$
$$\csc t \neq 0$$
$$1 - \sqrt{2}\cos t = 0 \text{ when } t = \frac{\pi}{4}$$

Comparing,

$$f\left(\frac{\pi}{4}\right) = -1$$
$$\lim_{t\to 0} f(t) = \lim_{t\to\pi} f(t) = -\infty$$

So, the maximum value of the function is -1 and there is no minimum value.

21. $B(t) = 0.5 - 0.4\cos\left(\frac{\pi t}{1.75}\right)$
To find the maximum and minimum volumes, find the critical point(s) from the derivative.

$$B'(t) = 0.4\left[\sin\left(\frac{\pi t}{1.75}\right)\right]\left(\frac{\pi}{1.75}\right)$$
$$0 = \frac{0.4\pi}{1.75}\sin\left(\frac{\pi t}{1.75}\right)$$
$$\sin\left(\frac{\pi t}{1.75}\right) = 0$$
$$\frac{\pi t}{1.75} = k\pi, \text{ where } k \text{ is an integer}$$
$$t = 1.75k$$

When k is even,

$$B(t) = 0.5 - 0.4(1) = 0.1$$

When k is odd,

$$B(t) = 0.5 - 0.4(-1) = 0.9$$

(a) So, the maximum volume is 0.9 liters and occurs when k is an odd value which can be renamed as $2n + 1$, where n is an integer, or $t = (2n + 1)1.75$

(b) The minimum volume is 0.1 liters and occurs when k is an even value, which can be renamed as $2n$, where n is an integer, or $t = (2n)1.75$

(c)

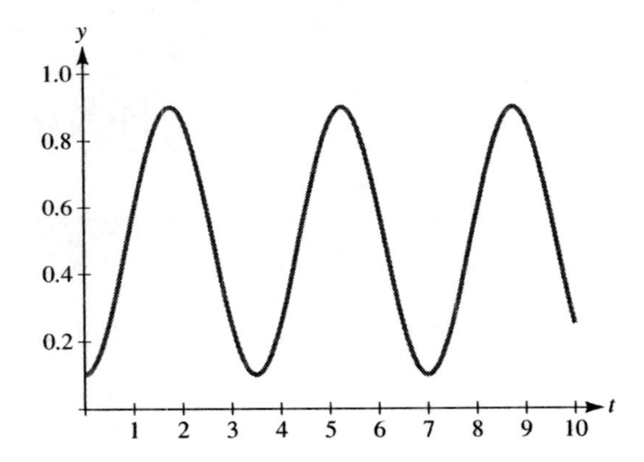

23. $\frac{dP}{dt} = (b + a\cos 2\pi t)P$
Separating the variables,

$$\int \frac{dP}{P} = \int (b + a\cos 2\pi t)\, dt$$
$$\ln|P| = bt + \frac{a}{2\pi}\sin 2\pi t + C_1$$
$$P = e^{bt + \frac{a}{2\pi}\sin 2\pi t} \cdot e^{C_1}$$
$$P(t) = Ce^{bt + \frac{a}{2\pi}\sin 2\pi t}$$

When $t = 0$, $P(0) = P_0$ so

$$P_0 = Ce^0, \text{ or } C = P_0$$

and $P(t) = P_0 e^{bt + \frac{a}{2\pi}\sin 2\pi t}$

25. **(a)** The equilibrium price occurs when $D(q) = S(q)$

$$9 - 6\sin\left(\frac{\pi q}{20}\right) = 5 + 2\sin\left(\frac{\pi q}{20}\right)$$

$$4 = 8\sin\left(\frac{\pi q}{20}\right)$$
$$0.5 = \sin\left(\frac{\pi q}{20}\right)$$
$$\frac{\pi q}{20} \approx 0.52360$$
$$q_e \approx 3.\bar{3} = \frac{10}{3} \text{ hundred units}$$

$$P_e = S\left(\frac{10}{3}\right) = \$6$$

(b) $$CS = \int_0^{10/3}\left[9 - 6\sin\left(\frac{\pi q}{20}\right)\right]dq - (6)\left(\frac{10}{3}\right)$$
$$= \int_0^{10/3} 9dq - 6\int_0^{10/3}\sin\left(\frac{\pi q}{20}\right)dq - 20$$

Using substitution for the second integral with $u = \frac{\pi q}{20}$ and $\frac{20}{\pi}du = dq$,

$$= \int_0^{10/3} 9\,dq - \frac{120}{\pi}\int_0^{\pi/6}\sin u\,du - 20$$
$$= 9q\Big|_0^{10/3} - \frac{120}{\pi}\left(-\cos u\Big|_0^{\pi/6}\right) - 20$$
$$= 9q\Big|_0^{10/3} + \frac{120}{\pi}\left(\cos u\Big|_0^{\pi/6}\right) - 20$$
$$= 30 + \frac{120}{\pi}\left(\frac{\sqrt{3}}{2} - 1\right) - 20 \approx 4.883$$

or $CS \approx 488$ units

$$PS = (6)\left(\frac{10}{3}\right) - \int_0^{10/3}\left[5 + 2\sin\left(\frac{\pi q}{20}\right)\right]dq$$
$$= 20 - \left[\int_0^{10/3} 5dq + 2\int_0^{10/3}\sin\left(\frac{\pi q}{20}\right)dq\right]$$

Using substitution for the second integral with $u = \frac{\pi q}{20}$ and $\frac{20}{\pi}du = dq$,

$$= 20 - \left[\int_0^{10/3} 5\,dq + \frac{40}{\pi}\int_0^{\pi/6}\sin u\,du\right]$$
$$= 20 - \left[5q\Big|_0^{10/3} + \frac{40}{\pi}\left(-\cos u\Big|_0^{\pi/6}\right)\right]$$
$$= 20 - \left[\frac{50}{3} - \frac{40}{\pi}\left(\cos u\Big|_0^{11/6}\right)\right]$$
$$= 20 - \left[\frac{50}{3} - \frac{40}{\pi}\left(\frac{\sqrt{3}}{2} - 1\right)\right] \approx 1.628$$

or $PS \approx 163$ units

27. Let x be the distance the bird flies over water. Then,

$$\cos\theta = \frac{S_1}{x}, \text{ or } x = \frac{S_1}{\cos\theta}$$

Let y be the distance on land opposite θ.Then,

$$\tan\theta = \frac{y}{s_1}, \text{ or } y = s_1\tan\theta$$

and the distance the bird flies over water is $s_2 - s_1\tan\theta$. The energy expended by the bird will be

$$E(\theta) = E_w\left(\frac{s_1}{\cos\theta}\right) + E_l(s_2 - s_1\tan\theta)$$
$$E(\theta) = E_w s_1(\cos\theta)^{-1} + E_l s_2 - E_l s_2\tan\theta$$

To minimize the energy expended, find the critical point(s) from the derivative.

$$E'(\theta) = -\frac{E_w s_1}{\cos^2\theta}(-\sin\theta) + 0 - E_l s_1\sec^2\theta$$
$$0 = E_w s_1\sec^2\theta\sin\theta - E_l s_1\sec^2\theta$$
$$0 = s_1\sec^2\theta(E_w\sin\theta - E_l)$$
$$E_w\sin\theta - E_l = 0$$
$$\sin\theta = \frac{E_l}{E_w}$$

29. Separate the area of the room into the areas of the matching triangular ends and the area of the rectangle.
Since $\sin\theta = \frac{h}{20}$, or $h = 20\sin\theta$, and $\cos\theta = \frac{b}{20}$, or $b = 20\cos\theta$, the areas are

$$2\left(\frac{1}{2}bh\right) = (20\cos\theta)(20\sin\theta)$$

$$lw = (20)(20\sin\theta)$$

The total area is

$$A(\theta) = 400\sin\theta\cos\theta + 400\sin\theta$$

To maximize the area for $0 \le \theta \le \pi$, find the critical point(s) from the derivative.

$$A'(\theta) = 400\,[(\sin\theta)(-\sin\theta) + (\cos\theta)(\cos\theta)] + 400\cos\theta$$

$$A'(\theta) = 400[\cos^2\theta - \sin^2\theta + \cos\theta]$$

$$0 = 400\left[\cos^2\theta - (1-\cos^2\theta) + \cos\theta\right]$$

$$0 = 400[2\cos^2\theta + \cos\theta - 1]$$

$$0 = 400(2\cos\theta - 1)(\cos\theta + 1)$$

$$\cos\theta = \frac{1}{2},\ \cos\theta = -1$$

$$\theta = \frac{\pi}{3},\ \pi$$

Comparing,

$$A(0) = 0$$

$$A\left(\frac{\pi}{3}\right) = 400\left(\frac{\sqrt{3}}{2}\right)\left(\frac{1}{2}\right) + 400\left(\frac{\sqrt{3}}{2}\right) = 300\sqrt{3}$$

$$A(\pi) = 0$$

So, the maximum area is $300\sqrt{3}$ square feet. The length of the fourth wall should be

$$20 + 2b = 20 + 2\left(20\cos\frac{\pi}{3}\right)$$

$$= 20 + 40\left(\frac{1}{2}\right)$$

$$= 40 \text{ feet}$$

31.

$$\frac{dP}{dt} = 0.03P + 0.004\cos 2\pi t$$

$$\frac{dP}{dt} - 0.03P = 0.004\cos 2\pi t$$

The integrating factor is

$$e^{\int -0.03dt} = e^{-0.03t}$$

$$P = e^{0.03t}\left(\int e^{-0.03t}\cdot 0.004\cos 2\pi t + C_1\right)$$

$$= e^{0.03t}\left(0.004\int e^{-0.03t}\cos 2\pi t\,dt + C_1\right)$$

$$= e^{0.03t}\left(0.004\left[\frac{e^{-0.03t}}{(-0.03)^2 + (2\pi)^2}(-0.03\cos 2\pi t + 2\pi\sin 2\pi t)\right] + C_2\right)$$

$$P(t) \approx \frac{0.004}{39.47932}(-0.03\cos 2\pi t + 2\pi\sin 2\pi t) + C_2 e^{0.03t}$$

When $t = 0$, $P(0) = 10{,}000$ so

$$10{,}000 \approx \frac{0.004}{39.47932}(-0.03) + C_2 e^0$$

or, $C_2 \approx 10{,}000$ and

$$P(10) \approx \frac{0.004}{39.47932}(-0.03\cos 20\pi + 2\pi\sin 20\pi) + 10{,}000e^{0.3}$$

$$\approx \frac{0.004}{39.47932}(-0.03) + 10{,}000(1.3499)$$

$$\approx 13{,}499 \text{ people}$$

33. **(a)** This corresponds to the period of V', which is $\dfrac{2\pi}{0.65} \approx 9.67$ seconds.

(b) The time of inhalation is $\frac{9.67}{2} = 4.835$, so

$$V(t) = \int_0^{4.835} 0.87\sin(0.65t)\,dt$$

$$= -\frac{0.87}{0.65}\left[\cos(0.65t)\Big|_0^{4.835}\right]$$

$$= -1.3385[-1 - 1] \approx 2.68 \text{ liters}$$

Yes. Note that integrating from $t = 0$ to $t = 9.67$ gives an answer of zero, as the volume inhaled and exhaled cancel out.

(c) The period of V' is still $\dfrac{2\pi}{0.65} \approx 9.67$ seconds.

$$\int_0^{4.835} 0.87t\sin(0.65t)\,dt$$

Using integration by parts with $u = 0.87t$ and $dV = \sin(0.65t)\,dt$

$$V(t) = (0.87t)\left[-\frac{1}{0.65}\cos(0.65t)\right]\Big|_0^{4.835}$$
$$-\int_0^{4.835} 0.87\left[-\frac{1}{0.65}\cos(0.65t)\right]dt$$
$$= -\frac{87}{65}t\cos(0.65t)\Big|_0^{4.835} + \frac{87}{65}\int_0^{4.835}\cos(0.65t)\,dt$$
$$= \left[-\frac{87}{65}t\cos(0.65t) + \frac{87}{65(0.65)}\sin(0.65t)\right]\Big|_0^{4.835}$$
$$= \frac{87}{65}\left[-t\cos(0.65t) + \frac{20}{13}\sin(0.65t)\right]\Big|_0^{4.835}$$
$$\approx \frac{87}{65}[(4.835+0)-(0+0)]$$
$$\approx 6.47 \text{ liters}$$

No, this time integrating from $t = 0$ to $t = 9.67$ does not give an answer of zero.

$$\int_0^{9.67} 0.87t\sin(0.65t)\,dt$$
$$= \frac{87}{65}\left[-t\cos(0.65t) + \frac{20}{13}\sin(0.65t)\right]\Big|_0^{9.67}$$
$$= \frac{87}{65}[(-9.67+0)-(0+0)] \approx -12.94$$

35. Let V represent the volume of the trough. Then $V =$ (area triangular end) (length of trough)
To find the height of the triangle, use

$$\cos\frac{\theta}{2} = \frac{h}{3}, \text{ or}$$
$$h = 3\cos\frac{\theta}{2}$$

To find the base of the triangle, use

$$\sin\frac{\theta}{2} = \frac{\frac{1}{2}b}{3}, \text{ or}$$
$$b = 6\sin\frac{\theta}{2}$$

So,

$$V(\theta) = \left[\frac{1}{2}\left(6\sin\frac{\theta}{2}\right)\left(3\cos\frac{\theta}{2}\right)\right](20)$$

Using the double-angle formula for sine,

$$\sin 2\left(\frac{\theta}{2}\right) = 2\sin\frac{\theta}{2}\cos\frac{\theta}{2}$$
$$\sin\theta = 2\sin\frac{\theta}{2}\cos\frac{\theta}{2}$$

and

$$V(\theta) = 90\sin\theta$$

To maximize V for $0 \le \theta \le \pi$, find the critical point(s) from the derivative.

$$V'(\theta) = 90\cos\theta$$
$$0 = 90\cos\theta, \text{ or}$$
$$\theta = \frac{\pi}{2}$$

Comparing,

$$V(0) = 0$$
$$V\left(\frac{\pi}{2}\right) = 90$$
$$V(\pi) = 0$$

So, the volume is maximized when $\theta = \frac{\pi}{2}$ radians.

37. Let x be the distance along the surface of the water opposite the angle at the source A. Then, the distance the light travels underwater is $\sqrt{a^2+x^2}$. Similarly, let $d - x$ be the distance along the surface of the water opposite the angle at observer B. Then, the distance the light travels in air is $\sqrt{b^2+(d-x)^2}$. Since $d = rt$, or $t = \frac{d}{r}$, the total time for the light to travel is

$$T(x) = \frac{\sqrt{a^2+x^2}}{V_1} + \frac{\sqrt{b^2+(d-x)^2}}{V_2}$$

To minimize the time, find the critical point(s) from the derivative.

$$T'(x) = \frac{1}{2v_1}(a^2+x^2)^{-1/2}(2x)$$
$$+\frac{1}{2v_2}\left[b^2+(d-x)^2\right]^{\frac{1}{2}} 2(d-x)(-1)$$
$$0 = \frac{x}{v_1(a^2+x^2)^{-1/2}} - \frac{d-x}{v_2[b^2+(d-x)^2]^{1/2}}$$

Note that, using alternate interior angles,

$$\sin\theta_1 = \frac{x}{\sqrt{a^2+x^2}}$$
$$\sin\theta_2 = \frac{d-x}{\sqrt{b^2+(d-x)^2}}$$

So,

$$\frac{\sin\theta_1}{v_1} = \frac{\sin\theta_2}{v_2}$$
$$\frac{\sin\theta_1}{\sin\theta_2} = \frac{v_1}{v_2}$$

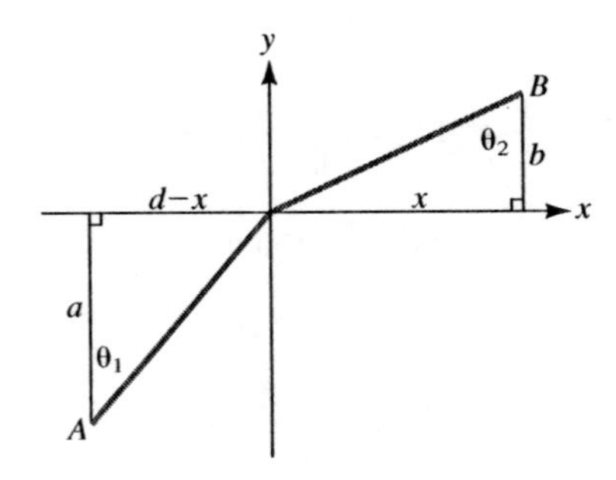

39. If $s_0 = 100$, $v_0 = 50$, and $\theta = 0$, then $x = (50\cos 0)t$ and $y = -\frac{1}{2}(32)t^2 + (50\sin 0)t + 100$. Simplifying,

$$x = 50t$$
$$y = -16t^2 + 100$$

From the first equation, $t = \frac{x}{50}$. Substituting, $y = -16\left(\frac{x}{50}\right)^2 + 100$ Press [y =]. Enter $-16 * (x/50)^2 + 100$ for $y_1 =$.
If $s_0 = 100$, $v_0 = 150$, and $\theta = \frac{\pi}{6}$, then $x = (150\cos\pi/6)t$ and $y = -16^2 + (150\sin\pi/6)t + 100$
Simplifying,

$$x = 75\sqrt{3}t$$
$$y = -16t^2 + 75t + 100$$

Substituting $t = \frac{x}{75\sqrt{3}}$ into the second equation, we obtain $y = -16\left(\frac{x}{75\sqrt{3}}\right)^2 + \frac{x}{\sqrt{3}} + 100$.
Enter this express for $y_2 =$.
If $s_0 = 0$, $v_0 = 50$, and $\theta = \pi/3$. then $x = \left(50\cos\frac{\pi}{3}\right)t$ and
$y = -16t^2 + \left(50\sin\frac{\pi}{3}\right)t$

Simplifying,

$$x = 25t$$
$$y = -16t^2 + 25\sqrt{3}t$$

Substituting $t = \frac{x}{25}$ into the second equation, we obtain $y = -16\left(\frac{x}{25}\right)^2 + \sqrt{3}x$.
Enter this expression for $y_3 =$.
If $s_0 = 0$, $v_0 = 200$ and $\theta = \frac{\pi}{8}$, then $x = \left(200\cos\frac{\pi}{8}\right)t$ and
$y = -16t^2 + \left(200\sin\frac{\pi}{8}\right)t$
Substituting $t = \frac{x}{200\cos\frac{\pi}{8}}$ into the second equation, we obtain $y = -16\left(\frac{x}{200\cos\frac{\pi}{8}}\right)^2 + \frac{\sin\frac{\pi}{8}x}{\cos\frac{\pi}{8}}$. Enter this expression for $y_4 =$.
Use window dimensions [0, 1000]200 by [0, 200]50.
Press [graph].

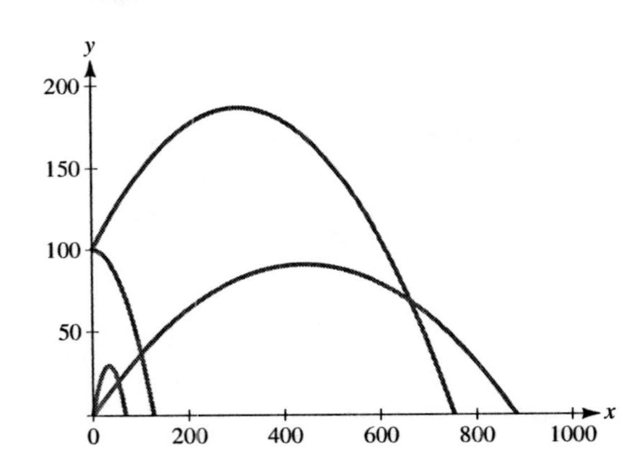

(a) Parabola

(b) Press [y =]. Highlight $y_1 =$ only. Press [graph]. Use the maximum function under the calc menu to find the maximum value is 100. Use the zero function under the calc menu to find the point of impact is 125. Repeat this procedure for $y_2 =$, $y_3 =$, and $y_4 =$. Given below is the table of results:

	y_1	y_2	y_3	y_4
highest point	100	187.89	29.30	91.53
point of impact	125	749.62	67.66	883.88

41. **(a)** Letting $s_0 = 3$, want to know when $y = 0$.

$$0 = -16t^2 + (40 \sin 12°)t + 3$$
$$16t^2 - 40 \sin 12°t - 3 = 0$$

$y = 0$ when

$$t = \frac{40 \sin 12° + \sqrt{(-40 \sin 12°)^2 + 4(16)(3)}}{2(16)}$$
$$t \approx 0.7649 \text{ seconds}$$

The x-value for this time will be the distance from the ramp.

$$x = (40 \cos 12°)(0.7649)$$
$$\approx 29.93 \text{ feet}$$

(b) Want $x = 39.93$ feet when $y = 0$. So,

$$39.93 = (v_0 \cos 12°)t$$
$$t = \frac{39.93}{v_0 \cos 12°}$$

and $y = 0$ when

$$0 = -16t^2 + (v_0 \sin 12°)t + 3$$
$$0 = -16\left(\frac{39.93}{v_0 \cos 12°}\right)^2 + (v_0 \sin 12°)\left(\frac{39.93}{v_0 \cos 12°}\right) + 3$$
$$16\left(\frac{39.93}{v_0 \cos 12°}\right)^2 = 39.93 \tan 12° + 3$$
$$\frac{26{,}663}{v_0^2} \approx 11.4874$$
$$v_0^2 \approx 2321$$
$$v_0 \approx 48.18 \text{ ft/sec}$$

43. (a)

$$F\left(\frac{\pi}{8}\right) = \frac{(0.4)(35)(\sec \frac{\pi}{8})}{1 + (0.4)\left(\tan \frac{\pi}{8}\right)} \approx 13.00$$
$$F\left(\frac{\pi}{6}\right) = \frac{(0.4)(35)(\sec \frac{\pi}{6})}{1 + (0.4)\left(\tan \frac{\pi}{6}\right)} \approx 13.13$$
$$F\left(\frac{\pi}{4}\right) = \frac{(0.4)(35)(\sec \frac{\pi}{4})}{1 + (0.4)\left(\tan \frac{\pi}{4}\right)} \approx 14.14$$
$$F(1) = \frac{(0.4)(35)(\sec 1)}{1 + (0.4)(\tan 1)} \approx 15.97$$
$$F(1.25) = \frac{(0.4)(35)(\sec 1.25)}{1 + (0.4)(\tan 1.25)} \approx 20.15$$
$$F(1.5) = \frac{(0.4)(35)(\sec 1.5)}{1 + (0.4)(\tan 1.5)} \approx 29.80$$

(b) As written, $F\left(\frac{\pi}{2}\right)$ is undefined, since $\tan \frac{\pi}{2}$ is undefined. However, rewriting as

$$F(\theta) = \frac{\mu W \left(\frac{1}{\cos \theta}\right)}{1 + \mu \left(\frac{\sin \theta}{\cos \theta}\right)} = \frac{\mu W}{\left[1 + \mu \left(\frac{\sin \theta}{\cos \theta}\right)\right] \cos \theta} = \frac{\mu W}{\cos \theta + \mu \sin \theta}$$

For $\theta = \frac{\pi}{2}$,

$$F\left(\frac{\pi}{2}\right) = \lim_{\theta \to \frac{\pi}{2}} \frac{\mu W}{\cos \theta + \mu \sin \theta} = \frac{\mu W}{0 + \mu \cdot 1} = W$$

(c) We want to sketch the graph of $F = \dfrac{(0.4)(35) \sec \theta}{1 + 0.4 \tan \theta}$ or

$$F = \frac{14 \sec \theta}{1 + 0.4 \tan \theta} = \frac{\frac{14}{\cos \theta}}{1 + 0.4\left(\frac{\sin \theta}{\cos \theta}\right)} = \frac{14}{\cos \theta + 0.4 \sin \theta}.$$

Press [y=]. Enter $14/(\cos(x) + 0.4 \sin(x))$ for$y_1 =$.
Use window dimensions $[0, \pi/2]\pi/12$ by $[0, 35]15$. Press [graph].

To find the angle at which F is minimized, use the minimum function under the calc menu with $x = 0.2$ as the left bound, $x = 0.5$ as the right bound and $x = 0.4$ as the guess. We see that F is minimized for $\theta \approx 0.38$.

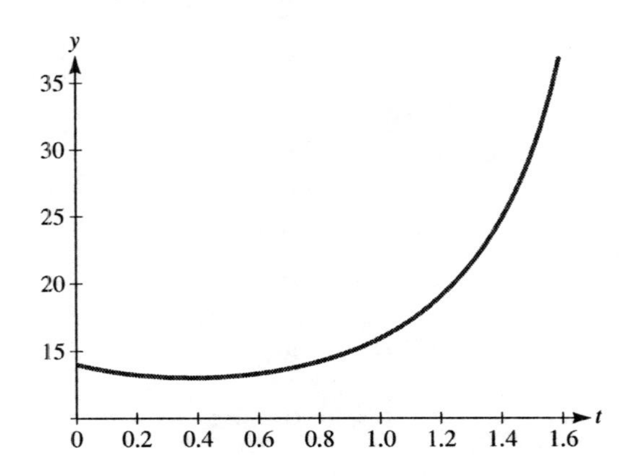

(d) To minimize F, find the critical point(s) from the derivative.

$$F'(\theta) = \frac{(1+\mu\tan\theta)(\mu W\sec\theta\tan\theta) - (\mu W\sec\theta)(\mu\sec^2\theta)}{(1+\mu\tan\theta)^2}$$

$$= \frac{\mu W\sec\theta\left[(1+\mu\tan\theta)\tan\theta - \mu\sec^2\theta\right]}{(1+\mu\tan\theta)^2}$$

$F'(\theta) = 0$ when, noting $\sec\theta \neq 0$,

$$(1+\mu\tan\theta)\tan\theta = \mu\sec^2\theta$$

$$\mu\tan^2\theta + \tan\theta - \mu\sec^2\theta = 0$$

$$\mu\tan^2\theta + \tan\theta - \mu(\tan^2\theta + 1) = 0$$

$$\tan\theta - \mu = 0$$

45. $$T(t) = 37.29 + 0.46\cos\left[\frac{\pi(t-16.37)}{12}\right]$$

(a) $$T'(t) = 0 + 0.46\left(\frac{\pi}{12}\right)\left(-\sin\left[\frac{\pi(t-16.37)}{12}\right]\right)$$

$$= -\frac{0.46\pi}{12}\sin\left[\frac{\pi(t-16.37)}{12}\right]$$

(b) To find the maximum body temperature, find the critical point(s) from the derivative. $T'(t) = 0$ when

$$\frac{\pi(t-16.37)}{12} = \pi k, \text{ where } k \text{ is any integer}$$

$$t - 16.37 = 12k$$

$$t = 12k + 16.37$$

Since $0 \le t \le 24$, $T'(t) = 0$ when $t = 4.37$ and when $t = 16.37$

Comparing

$$T(4.37) = 36.83\ °C$$

$$T(16.37) = 37.75\ °C$$

$$T(0) \approx 37.10\ °C$$

$$T(24) \approx 37.10\ °C$$

So, the maximum body temperature is 37.75 °C and it occurs when $t = 16.37$.

(c) From above, the minimum body temperature is 36.83 °C and it occurs when $t = 4.37$.

(d) $$T_{av} = \frac{1}{22-7}\int_7^{22} 37.29 + 0.46\cos\left[\frac{\pi(t-16.37)}{12}\right]dt$$

$$= \frac{1}{15}\left(37.29t + 0.46\left(\frac{12}{\pi}\right)\sin\left[\frac{\pi(t-16.37)}{12}\right]\right)\Big|_7^{22}$$

$$\approx \frac{1}{15}[(820.38 + 1.7488) - (261.03 - 1.1165)]$$

≈ 37.48 °C while awake

47. $$F'(\theta) = kh\csc\theta\left(\frac{\csc\theta}{R^4} - \frac{\cot\theta}{r^4}\right)$$

$$F'(\theta) = 0 \text{ when } \cos\theta_0 = \frac{r^4}{R^4}, \text{ for } 0 < \theta_0 < \frac{\pi}{2}$$

Rewriting,

$$F'(\theta) = \frac{kh}{\sin\theta}\left(\frac{1}{R^4\sin\theta} - \frac{\cos\theta}{r^4\sin\theta}\right)$$

$$= \frac{kh}{\sin^2\theta}\left(\frac{1}{R^4} - \frac{\cos\theta}{r^4}\right)$$

Since $k > 0$, $h > 0$, $\sin^2\theta > 0$, need to find sign of

$$\frac{1}{R^4} - \frac{\cos\theta}{r^4}$$

Now, $\cos\theta_0 = \frac{r^4}{R_4}$ and

$$\frac{1}{R_4} - \frac{\cos\theta_0}{r^4} = 0$$

Since the cosine function is strictly decreasing for $0 < \theta < \frac{\pi}{2}$, $\cos\theta > \cos\theta_0$ and

$$\frac{1}{R^4} - \frac{\cos\theta}{r^4} < 0$$

So, $F'(\theta) < 0$ when $0 < \theta < \theta_0$.
Similarly, when $\theta_0 < \theta < \frac{\pi}{2}$, $\cos\theta < \cos\theta_0$ and

$$\frac{1}{R_4} - \frac{\cos\theta}{r^4} > 0$$

So, $F'(\theta) > 0$ when $\theta_0 < \theta < \frac{\pi}{2}$.
Since there are no other critical values on interval $0 < \theta < \frac{\pi}{2}$, F decreases for $0 < \theta < \theta_0$ and F increases for $\theta_0 < \theta < \frac{\pi}{2}$, there is an absolute minimum at θ_0.

49. **(a)**

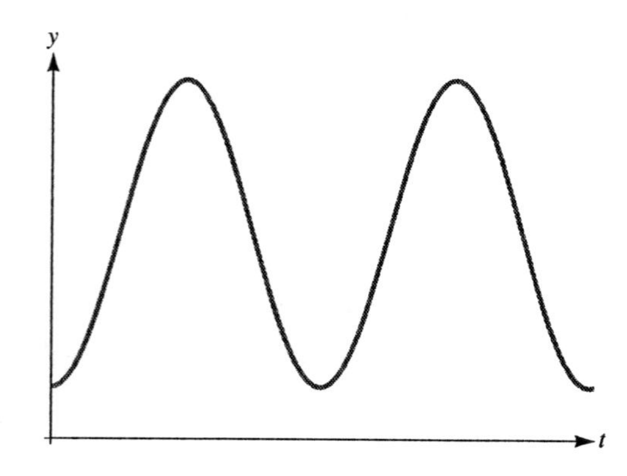

$$\beta(t) = \frac{\alpha\cos\gamma}{1-\sin^2\gamma\sin^2(\alpha t)}$$

To find the maximum and minimum, find the critical point(s) from the derivative.

$$\beta'(t) = \alpha\cos\gamma\left(-\left[1-\sin^2\gamma\sin^2(\alpha t)\right]^{-2}\left[-\sin^2\gamma\cdot 2\alpha\sin(\alpha t)\cos(\alpha t)\right]\right)$$

$$= \frac{2\alpha^2\cos\gamma\cos(\alpha t)\sin^2\gamma\sin(\alpha t)}{\left[1-\sin^2\gamma\sin^2(\alpha t)\right]^2}$$

$\beta'(t) = 0$ when

$$0 = 2\alpha^2\cos\gamma\cos(\alpha t)\sin^2\gamma\sin(\alpha t)$$

$$\cos(\alpha t) = 0 \text{ or } \sin(\alpha t) = 0$$

$$\alpha t = \frac{\pi}{2} + \pi k \text{ or } \alpha t = \pi k$$

where k is any integer

$$t = \frac{\pi}{2\alpha} + \frac{\pi}{\alpha}k \text{ or } t = \frac{\pi}{\alpha}k$$

$$\beta\left(\frac{\pi}{2\alpha}\right) = \frac{\alpha\cos\gamma}{1-\sin^2\gamma\sin^2\left(\alpha\cdot\frac{\pi}{2\alpha}\right)} = \frac{\alpha\cos\gamma}{1-\sin^2\gamma}$$

is the maximum value of β

$$\beta\left(\frac{\pi}{\alpha}\right) = \frac{\alpha\cos\gamma}{1-\sin^2\gamma\sin^2\left(\alpha\cdot\frac{\pi}{\alpha}\right)} = \alpha\cos\gamma$$

is the minimum value of β

(b) Since the time for the driven shaft to make one revolution is the same as the time for the driving shaft, the time is

$$T = \frac{2\pi}{\alpha}$$

$$\beta_{av} = \frac{1}{2\frac{\pi}{\alpha} - 0}\int_0^{2\pi/\alpha}\frac{\alpha\cos\gamma}{1-\sin^2\gamma\sin^2(\alpha t)}\,dt$$

(c) For $\alpha = 3$ and $\gamma = \pi/12$, then we wish to evaluate

$$\int_0^{\frac{2\pi}{3}}\frac{3\cos\frac{\pi}{12}}{1-\sin^2\frac{\pi}{12}\sin^2(3t)}\,dt$$

Enter this expression for $\boxed{y_1=}$. Use the $\int f(x)\,dx$ function under the calc menu with 0 as the lower limit and $2\pi/3$ as the upper limit to find $\int f(x)\,dx = 6.28\ldots$. Then multiply

this result by $\frac{1}{2\pi/3}$ to find the average output velocity A is 3.
Repeat this process changing the appropriate values of α and γ.
Given below is the table of results:

Input angular velocity α	3	5	10	12.7
Angle between shafts γ	$\frac{\pi}{12}$	$\frac{\pi}{6}$	0.4	0.26
Average output velocity A	3	5	10	12.7

(d) The average output angular velocity will be α, since it takes the same amount of time for the driven shaft to complete one revolution.

(e) Writing Exercise—Answers will vary.

51. (a) year 2000 $= P(0) \approx 460$ deer ticks
year 2005 $= P(5) \approx 2{,}483$ deer ticks
year 2010 $= P(10) \approx 13{,}211$ deer ticks

(b)

$$P'(t) = \left(1{,}250e^{0.3t}\right)\left(-1.38\left(\frac{2\pi}{365}\right)\sin\left[\frac{2\pi}{365}(t-145)\right]\right) + \left(1.47 + 1.38\cos\left[\frac{2\pi}{365}(t-145)\right]\right)1{,}250(0.3e^{0.3t})$$

year 2000, $P'(0) \approx 156$ deer ticks per year
year 2005, $P'(5) \approx 834$ deer ticks per year
year 2010, $P'(10) \approx 4{,}398$ deer ticks per year

(c)

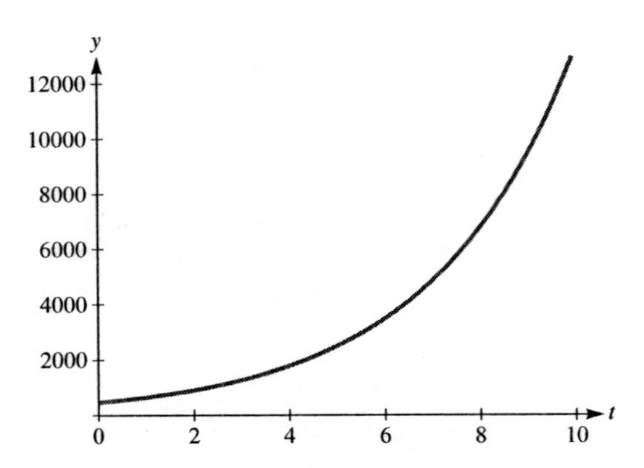

(d) Writing Exercise—Answers will vary.

53. $H(t) = 25 + 22\cos\left[\frac{2\pi}{365}(t-35)\right]$

(a) To find the maximum, find the critical point(s) of the derivative.

$$H'(t) = 22\left(-\frac{2\pi}{365}\right)\sin\left[\frac{2\pi}{365}(t-35)\right]$$

$H'(t) = 0$ when

$$\sin\left[\frac{2\pi}{365}(t-35)\right] = 0$$

$$\frac{2\pi}{365}(t-35) = \pi k$$

where k is any integer

$$t = \frac{365}{2}k + 35$$

$0 \le t \le 365$, $t = 35, 217.5$
Comparing,

$$H(35) = 47$$
$$H(217.5) = 3$$
$$H(0) \approx 43.1$$
$$H(365) \approx 43.1$$

The maximum is 47 hdd and occurs at the end of the 35th day of the year, namely February 5th.

(b) From above the minimum is 3 hdd and occurs in the middle of the 218th day of the year, namely August 6th.

(c) $H'(1) \approx 0.21$
Since $H'(0)$ is positive, H is increasing at this time.

$$H'(91) \approx -0.31$$

Since $H'(91)$ is negative, H is decreasing at this time.

$$H'(152) \approx -0.34$$

Since $H'(152)$ is negative, H is decreasing at this time.

(d)

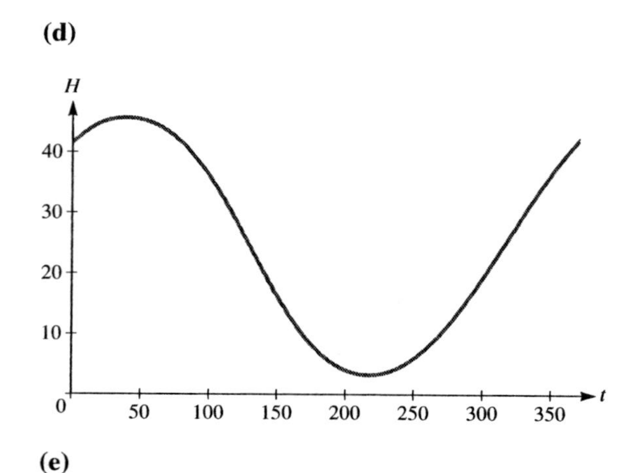

(e)

$$H_{av} = \frac{1}{90-0}\int_0^{90} 25 + 22\cos\left[\frac{2\pi}{365}(t-35)\right] dt$$

$$= \frac{1}{90}\left(25t + 22\left(\frac{365}{2\pi}\right)\sin\left[\frac{2\pi}{365}(t-35)\right]\right)\Big|_0^{90}$$

$$\approx \frac{1}{90}\left[(2{,}250 + 1{,}037.16) - (0 - 724.25)\right] \approx 44.6\ hdd$$

Checkup for Chapter 11

1. **(a)** $\sin\left(\frac{5\pi}{6}\right) = \frac{1}{2}$

(b) $\cos\left(-\frac{4\pi}{3}\right) = \cos\left(-\frac{4\pi}{3} + 2\pi\right) = \cos\left(\frac{2\pi}{3}\right) = -\frac{1}{2}$

(c) $\tan\left(\frac{3\pi}{4}\right) = \dfrac{\sin\left(\frac{3\pi}{4}\right)}{\cos\left(\frac{3\pi}{4}\right)} = \dfrac{\sqrt{2}/2}{-\sqrt{2}/2} = -1$

(d) $\sec\left(-\frac{5\pi}{6}\right) = \dfrac{1}{\cos\left(-\frac{5\pi}{6}\right)} = \dfrac{1}{-\cos\frac{5\pi}{6}} = \dfrac{1}{-\sqrt{3}/2} = -\dfrac{2}{\sqrt{3}} = -\dfrac{2\sqrt{3}}{3}$

2. $\tan\theta = \frac{2}{3} = \dfrac{\text{opposite}}{\text{adjacent}}$
Using a right triangle having an opposite side of length 2, an adjacent side of length 3, and a hypotenuse of length $\sqrt{4+9} = \sqrt{13}$,

$$\cos\theta = \frac{\text{adjacent}}{\text{hypotenuse}} = \frac{3}{\sqrt{13}} = \frac{3\sqrt{13}}{13}$$

3. **(a)**
$$f(x) = \sin(2x)$$
$$f'(x) = 2\cos(2x)$$

(b) $f(x) = x\cos x$
$f'(x) = (x)(-\sin x) + (\cos x)(1) = \cos x - x\sin x$

(c) $f(x) = \dfrac{\tan x}{x}$

$$f'(x) = \frac{(x)(\sec^2 x) - (\tan x)(1)}{x^2} = \frac{x\sec^2 x - \tan x}{x^2}$$

(d) $f(x) = x\sin(x^2)$
$f'(x) = (x)(2x\cos x^2) + (\sin x^2)(1)$
$= 2x^2\cos(x^2) + \sin(x^2)$

4. **(a)** $\int \sin(2x)\,dx$
Using substitution with $u = 2x$ and $\frac{1}{2}\,du = dx$,

$$= \frac{1}{2}\int \sin u\,du = -\frac{1}{2}\cos(2x) + C$$

(b) $\int x\cos(x^2)\,dx$
Using substitution with $u = x^2$ and $\frac{1}{2}\,du = x\,dx$,

$$= \frac{1}{2}\int \cos u\,du = \frac{1}{2}\sin(x^2) + C$$

(c) $\int_0^{\pi/2} \sec^2(2t)\,dt$
This integral represesnts the area under the curve $y = \sec^2(2t)$ from 0 to $\frac{\pi}{2}$. However, there is a vertical asymptote at $t = \frac{\pi}{4}$ and the area is infinite.

(d) $\int_0^{\pi/6} \sin^2 t\cos t\,dt$
Using substitution with $u = \sin t$ and $du = \cos t\,dt$,

$$= \int_0^{1/2} u^2\,du = \frac{u^3}{3}\Big|_0^{1/2} = \frac{1}{24}$$

5.

$$\sin(2\theta) = \cos\theta$$
$$2\sin\theta\cos\theta = \cos\theta$$
$$2\sin\theta\cos\theta - \cos\theta = 0$$
$$\cos\theta(2\sin\theta - 1) = 0$$
$$\cos\theta = 0, \quad 2\sin\theta - 1 = 0$$
$$\sin\theta = \frac{1}{2}$$

$\theta = \frac{\pi}{2}$ or $\theta = \frac{\pi}{6}, \frac{5\pi}{6}$

6. For the upper bound of the integral, find the intersection of the curves

$$\cos x = \sin x,$$
$$\text{or } x = \frac{\pi}{4}$$

$$\text{Area} = \int_0^{\pi/4} \cos x - \sin x\, dx$$
$$= (\sin x + \cos x)\Big|_0^{\pi/4}$$
$$= \left(\sin\frac{\pi}{4} + \cos\frac{\pi}{4}\right) - (\sin 0 + \cos 0)$$
$$= \frac{\sqrt{2}}{2} + \frac{\sqrt{2}}{2} - 0 - 1 = \sqrt{2} - 1$$

7.

$$\sin^2 x + \cos^2 x = 1$$
$$\frac{\sin^2 x}{\sin^2 x} + \frac{\cos^2 x}{\sin^2 x} = \frac{1}{\sin^2 x}$$
$$1 + \cot^2 x = \csc^2 x$$

8. $S(t) = 19.3 + 9.4\sin\left(\frac{\pi t}{6}\right) + 3.4\cos\left(\frac{\pi t}{3}\right)$

(a)
$$S'(t) = 9.4\left(\frac{\pi}{6}\right)\cos\left(\frac{\pi t}{6}\right) + 3.4\left(\frac{\pi}{3}\right) - \sin\left(\frac{\pi t}{3}\right)$$
$$= \frac{4.7\pi}{3}\cos\left(\frac{\pi t}{6}\right) - \frac{3.4\pi}{3}\sin\left(\frac{\pi t}{3}\right)$$
$$S'(6) = \frac{4.7\pi}{3}(-1) - \frac{3.4\pi}{3}(0)$$
$$\approx -4.92 \text{ thousand units per month}$$

Since $S'(6)$ is negative, S is decreasing.

(b) $S'(t) = 0$ when

$$0 = \frac{4.7\pi}{3}\cos\left(\frac{\pi t}{6}\right) - \frac{3.4\pi}{3}\sin\left(\frac{\pi t}{3}\right)$$
$$0 = 4.7\cos\left(\frac{\pi t}{6}\right) - 3.4\sin\left(\frac{\pi t}{3}\right)$$

Using the hint, with $\frac{\pi t}{3} = 2\left(\frac{\pi t}{6}\right)$,

$$0 = 4.7\cos\left(\frac{\pi t}{6}\right) - 6.8\sin\left(\frac{\pi t}{6}\right)\cos\left(\frac{\pi t}{6}\right)$$
$$0 = \cos\left(\frac{\pi t}{6}\right)\left[4.7 - 6.8\sin\left(\frac{\pi t}{6}\right)\right]$$
$$\cos\left(\frac{\pi t}{6}\right) = 0$$
$$\frac{\pi t}{6} = \frac{\pi}{2} + k\pi$$

where k is any intetger.
Since $0 \le t \le 12$, the solutions are $t = 3$ and $t = 9$.

$$4.7 - 6.8\sin\left(\frac{\pi t}{6}\right) = 0$$
$$\sin\left(\frac{\pi t}{6}\right) = \frac{47}{68}$$
$$\frac{\pi t}{6} \approx 0.7631 + 2\pi k$$
$$\text{or} \quad \frac{\pi t}{6} \approx 2.3785 + 2\pi k$$

Since $0 \le t \le 12$, the solutions are approximately $t = 1.4574$ and $t = 4.5426$.
Comparing $S(1.4574) \approx 25.95$
$S(3) = 25.3$
$S(4.5426) \approx 25.95$
$S(9) = 6.5$
$S(12) = 22.7$
So, largest sales are 25.95 thousand units in the months of February and May. Smallest sales are 6.5 thousand units at the end of the month of September.

9. $C'(x) = 3x + 2.5\sin(2\pi x)$

(a) $C(5) - C(0) = \int_0^5 C'(x)\, dx$

$$= \int_0^5 3x + 2.5\sin(2\pi x)\,dx$$
$$= \left[\frac{3}{2}x^2 + \left(\frac{2.5}{2\pi}\right) - \cos(2\pi x)\right]\Big|_0^5$$
$$= \left(\frac{75}{2} - \frac{2.5}{2\pi}\right) - \left(0 - \frac{2.5}{2\pi}\right) = 37.5$$

The net cost is 37.5 hundred, or \$3,750.

(b) $C'(5) = 3(5) + 2.5\sin(10x) = 15$ or, \$1,500 per unit

$$C''(x) = 3 + 2.5\,[\cos(2\pi x)]\,(2\pi)$$
$$= 3 + 5\pi\cos(2\pi x)$$
$$C''(5) = 3 + 5\pi \approx 18.7$$

The marginal cost is increasing at a rate of approximately \$1,870 per unit per unit.

10.

$$T(t) = 98.3 + 0.3\cos\left[\frac{2\pi(t-15)}{28}\right] + 0.2\cos\,[2\pi(t-0.6)]$$

(a)

$$T'(t) = 0 - 0.3\left(\frac{2\pi}{28}\right)\sin\left[\frac{2\pi(t-15)}{28}\right] - 0.2(2\pi)\sin\,[2\pi(t-0.6)]$$

When $t = 15$,

$$T'(15) = -0.3\left(\frac{\pi}{14}\right)\sin 0 - 0.2(2\pi)\sin\,[2\pi(14.4)]$$
$$\approx -0.7386$$

When $t = 15$, her body temperature is decreasing at the approximate rate of 0.74 degrees per day.

(b)

$$T_{av} = \frac{1}{30-0}\int_0^{30} 98.3 + 0.3\cos\left[\frac{2\pi(t-15)}{28}\right] + 0.2\cos\,[2\pi(t-0.6)]\,dt$$
$$= \frac{1}{30}\left(98.3t + 0.3\left(\frac{14}{\pi}\right)\sin\left[\frac{2\pi(t-15)}{28}\right] + 0.2\left(\frac{1}{2\pi}\right)\sin\,[2\pi(t-0.6)]\right)\Big|_0^{30}$$
$$= \frac{1}{30}\left(\left[98.3(30) + \frac{4.2}{\pi}\sin\frac{15\pi}{14} + \frac{0.1}{\pi}\sin 58.8\pi\right] - (\right.$$
$$\approx 98.29\ °\mathrm{F}$$

11. $B(t) = 105 + 31\cos kt$

(a) The maximum blood pressure occurs every 2π, so

$$kt = 2\pi$$
$$t = \frac{2\pi}{k}$$

Since the pulse rate is 80 heartbeats per minute,

$$\frac{80}{60} = \frac{1}{\frac{2\pi}{k}}$$
$$k = \frac{8\pi}{3} \approx 8.3776$$

(b)

$$B(t) = 105 + 31\cos\left(\frac{8\pi t}{3}\right)$$
$$= 105 + 31\cos\left[\frac{2\pi}{3/4}(t-0)\right]$$

period $= \frac{3}{4}$
amplitude $= 31$
phase shift $= 0$
vertical shift $= 105$

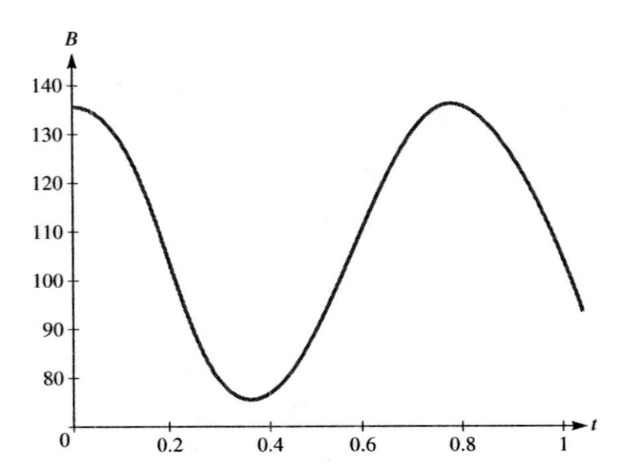

(c) The systolic pressure occurs when t is any multiple of 2π, so

$$B(2\pi) = 105 + 31(1) = 136$$

The distolic pressure occurs when t is π plus any multiple of 2π, so

$$B(\pi) = 105 + 31(-1) = 74$$

(d)
$$B'(t) = 31\left(\frac{8\pi}{3}\right)\left(-\sin\frac{8\pi t}{3}\right)$$
$$B''(t) = \frac{248\pi}{3}\left(-\frac{8\pi}{3}\right)\left(\cos\frac{8\pi t}{3}\right)$$

$B''(t) = 0$ when
$$\frac{8\pi t}{3} = \frac{\pi}{2} + n\pi$$

where n is any integer
Choosing $n = 0$,

$$t = \frac{3}{16}$$

$$B\left(\frac{3}{16}\right) = 105 + 31\cos\left[\frac{8\pi}{3}\left(\frac{3}{16}\right)\right] = 105 + 31(0) = 105$$

Review Problems

1. **(a)** The angle is $90° + \frac{1}{3}(90°) = 120°$.
In radians, is $\frac{\pi}{2} + \frac{1}{3}\left(\frac{\pi}{2}\right) = \frac{2\pi}{3}$ radians.
(b) The angle is $-180° + \frac{1}{2}(-90°) = -225°$.
In radians, is $-\pi + \frac{1}{2}\left(-\frac{\pi}{2}\right) = -\frac{5\pi}{4}$ radians.

3. **(a)** 0.25 radians $\left(\frac{180°}{\pi\,\text{rad}}\right) \approx 14.324°$
(b) 1 radian $\left(\frac{180°}{\pi\,\text{rad}}\right) \approx 57.296°$
(c) -1.5 radians $\left(\frac{180°}{\pi\,\text{rad}}\right) \approx 85.944°$

5. $\sin\theta = \frac{4}{5} = \dfrac{\text{opposite}}{\text{hypotenuse}}$
Using a right triangle having an opposite side of length 4, a hypotenuse of length 5, and an adjacent side of length $\sqrt{25 - 16} = 3$

$$\tan\theta = \frac{\text{opposite}}{\text{adjacent}} = \frac{4}{3}$$

7. $2\cos\theta + \sin 2\theta = 0,\ 0 \le \theta \le 2\pi$
$2\cos\theta + 2\sin\theta\cos\theta = 0$
$2\cos\theta(1 + \sin\theta) = 0$
$2\cos\theta = 0$ or $1 + \sin\theta = 0$
$\theta = \frac{\pi}{2}, \frac{3\pi}{2}$ or $\theta = \frac{3\pi}{2}$
So, the solutions are $\theta = \frac{\pi}{2}$ or $\theta = \frac{3\pi}{2}$.

9. $2\sin^2\theta = \cos 2\theta;\ 0 \le \theta \le \pi$
$2\sin^2\theta = \cos^2\theta - \sin^2\theta$
$2\sin^2\theta = (1 - \sin^2\theta) - \sin^2\theta$
$2\sin^2\theta = 1 - 2\sin^2\theta$
$4\sin^2\theta - 1 = 0$
$(2\sin\theta + 1)(2\sin\theta - 1) = 0$
$2\sin\theta + 1 = 0$ or $2\sin\theta - 1 = 0$
$\sin\theta = \pm\frac{1}{2}$

$$\theta = \frac{\pi}{6}, \frac{5\pi}{6}, \frac{7\pi}{6}, \frac{11\pi}{6}$$

11. $\cos^2\theta - \sin^2\theta = \cos(2\theta)$
$\cos^2\theta - (1 - \cos^2\theta) = \cos(2\theta)$
$2\cos^2\theta - 1 = \cos(2\theta)$
$\cos^2\theta = \frac{1}{2}(1 + \cos 2\theta)$
$\sin^2\theta + \cos^2\theta = 1$

$$\sin^2\theta = 1 - \cos^2\theta$$
$$\sin^2\theta = 1 - \left[\tfrac{1}{2}(1+\cos 2\theta)\right]$$
$$\sin^2\theta = 1 - \tfrac{1}{2} - \tfrac{1}{2}\cos 2\theta$$
$$\sin^2\theta = \tfrac{1}{2} - \tfrac{1}{2}\cos 2\theta$$
$$\sin^2\theta = \tfrac{1}{2}(1 - \cos 2\theta)$$

13. (a)

$$\cos\left(\frac{\pi}{2}+\theta\right) = \cos\frac{\pi}{2}\cos\theta - \sin\frac{\pi}{2}\sin\theta$$
$$= 0 - (1)\sin\theta$$
$$= -\sin\theta$$

$$\sin\left(\frac{\pi}{2}+\theta\right) = \sin\frac{\pi}{2}\cos\theta + \sin\theta\cos\frac{\pi}{2}$$
$$= (1)\cos\theta + 0$$
$$= \cos\theta$$

(b)

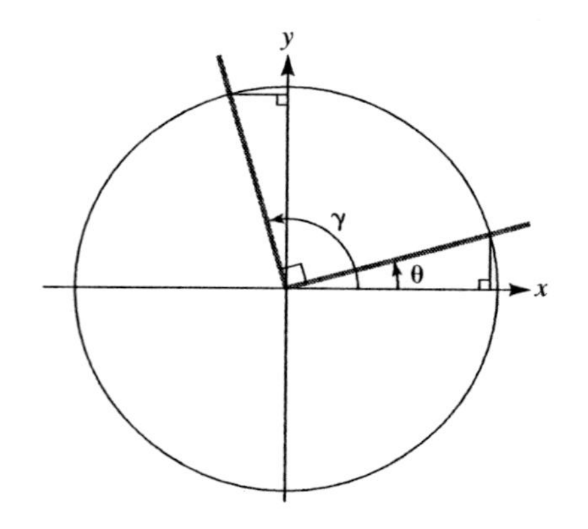

Adding $\frac{\pi}{2}$ to θ effectively interchanges the x and y axes, so that if $\gamma = \theta + \frac{\pi}{2}$, then

$$|\cos\gamma| = |\sin\theta| \text{ and } |\sin\gamma| = |\cos\theta|$$

Adding $\frac{\pi}{2}$ to θ moves the angle to the next quadrant. Noting
Quadrant I: $\sin x > 0$, $\cos x > 0$
Quadrant II: $\sin x > 0$, $\cos x < 0$
Quadrant III: $\sin x < 0$, $\cos x < 0$
Quadrant IV: $\sin x < 0$, $\cos x > 0$
So, $\cos\left(\frac{\pi}{2}+\theta\right) = -\sin\theta$ and $\sin\left(\frac{\pi}{2}+\theta\right) = \cos\theta$.

15.

$$f(x) = \cos^2 x = (\cos x)^2$$
$$f'(x) = 2(\cos x)(-\sin x)$$
$$= -2\cos x\sin x$$

17.

$$f(x) = \tan^2(3x+1) = [\tan(3x+1)]^2$$
$$f'(x) = 2\,[\tan(3x+1)]\left[\sec^2(3x+1)\right](3)$$
$$= 6\tan(3x+1)\sec^2(3x+1)$$

19.

$$f(x) = \ln(\cos^2 x) = \ln\left[(\cos x)^2\right]$$
$$= 2\ln(\cos x)$$
$$f'(x) = 2\frac{-\sin x}{\cos x}$$
$$= -2\tan x$$

21. $\int \sin 2t\,dt$

Using substitution with $u = 2t$ and $\frac{1}{2}\,du = dt$,

$$= \frac{1}{2}\int \sin u\,du$$
$$= -\frac{1}{2}\cos 2t + C$$

23. $\int \sin x\cos x\,dx$ Using substitution with $u = \sin x$ and $du = \cos x\,dx$,

$$= \int u\,du = \frac{1}{2}\sin^2 x + C$$

25. $\int \frac{\sec^2 t}{\tan t}\,dt$

Using substitution with $u = \tan t$ and $du = \sec^2 t\,dt$,

$$= \int \frac{1}{u}\,du = \ln|\tan t| + C$$

27. $\sin A = -\frac{\sqrt{2.5}}{2} \approx -0.79057$

So, the reference angle is

$$A \approx -52.2389°$$

Converting,

$$0.2389° \left(\frac{60'}{10}\right) \approx 14.334'$$

$$0.334' \left(\frac{60''}{1'}\right) \approx 20.04''$$

So, $A \approx -52°14'20''$.
The smallest angle such that $\csc\theta = \csc A$ would be in the third quadrant, or

$$\theta = 180° + 52°14'20'' \\ = 232°14'20''$$

29. We wish to solve $2\tan 3x - 5.87 - 2\sin 2x = 0$ for $0 \le x \le \frac{\pi}{2}$
Press [$y_1=$]. Enter $2\tan(3x) - 5.87 - 2\sin(2x)$ for $y_1 =$. Use window dimensions $[0, \pi/2]\pi/6$ by $[-20, 20]5$. Press [graph]. Use the zero function under the calc menu to find the x-intercepts in this interval. The x-intercepts are $x \approx 0.436$ and $x \approx 1.468$.

31. An element of volume is a disk, having a radius of y and a thickness of dx. So,

$$dV = \pi(y)^2\,dx \\ = \pi(\cos x + \sin x)^2\,dx$$

The volume of the solid is

$$\int dV = \int_{-\pi/2}^{\pi/6} \pi(\cos x + \sin x)^2\,dx$$

$$= \pi \int_{-\pi/2}^{\pi/6} (\cos^2 x + 2\sin x\cos x + \sin^2 x)\,dx$$

$$= \pi \int_{-\pi/2}^{\pi/6} 1 + 2\sin x\cos x\,dx$$

Using substitution with $u = \sin x$ and $du = \cos x\,dx$,

$$= \pi\left[\int_{-\pi/2}^{\pi/6} 1\,dx + 2\int_{-1}^{1/2} u\,du\right]$$

$$= \pi\left[x\Big|_{-\pi/2}^{\pi/6} + u^2\Big|_{-1}^{1/2}\right] = \pi\left[\left(\frac{\pi}{6} + \frac{\pi}{2}\right) + \left(\frac{1}{4} - 1\right)\right]$$

$$= \frac{2\pi^2}{3} - \frac{3\pi}{4}$$

33. **(a)** Using $f(x) = a + b\cos\left[\frac{2\pi}{p}(x - d)\right]$

period = 25
amplitude = 27
phase shift = 11
vertical shift = 33

(b)

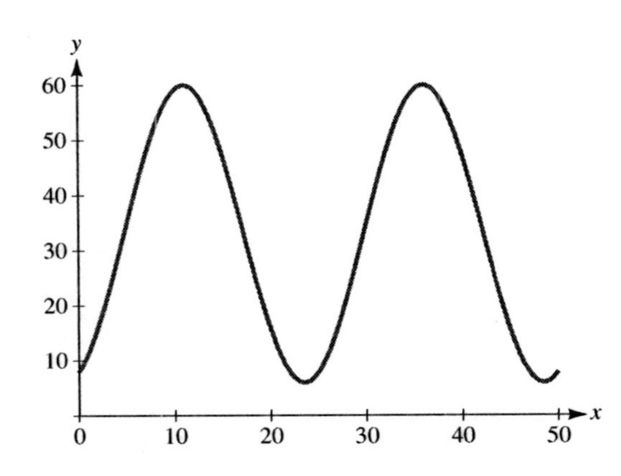

35. Let x be the distance on land from directly across from the power plant to point P. Let y be the distance underwater the cable will run. Then,

$$\sin\theta = \frac{900}{y} \text{ and } \tan\theta = \frac{900}{x}$$

$$y = \frac{900}{\sin\theta} \text{ and } x = \frac{900}{\tan\theta}$$

$$y = 900\csc\theta \text{ and } x = 900\cot\theta$$

$$C(\theta) = 5(900\csc\theta) + 4(3{,}000 - 900\cot\theta) \\ = 4500\csc\theta + 12{,}000 - 3{,}600\cot\theta$$

To find the minimum cos t, find the critical point(s) from the derivative (see problem 54 in Section 11.2).

$$C'(\theta) = -4{,}500\csc\theta\cot\theta + 3{,}600\csc^2\theta$$

$C'(\theta) = 0$ when

$$0 = -4{,}500\csc\theta\cot\theta + 3{,}600\csc^2\theta$$

$$0 = -4{,}500\csc^2\theta\cos\theta + 3{,}600\csc^2\theta$$

$$0 = 900\csc^2\theta(-5\cos\theta + 4)$$

Since $\csc\theta \neq 0$,

$$0 = -5\cos\theta + 4$$

$$\cos\theta = \frac{4}{5}$$

37. **(a)** $P(t) = 500e^{7\sin(2\pi t)}$
$P(0.125) \approx 70{,}570$ mosquitos
$P(0.25) \approx 548{,}317$ mosquitos
$P(0.325) \approx 255{,}673$ mosquitos
$P(0.5) \approx 500$ mosquitos

(b) $P'(t) = 500[14\pi\cos(2\pi t)][e^{7\sin(2\pi t)}]$
$P'(0.125) \approx 2{,}194{,}729$
$P'(0.25) = 0$
$P'(0.325) \approx -5{,}105{,}158$
$P'(0.5) \approx -21{,}991$

(c)

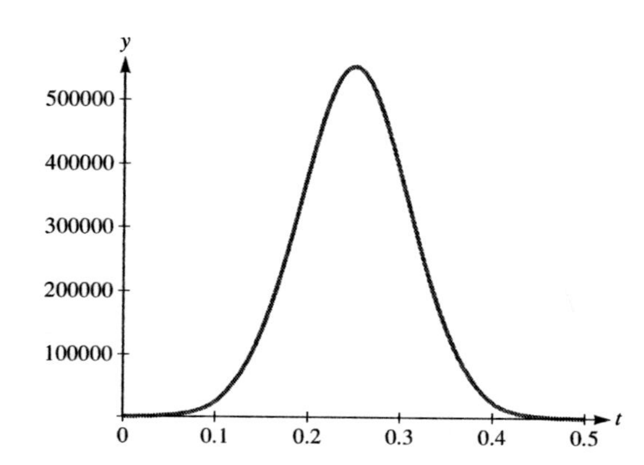

39. **(a)** $C(12) - C(0) - \int_0^{12} C'(t)\,dt$

$$= \int_0^{12} 275 + 275\cos\left[\frac{\pi}{6}(t-3)\right] dt$$

$$= 275\int_0^{12} 1 + \cos\left[\frac{\pi}{6}(t-3)\right] dt$$

Using substitution with $u = \frac{\pi}{6}(t-3)$ and $\frac{6}{\pi}\,du = dt$,

$$= 275\left[\int_0^{12} 1\,dt + \frac{6}{\pi}\int_{-\pi/2}^{3\pi/2} \cos u\,du\right]$$

$$= 275\left[t\Big|_0^{12} + \frac{6}{\pi}\left(\sin u\Big|_{-\pi/2}^{3\pi/2}\right)\right]$$

$$= 275\left[(12-0) + \frac{6}{\pi}(-1+1)\right] = 3{,}300 \text{ caribou}$$

$$\int_{12}^{24} C'(t)\,dt$$

$$= 275\left[t\Big|_{12}^{24} + \frac{6}{\pi}\left(\sin u\Big|_{3\pi/2}^{7\pi/2}\right)\right]$$

$$= 275\left[(24-12) + \frac{6}{\pi}(-1+1)\right] = 3{,}300 \text{ c}$$

(b) To find the maximum and minimum populations, find the critical point(s) from the derivative. $C'(t) = 0$ when

$$0 = 275 + 275\cos\left[\frac{\pi}{6}(t-3)\right]$$

$$\cos\left[\frac{\pi}{6}(t-3)\right] = -1$$

$$\frac{\pi}{6}(t-3) = k\pi$$

$$t = 6k + 3$$

where k is any odd integer
Since the period of C is 12, the maximum and minimum occur the same months of each year. Now,

$$C(t) = 275\left(t + \frac{6}{\pi}\sin\left[\frac{\pi}{6}(t-3)\right]\right) + C$$

Using the first year,

$$C'(t) = 0 \text{ when } t = 9$$

When $0 \le t < 9$, $C'(t) > 0$ so C is increasing. When $9 < t \le 12$, $C'(t) > 0$ so C is again increasing. So, the minimum occurs during the first month and the maximum occurs during the 12th month.

41. $$D(t) = 12.2 + 3.09\cos\left[\frac{2\pi}{365}(t-185)\right]$$

(a) $D(1) \approx 9.11$ hours on January 1st
$D(74) \approx 11.17$ hours on March 15th
$D(172) \approx 15.21$ hours on June 21st

(b) To find the maximum and minimum numbers of daylight hours, find the critical point(s) from the derivative.

$$D'(t) = -3.09\left(\frac{2\pi}{365}\right)\sin\left[\frac{2\pi}{365}(t-185)\right]$$

$D'(t) = 0$ when

$$\sin\left[\frac{2\pi}{365}(t-185)\right] = 0$$
$$\frac{2\pi}{365}(t-185) = k\pi$$
$$t = \frac{365}{2}k + 185$$

for any integer k
For $1 \le t \le 365$,
$D'(t) = 0$ when $t = 2.5$ or 185
Comparing,
$D(1) \approx 9.11$ hours
$D(2.5) \approx 9.11$ hours
$D(185) \approx 15.29$ hours
$D(365) \approx 9.11$ hours
So, the maximum number of daylight hours occurs on day 185, or July 4th. The minimum number of daylight hours occurs on days 1, 3 and 365, or January 1st, January 3rd, and December 31st.

(c)
$$D_{av} = \frac{1}{365}\int_0^{365} 12.2 + 3.09\cos\left[\frac{2\pi}{365}(1-185)\right] dt$$
$$= \frac{1}{365}\left(12.2t + 3.09\left(\frac{365}{2\pi}\right)\sin\left[\frac{2\pi}{365}(t-185)\right]\right)\Big|_0^{365}$$
$$\approx \frac{1}{365}(4{,}460.72 - 7.72) \approx 12.2 \text{ hours}$$

43. Let x be the height of the ball. Then,

$$\tan\theta = \frac{x}{600}$$
$$\sec^2\cdot\frac{d\theta}{dt} = \frac{1}{600}\cdot\frac{dx}{dt}$$
$$\frac{d\theta}{dt} = \frac{1}{600}\cdot\frac{dx}{dt}\cos^2\theta$$

When $x = 800$, $\cos\theta = \dfrac{600}{\sqrt{600^2+800^2}}$. Further,

$\dfrac{dx}{dt} = -20$ and

$$\frac{d\theta}{dt} = \frac{1}{600}(-20)\left(\frac{3{,}600}{1{,}000{,}000}\right)$$
$$\frac{d\theta}{dt} = -\frac{3}{250} \text{ radians per minute}$$

45. (a) Consider the small triangle on the bottom left, having hypotenuse $= x$. For that triangle,

$$\cos(\pi - 2\theta) = \frac{8.5-x}{x}$$
$$\cos(\pi + -2\theta) = \frac{8.5-x}{x}$$
$$\cos\pi\cos(-2\theta) - \sin\pi\sin(-2\theta) = \frac{8.5-x}{x}$$
$$\cos\pi\cos 2\theta = \frac{8.5-x}{x}$$
$$-\cos 2\theta = \frac{8.5-x}{x}$$
$$-(\cos^2\theta - \sin^2\theta) = \frac{8.5-x}{x}$$
$$\sin^2\theta - \cos^2\theta = \frac{8.5-x}{x}$$
$$\sin^2\theta - (1-\sin^2\theta) = \frac{8.5-x}{x}$$
$$2\sin^2\theta - 1 = \frac{8.5-x}{x}$$

From larger triangle, having hypotenuse $= L$,

$$\cos\theta = \frac{x}{L}$$
$$x = L\cos\theta$$

substituting,

$$2\sin^2\theta - 1 = \frac{8.5 - L\cos\theta}{L\cos\theta}$$
$$2\sin^2\theta - 1 = \frac{8.5}{L\cos\theta} - 1$$
$$2\sin^2\theta = \frac{8.5}{L\cos\theta}$$
$$L(\theta) = \frac{4.25}{\cos\theta\sin^2\theta}$$

To find the minimum value of L, find the critical point(s) from the derivative.

$$L(\theta) = 4.25 \sec\theta \csc^2\theta$$
$$L'(\theta) = 4.25\left[(\sec\theta)(2\csc\theta)(-\csc\theta\cot\theta) + (\csc^2\theta)(\sec\theta\tan\theta)\right]$$
$$= 4.25\left[-2\csc^3\theta + \csc\theta\sec^2\theta\right]$$

$L'(\theta) = 0$ when

$$0 = -2\csc^3\theta + \csc\theta\sec^2\theta$$
$$0 = \csc\theta(\sec^2\theta - 2\csc^2\theta)$$

Since $\csc\theta \neq 0$,

$$0 = \sec^2\theta - 2\csc^2\theta$$

Multiplying both sides by $\sin^2\theta$,

$$0 = \tan^2\theta - 2$$
$$\tan\theta = \pm\sqrt{2}$$

Since $0 < \theta < \frac{\pi}{2}$,

$$\tan\theta = \sqrt{2}$$
$$\theta \approx 0.955 \text{ radians, or } 54.74°$$

The minimum length is

$$L(0.955) = \frac{4.25}{(\cos 0.955)(\sin^2 0.955)}$$
$$\approx 11.04 \text{ inches}$$

(b)

$$\text{area} = \frac{1}{2}bh$$
$$A = \frac{1}{2}x(\sqrt{L^2 - x^2})$$
$$= \frac{1}{2}L\cos\theta\sqrt{L^2 - (L\cos\theta)^2}$$
$$= \frac{1}{2}L\cos\theta\sqrt{L^2(1 - \cos^2\theta)}$$
$$= \frac{1}{2}L^2\cos\theta\sqrt{\sin^2\theta}$$
$$= \frac{1}{2}L^2\cos\theta\sin\theta$$

To minimize the area, find the critical point(s) from the derivative.

$$\frac{dA}{d\theta} = \frac{1}{2}L^2(\cos^2\theta - \sin^2\theta)$$

$\frac{dA}{d\theta} = 0$ when

$$\cos^2 - \sin^2\theta = 0$$
$$\cos^2\theta - (1 - \cos^2\theta) = 0$$
$$2\cos^2\theta - 1 = 0$$
$$\cos^2\theta = \frac{1}{2}$$

Since θ cannot be larger than $\frac{\pi}{2}$, $\theta = \frac{\pi}{3}$. When $\theta = \frac{\pi}{3}$,

$$L = \frac{4.25}{\cos\frac{\pi}{3}\left(\sin\frac{\pi}{3}\right)^2} = \frac{34}{3}$$
$$A \approx \frac{1}{2}\left(\frac{34}{3}\right)^2 \cos\frac{\pi}{3}\sin\frac{\pi}{3}$$
$$\approx 27.81 \text{ square inches}$$

47.

$$F(t) = A + B\sin\left(\frac{\pi t}{12}\right)$$
$$R(t) = C + D\cos\left(\frac{\pi t}{12}\right)$$

(a)

$$F'(t) = \frac{B\pi}{12}\cos\left(\frac{\pi t}{12}\right)$$

$F'(t) = 0$ when

$$0 = \frac{B\pi}{12}\cos\left(\frac{\pi t}{12}\right)$$
$$0 = \cos\left(\frac{\pi t}{12}\right)$$
$$\frac{\pi t}{12} = k\pi + \frac{\pi}{2}, \text{ for } k = 0, 1, 2, \ldots$$
$$\frac{t}{12} = k + \frac{1}{2}$$
$$t = 12k + 6$$

or, $t = 6, 18, 30, \ldots$

$$F''(t) = -\frac{B\pi^2}{144}\sin\left(\frac{\pi t}{12}\right)$$

When $t = 6$, $F''(6) < 0$, so there is a maximum when $t = 6$.
When $t = 18$, $F''(18) > 0$, so there is a minimum when $t = 18$.
(Note, due to the cyclic nature of the sine function, all other values cycle as maximums and minimums.)
$F(6) = 70$, or $A + B \sin \frac{\pi}{2} = 70$
$F(18) = 22$, or $A + B \sin \frac{3\pi}{2} = 22$
So,
$A + B = 70$
$A - B = 22$
or, $A = 46$ and $B = 24$.
Similarly,

$$R'(t) = -\frac{D\pi}{12} \sin\left(\frac{\pi t}{12}\right)$$

$R'(t) = 0$ when

$$0 = -\frac{D\pi}{12} \sin\left(\frac{\pi t}{12}\right)$$

$$0 = \sin\left(\frac{\pi t}{12}\right)$$

$$\frac{\pi t}{12} = k\pi, \text{ for } k = 0, 1, 2, \ldots$$

$$t = 12k$$

or, $t = 0, 12, 24, \ldots$

$$R''(t) = -\frac{D\pi^2}{144} \cos\left(\frac{\pi t}{12}\right)$$

When $t = 0$, $R''(0) < 0$, so there is a maximum when $t = 0$.
When $t = 12$, $R''12) > 0$, so there is a minimum when $t = 12$.
$R(0) = 635$, or $C + D \cos 0 = 635$
$R(12) = 227$, or $C + D \cos \pi = 227$
So,
$C + D = 635$
$C - D = 227$
or, $C = 431$ and $D = 204$.
The period for each population is $\left|\frac{2\pi}{b}\right|$.
For the fox population,

$$= \left|\frac{2\pi}{\frac{\pi}{12}}\right| = 24 \text{ years}$$

For the rabbit population,

$$= \left|\frac{2\pi}{\frac{\pi}{12}}\right| = 24 \text{ years}$$

(b) The fox population is the largest when $t = 6$ (and cyclical values). At this time, the rabbit population is

$$R(6) = 431 + 204 \cos \frac{\pi}{2}$$
$$= 431 \text{ rabbits}$$

The fox population is the smallest when $t = 18$ (and cyclical values). At this time, the rabbit population is

$$R(18) = 431 + 204 \cos \frac{3\pi}{2}$$
$$= 431 \text{ rabbits}$$

(c) The rabbit population is the largest when $t = 12$ (and cyclical values). At this time, the fox population is
$F(12) = 46 + 24 \sin \pi = 46$ foxes.
The rabbit population is the smallest when $t = 0$ (and cyclical values). At this time, the fox population is
$F(0) = 46 + 24 \sin 0 = 46$ foxes

(d) To graph $F(t) = 46 + 24 \sin \frac{\pi t}{12}$ and $R(t) = 431 + 204 \cos \frac{\pi t}{12}$,
press [y=]. Input $F(t)$ for $y_1 =$ and $R(t)$ for $y_2 =$.
Use window dimensions [0, 24]2 by [0, 650] 100
Press [graph].

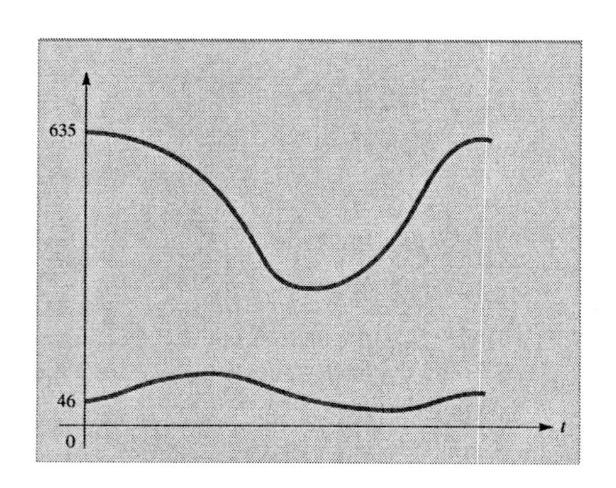

49. $$\frac{dy}{dx} + \frac{1}{x}y = \sin x$$

The integrating factor is

$$I = e^{\int \frac{1}{x}\,dx} = e^{\ln x} = x$$

$$y = \frac{1}{x}\left[\int x \cdot \sin x\,dx + C\right]$$

Using integration by parts with $u = x$ and $dV = \sin x\,dx$,

$$y = \frac{1}{x}\left[-x\cos x - \int 1 \cdot -\cos x\,dx + C\right]$$

$$y = \frac{1}{x}\left[-x\cos x + \sin x + C\right]$$

$$y = -\cos x + \frac{\sin x}{x} + \frac{C}{x}$$

Since $y = 1$ when $x = \pi$,

$$1 = 1 + 0 + \frac{C}{\pi}, \text{ or } C = 0$$

So,

$$y = -\cos x + \frac{\sin x}{x}$$

51. (a) $$\int_0^{\pi/2} \sin(x^2)\,dx;\, n = 10$$

Since $\Delta x = \dfrac{\frac{\pi}{2} - 0}{10} = \frac{\pi}{20}$,

$$= \frac{\pi}{60}\left[\sin 0 + 4\sin\left(\frac{\pi}{20}\right)^2 + 2\sin\left(\frac{2\pi}{20}\right)^2 + 4\sin\left(\frac{3\pi}{20}\right)^2 + 2\sin\left(\frac{4\pi}{20}\right)^2 + 4\sin\left(\frac{5\pi}{20}\right)^2 + 2\sin\left(\frac{6\pi}{20}\right)^2 + 4\sin\left(\frac{7\pi}{20}\right)^2 + 2\sin\left(\frac{8\pi}{20}\right)^2 + 4\sin\left(\frac{9\pi}{20}\right)^2 + \sin\left(\frac{10\pi}{20}\right)^2\right]$$

$$\approx \frac{\pi}{60}[0 + 0.098686 + 0.1970718 + 0.8809818 + 0.769218 + 2.3138752 + 1.5519563 + 3.7410876 + 1.9999304 + 3.639525 + 0.624266]$$

$$\approx 0.8282$$

(b) Press [y=]. Enter $\sin(x^2)$ for $y_1 =$.
Use window dimensions $[0, \pi]\pi/4$ by $[-5, 5]1$.
Press [graph]. Use the $\int f(x)\,dx$ function under the calc menu with $x = 0$ as the lower limit and $x = \pi/2$ as the upper limit. We see $\int_0^{\pi/2} \sin(x^2)\,dx \approx 0.8281$.

53. $$\int \tan x\,dx$$

$$= \int \frac{\sin x}{\cos x}\,dx$$

Using substitution with $u = \cos x$ and $du = -\sin x\,dx$,

$$= -\int \frac{1}{u}\,du = -\ln|\cos x| + C$$

55. $f(A, B, C) = \sin A \sin B \sin C$
Since $C = \pi - (A + B)$,

$$\begin{aligned}\sin C &= \sin[\pi - (A+B)]\\ &= \sin[\pi + -(A+B)]\\ &= \sin\pi\cos[-(A+B)] + \cos\pi\sin[-(A+B)]\\ &= -\sin[-(A+B)]\\ &= \sin(A+B)\end{aligned}$$

So,

$$f(A, B) = \sin A \sin B \sin(A+B)$$

$$\begin{aligned}\frac{\partial f}{\partial A} &= \sin B\,[\sin A\cos(A+B) + \sin(A+B)\cos A]\\ &= \sin B\,[\sin(2A+B)]\\ \frac{\partial f}{\partial B} &= \sin A\,[\sin B\cos(A+B) + \sin(A+B)\cos B]\\ &= \sin B\,[\sin(A+2B)]\end{aligned}$$

Want $\dfrac{\partial f}{\partial A} = 0 = \dfrac{\partial f}{\partial B}$, so

$$\sin B\,[\sin(2A+B)] = \sin B\,[\sin(A+2B)]$$

(rejecting $B = 0, \pi$)

$$\begin{aligned}\sin(2A+B) &= \sin(A+2B)\\ 2A+B &= A+2B\\ A &= B\end{aligned}$$

Since

$$\begin{aligned}\sin(A+2B) &= 0\\ A+2B &= \pi\\ A &= \pi - 2B\end{aligned}$$

Further,

$$\begin{aligned}C &= \pi - (\pi - 2B + B)\\ C &= B\end{aligned}$$

So, $A = B = C = \frac{\pi}{3}$ and

$$f\left(\frac{\pi}{3}, \frac{\pi}{3}, \frac{\pi}{3}\right) = \left(\sin\frac{\pi}{3}\right)^3 = \left(\frac{\sqrt{3}}{2}\right)^3 = \frac{3\sqrt{3}}{8}$$

57. $u_t = c^2 u_{xx}$

(a) When $u = e^{-c^2k^2t}\sin kx$,

$$\begin{aligned}u_t &= \sin kx\left(-c^2k^2e^{-c^2k^2t}\right)\\ u_x &= e^{-c^2k^2t}(k\cos kx)\\ u_{xx} &= ke^{-c^2k^2t}(-k\sin kx)\end{aligned}$$

So,

$$-c^2k^2e^{-c^2k^2t}\sin kx = c^2\left(-k^2e^{-c^2k^2t}\sin kx\right)$$

which shows the function satisfies the diffusion equation.

(b) Writing Exercise—Answers will vary.

59. $f(x, y) = \cos x + \cos y$

$$\begin{aligned}g(x, y) &= y - x\\ f_x &= -\sin x\\ f_y &= -\sin y\\ g_x &= -1\\ g_y &= 1\\ -\sin x &= -\lambda\\ -\sin y &= \lambda\\ \sin y &= -\sin x\\ \sin y &= \sin(-x)\\ y &= -x\end{aligned}$$

Now,

$$\begin{aligned}y - x &= \frac{\pi}{4}\\ -2x &= \frac{\pi}{4}\\ x &= -\frac{\pi}{8}\\ y &= \frac{\pi}{8}\end{aligned}$$

$$\begin{aligned}f\left(-\frac{\pi}{8}, \frac{\pi}{8}\right) &= \cos\left(-\frac{\pi}{8}\right)\cos\left(\frac{\pi}{8}\right)\\ &\approx 1.8478\end{aligned}$$